Fermentation and Biochemical Engineering Handbook

Principles, Process Design, and Equipment

Third Edition

Dedication

For Mother
For Walter, Christian, Brandon
For David, Kathy, David

Fermentation and Biochemical Engineering Handbook

Principles, Process Design, and Equipment

Third Edition

Edited by

Henry C. Vogel
Chapel Hill, WC

Celeste M. Todaro
CelesTech Inc., Haddonfield, New Jersey

ELSEVIER

AMSTERDAM • BOSTON • HEIDELBERG • LONDON • NEW YORK • OXFORD • PARIS
SAN DIEGO • SAN FRANCISCO • SINGAPORE • SYDNEY • TOKYO

William Andrew is an imprint of Elsevier

William Andrew is an imprint of Elsevier
225 Wyman Street, Waltham, MA 02451, USA
The Boulevard, Langford Lane, Kidlington, Oxford OX5 1GB, UK

First Edition 1983
Second Edition 1996
Third Edition 2014

Library of Congress Cataloging-in-Publication Data
A catalog record for this book is available from the Library of Congress

British Library Cataloguing-in-Publication Data
A catalogue record for this book is available from the British Library

ISBN: 978-1-4557-2553-3

For information on all Elsevier publications
visit our website at http://elsevierdirect.com

Contents

Part II
Equipment Design

Preface to the Third Edition

As we release the third edition, we do so without Henry. We honor him as father to grandfather, friend, founder of Fermentation and BioChemical Engineering Handbook and someone who worked diligently towards making penicillin commercially available.

Fermentation and Bio-Chemical Engineering Handbook started to relay the story of antibiotic drug production or what today is considered classic biochemical engineering or fermentation.

Classic industrial fermentation processes began with microorganisms and conditioning with nutrient concentration. The products historically are alcohols for consumption and not, glycerol, and carbon dioxide from yeast fermentation of various sugars to acetone, lactic acid, monosodium glutamate, and acetic acid from various bacteria—acids such as citric, gluconic acid, antibiotics, vitamins B12, and riboflavin from mold fermentation.

In the 1970's new discoveries in molecular and cell biology enabled the first *genetically engineering bacteria* to produce *human insulin.* Other protein—synthesising DNA fragments have been place in bacteria and yeast cells to allow commercialization of drugs, antibodies and other therapeutics. These developments in biotechnology brought advances in agricultural, food, pharmaceuticals, fine chemical, marine, biofuels, and green chemistry.

Bio-synthetized products such as peptides to make human insulin, anti-malaria drugs and ethanol-based biofuels to algal-derived jet fuel are all based upon fermentation.

And so this 3rd edition, bridges the Classic to Modern Biochemical Engineering or bio-synthesis. We have included classic theory, and processing equipment as they are used in early development or startups and as well as new bio-processing productions.

As we look forward to the continual advances in biosynthesis, it is worth a look back to not so long ago when the greatest medical development of the 20 century came about. I would like to share a Penicillin Story from my Mother, Charlotte Todaro:

"I graduated from a Newburgh, NY high school in June of 1948 with a College Entrance Degree and, in September of 1948 at the age of 17, entered the Methodist Hospital School of Nursing Program in Brooklyn for the next three years. Little did I realize back then I was to become one of the "human tools" to be implemented in the historical process of proving the biggest medical discovery accomplishment of that time.... The therapeutic value of Penicillin in combating infections.

When I along with the other student nurses in my class, 51 to be exact, received our third stripe on our student uniform, we were assigned mainly to the evening and night shifts on the nursing units to care for our patients. The staff consisted of the student nurse and a nurse's aide for the whole floor. We had the entire floor to give meds to and my recollection of my evenings and nights on the Surgical Halls 3 Unit was forever imbedded in my memory. That was 1950 and we were on the brink of the treatment of patients with the aqueous penicillin injections. Our trays every three hours were prepared with twenty or more syringes to be injected into the prescribed patients. Back then there were no neat little pre-packaged syringes to be loaded. We had to wash the syringes and needles for reuse every time, then sterilize by boiling them, and assemble them with the solution we prepared sent from the pharmacy. It came in large vials with a white powder to be diluted with a sterile solution. We went through vials and vials of these, administering shot after shot, 8 to each patient in a 24 hour period. The syringes were arranged in rows and we proceeded around to all the patients injecting them one after the other. We were not welcomed, but one after another they improved and were subsequently discharged to their homes, sore as they were.

My "husband to be" was one of my patients at that time. He had suffered continuously, time after time, with abscesses caused by osteomyelitis beginning at the age of 9 after a leg injury and had to be admitted to hospital where he underwent many surgeries ultimately losing his leg.

At the age of 23, after the first session of penicillin for several days, he was discharged. The inflammation and fever had subsided, and he was considered cured. Never again in his lifetime did he develop an abscess from the osteomyelitis that had constantly plagued him. Penicillin had saved his life. Mickey Mantle shortly thereafter too was saved from osteomyelitis by Penicillin."

Charlotte Todaro

Today's medicine, stem cell transplants, bone marrow transplantation, cancer chemotherapy would not be possible as infections are generated by these and antibiotics are needed. Unfortunately, there are so still so many people in the world who do not have access to antibiotics. Some of the proceed from this book will go towards that effort.

In the spirit of Classic to Modern Biochemical Engineering, we offer this book as part of this ever-evolving science.

I would like to give a heartfelt thank you to Steve Licht, a highly accomplished chemical process engineer, with incredible depth and breadth of knowledge and experience, for his support of this effort. He updated a number of chapters in the book and brought exciting new chapters, Biofuels and Distillation. His advice and support of this work are greatly appreciated and helped us bridge classic fermentation to modern biotechnology processes. I have not met a more benevolent chemical engineer.

The editor would also like to thank John Sheridan, a long time colleague and friend for his review and expertise in Environmental work and giving his time for editing several parts of this book. Thank you Jacqui Licht for creating incredibly detailed and beautiful colorized charts, figures and graphs to update this edition. Thank you Mother for sharing your story and always being there for me.

I am grateful to David Jackson of Elsevier for his wise counsel and unwavering commitment and patience to delivering this edition. Thank you to Sally Mortimore for the wonderful original groundwork as we began the third edition. Special thank you to Susan Li for a meticulous review to finalize this edition.

Celeste M. Todaro

Preface to the Second Edition

The second edition of the *Fermentation and Biochemical Engineering Handbook*, like the previous edition, is intended to assist the development, design and production engineer who is engaged in the fermentation industry. Particular emphasis is given to those unit operations most frequently encountered in the commercial production of chemicals and pharmaceuticals via fermentation, separation, and purification.

Some theory is included to provide the necessary insight into the unit operation but is not emphasized. Rather, the emphasis is placed on the practical aspects of development, design and operation—how one goes about collecting design data, what are the scale-up parameters, how to select the right piece of equipment, where operating problems arise, and how to troubleshoot.

The text is written from a practical and operating viewpoint, and all of the contributing authors have been chosen because of their industrial background and orientation. Several of the chapters which were in the first edition have been either deleted or replaced by other chapters which are more germane to current fermentation practice. Those chapters which were retained have been updated or have been rewritten to reflect current practice. Several new chapters were introduced to reflect current emphasis on cell cultures, nutritional requirements, statistical methods for fermentation optimization, cross-flow filtration, environmental concerns, and plant design.

The editors wish to express their gratitude to Mrs. Connie Gaskill of Heinkel Filtering Systems, Inc., for the wordprocessing assistance she gave to this edition.

Henry C. Vogel
Scotch Plains, New Jersey

Celeste M. Todaro
Bridgeport, New Jersey

September, 1996

Preface to the First Edition

This book is intended to assist the development, design and production engineer who is engaged in the fermentation industry. Particular emphasis is given to those unit operations most frequently encountered in the commercial production of chemicals and pharmaceuticals via fermentation, separation, and purification.

Some theory is included to provide the necessary insight into the unit operation but is not emphasized. Rather, the emphasis is placed on the practical aspects of development, design and operation—how one goes about collecting design data, what are the scale-up parameters, how to select the right piece of equipment, where operating problems arise and how to troubleshoot.

The text is written from a practical and operating viewpoint, and all of the contributing authors have been chosen because of their industrial background and orientation. Since the handbook concerns fermentation and often the engineers involved in fermentation are not versed in microbiology, it was thought advisable to introduce this subject at the beginning of the book.

Similarly, since much of fermentation deals with the production of antibiotics, it was deemed advisable to include some chapters specifically oriented to the production of sterile products.

The engineering using this handbook may wish that other unit operations or different pieces of equipment had been included other than those selected. The selection was based on the individual contributors and my own experience, over many years of work in the field, with unit operations and pieces of equipment that have been the backbone and workhorses of the industry.

The editor wished to express his thanks to Mr. Stanley Grossel of HofEnann-La Roche and Mr. John Carney of Davy McKee Corporation for reviewing and editing the draft copies. He also thanks Miss Mary Watson of Davy McKee Corporation for typing assistance, and Mr. Michael Garze of Davy McKee Corporation for his help in producing many of the graphs and illustrations. Dr. Sol Barer, the author of the microbiology chapter acknowledges the valuable input to the Celanese Biotechnology Department, and especially thanks Miss Maria Guerra for her patience in typing and retyping the manuscript.

Henry C. Vogel
Berkeley Heights, New Jersey
June 1983

In Memory of Henry C. Vogel

Henry C. Vogel (1919–2012) enjoyed a fruitful career as a Senior Staff Engineer at Davy McKee Corporation, a Manager of the Process Engineering Department at GAF, and an Assistant Director of the Task Force for Modernization at Merck & Co., Inc.

Upon graduating MIT in 1941, Vogel enlisted and served in the India-Burma Theater WWII. Upon his return, Vogel joined Merck & Co., Inc. He worked on the early penicillin process at Merck when the drug was first made commercially available. In 1952, Vogel graduated with an MS from Columbia University and received a patent from the USPTO for his process of recovering osmium.

Vogel was also active in professional organizations, co-chairing the New Jersey Chapter of the American Institute of Chemical Engineers and the New Jersey Lecture Series on Fermentation.

List of Contributors

Michael J. Akers, Baxter BioPharma, Bloomington, Indiana

Giovanni Bellini, 3 V, Cogeim

Ramesh R. Bhave, Oak Ridge National Laboratory, Oakridge, TN

Frederick J. Dechow, Mediquest Therapeutics, Bothell, WA

Barry Fox, Mendel Co., East Hanover, New Jersey

Howard L. Freese, Allvac, Monroe, NC

Edwin O. Geiger, Pfizer Inc., Groton, CT

Stephen M. Glasgow, Union Carbide, South Charleston, WV

Elliott Goldberg, Consultant

Yujiro Harada, K. F. Engineering Co., Ltd., Tokyo, Japan

Willem H. Kampen, Louisiana State University, Agriculture Center, Baton Rouge, LA

Mark Keyashian, Medarex, San Francisco, CA

John P. King, Foxboro Company, Rahway, NJ

Steven Licht, Solazyme Inc., San Francisco, CA

Maung K. Min, Consultant

James Y. Oldshue(deceased), Mixing Equipment Co., Inc., Rochester, NY

Laura Pellegrini, Politechnical di Milano

Kuniaki Sakata, Kyowa Hakko Kogyo Co., Ltd., Tokyo, Japan

Seijo Sato, Kyowa Medex Co., Ltd., Sunto-gun, Shizuoka Pref., Japan

Allan C. Soderberg, Fort Collins, CO

Curtis S. Strother, BioConvergence, Indianapolis, Indiana

Shinsaku Takayama, Tokai University, Numazu, Shizuoka Pref., Japan

Celeste M. Todaro, CelesTech Inc., Haddonfield, New Jersey

David B. Todd(deceased), Todd Engineering, Princeton, New Jersey

Mark R. Walden, Eli Lilly, Indianapolis, Indiana

Part I

Fermentation

Chapter 1

Fermentation Pilot Plant

Yujiro Harada, Kuniaki Sakata, Seiji Sato and Shinsaku Takayama

PROLOGUE

Yujiro Harada

The rapid development of biotechnology has impacted diverse sectors of the economy. Many industries are affected, including agricultural, bio-based chemicals, food processing, biological medicines, nutraceuticals, and biofuels. In order for current biotechnology research to continue revolutionizing industries, new processes must be developed to transform current research into viable market products. Specifically, attention must be directed toward the industrial processes of cultivation of cells, tissues, and microorganisms. Although several such processes already exist (e.g., r-DNA and cell fusion), more are needed and it is not even obvious which of the existing processes is best.

To develop the most cost-efficient process, scale-up data must be collected by repeating experiments at the bench and pilot scale level. These data must be extensive. Unfortunately, the collection is far more difficult than it would be in the chemical and petrochemical industries. The nature of working with living material makes contamination commonplace and reproducibility of data difficult to achieve. Such problems quickly distort the relevant scale-up factors.

In this chapter, three research scientists from Kyowa Kogyo Co. Ltd. (now Kyowa Hakko Bio Co. Ltd.) have addressed the problems of experimentation and pilot scale-up for microorganisms, mammalian cells, plant cells, and tissue. It is our sincere hope that the reader will find this chapter helpful in determining the best conditions for cultivation and the collection of scale-up data. Hopefully, this knowledge will, in turn, facilitate the transformation of worthwhile research programs into commercially viable processes.

1.0 MICROBIAL FERMENTATION

Kuniaki Sakato

Chemical engineers are still faced with problems regarding scale-up and microbial contamination in the fermentation of aerobic submerged cultures. Despite many advances in biochemical engineering to address these problems, the problems nevertheless persist. Recently, many advances have been made in the area of recombinant DNA, which themselves have spun off new and lucrative fields in the production of plant and animal pharmaceuticals. A careful study of this technology is therefore necessary, not only for the implementation of efficient fermentation processes, but also for compliance with official regulatory bodies.

There are several major topics to consider in scaling up laboratory processes to the industrial level. In general, scale-up is accomplished for a discrete system through laboratory and pilot scale operations. The steps involved can be broken down into seven topics that require some elaboration:

1. Strain improvements
2. Optimization of medium composition and cultural conditions such as pH and temperature
3. Oxygen supply required by cells to achieve the proper metabolic activities
4. Selection of an operative mode for culture process
5. Measurement of rheological properties of cultural broth
6. Modelling and formulation of process control strategies
7. Manufacturing sensors, bioreactors, and other peripheral equipment

Items 1 and 2 should be determined in the laboratory using shake flasks or small jar fermenters. Items 3–7 are usually determined in the pilot plant. The importance of the pilot plant is, however, not limited to steps 3–7. The pilot plant also provides the cultured broths needed for downstream processing and can generate information to determine the optimal cost structure in manufacturing and energy consumption as well as the testing of various raw materials in the medium.

1.1 Fermentation Pilot Plant

Microorganisms such as bacteria, yeast, fungi, or actinomycete have manufactured amino acids, nucleic acids, enzymes, organic acids, alcohols and physiologically active substances on an industrial scale. The "New Biotechnology" is making it increasingly possible to use recombinant DNA techniques to produce many kinds of physiologically active substances such as interferons, insulin, and salmon growth hormone which now only exist in small amounts in plants and animals.

This section will discuss the general problems that arise in pilot plant, fermentation and scale-up. The section will focus on three main topics: *(i)* bioreactors and culture techniques, *(ii)* the application of computer and sensing technologies to fermentation, and *(iii)* the scale-up itself.

1.2 Bioreactors and Culture Techniques for Microbial Processes

Current bioreactors are grouped into either culture vessels, or reactors using *biocatalysts* (e.g., immobilized enzymes/microorganisms) or plant and animal tissues.

Table 1.1 shows a number of aerobic fermentation systems which are schematically classified into *(i)* internal mechanical agitation reactors, *(ii)* external circulation reactors, and *(iii)* bubble column and air-lift loop reactors. This classification is based on both agitation and aeration as it relates to oxygen supply. In this table, reactor 1 is often used at the industrial level and reactors (a)2, (b)2, (c)2, and (c)3, can be fitted with draught tubes to improve both mixing and oxygen supply efficiencies.

Culture techniques can be classified into batch, fed-batch, and continuous operation (Table 1.2). In batch processes, all the nutrients required for cell growth and product formation are present in the medium prior to cultivation. Oxygen is supplied by aeration. The cessation of growth reflects the exhaustion of the limiting substrate in the medium. For fed-batch processes, the usual fed-batch and the repeated fed-batch operations are listed in Table 1.2.

TABLE 1.1 Classification of Aerobic Fermentation Systems

(a) Internal mechanical agitation reactors
  1. Turbine-stirring installation
  2. Stirred vessel with draft tube
  3. Stirred vessel with suction tube

(b) External circulation reactors
  1. Water jet aerator
  2. Forced water jet aerator
  3. Recycling aerator with fritted disc

(c) Bubble column and air-loop reactors
  1. Bubble column with fritted disc
  2. Bubble column with a draft tube for gyration flow
  3. Air lift reactor
  4. Pressure cycle reactor
  5. Sieve plate cascade system

A fed-batch operation is that operation in which one or more nutrients are added continuously or intermittently to the initial medium after the start of cultivation or from the halfway point through the batch process. Details of fed-batch operation are summarized in Table 1.3. In the table the fed-batch operation is divided into two basic models, one without feedback control and the other with feedback control. Fed-batch processes have been utilized to avoid substrate inhibition, glucose effect, and catabolite repression, as well as for auxotrophic mutants.

The continuous operations of Table 1.2 are elaborated in Table 1.4 as three types of operations. In a chemostat without feedback control, the feed medium containing all the nutrients is continuously fed at a constant rate (dilution rate) and the cultured broth is simultaneously removed from the fermenter at the same rate. A typical chemostat is shown in Fig. 1.1 The *chemostat* is quite useful in the optimization of media formulation and to

TABLE 1.2 Classification of Fermentation Processes

1. Batch process
2. Fed-batch process (semi-batch process)
3. Repeated fed-batch process (cyclic fed-batch process)
4. Repeated fed-batch process (semi-continuous process or cyclic batch process)
5. Continuous process

TABLE 1.3 Classification of Fed-Batch Processes in Fermentation

1. Without feedback control
   a. Intermittent fed-batch
   b. Constant rate fed-batch
   c. Exponentially fed-batch
   d. Optimized fed-batch
2. With feedback control
   a. Indirect control
   b. Direct control
      - Setpoint control (constant value control)
      - Program control
      - Optimal control

TABLE 1.4 Classification of Continuous Fermentation Processes

1. Without feedback control
   a. Chemostat
2. With feedback control
   a. Turbidostat
   b. Nutristat
   c. Phauxostat

investigate the physiological state of the microorganism. A *turbidostat* with feedback control is a continuous process to maintain the cell concentration at a constant level by controlling the medium feeding rate. A *nutristat* with feedback control is a cultivation technique to maintain a nutrient concentration at a constant level. A *phauxostat* is an extended nutristat which maintains the pH value of the medium in the fermenter at a preset value. Figure 1.1 is an example of chemostat equipment that we call a *single-stage continuous culture*. Typical homogeneous continuous culture systems are shown in Fig. 1.2.

1.3 Application of Computer Control and Sensing Technologies for Fermentation Process

The application of direct digital control of fermentation processes began in the 1960's. Since then, many corporations have developed computer-aided fermentation in both pilot and commercial plants. Unfortunately, these proprietary processes have almost never been published, due to corporate secrecy. Nevertheless, recent advances in computer and sensing technologies do provide us with a great deal of information on fermentation. This information can be used to design optimal and adaptive process controls.

In commercial plants, programmable logic controllers and process computers enable both process automation and labor-savings. The present and likely future uses of computer applications to fermentation processes in pilot and industrial plants are summarized in Table 1.5. In the table, *open circles* indicate items that have already been discussed in other reports while the *open triangles* are those topics to be elaborated here.

The acquisition of data and the estimation of state parameters on commercial scales will undoubtedly become increasingly significant. Unfortunately, the advanced control involving adaptive and optimized controls have not yet been sufficiently investigated in either the pilot or industrial scale.

Adaptive control is of great importance for self-optimization of fermentation processes, even on a commercial scale, because in ordinary fermentation the process includes several variables regarding culture conditions and raw materials. We are sometimes faced with difficulties in the mathematical modelling of fermentation processes because of the complex reaction kinetics involving cellular metabolism. The knowledge-based controls using fuzzy theory or neural networks have been found very useful for what we call the "black box" processes. Although the complexity of the process and the number of control parameters make control problems in fermentation very difficult to solve, the solution of adaptive optimization strategies is worthwhile and can contribute greatly to total profits. In order to establish such investigations, many fermentation corporations have been building pilot fermentation systems that consist of highly instrumented fermenters coupled to a distributed hierarchical computer network for on-and off-line data acquisition, data analysis, control and modelling. An example of the hierarchical computer system that is shown in Fig. 1.3 has become as common in the installation of large fermentation plants as it is elsewhere in the chemical industry. Figure 1.4 shows the details of a computer communication network and hardware.

As seen in Fig. 1.3, the system is mainly divided into three different functional levels. The first level has the

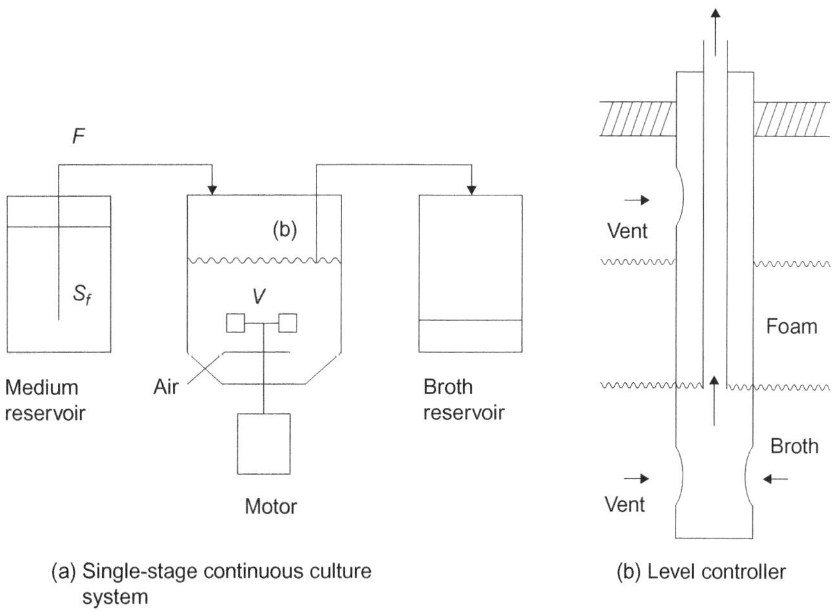

FIGURE 1.1 Chemostat System. V: Operation volume. F: Feed rate of medium. S_f: Concentration of limiting substrate.

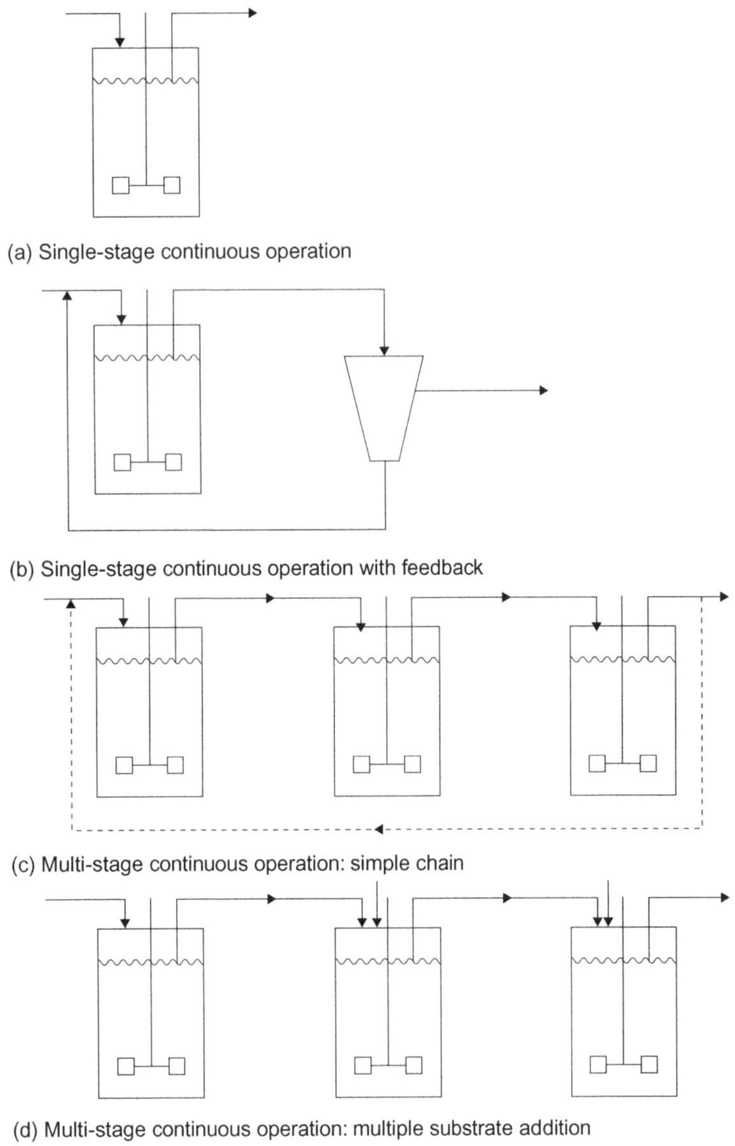

FIGURE 1.2 Homogeneous systems for continuous fermentation.

TABLE 1.5 Computer Applications to Fermentation Plants

	Pilot Scale		Production Scale	
	Present	Future	Present	Future
Sequence control	○	○	○	○
Feedback control	○	○	○	○
Data acquisition	○	○	Δ	○
Estimation of state variables	○	○	Δ	○
Advanced control	○	○	A few cases	Δ
Optimized control	○	○		Δ
Modelling	○	○		Δ
Scheduling			Δ	Δ

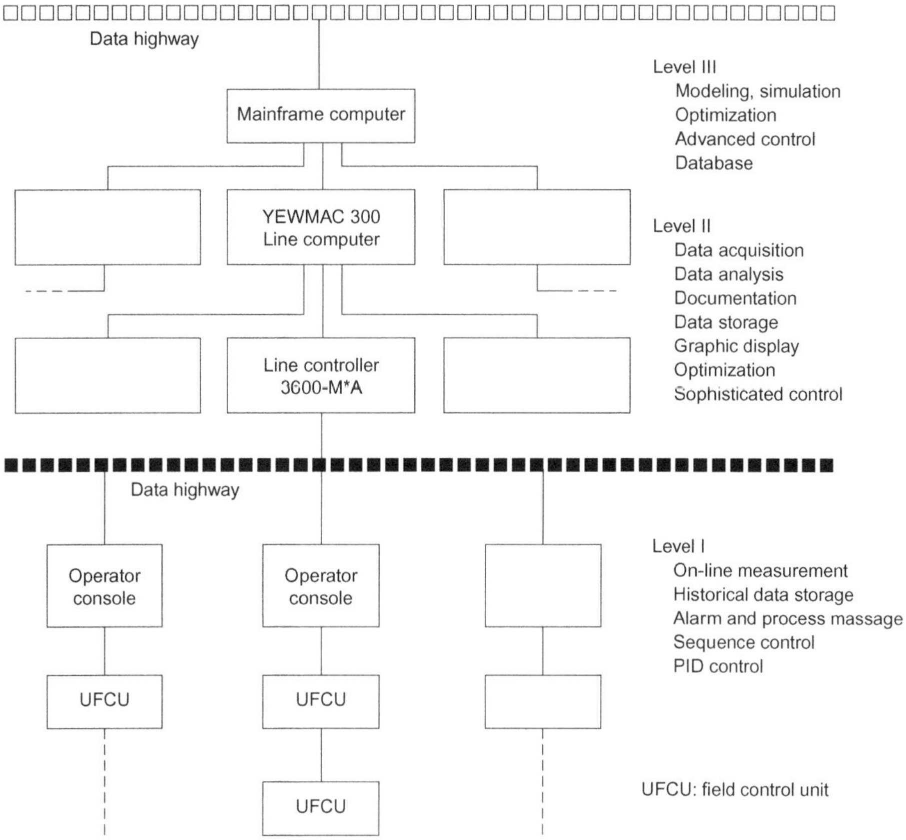

FIGURE 1.3 Configuration of distributed hierarchical computer system for fermentation pilot plant.

YEWPACK package instrumentation systems (Yokogawa Electric Corporation, Tokyo), which may consist of an operator's console (UOPC or UOPS) and several field control units (UFCU or UFCH) which are used mainly for on-line measurement, alarm, sequence control, and various types of proportional-integral-derivative (PID) controls. Each of the field control units interfaces directly with input/output signals from the instruments of fermenters via program controllers and signal conditioners. In the second level, YEWMAC line computer systems (Yokogawa Electric Corporation, Tokyo) are dedicated to the acquisition, storage, and analysis of data as well as to documentation, graphics, optimization, and advanced control. A line computer and several line controllers constitute a YEWMAC. The line controller also governs the local area network formed with some lower-level process computers using the BSC multipoint system. On the third level, a computer is reserved for modelling, development of advanced control, and the building of a database. Finally, the fermentation control system computer communicates with other business or R&D computers via a data highway or LAN. The run-time information is used for decision-making, planning, and other managerial functions. The lower-level computer, shown as the first level in Fig. 1.3, is directly interfaced to some highly instrumented fermenters. Figure 1.5 illustrates a brand new fermenter for fed-batch operation. Control is originally confined to pH, temperature, defoaming, airflow rate, agitation speed, backpressure, and medium feed rate. Analog signals from various sensors are sent to a multiplexer and A/D converters. After the computer stores the data and analyzes it on the basis of algorithms, the computer sends the control signals to the corresponding controllers to control the fermentation process.

Sensing in the fermentation area tends to lack the standard of reliability common to the chemical industry. Steam sterilization to achieve aseptic needs in fermentation is crucial for most sensors such as specific enzyme sensors. The various sensors that can be used in fermentation are summarized in Table 1.6. As in the chemical industry, almost all the physical measurements can be monitored online using sensors, although an accurate measurement device, such as a flow meter, is not yet available. The chemical sensors listed in Table 1.6 reflect the measurement of extracellular environmental conditions. The concentration of various compounds in the media are currently determined off-line following a manual sampling operation except for dissolved gas and exhaust gas concentration. Exhaust gas analysis can provide significant information about the respiratory activity, which is closely related to cellular metabolism and cell growth. This analysis is what is called *gateway sensor* and is shown schematically in Fig. 1.6.

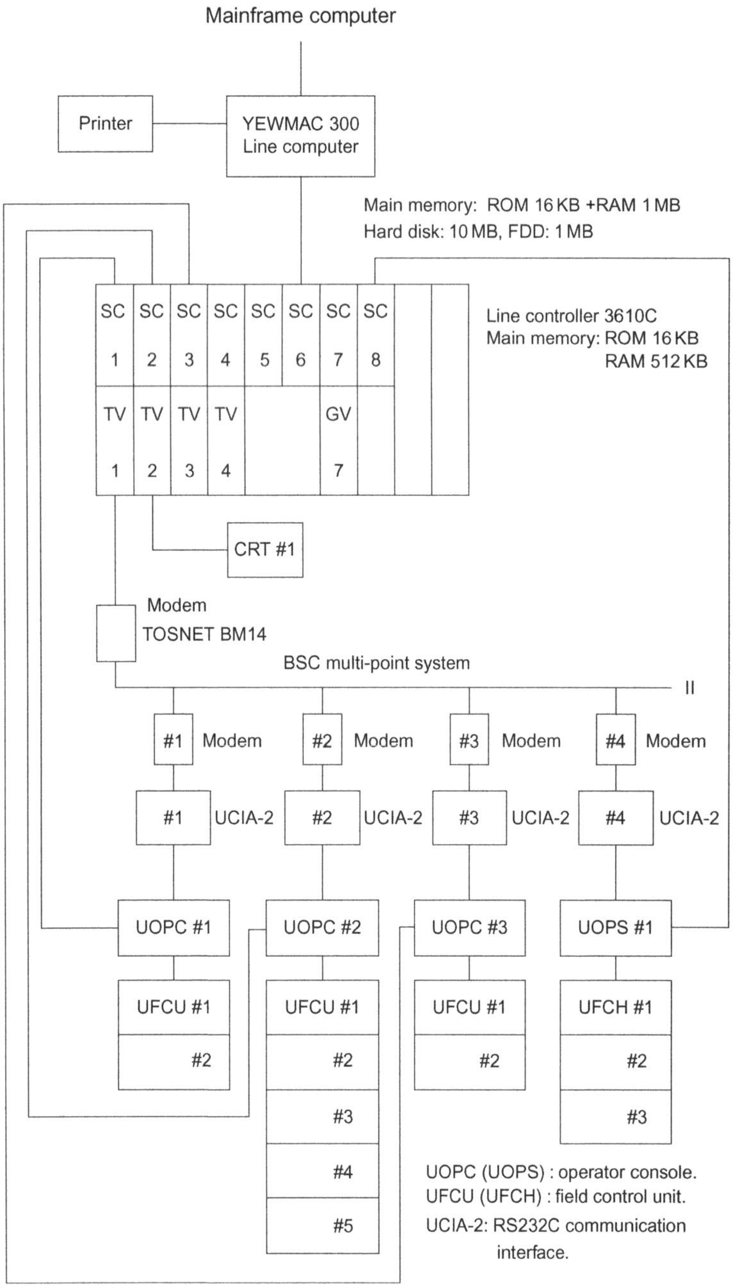

FIGURE 1.4 Local area network (LAN) for pilot fermentation plant.

The data analysis scheme of Fig. 1.6 includes the steady-state oxygen balance method and the carbon balancing method. In addition, the system can provide the oxygen supply conditions that relate to volumetric oxygen transfer coefficient (k_La), oxidation-reduction potential (ORP) and degree of oxygen saturation $Qo_2X/(Qo_2X)_{max}$. For the data analysis scheme of Fig. 1.6, the most significant advances in the fermentation field have been the development of steam sterilization, dissolved oxygen electrodes and the application of mass spectrometry to the exhaust gas analysis. Dissolved oxygen probes can be classified as either potentiometric (galvanic) or amperometric (polarographic). These electrodes are covered with a gas-permeable membrane; an electrolyte is included between the membrane and the cathode. It should be noted that these probes can measure the oxygen tension but not the concentration. The signal from both models of electrodes often drifts with time for long continuous measurements. Calibration then becomes difficult because of possible contamination. Most commercial probes have a vent to balance the pressure between the inside and outside of the probe. Often, the broth and electrolyte mix

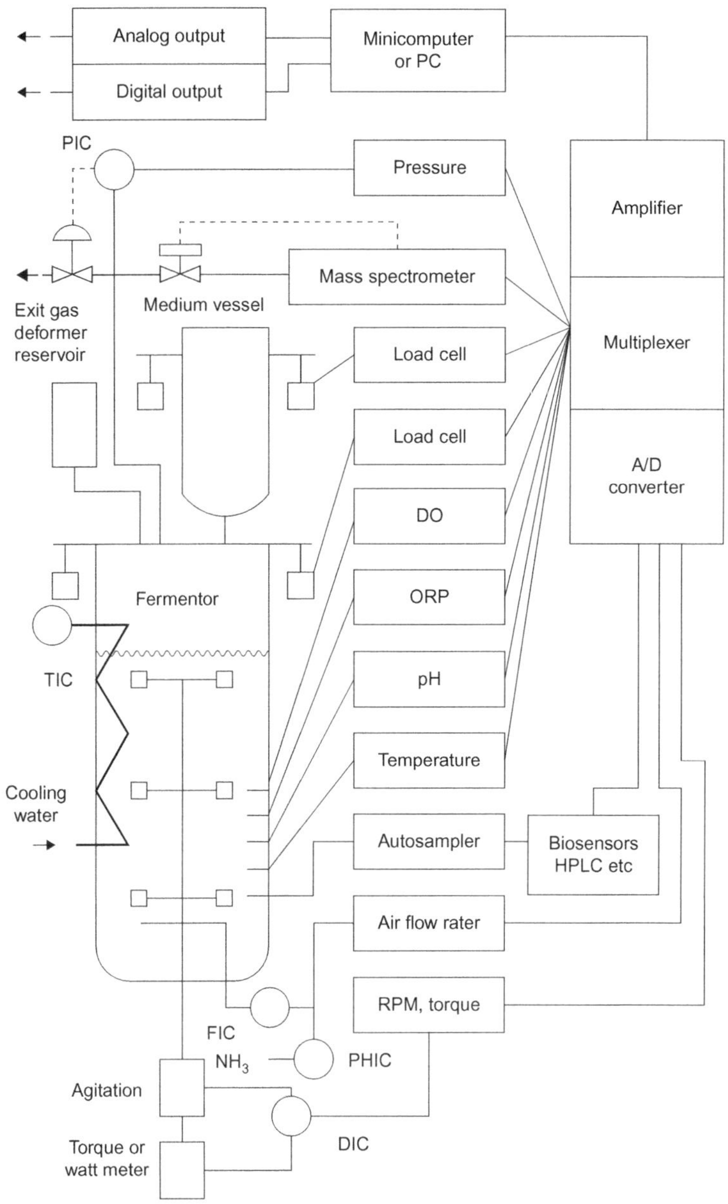

FIGURE 1.5 Highly instrumented pilot fermentor for fed-batch operations.

through the vent causing signal drift and rapid reduction in probe life. Therefore, fiber-optic chemical sensors such as pH, dissolved oxygen and carbon dioxide electrodes which need pressure compensation interference by medium components, drift and so on. This type of sensor is based on the interaction of light with a selective indicator at the waveguide surface of optical fiber. Fiber-optic sensors do not suffer from electromagnetic interferences. Also, these can be miniaturized and multiplexed, internally calibrated, steam-sterilized and can transmit light over long distances with actually no signal loss as well as no delayed time of the response. At present, a key factor for these sensors is to avoid the photodecomposition of the dyes during longtime measurements. Generally, the majority of measurements on oxygen uptake (Qo_2X) have been made with a paramagnetic oxygen analyzer while those on carbon dioxide evolution rate (Qco_2X) have been made with an infrared carbon dioxide analyzer.

Gateway sensors have become quite widespread in use in fermentation processes at both the pilot and plant levels. The sample's gas has to be dried by passing through a condenser prior to the exhaust gas analysis to avoid the influence of water vapor on the analyzers. Except for bakers' yeast production, few studies have been reported documenting the application of the steady-state oxygen balance method to the process control of fermentation processes in pilot and production plants. Recently the industrial use of this method has been published for the fed-batch process of glutathione fermentation. Based on the overall oxygen uptake rate Qo_2XV

TABLE 1.6 Sensors for Fermentation Processes

Physical
Temperature
Pressure
Shaft speed
Heat transfer rate
Heat production rate
Foam
Gas flow rate
Liquid flow rate*
Broth volume or weight
Turbidity*
Rheology or viscosity*
Biochemical
Viable cell concentration*
NAD/NADH level*
ATP/ADP/AMP/level*
Enzyme Activity*
Broth composition*
Chemical
pH
ORP
Ionic strength
Gaseous O_2 concentration
Gaseous CO_2 concentration
Dissolved O_2 concentration
Dissolved CO_2 concentration
Carbon source concentration
Nitrogen source concentration*
Metabolic product concentration*
Minor metal concentration*
Nutrient concentration*

Reliable sensors are not available.

and the exit ethanol concentration, the feed-forward/feedback control system of sugar feed rate has been developed to successfully attain the maximum accumulation of glutathione in the broth on the production scale (Fig. 1.7). In the figure, the feed-forward control of sugar cane molasses feeding was made with total oxygen uptake rate Qo_2XV and the sugar supply model which is based on the oxygen balance for both sugar and ethanol consumptions. In this system, oxygen, carbon dioxide and ethanol in outlet gas were measured on-line with a paramagnetic oxygen analyzer and two infrared gas analyzers as "gate way" sensors for a 120-kL production fermenter. Oxygen and ethanol concentration in outlet gas at the pilot level was continuously monitored with the sensor system consisting of two semiconductors. For the feedback control, a PID controller was used to compensate for a deviation, e, from a present ethanol concentration, E_{set}, calculated by the ethanol consumption rate model. Based on the deviation e, a deviation ΔF from the setpoint feed rate F can be calculated as shown in Fig. 1.7. The performance of this system was found to be very good using a YEWPACK Package Instrumentation System (Yokogawa Electric Corporation, Tokyo) and a 120 kL production fermenter (Fig. 1.8). The results, an average of 40% improvement of glutathione accumulation in the broth was attained, were compared with a conventionally exponential feeding of sugar cane molasses.

Recent research using mass spectrometry has made it possible to almost continuously measure not only oxygen and carbon dioxide concentrations but also many other volatiles at the same time. The increased reliability, freedom of calibration, and rapid analysis with a mass spectrometer has allowed the accurate on-line evaluation of steady-state variables in Fig. 1.8 for process control and scale-up. Figure 1.9 shows schematically the instrumentation system using a membrane on the inlet side for analyzing the exhaust gas from the fermenter. In Fig. 1.9, the left part is the gas sampling system that consists of a knockout pot, preventing the broth from flowing into the mass spectrometer, a filter and a pump, for sampling.

As shown in the right side of Fig. 1.9, a quadruple mass spectrometer, MSG 300, with a gas-tight ion source, secondary electron multiplier, direction detector, and a turbo-molecular pump (TURBOVAC 150) is equipped with a membrane inlet (all from Nippon Shinku, Tokyo). The resolution scale is 300. Mass spectrometry can also be used for the measurement of dissolved gases in a liquid phase using a steam sterilizable membrane probe. Recently, the application of the mass spectrometer to fermentation processes has increased markedly.

A laser turbidimeter has been developed for the on-line measurement of cell concentration, which is correlated to the turbidity of the cultured broth. However, the application of this turbidimeter to the continuous monitoring of cell growth might be limited to the lower range of cell concentration even in the highly transparent broths compared to the production media containing solid materials such as cane sugar molasses and corn steep liquor.

As indicated in Table 1.6, the biochemical sensor can be used for intracellular activities, which are closely

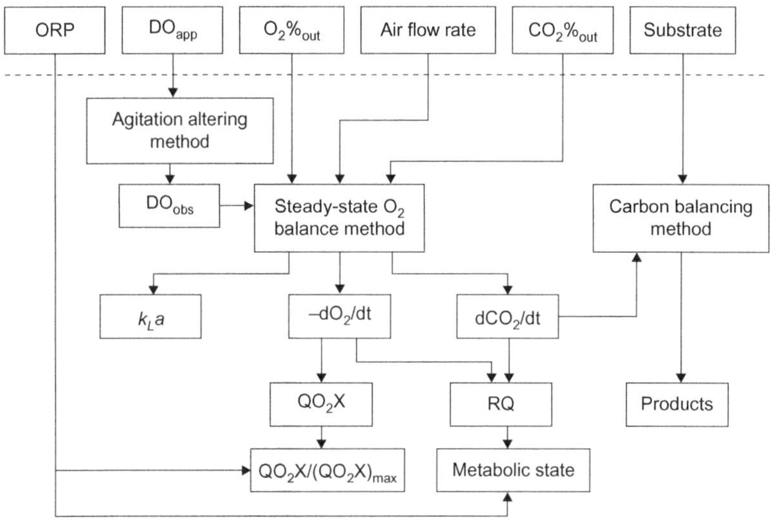

FIGURE 1.6 Estimation of metabolic parameters using *gateway sensor.*

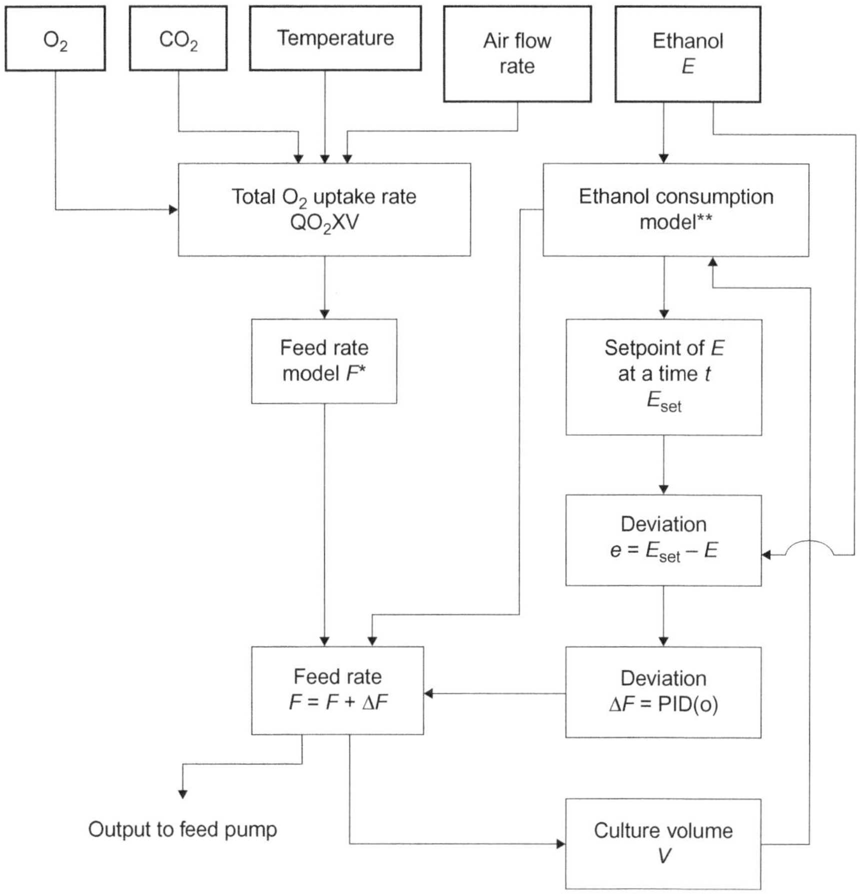

FIGURE 1.7 Configuration of process control system for glutathione fermentation.
*The feed rate F can be calculated from the oxygen balance for sugar and ethanol consumption in the broth.
**The optimal ethanol consumption profile is obtained for a constant consumption rate.

related to the level of key intermediates such as NAD/NADH and ATP/ADP/AMP. ATP is adenosine triphosphate; a nucleotide. It is the major source of energy for cellular reactions, this energy being released during its conversion to ADP. Formula: C10H16N5O13P3. Adenosine-5′-triphosphate (ATP) is an adenine ring, a ribose sugar, and three phosphate groups is used for energy transfer in plant and animal cells. ATP synthase produces ATP from ADP or AMP + Pi in water. ATP has many uses. It is used as a coenzyme, in glycolysis, for example. ATP is also found in nucleic acids in the processes of DNA replication and transcription. The high energy is from the two high-energy phosphoanhydride bonds. Nicotinamide adenine dinucleotide(NAD) accepts electrons to form NADH.

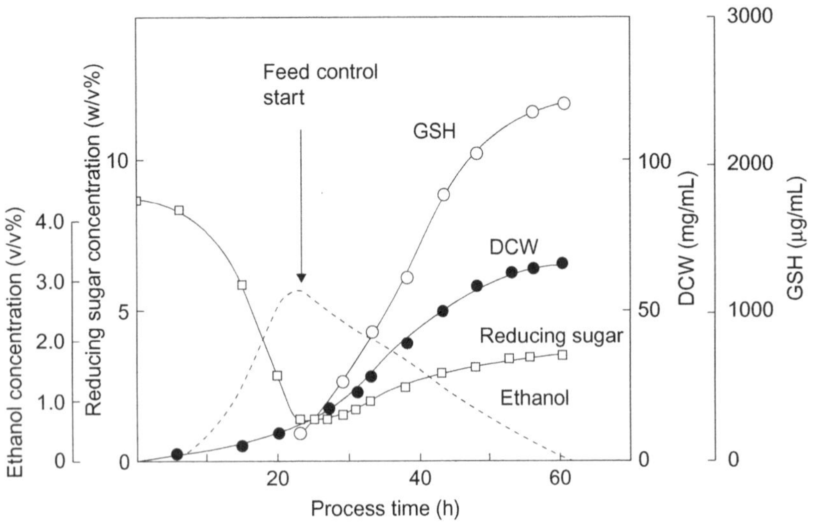

FIGURE 1.8 Trends of glutathione, reducing sugar, dry cell weight (DCW) and ethanol concentration in the broth during the glutathione fermentation in 120-kL fermenter using the feed-forward/feedback control system.

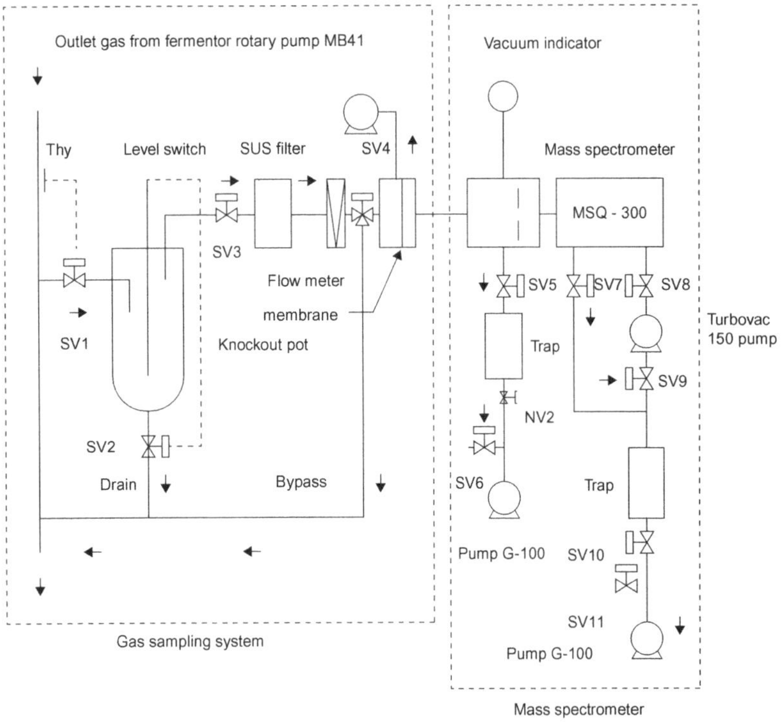

FIGURE 1.9 Schematic representation of analytical system for outlet gas from fermenter. (SV) solenoid valve; (NV) needle valve; (Thy) thermistor.

It is a coenzyme found in all living cells. It is used in cellular processes, most importantly as a substrate of enzymes that add or remove chemical groups from protein. The enzymes involved in NAD^+metabolism are targets for drug discovery. Sensors for monitoring on-line NADH on the intracellular level are commercially available. The *fluorometer* sensor can measure continuously the culture fluorescence, which is based on the fluorescence of NADH at an emission wavelength of 460 nm when excited with light at 360 nm. The sensor response corresponds to the number of viable cells in the lower range of the cell concentration. It should be especially noted that the sensor reflects the metabolic state of microorganisms.

Other useful sensors are the Fourier transform infrared spectrometer (FTIR) and the near-infrared (NIR) spectrometer for the on-line measurement of composition changes in complex media during cultivation. The FTIR measurements are based on the type and quantities of infrared radiation that a molecule absorbs. The NIR measurements are based on the absorption spectra following the multi-regression analyses. These sensors are available for fermentation processes.

1.4 Scale-Up

The supply of oxygen by aeration-agitation conditions are closely related to the following parameters:

1. Gas/liquid interfacial area
2. Bubble retention time ("hold-up")
3. Thickness of liquid film at the gas/liquid interface

Based on these three parameters, the four scale-up methods have been investigated keeping each parameter constant from laboratory to industrial scale. The parameters for scale-up are the following:

1. Volumetric oxygen transfer coefficient (k_La)
2. Power consumption volume
3. Impeller tip velocity
4. Mixing time

Even for the simple stirred, aerated fermenter, there is no one single solution for the scale-up of aeration-agitation which can be applied with high probability of success for all fermentation processes. Scale-up methods based on aeration efficiency (k_La) or power consumption/unit volume have become the standard practice in the fermentation field.

Scale-up based on impeller tip velocity may be applicable to the case where an organism sensitive to mechanical damage was employed with culture broths showing non-Newtonian viscosity. Furthermore, scale-up based on constant mixing time cannot be applied in practice because of the lack of any correlation between mixing time and aeration efficiency. It might be interesting and more useful to obtain information on either mixing time or impeller tip velocity in non-Newtonian viscous systems.

The degree of oxygen saturation $Qo_2/(Qo_2)_{max}$ and oxidation-reduction potential (ORP) have already been found to be very effective for the scale-up of fermentation processes for amino acids, nucleic acids, and coenzyme Q_{10}. The successful scale-up of many aerobic fermentations suggests that the dissolved oxygen concentration level can be regarded as an oxygen. Measurements using conventional dissolved oxygen probes are not always adequate to detect the dissolved oxygen level below 0.01 atm. Even 0.01 atm is rather high compared to the critical dissolved oxygen level for most bacterial respirations. Due to the lower detection limit of dissolved oxygen probes, oxidation-reduction potential (ORP) was introduced as an oxygen supply index, which is closely connected to the degree of oxygen saturation.

The ORP value E_h in a non-biological system at a constant temperature is given in the following equation:

$$E_h = 454.7 - 59.1 + \log(P_L) \tag{1.1}$$

where

P_L = the dissolved oxygen tension = (atm)
E_h = the potential vs hydrogen electrode

In microbial culture systems, the ORP value E can be expressed as follows:

$$E = E_{DO} + E_{pH} + E_t + E_{md} + E_{cm} \tag{1.2}$$

where

E_{DO} = the dissolved oxygen
E_{pH} = the pH
E_t = the temperature
E_{md} = the medium
E_{cm} = all metabolic activity to the whole ORP E

For most aerobic fermentations at constant pH and temperature, Eq. (1.2) can be simplified to the following,

$$E = E_{DO} \tag{1.3}$$

As a result, we can generally use the culture ORP to evaluate the dissolved oxygen probe.

An example using the ORP as a scale-up parameter has been reported for the coenzyme Q_{10} fermentation using *Rhodopseudomonas spheroides*. In this case, coenzyme Q_{10} production occurred under a limited oxygen supply where the dissolved oxygen level in the broth was below a detection limit of conventional dissolved oxygen probes. Therefore, the oxidation-reduction potential (ORP) was used as a scale-up parameter representing the dissolved oxygen level. As a result, the maximum coenzyme Q_{10} production was attained, being kept the minimum ORP around 200 mV in the last phase of culture (Fig. 1.10).

In the scale-up of ordinary aerobic processes, oxygen transfer conditions have been adjusted to the maximum oxygen requirement of the fermentation beer during the whole culture period. However, the excess oxygen supply occurs in the early growth due to the lower cell concentration under these conditions. It should be noted that such excess supply of oxygen sometimes has the harmful effect of bioproducts formation. In other words, the oxygen supply should be altered according to the oxygen requirements of microorganisms in various culture phases.

1.5 Bioreactors for Recombinant DNA Technology

There are many microorganisms used widely in industry today that have been manipulated through recombinant DNA technology. To assure safety in the manufacture of amino acids, enzymes, biopharmaceuticals such as interferons, and other chemicals using altered microorganisms, guidelines have existed for their industrial application. At the time of the second edition of this handbook (1996), more than 3,000 experiments using recombinant DNA technology had been made in Japan, while the industrial applications were around 500. At the time of the third edition (2014) such technology is commonplace. In most of

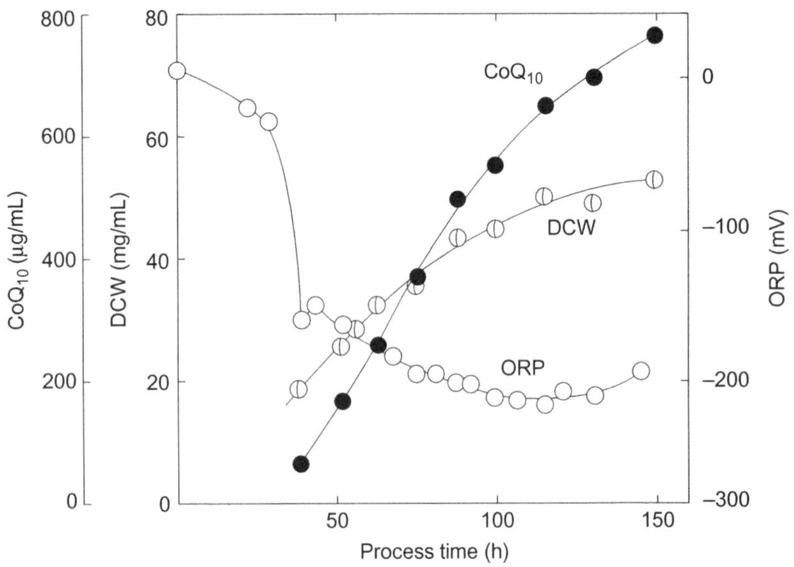

FIGURE 1.10 Coenzyme Q_{10} fermentation under an optimal aeration-agitation condition using 30 liter jar fermenter and the constant rate fed-batch culture. DCW: dry cell weight, ORP: oxidation-reduction potential.

the OECD countries, large-scale fermentation processes can be regarded as those including cultured broths over 10 liters. Organizations which have pilot plants employing recombinant DNA organisms must evaluate the safety of the microorganism and process based on the safety of a recipient microorganism and assign it to one of the following categories: GILSP (Good Industrial Large-Scale Practice), Categories 1, 2, and 3 or a special category.

This classification is quoted from *Guideline for Industrial Application of Recombinant DNA Technology* which has been published by the Ministry of International Trade and Industry in Japan. This guideline can be applied to the manufacturing of chemicals. There are also two major guidelines for pharmaceuticals and foods by the Ministry of Health and Welfare, and for the agricultural and marine field by the Ministry of Agricultural, Forestry and Fishery.

Regulatory guidelines for industrial applications of recombinant DNA technology, even though there are differences in each country, are primarily based on "Recombinant DNA Safety Considerations" following the "Recommendation of the Council," which have been recommended to the member nations of OECD in 1986.

GILSP (Good Industrial Large-Scale Practice). A recipient organism should be nonpathogenic, should not include such organisms as pathogenic viruses, phages, and plasmids; it should also have a long-term and safe history of industrial uses, or have environmental limitations that allow optimum growth in an industrial setting, but limited survival without adverse consequences in the environment.

Category 1. A nonpathogenic recipient organism which is not included in the above GILSP.

Category 2. A recipient organism having undeniable pathogenicity to humans that might cause infection when directly handled. However, the infection will probably not result in a serious outbreak in cases where effective preventive and therapeutic methods are known.

Category 3. A recipient organism capable of resulting in disease and not included in Category 2 above. It shall be carefully handled, but there are known effective preventive and therapeutic methods for said disease. A recipient organism which, whether directly handled or not, might be significantly harmful to human health and result in a disease for which no effective preventive nor therapeutic method is known, shall be assigned a classification separate from Category 3 and treated in a special manner.

Based on the Category mentioned above, the organization should take account of "Physical Containment." Physical containment involves three elements of containment: equipment, operating practices/techniques, and facilities. Physical containment at each Category for the GILSP level is given in "Guideline for Industrial Application of Recombinant DNA Technology" in Japan. Using appropriate equipment, safe operating procedures, and facility design, personnel and the external environment can be protected from microorganisms modified by recombinant DNA technology.

FURTHER READING

[1] S. Aiba, A.E. Humphery, N.F. Mills, Biochemical Engineering, second ed., Academic Press, New York, 1973.

[2] G.T. Banks, Scale-up of fermentation process, Top. Enzyme Ferment. Technol. 3 (1979) 170.

[3] H.W. Blanch, S.M. Bhabaraju, Non-Newtonian Fermentation Broths: Rheology and Mass Transfer, Biotechnol. Bioeng. 28 (1976) 745.

[4] Committee on the Introduction of Genetically Engineered Organisms into the Environment, Introduction of Recombinant DNA-Engineered Organisms into the Environment: Key Issues, National Academy of Science, Washington, 1987.

[5] E. Heinzle, H. Kramer, I.J. Dunn, State analysis of fermentation using a mass spectrometer with membrane probe, Biotechnol. Bioeng. 27 (1985) 238.

[6] A.E. Humphrey, Algorithmic monitors for computer control of fermentations, in: S. Aiba (Ed.), Horizons of Biochemical Engineering, Tokyo Press, Tokyo, 1987, p. 203.

[7] J.F. Kenny, C.A White, Principles of immobilization of enzymes, Handbook of Enzyme Biotechnology, second ed., Ellis Howood, Chichester, 1985.

[8] K.B. Konstantinov, T. Yoshida, Knowledge-based control of fermentation processes, Biotechnol. Bioeng. 39 (1992) 479–486.

[9] G.A Martin, W.P. Hempfling, A method for the regulation of microbial population density during continuous culture at high growth rates, Arch. Microbiol. 107 (1976) 41–47.

[10] Organization for Economic Co-operations and Development: Recombinant DNA Safety Considerations-Safety Considerations for Industrial, Agricultural Environmental Applications of Organisms derived by Recombinant DNA Techniques, OECD, Paris, 1986.

[11] Organization for Economic Co-operations and Development: Recommendation of the Council-Concerning Safety Considerations for Applications of Recombinant DNA Organisms in Industry, Agriculture and Environment, OECD, Paris, 1986.

[12] M.J. Rolf, H.C. Lim, Adaptive on-line optimization for continuous bioreactors, Chem. Eng. Commun. 29 (1984) 229.

[13] K. Sakato, H. Tanaka, Advanced control of gluthathione fermentation process, Biotechnol. Bioeng. 40 (1992) 904.

[14] K. Sakato, H. Tanaka, S. Shibata, Y. Kuratsu, Agitation-aeration studies on coenzyme Q10 production using rhodopseudomonas spheroides, Biotechnol. Appl. Biochem. 16 (1992) 19.

[15] N.S. Wang, G.N. Stephanopoulus, Computer applications to fermentation processes in CRC critical reviews, Biotechnology 2 (1984) 1.

[16] D.W. Tempest, J.T.M. Wouters, Properties and performance of microorganisms in chemostat culture, Enzyme Microb. Technol. 3 (1981) 283.

[17] T. Yamane, S. Shimizu, Fed-batch techniques in microbial processes, Ad. Biochem. Eng 30. (1984) 148.

[18] I.A. Venetka, D.A. Walt, Fiber-optic sensor for continuous monitoring of fermentation pH, Bio/Technology 11 (1993) 726–729.

Chapter 2

Mammalian Cell Culture System

Seijo Sato

1.0 INTRODUCTION

The large-scale production of mammalian cell culture has become one of the most important technologies since the advent of genetic engineering in 1975. Interest in mammalian cell culture intensified with the development of interferons [1]. Suddenly, large amounts of human fibroblasts [2] and lymphocyte cells [3] were needed to run clinical trials and laboratory tests on the so-called "miracle drugs." The demand for large-scale reactors and systems resulted in rapid gains in the technology. At the same time, culture media, microcarriers [4] and hollow-fiber membranes [5] were also being improved.

Advances in genetic engineering, particularly in the 1996–2001 time period, generated interest in the large-scale cultivation of mammalian cells. Through genetic engineering the mass production of cells derived from proteins and peptides has real possibilities. Mammalian cells are not only useful sources of proteins and peptides for genetic engineering, but also serve as competent hosts capable of producing proteins containing sugar chains, large molecular proteins and complex proteins consisting of subunits and variegated proteins, such as monoclonal antibodies. Since monoclonal antibodies cannot be produced by bacterial hosts, mammalian cells must be used. Therefore, the demand for large scale production of high-density mammalian cells grew by large increments with the introduction of biological medicines in the 2001–2006 timeframe, and had continued to increase through the year of the third edition of this handbook (2014).

Industry responded quickly to develop new methods to meet this growing demand, as it had done in the past for industrial microbiology.

2.0 CULTURE MEDIA

Since a mammalian cell culture medium was first prepared [6] many different kinds of basal media have been established. For example, there are Eagle's minimum essential medium (MEM) [7], Duldecco's modified MEM (DME) [8], 199 medium [9], RPMI-1640 [10], L-15 [11], Hum F-10 and Hum F-12 [12], DM-160 and DM-170, etc [13]. The MIT group [14] created the High-GEM (High Growth Enhancement Medium) in which fructose replaces glucose as the energy source to achieve a 3- to 4-fold decrease in the accumulation of lactic acid. These basal media are now commercially available.

In order to generate useful proteins in very small amounts, the serum-free or chemically defined media are more useful than media containing serum. Yamane et al. [15] detected that the effective substances in albumin were oleic acid and linoleic acid; he then tried to formulate a serum-free medium containing those fatty acids as RITC-media. Barnes and Sato [16] hypothesized that the role of serum is not to supply nutrients for cells, but to supply hormones and growth factors. They then made up different kinds of serum-free media containing either peptide hormones or growth factors. The additive growth factors used for serum substituents were PDGF (platelet derived growth factor) [17], EGF (epidermal growth factor) [18], FGF (fibroblast growth factor) [19], IGF-I [20], IGF-II [21] (insulin-like growth factor I, II, or somatomedins), NGF (nerve growth factor) [22], TGF [23,24], (transforming growth factor -α, -β). IL-2 [25] or TCGF [25] (interleukin 2 or T-cell growth factor), IL-3 (interleukin-3 or muti-CSF) [26], IL-4 [27] or BCGF-1 (interleukin-4 or B-cell growth factor-1), IL-6 [28] or MGF (interleukin-6 or myeloma growth factor), M-, GM-, G-CSF [29] (macrophage-, macrophage-granulocyte-, granulocyte-colony stimulating factor), Epo (erythropoietin) [30], etc.

The way to create a serum-free culture is to adapt the cells to the serum free medium. In our laboratory, we tried to adapt a human lymphoblastoid cell line, Namalwa, from a medium containing 10% serum to serum-free. We were able to adapt Namalwa cell to a ITPSG serum-free medium which contained insulin, transferrin, sodium pyruvate, selenious acid and galactose in RPMI-1640 [31]. In the case of cell adaptation for production of autocrine growth factor, we were able to grow

Fermentation and Biochemical Engineering Handbook.

the cell line in serum- and protein-free media as well as in K562-K1(T1) which produces an autocrine growth factor, LGF-1 (leukemia derived growth factor-1) [32].

3.0 MICROCARRIER CULTURE AND GENERAL CONTROL PARAMETERS

The method for animal cell culture is chosen according to whether the cell type is anchorage dependent or independent. For anchorage dependent cells, the cells must adhere to suitable material such as a plastic or glass dish or plate. As shown in Table 2.1, several types of culture methods were developed for cell adherent substrates such as glass, plastic, ceramic and synthetic resins. Adherent reactors were made up to expand the cell adherent surfaces such as roller bottle, plastic bag, multi-dish, multi-tray, multi-plate, spiral-film, glass-beads propagator [34], Gyrogen [35] and so on. In 1967, van Welzel demonstrated the feasibility of growing cells on Sephadex or DEAE-cellulose beads kept in suspension by stirring [4]. The drawback for the anchorage-dependent cells has been overcome by the development of the microcarrier culture method. Using the microcarrier culture systems and anchorage-dependent cells, it is now possible to apply the suspension culture method on a commercial scale [5].

TABLE 2.1 Available Materials and Methods for Cell Culture

Anchored	Flat plate	Solid single trays and dishes
		Multi-plate
		Multi-tray
		Multi-dish
	Cylinder and tubes	Roller bottle
		Spiral film
		Gyrogen
	Membrane	Dialysis membrane
		Ultrafiltration membrane
		Hollow fiber
Suspended	Microcarrier	Polymer beads
		Glass beads [33]
	Microcapsule	Sodium alginate gel
	Soluble polymer	Serum (Serum albumin)
		Methylcellulose
		Pluronic F 68 (Pepol B188)
		Polyethyleneglycol
		Polyvinylpyrrolidone

The most important factor in this method is the selection of a suitable microcarrier for the cells. Microcarriers are made of materials such as dextran, polyacrylamide, polystyrene cellulose, gelatin and glass. They are coated with collagen or the negative charge of dimethylaminoethyl, diethylaminopropyl and trimethyl-2-hydroxyaminopropyl groups as shown in Table 2.2.

In scaling up batch culture systems, certain fundamental laws of microbial cell systems can be applied to mammalian cells where the suspension cultures contain the anchorage-dependent cells. This is not the case with animal cells which are sensitive to the effects of heavy metal ion concentration, shear force of impeller agitation or air sparging, and are dependent on serum or growth factors. For these reasons, the materials for construction of fermenters are 316 low carbon stainless steel, silicone and Teflon. Different agitation systems such as marine-blade impeller types, vibromixer and air-lift are recommended to mitigate the shear stress. The maximum cell growth for large-scale cell suspension using ajar fermenter is governed by several critical parameters listed in Table 2.3.

For each parameter, the pH, DO (dissolved oxygen), ORP (oxidation-reduction potential), temperature, agitation speed, culture volume and pressure can be measured with sensors located in the fermenter. The output of the individual sensors is accepted by the computer for the on-line, continuous and real-time data analysis. Information stored in the computer control system then regulates the gas flow valves and the motors to the feed pumps. A model of a computer control system is shown in Fig. 2.1. The computer control systems, like the batch systems for mammalian cell culture, seem to level out at a maximum cell density of 10^6 cells/ml. It may be impossible for the batch culture method to solve the several limiting factors (Table 2.4) that set into high density culture where the levels are less than 10^7 cells/ml [37].

4.0 PERFUSION CULTURE SYSTEMS AS A NEW HIGH DENSITY CULTURE TECHNOLOGY

In monolayer cultures, Knazeck et al. [33] have shown that an artificial capillary system can maintain high density cells using perfusion culture. The artificial capillary system is very important when cell densities approach those of in vivo values obtained via in vitro culture systems. Perfusion culture systems are continuous culture systems that are modelled after in vivo blood flow systems. In perfusion culture systems, a continuous flow of fresh medium supplies nutrients and dissolved oxygen to

TABLE 2.2 Microcarriers

TYPE	Name	Supplier	Material	Specific Gravity (SG)	Size (μ)	Specific Volume (SA) with Units (cm^3/g)
Negative charge	Biocarrir	Bio-Rad	Polyacrylamide	1.04	120–180	5000
	Superbeads	Flow Laboratorie	Dextran	–	135–205	5000–6000
	Cytodex 1	Pharmacia	Poly-acrylamide	1. 03	131–220	6000
	Cytodex 2	Pharmacia	Dextran	1.04	141–198	5500
	Dormaceil	Pfeir-Langen	Dextran	–	–	–
	DE-52	Whatman	Micro-celluiose	–	40–50 (L:80–400)	–
	DE-53	Whatman	Micro-celluiose		40–50 (L:80–400)	–
Collagen coated	Cytodex 3	Pharmacia	Dextran	1.04	133–215	4600
	Glass beads	Whatman	Glass	1.02–1.04	150–210 90–150	–
Collagen	Microsphere	Koken	Collagen	1.01–1.02	100–400	
Gelatin	Gel-Beads	KC-Bio.	Gelatine	–	235–115	3800
Tissue culture treated	Biosilon	Nunc.	Polystyrene	1.05	160–300	225
	Cytosphere	Lux	Polystyrene	1.04	160–230	250
Growth factor treated	MICA	Mulles-Lieheim	Oxiraneacryl	1.03	50–250	6300
Glass	Hollow glass	KMS Fusion	Class	1.04	100–150	385
	Bioglas	Solohill Eng.	Class	–	–	–

TABLE 2.3 Critical Parameters of General Cell Culture

1. Chemical parameters:
   - Decrease of general critical nutrients:
     - glutamine and glucose
   - Increase of inhibitory metabolites:
     - ammonium ions and lactic acid (pH control)
   - Oxidation–reduction potential:
     - gas sparging, chemically by adding cysteine, ascorbic acid and sodium thioglycollate, etc.
2. Physical parameters:
   - Decrease of dissolved oxygen:
     - aeration volume, agitation speed and oxygen contents of gas phase
   - Temperature and pressure:
     - optimum condition control.
   - Osmotic pressure:
     - control of additional ion concentration etc.
3. Physiological parameters:
   - Cell viability:
     - contamination of cytotoxic compounds
   - Cell density:
     - increase of inhibitory metabolites and chalone like substance, ratio of fresh medium and cell adhesive surface
   - Product concentration:
     - cell density and induction conditions, etc.

the cultivating cells. Inhibitory metabolites such as ammonium ions, methylglyoxal, lactate and high molecular chalone-like substances are then removed automatically. If the cells cultivated under continuous flow conditions can be held in the fermenter membranes, filters, etc., then the cells can grow into high density by the "concentrating culture." Thus, these perfusion culture systems may be able to solve some of the limiting factors associated with high density cell growth such as the mouse ascites level.

The perfusion culture systems are classified into two types by static and dynamic methods as shown in Fig. 2.2.

The most important technique for perfusion culture methods is to separate the concentrated cells and conditioned medium from the suspended culture broth. As noted above, the separation methods chiefly used are filtration with tubular and flat membranes as well as ceramic macro porous filters. These membrane reactors can be employed for both anchorage-dependent and suspension growing cells. Static maintenance type systems are commercially available for disposable reactors, and small size unit reactors from 80 ml to 1 liter are used for continuous production of monoclonal antibodies with hybridoma cells. The maintainable cell densities are about

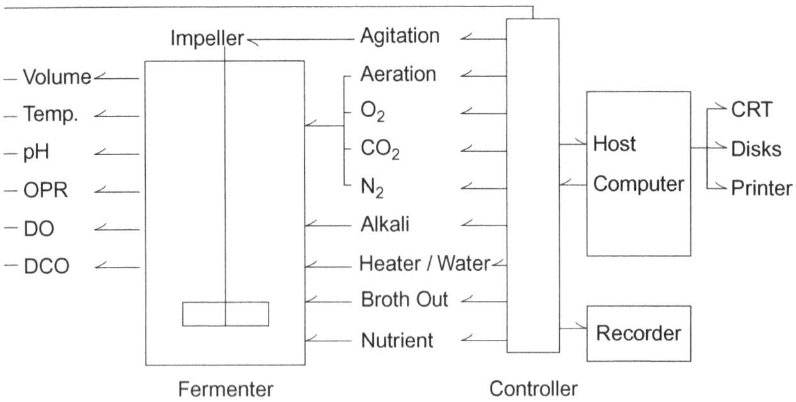

FIGURE 2.1 General control system of batch fermenter.

TABLE 2.4 Limiting Factors of High-Density Cell Cultivation [36]

Limiting factors:		
	Physical factors:	Shear force
	Physiological factors:	Accumulation of chalone-like substances
		Contact inhibition
Chemical factors:		Exhaustion of nutrient
		Decrease of dissolved oxygen
		Accumulation of ammonium ion
		Accumulation of lactate

10^7-10^8 cells/ml which is essentially mouse ascites level. However, in these systems, the cell numbers cannot be counted directly because the cells adhere to membranes or hollow fibers. Therefore, the measurement of cell density must use indirect methods. Such indirect methods include the assaying of the quantities of glucose consumption and the accumulation of lactate. The parameters of scale-up have not yet been established for these static methods.

Tolbert et al. [42] and Himmelfarb et al. [43] have obtained high-density cell growth using a rotating filter perfusion culture system. Lehmann et al. [44] used an agitator of hollow fiber unit for both perfusion and aeration. In our laboratory, we [46] constructed a membrane dialysis fermenter using a flat dialysis membrane. The small size system is well-suited for the cultivation of normal lymphocytes (Lymphokine activated killer cells) [47]. These cells are employed in adoptive immunotherapy due to their high activities for thirty or more days and their acceptance by the reactor cells.

To eliminate the use of a membrane and a filter, we have also tried to make a perfusion culture system using a sedimentation column [45].

5.0 SEDIMENTATION COLUMN PERFUSION SYSTEMS

We have developed several new perfusion systems which do not use filtration methods for cell propagation. When the flow rate of the continuous supplying medium is minimized, for example, when it is 1−3 times its working volume per day, the system has the ability to separate the suspended cells from the supernatant fluid. This is accomplished by means of an internal cell-sedimentation column in which the cells settle by gravity. The shape and length of the column are sufficient to ensure complete separation of cells from the medium. Cells remain in culture whereas the effluent medium is continuously withdrawn at a rate less than that of the cell sedimentation velocity. We experimented with several shapes for the sedimentation column and found that the cone and two jacketed types work best.

With the cone for a continuous flow rate of perfusion, the flow rate in the column is inversely proportional to the square of the radius of the cone at any given position. If the ratio of the radii of the inlet and outlet is 1:10 and the flow rate of the outlet is 1/100 of the inlet flow rate, then the separation efficiency of the supernatant fluid and suspended cells are improved. As shown in Fig. 2.3, the jacket type sedimentary system allows easy control of the temperature for separating the static supernatant from the cells. This jacket method was applied to an air-lift fermenter since it had not been done in an air-lift perfusion culture. According to Katinger et al. [48], air-lift methods have smaller shear forces than impeller type agitation. However, in perfusion culture, comparable maximum cell densities were obtained using all three types of fermenters.

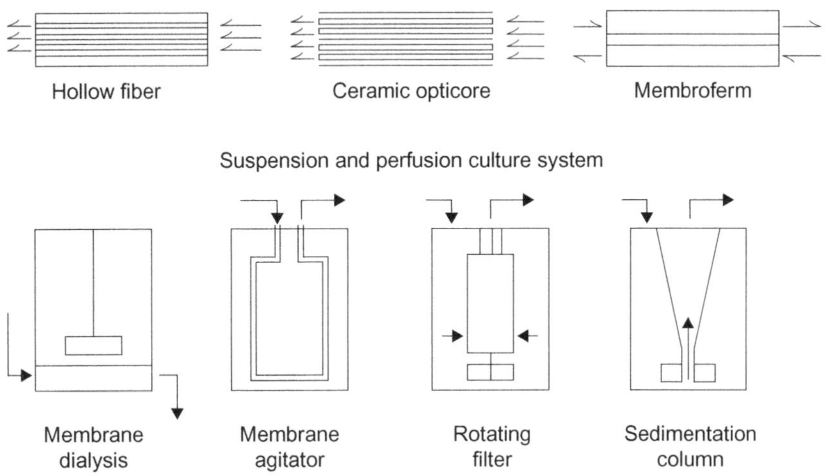

FIGURE 2.2 Static maintenance culture systems. Static maintenance type: hollow fiber [38], ceramic opticeli [39], membroferm [39], static maintenance systems [40]. Suspension culture type: membrane dialysis [41], rotating filter [42,43], membrane agitator [44], sedimentation column systems [45].

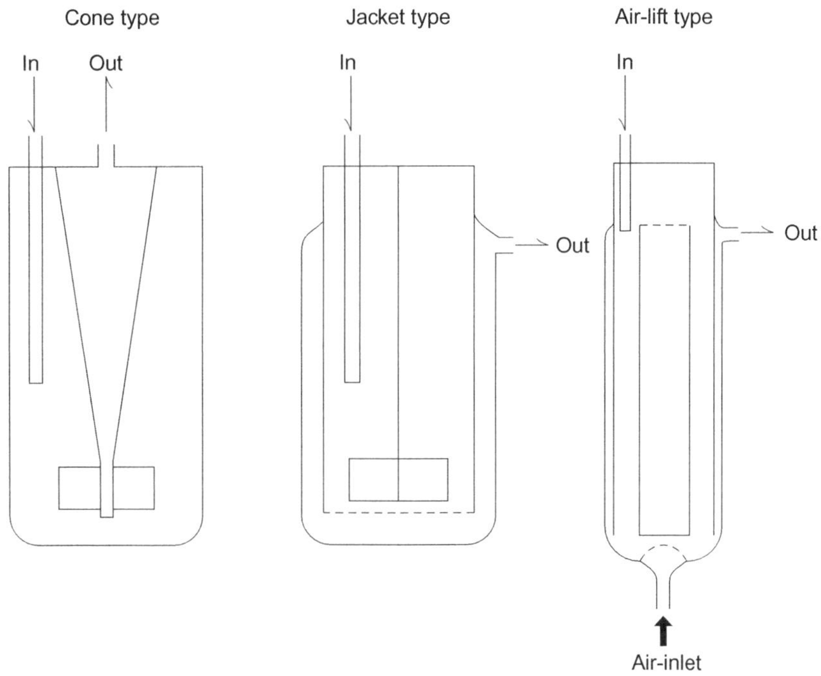

FIGURE 2.3 Sedimentation column perfusion system.

6.0 HIGH DENSITY CULTURE USING A PERFUSION CULTURE SYSTEM WITH SEDIMENTATION COLUMN

The specific standard methods of a new perfusion culture will now be described for growth and maintenance of mammalian cells in suspension cultures at high density. The bio fermenter was used for high-density culture of Namalwa cells with serum-free medium as the model. In 1980, the parent Namalwa cells was obtained from Mr. F. Klein of Frederick Cancer Research Center, Frederick, Maryland, U.S.A. In our laboratories, we were able to adapt the cells to a serum and albumin-free medium and named the cells KJM-1. ITPSG and ITPSG + F68 used a serum-free medium containing insulin, 3 g/ml; Transferrin, 5 g/ml; sodium pyruvate, 5 mM; selenious acid, 1.25×10^{-7} M; galactose, 1 mg/ml; and/or Pluronic F 68 (Pepol B-188) 0.1 mg/ml; in RPMI-1640 basal medium.

The bio fermenter BF-F500 system consisted of a 1.5 l culture vessel, 2 l medium reservoir and effluent bottle (2 l glass vessels) for fresh and expended media which were connected to the perfusion (culture) vessel by a peristaltic pump. As shown in Fig. 2.4, the fermenter systems have a conical shape sedimentation column in the center of the fermenter, and an impeller on the bottom of the sedimentation column. The Namalwa cells, KJM-1, were cultivated by continuous cultivation in the biofermenter. In Fig. 2.5, the culture has been inoculated at 1 to 2×10^6 cells/ml with an initial flow rate of approximately 10 ml/h, sufficient to support the population growth. At densities of $7 \times 10^6 - 1.5 \times 10^7$ cells/ml, we have used a nutrient flow rate of 1340 ml/h using ITPSG and ITPSG-F68 serum-free media. The flow rate of fresh media was increased stepwise from 240 to 960 ml/d in proportion to the increase in cell density. This resulted in an increase of 4 to 10 fold in cell density compared to the conventional batch culture systems. This system was then scaled up to a 45 l SUS316L unit mounted on an auto-sterilization sequence system with a medium reservoir and an effluent vessel of 90 l each (Fig. 2.6).

The system was agitated from below by a magnet impeller and was controlled and analyzed using a personal computer system. The system is shown in Fig. 2.7. In the 45 l perfusion fermenter, we were able to obtain high-density cell growth and duplicate the results of the small-scale fermenter system.

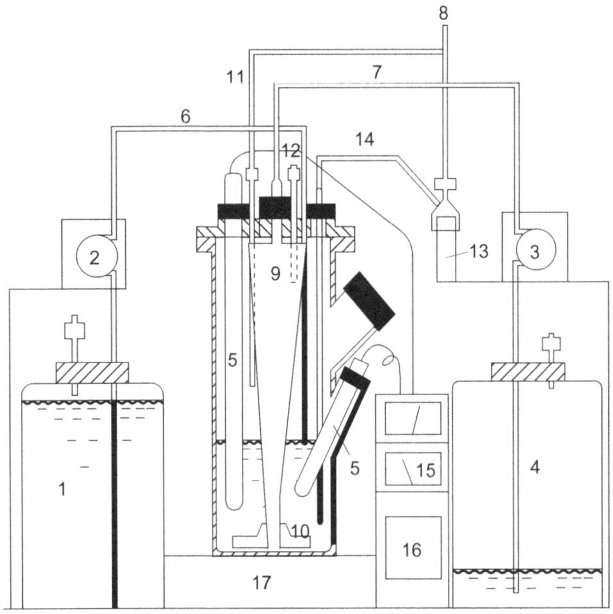

1. Fresh medium vessel
2. Feeding pump
3. Effluent pump
4. Effluent vessel
5. Sensor
6. Feeding line
7. Effluent line
8. Air line
9. Sedimentation column
10. Impeller
11. Air inlet
12. Air outlet
13. Sampling system
14. Sampling line
15. Detector
16. Recorder
17. Stirrer

FIGURE 2.4 Continuous culture system.

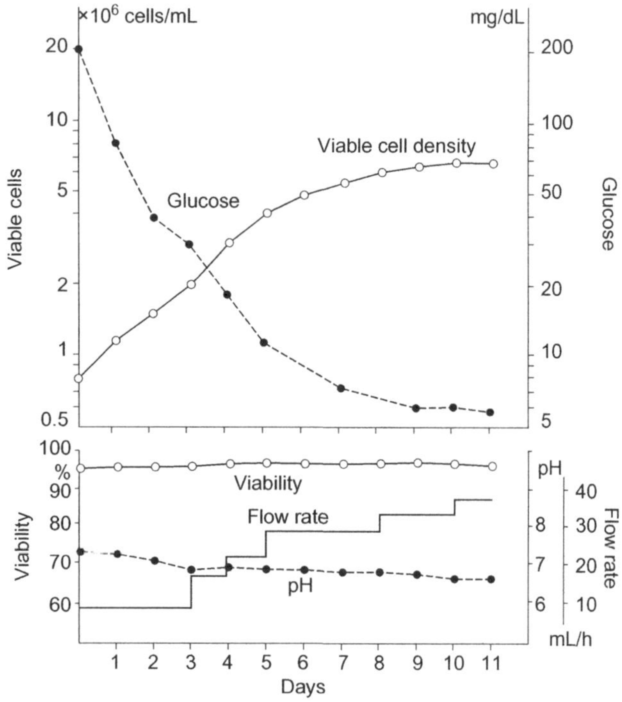

FIGURE 2.5 High density Namalwa cell culture in serum-free medium.

FIGURE 2.6 A 45L SUS 316L Unit.

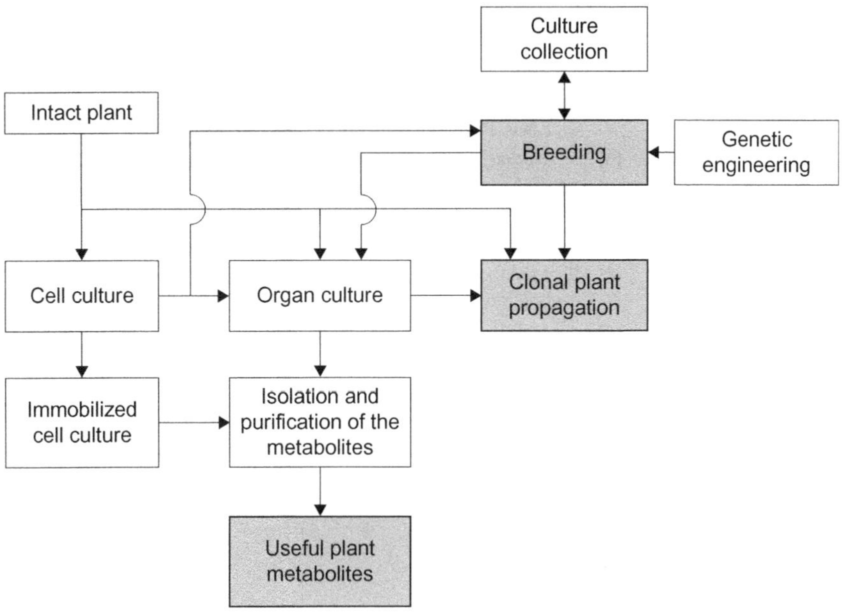

FIGURE 2.7 The area of plant cell, tissue and organ cultures.

ACKNOWLEDGMENT

This work was supported by funds obtained through the Research and Development Project of Basic Technologies for Future Industries from the Ministry of International Trade and Industry of Japan.

REFERENCES AND BIBLIOGRAPHY (SECTION 2)

[1] S. Reuveny, A Mizrahi, M. Hotler, A Freeman, Biotech. Bioeng. 25 (1983) 2969.
[2] K. Nilsson, K. Mosbach, FEBS Lett. 118 (1980) 145.
[3] L.K. Nyiri, in: R.T. Acton, J.D. Lynn (Eds.), Cell Culture and Its Application, 161, Academic Press, 1977.
[4] F. Klein, W.I. Jones, B.G. Mahlandt, R.E. Lincoln, Appl. Microbiol. 21 (1971) 265.
[5] K. Ulrich, G.E. Moore, Biotech. Bioeng. 23 (1981) 2117.
[6] H.W.D. Katinger, W. Schrirer, E. Kromer, Ger. Chem. Eng. 2 (1979) 31.
[7] S. Hosoi, H. Mioh, C. Anzai, S. Sato, N. Fujiyoshi, Cytotech. 1 (1988) 151–158.
[8] G.M. Toth, in: R.T. Actone, J.D. Lynn (Eds.), Cell Culture and Its Application, Academic Press Inc., 1977, p. 617.
[9] Y. Ishikawa, Bio. Ind. 2 (1985) 553.
[10] B.A. Blazar, L.M. Scttor, M. Strome, Cancer Res. 43 (1983) 4562.
[11] R.A Knazek, P.M. Gullino, P.O. Kohler, R.L. Dedrick, Science 178 (1972) 65.
[12] S. Sato, Bio. Ind. 1 (9) (1984) 20.
[13] K. Ku, M.J. Kuo, J. Delente, B.S. Wildi, J. Feder, Biotech. Bioeng. 23 (1981) 79.
[14] EA Bogner, G.G. Pugh, B.K. Lydersen, J. Tissue Culture Method 8 (1983) 147.
[15] N. Fujiyoshi, Hakko to Kogyo 45 (3) (1987) 198.
[16] W.I. Tolbert, in: J. Feder, W.I. Tolbert (Eds.), Large Scale Mammalian Cell Culture, Academic Press Inc., 1985, p. 97.
[17] S. Sato, Cell Technol. Suppl. 7 (1988) 35–42.
[18] W.I. Tolbert, J. Feder, R.C. Kimes, In Vitro 17 (1981) 885.
[19] P. Himmelfarb, P.S. Thayer, H.E. Martin, Science 164 (1969) 555–557.
[20] J. Lehmann, G.W. Piehl, R.S. Braunschweig, Intl. Magazine Biotech, Special Publication of BTF, 1985.
[21] Y. Kohgo, S. Kakamaki, Y. Kanisawa, S. Nojiri, T. Ueno, Y. Itoh, et al., Cytotechnology 2 (1989) 49.
[22] S.A Rosenberg, J. Natl. Cancer Inst. 75 (1985) 595.
[23] S. Sato, K. Kawamura, N. Fujiyoshi, J. Tissue Culture Method 8 (1983) 167.
[24] EA Havell, J. Vilcek, Antimicrob. Agents Chemother. 2 (1972) 476.
[25] E. Knight Jr., Proc. Natl. Acad. Sci., USA 73 (1976) 520.
[26] H. Strander, K.E Mogensen, K. Cantell, J. Clin. Microbiol. 1 (1975) 116.
[27] A.L. van Welzel, Nature 216 (1967) 64.
[28] R.A Knazek, P.M. Gullino, P.O. Kohler, R.L. Dedrick, Science 178 (1972) 65.
[29] V.J. Evans, J.C. Bryant, M.C. Fioramonti, W.T. McQuilkin, K.K. Sanford, W.R. Earle, Cancer Res. 16 (1956) 77.
[30] H. Eagle, Science 130 (1959) 432.
[31] R. Duldecco, G. Freeman, Virology 8 (1959) 396.
[32] J.F. Morgan, H.J. Morton, R.C. Perker, Proc. Soc. Exp. Biol. Med. 73 (1950) 1.
[33] T. Takaoka, K. Katsuta, Exp. Cell Res. 67 (1971) 295.
[34] G.E. Moore, R.E. Grener, H.A. Franlin, JAMA 199 (1967) 519.
[35] A Leibovitz, Am. J. Hyg. 78 (1963) 173.
[36] T. Imamura, C.L. Crespi, W.G. Thilly, H. Brunengraber, Anal. Biochem. 124 (1982) 353.
[37] R.G. Ham, Exp. Cell Res. 29 (1963) 515.

[38] I. Yamane, O. Murakami, M. Kato, Proc. Soc. Exp. Biol. Med. 149 (1975) 439.
[39] D. Barnes, G. Sato, Anal. Biochem. 102 (1980) 255.
[40] S. Chohen, J.M. Taylor, Recent Progr. Hormone Res. 30 (1974) 535.
[41] D. Gospodarowitz, J. Biol. Chem. 250 (1975) 2515.
[42] E. Rinderknecht, R.E. Humbel, Proc. Natl. Acad. Sci. USA 73 (1976) 2365.
[43] T.L. Blundell, R.E. Humbel, Nature 287 (1980) 781.
[44] H. Thoene, Y.A Brade, Physiol. Rev. 60 (1980) 1284.
[45] J.P. Tam, H. Marquardt, D.F. Rosenberger, T.W. Wong, G.J. Todaro, Nature 309 (1984) 376.
[46] A.B. Roberts, M.A Anzano, L.C. Lamb, J.M. Smith, M.B Sporn, Proc. Natl. Acad. Sci. USA 78 (1981) 5339.
[47] S. Gillis, J. Watson, J. Exp. Med. 152 (1980) 1709.
[48] J.C. Lee, J.N. Ihle, Nature (London) 209 (1981) 407.

FURTHER READING

[1] R. Ross, A Vogel, Cell 14 (1978) 203.
[2] A. O'Garra, D.J. Warren, M. Holman, A.N. Popham, C.L. Sanderson, G. Klaus, Proc. Natl. Acad. Sci., USA 83 (1986) 5228.
[3] T. Hirano, K. Yasukawa, H. Harada, T. Taga, Y. Watanabe, T. Mastuda, et al., Nature 324 (1986) 73.
[4] T. Okabe, H. Nomura, N. Sato, N. Ohsawa, J. Cell. Physiol. 110 (1982) 43.
[5] T. Miyake, K. Charles, H. Kung, E. Goldwasser, J. Biol. Chem. 252 (1977) 558.
[6] S. Sato, K. Kawamura, N. Fujiyoshi, Tissue Culture 9 (1983) 286.
[7] A Mihara, K. Fujiwara, S. Sato, T. Okabe, N. Fujiyishi, In Vitro 23 (1987) 317.
[8] M.W. Gladeen, Trends Biotechnol. 1 (1983) 102.
[9] H.G. Girard, M. Stutch, H. Erdom, I. Gurhan, Develop. Biol. Stand. 42 (1979) 127.

Chapter 3

Bioreactors for Plant Cell Tissue and Organ Cultures

Shinsaku Takayama

1.0 BACKGROUND OF THE TECHNIQUE—HISTORICAL OVERVIEW

Haberlandt [1] first reported plant cell, tissue, and organ cultures in 1902. He separated plant tissues and attempted to grow them in a simple nutrient medium. He was able to maintain these cells in a culture medium for 20 to 27 days. Although these cells increased eleven-fold in the best case, no cell division was observed. Gautheret [2] was the first to succeed in multiplying the cells from the culture in 1934. He used the cambial tissues of *Acer pseudoplatanus, Salix capraea, Sambucus nigra*. After 15 to 18 months in subculture, cell activity ceased. He reasoned that this inactiveness was due to the lack of essential substances for cell division. He suspected that auxin may have been one of the deficient substances. This compound was first reported in 1928 and was isolated by Kogel in the 1930's. Addition of auxin to the medium prompted plant cell growth. This finding was reported almost simultaneously by Gauthret [3] and White [4] in 1939. Plant cell tissue and organ culture techniques rapidly developed, and in the mid-1950's another important phytohormone, cytokinins, had been discovered (Miller, Skoog, Okumura, Von Saltza and Strong 1955) [5]. By 1962 Murashige and Skoog [6] had reported a completely defined medium which allowed the culture of most plant cells. Their medium has now become the mostly widely used medium in laboratories around the world.

After these initial discoveries and some significant improvements in media, scientific research on the cultivation of plant cell, tissue, and organs shifted to the area of basic physiological research. Industrial applications were also sought in the production of secondary metabolites, clonal plants, and the improvement of various plant tissues.

Plant cell, tissue, and organ culture can be performed by either solid or liquid culture methods, however, in order to scale up the culture to the level of industrial processes, the liquid culture method must be employed.

By the time of the second edition of this handbook (1996), pilot bioreactors as large as 20 kl had been constructed in the research laboratories of Japan Tobacco and Salt Co. and in those of Nitto Denko Co. Solid culture methods were used in large scale pilot experiments for the production of tobacco cells, and liquid culture methods were used in the production of Panax ginseng cells. An outstanding example of cell suspension culture in a pilot scale bioreactor (750 1) was the production of shikonins by Mitui Petrochemical Industries. In all these examples, various technologies have been used to improve the productivity of the metabolites. The technologies include: *(i)* selection of a high yielding cell strain, *(ii)* screening of the optimum culture condition for metabolite production, *(iii)* addition of precursor metabolites, *(iv)* immobilized cell culture, and *(v)* differentiated tissue and/or organ culture. The productivity of various metabolites such as ginsenoside, anthraquinones, rosmalinic acid, shikonins, ubiquinones, glutathione, tripdiolide, etc., reached or exceeded the amount produced by intact plants. To date, the production costs remain very high which is why most of the metabolites are still not produced on an industrial or pilot plant scale. Development of large-scale industrial culture systems and techniques for plant cell, tissue, and organs, and the selection of the target metabolites are the chief prerequisites for the establishment of the industrial production of plant metabolites.

2.0 MEDIA FORMULATIONS

The formulation of the medium for plant cell, tissue, and organ culture depend primarily on nutritional requirements. Intact plants grow photoautotrophically in the soil, (i.e., they use CO_2 as the principal carbon source and synthesize sugars by photosynthesis). In the case of aseptic cultures however, establishment of an autotrophic culture is not achieved so that heterotrophic or mixotrophic growth becomes the distinguishing characteristic.

Therefore, such cultures require the addition of carbon as an energy source. Given this fact, the culture medium must be formulated as a chemically defined mixture of mineral salts (macro- and microelements) in combination with a carbon source (usually sucrose). In addition to these constituents, organic constituents such as vitamins, amino acids, sugar alcohols, and plant growth regulators are usually added to the medium. Media commonly used are listed in Table 3.1.

3.0 GENERAL APPLICATIONS

The most important fields of research for industrial applications, plant cell tissue and organ cultures are clonal propagation and secondary metabolite production. Plants cultivated in vitro have great changes in their morphological features, from cell tissue to differentiated embryo, roots, shoots or plantlets.

TABLE 3.1 Formulations of Most Frequently Used Plant Tissue Culture Media

Ingredients (mg ℓ^{-1})	MS	B5	White	Heller
$(NH_4)_2SO_4$			134	
$(NH_4)NO_3$	1650			
$NaNO_3$				600
KNO_3	1900	2500	80	
$Ca(NO_3)_2$			300	
$CaCl_3 \cdot 2H_2O$	440	150		75
$MgSC_4 \cdot 7H_2O$	370	250	720	250
Na_2SO_4			200	
KH_2PO_4	170			125
$NaH_2PO_4 \cdot H_2O$		150	16.5	
KCl			65	750
$FeSC_4 \cdot 7H_2O$	27.8	27.8		
Na_2EDTA	37.3	37.3		
$FeCl_3 \cdot 6H_2O$				1.0
$Fe_2(SO_4)_3$			2.5	
$MnSC_4 \cdot 4H_2O$	22.3		7	0.01
$MnSC_4 \cdot H_2O$		10		
$ZnSC_4 \cdot 7H_2O$	8.6	2	3	1
H_3BO_3	6.2	3	1.5	1
KI	0.83	0.75	0.75	0.01
$Na_2MoO_4 \cdot 2H_2O$	0.25	0.25		
$CuSO_4 \cdot 5H_2O$	0.025	0.025		0.03
$CoCl_2 \cdot 6H_2O$	0.025	0.025		
$NiCl_2 \cdot 6H_2O$				0.03
$AlCl_3$				0.03
Myo-inositol	100	100		
Nicotinic acid	0.5	1.0	0.5	
Pyridoxine · HCl	0.5	1.0	0.1	
Thiamine · HCl	0.1	10.0	0.1	1.0
Glycine	2.0		3.0	
Ca D-pantothenic acid			1.0	
Sucrose	30,000	20,000	20,000	20,000
Kinetin	0.04–10	0.1		
2,4-D		0.1–1.0	6.0	
IAA	1.0–30			
pH	5.7–5.8	5.5	5.5	

Applications to Secondary Metabolite Production. Plant tissue culture is a potential method for producing secondary metabolites. Both shikonins (Fujita and Tabata 1987) [7] and ginseng saponins (Ushiyama et al., 1986) [8] have now been produced on a large scale by this method. However, the important secondary metabolites are usually produced by callus or cell suspension culture techniques. The amounts of some metabolites in the cell have exceeded the amounts of metabolites in the cells of the original plants grown in the soil. So it is expected that cell culturing may be applicable to industrial processes for the production of useful secondary metabolites. It is common knowledge that when a cell culture is initiated and then transferred, the productivity of the metabolite decreases (Kurz and Constabel, 1979) [9]. Once productivity decreases, it becomes very difficult to arrest or reverse the decrease. In order to avoid this phenomenon, many cell strains were screened to select those which would maintain metabolite productivity. Some metabolites such as anthocyanins, shikonins, vinca alkaloids, and ubiquinones have been reported to have increased their productivity significantly. Deus-Neumann and Zenk (1984) [10] have checked the stability of the productivity of the selected cell strains reported in the literature and noted that the production of some metabolites such as anthraquinone *(Morinda citrofolia)*, rosmalinic acid *(Colius blumei)*, visnagin *(Ammi visnaga)*, diosgenin *(Dioscorea deltoidea)*, etc., were stable after several subcultures, but some metabolites such as nicotine *(Nicotiana rustica)*, shikonin *(Lithospermum erythrorhizon)*, ajmalicine *(Catharanthus roseus)*, rotenoids *(Derris eliptica)*, anthocyan *(Daucus carota)*, etc., were shown to be unstable after several subcultures.

Clonal Plant Propagation. Plants are propagated clonally from vegetative tissue or organs via bypass sex. Conventional clonal propagation can be performed by leaf or stem cutting and layering or dividing of the plants, however the efficiency is very low. Recently, many plants were propagated efficiently through tissue culture. This technique was first reported in 1960 by G. Morel [11] for the propagation of orchids and since then, many plants have been propagated by tissue culture. Today there are many commercial tissue culture nurseries throughout the world. Most of these tissue culture nurseries are using flasks or bottles containing agar medium for commercial propagation, but the efficiency is also low. In order to improve the efficiency, use of a bioreactor is desirable. Using a small bioreactor (4 to 10 liters), the author has produced over 4,000 to 10,000 plantlets within 1 to 2 months. The bioreactor system allows the induction of somatic embryos from vegetative cells which then leads to the production of artificial seeds (Redenbaugh et al., 1987) [12].

4.0 BIOREACTORS—HARDWARE CONFIGURATION

The configuration of bioreactors most frequently used for plant cell, tissue, and organ cultures is fundamentally the same as that used for microbial or animal cell cultures. However, in plants, the cells, tissues, and organs are all susceptible to mechanical stresses by medium aeration and agitation. At times, the production of both cells mass and metabolites is repressed severely and the bioreactor must therefore have the characteristics of low shear stresses and efficient oxygen supply. For these reasons, different bioreactors (Fig. 3.1) have been investigated in order to select the most suitable design. Wagner and Vogelmann (1977) [13] have studied the comparison of different types of bioreactors for the yield and productivity of cell mass and anthraquinone (Fig. 3.2). Among different types of bioreactors, the yield of anthraquinones in the air-lift bioreactor was about double that found in those bioreactors with flat blade turbine impellers, perforated disk impellers, or draft tube bioreactors with Kaplan turbine impellers. It was also about 30% higher than that of a shake flask culture. Thus, the configuration of the bioreactor is very important and development efforts are underway for both bench scale and pilot scale bioreactors.

Aeration-Agitation Bioreactor. This type of bioreactor (Fig. 3.3) is popular and is fundamentally the same as that used with microbial cultures. For small scale experiments, the aeration-agitation type bioreactors is widely used. However, when the culture volume is increased, many problems arise. The following are some of the scale-up problems in large aeration-agitation bioreactors: *(i)* increasing mechanical stresses by impeller agitation and *(ii)* increasing foaming and adhesion of cells on the inner surface of the bioreactor. Despite these problems, a large scale pilot bioreactor (volume 20 kl) was

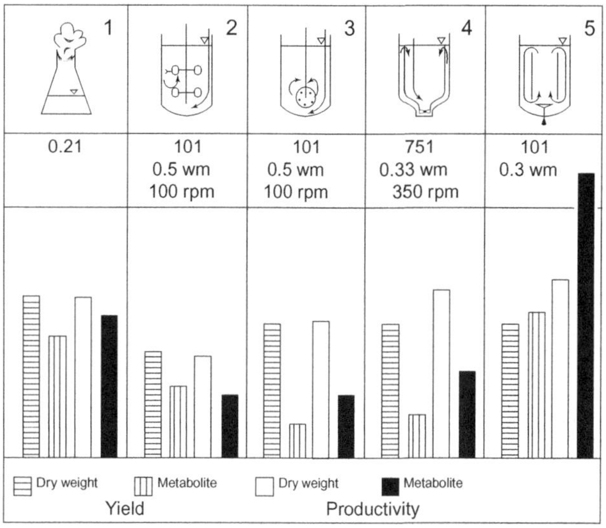

FIGURE 3.2 Comparison of yield and productivity for cell mass and anthraquinones in various bioreactor systems. *(1)* Shake Flask. *(2)* Flat Blade Turbine. *(3)* Perfolated Disk Impeller. *(4)* Draft Tube Bioreactor with Kaplan Turbine. *(5)* Air-lift Bioreactor.

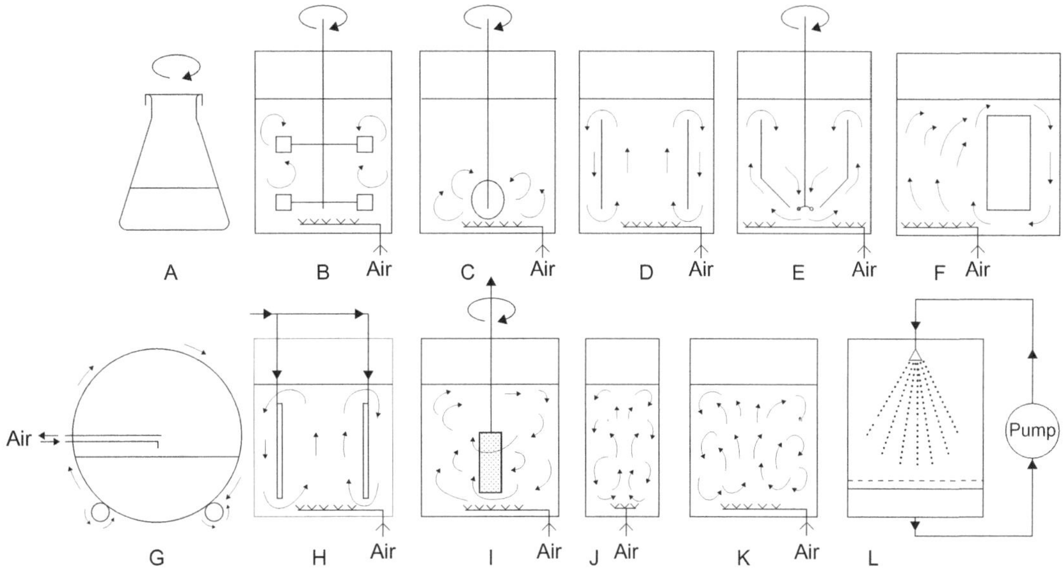

FIGURE 3.1 Different types of bioreactors for plant cells, tissues and organs. (A) Shake Flask. (B) Aeration-Agitation. (C) Percolated Impeller. (D) Draught Tube Air-lift. (E) Draft Tube with Kaplan Turbine. (F) Air-lift loop. (G) Rotating Drum. (H) Light Emitting Draught Tube. (I) Spin Filter. (J) Bubble Column. (K) Aeration. (L) Gaseous Phase.

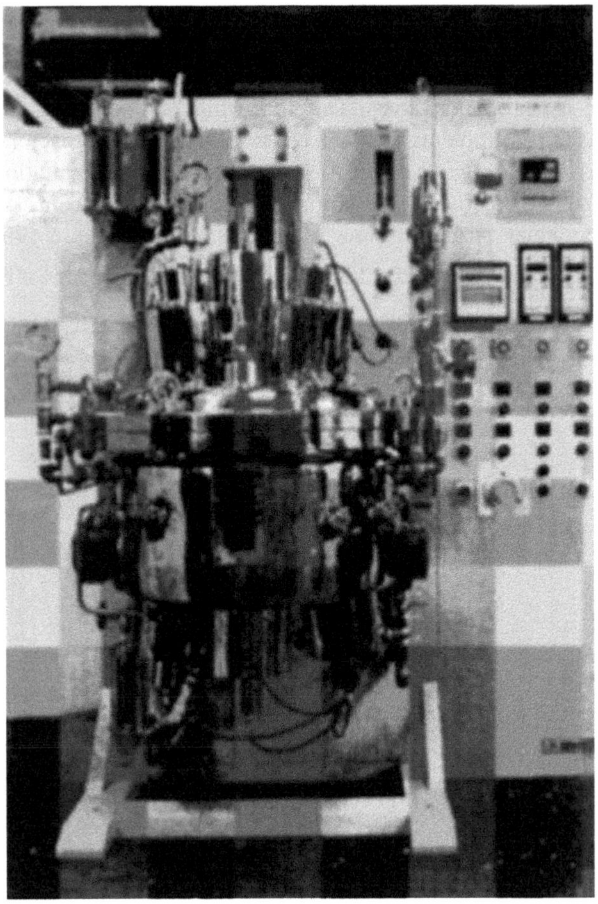

FIGURE 3.3 Ninety-five liter automated bioreactor for plant cell, tissue and organ cultures. *(Photo courtesy of K. F. Engineering Co., Ltd., Tokyo).*

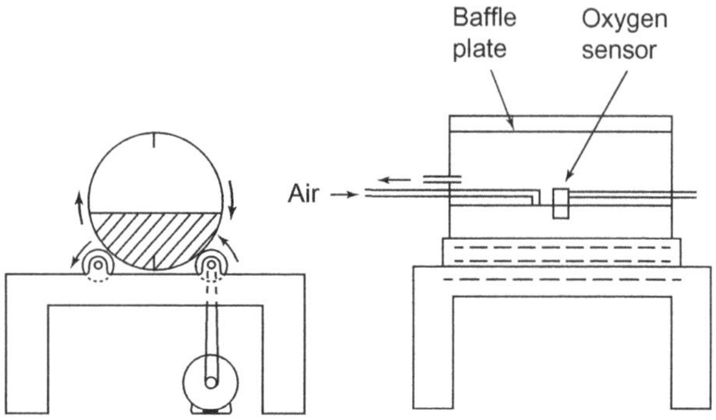

FIGURE 3.4 Schematic diagram of the rotating drum bioreactor (Tanaka, H., et al., 1983).

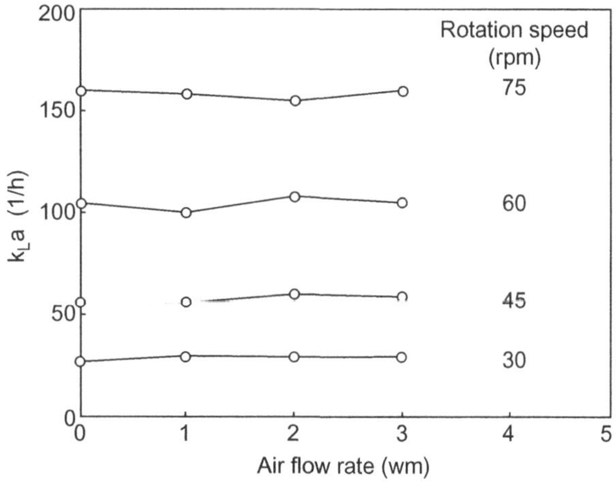

FIGURE 3.5 Effect of the airflow rate on k_La in rotating drum fermenter (Tanaka, H., et al., 1983).

constructed. It successfully produced both cell mass and metabolites. This bioreactor is therefore the most important type for bioreactor systems.

Air Driven Bioreactors. The simplest design is the air-driven bioreactor equipped with sparger at the bottom of the vessel. It is widely used for plant cell, tissue, and organ cultures. In cases where the cells grow rapidly and the cell mass occupies 40–60% of the reactor volume, the flow characteristics become non-Newtonian and the culture medium can no longer be agitated by simple aeration.

Rotating Drum Bioreactor. The rotating drum bioreactor (Fig. 3.4) turns on rollers and the oxygen supply mechanism is entirely different from either the mechanically agitated or the air-lift bioreactor. Tanaka et al. (1983) [14], reported that the oxygen transfer coefficient is affected by a change of airflow rate under all rotational speeds (Fig. 3.5). This characteristic is suitable not only for the growth of plant cell, tissue, and organs but also for the production of metabolites under high viscosity and high density cultures. It is superior to the cultures using either mechanically agitated or air-lift bioreactors since the cultures are supplied ample oxygen and are only weakly stressed. Recently a 1 kl bioreactor of this type was constructed and used for a pilot scale experiment (Tanaka, 1987) [15].

Spin Filter Bioreactor. This type of bioreactor (Styer, 1985) [16] is equipped with a filter driven by a magnetic coupling in the stir plate (Fig. 3.6). This spinning filter operates as a medium agitator without generating shear stress and also serves as an excellent filter for the removal of the medium from the bioreactor without the cells plugging it. The spin filter bioreactor will be most suitable for the continuous culture of plant cells. When a conventional bioreactor was used and the feeding rate of the medium was increased, the cell density was decreased because of washout. However, when a spin filter bioreactor was used, the cell density was maintained constant and half of the spent medium was effectively removed through the spin filter.

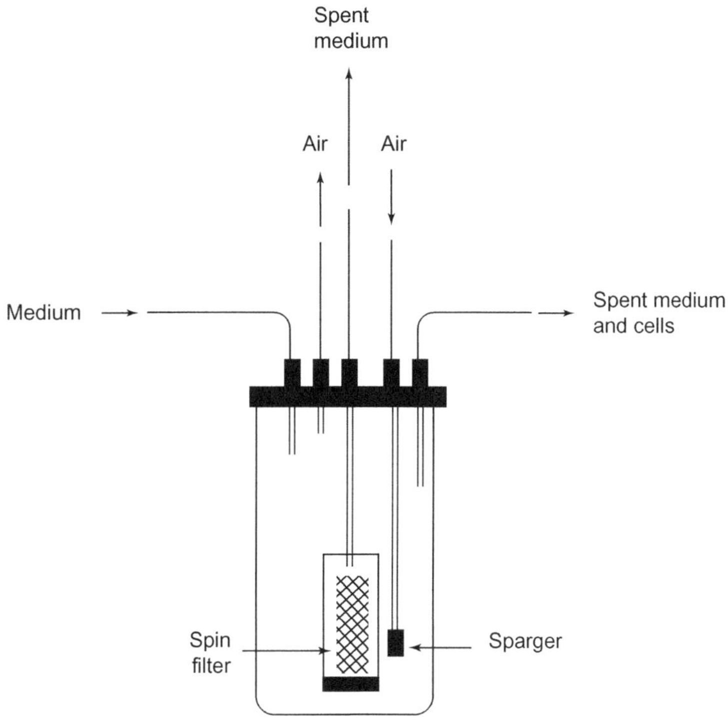

FIGURE 3.6 The spin filter bioreactor (Styer, 1985).

Gaseous Phase Bioreactor. As shown in Fig. 3.7, this type of bioreactor is equipped with filters on which the culture is supported and with a shower nozzle for spraying on the medium (Ushiyama et al., 1984 [17]; Ushiyama, 1988) [18]. Seed cultures are inoculated on the filters and the medium is supplied to the culture by spraying from a shower nozzle. The drained medium is collected on the bottom of the bioreactor. This type of bioreactor is excellent for plant cell, tissue, and organ cultures because there is no mechanical agitation (e.g., driven impeller, aerator) and, therefore, the growth rate and the secondary metabolite production are enhanced.

Light Introducing Bioreactor. Plants are susceptible to light irradiation and as a consequence various metabolic and/or physiological changes are generated. Some important reactions are: *(i)* photosynthesis, *(ii)* activation of specific enzymes such as phenylalanine ammonia lyase (PAL) and to induce the production of flavonoids or anthodyanins, *(iii)* photomorphogenesis such as development of leaves. For these reactions, the introduction of light into the bioreactor is required. Inoue (1984) [19] reported a bioreactor equipped with transparent pipes. The light was emitted from the surface of the pipe into the bioreactor. Ikeda (1985) [20] reported an air-lift bioreactor equipped with a photo introducing draft tube

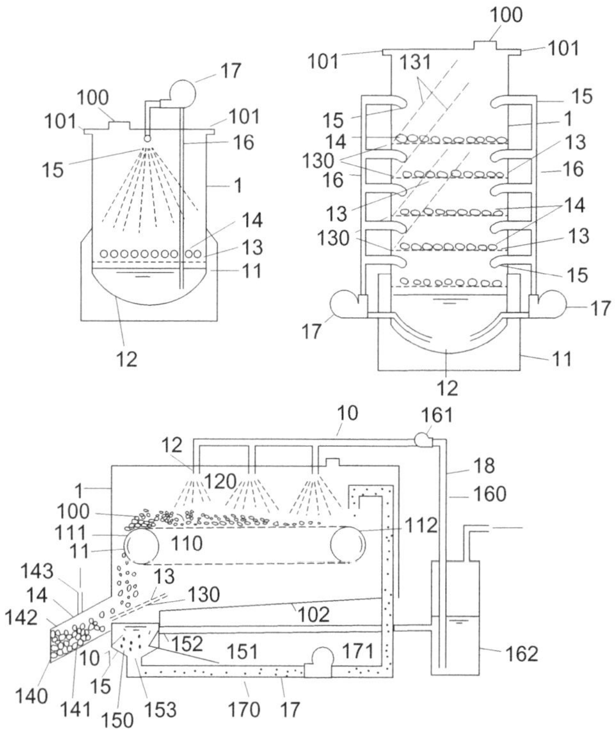

FIGURE 3.7 Gas phase bioreactor (Ushiyama, et al., 1984).

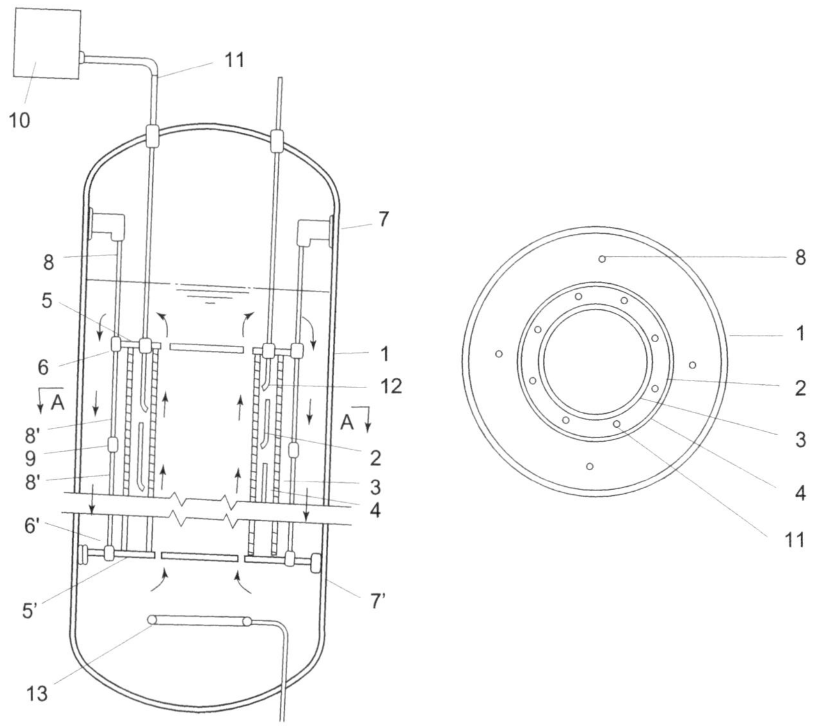

FIGURE 3.8 Photo introducing bioreactor (Jkeda, 1985).

(Fig. 3.8). The draft tube was constructed as an airtight tube which consisted of a transparent inner and outer tube. Within the center of the draft tube was a light introducing optical fiber. The light source was a sunlight collector system that operated automatically by computer control and the collected light was introduced into the bioreactor through the optical fibers. Introduction of light into the bioreactor will become an important technique for the production of specific plant metabolites.

Gas Permeable Membrane Aerator Bioreactor. This type of bioreactor has not yet been fully developed. Nevertheless, some information is available. For example, one bioreactor is equipped with an aerator composed of fine tubes made of polycarbonate, polypropylene, silicone gum, etc. This type of bioreactor should be valuable for immobilized plant cell cultures.

5.0 BIOREACTOR SIZE

For industrial production of secondary metabolites, large scale bioreactor systems (sometime over 100 kl) will be required. The 75 kl and 20 kl bioreactor systems used for pilot scale experimentation were at the time of the second edition of this handbook (1996) the largest in the world. They were at the DIVERSA Gesellschaft für Bio- und Verfahlenstechnik mbH in Germany and Nitto Denko Co., Ltd. in Japan. When there is a limited demand for a particular metabolite (e.g., a pigment), the production of commercial quantities can be done in the pilot scale bioreactor. Shikonin is produced this way. In 1983, Mitui Petrochemical Industries became the first to commercially produce a plant metabolite by using a 750 1 bioreactor. For routine experiments, the smaller bioreactors of vessel volume 1 to 100 liters are more widely used. Small bioreactors with volumes from 1 to 20 liters are used commercially for the production of clonal plants. These small bioreactors are valuable for the rapid propagation of large numbers of clonal plantlets. Through asexual embryogenesis, 10,000 to 1,000,000 embryos can be produced per liter and these embryos are then grown to plantlets. Using 2 to 10 1 bioreactors, it is also possible to produce 5,000 to 10,000 plantlets from plant tissue, which can then be transplanted directly into soil.

6.0 CULTURE PERIOD

The growth of plant cells, tissues, and organs is much slower than microbial organisms. The most rapid growth cell line reported in scientific journals is the bright yellow *Nicotiana tabacum cv.* (Noguchi et al., 1987) [21]. The doubling time of this cell strain was 15 h, and the duration to maximum growth was 80 h (3.3 days) when cultured in a 20 kl pilot scale bioreactor. In general, the growth of the cells of herbaceous annual plants is rapid and their doubling time is usually about 1 to 3 days (duration to maximum growth was 10 to 20 days), and that of woody plants or differentiated organs is slow (doubling time is about 2 to 10 days and the culture period is about 20 to 100 days).

7.0 AERATION AND AGITATION

The oxygen requirement of plant cells is quite low compared to microorganisms. Kato et al. (1975) [22] have examined the effect of k_La on biomass production (Fig. 3.9a). They observed that the volumetric oxygen transfer coefficient, k_La was constant after 10 h and the final biomass concentration became constant at 0.43 g cell dry weight/g sucrose. When k_La was set under 10 h, cell yield became dependent on k_La values. The effect of agitation speed on final cell mass concentration was also analyzed by Kato et al. (1975) [22] using a 15 l bioreactor (Fig. 3.9b). At lower agitation speeds (less than 150 rpm), cell mass concentration became constant. However, when the agitation speed exceeded 150 rpm, the cultures became bulky and started to foam profusely. An agitation speed of either 50 or 100 rpm seemed to be optimal for production of cell mass and also for avoiding the culture problems.

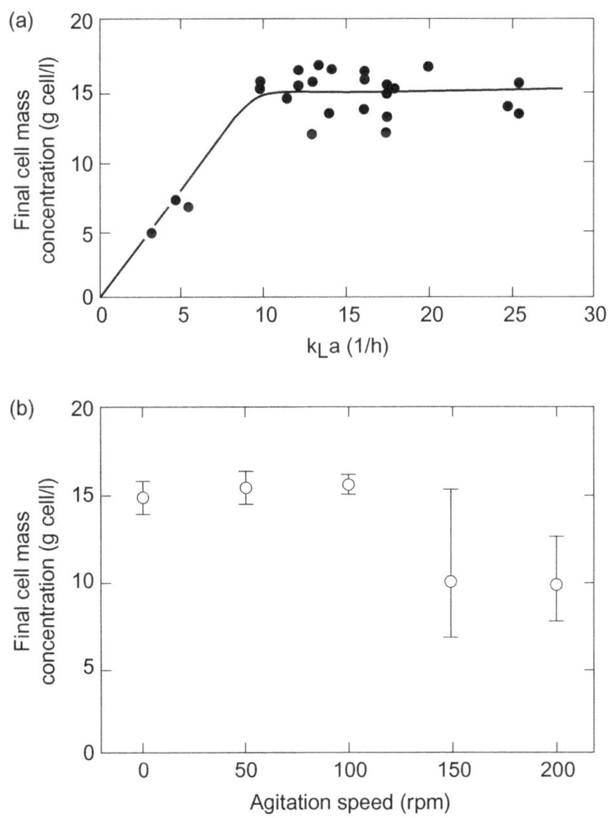

FIGURE 3.9 Effect of initial oxygen transfer coefficient and agitation by turbine impeller on cell growth (Kato, 1975).

8.0 MICROBIAL CONTAMINATION

According to Manfredini et al. (1982) [23], the most frequent factors causing microbial contamination include: *(i)* construction materials, *(ii)* seals and valves, *(iii)* complexity of the plant, *(iv)* operator error, *(v)* instrumentation failure, *(vi)* process air, *(vii)* transfer and feed lines, *(viii)* contamination from vegetative preculture, *(ix)* critical medium composition, *(x)* inadequate procedures. As the culture periods of plant cell, tissues, and organs are usually quite long, (particularly for continuous cultures), special designs and operations are necessary to avoid microbial contamination. For example, Hashimoto et al. (1982) [24], reduced contamination in their 20 kl bioreactor by using three air filters in a series; the third filter was a membrane of uniform pore size of 0.4 μm. The design specified aseptic seals in the agitator shaft and exit pipelines for sterilized air after steam sterilization.

9.0 CHARACTERISTICS

The special characteristics of plant cells that tend to hamper large-scale cultivation of the cells are described below.

Bubbling and Adhesion of Cells to the Inner Surface of the Bioreactor. Plant cell cultivation is usually performed by bubbling and agitation which cause foaming and adhesion of the cells to the surface of a bioreactor. Because of this phenomenon, cell growth is inhibited. The authors (Takayama et al., 1977) [25] have examined the possible causes and concluded that the adhesion of cells appeared to be the result of gel formation from pectin and calcium. By reducing the concentration of $CaCl_2 \cdot 2H_2O$ in the medium, the foaming and the number of cells that adhered to the walls was decreased markedly. The cells became easily removable from the inside wall of the fermenter and were returned to the medium. Cell destruction was measured by A660 values which also depend on the $CaCl_2 \cdot 2H_2O$ levels. A lower level of $CaCl_2 \cdot 2H_2O$ in the medium markedly inhibits cell destruction. These observations are particularly pertinent when large-scale cultivation is being considered.

Cell Morphology and Specific Gravity. According to Tanaka (1982) [26], plant cells have a tendency to grow in aggregates of different sizes. The size distribution of cell aggregates is different from one plant species to another (Tanaka, 1982) [27]. Specific gravity of these cells ranges from 1.002 to 1.028. If the diameter of the cell aggregate is less than 1 to 2 mm, the cells can be suspended and do not sink to the bottom of the bioreactor (Tanaka, 1982) [26], but, when the specific gravity of the cell is greater than 1.03, the diameter of the aggregate becomes 0.5 to 1.0 cm and the cells sink to the bottom of

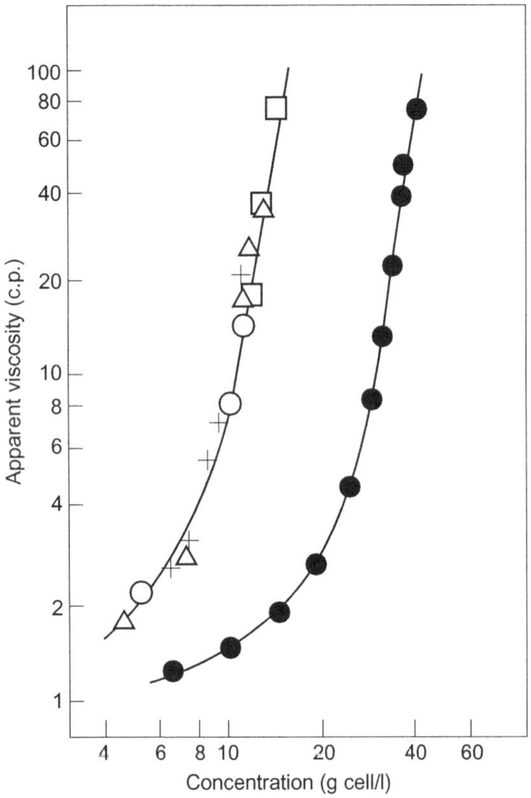

FIGURE 3.10 Relationship between apparent viscosity and concentration of cells and pseudocells in culture media (Tanaka, 1982). (□) *C. Roseus*, (○) *C. Tricupsidata B.*, (Δ) *N. tabacum L.*, (+) granulated sugar.

the bioreactor and cannot be suspended (Fig. 3.10). When the agitation is increased, the size of the cell aggregate becomes smaller (Tanaka, 1981) [28], but the growth of the cell is repressed. In order to separate the cells from the aggregate, the amount of calcium is decreased to suppress the gel formation of pectin which plays an important role in the cell, e.g., cementing plant cells, but has little effect on cell separation (Takayama, 1977) [25].

Viscosity, Fluidity, and Oxygen Supply. When plant cells grow well, they can occupy 40 to 60% of the whole culture volume, and the apparent viscosity becomes very high. Tanaka (1982) [26] examined the relationship between apparent viscosity and concentration of solids in suspension, and concluded that when the cell density exceeds 10 g/l, the slope of the apparent viscosity increases rapidly, and when cell density reaches 30 g/l, the culture medium becomes difficult to agitate and supply with oxygen.

10.0 MANIPULATION

Large Scale Batch Culture. Batch culture systems are in use worldwide and many experimental results have been reported using 10 1 to 20 kl bioreactors. Noguchi et al. (1987) [21] have examined the growth of tobacco BY-2 cells using a 20 kl aeration-agitation bioreactor with 15 kl medium. The medium used was Murashige and Skoog's inorganic nutrients with three times the normal amount of phosphate and 3% sucrose, incubated at 28°C and aerated at 0.3 vvm. The results revealed that the highest growth rate was observed from the incubation time of 45 to 70 hours with a doubling time of about 15 hours, which was almost the same as the growth in flask cultures. Ushiyama et al. (1986) [8] examined the growth of Panax ginseng root cultures in 30 1, 2 kl, and 20 kl aeration-agitation bioreactors. The productivity of the cultures in 2 kl and 20 kl bioreactors was 700 and 500 mg/l/day in dry weight, respectively. Building upon this basic research, large scale batch culture techniques have been developed for the industrial production of cell mass. However, culture conditions suitable for cell mass production are not always suitable for secondary metabolite production. Accordingly, in order to produce both cell mass and metabolites efficiently, two-stage culture techniques have been adopted. This technique uses two batch bioreactors and was first reported by Noguchi et al. (1987) [21] for the production of low nitrogen content tobacco cells. In the 1980's, this technique was widely used for secondary metabolite production such as shikonin (Fujita et al.) [27], rosmarinic acid (Ulbrich, 1985) [27] and digoxin production (Reinhard, 1980) [28].

For shikonin production by *Lithospermum erythrorhizon*, two-stage cell culture was used (see Fig. 3.11). The first stage culture was grown in a MG-5 medium which was suitable for cell mass production. It was then transferred to 2nd-stage culture where it was grown in an M-9 medium, modified by a higher Cu^{++} content and a decreased salt content.

Large Scale Continuous Culture. The growth rate of plant cells is usually low compared to that of microbial organisms. In order to enhance the productivity of cell mass and metabolites, continuous culture methods should be employed (Wilson, 1978 [29]; Fig. 3.12a). In the research laboratories of Japan Tobacco and Salt Co., a pilot plant (1500 1) and an industrial plant system (20 kl) have been used for developing continuous culture techniques (Hashimoto, et al., 1982 [24]; Azechi et al., 1983 [30]). A 20 kl bioreactor having a working volume of 6.34 kl, was used for the experiment (Fig. 3.12b) and ran for 66 days of continuous operation. The conditions were: aeration rate, 0.35–0.47 vvm; agitation speed, 27.5–35 rpm; dilution rate, 0.28–0.38 days. In this experiment, the residual sugar content was an important index of the operation and, at steady state, its value was maintained above 5 g/l. Other control parameters such as aeration, agitation, and dilution rates were changed gradually. The success of this experiment will soon lead to the

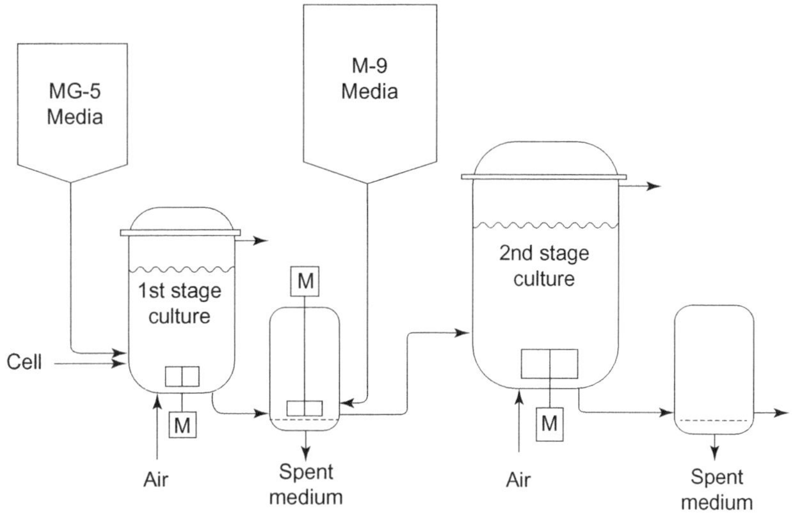

FIGURE 3.11 Two stage culture methods (Fujita, 1984).

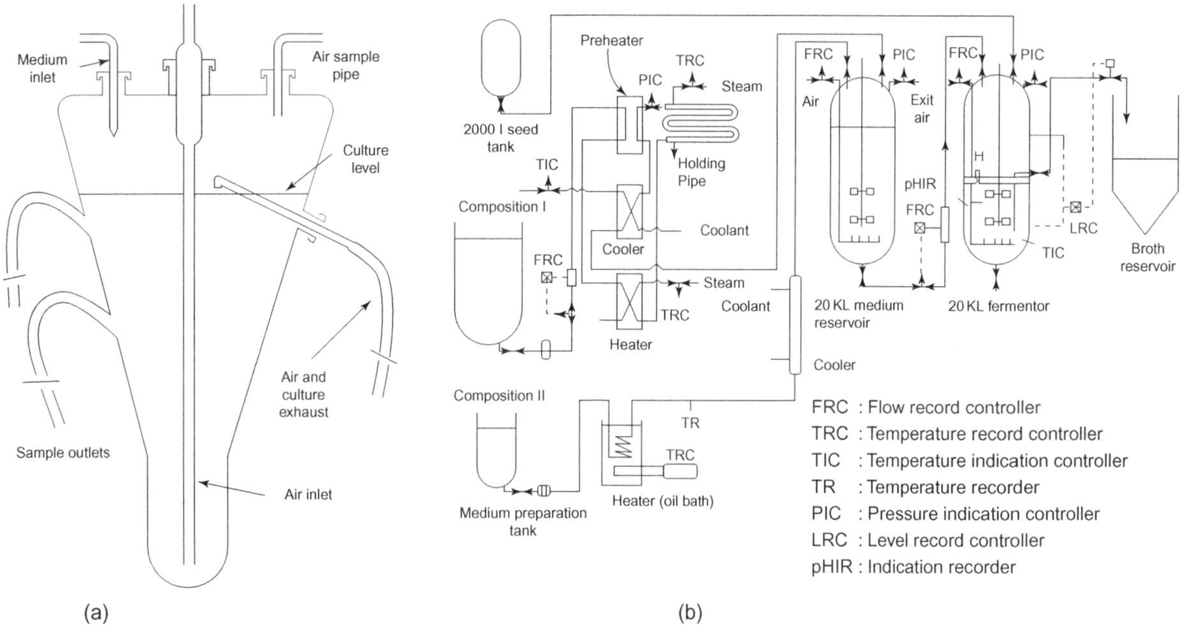

FIGURE 3.12 Continuous culture system. (a) Small glass vessel (Wilson, et al., 1976). (b) Pilot plant for continuous cultures with 20 kl bioreactor (Azechi, et al., 1983).

establishment of long-term industrial continuous culture systems for secondary metabolite production.

Immobilized Culture. Brodelius et al. first reported immobilization of plant cells in 1979 [31], and since then many reports have been published (Fig. 3.13). Unfortunately, an immobilized cell culture technique has not yet been established as an industrial process for secondary metabolite production. However, this technique has many excellent features and should be the subject of future development research.

11.0 SCALE-UP PROBLEMS

Scale-up techniques for plant cell cultures are not well understood. Some attempts have been made using tobacco cells and applying essentially the same parameters as those for the scale-up of microbial cultures (Azechi, 1985) [32]. The results showed that k_La values are useful as scale-up parameters, however, the situation for secondary metabolite production is quite different. The productivity of the metabolites decreased as the culture volume

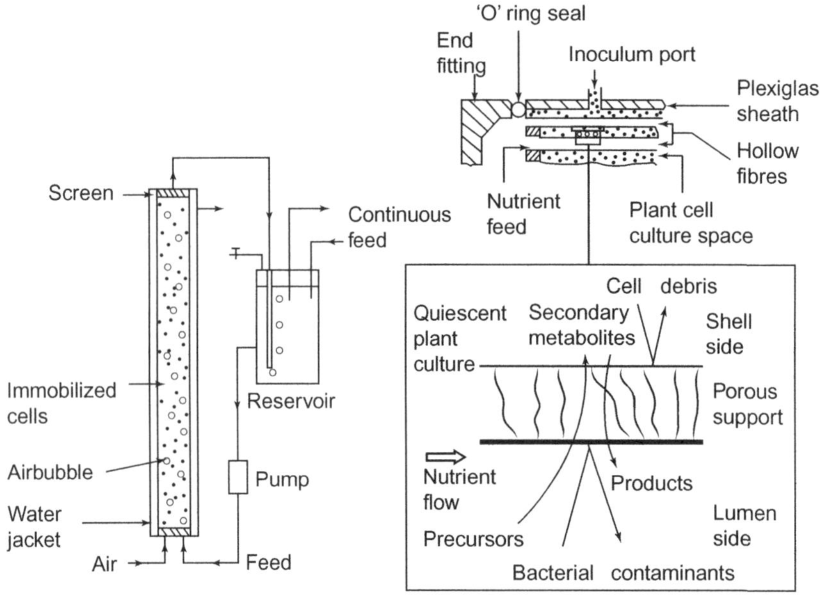

FIGURE 3.13 Immobilized plant cell cultures (Prenosil and Pederson, 1983).

increased. An example of this is the productivity of the indole alkaloid, a serpentine which declined significantly as the culture volume increased from 0.1 to 80 1 (Fowler, 1987) [33]. Possible reasons for the loss of product on scale-up are the following: *(i)* altered and inadequate mixing of the nutrient and cells at the high reactor volumes and *(ii)* lowered dissolved oxygen level (Breuling et al, 1985) [34]. Fujita and Tabata (1987) [1] used the scale-up of suspension cultures of *Lithospermum erythrorhizon* cells for their ability to produce shikonins as the criterion for comparing the aeration-agitation type bioreactor with a modified paddle impeller and the rotary drum type bioreactor. When the aeration-agitation type bioreactor was scaled up to a volume of 1000 1, the shikonin productivity decreased, but when the rotary drum bioreactor was scaled up to 1000 1, there was no decrease in the yield of shikonins (Fig. 3.14). Thus, in an industrial pilot plant for secondary metabolite production, mild agitation and oxygen supply will be important variables.

12.0 BIOPROCESS MEASUREMENT AND CONTROL

Bioprocess Measurement and Control in Large Scale Culture. Measurement and control systems used in the bioreactor for plants are essentially the same as those for microbial or animal cell cultures, but, in special cases, where the mineral components influence the productivity of secondary metabolites, the kind of salts used for the electrode must be taken into consideration.

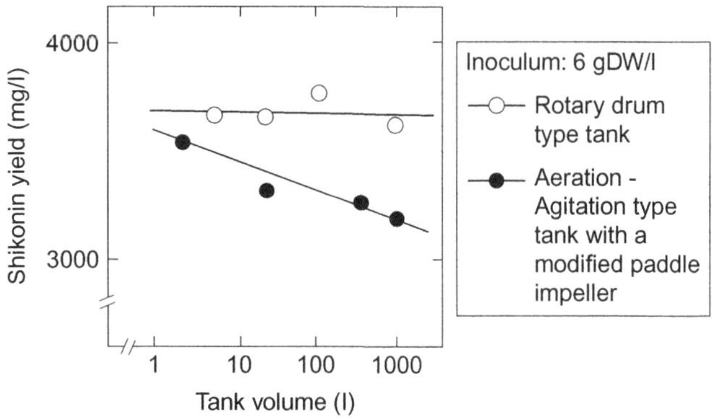

FIGURE 3.14 Scale up of suspension cultures for Lithospermum erythrorhizon cells (Fujita and Tabata, 1987).

Mass balance. The mass balance equation (Eq. 3.1) is generally used for heterotrophic organisms. Pareilleux and Chaubet (1980) [35] have applied the equation to apple cell cultures:

$$\frac{1}{X}\cdot\frac{dS}{dt}=\frac{1}{Y_G}\cdot(\mu+m) \qquad (3.1)$$

where

$\frac{1}{X}\cdot\frac{dS}{dt}$ = specific rate of sugar consumption (g sugar/g cell dry weight · h)
μ = specific growth rate (h)
Y_G = maximum growth yield (g cell weight/g sugar)
m = maintenance coefficient for sugar (g sugar/g cell dry weight/h)

From the equation, m and Y_G can be estimated from a plot of l/X (dS/dt) against μ. The values observed for Y_G and m are in good agreement with those reported previously by Kato and Nagai (1979) [36] calculated from tobacco cell cultures; i.e., m values were smaller and Y_G values were higher compared to the values reported for many microorganisms.

Measurement and Mathematical Fitting of Cell Growth. Richards (1960) used a generalized logistic curve for the mathematical fitting of the growth curve of plants. The equation is:

$$\ln\frac{A-W}{W-B}=f(t) \tag{3.2}$$

where

A = asympotic value of dry weight
B = inoculated value of dry weight
t = days culture in vitro
W = dry weight at any instant t (days or hours)
$f(t)$ = suitable function of time t (days or hours)

By eliminating the natural logarithm

$$W = B + (A-B)/[1+e^{f(t)}] \tag{3.3}$$

Differentiating with respect to t:

$$\frac{dW}{dt} = -\frac{(A-W)(W-B)}{A-B}f'(t) \tag{3.4}$$

The above equation is the growth rate. To calculate the specific growth rate, u, divide by the dry weight

$$\mu = \frac{1}{W}\cdot\frac{dW}{dt} \tag{3.5}$$

Doubling time, (T) becomes:

$$T = \frac{\ln(2)}{\mu} \tag{3.6}$$

REFERENCES (SECTION 3)

[1] G. Haberlandt, Kulturversuche mit isolierten Pfanzen, Sber. Akad. Wiss. Wien. 111 (1902) 69–92.

[2] R.J. Gautheret, Culture de tissue cambial, C. R. Hebd. Seanc. Acad. Sci., Paris 198 (1934) 2195–2196.

[3] R.J. Gauthret, Sur la possibilite de realiser la culture in definie des tissu de tubericule de carotte, C. R. Hebd. Seanc. Acad. Sci., Paris 208 (1939) 118–121.

[4] P.R. White, Potential unlimited growth of excised plant callus in an artificial medium, Amer. J. Bot. 26 (1939) 59–64.

[5] C.O. Miller, F. Skoog, F.S. Okumura, M.H. Von Saltza, F.M. Strong, Structure and synthesis of kinetin, J. Ameri. Chem. Soc. 77 (1955) 2662–2663.

[6] T. Murashige, F. Skoog, A revised medium for rapid growth and bioassays with tobacco tissue cultures, Physiol. Plant 15 (1962) 473–497.

[7] Y. Fujita, M. Tabata, Secondary metabolites from plant cells—pharmaceutical applications and progress in commercial production, Plant Tissue and Cell Culture, Alan R. Liss Inc., 1987.

[8] K. Ushiyama, Large scale culture techniques of plant cells and the secondary metabolite production, Hakko to Kogyo, 46 (1988) 7–11, in Japanese.

[9] W.G.W. Kurz, F. Constabel, Plant cell suspension cultures and their biosynthetic potential, Microbial. Technol. 1 (1979) 389–416.

[10] B. Deus-Neumann, M.H. Zenk, Instability of indole alkaloid production in Catharanthus roseus cells in suspension cultures, Planta Med. 50 (1984) 427–431.

[11] G. Morel, Producing virus-free cymbidium, Amer. Orchid Soc. Bull. 29 (1960) 495–497.

[12] K. Redenbaugh, P. Viss, D. Slade, J.A Fujil, Scale-up; artificikal kalin, Plant Tissue and Cell Cultures, 1987.

[13] F. Wagner, H. Vogelmann, Cultivation of plant tissue culture in bioreactors and formation of secondary metabolites, in: W. Barz, E. Reinhard, M.H. Zenk (Eds.), Plant Tissue and Its Biotechnological Applications, Springer Verlag, Berlin, 1987, pp. 245–252. ISBN 3-540-08227–1.

[14] H. Tanaka, F. Nishijima, M. Suwa, T. Iwamoto, Rotating drum fermentor for plant cell suspension cultures, Biotechnol. Bioeng. 25 (1983) 2359–2370.

[15] H. Tanaka, Large-scale cultivation of plant cells at high density: a review, Process Biochem. (1987) 106–113.

[16] D.J. Styer, in: R.R. Henke, K.W. Hughes, M.J. Constantin, A. Hollander (Eds.), Bioreactor Technology for Plant Propagation, Plenum Press, 1985, pp. 117–130. ISBN 0-306–41919.

[17] K. Ushiyama, H. Oda, Y. Miyamoto, Y. Ishida, Japan Patent, Kokai, 59–45879 (1984).

[18] K. Ushiyama, H. Oda, Y. Miyamoto, Large scale tissue culture of Panax ginseng root, Abst. VIth International Congress of Plant Tissue and Cell Culture, University of Minnesota, 1986. p. 252.

[19] H. Inoue, Culture vessel for Photo-requiring organisms, Japan Patent, Kokai 59–21682 (1984).

[20] H. Ikeda, Culture vessel for photoautotrophic culture, Japan Patent, Kokai 60–237984 (1985).

[21] M. Noguchi, T. Matumoto, Y. Hirata, Y. Kamamoto, A Akutu, A Kato, et al., Improvement of growth rates of plant cell cultures, Plant Tissue Culture and Its Bio-technological Application, Springer Verlag, Berlin, 1987. ISBN 3-540-18227–1.

[22] A Kato, Y. Shimizu, S. Noguchi, Effect of initial k_La on the growth of tobacco cells in batch culture, J. Ferment. Technol. 53 (1975) 744–751.

[23] R. Manfredini, L.G. Saporiti, V. Cavallera, Technological approach to industrial fermentation: limiting factors and practical solutions, La Chimica E l'Industria 64 (1982) 325–334.

[24] T. Hashimoto, S. Azechi, S. Sugita, K. Suzuki, Large scale production of tobacco cells by continuous cultivation, Plant Tissue Culture, Maruzen Co., Tokyo, 1982.

[25] S. Takayama, M. Misawa, K. Ko, T. Misato, Effect of cultural conditions on the growth of Agrostemma githago cells in suspension culture and the concomitant production of an anti-plant virus substance, Physiol. Plant 41 (1977) 313–320.

[26] H. Tanaka, Some properties of pseudo cells of plant cells, Biotechnol. Bioeng. 24 (1982) 2591–2596.

[27] H. Tanaka, Oxygen transfer in broth of plant cells at high density, Biotechnol. Bioeng. 24 (1982) 425–442.

[28] H. Tanaka, Technological problems in cultivation of plant cells at high density, Biotechnol. Bioeng. 23 (1981) 1203–1218.

[29] G. Wilson, Growth and product formation in large scale and continuous culture systems, Frontiers of Plant Tissue Culture, University of Calgary, Canada, 1978.

[30] S. Azechi, T. Hashimoto, T. Yuyama, S. Nagatsuka, M. Nakashizuka, T. Nishiyama, et al., Continuous cultivation of tobacco plant cells in an industrial scale plant, Hakkokogaku 60 (1983) 117–128.

[31] P. Brodelius, B Deus, K. Mosbach, M.H. Zenk, Immobilized plant cells for the production and transformation of natural products, FEBS Lett. 103 (1979) 93–97.

[32] S. Azechi, Large scale culture of plant cells, Bio-Engineering, Nikkan Kougyou Shinbunsya, 1985.

[33] D.D. Blesboer (Ed.), Plant Tissue and Cell Culture, Alan R. Liss Inc., 1987. ISBN 0-8451-1802–1.

[34] M. Breuling, A.W. Alfermann, E. Reinhard, Cultivation of cell cultures of Berberis wilsonde in 20–1 air-lift bioreactor, Plant Cell Rep. 4 (1985) 220–223.

[35] A. Pareilleux, N. Chaubet, Growth kinetics of apple plant cell cultures, Biotechnol. Lett. 2 (1980) 291–296.

[36] A. Kato, S. Nagai, Energetics of tobacco cells, Nicotiana tabacum L. growing on sucrose medium, Europ. J. Appl. Microbiol. Biotechnol. 7 (1979) 219–255.

FURTHER READING

[1] J.E. Prenosil, H. Pederson, Immunobilized plant cell reactors, Enzyme Microtechnol. 5 (1983) 323–331.

[2] M.H. Fowler, Process systems and approaches for large scale plant cell cultures, in: C.E. Green, D.A Somers, W.P. Hackett, Y. Fujita (Eds.), Plant Tissue Culture and Fine Chemicals, CMC Press Inc., Tokyo, 1984, pp. 191–197.

[3] E. Reinhard, A.W. Alfermann, Biotransformating by plant cell cultures, Adv. Biochem. Eng. 16 (1980) 49–83.

[4] B. Ulbrich, W. Welsneer, H. Arens, Large-scale production of rosmarinic acid from plant cell cultures of Coleus blumei Benth, in: K.H. Neumann, et al. (Eds.), Primary and Secondary Metabolism of Plan Cell Cultures, Springer-Verlag, Berlin, 1985, pp. 293–303.

Chapter 4

Nutritional Requirements in Fermentation Processes

Willem H. Kampen

1.0 INTRODUCTION

Specific nutritional requirements of microorganisms used in industrial fermentation processes are as complex and varied as the microorganisms in question. Not only are the types of microorganisms diverse (bacteria, molds and yeast, normally), but also the species and strains become very specific as to their requirements. Microorganisms obtain energy for support of biosynthesis and growth from their environment in a variety of ways. The following quotation is reprinted by permission of Prentice-Hall, Incorporated, Englewood Cliffs, New Jersey.

The most useful and relatively simple primary classification of nutritional categories is one that takes into account two parameters: The nature of the energy source and the nature of the principal carbon source, disregarding requirements for specific growth factors. Phototrophs use light as an energy source and chemotrophs use chemical energy sources.

Organisms that use CO_2 as the principal carbon source are defined as autotrophic; organisms that use organic compounds as the principal carbon source are defined as heterotrophic. A combination of these two criteria leads to the establishment of four principal categories: *(i)* photoautotrophic, *(ii)* photoheterotrophic, *(iii)* chemoautotrophic and *(iv)* chemoheterotrophic organisms.

Photoautotrophic organisms are dependent on light as an energy source and employ CO_2 as the principal carbon source. This category includes higher plants, eucaryotic algae, blue green algae, and certain photosynthetic bacteria (the purple and green sulfur bacteria).

Photoheterotrophic organisms are also dependent on the light as an energy source and employ organic compounds as the principal carbon source. The principal representatives of this category are a group of photosynthetic bacteria known as the purple non-sulfur bacteria; a few eucaryotic algae also belong to it.

Chemoautotrophic organisms depend on chemical energy sources and employ CO_2 as a principal carbon source. The use of CO_2 as a principle carbon source by chemotrophs is always associated with the ability to use reduced inorganic compounds as energy sources. This ability is confined to bacteria and occurs in a number of specialized groups that can use reduced nitrogen compounds (NH_3, NO_2), ferrous iron, reduced sulfur compounds (H_2S, S, $S_2O_3^{2-}$), or H_2 as oxidizable energy sources.

Chemoheterotrophic organisms are also dependent on chemical energy sources and employ organic compounds as the principle carbon source. It is characteristic of this category that both energy and carbon requirements are supplied at the expense of an organic compound. Its members are numerous and diverse, including fungi and the great majority of the bacteria.

The chemoheterotrophs are of great commercial importance. This category may be subdivided into respiratory organisms, which couple the oxidation of organic substrates with the reduction of an inorganic oxidizing agent (electron acceptor, usually O_2), and fermentative organisms, in which the energy-yielding metabolism of organic substrates is not so coupled. In addition to an energy source and a carbon source, the microorganisms require nutritional factors coupled with essential and trace elements that combine in various ways to form cellular material and products.

Since photosynthetic organisms (and chemoautotrophs) are the only net producers of organic matter on earth, it is they that ultimately provide, either directly or indirectly, the organic forms of energy required by all other organisms [1].

Compounds that serve as energy carriers for the chemotropism, linking catabolic and biosynthetic phases of metabolism, are adenosine phosphate and reduced pyridine nucleotides (such as nicotinamide dinucleotide or NAD). The structure of adenosine triphosphate (ATP) is shown in Figure 4.1. It contains two energy-rich bonds, which upon hydrolysis, yield nearly eight kcal/mole for each bond broken. ATP is thus reduced to the diphosphate (ADP) or the monophosphate (AMP) form.

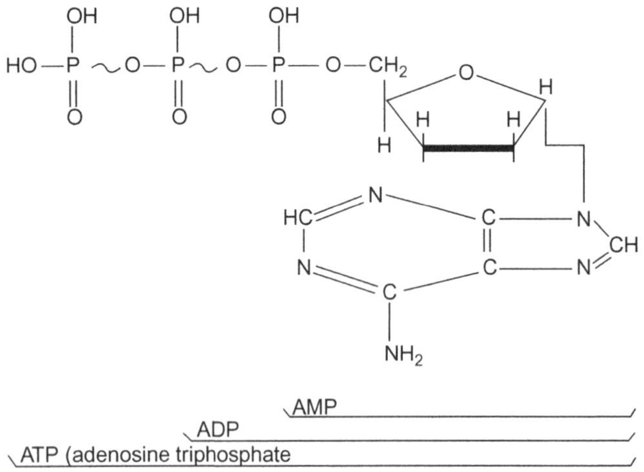

FIGURE 4.1 Chemical structure of ATP, which contains two energy-rich bonds. When ATP yields ADP, the Gibbs free energy change is −7.3 kcal/kg at 37°C and pH 7.

Plants and animals can use the conserved energy of ATP and other substances to carry out their energy requiring processes, i.e., skeletal muscle contractions, etc. When the energy in ATP is used, a coupled reaction occurs. ATP is thus hydrolyzed.

Biochemically, energetic coupling is achieved by the transfer of one or both of the terminal phosphate groups of AMP to an acceptor molecule, most of the bond energy being preserved in the newly formed molecule, e.g., glucose + ATP → glucose-6-phosphate + ADP [1].

Mammalian skeleton muscle at rest contains 350–400 mg ATP per 100 g. ATP inhibits enzymatic browning of raw edible plant materials, such as sliced apples, potatoes, etc.

2.0 NUTRITIONAL REQUIREMENTS OF THE CELL

Besides a source of energy, organisms require a source of materials for biosynthesis of cellular matter and products in cell operation, maintenance and reproduction. These materials must supply all the elements necessary to accomplish this. Some microorganisms utilize elements in the form of simple compounds, others require more complex compounds, usually related to the form in which they ultimately will be incorporated in the cellular material. The four predominant types of polymeric cell compounds are the lipids (fats), the polysaccharides (starch, cellulose, etc.), the information-encoded polydeoxyribonucleic acid and polyribonucleic acids (DNA and RNA), and proteins. Lipids are essentially insoluble in water and can thus be found in the nonaqueous biological phases, especially the plasma and organelle membranes. Lipids also constitute portions of more complex molecules, such as lipoproteins and liposaccharides. Lipids also serve as the polymeric biological fuel storage.

Natural membranes are normally impermeable to highly charged chemical species such as phosphorylated compounds. This allows the cell to contain a reservoir of charged nutrients and metabolic intermediates, as well as maintaining a considerable difference between the internal and external concentrations of small cations, such as H^+, K^+ and Na^+. Vitamins A, E, K and D are fat-soluble and water-insoluble. Sometimes they are also classified as lipids.

DNA contains all the cell's hereditary information. Upon cell division, each new cell receives a complete copy of its parents' DNA. The sequence of the subunit nucleotides along the polymer chain holds this information. Nucleotides are made up of deoxyribose, phosphoric acid, and a purine or pyrimidine nitrogenous base. RNA is a polymer of ribose-containing nucleotides. Of the nitrogenous bases, adenine, guanine, and cytosine are common to both DNA and RNA. Thymine is found only in DNA and uracil only in RNA [1]. Prokaryotes contain one DNA molecule with a molecular weight on the order of 2×10^9. This one molecule contains all the hereditary information. Eukaryotes contain a nucleus with several larger DNA molecules. The negative charges on DNA are balanced by divalent ions in the case of prokaryotes or basic amino acids in the case of eukaryotes. Messenger RNA-molecules carry messages from DNA to another part of the cell. The message is read in the ribosomes. Transfer RNA is found in the cytoplasm and assists in the translation of the genetic code at the ribosome.

Typically 30–70% of the cell's dry weight is protein. All proteins contain C, H, N, and O. Sulfur contributes to the three-dimensional stabilization of almost all proteins. Proteins show great diversity of biological functions. The building blocks of proteins are the amino acids. The predominant chemical elements in living matter are: C, H, O, and N, and they constitute approximately 99% of the atoms in most organisms. Carbon, an element of

TABLE 4.1 Physiological Functions of the Principal Elements [1,3]

Element	Symbol	Atomic	Physiological Function
Hydrogen	H	1	Constituent of cellular water and organic cell materials
Carbon	C	6	Constituent of organic cell materials
Nitrogen	N	7	Constituent of proteins, nucleic acids and coenzymes
Oxygen	O	8	Constituent of cellular water and organic materials, as O_2 electron acceptor in respiration of aerobes
Sodium	Na	11	Principal extracellular cation.
Magnesium	Mg	12	Important divalent cellular cation, inorganic cofactor for many enzymatic reactions, incl. those involving ATP; functions in binding enzymes to substrates and present in chlorophylls
Phosphorus	P	15	Constituent of phospholipids, coenzymes and nucleic acids
Sulfur	S	16	Constituent of cysteine, cystine, methionine and proteins as well as some coenzymes as CoA and cocarboxylase
Chlorine	Cl	17	Principal intracellular and extracellular anion
Potassium	K	19	Principal intracellular cation, cofactor for some enzymes
Calcium	Ca	20	Important cellular cation, cofactor for enzymes as proteinases
Manganese	Mn	25	Inorganic cofactor cation, cofactor for enzymes as proteinases
Iron	Fe	26	Constituent of cytochromes and other heme or non-heme proteins, cofactor for a number of enzymes
Cobalt	Co	27	Constituent of vitamin B_{12} and its coenzyme derivatives
Copper	Cu	29	
Zinc	Zn	30	Inorganic constituents of special enzymes
Molybdenum	Mo	42	

prehistoric discovery, is widely distributed in nature. Carbon is unique among the elements in the vast number and variety of compounds it can form. There are upwards of a million or more known carbon compounds, many thousands of which are vital to organic and life processes [2]. Hydrogen is the most abundant of all elements in the universe, and it is thought that the heavier elements were, and still are, being built from hydrogen and helium. It has been estimated that hydrogen makes up more than 90% of all the atoms or three quarters of the mass of the universe [2]. Oxygen makes up 21 and nitrogen 78 volume percent of the air. These elements are the smallest ones in the periodic system that can achieve stable electronic configurations by adding one, two, three or four electrons respectively [1,3]. This ability to add electrons, by sharing them with other atoms, is the first step in forming chemical bonds, and thus, molecules. Atomic smallness increases the stability of molecular bonds and also enhances the formation of stable multiple bonds.

The biological significance of the main chemical elements in microorganisms is given in Table 4.1[1,3]. Ash composes approximately 5 percent of the dry weight of biomass with phosphorus and sulfur accounting, for respectively 60 and 20 percent. The remainder is usually made up of Mg, K, Na, Ca, Fe, Mn, Cu, Mo, Co, Zn and Cl [1].

The predominant atomic constituents of organisms, C, H, N, O, P, and S, go into making up the molecules of living matter. All living cells on earth contain water as their predominant constituent. The remainder of the cell consists largely of proteins, nucleic acids, lipids, and carbohydrates, along with a few common salts. A few smaller compounds are very ubiquitous and function universally in bioenergetics, e.g., ATP for energy capture and transfer, and NAD in biochemical dehydrogenation. Microorganisms share similar chemical compositions and universal pathways. They all have to accomplish energy transfer and conversion, as well as synthesis of specific and patterned chemical structures [1].

The microbial environment is largely determined by the composition of the growth medium. Using pure compounds in precisely defined proportions yields a defined or synthetic medium. This is usually preferred for researching specific requirements for growth and product formation by systematically adding or eliminating chemical species from the formulation. Defined media can be easily reproduced, have low foaming tendency, show translucency and allow easy product recovery and purification.

Complex or natural media such as molasses, corn steep liquor, meat extracts, etc., are not completely defined chemically, however, they are the media of choice in industrial fermentations.

In many cases the complex or natural media have to be supplemented with mainly inorganic nutrients to satisfy the requirements of the fermenting organism. The objective in media formulation is to blend ingredients rich in some nutrients and deficient in others with materials possessing other profiles to achieve the proper chemical balance at the lowest cost and still allow easy processing [4]. Fermentation nutrients are generally classified as: sources of carbon, nitrogen and sulfur, minerals and vitamins.

3.0 THE CARBON SOURCE

Biomass is typically 50% carbon on a dry weight basis, an indication of how important it is. Since organic substances are at the same general oxidation level as organic cell constituents, they do not have to undergo a primary reduction to serve as sources of cell carbon. They also serve as an energy source. Consequently, much of this carbon enters the pathways of energy-yielding metabolism and is eventually secreted from the cell as CO_2 (the major product of energy-yielding respiratory metabolism or as a mixture of CO_2 and organic compounds, the typical end-products of fermentation metabolism). Many microorganisms can use a single organic compound to supply both carbon and energy needs. Others need a variable number of additional organic compounds as nutrients. These additional organic nutrients are called growth factors and have a purely biosynthetic function, being required as precursors of certain organic cell constituents that the organism is unable to synthesize. Most microorganisms that depend on organic carbon sources also require CO_2 as a nutrient in very small amounts [1]. In the fermentation of beet molasses to ethanol and glycerol, it was found that by manipulating several fermentation parameters, the ethanol yield (90.6%) and concentration (8.5% v/v) remained essentially the same, while the glycerol concentration went from 8.3 g/l to 11.9 g/l. The CO_2 formation, however, was reduced! With glycerol levels over 12 g/l, the ethanol yield and concentration reduced with the CO_2-formation near normal again [5]. In fermentations, the carbon source on a unit of weight basis may be the least expensive raw material, however, quite often represents the largest single cost for raw material due to the levels required. Facultative organisms incorporate roughly 10% of substrate carbon in cell material, when metabolizing anaerobically, but 50–55% of substrate carbon is converted to cells with fully aerobic metabolism. Hence, if 80 grams per liter of dry weight of cells are required in an aerobic fermentation, then the carbon required in that fermentation equals $(80/2)\ (100/50) = 80$ grams of carbon. If this is supplied as the hexose glucose, with molecular weight 180 and carbon weight 72, then $(80)\ (180)/72 = 200$ gram per liter of glucose are required.

Carbohydrates are excellent sources of carbon, oxygen, hydrogen, and metabolic energy. They are frequently present in the media in concentrations higher than other nutrients and are generally used in the range of 0.2–25%. The availability of the carbohydrate to the microorganism normally depends upon the complexity of the molecule. It generally may be ranked as:

hexose > disaccharides > pentoses > polysaccharides

Carbohydrates have the chemical structure of either polyhydroxyaldehydes or polyhydroxyketones. In general, they can be divided into three broad classes: monosaccharides, disaccharides and polysaccharides. Carbohydrates have a central role in biological energetics, the production of ATP. The progressive breakdown of polysaccharides and disaccharides to simpler sugars is a major source of energy-rich compounds [1]. During catabolism, glucose, as an example, is converted to carbon dioxide, water and energy. Enzymes catalyze the conversion from complex to simpler sugars. Three major interrelated pathways control carbohydrate metabolism:

- The Embden-Meyerhof pathway (EMP)
- The Krebs or tricarboxylic acid cycle (TCA)
- The pentose-phosphate pathway (PPP)

In the EMP, glucose is anaerobically converted to pyruvic acid and on to either ethanol or lactic acid. From pyruvic acid it may also enter the oxidative TCA pathway. Per mole of glucose broken down, a net gain of 2 moles of ATP is being obtained in the EMP. The EMP is also the entrance for glucose, fructose, and galactose into the aerobic metabolic pathways, such as the TCA-cycle. In cells containing the additional aerobic pathways, the $NADH_2$ that forms in the EMP where glyceraldehyde-3-phosphate is converted into 3-phosphoglyceric acid, enters the oxidative phosphorylation scheme and results in ATP generation [3]. In fermentative organisms the pyruvic acid formed in the EMP pathway may be the precursor to many products, such as ethanol, lactic acid, butyric acid (butanol), acetone and isopropanol [1].

The TCA-cycle functions to convert pyruvic and lactic acids, the end products of anaerobic glycolysis (EMP), to CO_2 and H_2O. It also is a common channel for the ultimate oxidation of fatty acids and the carbon skeletons of many amino acids. The overall reaction is:

$$2C_3H_4O_3 + 5O_2 + 30\ ADP + 30\ P_i \rightarrow 6CO_2 + 4H_2O + 30\ ATP$$

for pyruvic acid as the starting material [3]. Obviously, the EMP-pathway and TCA-cycle are the major sources of ATP energy, while they also provide intermediates for lipid and amino acid synthesis.

The PPP handles pentoses and is important for nucleotide (ribose-5-phosphate) and fatty acid biosynthesis ($NADPH_2$). The Entner-Doudoroff pathway catabolizes glucose into pyruvate and glyceraldehyde-3-phosphate. It is important primarily in Gram negative prokaryotes [6].

The yeast *Saccharomyces cerevisiae* will ferment glucose, fructose and sucrose without any difficulties, as long as the minimal nutritional requirements of niacin (for NAD), inorganic phosphorus (for phosphate groups in 1, 3-diphosphoglyceric acid and ATP) and magnesium (catalyzes, with hexokinase and phosphofructokinase, the conversion of glucose to glucose-6-phosphate and fructose-6-phosphate to fructose-1, 6-diphosphate) are met. Table 4.2 lists some of the important biological molecules involved in catabolism and anabolism [1,3]. *S. cerevisiae* ferments galactose and maltose occasionally,

TABLE 4.2 Fundamental Biological Molecules [1,3]

Simple Molecule	Constituent Atoms	Derived Macro Molecules
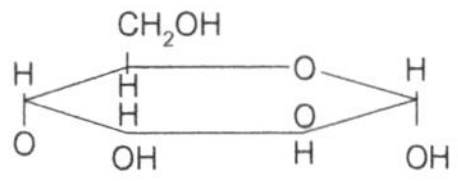 Glucose (carbohydrate)	C,H,O	glycogen, starch, cellulose
$HSCH_2CH(NH_2)COOH$ Cysteine (amino acid)	C,N,H,O,S	proteins
$NH_2(CH_2)_4\ CH(NH_2)COOH$ Lysine (basic amino acid)	C,N,H,O	
$CH_3(CH_2)_{14}COOH$ Palmitic acid (fatty acid)	C,H,O	fats and oils
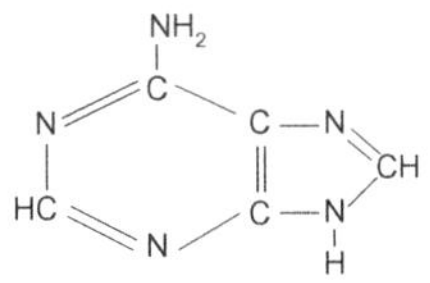 Adenine (purine)	C,N,H,O	nucleotides (nucleic acids, DNA and RNA)
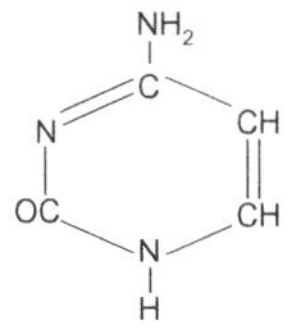 Cytosine (pyrimidine)	C,N,H,O	

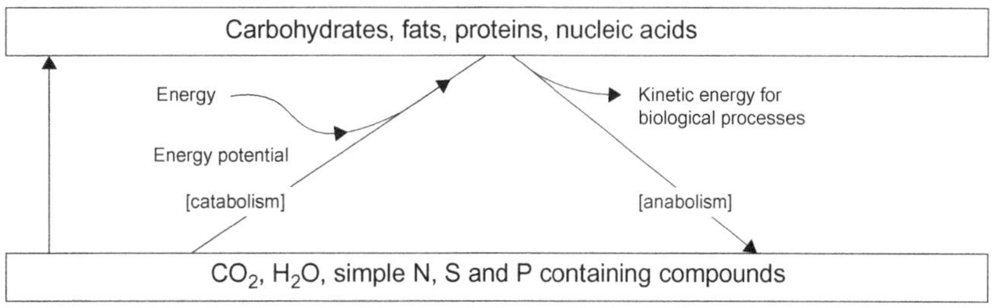

but slowly; inulin very poorly; raffinose only to the extent of one third and melibiose and lactose it will not ferment. *S. cerevisiae* follows the Embden-Meyerhof pathway and produces beside ethanol, 2 moles of ATP per mole of glucose.

Catabolite repression, transient repression and catabolite inhibition regulate the utilization of many carbohydrates [1]. Catabolite repression is a reduction in the rate of synthesis of certain enzymes in the presence of glucose or other easily metabolized carbon sources. In addition to this repression during steady-state growth in glucose, a period of more intense repression may occur immediately after the cells have been exposed to very high levels of glucose. This effect may last up to one generation or until glucose levels have been reduced to more acceptable levels. This is transient repression. Catabolite inhibition is a control exerted by glucose on enzyme activity rather than on enzyme formation, analogous to the feedback inhibition in biosynthetic pathways. Enzymes involved in the utilization of other carbohydrates are inhibited by glucose.

Simple sugars are available in powder or in liquid form and in a variety of purities. Glucose is usually made from corn starch through hydrolysis and sucrose from sugar cane or sugar beets. Sucrose is most often purchased in the form of molasses. Sugar beet molasses is the main by-product of table sugar production. Blackstrap molasses is the remaining by-product of raw sugar production from sugar cane, it is the prevailing type of cane molasses. High test molasses or inverted cane syrup is a by-product of the refineries in which raw sugar is refined into white or table sugar. Both blackstrap and beet molasses are widely used in the fermentation industry. Their approximate composition differs considerably, as indicated in Table 4.3 [5]. The data on sugar beet molasses are averages from two (2) samples each of Dutch and French molasses from the 1990 campaign (column A). The US beet molasses data are averages from five (5) factories belonging to American Crystal Sugar over the 1991 season (Column B). The blackstrap molasses data are averages from several samples from Brazil, Dominican Republic and Haiti, over the period 1975–1983. Table 4.4 is an indication of how complex a medium blackstrap molasses is. Upon diluting to 25%, it was stripped under reduced pressure (40 mm Hg) at 38°C. The distillate was extracted with an ether/pentane (1:1) mixture. Separation and identification was done with a capillary column and mass spectrometer [5].

Molasses is produced through nonsugar accumulation during the sugar production process and the accompanying increased solubility of sucrose. Of the non-sugars, the mineral salts have a much greater influence on sucrose solubility than the organic compounds. As a rule of thumb, one gram of mineral salts present in normal molasses will retain five grams of non-crystallizable sucrose. Through chromatographic separation processes, it is possible to recover up to approximately 85 percent of this sucrose.

The composition of molasses varies from year to year, since it depends on many factors, such as variety of sugar cane or beet, soil type, climatic conditions (rainfall, sunshine), time of harvesting, process conditions, etc.

Beet molasses contains approximately 1.9% N of which roughly 1.2% consists of betaine, 0.6% amine-N and 0.025% ammoniacal-N. Cane molasses does not contain betaine and has less than 50% of the organic nitrogen content of beet molasses. Beet molasses contains a relatively high amount of protein, while cane molasses contains high levels of gums and pectins. Also present are: hemicellulose, reversible and irreversible colloids, pigments, inositol, etc. Both types of molasses also contain trace elements, vitamins and growth factors, however, cane molasses usually contains more than beet molasses. Beet molasses has a characteristic, often unpleasant smell, and a pH around 8.0; while blackstrap molasses usually has a fruity, pleasant, mildly acidic smell and a pH value below 7.0.

While sugar alcohols are not common, large scale, they may be used in bioconversions such as from glycerol. Methane, methanol and n-alkanes have been used in biomass production.

Fatty acids may be converted by fungi after hydrolysis by lipase. Other organic acid carbon sources would be oleic, linoleic and linolenic acids. These might also serve as foam control agents. Carbon dioxide is a possible carbon source in nature, but is not practical commercially due to low growth rates.

The most economically important and most widely used carbon sources are the carbohydrates. They are commonly found and most are economically priced.

4.0 THE NITROGEN AND SULFUR SOURCE

Following the carbon source, the nitrogen source is generally the next most plentiful substance in the fermentation media. A few organisms can also use the nitrogen source as the energy source. Nitrogen and sulfur occur in the organic compounds of the cell, principally in reduced form, as amino and sulfhydryl groups, respectively. Most photosynthetic organisms assimilate these two elements in the oxidized inorganic state, as nitrates and sulfates; their biosynthetic utilization thus involves a preliminary reduction. Many nonphotosynthetic bacteria and fungi can also meet the needs for nitrogen and sulfur from nitrates and sulfates. Some microorganisms are unable to bring about a reduction of one or both of these anions and must be supplied with the elements in the reduced form.

TABLE 4.3 Average Composition (%) of European (A) and U.S. (B) Beet Molasses versus Brazilian/Caribbean Blackstrap Molasses Samples [5]

Component	Beet Molasses		Blackstrap
	Column A	Column B	
water	16.5	19.2	18.0
sucrose	51.0	48.9	32.0
glucose + fructose	1.0	0.5	27.0
raffinose	1.0	1.3	—
organic non-sugars	19.0	18.0	14.0
ash	11.5	12.1	9.0
Ash components:			
SiO_2	0.1	—	0.7
K_2O	3.9	6.4	3.5
CaO	0.26	0.21	1.9
MgO	0.16	0.12	0.1
P_2O_5	0.06	0.03	0.2
Na_2O	1.3	1.6	—
Fe_2O_3	0.02	0.03	0.4
Al_2O_3	0.07	—	—
CO_3	3.5	—	—
Sulfates as SO_3	0.55	0.74	1.8
Cl	1.6	0.8	0.4
	11.5		9.0
Vitamins (mg/100 g):			
Thiamine (B_1)	1.3	0.01	8.3
Riboflavin (B_2)	0.4	1.1	2.5
Nicotinic acid	51.0	8.0	21.0
Ca-pantothenate (B_3)	1.3	0.7	21.4
Folic acid	2.1	0.025	0.04
Pyridoxine-HCl (B_2)	5.4	—	6.5
Biotin	0.05	—	1.2

The average viscosity of sample B was 1,062 cP at 45°C.

In these cases, nitrogen may be supplied as ammonia salts and sulfur as a sulfide or an organic compound like cysteine, which contains a sulfhydryl group. Organic nutrients as amino acids and more complex protein degradation products as peptones may also supply nitrogen and sulfur in the reduced form as well as carbon and energy. Several prokaryotic groups can also utilize the most abundant natural nitrogen source, N_2, which is unavailable to eukaryotes. This process of nitrogen assimilation is termed *nitrogen fixation* and involves a preliminary reduction of N_2 to ammonia [1]. Nitrogen is used for the anabolic synthesis of nitrogen-containing cellular substances, such as amino acids, purines, DNA, and RNA. Many algae and fungi use ammonium nitrate and sodium

TABLE 4.4 Isolation of Some Volatile Compounds in Blackstrap Molasses [5]

methanol	acetic acid
methylformate	propionic acid
2-methyl-furamidon-3	isobutyric acid
furfurylacetate	n-butyric acid
methylfurfural	isovaleric acid
2-acetylfuran	n-valeric acid
phenol	isocaproic acid
quaiacol	n-caproic acid
benzaldehyde	alanine
2.5-dimethylpyrazine (1, 4)	aspartic acid
2, 6-dimethylpyrazine (1, 4)	glutamic acid
2-methyl-6-ethylpyrazine (1, 4)	leucine
2-methyl-5-ethylpyrazine (1, 4)	isoleucine
trimethylpyrazine (1, 4)	glycine
syringic acid	methionine
vanillic acid	asparagine
p-hydroxy benzoic acid	glutamine
p-coumaric acid (trans-)	valine
p-coumaric acid (cis-)	tyrosine
p-hydroxphenylacetic acid	

nitrate as nitrogen sources, however, yeasts and bacteria have problems utilizing nitrogen in this form [1]. Few organisms are able to assimilate nitrites.

Organic sources of nitrogen in synthetic media are specific amino acids, purines, pyrimidines, and urea. Urea, depending upon the buffer capacity of the system, will raise the pH-value of the medium. Organic urea is also formed in the urea cycle reaction, starting with ammonia:

$$NH_3 + ATP + CO_2 \rightarrow H_2NCOOP + ADP$$

Ammonium sulfate produces acidic conditions because the ammonia is rapidly utilized and free acid is then liberated.

Many commercial fermentations use complex organic nitrogen sources, which are by-products of the agricultural and food processing industries, such as: corn steep liquor, dried distillers solubles, yeast, fish or bone meal, corn germ or gluten meal, protein peptones, hydrolysates and digests from casein, yeast, cottonseed, milk proteins, etc. These sources of nitrogen provide many other nutrients and are usually reasonably priced. The composition of some of these products is given in Tables 4.5–4.7.

Industrial fermentations are generally more rapid and efficient when these materials are used, since they reduce the number of compounds which the cells would otherwise have to synthesize "de novo" [4]. The availability of nitrogen as well as the concentration in the media has to be considered in each case. Proteins can only be assimilated by microorganisms that secrete extracellular proteases, which enzymatically hydrolyze the proteins to amino acids. Microorganisms without this ability require protein hydrolysates, peptones, or digests composed of free amino acids prepared by hydrolyzing proteinaceous materials with acids or enzymes.

5.0 THE SOURCE OF TRACE AND ESSENTIAL ELEMENTS

Minerals supply the necessary elements to cells during their cultivation. Typical biological functions of the main elements were listed in Table 4.1. Table 4.8 shows the trace element composition in samples of Puerto Rican blackstrap and Dutch beet molasses in the year 1986. Phosphorous occurs principally in the form of sugar-phosphates, such as the nucleotides which compose DNA, RNA, and ATP. Phosphorus is assimilated in its inorganic form where the phosphate ion is esterified. The P-atom does not change in valence and remains as part of a phosphate group. Upon the death of the cell, it is again liberated as inorganic phosphorus through hydrolysis. Sulfur is present to the greatest extent in the amino acids cysteine and methionine. It is also commonly supplied as H_2SO_4 for pH adjustment, and as ammonium sulfate and potassium bisulfate. Many of the other elements are found complexed with enzymes: e.g., Mg^{2+} with phosphohydrolase and phosphotransferase, K^+ with pyruvate phosphokinase (and Mg^{2+}), and Na^+ with plasma membrane ATP-ase (and K^+ and Mg^{2+}) [1].

Requirements for trace elements may include iron (Fe^{2+} and Fe^{3+}), zinc (Zn^{2+}), manganese (Mn^{2+}), molybdenum (Mo^{2+}), cobalt (Co^{2+}), copper (Cu^{2+}), and calcium (Ca^{2+}). The functions of each vary from serving in coenzyme functions to catalyze many reactions, vitamin synthesis, and cell wall transport. The requirements are generally in very low levels and can sometimes even be supplied from quantities occurring in water or from leachates from equipment. Trace elements may contribute to both primary or secondary metabolite production. Manganese can influence enzyme production. Iron and zinc have been found to influence antibiotic production. Primary metabolite production is usually not very sensitive to trace element concentration, however, this is a different matter for secondary metabolite production.

TABLE 4.5 Typical Chemical Composition of Pharmamedia*; [1] a Nitrogen and Energy Source from the Cottonseed Embryo; Soy Bean Meal (expeller) [2]; Meat and Bone Meal [3]; and Peanut Meal and Hulls [4]

	[1]	[2]	[3]	[4]
dry matter	99.0	90.0	92.0	90.5
protein, %	59.2	42.0	50.0	45.0
carbohydrates,%	24.1	29.9	0.0	23.0
fat, %	4.0	4.0	8.0	5.0
fiber, %	2.6	6.0	3.0	12.0
ash, %	6.7	6.5	31.0	5.5
calcium, %	0.25	0.25	8.9	0.15
magnesium, %	0.74	0.25	1.1	0.32
phosphorus, %	1.31	0.63	4.4	0.55
available P, %	0.31	0.16	4.4	0.2
potassium, %	1.72	1.75	1.46	1.12
sulfur, %	0.6	0.32	0.26	0.28
biotin, mg/kg	1.52	—	—	—
choline, mg/kg	3270	2420	1914	1672
niacin, mg/kg	83.3	30.4	55	167.2
panthotenic acid, mg/kg	12.4	14.1	8.8	48.4
pyridoxine, mg/kg	16.4	—	—	—
riboflavin, mg/kg	4.82	3.1	4.4	5.3
thiamine, mg/kg	3.99	—	1.1	7.26
arginine, %	12.28	2.9	4.0	4.6
cystine, %	1.52	0.62	1.4	0.7
glycine, %	3.78	—	6.6	3.0
histidine, %	2.96	—	0.9	1.0
isoleucine, %	3.29	—	1.7	2.0
leucine, %	6.11	—	3.1	3.1
lysine, %	4.49	2.8	3.5	1.3
methionine, %	1.52	0.59	0.7	0.6
phenylalanine, %	5.92	—	1.8	2.3
threonine, %	3.31	1.72	1.8	1.4
tryptophan, %	0.95	0.59	0.2	0.5
tyrosine, % valine, %	3.42	—	1.22	—
	4.57	—	2.4	2.2

*Trademark.

TABLE 4.6 Average Composition of Corn Steep Liquor [4]

Total solids, %	54.0
pH	4.2
Ash (oxide), % dry basis	17
Crude protein (N × 6.25)	47
Fat	0.4
Total acids as lactic acid	2.6
Nitrogen	7.5
Phytic acid	7.8
Reducing sugars as glucose	2.5
Ash constituents. % dry basis:	
calcium	0.06
chlorine	0.70
magnesium	1.5
total phosphorus	3.3
-phytin phosphorus	2.2
-inorganic phosphorus by difference	1.1
potassium	4.5
sodium	0.2
total sulfur	0.58
-sulfate sulfur	0.25
-sulfite sulfur	0.01
boron (ppm, dry basis)	30
copper	25
iron	300
manganese	50
molybdenum	2
strontium	2.5
zinc	175
Vitamins, ppm dry basis:	
biotin	0.1
choline	5600
folic acid	0.5
inositol	5000
Niacin	160
Panthotenic acid	25
Pyridoxine	20
Riboflavin	10
Thiamine	5

Bacillus licheniformis produces the secondary metabolite bacitracin [7]. A manganese concentration of 0.07×10^{-5} M is required, but at a concentration of 4.0×10^{-5} M manganese becomes an inhibitor. *Streptomyces griseus* produces streptomycin as a secondary metabolite. For maximum growth it requires a concentration of 1.0×10^{-5} M of iron and 0.3×10^{-5} zinc, while a zinc concentration of 20×10^{-5} M becomes an inhibitor. *Aspergillus niger* produces citric acid as a primary metabolite. Concentrations of 2.0×10^{-5} M of zinc, 6.0×10^{-5} M of iron and/or 0.02×10^{-5} M of manganese act as inhibitors [8]. It produces only citric acid from glucose or sucrose during an iron deficiency and/or proper Cu/Fe ratio in the fermentation media. Raw materials such as molasses may have to be treated to remove iron. Manganese enhances longevity in cultures of *Bacillus sp.*, iron in *Escherichia sp.*, while zinc suppresses longevity of *Torulopsis sp* [8].

Cells are 80% or more water and in quantitative terms this is the major essential nutrient. Water is the solvent within the cell and it has some unusual properties, like a high dielectric constant, high specific heat and high heat of vaporization. It furthermore ionizes into acid and base, and has a propensity for hydrogen bonding. In most fermentations, microorganisms inhabit hypotonic environments in which the concentration of water is higher than it is within the cell. The cell walls are freely permeable to water, but not to many solutes. Water tends to enter the cell to equalize the internal and external water concentrations. Many eukaryotes and nearly all prokaryotes have a rigid wall enclosing the cell, which mechanically prevents it from swelling too much and undergoing osmotic lysis. The product of osmotic pressure and the volume containing one gram-molecule of solute is a constant. Thus, osmotic pressure is directly proportional to the concentration and

$$P = (0.0821)(T)/V$$

for a substance, where, P is measured in atmospheres, V in liters, and T in degrees Kelvin. A solution of 180 g/L of glucose (MW = 180) at 30°C, contains 1 gram-molecule per liter. Thus, $V = 1$ liter and $K = 303$ K. Hence, the osmotic pressure P = 24.9 atm. (366 psi).

All the required metallic elements can be supplied as nutrients in the form of the cations of inorganic salts. K, Mg, Ca and Fe are normally required in relatively large amounts and should normally always be included as salts in culture media. Table 4.9 shows which salts are soluble and which are insoluble in water [9], as well as commonly used inorganic and trace elements and concentration ranges.

Oxygen is always provided in water. Some organisms require molecular oxygen as terminal oxidizing agents to fulfill their energetic needs through aerobic respiration.

TABLE 4.7 Analysis of Microbiological Media Prepared Proteins *(Difco Manual, Ninth Edition)*

Constituent	Peptone	Tryptone	Casamino Acids	Yeast Extract
Ash %	3.53	7.28	3.64	10.1
Soluble Extract %	0.37	0.30		
Total Nitrogen %	16.16	13.14	11.15	9.18
Ammonia Nitrogen %	0.04	0.02		
Free Amino Nitrogen %	3.20	4.73		
Arginine %	8.0	3.3	3.8	0.78
Aspartic Acid %	5.9	6.1	0.49	5.1
Cystine %	0.22	0.19		
Glutamic Acid %	11.0	18.9	5.1	6.5
Glycine %	23.0	2.4	1.1	2.4
Histidine %	0.96	2.0	2.3	0.94
Isoleucine %	2.0	4.8	4.6	2.9
Leucine %	3.5	3.5	9.9	3.6
Lysine %	4.3	6.8	6.7	4.0
Methionine %	0.83	2.4	2.2	0.79
Phenylalanine %	2.3	4.1	4.0	2.2
Threonine %	1.6	3.1	3.9	3.4
Tryptophan %	0.42	1.45	0.8	0.88
Tyrosine %	2.3	7.1	1.9	0.60
Valine %	3.2	6.3	7.2	3.4
Organic Sulfur %	0.33	0.53		
Inorganic Sulfur %	0.29	0.04		
Phosphorous %	0.07	0.75	0.35	0.29
Potassium %	0.22	0.30	0.88	0.04
Sodium %	1.08	2.69	0.77	0.32
Magnesium %	0.056	0.045	0.0032	0.030
Calcium %	0.058	0.096	0.0025	0.040
Chloride %	0.27	0.29	11.2	0.190
Manganese (mg/L)	8.6	13.2	7.6	7.8
Copper (mg/L)	17.00	16.00	10.00	19.00
Zinc (mg/L)	18.00	30.00	8.00	88.00
Biotin μg/gm)	0.32	0.36	0.102	1.4
Thiamine μg/gm)	0.50	0.33	0.12	3.2
Riboflavin μg/gm)	4.00	0.18	0.03	19.00

TABLE 4.8 Average Composition of (trace) Elements in Puerto Rican Blackstrap Molasses in 1986 and One European Beet Molasses Sample of the Same Year [5]

Element	Blackstrap	Beet Molasses
B	0.041	0.0003
Ca	0.86	0.42
Co	0.000054	0.00006
Cu	0.0028	0.0005
P	0.071	0.012
Fe	0.0158	0.0115
Mg	1.14	—
Mn	0.0057	0.0018
Ni	0.000123	—
Na	0.058	0.083
Pb	0.75	0.78
K	2.68	3.39
Sr	—	0.004
Zn	0.011	0.0034

TABLE 4.9 Solubility of the Common Salts [5] and Commonly used Inorganic and Trace Elements and Concentration Ranges

Salt	Soluble	Insoluble
Nitrates	All	
Sulfides	Na, K, Ca, Ba	All others
Chlorides	All others	Ag, Hg, Pb
Carbonates	Na, K	All others
Sulfates	All others	Pb, Ca, Ba, Sr
Phosphates	Na, K	All others
Silicates	Na, K	All others
Acetates	All	
Oxalates	All others	Ca (depends upon concentration)
Source		Quantity (g/L)
KH_2PO_4		1.0–5.0
$MgSO_4 \cdot 7H_2O$		0.1–3.0
KCl		0.5–12.0
$CaCO_3$		5.0–17.2
$FeSO_4 \cdot 4H_2O$		0.01–0.1
$ZnSO_4 \cdot 8H_2O$		0.1–1.0
$MnSO_4 \cdot H_2O$		0.01–0.1
$CuSO_4 \cdot 5H_2O$		0.003–0.01
$Na_2MoO_4 \cdot 2H_2O$		0.01–0.1

(from Stanbury & Whitaker, 1984).

These organisms are obligately aerobic. For obligate anaerobes molecular O_2 is a toxic substance. Some organisms are facultative anaerobes and can grow with or without molecular O_2. Lactic acid bacteria have an exclusive fermentative energy-yielding metabolism, but are not sensitive to the presence of oxygen [4]. *Saccharomyces cerevisiae* produces ethanol anaerobically and cell mass aerobically, and it can shift from a respiratory to a fermentative mode of metabolism.

Sodium and chloride ions are respectively the principal extracellular cations and anions in animals and plants. Potassium is the principal intracellular cation. *Candida intermedia*, (SCP) single cell protein, grows better on normal alkanes with sources of N, O, and P, if small amounts of $ZnSO_4 \cdot 7H_2O$ are added [7]. Takeda, et al., reported that yeast can be grown by continuously feeding a medium consisting of hydrocarbon fractions boiling at 200° to 360°C, small amounts of inorganic nitrogen, inorganic salts and organic nitrogen to which the fermentation waste liquor previously used or CSL and ethanol are added [7]. This suggests that SCP production on hydrocarbons may be an outlet for stillage from ethanol-from-beet molasses plants or whey. Table 4.10 shows the approximate composition of such concentrated stillage from a European producer [5].

The combination of minerals is also important in regulating the electrolytic and osmotic properties of the cell interior. In most cases, the complex industrial carbon and/or nitrogen sources supply sufficient minerals for proper fermentation.

6.0 THE VITAMIN SOURCE AND OTHER GROWTH FACTORS

Vitamins are growth factors that fulfill specific catalytic needs in biosynthesis and are required in only small amounts. They are organic compounds that function as coenzymes or parts of coenzymes to catalyze many reactions. Table 4.11 itemizes the vitamins together with their active forms, catalytic function, molecular precursors, and raw material sources [10].

The vitamins most frequently required are thiamin and biotin. Required in the greatest amounts are

TABLE 4.10 Composition of a Concentrated French Stillage (Ethanol-from-beet molasses)

Moisture	31.4%
True protein	4.7%
Betaine	10.6%
L-pyroglutamic acid	7.2%
Lactic acid	4.1%
Acetic acid	1.3%
Butyric acid	1.3%
Other organic acids	1.8%
Glycero	l 5.4%
Raffinose	0.3%
Glucose + fructose	0.9%
Melibiose	1.4%
Inositol	0.7%
k^+	8.1%
Na^+	1.7%
Ca^{2+}	0.4%
Mg^{2+}	0.03%
Fe^{3+}	0.02%
P_2O_5	0.34%

usually niacin, pantothenate, riboflavin, and some (folic derivatives, biotin, vitamin B_{12} and lipoic acid) are required in smaller amounts [4]. In industrial fermentations, the correct vitamin balance can be achieved by the proper blending of complex materials and, if required, through the addition of pure vitamins. A satisfactory growth medium for baker's yeast, for example, can be achieved by mixing cane molasses, rich in biotin, with beet molasses, rich in the B-group vitamins. The production of glutamic acid by *Corynebacterium glutamicum* is a function of the concentration of biotin in the medium, which must be maintained in the range of 2–5 μg/l [4].

Organic growth factors are: vitamins, amino acids, purines and pyrimidines. Some twenty-two amino acids enter into the composition of proteins, so the need for any specific amino acid that the cell is unable to synthesize is obviously not large. The same argument applies to the specific need for a purine or pyrimidine: five different compounds enter into the structure of the nucleic acids. An often cited example of the importance of nutritional quality of natural sources occurred during the early phases of the development of the penicillin process [4]. A fivefold improvement in antibiotic yield was obtained when CSL was added to the fermentation medium for *Penicillium chrysogenum.* It was later found that the CSL contained phenylalanine and phenylethylamine, which are precursors for penicillin G. Today, any one of several nitrogen sources are used in conjunction with continuous additions of another precursor for penicillin G, phenylacetic acid [4].

Often, cells of a single microbial strain can synthesize more than one member of a chemical family. The final yields of the various members can be shifted by appropriate precursor pressure. The absence or presence of certain growth factors may accomplish this. In the absence of either exogenous phenylalanine or tryptophan, the ratio of tyrocidines A:B:C synthesized by *Bacillus brevis* is 1:3:7. If either L- or D-phenylalanine is provided, the main component formed is tyrocidine A. If L- or D-tryptophan is furnished, component D predominates; when both phenylalanine and tryptophan are supplied, each of the four components is synthesized [9].

Metabolic modifiers are added to the fermentation media to force the biosynthetic apparatus of the cell in a certain direction [11]. Most metabolic blocks of commercial importance, however, are created by genetic manipulation. In screening media formulations in a relatively short period of time, banks of shake flask cultivations as well as continuous culture methods are most useful.

Betaine or 1-carboxy-N, N, N-trimethylmethanammonium hydroxide inner salt is a relatively new and very interesting compound from a nutritional point of view. Its formula is

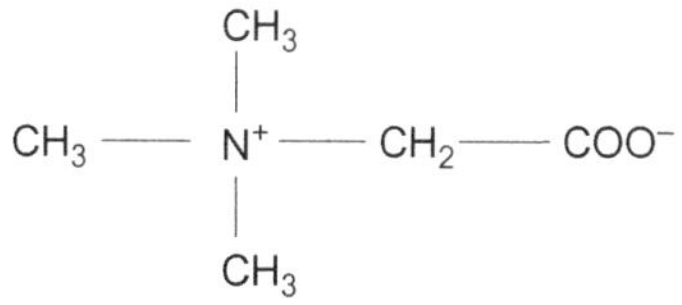

Betaine is important in both catabolic and anabolic pathways [12]. It is a methyl donor in the body synthesis of such essential compounds as methionine, carnitine and creatinine. Betaine is a very important regulator of osmotic pressure in different plants, bacteria and marine animals. It regulates the osmolality (number of moles of solute per 1000 gram of solvent) by acting as a nonpolar salt in the cells by adjusting the concentration of salts and/or loss of water. Betaine is a growth factor in many fermentation processes, stimulating the overproduction of desired metabolites. It functions as an osmoprotectant, as a metabolic regulator and as a precursor or intermediate. Betaine or trimethylglycine enhances biochemical reactions, not only in bacteria and fungi, but also in plants and in mammals. Tissue has only a limited capacity for

TABLE 4.11 Vitamins: their Sources and Metabolic Functions [10]

Accessory Growth Factor	Active Form	Chemical Group Transferred	Substance Needed to Fulfil Metabolic Requirement	Raw Material Source of Growth Factor
Thiamin (vitamin B_1)	Thiaminepyro-phosphate	Decarboxylation and aldehyde groups	i. Pyrimidine ii. Thiazole iii. Pyrimidine + Thiazole iv. Thiamine	Rice polishings Wheat germ Yeast
Riboflavin (vitamin B_2)	i. Flavin mononucleotide ii. Flavin adenine dinucleotide	Hydrogen Hydrogen	i. Riboflavin	Cereals Cornsteep liquor Cottonseed flour
Pyridoxal (vitamin B_6)	Pyridoxal phosphate	Amino group and decarboxylation	i. Pyridoxine ii. Pyridoxamine or pyridoxal iii. Pyridoxal phosphate	*Penicillium* spent mycelium Yeast *Rice polishing* Cereals Wheat seeds Maize seeds Cornsteep liquor Cottonseed flour
Nicotinic acid or nicotinamide	i. Nicotinamide adenine dinucleotide ii. Nicotinamide adenine dinucleotide phosphate iii. Nicotinamide mononucleotide	Hydrogen Hydrogen Hydrogen	i. Nicotinic acid or nicotinamide ii. Nucleotides of nicotinamide	*Penicillium* spent mycelium Wheat seeds Live
Pantothenic acid	Coenzyme A	Acyl group	i. Pantothenic acid	Beet molasses *Penicillium* spent mycelium Cornsteep liquor Cottonseed flour
Cyanocobalamin (Vitamin B_{12})		Carboxyl displacement Methyl group synthesis	i. Cyanocobalamin ii. Other cobalamins	Activated sewage sludge Liver Cow dung *Streptomyces griseus* mycelium Silage Meat
Folic acid	Tetrahydrofolic acid	Formyl group	i. Folic acid ii. Para-amino benzoic acid	*Penicillium* spent mycelium Spinach Liver Cottonseed flour
Biotin	Biotin	CO_2 fixation	i. Biotin	High test cane molasses Corn steep liquor *Penicillium* spent mycelium Cottonseed flour
Lipoic acid	Lipoic acid	Hydrogen and acyl groups	i. a-Lipoic acid or thioctic acid	Liver
Purines	Purine' nucleotides		i. Purines ii. Nucleotides derived from purines	Meat Dried blood
Pyrimidines	Pyrimidine nucleotides		i. Pyrimidines ii. Nucleotides derived from pyrimidines	Meat
Inositol	Phosphatides		Inositol	Cornsteep liquor
Choline	Phosphatides		Choline	Egg yolk Hops
Hemins	Cell hemins	Electrons	Hemins	Blood

(*With permission from* [10].

betaine synthesis and during conditions such as stress, the need for additional betaine arises. Cell organs involved in energy metabolism, such as mitochondria and chloroplast, contain high levels of betaine.

These are the respective sites of photosynthetic and respiratory function in eukaryotic cells. Sugar beets are high in betaine. It ends up in the molasses during the recovery of sucrose. When fermenting beet molasses to ethanol, the betaine ends up in the stillage.

7.0 PHYSICAL AND IONIC REQUIREMENTS

Each reaction that occurs within the cell has its own optimum (range of) conditions. For instance, although a given medium may be suitable for the initiation of growth, the subsequent development of a bacterial strain may be severely limited by chemical changes that are brought about by the growth and metabolism of the microorganisms themselves. In the case of glucose-containing media, organic acids that may be produced as a result of fermentation may become inhibitory to growth. In contrast, the microbial decomposition or utilization of anionic components of a medium tends to make the medium more alkaline [4]. The oxidation of a molecule of sodium succinate liberates two sodium ions in the form of the very alkaline salt, sodium carbonate. The decomposition of amino acids and proteins may also make a medium alkaline as a result of ammonia production. To prevent excessive changes in the hydrogen ion concentration, either buffers or insoluble carbonates are often added to the medium. The phosphate buffers, which consist of mixtures of mono-hydrogen and dihydrogen phosphates (e.g., K_2HPO_4 and KH_2PO_4), are the most useful ones. KH_2PO_4 is a weakly acidic salt, whereas KH_2HPO_4 is slightly basic, so that an equimolar solution of the two is nearly neutral, having a pH of 6.8. If a limited amount of strong acid is added to such a solution, part of the basic salt is converted to the weakly acidic one:

$$K_2HPO_4 + HCl \rightarrow KH_2PO_4 + KCl$$

If, however, a strong base is added, the opposite conversion occurs:

$$KH_2PO_4 + KOH \rightarrow K_2HPO_4 + H_2O$$

Thus, the solution acts as a buffer by resisting radical changes in hydrogen ion concentration (pH) when acid or alkali is produced in the medium. Different ratios of acidic and basic phosphates may be used to obtain pH-values from approximately 6.0–7.6. Good buffering action, however, is obtained only in the range of pH 6.4–7.2 because the capacity of a buffer solution is limited by the amounts of its basic and acidic ingredients. Bacteria and fungi can generally tolerate up to 5 g/L of potassium phosphates. When a great deal of acid is produced by a culture, the limited amounts of phosphate buffer that may be used become insufficient for the maintenance of a suitable pH. In such cases, carbonates may be added to media as "reserve alkali" to neutralize the acids as they are formed. By adding finely powdered $CaCO_3$ to media, it will react with hydrogen ions to form bicarbonate, which in turn is converted to carbonic acid, which decomposes to CO_2 and H_2O in a sequence of freely reversible reactions [1]. Several fermentations are run on pH control through addition on demand of acid or alkali. The pH of the medium affects the ionic states of the components in the medium and on the cellular exterior surface. Shifts in pH probably affect growth by influencing the activity of permease enzymes in the cytoplasmic membrane or enzymes associated with enzymes in the cell wall [4]. The pH affects solubility; proteins will coagulate and precipitate (salting-out) at their isoelectric point.

In the preparation of synthetic media, sometimes precipitates form upon sterilization, particularly if the medium has a relatively high phosphate concentration. The precipitate results from the formation of insoluble complexes between phosphates and mainly calcium and iron cations. By sterilizing the calcium and iron salts separately and then adding them to the sterilized and cooled medium, the problem can be avoided [4].

Alternatively, a chelating agent such as EDTA (ethylenediamine tetraacetic acid), may be added at a concentration of approximately 0.01%, to form a soluble complex with these metals. When two or more microorganisms are placed in a medium, their combined metabolic activities may differ, either quantitatively or qualitatively, from the sum of the activities of the individual members growing in isolation in the same medium. Such phenomena result from nutritional or metabolic interactions and are collectively termed *synergistic* effects [13].

Typical difficulties in scaleup can occur due to ionic strength. A laboratory fermenter or shake flask sterilized in an autoclave will have a higher nutrient concentration after sterilization due to evaporation. A large fermenter, sterilized in part by direct steam injection, will have a lower concentration, due to steam condensate pickup. These differences usually do not prevent growth, but can certainly alter yield of batches. The range of pH tolerated by most microorganisms can be as broad as 3 to 5 pH units. Rapid growth and/or reaction rates are normally in a much more narrow range of 1 pH unit or less. In small-scale experiments, it seems to be common to use NaOH for pH control, but it may only contribute problems in scaleup.

8.0 MEDIA DEVELOPMENT

Factors that must be considered in developing a medium for large-scale fermentations are:

1. The nutrient requirements of the selected microorganism
2. The composition of available industrial nutrients
3. The nutrient properties in relation to storage and handling, pasteurization or sterilization, processing and product purification
4. Cost of the ingredients

In the calculation of the cost of the medium, all costs have to be recognized. Thus, in addition to the purchase price, which is obvious, material handling and storage, labor and analytical requirements must be included. Dilute nutrients require greater storage volumes than concentrated sources. The stability of nutritional requirements is important, refrigeration or heating may be required. High volatility (alcohols), corrosiveness (acids and alkalis) and explosive characteristics (starch powders) pose certain environmental and safety risks. Pretreatment costs for certain raw materials, such as starch liquefaction and saccharification, may be substantial. The rheological properties of the medium may affect such items as mixing, aeration and/or temperature control. The surface tension has an effect on the foaming tendency of the broth. Finally, the solids concentration, odor, color, etc., are pivotal in determining the costs of product recovery and purification.

Product concentration, yield, and productivity are among the most important process variables in determining conversion costs [4]. The concentration of the product influences its recovery and refining costs. Raw material costs are affected by the yield. Productivity, or the rate of product formation per unit of process capacity, helps determine the amount of capital, labor, and indirect costs assignable to the product. The influence of the medium on the interplay of these three variables cannot be ignored.

Raw material costs in fermentations may vary from 15 to 60% of the total manufacturing cost. Simply trying to cut manufacturing costs by substituting raw materials with cheaper ones may not be the answer. If carbohydrate costs represent, for example, 10% of the total manufacturing cost, it requires a 50% reduction in the carbon source to effect a 5% reduction in manufacturing cost. The question is then how the new raw material effects the multiple interactions of a complex medium.

A better approach would be to explore how the impact of a change in raw material would impact the product yield and purity. This could have a far greater influence on the final cost than a cheaper carbohydrate source. Performing multivariable experiments would be the most effective way. Interacting variables used are nutrients, pH, aeration, temperatures, etc. This allows the determination of optimum levels for a given process. As the number of variables increases it becomes impractical to investigate all combinations. An evaluation of five nutrients at only three concentration levels yields 3^5 or 243 combinations and possible trials. A statistical approach may be taken to deal with the complexity. Several computer programs for statistical experimental design are available. This allows for a three dimensional view of interactions between key variables through response surface methodology techniques. These techniques not only allow for optimized process conditions, but also lend insight into process requirements that will exceed simple optimization trials. Geiger discusses the statistical approach in greater detail in Ch. 4.

Most large-scale fermentations, i.e., batch size in excess of 50,000 liters, use inexpensive raw materials in large volume. Here materials such as molasses and corn steep liquor are normally purchased in truckload quantities. The carbohydrate source, because of its large volume, is the only fermentation raw material, which has any influence on plant location. In order to maximize production rates one may operate a continuous fermentation process if the product is associated with growth of the organism cells or an intracellular product. Low sensitivity to contamination, such as with a thermophile or production at a reduced pH, are of importance. It is impractical to consider anything but batch fermentation when the product is an extracellular metabolite that is sensitive to culture and medium balance. Similarly, batch fermentation is preferred when the culture can undergo mutation at extended operating time and is sensitive to contamination.

There are many and varied conditions worldwide which impact on the cost of fermentation raw materials. These can be climatic, e.g., drought or floods, or political, e.g., government subsidies for, or restriction on, farm products or a national ethanol fuel program. These conditions greatly affect the world price of sugar, molasses and corn and are responsible for much of the variability. The rise and fall of sugar prices affects all sources of carbohydrates. To ensure against the effects of wide swings, a prudent course of action would be to develop processes that permit alternate sources. The demands of the final product may have an important bearing on the selection of the fermentation ingredients. Odor and color on the one hand may play a role, on the other product purity specifications which are very demanding, as is the case for vaccines, require extremely pure nutrients.

Some information on industrial protein sources is given in Tables 4.12–4.14. These were taken from *Traders Guide to Fermentation Media Formulations*, which can serve as an excellent reference [4].

TABLE 4.12 Composition of Yeast Extract (Standard Grade)

Constituent	%w/w
Moisture	>5 as loss on drying
Total nitrogen	<7.0
Ash	10.1
pH of 2% solution 6.7 ± 0.2 at 25°C (autoclaved for 2 min.)	
Coagulable protein	no precipitate in 5% solution, boiling
Chloride	0.19
Phosphorus	9.89
Sodium	0.32
Potassium	0.042
Iron	0.028
Calcium	0.04
Magnesium	0.030
Silicon dioxide	0.52
Manganese (ppm)	7.8
Lead (ppm)	16
Arsenic (ppm)	0.11
Copper (ppm)	19.0
Zinc (ppm)	8.8
AMINO ACIDS	
Lysine	4.0
Tryptophan	0.88
Phenylalanine	2.2
Methionine	0.79
Threonine	3.4
Leucine	3.6
Isoleucine	2.9
Valine	3.4
Arginine	0.78
Tyrosine	0.6
Aspartic	5.1
Glutamine	6.5
Glycine	2.4
Histidine	0.94

These materials would primarily be used in large-scale fermentations. The economic implications of a fermentation medium on the profitability of a process have to be considered before fermentation process design can be started (Table 4.15).

Grade and quality information can be obtained for many materials from written sources such as the United States Pharmacopeia, Food Chemical Codex, The Merck Index, and suppliers' catalogs.

The usual fermentation process classification was in high-volume-low-cost products and low-volume-high-cost products. The first were carried out in fermenters with working volumes of 50,000 liters and up, the latter in fermenters of less than 50,000 liter working volumes. With the advent of biotechnology, the extremely-low-volume-extremely-high-price products entered the market. Vaccines, hormones, specialty enzymes and antibodies fall into this class. A working volume of 500 liters is considered large. Selections of raw materials are based upon these definitions of scale. Commercially prepared microbiological media are available for these fermentations. The variability and consistency of these prepared media are, of course, excellent and unlike that of the industrial raw materials.

9.0 EFFECT OF NUTRIENT CONCENTRATION ON GROWTH RATE

When inoculating a fresh medium, the cells encounter an environmental shock, which results in a lag phase. The length of this phase depends upon the type of organism, the age and size of the inoculum, any changes in nutrient composition, pH and temperature. When presented with a new nutrient the cell adapts itself to its new environment and normally produces the required enzyme.

Essentially all nutrients can limit the fermentation rate by being present in concentrations that are either too low or too high. At low concentrations, the growth rate is roughly proportional to concentration, but as the concentration increases, the growth rate rises rapidly to a maximum value, which is maintained until the nutrient concentration reaches an inhibitory level, at which point the growth rate begins to fall again. The same type of hyperbolic curve will be obtained for all essential nutrients as the rate-limiting nutrient. The effects of different nutrients on growth rate can best be compared in terms of the concentrations that support a half-maximal rate of growth, this is the saturation constant (K_s) [13]. For carbon and energy sources this concentration is usually on the order of 10^{-5} to 10^{-6} M, which corresponds for

TABLE 4.13 Composition of Various Hydrolyzed Proteins (%)

Constituent	Blood	Meat Peptone	Meat Protein	Casein	Cottonseed Protein
Total nitrogen	10.0	9.5	9.7	13.5	8.6
Amino nitrogen (as % of TN-Sorensen)	8.0	8.6	10.5	30.0	31.0
AMINO ACIDS					
alanine	2.3	2.5	—	2.6	2.2
arginine	2.3	4.3	3.8	3.7	4.0
aspartic acid	2.5	3.7	3.9	5.7	4.2
cystine	0.5	0.2	0.4	0.3	0.8
glutamic acid	2.1	5.5	5.7	20.1	9.9
glycine	0.7	3.4	—	1.0	2.1
hydroxyproline	0.5	4.1	—	—	—
histidine	4.0	0.4	1.2	2.2	1.2
leucine	16.7	2.2	3.8	9.4	3.2
isoleucine	—	1.5	2.5	4.8	1.6
lysine	5.2	2.7	4.2	6.8	1.8
methionine	—	0.6	1.1	2.8	0.9
phenylalanine	3.0	1.2	2.6	5.5	2.8
proline	1.5	3.0	—	9.7	2.2
serine	0.3	1.7	—	5.6	2.3
threonine	—	1.0	2.1	4.3	1.6
tyrosine	2.5	0.8	—	4.4	1.7
tryptophan	1.0	0.2	0.2	1.2	0.3
valine	—	1.8	2.6	6.2	2.2
Form of material	65% solids solution	60% solids solution	60% solids solution	dry powder	dry powder

glucose to a concentration between 20 and 200 mg/L. In general, K_s for respiratory enzymes, those associated with sugar metabolism, is lower than K_s for the hydrolytic enzymes, those associated with primary substrate attack.

The Monod equation is frequently used to describe the stimulation of growth by the concentration of nutrients as given by:

$$\mu = \mu_{max} S/(K_s + S) \quad (4.1)$$

where,

μ = specific growth rate, h^{-1} defined as (1/X) (dx/dt)
μ_{max} = maximum value of μ, h^{-1}
K_S = saturation constant, gL^{-1} at $\mu_{max}/2$
S = substrate concentration, gl^{-1}
t = time, h

The saturation constant K_s for *Saccharomyces cerevisiae* on glucose is 25 mg/L, for *Escherichia* on lactose: 20 mg/L, and for *Pseudomonas* on methanol: 0.7 mg/L. Here, μ_{max} is the maximum growth rate achievable when $S \gg K_s$ and the concentration of all other essential nutrients is unchanged. The saturation constant K_s is the approximate division between the lower concentration range where μ is essentially linearly related to S and the higher rate where μ becomes independent of S [9].

TABLE 4.14 Typical Nutrient Composition of Distillers Feeds (Corn)

From Corn	
Moisture %	4.5
Protein %	28.5
Lipid %	9.0
Fiber %	4.0
Ash %	7.0
AMINO ACIDS mg/g	
Lysine	0.95
Methionine	0.50
Phenylalanine	1.30
Cystine	0.40
Histidine	0.63
Tryptophan	0.30
Arginine	1.15
MINERALS	
K%	2.10
Na%	0.15
Ca%	0.30
Mg%	0.60
P%	200.0
Fe, ppm	60.0
Mn, ppm	100.0
Zn, ppm	55.0
Cu, ppm	1.60

TABLE 4.15 Nutrient Sources for Industrial Fermentation

Nutrient	Raw Material
Carbon	
Glucose	corn sugar, starch, cellulose
Sucrose	sugarcane, sugar beet molasses
Glycerol	
Starch	
Maltodextrine	
Lactose	milk whey
Fats	vegetable oils
Hydrocarbons	petroleum fractions
Nitrogen	
Protein	soybean meal, corn steep liquor, distillers' solubles
Ammonia	pure ammonia or ammonium salts urea
Nitrate	nitrate salts
Phosphorus source	phosphate salts
Vitamins and growth factors	
	Yeast, Yeast extract
	Wheat germ meal, cotton seed meal
	Beef extract
	Corn steep liquor

Growth media are required for industrial fermentation, since any microbe requires water, (oxygen), an energy source, a carbon source, a nitrogen source and micronutrients for growth.
Carbon & energy source + nitrogen source + O_2 + other requirements → Biomass + Product + byproducts + CO_2 + H_2O + heat
http://en.wikipedia.org/wiki/Industrial_fermentation

The effect of excessive nutrient or product concentrations on growth is often expressed empirically as:

$$\mu = \mu_{max} K_i/(K_i + I) \tag{4.2}$$

where,

K_i = inhibition, constant, gl^{-1}
I = concentration of inhibitor, gl^{-1}

Equations (4.1) and (4.2) can be combined to illustrate the characteristics common to many substrates:

$$\mu = \mu_{max}\left(\frac{S}{K_S + S}\right)\left(\frac{K_i}{K_i + S}\right) \tag{4.3}$$

The kinetic models of Eqs. (4.1)–(4.3) are illustrated in dimensionless form in Figure 4.2. It can be seen that the adding of large amounts of substrate to provide high concentrations of product(s) and to overcome the rate-limiting effects of Eq. (4.1) can result in concentrations that are so high that the fermentation is limited by the effects of Eq. (4.2). The ideal operating range would be $1 < S/K_s < 2$ where the growth rate is near its maximum and is relatively insensitive to substrate concentration [4]. The concentration ranges which enhance or inhibit fermentation activity vary with each microorganism, chemical species, and growth conditions.

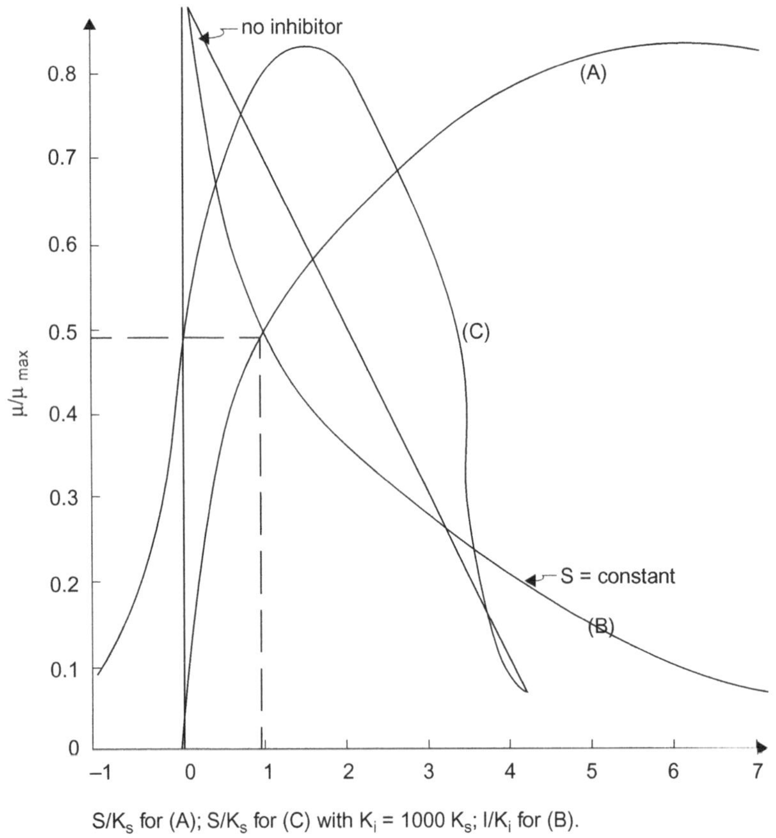

FIGURE 4.2 *(A)* Monod Growth Model; *(B)* model for growth inhibition; and *(C)* model for substrate activation and inhibition of growth.

Product formation is related to the substrate consumption as follows:

$$\Delta P = Y_{p/s} \tag{4.4}$$

where,

ΔP = product concentration - initial product concentration in gl^{-1}.
ΔS = substrate concentration - initial substrate concentration in gl^{-1}
$Y_{p/s}$ = product yield, $g\text{-}P.g\text{-}S^{-1}$

This equation is especially useful when the substrate is a precursor for the product. Many other models are available. When calculating a material balance for medium formulation, the choice of models depends upon the data that are available.

Ethanol-from-biomass fermentation is an example of product inhibition, while high fermentable carbohydrate concentrations also can become inhibitive. Aiba and Shoda developed a mathematical model for the anaerobic fermentation, where nine constants are required to describe the model [13].

$$\frac{dS}{dt} = \frac{1}{Y_G}\frac{dX}{dt} + \frac{1}{Y_P}\frac{dP}{dt} + MX$$

$$\frac{dX}{dt} = \frac{\mu_o}{1 + \frac{P}{K_P}} \frac{S}{K_s + S} X$$

$$\frac{dP}{dt} = \frac{q_o}{1 + \frac{P}{K_P}} \frac{S}{K'_s + S} X$$

where,

S, P, X = concentration of substrate, product and cell mass, gl^{-1}
M = maintenance constant, $g\text{-}S, g\text{-}X^{-1}.h^{-1}$
K_s, K'_s, = saturation constants, gl^{-1}
K_p, K'_p, = inhibition constants, gl^{-L}
q_o = maximum specific product formation at $P = 0$, $g\text{-}P.g\text{-}s^{-1}$
t = time, h
μ_o = maximum specific growth rate at $P = 0$, h^{-1}
Y_G = true growth constant, $g\text{-}X.g\text{-}S^{-1}$
Y_P = product yield, $g\text{-}P.g\text{-}X^{-1}$

It is doubtful that cellular models of any greater complexity will have much utility in soluble substrate fermentations [9]. However, another level of complexity may be warranted in insoluble substrate fermentations such as hydrocarbons and cellulose.

Immobilized microorganisms and enzymes are becoming commercially available. One example is co-immobilized yeast and glucoamylase (Gist Brocades NV), which is used to simultaneously saccharify starch dextrins into glucose and ferment this to ethanol in fluidized bed reactors.

In these fluidized bed fermenters, the "reaction rate" is controlled by the superficial flow velocity and its effects on the diffusion of substrate from the bulk of the medium to the enzymatically active surface, by the enzymatic reaction at the surface, or by diffusion of the reactant products back into the bulk of the medium being fermented.

REFERENCES

[1] R.Y. Stanier, M. Doudoroff, E.A. Adelberg, The Microbial World, third ed., Prentice Hall, Englewood Cliffs, NJ, 1976.

[2] R.C. West, Handbook of Chemistry and Physics, sixtieth ed., CRC Press, Boca Raton, Florida, 1979.

[3] T.P. Bennett, E. Frieden, Modern Topics in Biochemistry, The MacMillan Company, New York, 1967.

[4] D.W. Zabriski, et al., Trader's Guide to Fermentation Media Formulation, Trader's Oil Mill Co., Ft. Worth, TX, 1980.

[5] W.H. Kampen, Private Work or Correspondence.

[6] Henry C. Vogel (Ed.), Fermentation and Biochemical Engineering Handbook, Noyes Publications, Park Ridge, NJ, 1983.

[7] S. Gutcho, Proteins from Hydrocarbons, Noyes Data Corporation, Park Ridge, NJ, 1973.

[8] A.H. Rose (Ed.), Advances in Microbial Physiology, vol. 4, Academic Press, New York, 1970.

[9] H.J. Peppler, D. Perlman, second ed., Microbial Technology, vol. I, Academic Press, New York, 1979.

[10] A. Rhodes, D.L. Fletcher, Principals of Industrial Microbiology, Ch. 6 Pergamon, New York, 1966

[11] A. Dinsmoor Webb (Ed.), Advances in Chemistry Series 137, ACS, Washington, DC, 1974.

[12] S. Budavari, et al., The Merck Index, An Encyclopedia of Chemicals Drugs and Biologicals, eleventh ed., Merck and Company, Inc., Rahway, New Jersey, 1989.

[13] J.E. Bailey, D.F. Ollis, Biochemical Engineering Fundamentals, McGraw-Hill Book Co., New York, 1977.

FURTHER READING

[1] P.F. Stansbury, A. Whitaker, S.J. Hall, Principles of Fermentation Technology, 1997.

[2] J.E. Bailey, P.F. Ollis, Biochemical Engineering Fundamentals, McGraw Hill Publication.

Chapter 5

Fermentation for Biofuels and Bio-Based Chemicals

Steve Licht

1.0 INTRODUCTION, HISTORY, AND THEORY

1.1 The Biofuel and Bio-Based Chemical Industries Originated with Natural Fermentation Products

Beer, wine, vinegar.

Not just menu items. They are all products of naturally occurring fermentation processes that were in beneficial use at commercial scale, long before humankind discovered or understood microbes—the agents that transform raw agricultural products into these savory and sour treats.

Boiling and condensing vapor from fermented brews made from locally available grains, fruits, tubers, or sugar led to the production of "distilled spirits" (note the non-material connotation of "spirits") by people in many parts of the world. Whatever name they were given by their makers—whisky, rum, brandy, vodka, grappa—they all had the same intoxicating quality when imbibed by people or animals. And their active ingredient was the same: ethyl alcohol, a name that was shortened to "ethanol" in 1957 with adaption of the IUPAC (International Union of Pure and Applied Chemistry) system for naming organic chemical compounds.

If brews made in fermentation vessels were left open to air for too long a period before being distilled, the intoxicating quality was replaced by a strong sour quality, in gradients proportional to the time the brew was exposed to air. The distilled product from these over-aged brews was called "vinegar." It was the first bio-based chemical, not intended as a beverage, in production. The roots of its usage, and of practical methods for its production at a commercial scale, lead deep into the past. Vinegar was known, used, and traded internationally before the time of the Roman Empire.

Let us return now to the story of the better alcoholic fermentation brews, the ones that were not aged until the alcohol turned into vinegar.

After strongly or repeatedly distilling, the distilled spirits were known to acquire the ability to burn, accidentally or on purpose. Strong spirits were especially easy to ignite when soaked into cloth, or placed in an oil lamp with a wick. It became known that under just the right conditions, or the wrong conditions (depending on your intention), the vapor from distilled spirits, dispersed in air, sometimes would explode, with major force.

Biofuels are not new. No doubt, the idea of using distilled spirits deliberately as fuel occurred to many people in many places, in the centuries before ours. Thus, the oldest known fermented biofuel involved firing wicked alcohol burners in alchemists' laboratories and in the butlers' pantries of the rich, long before the discovery of microorganisms or of the science behind fermentation.

The same varieties of locally made distilled spirits that could be used as intoxicating beverages were also found to be useful in many of the practical arts. Distilled spirits of varying strengths were useful for thinning the vegetable-based or animal-derived oils used as bases for making paint, ink, lacquer, or shellac (made by dissolving beetle shells in alcohol, hence the name "shell-lac") for coating wood, plaster, leather, and metal; as solvents for cleaning grease and oil stains from butchery and tanning; and for many other home, farm, and commercial uses.

Distilled spirits intended for these practical uses were available for sale in hardware stores and medicine shops, in bottles, jugs, cans, and barrels, when the internal combustion engine was developed and patented by Otto in the second half of the 19th century.

In the beginning, internal combustion engines were necessarily stationary. They were powered by the explosion of air mixed with gases generated from coal, for the purpose of providing town lighting, and this gas was delivered in pipes.

In an effort to make such engines usable away from a permanent base installation, without being attached by hose to fixed gas pipes, and movable from place to place—or perhaps even *mobile*, able to power a boat or a

Fermentation and Biochemical Engineering Handbook.

railcar—various types of flammable liquids then known were tested by stationary engine makers such as Otto, and some were found to be suitable.

Among the most suitable fuels for mobile internal combustion engines were benzine, a coal tar distillate, containing benzene along with other hydrocarbons; turpentine, consisting chiefly of terpenes, volatile unsaturated hydrocarbons found naturally in plants, and made by boiling pine tree stumps and water in enclosed vats and decanting the oily turpentine from the surface of the condensate; and, of course, distilled spirits containing ethyl alcohol.

From the time of first invention in 1885 onward for about ten years, the first prototype automobiles, truly "horseless" carriages, powered by internal combustion engines using the 4-stroke Otto cycle, were operated on the streets of Europe in the absence of a pre-existing network of fueling stations. In order to procure fuel, their drivers could (and did) stop at just about any hardware store or pharmacy to purchase and refuel with benzine intended for service as a paint thinner, or with distilled spirits as a viable biofuel alternative.

It may not be commonly known in our day that ethanol was one of several viable fuels commonly used in automobiles of the pioneering era, 125 years ago. At least two of the viable fuels of that era were derived by distillation from plant material or fermented plant material: distilled spirits and turpentine. They both fit the definition of what we classify as "biofuels" today, in the 21st century.

Not until the closing years of the 19th century, when creative businesspeople got the idea to make money by recovering and broadly distributing petrol or gasoline (then a waste product from the recently commercialized distillation of crude oil to make kerosene for lighting) for use as an Otto cycle motor fuel, did the economic balance between the biofuel choices (ethanol and turpentine) and the mineral alternatives (coal–gas condensates and petroleum distillates) tilt heavily in favor of the petroleum products.

1.2 Theory: Principles of Chemistry and Biology Guide the Selection of Fermentation Products, Substrates, and Organisms

Each micro-organism, be it naturally occurring or modified by genetic engineering, generates a suite of metabolic enzymes, with many alternative metabolic pathways that begin with different prospective nutrients and end with different fermentation products and byproducts.

The limits of possibility for suitable substrates and feasible fermentation products are written in the map of enzyme-enabled chemical transformations that are expressed in any given organism strain. It is usual for any one strain to be able to use any one or more of a selection of substrates for its growth and energy source. Related organisms (such as bacteria of the same genus) typically share similar metabolic pathways and can grow on a similar set of substrates. More distantly related organisms (such as one species of fungus arbitrarily paired with one species of bacteria) usually share at least half of the same metabolic pathways. Yet each make some enzymes not made by the other species and therefore are capable of carrying out biochemical reactions that the other one cannot.

1.2.1 One Example of the Way that One Organism's Metabolic Pathways Enable it to Process a Variety of Substrates into a Selection of Several Exported Fermentation Products

Here we will use one anaerobic bacteria as an example of the theoretical evaluations that can be made regarding selection of a fermentation organism and pairing it with substrates and fermentation products.

The identity of several fermentation products that may be produced by natural biological pathways in a well-studied bacteria, *Clostridium ljungdahlii*, recently recruited for use in biofuel production from sugars or synthesis gases containing carbon dioxide and carbon monoxide [1], is represented in Fig. 5.1 and is also found at the Internet address given in Reference [2]:

The potential substrates and products are shown around the edges of the metabolic diagram. In the upper left corner is the most common metabolic pathway: glycolysis/gluconeogenesis, which digests glucose and similar common C6 sugars (this pathway is essentially universal in heterotrophic organisms, including bacteria, fungi, and animals). To the right of this is the pentose phosphatase pathway, which is found in a minority of heterotrophic organisms and which starts the digestion of C5 sugars before passing intermediates to the C6 glycolysis pathway. In the lower left of the diagram is a metabolic pathway unique to homoacetogenic bacteria: the Wood–Ljungdahl pathway, which is named after its discoverers. This pathway enables microbes so equipped to take in CO_2, CO, and/or H_2 as substrates, or "food sources"; no other known eukaryote or bacteria commonly studied and propagated in laboratories is able to do so.

As seen in Fig. 5.1, the substrates taken in by *Clostridium ljungdahlii* to be transforrmed into desired products and byproducts may be any of the following molecular species:

- Hexoses, six-carbon (C6) sugars:
  - Glucose
  - Fructose
  - Mannose
  - Rhaminose
- Pentoses, five-carbon (C5) sugars:
  - Xylose
  - Arabinose
  - Ribose

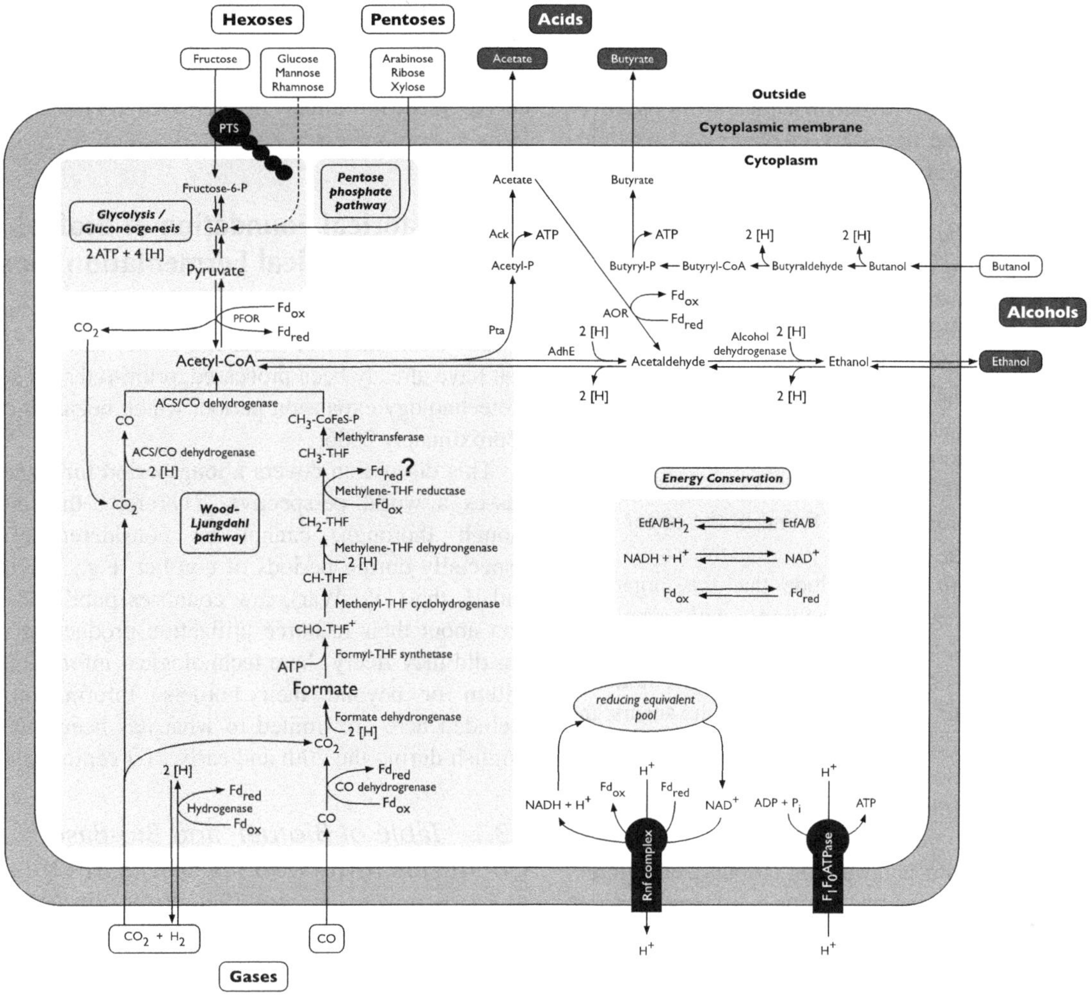

FIGURE 5.1 Metabolic Diagram of *Clostridium ljungdahlii.*

- Alcohols:
  - Ethanol
  - *N*-Butanol
- Organic acids:
  - Acetic acid
  - Butyric acid
- Three gaseous compounds found in combustion gases, or produced from solid hydrocarbons or biomass at high temperature under oxidizing or reducing conditions in a gasifier:
  - Carbon dioxide (CO_2)
  - Carbon monoxide (CO)
  - Hydrogen (H_2)

The products that are known to be made by *Clostridium ljungdahlii* and exported out of the cell are shown outside the borders of the diagram in *Fig. 5.1.* They consist of the following:

- Acetate
- Butyrate
- Ethanol
- Carbon dioxide
- Water
- Hydrogen gas (H_2)

These compounds are produced by the cell in addition to the many molecules made by the bacteria that are necessary for cellular life (e.g., amino acids, purines, pyridines, fatty acids) and that are retained inside the cell in its essential structures and mechanisms. Fermentation is concerned with the economic products produced by cells in excess of their own needs and exported. Therefore, for simplicity in describing the technology of biofuel- and bio-based chemical production, the multitude of other cellular chemical pathways are excluded from this discussion.

The metabolic steps that begin with sugars as substrates (starting at the top left of Fig. 5.1) all converge with the production of pyruvate (shown at the center of Fig. 5.1). Those that begin with CO_2, CO, and H_2 gases (starting at the bottom left) converge with the production of acetyl-CoA (in the center of Fig. 5.1). Pyruvate and acetyl-CoA are inter-convertible; this inter-convertibility makes it possible for reactions that run from one substrate to these two compounds to run in either the forward or reverse direction if the necessary enzymes and reactants are present at sufficient concentration, in unbound states, in the cell.

For *Clostridium ljungdahlii*, examination of the metabolic pathways shows that ethanol can be produced from acetyl-CoA (see the arrow exiting the left edge of the diagram). Acetate can also be made from acetyl-CoA (arrow going toward the top center); and acetate can be converted to ethanol (diagonal pathway from top center to right edge).

Alternatively, ethanol can be processed as a substrate (arrow leading in from the left edge) and converted to acetyl-CoA. Additionally, *n*-butanol can be oxidized to butyrate, in the reaction pathway found in the upper right quarter of the diagram.

From this diagram, we conclude that this organism, under the right fermentation conditions, with supply of appropriate substrates, could be used to produce ethanol or acetate from a variety of sugars or $CO/CO_2/H_2$ gas mixtures. It also could be used to produce acetate from the sugars, the gases, or the ethanol. Additionally, on an unrelated pathway, this organism can convert butanol to butyrate.

Genetic engineering is employed if it is desired to increase the production rate along one pathway, reduce the reaction rate of another pathway, or introduce a different chemical reaction (allowing the use of a different substrate compound, or leading to a different product) somewhere in the metabolic pathway. Those are the primary motivations inspiring genetic modifications to organisms intended for development in biofuel- or bio-based chemical production fermentation processes.

1.3 Historical Foundation of Biofuel- and Bio-Based Chemical Fermentation Industries

The following is a broad review of products that have already been made at industrial scale, and the substrates that have already been processed, prior to the present major biotechnology expansion period, which began in earnest in approximately 2006.

This discussion covers a long period time and encompasses a world perspective. Therefore, this discussion, though thorough, cannot be considered exhaustive. Especially during periods of conflict (e.g., World Wars I and II, the Cold War), few countries published information about their resource utilization production facilities; nor did they freely share technological information, either within or beyond their borders. Information sources included here are limited to what has been published in English during the 20th and early 21st centuries.

1.3.1 Table of Biofuel- and Bio-Based Chemical Production Processes in Operation at Commercial Scale, 1910 through 2013

Year Begun	Fermented Product	Substrate	Fermentor Type & Organism	Location	Scale	Additional Information	Ref.
Before 1900	Vinegar: 8%–10% acetic acid, with dissolved solids remaining from substrate	Cane sugar, molasses, crushed fruit, ethanol	Batch or fed-batch fermentation with aerobic bacteria, *Acetobacter aceti* or *Bacterium aceti*	Worldwide	Up to 1 million liters/yr. per site	8% Acetic acid limit in batch fermentation is due to starting high sugar osmotic inhibition. The 10% limit in fed-batch mode is due to the organism	[3, pp. 670–672]
Before 1900	Distilled vinegar, 3% acetic acid	Cane sugar, molasses, crushed fruit, ethanol	Batch or fed-batch fermentation with aerobic bacteria, *Acetobacter aceti* or *Bacterium aceti*	Worldwide	Up to 1 million liters/yr. per site	Batch fermentation to 8% followed by water dilution and distillation to low-boiling azeotrope (3%)	[3, pp. 670–672]

Year Begun	Fermented Product	Substrate	Fermentor Type & Organism	Location	Scale	Additional Information	Ref.
1915	Acetone	Cane sugar, molasses, corn	Batch fermentation of *Clostridium acetobutylicum* and *Clostridium saccharo-butylacetonicum liquifaciens*, anaerobic, at 50,000 gallon scale by 1940	Developed in England by Chaim Weizmann. Manufactured in the USA in Indiana (IN) and Illinois (IL)		Commercial Solvents Corporation was founded in the USA in 1920 to use the US & British Patents of 1915	[3, pp. 666–670]
1915	Butyl alcohol (Butanol)	Cane sugar, molasses, corn	Butanol is co-produced with acetone and ethanol by the above organisms at a mass ratio of about 60% butanol, 30% acetone, 10% ethanol	In England and in the USA at Terra Haute, IN and Peoria, IL		Commercial Solvents Corporation started production in the USA in 1920	[3, pp. 666–670]
Before 1900	Ethanol used as solvents, undenatured, 70% to 95% v/v	Made from cane sugar, beet sugar, potatoes, grapes, or practically any sugar or starch source	Batch anaerobic fermentation, following batch aerobic cell-growth stages, with *Saccharomyces cerevisiae* at fermentor sizes up to 200,000 liters by 1940	Worldwide	Up to 10 million liters/yr. per site. 20 million liters total U.S. production in 1940	70% Ethanol used as medical "rubbing alcohol"	[3, pp. 646–660]
By 1920	Denatured ethanol solvents, 90% to 100% v/v. Called "absolute alcohol" if 99100%	Same as undenatured ethanol	Same production lines as undenatured ethanol, using *Saccharomyces cerevisiae*, known in the beverage alcohol industry as "ale yeast" or "top-fermenting brewers' yeast" used for warm fermentations, as opposed to cold-fermenting *Saccharomyces pastorianus*, "lager yeast" or "bottom-fermenting yeast"	Worldwide	Up to 10 million liters/ yr. per site. 450 million liters total US production in 1940	"Denaturing" adds a poisonous compound like methanol to discourage drinking	[3, pp. 646–660]
From circa 1916 (start of WWI) through 1945 (end of WWII)	Motor fuel ethanol blends, up to 20%, in gasoline /petrol	Made by blending anhydrous denatured ethanol into gasoline / petrol	Same anaerobic fermentors and *Saccharomyces cerevisiae* organisms as in industrial ethanol production, given above	Prevalent in Spain (wine-based ethanol) and Sweden (grain- or potato based). Also practiced in Caribbean sugarcane-growing nations	Up to 20% of Sweden's national petrol supply in late 1930s through 1945	Blending ethanol into gasoline/petrol increases the octane rating and improves poor octane fuel engine performance	[3, pp. 646–660]

Year Begun	Fermented Product	Substrate	Fermentor Type & Organism	Location	Scale	Additional Information	Ref.
Before 1900 (Batch Process Using Whey) and by 1931 Continuous Fermentation	Lactic acid	Cane or beet sugar (sucrose), corn syrup (glucose), or lactose from concentrated whey	Initially, anaerobic fermentation of milk whey, by *Streptococcus lactis.* Later, anaerobic fermentation of *Lactobacillus delbrucki* with non-lactose sugars, or *Lactobacillus bulgaricus* with lactose. It is noteworthy that ***continuous fermentation processes were in use by 1931***	Initially, global production of lactic acid was used in leather tanning processes worldwide Specialized fermentation plants developed mainly in the USA and Europe in the 1930s, and later all over the globe	2,200 tonnes per year by 1940	In the early 2000s, lactic acid production using genetically modified bacteria with high production rate was used to feed one of the first fermented bio-based polymer chemical production operations, which continue today in 2013	[3, pp. 673–674]
1917	Citric Acid	Cane or Beet Sugar (Sucrose) or Corn Syrup (Glucose)	Aerobic fermentation of the mold *Aspergillus niger*, initially in shallow pans, later in agitated and air-life fermentors	Initially in the USA; production expanded from the 1970's in the USA, Ireland (Pfizer / ADM), and Worldwide	Production reached 4,000 tonnes per year by 1939, and grew into the 10's of thousands by 1990	The production of citric acid cheaply by fermentation permitted many new industrial (non-food) uses	[3, pp. 672–673]
1977–1981 Began in Response to the Oil Embargos of 1974 and 1978, against the USA	Fuel grade ethanol blended up to 10% with gasoline to make "gasohol" and thus extend the gasoline / petrol supply without importing more crude oil	Corn syrup (glucose only, not isomerized to fructose) in corn wet mills	Batch or continuous multi-stage fermentation of *Saccharomyces cerevisiae*, following batch or continuous aerobic propagation, at fermentor sizes up to 2,000,000 liters by 1981. Continuous fermentation with 5 or more stages was used sometimes, representing <10% of the US production capacity	Corn-based in the USA (first in Illinois, Iowa, Tennessee, Ohio); and cane-sugar based in Brazil and Caribbean nations	Typically 3 to 50 million US gallons (11 to 180 million liters) per year, at each site	Fuel grade ethanol often was made in "corn wet mills" to offset seasonal demand cycles for high fructose sweeteners for beverage use, and also by sugar cane refineries to balance harvest cycles and purified sucrose product demand	
Mainly built from 1974 to 1985, with only a few built between 1985 and 1999	Fuel grade ethanol	Whole kernel #2 yellow corn in "dry mills." A small number of plants used whole kernel wheat, milo, or barley instead of corn	Batch anaerobic fermentation, following batch aerobic propagation of *Saccharomyces cerevisiae*, at fermentor sizes up to 1,200,000 liters by 1999. Continuous fermentation with 5 or more stages was used sometimes, representing <10% of the U.S. production capacity	These earliest dedicated fuel-grade ethanol plants were built in many regions of the USA—anywhere corn was grown	10,000 to 50 million U.S. gallons (40,000 to 180 million liters) per year (small plants located on farms to dedicated industrial-scale plants)	These were known as "corn dry mills" because they did not steep the corn kernels in hot water before milling, to separate the starch from other saleable products, as did corn wet mills	

Year Begun	Fermented Product	Substrate	Fermentor Type & Organism	Location	Scale	Additional Information	Ref.
2000 through 2010	Fuel grade ethanol	Whole kernel #2 yellow corn	Batch anaerobic fermentation, following batch or continuous aerobic propagation of *Saccharomyces cerevisiae*, at fermentor sizes up to 4,000,000 liters by 2006. Continuous fermentation rarely was used and sometimes comprised <3% of the U.S. production capacity	Built in all corn-growing Midwestern states in the USA, plus some western states using corn delivered by rail	This generation of corn dry mills were sized from 10 to 40 to 500 million liters per year. The total USA capacity reached 48 billion liters/yr. in 2010	An increasing number of plants were built annually, with a major spike from 2006-2008 dropping to almost no construction activity by 2009	
1978-2013	Fuel Grade Ethanol, hydrous (with up to 5% water) in the warmest countries, denatured anhydrous ethanol (<1% water, same as grain-based fuel ethanol) in cooler countries	Cane and Beet Sugar	Rapid (<<24-hour) batch or continuous multi-stage fermentation, first with the common brewers' yeast *Saccharomyces cerevisiae*, and later with heat resistant, highly productive *Schizosaccharomyces pombe* fission yeast, following batch or continuous aerobic propagation, at fermentor sizes up to 4,000,000 liters by 2010	Largest industrial output is in Brazil. Also in Columbia, Guatemala, India, Thailand, and other sugar-producing nations	Global capacity exceeded 70 billion liters per year in 2012	Hydrous (95%) sugar cane ethanol sold for use in autos in petrol stations, *blended in any proportion* or not blended with petrol, achieved great success in displacing petrol in Brazil by 2006 due to deployment of "flex fuel" vehicles	
1939–1978 WWII and Cold War Years	Motor fuel ethanol blends, up to 20%, in gasoline / petrol; sometimes used at 85%–100% ethanol concentration in modified vehicles when standard motor fuel was not available due to war or remote locations (especially in Siberia)	Early cellulosic ethanol production scale used any available biomass, especially side streams from pulping operations, logging residues, and crop residues	Glucose produced by acid hydrolysis was fed to batch anaerobic fermentation using *Saccharomyces cerevisiae*, following batch aerobic cell-yeast growth / conditioning	Strong sulfuric acid cellulose hydrolysis process plants were used in Germany from the early WWII years. The US built a similar plant on the West Coast during WWII. Russia constructed other strong acid plants based on improvements to the WWII German know-how	Individual plant scale was typically about 10 to 40 million liters per year. Production total figures were national secrets not released to public sources	World peace permitted secure global sea trade of crude oil and finished petroleum products, compared to the WWII period. Major American, Canadian, and Persian Gulf oil field discoveries in the 1950s resulted in inexpensive crude oil availble worldwide. Cellulosic ethanol could not compete economically	

Year Begun	Fermented Product	Substrate	Fermentor Type & Organism	Location	Scale	Additional Information	Ref.
1978–2013 R&D and Demo Projects for Advanced or Cellulosic Biofuels	The intended fermentation product usually was fuel grade ethanol, meeting the same product specs as sugar-based and grain-based fuel ethanol. The final form could be hydrous (with up to 5% water) in the warmest countries, or anhydrous ethanol (<1% water) in cooler countries. Fermented *N*-butanol was also introduced for blending with gasoline/petrol	The fermentation substrates were usually the natural C6 sugars, plus unconventional C6 and C5 sugars released during the biomass pretreatment and hydrolysis processes. An alternative technology pathway gasified biomass or municipal solid waste to $CO-CO_2-H_2$ synthesis gases, and these were directly used as substrates by *Clostridium* or other bacteria	There were many variations in the biomass pretreatment processes and hydrolysis to sugar, but not as many in fermentation. Processes used batch or continuous multi-stage fermentation of *Saccharomyces cerevisiae* or certain ethanol-producing bacteria (especially *Clostridium* strains), following batch or continuous aerobic (yeast) or anaerobic (bacterial) propagation. Yeast and bacteria were genetically modified to use unconventional C6 and C5 sugars as substrates, or to produce digestive enzymes. The greatest new technology advance was in the ability to sustain a fermentation on gaseous CO and CO_2 substrates	R&D was a global effort, especially strong in the USA, Brazil, China, Denmark, Sweden, Finland, Holland, Canada, New Zealand, India, and South Africa	Demonstration scale projects typically ranged from 50 thousand to 12 million liters per year annual capacity, on the basis of 24 hour operation and 90% utilization. The first commercial projects varied from 30 to 95 million liters per year of ethanol production	World chemists continued to research the production of cellulosic ethanol throughout the 1940s and 1950s. Cellulosic feedstock implementation for ethanol. Modern process developments from circa 1990 through 2013. New commercial-scale production constructed and coming on-line 2013	

2.0 FERMENTATION ORGANISM DEVELOPMENT FOR A BIOFUEL- OR BIO-BASED CHEMICAL PRODUCT

This section navigates the steps in a typical development program for a new biofuel- or bio-based chemical production process and emphasizes the development of the fermentation step of that process.

Industrial experience with the development of fermentation processes and organisms has led many practitioners, over time, to converge on a common set of scientifically valid investigatory and developmental steps that (if successful) lead to the creation of one or more technically viable production processes, at lab scale. These steps, outlined here, are not the only way to arrive at the end goal. They are presented as a project planning guideline, useful if no program-specific or company-specific development program exists.

The focus throughout is on how fermentation process development decisions affect the new biofuel- or bio-based chemical project's outcome. The goal of this focus is to provide the type of technical-minded people expected to use this handbook with a clear picture of why their work is important to the entire program and how they might work most wisely in the earlier stages to produce the best outcome in the later stages.

2.1 Native Strain Screening, Selection, and Genetic Modification

The development of a new production process for a fermented biofuel- or bio-based chemical often follows extensive laboratory and field studies for identifying one or more native microorganism strains with desirable qualities that suggest the possibility of developing a commercially viable fermentation process using those strains.

It would be unusual to find that a newly discovered native strain was able to produce the desired fermentation product at a high rate, with high yield, high titer, and high product specificity relative to alternate by-products.

It is standard practice for the most promising native strains to be taken into a microbiology laboratory to undergo a series of preliminary screening evaluations, followed by stress-based transformations and selections, before serious work begins on developing a fermentation process using any of the native organisms or lab-selected strains.

It is also common for the most promising native strains to be subjected to genetic engineering to improve the economics of production, by processing additional or different nutrients than the native strain (such as incorporating the ability to ferment C5 sugars, or isomers of C6 sugars other than common glucose and fructose) or altering metabolic pathways to favor products not naturally produced in significant quantities by the natural organism (a very common goal). The metabolic pathways are altered by the molecular biology or genetic engineering team knocking out one or more genes; multiplying genes to higher copy numbers; crossing over diploid gene sequences followed by meiosis that will yield haploid cells with differing traits; or introducing new genes (from other organisms) for enzymes that catalyze biochemical reaction steps that lead to the desired product.

The methods for these evaluations and selections are not unique for biofuels. Rather, they are the same methods used for other microbial fermentations (see Chapter 1 for a discussion of those methods). Detailed coverage of these microbiological methods, especially those of genetic engineering, goes beyond the scope of this work. Broadly stated, the evaluation tools and the selection methods have produced many major transformations that are in profitable commercial use today. Scientists are adding to and evolving these advanced methods at a rapid pace. The speed and utility of the methods, and the spectrum of productive organism stains they produce, are likely to continue to see strong gains in the near future.

Biofuels and bio-based commodity chemicals must sell for low commodity prices to induce industrial users, wholesalers, and distributors to substitute them for conventional petroleum-based fuels and petrochemicals. There is greater pressure to reach a higher fermentation production rate, higher yield, higher titer, and higher product specificity than would be necessary to justify the production of highly valued products, such as nutritionals, beverages, flavors, fragrances, pharmaceuticals, and cosmetics. Thus, the importance of optimizing the product cycle to minimize the product costs before initial market entry cannot be overstated. This product cycle optimization is a central distinguishing characteristic of the development of biofuels and bio-based chemicals as compared to non-commodity products.

2.2 Adaption to the Fermentation Background and Lowest Cost Nutrient Mix, to Overcome Inhibition and Minimize Operating Costs

The early stages of biofuel- or bio-based chemical fermentation process development (the organism strain development and screening stages) typically use standardized fermentation substrates as the main "food" sources, with generous provision of trace nutrients. This is done to reduce the possibility that poor organism performance during these selection stages is caused by insufficient nutrition rather than insufficient phenotypic capability. No organism, be it a yeast cell, a fish, or a human, will perform at its best if underfed.

Growth testing of microbes during the screening processes usually takes place in clean laboratory ware, with filtered distilled water used to make a standardized and controlled growth medium. Variations in media or nutrients are rarely allowed to occur during this stage, so that differences in microbe performance can reasonably be ascribed to their strain identity instead of to other variables.

The main energy source during screening usually is a standardized, purified, and controlled grade of monosaccharaide (usually glucose derived from corn or wheat starch), or a disaccharide such as sucrose (from cane sugar or beet sugar). In cases where glucose is not the dominant food for the anticipated commercial production process, another similarly controlled and standardized substrate (such as common laboratory filter paper, or Avicel™, for cellulose-based processes) is used instead of a simple sugar.

Trace nutrients provided during screening also come from controlled sources. Common sources include various grades of molasses, corn steep liquor, and yeast extracts from spent brewers' yeast. Phosphorous and nitrogen sources are also required and are often provided by the addition of di- or mono-ammonium phosphate, urea, or other inorganic salts to the primary nutrient solution, in bulk, before use.

All of these nutrient sources must be thoroughly sterilized before use, usually by a HTST (High Temperature, Short Time) process step, which can take place batchwise, or with continuous flow through a Pasteurization unit that includes heating, holding, and cooling steps.

After the microbial strain screening process has led to the selection of a number of candidate strains that are able to make the desired product under reasonable process conditions, it is time to work on minimizing the operating costs by reducing the amount of nutrients provided.

It is also time to adapt the strains to conditions closer to those that will be found inside commercial fermentation vessels during fermentation of real substrate to maximum titer. During the course of fermentation, the microbes will encounter steadily increasing concentrations of a multitude of fermentation products and byproducts, any of which may be potentially inhibitory, plus inert or detrimental trace components that may be in the fermentation broth due to their presence in the commercial-grade main food source or trace nutrient reagents. This poorly defined group of chemical species is conventionally called the fermentation "background".

No strain is ready for service in production operations until it has been adapted to living with minimal nutrition in the presence of a fermentation background that is representative of what will be encountered in the real-world situation.

A recommended approach to defining an experimental program for satisfying the nutrition and inhibition challenges to microbial growth is first to set up two different experimental programs, which work on one of the two variables while keeping the other variable steady.

One test matrix will study the ability of candidate strains to multiply and produce the desired product, as a function of variations in nutrient identity and quantity. The second test matrix will study the effects of fermentation broth background compounds on the candidate strains, with concentrations increasing by gradients up to that expected in the worst case commercial application.

Acclimated strains that come out of one of these test matrices should be subjected to the challenges of the other test program.

Acclimated strains are the descendants of candidate strains selected in initial screening that, over time, have adapted to their new food or new inhibitors (typically by induced changes in gene expression, or by epigenetic transformations that are passed on to daughter cells in mitosis—not by any genetic changes) that perform well in one of the tests should subsequently be subjected to the other.

Strains that have acclimated to minimal cost nutrition should be challenged to adapt to the fermentation inhibition background, and vice versa. This process may be repeated, iteratively, until a short list of strains that have done best in both challenges is selected for going forward to fermentation *process development*. That stage will rely on strains reliably performing consistently, batch after batch. Thus, it is necessary to fully adapt and stabilize strains before larger scale fermentation process development can become fruitful.

After the candidate strains for fermentation process development are selected, it is common to continue strain development from the early stages, as a parallel program, and then move any majorly improved strains discovered quickly through the nutrient optimization and background inhibition adaption stages. If the resulting strain from that work promises to be better than one currently used in the later stage fermentation process development, it may be rolled out to the later stage development program after a cost/benefit evaluation of its impact on the overall program.

3.0 BIOFUEL- OR BIO-BASED CHEMICAL FERMENTATION PROCESS DEVELOPMENT AND DESIGN

This section outlines one time-tested typical approach to scaling up a candidate process from small lab scale, through several pilot plant development stages, to a full-scale commercial design.

Emphasis is on the fermentation steps, but with recognition that these steps are imbedded in a larger production cycle. The long-term demonstration of the *full-cycle economics* at pilot plant scale, and the development of accurate capital cost estimates for the scaled-up production equipment, will be most important in determining whether a technically viable production process is worth implementing commercially.

Outline of a Biofuel Fermentation Process Development Program

- **3.1 Small Scale Lab Fermentation to Select Operating Parameter Working Ranges**
  - Small lab scale screening set initial fermentation time/temperature/concentration parameters.
- **3.2 Fermentation Process Parameter Improvement Studies within Working Ranges**
  - Fermentation process condition improvement studies are made, with many parallel fermentations conducted at the same time in shake flasks and roller bottles.
- **3.3 Fermentation Cycle Development in Bench-Top Fermenters**
  - Fermentation process cycle development in instrumented 0.25 to 5 liter bench-top fermenters.
- **3.4 Nutrient Option Studies to Improve Production Economics**
  - Nutrient option studies conducted with economic factors (not scientific goals) as targets.
- **3.5 Product Recovery Method and Quality Control Test Method Development**
  - Product recovery method testing and product quality testing (in accordance with commercial goals).
- **3.6 Lab Fermentation Scale-up Studies**
  - Lab fermentation scale-up studies (to approximately 50 liters upper limit) to study mass transfer-related production constraints.
- **3.7 Pilot Plant Fermentation Scale-up Studies to Establish Commercial Fermenter Design**
  - Pilot plant fermentations in successively larger fermenters configured like commercial equipment—10× typical scale-up limit per experiment stage, except for well-known organisms.

- **3.8 Completing the Production Cycle: Modeling of Recycle Loops and Effluent Treatment**
  - Modeling of recycle loops (especially the water cycle) at lab and pilot plant scale, through at least 5 cycles, with tracking of inhibitor and waste product concentrations; and establishing proof of stability.
- **3.9 Long-Term Fermentation Demonstration Runs (With Recycle) to Support Commercialization**
  - 1000-Hour demonstration runs (measured as cumulative fermentation time) are the norm for establishing ranges for long-term stable production rate, yield, titer, and other production cycle parameters.

4.0 PRACTICAL APPLICATIONS

After the fermentation demonstration runs have established product yield data for each given set of microbe strains, nutrient mixes, substrate types, and process operating parameters, it is time to begin the scale-up process. The next step is to prepare preliminary designs for commercial production equipment and computing plant economics, based on different combinations of these possible "inputs".

The desired "output" from this evaluation of practical applications of the technology just demonstrated will be a production process that yields the maximum return on investment (ROI) when evaluated as a business proposal, over the next 5 to 10 years, including the capital investment in the construction of the plant and commissioning of time and expenses.

4.1 Selection

- **4.1.1 Selection of Organism Strain and Fermentation Cycle Parameters**
  - Full cycle economics are key decision drivers for organism strain and cycle parameter selection.
- **4.1.2 Intellectual Property (IP) Rights Set Boundaries on Strain Selection and Cycle Parameters**
  - IP considerations (patent restrictions, licenses, and royalties) constrain and guide selection.
- **4.1.3 Selection of Fermentation Feedstock and Nutrient Sources and Quality, and Product/Byproduct Forms and Sales Specifications**
  - Feedstock and byproduct choices are guided by regional and seasonal availability.
- **4.1.4 Production Site Selection**
  - Site selection is influenced by production and transportation factors, government permit requirements, and local political views regarding the proposed facility.

4.2 Critical Factors

- **4.2.1 Economical Selection of Full-Plant and Individual Fermenter Production Capacity**
  - Production scale for individual fermentors and overall plant capacity should be subjected to parametric study with economics.
- **4.2.2 Supply and Off-Take Agreement Form and Pricing Structure**
  - Supply and off-take agreement form, pricing, and exclusivity are key inputs to the economics and require the involvement of business representatives in addition to the technical staff; without this involvement, misguided decisions based on preliminary price information might result.
- **4.2.3 License Agreements Can Constrain Production Scale and Available Customer Base**
- **4.2.4 Risk Analysis**
  - Perform risk analysis across all aspects of the project before authorizing commercial go-ahead.
- **4.2.5 Return on Investment (ROI) Sensitivity Studies**
  - Return on investment (ROI) sensitivity studies (often called "tornado charts") should be prepared. These charts aid in studying the effect of time delays, changes in capital or operating costs, and market price reactions before authorizing commercial go-ahead.

4.3 Troubleshooting

- When faced with production trouble, such as low rate, yield, or titer, or infection, first identify the symptoms of malfunction, the time of their first appearance, and their magnitude over time. (With a new production process, new product, new feedstock, or new organism, troubleshooting must begin *without* an established guideline for possible causes of any given undesirable effect. Thus, the focus is on "What happened *at that time*?")
- Control system historical data and QC lab management system data should be merged and plotted to identify temporal relationships between mal-operation and other anomalies (such as temperature, pH, nutrient source, equipment opening, starting or stopping). These anomalies present opportunities for investigation.
- Causes and effects occurring at the same time, even in seemingly unrelated plant systems, could be related. All pairings of changes that happen at the same time should be investigated.
- If causes of production difficulties are not found by studying the historical plant and lab data, then small-scale lab screening studies (varying the parameters that varied in the production plant) should be proposed and performed quickly.

5.0 PLANT OPERATIONS

The major tasks involved in taking a new biofuel- or bio-based chemical plant from an inanimate, cold, completed facility to a warm and active production facility are the same as for other fermentation products. A short but comprehensive outline of these tasks is as follows:

- 5.1 Conduct plant operations planning tasks during the design/build phase of the project:
  - Write operating instructions.
  - Hire operations, maintenance, and site support staff.
  - Train the staff during plant system commissioning.
- 5.2 Prepare and approve quality assurance and quality control (QA/QC) plans.
- 5.3 Make and approve a list of critical production parameters, with target values and allowable ranges with upper and lower bounds.
- 5.4 Conduct initial process and utility system check-outs with water, steam, air, and other non-hazardous materials.
- 5.5 Sterilize fermenters and connected systems, and prove sterility.
- 5.6 Schedule the first production runs as "engineering runs" with close supervision, extra data recording, and extra sampling for analysis.
- 5.7 Analyze engineering run data, and evaluate according to performance test parameters. When acceptable, move on to the performance test.
- 5.8 Perform plant performance tests as dictated by business and financial contracts. When the tests are passed, move on to normal production operations.
- 5.9 Gradually reduce technical support while establishing a regular operating cycle, with "factory mentality" replacing the R&D mentality that inspired and enlightened the project prior to performance testing and acceptance.

6.0 REFERENCES

[1] M. Kopke, et al., Clostridium ljungdahlii represents a microbial production platform based on syngas, Proc. Natl. Acad. Sci. USA. 107 (29) (2010) 13087–13092 Published online 2010 July 2. doi: 10.1073/pnas.1004716107.

[2] <http://www.ncbi.nlm.nih.gov/core/lw/2.0/html/tileshop_pmc/tileshop_pmc_inline.html?title = Click%20on%20image%20to%20zoom&p = PMC3&id = 2919952_pnas.1004716107fig01.jpg>.

[3] R.N. Shreve, The Chemical Process Industries, McGraw-Hill, New York, 1945, pp. 670–672, 666–670, 646–660, 673–674, 672–673.

7.0 EXAMPLES OF BIOFUEL- AND BIO-BASED CHEMICAL INDUSTRIAL PRODUCTION PROCESSES

7.1 Current Biofuels Technologies: Overview of the Production Processes for Fuel Grade Ethanol from Corn, Wheat, or Barley in the United States, Canada, and Europe

Ethanol production requires only the starch portion of a corn kernel. The remaining protein, fat, fiber, and other nutrients are returned to the global livestock and poultry feed markets. Every bushel of corn processed by an ethanol plant yields approximately 30% of the incoming dry matter available in other products, for human food (in the wet mill plants) or for animal feed. Wheat and barley contain less starch than corn, so the remaining percentage not fermented is greater than 30% for these grains.

7.1.1 Two Different Ethanol Production Processes: Wet Milling and Dry Milling

Ethanol is commercially produced using either a wet mill or dry mill process. Wet milling involves separating the grain kernel into its component parts (germ, fiber, protein, and starch) prior to fermentation. We will first discuss wet milling, and then proceed to discuss drying milling.

7.1.2 The Ethanol Production Process: Wet Milling

In the wet milling process depicted in Figure 5.2, above, the grain is soaked or "steeped" in water and dilute sulfurous acid for 24 to 48 hours. This steeping facilitates the separation of the grain into its many component parts.

After steeping, the corn slurry is processed through a series of grinders to separate the corn germ. The corn oil from the germ is either extracted on-site or sold to crushers who extract the corn oil. The remaining fiber, gluten, and starch components are further segregated using centrifugal, screen, and hydro-clone separators.

The steeping liquor is concentrated in an evaporator. This concentrated product, heavy steep water, is co-dried with the fiber component and is then sold as corn gluten feed to the livestock industry. Heavy steep water is also sold by itself as a feed ingredient and is used as a component in Ice Ban™, an environmentally friendly alternative to salt for removing ice from roads.

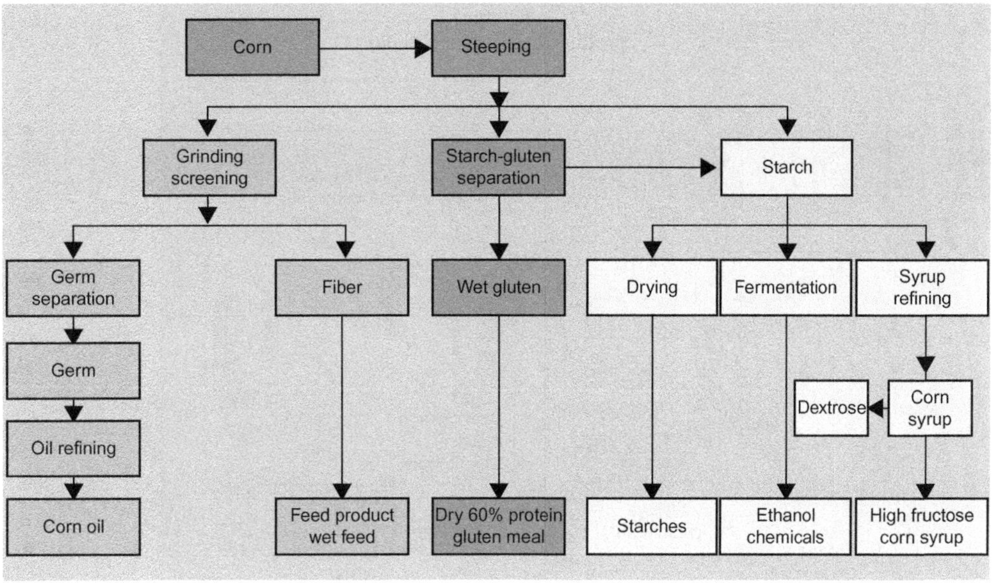

FIGURE 5.2 Flow diagram courtesy of the Renewable Fuels Association.

The gluten component (protein) is filtered and dried to produce the corn gluten meal co-product. This product is highly sought after as a feed ingredient in poultry broiler operations.

The starch and any remaining water from the mash can then be processed in one of three ways: fermented into ethanol, dried and sold as dried or modified corn starch, or processed into corn syrup.

The fermentation process for ethanol from the starch produced in dry mills is very similar to the dry mill process described below; the same enzymatic digestion steps are used to reduce the starch to simple sugars because yeast cannot directly metabolize starch.

7.1.3 Dry Milling Ethanol Production Process

This section was provided by ICM Inc. of Colwich, Kansas, the largest designer of dry milling corn ethanol plants in the world. An interactive version of this section is available at the ICM website http://icminc.com/ via the following link:

http://icminc.com/innovation/ethanol/ethanol-production-process.html

There the reader can access an interactive process flow diagram for a dry mill corn ethanol plant and learn more about the dry mill process by guiding the cursor over the steps.

ICM-designed plants utilize the dry mill process, where the entire grain kernel is ground into flour. The starch in the flour is converted to ethanol during the fermentation process, creating carbon dioxide and distillers' grain.

For readers unable to access the aforementioned link, the following process flow diagram provided by the U.S. Department of Energy (DOE) Renewable Energy Lab is similar to that provided by ICM and may be used in lieu of the interactive version.

Grain Delivery/Storage

Grain is delivered by truck or rail to the ethanol plant, where it's loaded in storage bins designed to hold enough grain to supply the plant for 7–10 days.

Milling

The grain is screened to remove debris and is then ground into coarse flour.

Cooking (Hot Slurry, Primary Liquefaction, and Secondary Liquefaction)

During the cooking process, the starch in the flour is physically and chemically prepared for fermentation.

Hot Slurry

The milled grain is mixed with process water, the pH is adjusted to about 5.8, and an alpha-amylase enzyme is added. The slurry is heated to 180–190°F for 30–45 minutes to reduce viscosity.

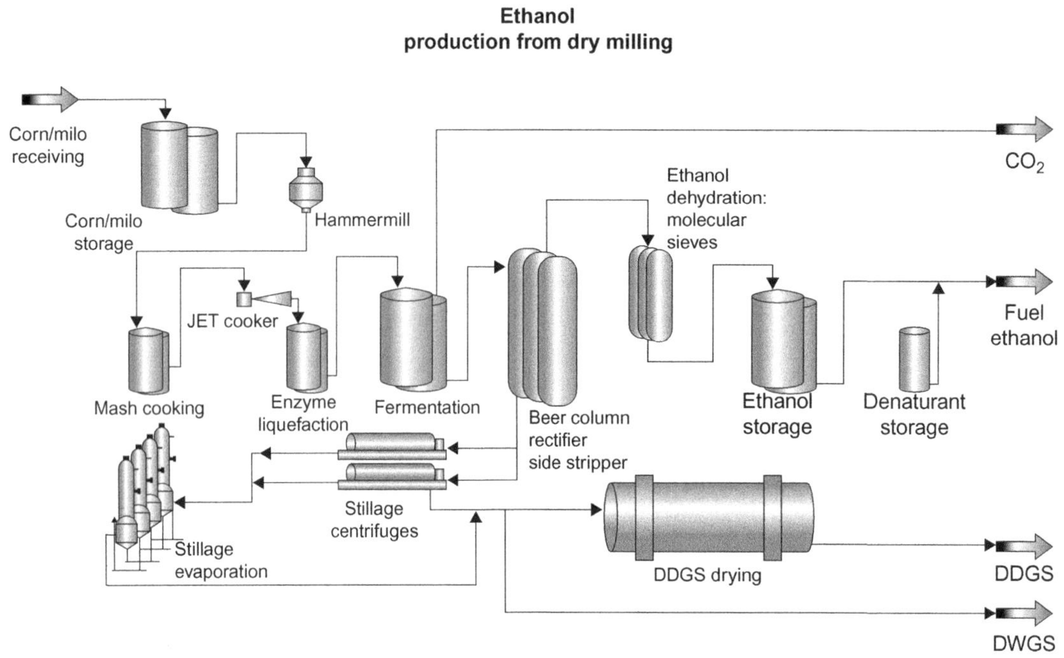

Primary Liquefaction

The slurry is then pumped through a pressurized jet cooker at 221°F and held for 5 minutes. The mixture is then cooled by an atmospheric or vacuum flash condenser.

Secondary Liquefaction

After the flash condensation cooling, the mixture is held for 1–2 hours at 180–190°F to give the alpha-amylase enzyme time to break down the starch into short chain dextrins. After pH and temperature adjustment, a second enzyme, glucoamylase, is added as the mixture is pumped into the fermentation tanks.

Simultaneous Saccharification and Fermentation

Once inside the fermentation tanks, the mixture is referred to as mash. The glucoamylase enzyme breaks down the dextrins to form simple sugars. Yeast is added to convert the sugar to ethanol and carbon dioxide. The mash is then allowed to ferment for 50–60 hours, resulting in a mixture that contains about 15% ethanol as well as the solids from the grain and added yeast.

Distillation

The fermented mash is pumped into a multi-column distillation system, where additional heat is added. The columns utilize the differences in the boiling points of ethanol and water to boil off and separate the ethanol. By the time the product stream is ready to leave the distillation columns, it contains about 95% ethanol by volume (190 proof). The residue from this process, called stillage, contains non-fermentable solids and water and is pumped out from the bottom of the columns into the centrifuges.

Molecular Sieves

The 190-proof product stream is pumped into the molecular sieve system. These specialized tanks contain molecular sieve beads that adsorb water molecules from the process stream while ethanol molecules pass through unaffected. When the product stream leaves the molecular sieves, it contains approximately 99% ethanol by volume (200 proof).

Ethanol Storage and Loadout

The 200-proof ethanol is pumped to on-site storage tanks, where it is denatured and stored until it is ready to be shipped by tanker truck or rail.

Liquid/Solid Separation

The stillage from the distillation system is pumped into centrifuges to separate the majority of the solid matter from the solution. This creates two products: A semi-solid product called wet cake is removed and conveyed to rotary dryers. A mostly water process stream, called thin stillage, is pumped to the evaporation system.

Evaporation

The thin stillage from the centrifuges is pumped into a series of evaporators where a majority of the water in the solution is removed. The resulting product stream is called syrup. The syrup can be sold as a stand-alone product or added to the wet cake before moving into the dryer system.

DDGS Drying

The wet cake is conveyed to dryers where it is converted into a low-moisture (10–12%) product called dried distillers grains with solubles (DDGSs).

7.2 The Emerging Next-Generation Biofuel Industry: Examples of 20 Companies Developing New Bio-Based Products and/or Working with New Substrates

The 20 companies listed in the following table are the top 20 among a list of 50 compiled by the *Biofuels Digest*, **www.biofuelsdigest.com**, an on-line news source for the biofuels- and bio-based chemicals industries. The 20 companies in the list were voted to be the top 20 out of the "50 Hottest Companies in Bioenergy 2012–2013" by readers of the digest. There were more than 200 companies known to be operating in this industry in 2013. Although the rankings are necessarily subjective, this list does represent a cross section of the types of products, substrates, and projects actively under development at this time.

Readers of the electronic book version of this Handbook may open the following hyperlinks to articles in the *Biofuels Digest*, www.biofuelsdigest.com, by clicking on their names in the last column while on-line.

Company	Type of Bio Technology & Substrate	Bio-Product	Company Website	Link to More Information
1. Solazyme	Fermentation of glucose, sucrose from any source	Renewable oils for many uses made by genetically modified or unmodified heterotrophic microalgae	http://solazyme.com/	1. Solazyme
2. KiOR	Not biological or fermentation: destructive distillation of wood chips in an oil refinery-type FCC process and hydro-treating	No organism. Broad boiling point range hydrocarbon liquid sent to oil refineries as a substitute for regular crude oil	http://www.kior.com/	2. KiOR
3. LanzaTech	Fermentation of CO and CO_2 gases without requiring H_2	Ethanol is the first product, made by gas fermenting bacteria. Bio-based chemicals will follow.	http://www.lanzatech.com/	3. LanzaTech
4. Novozymes	Supports many technologies that use enzymatic hydrolysis	Novozymes is a producer of enzymes for grain and cellulosic ethanol producers	http://www.novozymes.com/	4. Novozymes
5. POET	Corn ethanol production, and corn stover–based cellulosic ethanol	All the fuel ethanol production comes from fermentation of *Saccharomyces cerevisiae*	http://www.poet.com/	5. POET
6. DuPont	DuPont subsidiaries and joint ventures produce enzymes for grain ethanol and cellulosic ethanol, and also have developed organisms for Butanol production in a JV with BP (Butamax)	Fuel grade ethanol work uses commercially available yeast. Butanol production comes from proprietary bacteria that are the subject of several patents. These patents have been challenged by their rival, Gevo.	http://www2.dupont.com/home/en-us/index.html	6. DuPont
7. Gevo	Bacterial fermentation of glucose from corn or sucrose to Butanol	Butanol production comes from proprietary bacteria that are the subject of several patents. These patents have been challenged by their rival, Butamax.	http://gevo.com/	7. Gevo

Company	Type of Bio Technology & Substrate	Bio-Product	Company Website	Link to More Information
8. Sapphire Energy	Photosynthesis of algae using sunlight and CO_2	Drop-in fuels—algae-derived gasoline/petrol, jet fuel, and diesel	http://www.sapphireenergy.com/	8. Sapphire Energy
9. Joule Unlimited	The company's *Helioculture*™ platform uses photosynthetic microorganisms as living catalysts to *produce fuel*, not as intermediates to produce lipids or sugars that are *subsequently converted to fuel*.	The entire process takes place in a novel *SolarConverter*® system, from photon capture to product creation and initial separation, with no requirement for arable land, fresh water, or crops.	http://www.jouleunlimited.com/	9. Joule Unlimited
10. ZeaChem	ZeaChem utilizes a hybrid of biochemical and thermochemical processing of cellulosic biomass—initially hybrid poplar trees. Xylose [C_5] and glucose [C_6]) are sent to fermentation where an acetogenic process using ferments the sugars to acetic acid without CO_2 as a by-product.	Products from fermentation with conventional microbes are acetic acid, ethyl acetate, and ethanol. Acetic acid is converted to an ester and then reacted with hydrogen to make ethanol. The hydrogen is generated by gasifying the lignin residue from biomass pretreatment and hydrolysis.	http://zeachem.com/	10. ZeaChem
11. Honeywell's UOP	All processes are non-biological. They use thermal pyrolysis and/or hydrogenation to produce renewable diesel and jet fuel.	Drop-in replacement fuels: "green" renewable diesel and jet fuel	http://www.uop.com/processing-solutions/biofuels/	11. Honeywell's UOP
12. BP Biofuels	BP has two technology famililies: enzymatic conversion and fermentation of pretreated cellulosic biomass (bagasse), and sugar cane or corn-based Butanol production (Butamax)	Ethanol is the product from cellulosic ventures based on sugar cane bagasse. Butanol is the product of bacterial fermentation of sugar derived from corn or sugar cane in Butamax, a JV with DuPont (see above).	http://www.bp.com	12. BP Biofuels
13. LS9	LS9 technology uses biological catalysts, which have been engineered to selectively convert abundant renewable feedstocks directly to drop-in and differentiated products through simple scalable processes. LS9 applies synthetic biology to combine the highly efficient metabolism of microorganisms with new biocatalytic capabilities engineered into each cell.	The resulting catalysts selectively and efficiently convert a diversity of feedstocks to a final product in a single-step fermentation process. The product is secreted from the cell where it conveniently forms a non-toxic, light organic phase that is easily recoverable. Drop-in fuels planned include renewable gasoline/petrol, jet fuel, and diesel.	http://ls9.com/	13. LS9
14. Beta Renewables	Cellulosic biomass (straw, grasses, bagasse, wood waste) is pretreated and enzymatically hydrolyzed to sugars, followed by conventional fermentation. Beta Renewables calls their configuration "The PROESA™ Process".	Cellulosic fuel grade ethanol is the initial product of commercially available C6/C5 yeast fermentation in the first Beta Renewables commercial plant in Crescentino, Italy, operational before the end of 2012.	http://betarenewables.com/	14. Beta Renewables
15. Amyris	Amyris uses synthetic biology methods to ferment conventional sugar (glucose or sucrose) to produce Biofene, Amyris's brand	Amyris's initial portfolio of commercial products are derivatives of Biofene, Amyris's brand of farnesene, manufactured	http://www.amyrisbiotech.com/	15. Amyris

Company	Type of Bio Technology & Substrate	Bio-Product	Company Website	Link to More Information
	of a long-chain, branched hydrocarbon molecule called farnesene (trans-ß-farnesene). This molecule forms the basis for a wide range of products varying from drop-in fuels to specialty products.	using proprietary engineered microbes in fermentation. Products address these market segments: Cosmetics, Flavors & Fragrances, Polymers & Plastic Additives, Home & Personal Care, Lubricants, Fuels.		
16. Virent	Not biological technology. Virent uses patented BioForming® technology featuring catalytic chemistry, which converts plant-based sugars into a full range of hydrocarbon products, including gasoline, diesel, jet fuel, and chemicals for plastics and fibers.	Virent products are "drop-in" replacements for petrochemicals and fuels, made from renewable resources. No organism is used, and no biotechnology is employed.	http://www.virent.com/	16. Virent
17. INEOS Bio	Combination of thermal and biological processing. Gasification of any biomass (initially, municipal solid waste) to syngas followed by fermentation to bioethanol.	Ethanol is produced from syngas by proprietary strains of naturally occurring bacteria, which have evolved to efficiently convert carbon monoxide and hydrogen gases. INEOS Bio bacteria are highly selective to ethanol.	http://www.ineos.com/businesses/INEOS-Bio/	17. INEOS Bio
18. Enerkem	Enerkem's technology platform is a 4-step thermochemical process that consists of: feedstock preparation, gasification, cleaning and conditioning of syngas, and catalytic synthesis. Enerkem converts mixed waste and residues into a pure synthesis gas (or syngas), which is suitable for the production of biofuels and chemicals using proven, well-established, and commercially available catalysts.	No biological processes are used. Enerkem's primary focus is the commercial production of cellulosic ethanol. Its unique process first requires the production of methanol as a chemical building block for the production of ethanol. Enerkem can also sell its methanol as an end product, or use it as a key intermediate to produce other renewable chemicals.	http://enerkem.com/en/home.html	18. Enerkem
19. Abengoa Bioenergy	Abengoa produces conventional first-generation fuel ethanol from corn and other grains in the USA and Spain, and from cane sugar in Brazil. Abengoa also has completed a cellulosic ethanol demonstration facility in Spain and is constructing a corn stover–based cellulosic ethanol facility in Kansas in the USA.	To make cellulosic ethanol, commercially available enzymes are used for hydrolysis, and yeast with C6/C5 fermentation ability is used in batch fermentation.	http://www.abengoabioenergy.com/web/en/index.html	19. Abengoa Bioenergy
20. Ceres	Ceres provides "upstream" technology for biofuels and bio-based chemicals. Its technology makes biofuels, bio-based chemicals, and biopower more economical and scalable. Ceres starts with improving the crops themselves; a process as old as farming, but now coupled with the latest genomics-based breakthroughs.	Ceres research starts by sequencing plant DNA, discovering genes and their function, and determining their potential use. Biochemistry experts study crop composition and metabolism to improve fuel yields and facilitate refining. Molecular markers speed product development and allow working with multiple traits in a simple, efficient way.	http://www.ceres.net/	20. Ceres

7.3 OPX Bio Organism Advanced Rapid Development Method Overview

An example of a modern methodology for rapid production in genetic modification and screening was recently developed by OPX Biotechnologies (OPXBIO) of Boulder, Colorado. OPXBIO provides the following non-confidential summary of its core technology, the EDGE method, on its website, www.opxbio.com:

> OPXBIO's patented Efficiency Directed Genome Engineering (EDGE) technology platform has proven to be a big advantage over traditional bioengineering approaches. EDGE enables OPXBIO to more rapidly, rationally, and robustly engineer microbes and develop cost-effective bioprocesses.
>
> The EDGE technology is rapid—up to 5,000 times faster than conventional bioengineering methods for redesigning the genetic code of microbes. EDGE is also rational—meaning that we first determine and then purposefully program the specific, optimum genetic code in a microbe for bio-based production of our products rather than counting on random genetic changes to evolve a microbe that is only slightly improved.
>
> EDGE is also robust it enables OPXBIO to construct the optimum production pathway in the microbe while also making the genetic changes that increase the microbe's vitality and productivity [and lead] to consistently better performance and cost.
>
> EDGE creates a strong connection between microbe design and overall production process design so that the entire process can be optimized for best performance and cost.
>
> As an example, with EDGE, we developed the microbe to make BioAcrylic at commercially viable rates in just two years, perhaps up to 5 years faster than would have been possible with conventional technologies.
>
> **How EDGE Works**
>
> - With EDGE, we change the genetic code of millions of microbes. Then we run lab experiments to simultaneously test the ability of all the microbes to make our product.
> - Unlike conventional approaches, with EDGE, we track every genetic change we make so we can relate specific genetic changes to microbe performance in our lab experiments. We can determine WHICH genetic changes affect performance in WHAT way.
> - We then analyze all the data from our lab evaluations to determine how best to redesign a microbe's genetic code to achieve the best possible production of our desired product.
> - Even in the later stages of process development and scale-up we can still use EDGE to rapidly adjust the genetic code of the microbe to further improve process performance and cost.
> - The EDGE platform is flexible, enabling us to redesign microbes to make a wide variety of products using a range of feedstocks.

OPXBIO is currently using EDGE to develop microbes and bioprocesses to make BioAcrylic—a more sustainable ingredient for diapers, detergents, paints, and adhesives. OPXBIO is using EDGE to make fatty acids for cleaning products and lubricants and for conversion into renewable diesel and jet fuels.

7.4 ICM Cellulosic Ethanol Example: Improving Corn Ethanol Plant Yield with Cellulosic Bolt-On Technology

The following is an extract from an official press release of ICM Corporation, describing cellulosic ethanol technology available for improving the yield of corn ethanol plants.

(Colwich, Kan.–November 29, 2012)—ICM, Inc. announces that it successfully completed its 1,000-hour run of an integrated fiber campaign conducted at its pilot plant in St. Joseph, Missouri. ICM has developed and validated its proprietary Generation 1.5 Integrated Fiber to Cellulosic Ethanol Technology for the production of cellulosic ethanol at existing grain ethanol plants. Through its Generation 1.5 model, ICM has proved substantial operating and capital expense cost savings over a traditional approach of cellulosic ethanol production. To ICM's knowledge, this run is the largest-scale fermentation integration of grain fiber to cellulosic ethanol technology to date.

The 1,000 hours of continuous production is a significant achievement, as it qualifies these data for consideration of federal loan guarantee programs which can be utilized in the financing of new, advanced generation renewable energy technologies. ICM's completion of the 1,000-hour run was achieved through the sequential completion of (24) 15,000 gallon pilot fermenters and (5) 585,000 gallon commercial scale reactors. Another critical achievement from the 1,000-hour run was to demonstrate that the dried distillers grains (DDGs) co-product of ICM's integrated fiber cellulosic process has a significant concentration of protein-fat amounts that accompany the integrated fiber to ethanol process. Additionally, the standard DDGs market has indicated acceptance of the feed ingredient which enables diversification of the co-product and expansion into new feed markets.

From the mechanical operations and process perspectives, the 1,000-hour run performed continuously and exceptionally well as would be required in a commercial operation. Corn fiber yields greater than 100 gallons per ton were performed up to the 585,000 gallon fermentation scale with all inputs (enzymes, chemicals, organisms) utilized at an economically-feasible range. ICM believes that its success with the integrated fiber 1,000-hour run is because of dedicated individuals and extensive testing of

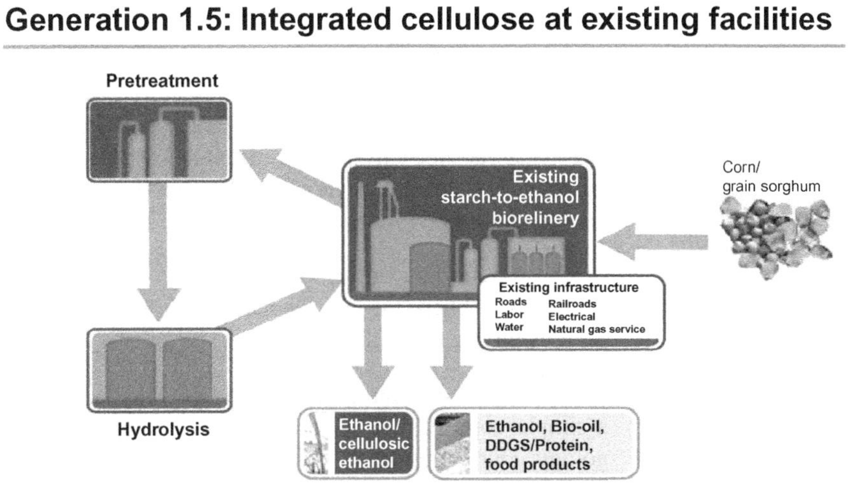

various feedstocks at the pilot scale for next generation conversion technology including: corn stover, corn fiber, wheat fiber, barley fiber, switchgrass, energy sorghum, and bagasse.

Commercial target achievements and competitive advantages in ICM's Integrated Fiber cellulosic process testing to date include:

- An additional 7–10% increase of ethanol yield/bushel showing an approximate 3.1 gallons per bushel equivalent. Production previously achieved at laboratory scale has been replicated and validated at both pilot and commercial scales in each reactor run.
- >90% conversion of C6 sugars and >80% conversion of C5 sugars—establishing a pathway to Generation 2—and ability to produce clarified starch and cellulosic sugars to ethanol production and alternative chemicals.
- Fermentation yields >90% (complete integrated process from feedstock to distillation, including backset, wastewater treatment, co-product recovery).
- Co-product market diversification capabilities by delivering substantially higher oil recovery rates and protein concentration.
- Reduced use of the energy center (fewer post-fermentation insolubles, reduced greenhouse gas emissions, and reduced carbon emissions).
- Ability to co-distill, utilize existing distillation capabilities.
- The same quantity of ethanol can be produced with 10% less bushels, creating more efficiency per bushel. With an ethanol market capacity of 14 billion U.S. gallons per year (BGY), approximately 5 billion bushels of corn is utilized; not only can 1.4 BGY of cellulosic ethanol be produced, but the same production amount could be produced with 10% less grain.

Over the course of the past eight weeks, ICM has demonstrated that its proprietary integrated fiber process design capably and efficiently produces fuel ethanol from cellulosic material as a bolt-on technology to a Generation 1 plant. It also reinforced the significance of incentives to exist for GHG reduction per gallon/ethanol, a more efficient use of Generation 1 feedstocks, and enabled a progression to Generation 2 production capabilities. The U.S. Environmental Protection Agency's (EPA's) denial of the waiver requests it received from governors of nine states on November 16 strongly affirms the critical importance of the next generations of advanced biofuels in meeting the Renewable Fuels Standards production requirements. Recent studies released by the Union of Concerned Scientists and the U.S. Department of Energy's (DOE) most updated iteration of its "Billion Ton Study" both demonstrate that the U.S. has plenty of available biomass feedstocks (54 billion gallons of biofuels capacity) to satisfy a substantial part of its energy needs.

About ICM, Inc.:

Established in 1995 and headquartered in Colwich, Kan., ICM, Inc., provides innovative technologies, solutions, and services to sustain agriculture and advance renewable energy, including food and feed technologies that will increase the supply of world protein. By providing proprietary process technology to 102 facilities with a combined production capacity of approximately 7 billion gallons of annual ethanol production, ICM has become a world leader in biorefining technology. The full-service provider also offers a comprehensive line of more than 100 products and services tailored to make biofuels production more efficient and more profitable. ICM is further upholding its responsibility as an industry leader by heavily investing in the continued advancement of renewable energy technologies.

In an effort to speed that advance, ICM has been conducting research and testing at its state-of-the-art research facility in St. Joseph, MO, in conjunction with a growing list of strategic partners spanning multiple industries.

7.5 INEOS BIO: An Example of Non-Conventional Fermentation Process with Synthesis Gases (CO, H_2, CO_2) Generated from Municipal Solid Waste as Substrate for Bacterial Fermentation to Ethanol

INEOS Bio's project to convert municipal solid waste to fuel grade ethanol and exportable electricity represents an example of a commercially novel approach to biofuel production. The process consists of gasification of municipal solid waste (preferably green waste from yard trimmings) to a carbon monoxide (CO) based synthesis gas, followed by fermentation by bacteria in a liquid-gas contact environment. This information is provided by INEOS Bio at its website, www.ineos.com/BUSINESSES/INEOS-bio/ij8.

The INEOS Bio Technology Platform Summary

The following is an extract from information about new fermentation technology developed by INEOS Bio. The novelty is that the substrate for fermentation is gaseous, not a dissolved solid or liquid. This and further information may be found at the INEOS Bio company website, http://www.ineos.com/en/businesses/INEOS-Bio/.

Bioethanol is a renewable fuel. It reduces greenhouse gas emissions compared to gasoline (petrol). When bioethanol is combusted (burnt), it effectively returns carbon dioxide back to the atmosphere, which had been taken out relatively recently as the biomass grew. Conversely, burning traditional gasoline (petrol) releases carbon dioxide into the atmosphere from carbon that has been stored in the form of oil over a significantly longer time period.

There are three main approaches to producing bioethanol from biomass materials:

1. Conventional fermentation of sugars obtained from sugar and starch crops
2. Hydrolysis of cellulose to sugars using acid or enzymes followed by fermentation of the sugars
3. Gasification of any biomass to syngas followed by conversion to bioethanol

The INEOS Bio process falls into the third category. It is an advanced thermochemical and biochemical bioethanol technology, which can convert cellulose, hemi-cellulose, and lignin efficiently to clean bioethanol. It delivers more than 90% greenhouse gas savings compared to gasoline (petrol), is cost competitive, and is independent of food production. The INEOS Bio bioethanol is expected to make a significant contribution to the growth in the European and North American bioethanol markets.

Central to the INEOS Bio technology platform is a highly selective, highly efficient, proprietary biochemical conversion of synthesis gas to ethanol. The inspired use of a bacterial biocatalyst is the essential key to unlocking the value chains from a variety of low-cost, waste carbon materials through synthesis gas to:

1. Clean renewable advanced biofuel
2. Valuable chemical intermediates such as ethylene for polyethylene manufacture

A process of continuous innovation has transformed a simple concept into a world leading commercial technology. Our proprietary biocatalyst far out-performs any known, conventional catalyst for the transformation of synthesis gas to fuels.

INEOS Bio Ethanol Technology Core Process Sequence

1. **Gasification**

   The prepared organic carbon material is gasified using a controlled amount of oxygen to produce synthesis gas, a mixture of principally carbon monoxide and hydrogen. The gasifier design and operating conditions have been carefully chosen to inhibit the formation of dioxins and furans and to suppress the carry-over of volatile metals. The hot synthesis gas is quenched and cleaned. Heat is recovered to generate renewable power for use in the process.
2. **Fermentation**

   The cleaned, cooled synthesis gas is passed into a patented fermentation process, where it is converted selectively into ethanol by naturally occurring anaerobic bacteria (the biocatalyst). The fermentation environment, containing the right quantity and type of nutrients, is maintained at carefully controlled conditions. The bacteria, in this healthy state, achieve a high selectivity to ethanol and high yield of ethanol. The high selectivity and yield translate to outstanding process efficiencies. The off-gas from the fermenter is used to generate additional power and heat.
3. **Purification**

   The ethanol solution is purified and refined to make anhydrous ethanol (>99.7% ethanol). This can be blended into gasoline as required for the local, renewable road transport fuel market.

Advantages of a Biocatalyst Compared to Non-Biological Chemical Catalysts

At the heart of the INEOS Bio process are patented bacteria. The bacteria are naturally occurring micro-organisms that have evolved to efficiently convert carbon monoxide and hydrogen gases generated in their natural environment to ethanol. The bacteria are harmless to people and to the environment and are totally safe to handle in an industrial process.

The INEOS Bio proprietary bacteria out-perform any known conventional catalyst. The bacteria are highly selective to ethanol, are inexpensive compared to conventional catalysts, continually regenerate themselves, are tolerant to most impurities, and operate at low temperatures and pressures.

By contrast, conventional catalysts have relatively low selectivity to the desired product, which means that the various by-products need to be separated out and purified or recycled. They need to operate at high temperatures and pressures to convert syngas to fuels. They are often costly to synthesize, are easily poisoned, and need to be regularly removed and replaced. This results in higher production costs.

Gasification

Gasification is a process whereby carbon materials are heated in a controlled, partial oxygen environment. The aim is to convert as much of the carbon in the feed material as possible to carbon monoxide, which has valuable chemical energy. Incineration, by contrast, is a process in which wastes are fully combusted using excess air to produce carbon dioxide.

The oxygen-blown gasification technology used in the INEOS Bio process suppresses the formation of dioxins and furans thanks to the reducing environment in the gasifier, and destroys any dioxins and furans formed through exposure to high temperature and residence time in the upper part of the gasifier. In addition, the carry-over of volatile metals into the syngas is minimized. Our design reduces the burden on, and cost of, the syngas clean-up stage.

It is the use of the thermo-chemical gasification step that gives the whole process its feed flexibility. This feed flexibility enables many biomass materials and wastes to be converted to syngas for onward, high-yield conversion by the INEOS Bio biocatalyst to bioethanol.

Process Benefits of Gasification Followed by CO-Fed Bacterial Fermentation

Unlike other technologies that rely on one primary source of feedstock, the INEOS Bio process can produce ethanol and renewable energy from numerous feedstocks, including municipal solid waste, yard waste, forestry waste, and agricultural waste while breaking the link between food crops and ethanol production. This flexibility allows facilities, like the Florida BioEnergy Center, to be built anywhere in the world, wherever there is biomass waste, and thereby helps to provide jobs and locally sourced renewable energy for urban and rural communities.

Leading Economics

The outstanding energy efficiency, selectivity to ethanol, and feedstock flexibility of the INEOS Bio advanced bioethanol process result in highly cost-competitive production of ethanol.

Feedstock Flexibility

Unlike conventional bioethanol technologies, which use food crops, or even the emerging cellulosic fermentation technologies, which can convert cellulose and hemicellulose but not lignin, the INEOS Bio process can convert all ligno-cellulosic materials as well as other carbon materials into ethanol.

The range of organic materials that can be used includes, but is not limited to:

- The biogenic portion of municipal solid waste (MSW), which includes garden and food wastes
- Organic commercial and industrial wastes
- Wood waste
- Forestry residues and products (e.g., brash, bark, saw dust, wood chippings)
- Agricultural residues (e.g., sugar cane bagasse, corn stover, straw)
- Ligno-cellulose energy crops (e.g., trees, coppice, miscanthus and switch grass)

High Energy Efficiency

The winning combination of a thermo-chemical process (gasification) and a low temperature, low pressure biochemical process (bacterial fermentation) with excellent heat integration throughout delivers high energy efficiency.

Biocatalyst Advantages

Conventional catalysts used for the conversion of syngas to transport fuels can be expensive to prepare, are easily poisoned, need to be replaced regularly, require high temperatures and pressures, and can have relatively poor selectivity to the target fuel product (e.g., ethanol or diesel). In contrast, the proprietary INEOS biocatalyst is low cost, regenerates itself, is tolerant to impurities in the syngas, has extremely high selectivity to ethanol, and operates at low temperature and pressure.

Security of Energy Supply

The feedstock flexibility of the INEOS Bio technology enables almost any country in the world to manufacture a portion of its own road transportation fuel from the biomass and wastes it generates. Wherever there are people, wastes are generated and transportation fuels are needed.

Every major city in the world could potentially have its own INEOS Bio plant or plants.

Climate Friendly

When the bioethanol is used as a road transport fuel, the greenhouse gas (GHG) emissions are reduced by at least 90% compared to gasoline (petrol). This outstanding GHG performance puts clear water between INEOS Bioethanol and bioethanol made from sugar and starch crops and from emerging cellulosic fermentation technologies.

Positive Impact on Society

A localized approach that reduces wastes going to landfill and air pollution, creates jobs, generates tax revenue, and safely produces renewable fuel and clean energy.

Reduced Pressure on Land Use

Significant greenhouse gas (GHG) emission savings can be achieved from the road transport sector without increasing pressure on land use, or increasing agrichemical use (e.g., fertilizers and pesticides). The diversion of organic wastes from landfill avoids harmful effects of landfill.

7.6 A Representative List of Cellulosic and Non-Traditional Biofuel Production Processes, Under Development or Entering Commercial Demonstrations, 2000–2013:

The DOE Bioenergy Fact Sheets are a comprehensive set of free, publically available research reports and technology development reports covering all of the biofuel and bio-based chemical production programs supported by the U.S. DOE, effective May 2013. All of these may be obtained at no cost by downloading from their links (e-book format only), and also from the U.S. DOE web page on biofuels: http://energy.gov/public-services/vehicles/biofuels

DOE Bioenergy Technology Office Project Fact Sheets

There are numerous facets to biomass research. The U.S. Department of Energy (DOE) Bioenergy Technologies Office consists of a large array of related but discrete projects covering various aspects of biomass feedstock generation and conversion technologies. DOE national laboratories conduct some projects; universities or other outside research organizations carry out others. Project selection also varies over time, with some projects taking only a year and others continuing for several.

The following fact sheets summarize these various internal, external, and joint projects that make up the Bioenergy Technologies Office's project portfolio. For a listing and summaries of major fact sheets and publications, see the DOE Publications Listing page. The web address of that page (for readers without the electronic book's hyperlinks) is https://www1.eere.energy.gov/biomass/key_publications/html

Research and Development Projects

- Feedstocks
- Biochemical Platform Technology
- Thermochemical Platform Technology
- Integrated Biorefineries
- Biobased Products
- Congressionally Directed Projects

Feedstock Research Projects

- Active
  - Feedstock Infrastructure
  - Feedstock Supply Logistics
  - Multi-Component Harvesting of Wheat Straw
- Complete
  - Harvesting and Collection of Biomass
  - Selective Harvest of Wheat Straw
  - Biomass Bulk Processing and Storage
  - Feedstock Supply Forecasts and Analysis
  - Feedstock Supply System Logistics

Biochemical Platform Research Projects

- Active
  - Biochemical Platform Analysis
  - Biochemical Platform Processing Integration
  - Feedstock Sugar Interface
  - Integration of Biomass Pretreatment Technologies With Enzymatic Digestion and Hydrolyzate Fermentation
  - Pretreatment and Enzymatic Hydrolysis
  - Targeted Conversion Research
- Complete
  - Biological Fractionation Fundamentals
  - Biomass Structure
  - Biomass Surface Characterization
  - Chemical Processing Fundamentals
  - Development of Improved Ethanol Fermentation Microorganisms
  - Development of New Sugar Hydrolysis Enzymes: Genencor International
  - Development of New Sugar Hydrolysis Enzymes: Novozymes Biotech, Inc.
  - Emerging Pretreatment Options
  - Enzymatic Biomass Conversion Process Integration
  - Enzyme-Mediated Pretreatment
  - Fundamental Study on Kinetics and Transport Phenomena in Low-Water, Dilute-Acid Total Hydrolysis of Cellulosic Biomass

- Fundamentals of Biomass Hydrolysis at Low pH
- Pretreatment of Applications for Corn Dry Mills

Thermochemical Platform Research Projects

- Active
  - Advancement of High-Temperature Black Liquor Gasification Technology
  - Biomass Gas Clean-Up Using a Therminator
  - Catalytic Hydrothermal Gasification of Wet Biomass Feedstocks
  - Causticizing for Black Liquor Gasifiers
  - Cost-Benefit Analysis of Fuel and Chemical Production at Pulp Mills
  - Direct Causticization for Black Liquor Gasification
  - Evaluation of Membrane Systems for Biomass Gasifiers
  - Feed Processing, Handling, and Gasification
  - Evaluation of Sulfur in Syngas
  - Fuel Chemistry and Bed Performance in a Black Liquor Steam Reformer
  - Gasification of Biorefinery Residues
  - Mixed Alcohols Production From Syngas
  - New Catalysts for In-Process Elimination of Tars
  - Pressurized Entrained-Flow Draft Black Liquor Gasification
  - Pyrolysis Oil Upgrading
  - Refractory for Black Liquor Gasifiers
  - Syngas Clean-Up and Conditioning
  - Thermochemical Conversion Analysis
  - Thermochemical Conversion of Corn Stover
  - Trace Metal Scavenging From Biomass Syngas Using Novel Sorbents
- Complete
  - Catalytic Cleanup Conditioning Fundamentals
  - Evaluation of RVS-1 Sorbent for Removal of Sulfur from Black Liquor Gasification
  - Fundamentals of Biomass Gasification
  - Gasification of Wet Biomass
  - Materials Evaluation for Black Liquor and Biomass Gasifiers
  - Pulsed Black Liquor Reformer Materials Evaluation
  - Syngas Particulate Cleanup
  - Thermochemical Process Modeling and Optimization

Integrated Biorefineries

- Recovery Act—Active
  - ADM Recovery Act Biorefinery Project
  - Algenol Recovery Act Biorefinery Project
  - Amyris Recovery Act Biorefinery Project
  - API Recovery Act Biorefinery Project
  - Bluefire Recovery Act Biorefinery Project
  - Clear Fuels Recovery Act Biorefinery Project
  - Enerkem Recovery Act Biorefinery Project
  - Haldor Topsoe Recovery Act Biorefinery Project
  - ICM Recovery Act Biorefinery Project
  - INEOS Bio Recovery Act Biorefinery Project
  - Logos Recovery Act Biorefinery Project
  - Myriant Recovery Act Biorefinery Project
  - REII Recovery Act Biorefinery Project
  - Sapphire Recovery Act Biorefinery Project
  - Solazyme Recovery Act Biorefinery Project
  - UOP Recovery Act Biorefinery Project
  - ZeaChem Recovery Act Biorefinery Project
- Active
  - Abengoa Commercial Scale Biorefinery
  - Poet Commercial Scale Biorefinery
  - Mascoma Demonstration Scale Biorefinery
  - RSA Demonstration Scale Biorefinery
- Complete
  - Making Industrial Biorefining Happen
  - Integrated Corn-Based Biorefinery
  - Advancing Biorefining of Distiller's Grain and Corn Stover Blends
  - Big Island Demonstration Project - Black Liquor
  - Collection, Commercial Processing, and Utilization of Corn Stover
  - Enhancement of Co-Products From Bioconversion of Municipal Solid Waste
  - A New Biorefinery Platform Intermediate
  - A Second-Generation Dry Mill Biorefinery
  - Separation of Corn Fiber and Conversion to Fuels and Chemicals: Pilot-Scale Operation
  - GTI Recovery Act Biorefinery Project
  - Verenium Biofuels Corporation Demonstration Scale Biorefinery
  - Elevance Recovery Act Biorefinery Project
- Inactive
  - Range Fuels Commercial Scale Biorefinery
  - Flambeau River Biofuels LLC Demonstration Scale Biorefinery
  - NewPage Corp. Demonstration Scale Biorefinery
  - Lignol Innovations, Inc Demonstration Scale Biorefinery

Biobased Products

- Active
  - Catalytic Upgrading of Glycerol to Propylene Glycol
  - Chemicals from Oilseeds
  - Engineering Thermotolerant Biocatalysts for Biomass Conversion to Products
  - Fungal Genomics
  - New Sustainable Chemistry for Adhesives, Elastomers, and Foams

  - Opportunity Analysis for Bio-Based Products
  - Platform Chemicals from an Oilseed Biorefinery
  - Production of Polyhydroxyalkanoate Polymers
  - Separative Bioreactor
  - Top Value-Added Chemicals from Biomass: Sugar, Syngas, and Lignin Pathways
  - Value-Added Products From Hemicellulose Utilization in Dry Mill Ethanol Plants
- Complete
  - Arabinose Yeast Cooperative Research and Development
  - Biocatalyst Development for Chemicals
  - Evaluation of Performance and Emission Benefits of Renewable Diesel Transportation Fuel
  - Functionalized Vegetable Oils for Use as Polymer Building Blocks
  - Pyrolysis Oil Upgrading for Products
  - Wood Adhesives from Bark-Derived Phenols

Congressionally Directed Projects

- Active
  - Agricultural Mixed Waste Biorefinery Using Thermal Conversion Process
  - Biomass-Derived Hydrogen from a Thermally Ballasted Gasifier
  - Energy from Biomass Research and Technology Transfer Program
  - Gasification-Based Fuel Conversion and Electricity Production in the Forest Products Industry
  - Gasification Research Center
  - Hybrid Poplar Research
  - Indiana Corn Dry Mill
  - Integration of Succinic Acid Production in a Dry Mill Ethanol Facility
  - Iowa Switchgrass Project
  - Kentucky Rural Energy Supply Program
  - Micro-Modular Biopower System for Cooling, Heating, and Power
  - Mississippi Ethanol Gasification Project
  - Native Flora Utilization
  - North Central Texas Dairy Waste Control Pilot Project
  - On-Farm Small-Scale Waste Energy Demonstration
  - Southeast Alaska Wood Residue Biorefinery
- Complete
  - AgraPure Mississippi Biomass Project
  - Black Belt Bioenergy Demonstration Project
  - North Country Hospital Biomass Conversion and Cogeneration Project
  - Salix Consortium: Energy Crops for Power Production

Part II

Equipment Design

Chapter 6

Fermentation Design

Allan C. Soderberg

1.0 INTRODUCTION

Industrial scale fermentation technology tends to be a "proprietary science." The industries with submerged liquid fermentation processes as a "synthetic" step for producing a commercial product generally have developed their own technology and have not shared developments with their competitors, academe, or the public. If major fermentation industries decided to openly discuss the criteria of their procedures and processes for their fermentation departments, they would not agree on most systems and equipment, from culture storage methods to valves, from lab culture propagation to fermenter design, from scale-up to sterile filters, or from tank inoculation methods to continuous sterilizers. The experience of every author or speaker, though he may have years of practical knowledge, is probably regarded as inferior to the experience of the reader or listener. That is, the subjective analysis of the data by each company has resulted in different solutions to common problems, or each company has a customized plant suited to its procedures and products.

2.0 FERMENTATION DEPARTMENT, EQUIPMENT AND SPACE REQUIREMENTS

2.1 The Microbiological Laboratories

Isolation of organisms for new products normally does not occur in laboratories associated with production cultures, however, production (microbiological) laboratories frequently do mutation and isolation work to produce strains with higher yields, to suppress a by-product, to reduce the formation of a surfactant, to change the physical properties of the broth to facilitate the product recovery, etc. The experience, imagination and personal skill of the individual is fundamental for success. The results of mutation work have been of great economic value to the fermentation industry, therefore, the methods used remain closely guarded and are almost never published. Other on-going studies include new culture preservation techniques; improved culture storage methods; culture stability testing; new propagation procedures; media improvements; search for inducers, repressors, inhibitors, etc. Here again, the imagination of the researcher is essential to success because specific research methods are commonly nontraditional.

The highly developed production cultures must be preserved from degradation, contamination and loss of viability. Every conceivable method is being used and supported by experimental data—sand, soil, lyophils, spore and vegetative suspensions, slants and roux bottles, surface colonies under oil, etc. The temperature for culture storage varies from −196°C (liquid nitrogen) up to +2°C and above. The containers generally are glass, but vary from tubing, to test tubes, flasks (any shape and size), roux bottles, serum bottles, etc. A good argument can be made that the only important variable is to select the correct medium to grow the organism in or on before it is stored. Obviously, carbon, nitrogen, water and minerals are required for growth, but sometimes high concentrations of salts, polyols or other chemicals are needed to prevent a high loss of viability during storage. Frequently, a natural product (oat meal, tomato juice, etc.) is helpful for stability compared to a totally synthetic medium. Under the right conditions, procedures based on vegetative growth can be more stable than ones based on spores.

Submerged fermentation procedures are used almost exclusively today. A few surface fermentation processes (on liquids or solids) are still used. Cost comparisons of labor, air compression, infection, etc., can be made, but modern batch fed, highly instrumented and computerized submerged methods predominate. Submerged methods are also the predominant culture propagation technique. The general principle is to have the fewest possible transfers from the primary culture stock to the fermenter. This is based on the assumptions that transferring and media sterilization are the main infection risks. Generally, a lyophilized or frozen culture is used to inoculate a flask of liquid medium which is then shaken until sufficient cell mass has been produced. (Some prefer solid media, in which case a sterile solution must be added to suspend the culture in order to transfer the culture to the seed tank.)

Fermentation and Biochemical Engineering Handbook.

The medium in the seed flask frequently contains production raw materials rather than microbiological preparations used in research laboratories. (For a general description of various microbiological tasks performed in industry, see Peppler and Perlman [1].)

After the culture is grown, the flask (fitted with a hose and tank coupling device) is used to inoculate the seed fermenter. However, some transfer the culture from the seed flask to a sterile metal container (in the laboratory) which has a special attachment for the seed fermenter. This technique is usually abandoned in time. Ingenuity for the minimum transfers in the simplest manner will usually give the best results.

The space requirements and the equipment necessary for designing a culture maintenance lab vary so widely, from simple laminar flow hoods to air locked sterile rooms, that only each company can specify the details. The number of rooms and work areas depend upon the number of types of cultures maintained, as well as the variety of techniques for mutation, isolation and testing. Therefore, lab space and equipment might include:

1. *Glassware and Equipment Washing Area.* Washing and drying equipment, benches, carts.
2. *Media Preparation Area(s).* Space must be provided for large raw material lots, not only for growth in flasks, but testing of cultures in very small glass fermenters, large statistically designed shake flask experiments, serial growth experiments in Petri dishes for stability experiments and others. Equipment will be required to hydrolyze starch and proteins, to process molasses, in addition to kettles, homogenizers, centrifuges, sterilizers and large benches.
3. *Inoculation Rooms.* Frequently, separate rooms are used for work with bacteria, actinomycetes, molds, and sterility testing. High intensity UV lighting is commonly used when the rooms are unoccupied. These rooms generally have only work benches (or hoods) for easy cleaning.
4. *Incubator Areas.* Space is required for incubators (various temperatures), some of which could be the walk-in type, and/or floor cabinet models. Shaker cabinets at various temperatures are also needed.
5. *Office.* Record keeping and administration will require one or more offices, depending upon the size of the staff.
6. *Laboratories.* Depending upon the size of the facility, separate laboratories could be required for culture mutation, culture isolation, and testing in bench top fermenters. Space must be provided for microscopes, special analytical equipment for DNA, ATP, Coulter counters, water baths, pH and DO instruments, laminar flow hoods, balances, lyophilization equipment, etc.
7. *Other.* Space must be provided for refrigerators and freezers, which are the repositories of the production culture collection. Laboratory employee facilities are segregated from other areas.

The square feet of floor space per technician required for these laboratories will be four to eight times that required for the analytical laboratories of the fermentation department. The reason for this is cleanliness, and the rooms have specific purposes for which they may not be used every day. The work force moves from room to room depending upon the task scheduled. Also, the total work area depends upon the variety of microbiological tasks performed. A large plant may even have a pilot plant.

2.2 Analytical Support Laboratories

The functions of these laboratories usually are sterility testing of production samples, and chemical assays of: raw materials for approval to use in the processes, blends or batches of raw materials before sterilization, scheduled samples of production batches, fermenter feeds, waste streams and miscellaneous sources. In many instances the analytical work for the culture laboratories will also be performed.

Typical laboratories have Technicon Auto-analyzers for each of the common repetitive assays (the product of the fermentations, carbohydrates, phosphate, various ions, specific enzymes, etc.). Other equipment generally includes balances, gas chromatographs, high pressure liquid chromatographs, Kjeldahl equipment, titrimeters, UV/visible spectrophotometers, an atomic absorption spectrophotometer, pH meters, viscosimeter, refractometer, densitometer, etc. The cell mass is usually followed for its intrinsic value as well as to calculate specific uptake rates or production rates in the fermenter. Therefore, centrifuges and various types of ovens are required for drying in addition to ashing.

Fermenter sterility testing requires a room with a laminar flow hood to prepare plates, tubes and shake flasks. Space needs to be provided for incubators and microscopes. Since it is very important to identify when infection occurs in large scale production, microscopic examination of shake flasks is usually preferred because a large sample can be used, and it gives the fastest response. Similarly, stereo microscopes are used for reading spiral streaks on agar plates before the naked eye can see colonies.

Chemical and glassware storage, dish washing, sample refrigerators, glassware dryers, autoclaves for the preparation of sterile sample bottles for the plant, computer(s) for assay calculations, water baths, fume hoods, etc., are additional basic equipment items needed. Typical overall space requirements are 450 ft^2 of floor space per working chemical technician.

2.3 Production: Raw Material Storage

Raw material warehousing most often is a separate building from manufacturing. Its location should be on a rail siding (for large plants) and have easy access by twenty-ton trailers. The dimensions of the building should make it easy to stack a palletized forty-ton rail car's contents—two pallets wide and three or four pallets high, from the main aisle to the wall. In this manner, raw material lots can be easily identified and used when approved.

Large volume dry raw materials should be purchased in bulk (trucks or rail cars) and stored in silos. Pneumatic conveying from the silos to the mixing tanks can be controlled from the panel in the instrument control room after selecting the weight and positioning diverter valves. Wherever possible, liquid raw materials should be purchased in bulk and pumped. For safety and environmental reasons, drummed, liquid raw materials should be avoided, if possible, the silos and bulk liquid tanks can usually be placed close to the batching area, whereas the warehouse can be some distance away. Since large volume materials are pneumatically conveyed or pumped, the floor space of the batching area for storing miscellaneous materials can be relatively small.

The equipment needed in warehousing are fork lift trucks, floor-washing machines, etc. Special materials must be on hand to clean up spills quickly, according to federal regulations. Good housekeeping and pest control are essential.

2.4 Media Preparation or Batching Area

For good housekeeping, all equipment should be on or above the floor and no pits should be used. On the other hand, grated trenches make it easy to clean the floors, and minimize the number of floor drains.

The number, shape and volume of batching tanks that different companies use show personal preference and are not very important. Usually two or three different sized tanks are used; smaller batching tanks are for inoculum tanks and the larger tanks for feed and fermenter media preparation. The type of agitation varies widely. Batching tanks, 10,000 gallons and smaller, could be specified as 304 stainless steel, dished or flat bottom and heads, H/D ratio about 0.7 to keep a working platform low, a slow speed (60 to 90 rpm) top-entering agitator with airfoil type impellers, horsepower approximately 1.25 per 1000 gallons. The tanks need to be equipped with submerged (bottom) nozzles which are supplied with both steam and air. Hot and cold water are usually piped to the top. The hatch, with a removable grate of ½" S/S rod on 6" × 6" centers, should be as large as a 100 lb. bag of raw materials. A temperature recorder is the minimum instrumentation. The cyclone, with a rotary air lock valve to permit material additions from the bulk storage silos, is normally located above the tank(s). For tanks larger than 10,000 gallons, the bottom head should be dished, the H/D ratio made 1 to 2, and airfoil type agitators used.

The size and number of batching tanks depend upon whether the plant uses continuous sterilizers or batch sterilization. The difference is that in the latter case, the tanks can be large (50 to 80% of the size of the fermenter), and usually all the materials are mixed together. For continuous sterilizers, there is usually a minimum of four smaller tanks so that proteins, carbohydrates and salts can be batched and sterilized separately. In this case, the tanks are considerably smaller than the fermenter.

The media preparation area is also where hydrolysates of proteins, and starches, as well as special processing of steep liquor, molasses and other crude materials takes place. Very strict accuracy of weights, volumes, pH adjustments and processing instructions are the first step to reproducible fermentation results. A well-run batching area depends upon purchasing a uniform quality of raw materials, adequate equipment, detailed batching instructions and well trained, reliable personnel. Record keeping of batch quantities, lot numbers, pH, temperatures, etc. are necessary for quality and good manufacturing practices.

2.5 The Seed Fermenter Layout

Some companies prefer to locate all the seed fermenters in one area so that a group of workmen become specialists in batch sterilizing, inoculating, and coddling the first (plant) inoculum stage to maturity. Other companies locate the seed fermenters adjacent to the fermenters. Small plants cannot afford to isolate equipment and have a specialized work force, however, large plants do isolate groups of similar equipment, and specialize the work force, which often results in higher productivity.

The operation of fermenters is basically the same regardless of size, but seed fermenters usually do not have sterile anti-foam and nutrient feeds piped to the tanks as the main fermenters have. Therefore, foaming in the seed fermenters can lead to infection, which is one of the reasons they need more attention. Careful inoculation procedures, sampling and sterilizing the transfer lines from the seed fermenter require alert personnel. Careful attention to these details is more important than the proximity of the seed and main fermenters.

The number of inoculum stages or scale-up is traditional. The rule of a tenfold volume increase per stage is followed by some companies, but is not critical. The multiplication rate of an organism is constant after the lag phase so the amount of cell mass developed to inoculate the next stage, minus the starting amount, is a matter of time, providing, of course, there is sufficient substrate and environmental conditions are reasonable. After all, the theory is that one foreign organism or spore, if not killed during sterilization, will, in time, contaminate the fermenter. Larger cell masses of inoculum can shorten the growth phase of the next larger stage. Using this concept, some companies make the inoculum volume larger than a tenth of the fermenter volume so that the number of transfers from laboratory flask to the final fermenter is minimum. This also assumes there is a higher risk of infection during transfers as well as a certain viability loss. A higher inoculum cell mass may reduce the lag time in the fermenter. This, combined with using continuous sterilization for a short "turn around" time of the fermenter, can increase productivity for little or no cost.

2.6 The Main Fermenter Layout

For simplicity of piping, especially the utility piping, the fermenters are usually placed in a straight line, sometimes two or more parallel lines. In this manner the plant is easily expanded, and other tank layouts do not seem as convincing. It is desirable to have the working platform extend completely around the circumference of the top dish, and to have enough room between tanks for maintenance carts (1 to 1.5 meters). Good lighting and ventilation on the working platform should not be overlooked. Using water from hoses for cleaning is common so care must be taken to have nonskid floors with adequate drains, especially at the top of stairs. Open floor grating is not desirable. All structural steel should be well primed to prevent corrosion from the very humid atmosphere. Electronic instrumentation and computers must be placed in control rooms which run at constant (HVAC) temperature. Most fermenter buildings are between 40 and 100 feet high, making it possible to have one or more floors between the ground floor and the main fermenter working platform. The intermediate floors can be used for the utility and process piping, sterile air filters, the sterile anti-foam system, instrumentation sensors (temperature, pH, DO, etc.), heat exchangers, motor control center, laboratories and offices. Buildings 40 feet or more high frequently have elevators installed.

Fermenters can be located outdoors in most countries of the world. The working platforms usually are enclosed and heated in temperate zones, and only shaded in more tropical zones. In more populated areas, open fermenter buildings make too much noise for local residents. The environmental awareness, or the tolerance of the public, could preclude open fermenter buildings in the future. Odor is also offensive to the public. The environmental authorities are demanding that equipment be installed to eliminate the offensive odor of the off-gases. (Noise levels inside a fermenter building will be greater than 90 dBA if no preventive measures are taken.)

Harvest tanks can be justified as the responsibility of the fermentation or recovery department. They are economical (carbon or stainless steel) with a shape described by ($H/D \cong 1$) and should be insulated and equipped with cooling coils and agitator(s).

2.7 Nutrient Feed Tanks

Essential equipment to a productive fermentation department are sterilizable tanks for nutrient feeds. Multiproduct plants usually require several different sizes of feed tanks: *(i)* a small volume to be transferred once every 12 or 24 hours such as a nitrogen source; *(ii)* a large volume carbohydrate solution fed continuously, perhaps varying with the fermenter volume; *(iii)* a precursor feed, fed in small amounts relative to assay data; *(iv)* anti-foam. (Some companies prefer a separate anti-foam feed system for each fermenter. A continuously sterilizing system for anti-foam is discussed below which is capable of servicing all the fermenters.); *(v)* other tanks for acids, bases, salts, etc. Many companies prefer to batch sterilize a known quantity and transfer the entire contents quickly. Sometimes, the feeds require programming the addition rate to achieve high productivity. In this latter case, large volume tanks are used and the contents are presterilized (batch or continuous) or the feed is continuously sterilized between the feed tank and the fermenter. Usually feed tanks are not designed as fermenters, even though they are sterilizable, and there is no need for high volume air flow, but only sufficient air pressure for the transfer. For solvable nutrients the agitator and anti-foam system are not required. Since the air requirements are needed only to transfer the feed, the air piping design is different and the sterile air filter is proportionately smaller. Instrumentation is usually limited to temperature, pressure and volume. The H/D ratio of the vessel can be near one for economy and need not be designed for the aeration/agitation requirements of a fermenter.

2.8 Sterile Filters

Sterile air filtration is simple today with the commercial units readily available. However, some companies still design their own (see Aiba, Humphrey and Millis [2]) to use a variety of filter media such as carbon, cotton, glass staple, etc. (For recent papers about industrial applications of cartridge filters, see Bruno [3] and Perkowski [4].)

The essential method to obtain sterile air, whether packed-bed or cartridge filters are used, is to reduce the humidity of the air after compression so that the filter material always remains dry. The unsterilized compressed air must never reach 100% relative humidity. Larger plants install instrumentation with alarms set at about 85% relative humidity. Careful selection of the cartridge design or the design of packed-bed filters will result in units that can operate in excess of three years without replacement of filter media. If a fiber material is used in a packed-bed type filter, the finer the fiber diameter the shallower the bed depth needs to be for efficient filtration. Other filter media are less common and tend to have special problems and/or shorter life. The bed depth of filters is only 10 to 18 inches for fibers of less than 10 microns. These filters run "clean" for 2 weeks or longer before being resterilized (Fig. 6.1).

Some plants have a separate filter for each sterile vessel. Others place filters in a central group which feeds all the vessels. In this case, one filter, for example, might be taken out of service each day, sterilized and put back into service. If there were ten filters in the group, each one would be sterilized every tenth day. This system has the advantage that the filter can be blown dry after sterilization with sterile air before it is put into service again.

2.9 Air Compressors

It is ideal to have oil-free compressed air. Centrifugal machines generally are available up to 40,000 cfm. "Oil free" screw air compressors are available in smaller sizes. Regarding oil-free screw type compressors, it is necessary to read the fine print of the manufacturer. For example, one manufacturer uses no lubricant on the screws and another claims to be oil free, but does use a non-hydrocarbon liquid lubricant. Carbon ring reciprocating compressors are available and used, but maintenance is annoying.

For small plants, non-lubricated screw compressors with two-speed motors and constant pressure control will provide versatility. For large plants, centrifugal air compressors, driven by non-condensing steam turbines with 50 psig steam extraction for process requirements, are suitable. In all cases, extra considerations include locating the intake 20 feet or more above the ground level and installing filters on the intake to the compressors to prevent dirt accumulation on the sterile filters. Occasionally, the noise levels measured at the suction inlet exceed OSHA regulations and bother the neighbors of the plant. The air from the compressors requires heat exchangers to lower the air temperature below the dew point, plus additional heat exchangers to reheat and control the air to have the relative humidity at about 85%.

FIGURE 6.1 Domnick-Hunter sterile air filter.

2.10 Valves (to Maintain Sterility)

Most companies have tried gate, diaphragm, ball, and plug valves, to name a few. Some have designed and patented special valves for the bottom or sample positions. Some companies will disassemble all fermenter valves after an infected run. Manufacturers do not use threaded nipples or valves on a fermenter because the threads are a site of potential infection. In general, valves are less of a sterility problem when a continuous sterilizer is used for the substrate than fermenters which batch sterilize the substrate. This is because, in the former case, the vessel is sterilized empty, and all valves are opened and sterilized in an outward direction so that a steam plume can be seen. The temperature of the valves during sterilization can be checked with a Tempilstik™. Batch sterilizing requires all valves below the liquid level to be sterilized with steam passing through the valve into the substrate. This depends upon steam pressure and how much the valve is opened (which might affect the *P/T* conditions of sterilization). This is much more subject to human error and infection. Most plants drill and tap the body of the valve near the valve seat in order to drain the condensate away from all sections of pipe where a steam seal is required for sterility. In general, diaphragm and ball valves require considerable maintenance, but tend to be popular in batch sterilizing operations, while plug type valves are more typical on fermenters where continuous sterilizers are used. Plug or diaphragm valves are commonly used for inoculum transfer and sterile feed piping. All the process valves and piping today are 316 S/S. Utility piping remains carbon steel up to the first S/S valve on the fermenter. Valves used in non-process piping are selected for the best type of service and/or control. Butterfly valves have been used in applications where perfect closure is not essential, such as a vent valve.

In summary, the valves which maintain a sterile environment on one side and a non-sterile environment on the other side are the essential valves. They must be devoid of pockets, easily sterilized, maintained, and occasionally replaced.

2.11 Pumps

Apart from continuous sterilizers, pumps are a minor concern in the fermentation department. A simple way to transfer inoculum from a large laboratory flask to a seed fermenter, without removing the back pressure on the vessel, is to use a peristaltic pump. Connect the sterile adapter (which is attached to the flask) to the seed fermenter by sterile technique. Install the gum rubber tubing in the pump, open the hose clamp and start the pump.

Inoculum from seed fermenters and sterile feeds are transferred to the fermenter by air pressure. Centrifugal pumps (316 S/S) are used to pump non-sterile raw materials, slurries, harvested broth, etc. The centrifugal pumps and piping should be cleaned immediately after a transfer has been completed. Occasionally a specialty pump may be required.

2.12 Cooling Equipment

Cooling is required to cool media from sterilizing temperatures, to remove the exothermic heat of fermentation, to cool broth before harvesting, and to cool the compressed air. Some portion of the heat can be reclaimed to produce hot water for the preparation of new substrate, and for general cleaning of equipment, platforms and floors, however, the excess heat must be disposed to the environment. Cooling water is provided from cooling towers, but chilled water (5°–15°C) is produced by steam vacuum, or refrigeration units.

In any case, the fermentation department should always be concerned about its cooling water supply, i.e., the temperature and chloride content. Chloride ions above 150 ppm when stainless steel is above 80°C (while sterilizing) will cause stress corrosion cracking of stainless steel. A conductivity probe should be in the cooling water line. When the dissolved solids (salts) get too high, it may indicate a process leak, or that the salt level is too high and some water must be discharged and fresh water added. If cooling water is discharged to a stream, river, etc., an NPDES permit may be needed and special monitoring required. The chloride content should be determined analytically every two weeks to control the chloride to less than 100 ppm. Draining water from the cooling tower and adding fresh water does this.

2.13 Environmental Control

Stack odors have to be avoided. Certain raw materials smell when sterilized. Each fermentation process tends to have its own unique odor ranging from mild to strong and from almost pleasant to absolutely foul. Due to the high volume of air discharged from a large fermenter house, odor is neither easy nor cheap to eliminate. Carbon adsorption is impractical. Normally, more air is exhausted than required for steam production from the boilers which eliminates that route of disposal. Wet scrubbing towers with sodium hypochlorite are expensive, and discharge Na^+ and Cl_2 to the waste system which may preclude this method. Ozone treatment can be effective. A very tall exhaust stack for dilution of the off gas with the atmosphere before the odor reaches the ground is possible in some cases, but is not considered an acceptable solution by U.S. Authorities.

The fermentation department should monitor and control the COD/BOD of its liquid waste to the sewer. Procedures for cleaning up spills and reporting should be Standard Operating Procedure. A primary aeration basin will reduce the COD to 80–90 ppm. Secondary aeration lagoons will reduce the BOD to acceptable levels which have no odor.

Noise levels are very difficult to reduce to Federal standards. Hearing protection for employees is essential. The move towards greater automation has resulted in operators having less exposure to noisy work areas.

3.0 GENERAL DESIGN DATA

Most companies produce more than one product by fermentation simultaneously. It is not necessary to have separate fermenter buildings to isolate products. Well-designed fermenters which are operated properly, not only keep infection out, but also prevent cross contamination of products. Over the years, most fermentation plants have been enlarged by the addition of new fermenters despite major yield improvements. Therefore, as plants grow, the engineer must always keep in mind there will be a need for further expansions. The layout of labs, fermenter buildings, the media preparation area and warehousing must be able to be expanded. Utilities and utility piping must also be installed with spare capacity to handle average and peak loads as well as future growth.

Some guidelines for piping design are:

1. 50 to 150 psig steam	0.5 psi loss/100 ft
2. 100 psig instrument air	0.5 psi loss/100 ft
3. 50 psig fermenter air, (from compressors to sterile filters)	2.0 psi loss; total Δp
4. Water in schedule 40 steel pipes	6–10 ft/sec
5. Gravity flow sewers	2.5 ft/sec

The consumption of utilities in a fermentation department depends upon the fermenter cycles since most of the steam and water are used to clean, prepare, sterilize and cool each batch. The data presented below are based on a one-week (168 hours) cycle including turnaround time and 1000 gallons of fermenter installed capacity (abbreviated: 1000 I.c. Wk).

Steam	
1. 45 psig steam for media sterilization	1350 lb
2. 45 psig steam for equipment and piping-cleaning and sterilizing	3150 lb
Total steam	4500 lb

Water (in)	
1. Steam in *(1), (2),* above	540 gal
2. Media makeup	570 gal
3. Equipment cleaning	2880 gal
4. Cooling tower water (makeup)	550 gal
Total water	4540 gal

The fermentation department can consume up to 2/3 of the total plant electrical requirements (depending upon the recovery process), which includes mechanical agitation (usually 15 hp/1000 gal) and electrically driven air compressors.

There is no relationship between the cubic feet of compressed air for large fermenters and their installed capacity. The compressed air required for fermenters is calculated by linear velocity through the fermenter and the square feet of cross-sectional area of a vessel, not its volume. Therefore, if volume is constant, short squat vessels require more compressed air than tall slender vessels. More on this is discussed under fermenter design.

4.0 CONTINUOUS STERILIZERS

4.1 A Justification for Continuous Sterilization

The design of any fermentation plant begins with the annual capacity of product for sale, the yield of product isolation, and the productivity of the fermenters. The size of the fermenters should be the largest size possible consistent with the product degradation rate during isolation, the economy of isolation equipment, manpower and operating costs. Unfortunately, many companies have not built fermenters over a wide range of sizes, but have built new fermenters "just like the last one." One factor contributing to the reluctance to scale up is that small fermenters are batch sterilized, and there is a hesitancy to build and operate continuous sterilizers at the same time fermenters are scaled-up. Large fermenters and continuous sterilizers are economically sound. There are the same number of valves and operations on a small fermenter as on a large one, therefore, labor savings per kilo of product are made by making larger fermenters. A continuous sterilizer is economically advantageous at almost any industrial scale with five or six fermenters.

Reduced Fermenter Turn-Around Time. A fermenter can be productive only when fermenting. Emptying, cleaning, filling, batch sterilizing and cooling are nonproductive time. A continuous sterilizer will shorten the turnaround time leaving more time for production. The increased number of harvests per year for a fermenter is related to the fermentation cycle; e.g., using a 30,000 gallon fermenter and a 150 gpm continuous sterilizer, the increased capacity annually is illustrated in Table 6.1.

More Effective. Sterilization. The internal parts of a fermenter are sterilized easier with no liquid inside. A lower percentage of media contamination can be achieved with a continuous sterilizer than by batch sterilization.

Higher Fermentation Yields. With a continuous sterilizer, proteins can be sterilized separately from carbohydrates and salts. The residence time at high temperature is short. There is less interaction and degradation of raw materials, resulting in higher fermentation yields.

Reduced Agitator Cost. It is not necessary to buy a two-speed motor where the slow speed (low horsepower) is used for mixing during batch sterilization and high speed only during aeration.

4.2 Support Equipment for a Sterilizer

All continuous sterilizers have a heating section, a retention section, and a cooling section. However, before the design

TABLE 6.1 Increased Harvests per Year Due to a Continuous Sterilizer

Fermentation Time (hr)	Percent Increased Annual Harvest Volume
200	5
150	6
100	9
50	20

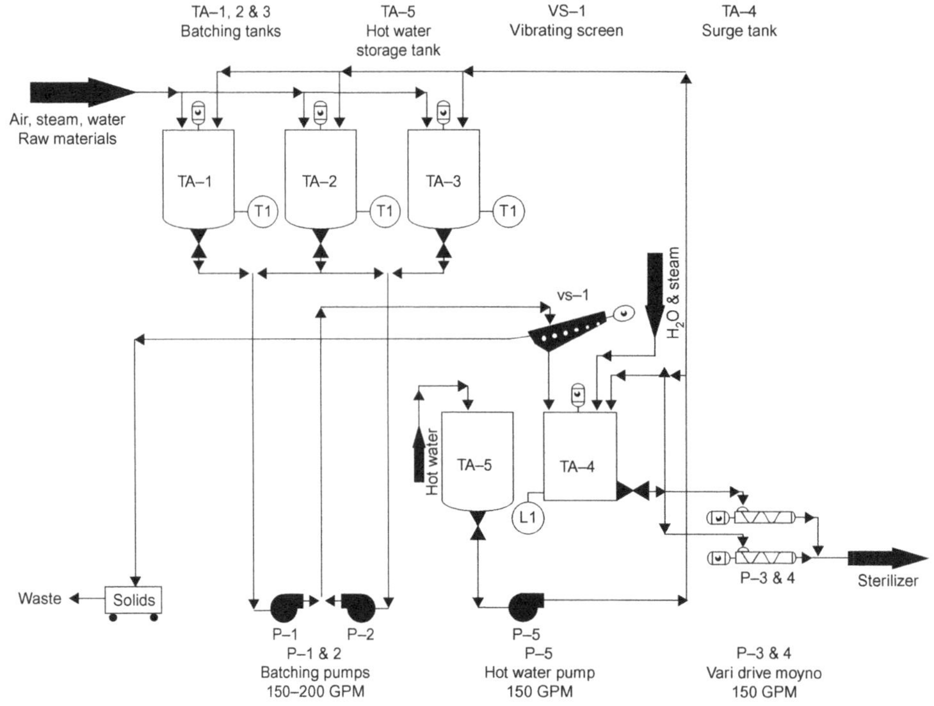

FIGURE 6.2 Batching area flow chart.

of the sterilizer is discussed, a brief review of batching equipment in support of the sterilizer is necessary.

Figure 6.2 is a flow diagram of batching equipment. Tanks 1, 2, and 3 illustrate that the proteins, carbohydrates and salts can be prepared and pumped separately to prevent interaction during sterilization. Notice that Tank 5 is for storage of hot water from the cooling section of the sterilizer and is used for media preparation, especially assisting in dissolving salts, sugars, etc. Omitted from Fig. 6.2 are the bulk storage and pneumatic conveying equipment of large volume of dry materials, the bulk liquid storage system, starch hydrolysis systems for dextrin and glucose, and other systems for economy and high volume handling.

After the raw materials are dissolved, suspended, and treated, they should be passed through a vibrating screen. The success of sterilization depends upon moist heat penetrating to the center of the suspended solids. This reaction is a function of time and temperature, and the time-temperature design basis of the sterilizer must be capable of the task. Therefore, to prevent long sterilization times, a screen size with openings of about 4 mm^2 is reasonable. Also, the non-dissolving raw materials must be a fine grind when purchased so that good dispersion in the batching tanks will be achieved.

Tank 4 is a surge tank for the operation of the sterilizer. Pump 1 (or 2) from the batching tanks must fill Tank 4 faster than Pump 3 (or 4) feeds the sterilizer. Figure 6.3 illustrates the piping and instrumentation of the surge tank. It is filled from the batching tanks sequentially so that there is a minimum mixing of the segregated raw materials before sterilizing.

The hot water from Tank 5 also supplies Tank 4 and Pump 3 (or 4). This is necessary to start and finish a batch through the sterilizer. For example, to start, the sterilizer is first steam sterilized (no liquid). At the end of this cycle, hot water from Tank 5 is started through the sterilizer to *set* or *balance* the instrumentation. When this is achieved, media is fed to Pump 3 (or 4) by remote operating valves. Similarly, after all the media has been pumped, it is necessary to pump water through the sterilizer until the fermenter volume is correct. If another fermenter is to be filled immediately, the sterile water is diverted to the awaiting empty (and sterile) fermenter, and then the new media for the second fermenter is pumped into the sterilizer.

The control room for the operation of a continuous sterilizer should be close to Tank 4, Pumps 3 and 4, the main steam valves and the valves of the sterilizer itself. This location is essential to sterilize the empty sterilizer and control the pumping of water and/or media.

Figure 6.4 is a block flow diagram of a sterilizer that is suitable for fermenter volumes of 20,000 to 60,000 gallons capacity. It is based on pumping 150 gpm of non-sterile media to the steam injector. Energy savings could be about 45% if the hot water storage capacity (Tank 5) were equal in volume to a fermenter. Using the excess hot water for other purposes in the plant (e.g., in crystallizers, vacuum evaporators, space heaters, cleaning, etc.) can provide additional energy savings.

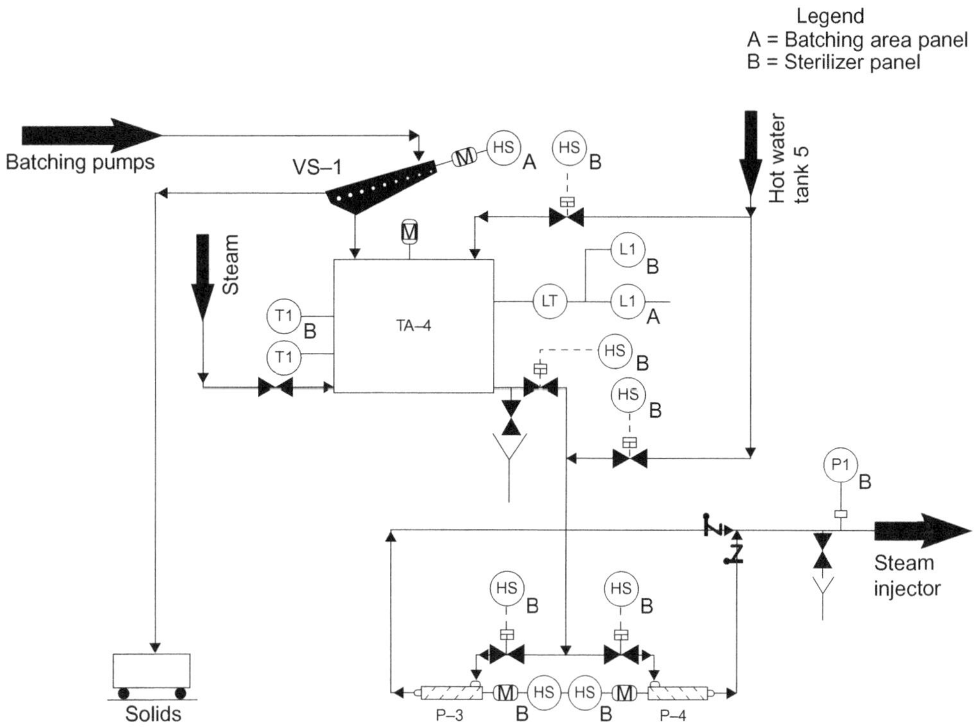

FIGURE 6.3 Surge tank and pumps to sterilizer.

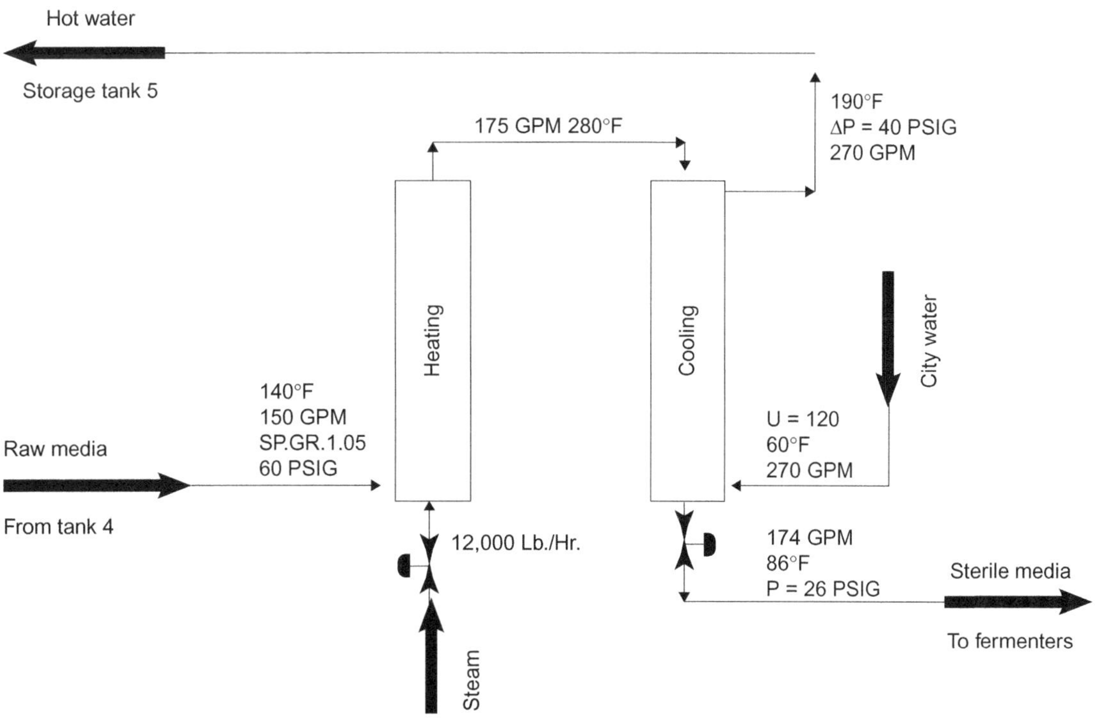

FIGURE 6.4 Material and energy balance of a sterilizer without an economizer.

Notice that the pressure in the sterilizer during operation is greater than the pressure of the cooling water. If any leak should occur in the inner pipe, media will pass into the non-sterile cooling water. In addition, the pressure maintained in the sterilizer is greater than the equilibrium boiling point in the heating section. This reduces the noise and hammering. Proper selection of the steam control valve will reduce noise also. However, there remains considerable noise at the steam injector, and it is good to locate it (and all the sterilizer) outdoors. The injector can be enclosed in an insulated "box" to reduce noise levels still further. One final remark: if the steam supply is directly from a boiler, non-volatile additives must be used. Biotech companies have chosen to use clean steam generators.

Figures 6.5 and 6.6 show more details of the piping and instrumentation of the sterilizer.

4.3 The Sterilizing Section

The hot section (Fig. 6.5) is controlled by a cascade loop which is based on a selected pumping rate (150 gpm) and sterilization temperature set in the TIC. Changes in the feed temperature are monitored at TT1 which will automatically override the steam supply to keep the temperature at setpoint. Steam flow rate is monitored (by FE) and flow is automatically compensated should a large draw down of steam occur elsewhere in the plant. Temperature is recorded at the beginning and end of the hot section. The hot section should be well insulated and special care should be given to the pipe supports for expansion. (Instrumentation symbols used here and in Figs. 6.3, 6.5, 6.6 and 6.7 conform to the standard symbols of the Instrument Society of America.)

The pumping rate, the pipe diameter and the length of the hot section of the sterilizer, fix an average retention time. The design basis of the retention time depends upon the bacterial spore count, the maximum particle size of the suspended solids, and the fluid velocity. For economy, the minimum velocity which gives turbulent flow should be used, i.e., a Reynold's number of about 3000 to keep the pipe short and the pressure drop low. The installation of (carefully selected) short static mixers can help in some cases to increase turbulence, reduce the velocity and the length of the hot section. Due to the source of raw materials normally used in fermentation media, bacterial counts can run very high, and some suspended solids can be almost hydrophobic. Based on the particle size which will pass through the screen stated above, three minutes retention time is borderline for sterilization. Five to six minutes retention time is often designed because, in time, inorganic scale will deposit on the wall of the hot section resulting in a smaller diameter and a higher fluid velocity or a shorter retention time. The hot section is easily cleaned once a year to remove the scale.

4.4 The Cooling Section

Most commercial fermentation processes use media with a high concentration of dissolved and suspended solids. Unless a uniform flow profile is maintained, solids may build up in the cooling section. The following are examples of types of heat exchangers to be considered for continuous sterilizers of fermentation media.

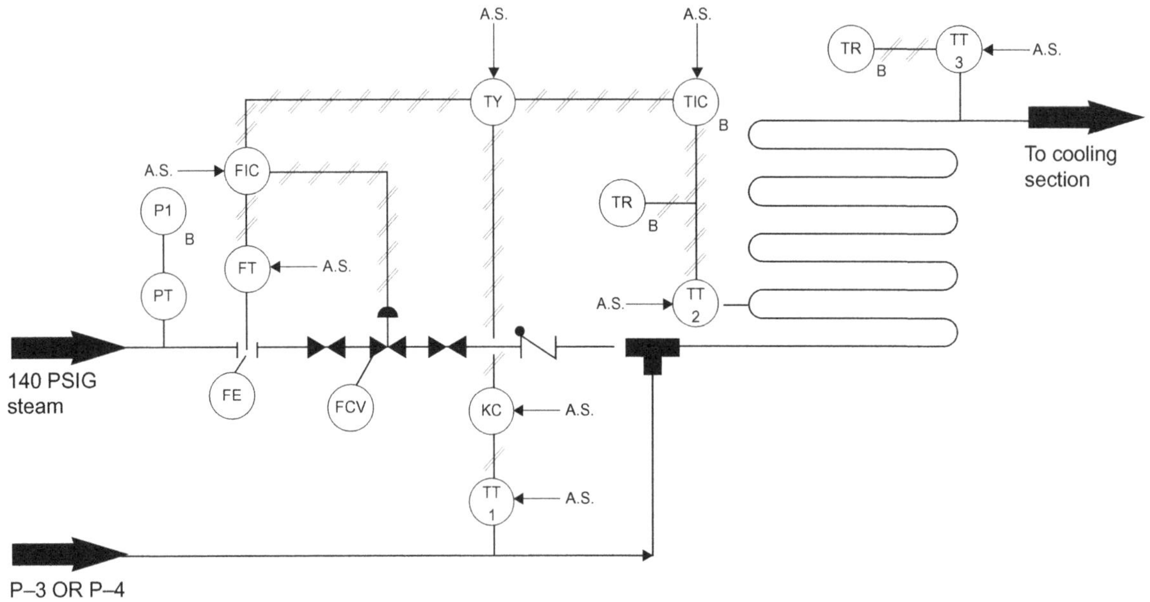

FIGURE 6.5 Piping and Insrumentation Drawing of the sterilizing section of a continuous sterilizer.

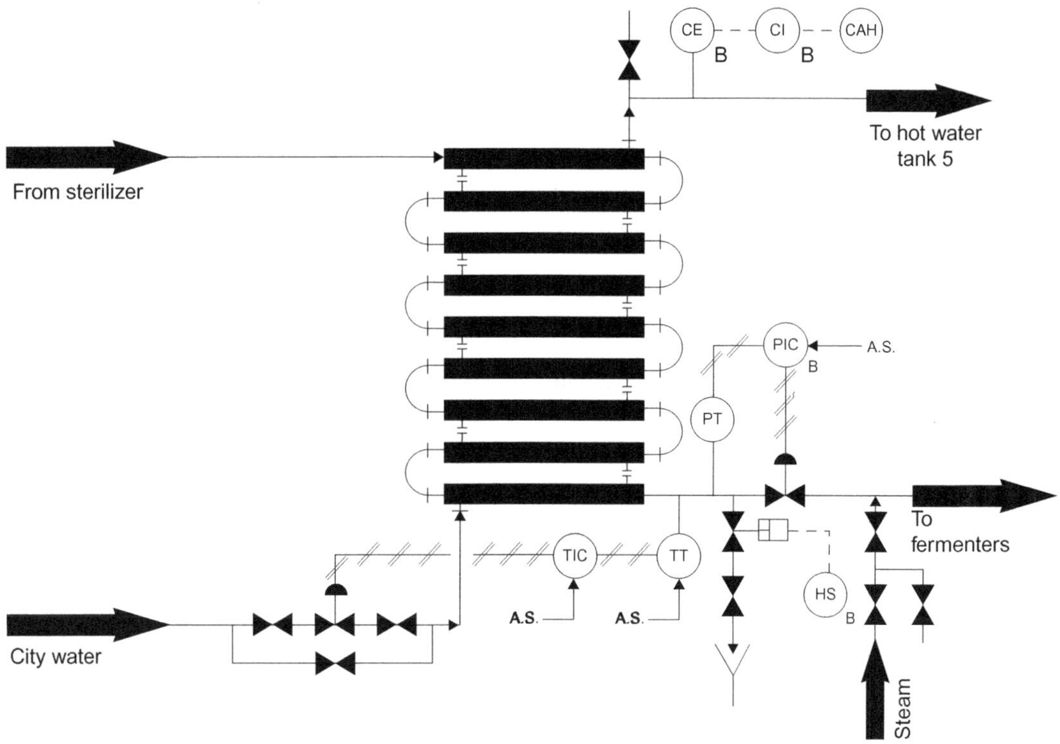

FIGURE 6.6 Piping and Instrumentation Drawing of the cooling section of a continuous sterilizer.

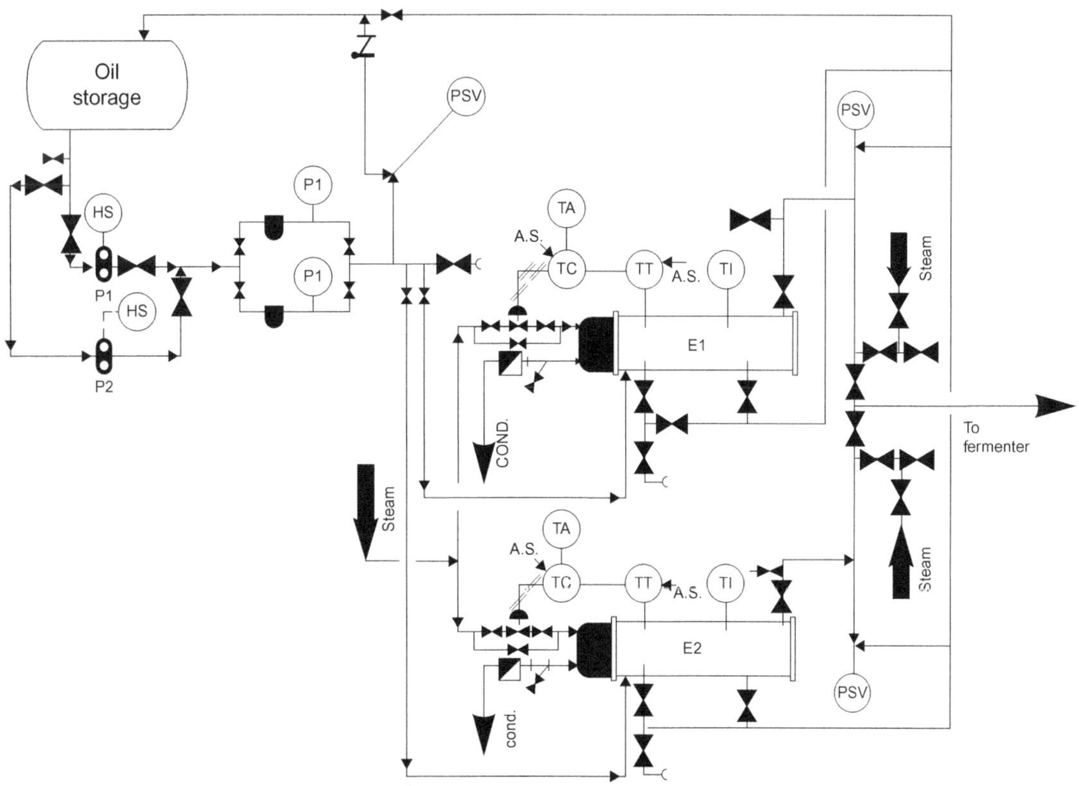

FIGURE 6.7 Piping and Instrumentation Drawing of a continuous oil (antifoam) sterilizer.

Concentric Double Pipe Heat Exchangers. This type of heat exchanger offers the most advantages for a continuous sterilizer with a range of flow rates suitable to the vast majority of commercial fermenters. (Wiseman states production fermenters are 25–1000 m^3 [5].)

- It is not limited by the flow ratio of the media and the cooling water
- It has the least crevices for corrosion.
- It requires the least cleaning and is cleaned relatively easily
- Scale in the cooling section is relatively minor.
- The velocity profile and pressure drop do not result in heat transfer difficulties
- It is easy to operate and instrument

The cooling section, Fig. 6.6, is of double pipe construction. Cooling water and sterile media pass countercurrently. The back pressure control valve (for sterilization) is located at the low point of the piping. A Masoneilian Camflex™ valve is a suitable design for this service. A steam bleed should be located on each side of this valve in order to sterilize the sterilizer forward from the steam injector and backward from the fermenter.

Notice also, there is no liquid metering device on the sterilizer. From a maintenance standpoint, it is much preferred to have dP cells on the fermenters for filling and controlling the volume than to measure the volume pumped through the sterilizer. The piping arrangement from the continuous sterilizer to the fermenters will depend somewhat upon the experience of the company as to the number, types, and locations of valves and steam bleeds. However, in general, the piping arrangements of fermenters filled by means of continuous sterilizers are more simplified than batch sterilized systems because all steam bleeding through valves is done in an outward direction. Other types of heat exchangers include those listed below.

Plate Heat Exchangers. The advantages are:

- Plate heat exchangers have a high film coefficient for heat transfer of certain classes of fluids
- The pressure drop across a unit for clear solutions is moderate

The disadvantages are:

- The velocity profile across each plate is not uniform by a factor of five due to the plate corrugations. The friction factors range from 10 to 400 times those in a single pipe with the same port flow rate and with the same surface area. The non-uniformity of flow rates causes suspended solids to accumulate between the plates creating problems of cleaning and sterilizing
- There is a pressure drop through the pressure ports causing an unequal distribution of flow through the plate stack. Solids then begin to accumulate in the plates with the lowest pressure drop until plugging results. Gaskets often leak or rupture
- Plate heat exchangers have the most feet of gasket material for any commercial heat exchanger. The crevices at the gasket have a high incidence of chloride corrosion. Although cooling water may have less than 50 ppm chloride, scale buildup in the gasket crevice usually is several times the concentration in the cooling water. Should the fermentation media contain chlorides as well, stress corrosion will occur from both sides simultaneously. Corrosion due to chlorides is serious when the concentration is above 150 ppm and 80°C. The first evidence of stress corrosion results in non-sterile media, rather than a visible leak or a major leak of water between the two fluids
- Operationally, the plate heat exchanger is more difficult to sterilize and put into operation without losing the back pressure and temperature in the heating section than the concentric pipe exchanger
- The optimum ratio of flow rates for the two fluids is 0.7 to 1.3. This constraint limits the range of media pumping rate

Spiral Heat Exchangers. Spiral heat exchangers have similar problems to the plate type when the gap is small. The velocity profile is better than the plate type. These types of exchangers can be used for media with low suspended solid concentrations and become more the exchanger of choice for continuous sterilizers with high volumetric throughput because the gap becomes larger.

The amount of gasketing material is less than for the plate type resulting in fewer problems.

Shell and Tube Heat Exchangers. The shell and tube exchanger is the least practical choice for cooling fermentation media with high suspended solids. It is very difficult to maintain sterility and cleanliness. It is the easiest to plug and foul.

There is an excellent application for a shell and tube heat exchanger, the continuous sterilization of anti-foam. In this case, the exchanger is not the cooler, but the heater. If the anti-foam liquid has no suspended solids or material which will foul the heating surface, only one exchanger is needed per fermentation building or plant. However, if a crude vegetable oil containing non-triglycerides is the anti-foam agent, then fouling will occur. Figure 6.7 shows one of the several possible systems for the continuous sterilization of crude vegetable oil. In this case, steam is supplied to the tubes. The main features of the system are two heat exchangers, each having the capacity in their shells to hold oil long enough to sterilize even though the supply pump should run continuously. One heat exchanger is in service while the spare, after being cleaned, is waiting to be put to service when the first can no longer maintain setpoint temperature.

With such an anti-foam sterilizer as Fig. 6.7, a fermentation facility can install a sterile, recirculating, anti-foam system. Commercial anti-foam probes are available and reliable. Frequently, a variable timer is placed in the circuit between the probe and a solenoid valve which permits anti-foam additions to the fermenter. In this manner, anti-foam can be programmed or fed by demand with the ability to change the volume of the addition. It is also possible to place a meter in the sterile anti-foam line of each fermenter in order to control and/or measure the volume added per run.

Small continuous sterilizers are used in fermentation pilot plants as well as for nutrient feeds to a single vessel or group of fermenters.

There are many references in the literature about the theory, design and application of continuous sterilization. For reference, see the following sources and their bibliographies: Peppler, H. J. [6]; Aiba, Humphrey, and Millis [2]; Lin, S. H. [7,8]. Ashley, M. H. J., and Mooyman, J. [9]; Wang, D. I. C., et al. [10].

5.0 FERMENTER COOLING

When designing a fermenter, one primary consideration is the removal of heat. There is a practical limit to the square feet of cooling surface that can be achieved from a tank jacket and the amount of coils that can be placed inside the tank. The three sources of heat to be removed are from the cooling of media after batch sterilization, from the exothermic fermentation process, and the mechanical agitation.

The preceding topic about the design of a continuous sterilizer emphasized reduced turnaround time, easier media sterilization, higher yields and one speed agitator motors. The reduced turnaround time is realized because the heat removal after broth sterilization is two to four times faster in a continuous sterilizer than from a fermenter after batch sterilization. The cooling section of a continuous sterilizer is a true countercurrent design. Cooling a fermenter after batch sterilization is more similar to a cocurrent heat exchanger.

Assuming that all modern large scale industrial fermentation plants sterilize media through a continuous sterilizer, the heat transfer design of the fermenter is only concerned with the removal of heat caused by the mechanical agitator (if there is one) and the heat of fermentation. These data can be obtained while running a full scale fermenter. The steps are as follows:

1. Heat Loss by Convection and Radiation
   - **a.** *Perry's Handbook* [14]:

$$\cup = 1.8\ \text{Btu/hr/}^\circ\text{F/ft}^2$$

   (No insulation; if tank is insulated determine proper constant.)

   - **b.** Calculate tank surface area = A
   - **c.** Temp, of Broth = T_1
   - **d.** Ambient Air Temp. = T_2

$$Q_1 = \cup A(T_1 - T_2) = \text{Btu/hr}$$

   Convection and radiation depend upon whether the tanks are insulated or not, and the ambient air temperature, especially during the winter. Measurements of convection and radiation heat losses are, on average, 5% or less of total heat of fermentation (winter and uninsulated tanks).

2. Heat Loss by Evaporation
   - **a.** If fermenters have level indicators, the average evaporation per hour is easily determined.
   - **b.** Calculate pounds of water/hour evaporated from psychometric charts based on the inlet volume and humidity of air used, and at the broth temperature. The exhaust air will be saturated. Determine heat of vaporization from steam tables at the temperature of the broth = H_{EV} = Btu/lb.

$$Q_2 = H_{EV} \times (\text{1b water evap/hr}) = \text{Btu/hr}$$

   Evaporation depends upon the relative humidity of the compressed air, temperature of the fermentation broth and the aeration rate. It is not uncommon that the loss of heat by evaporation is 15 to 25% of the heat of fermentation. Modern plants first cool the compressed air then reheat it to 70–80% relative humidity based on summertime air intake conditions. Consequently, in winter the air temperature and absolute humidity of raw air are very low and the sterile air supply will be much lower in relative humidity than summer conditions. Therefore, in the winter more water is evaporated from the fermenters than in the summer. (Water can be added to the fermenter or feeds can be made more dilute to keep the running volume equal to summer conditions and productivity in summer and winter equal.)

3. Heat Removed by Refrigerant
   - **a.** This is determined by cooling the broth as rapidly as possible 5°F below the normal running temperature and then shutting off all cooling. The time interval is then very carefully measured for the broth to heat up to running temperature (ΔT and time).
   - **b.** Assume specific heat of broth = 1.0 Btu/lb-°F
   - **c.** Volume of broth by level indicator (or best estimate) = gal

$$Q_3 = \text{Sp.Ht.} \times \text{broth vol.} \times 8.345 \times \Delta T \div \text{time(hr)}$$

$$Q_3 = \text{Btu/hr}$$

4. Heat Added by Mechanical Agitation
   a. Determine or assume motor and gear box efficiency (about 0.92)
   b. Measure kW of motor

$$Q_4 = \text{kW} \times 3415 \times \text{efficiency} = \text{Btu/hr}$$

5. Heat of Fermentation = ΔH_f

$$Q_1 + Q_2 + Q_3 - Q_4 = \Delta H_f$$

The heat of fermentation is not constant during the course of the fermentation. Peaks occur simultaneously with high metabolic activity. Commercial fermentation is not constant during the course of the fermentation. Commercial fermentations with a carbohydrate substrate may have peak loads of 120 Btu/hr/gal. The average ΔH_f for typical commercial fermentations is about 60 Btu/hr/gal. The average loss of heat due to evaporation from aeration is in the range of 10 to 25 Btu/hr/gal. Fermentations with a hydrocarbon substrate usually have a much higher ΔH_f than carbohydrate fermentations. Naturally, most companies determine the ΔH_f for each product, especially after each major medium revision. (Typically, data are collected every eight hours throughout a run to observe the growth phase and production phase. Three batches can be averaged for a reliable ΔH_f.) In this manner, the production department can give reliable data to the engineering department for plant expansions.

The following is how the heat transfer surface area could be designed for a small fermenter. The minimum heat transfer surface area has been calculated (based on the data below) and presented in Table 6.2. Assume:

S/S fermenter capacity	30,000 gal
Agitator	15 hp/1,000 gal
Heat of fermentation (peak)	100 Btu/hr/gal
Heat of agitation	38 Btu/hr/gal
Heat transfer, U coils	120 Btu/hr/sq ft
Heat transfer, U jacket	80 Btu/hr sq ft
Safety factor	No Btu lost in evaporation
Chilled water supply	50°F
Chilled water return	60°F
Broth temperature (28°C)	82°F

TABLE 6.2 The Heat Transfer

	Surface Area (ft^2) Required for Tank with:	
	Coils Only	Jacket Only
Mechanical agitation	200	5
Air agitation only	150	6

After the heat transfer surface area requirements are known, various shaped (height to diameter) tanks should be considered. Table 6.3 illustrates parameters of 30,000 gallon vessels of various H/D ratios.

Comparing Tables 6.1 and 6.2 shows that if mechanical agitation is used and a jacket is desired, then additional internal coils are required. The internal coils can be vertical, like baffles, or helical. Agitation experts state that helical coils can be used with radial turbines if the spaces between the coil loops are 1 to 1.5 pipe diameters. Once helical coils are accepted, why use a jacket at all? Reasons in favor of coils (in addition to the better heat transfer coefficient) are:

1. Should stress corrosion cracking occur (due to chlorides in the cooling water), the replacement of coils is cheaper than the tank wall and jacket.
2. The cost of a fermenter with helical coils is cheaper than a jacketed tank with internal coils.
3. Structurally, internal coils present no problems with continuous sterilization. However, if batch sterilization is insisted upon, vertical coils are one solution to avoiding the stress between the coil supports and tank wall created when cooling water enters the coils while the broth and tank wall are at 120°C. Notice that the method of media sterilization, batch or continuous, is related to the fermenter design and the capital cost.

6.0 THE DESIGN OF LARGE FERMENTERS (BASED ON AERATION)

6.1 Agitator Effectiveness

Laboratory scale work frequently reports aeration rates as the volume of air at standard conditions per volume of liquid per minute, or standard cubic feet of air per hour per gallon. Production engineers realized that the scale-up of aeration for a large range of vessel sizes was by

TABLE 6.3 Maximum Heat Transfer Surface Area Available (ft^2) on 80% of the Straight Side

H/D	F (ft)	D (ft)	Jacket	Coils*
2	27.3	13.7	938	1,245
2.5	31.7	12.7	1,010	1,340
3	35.8	11.9	1,070	1,400
3.5	39.7	11.3	1,130	1,455
4	43.4	10.8	1,180	1,150

Coil area is based on 3.5 inch o.d., 3.5 inch spacing between helical coils and 12 inches between the tank wall and the center line of the coil.

superficial linear velocity (SLV), or feet per second. Large scale fermenters, for energy savings in production equipment, use air-agitated fermenters. The cost savings are not apparent when comparing the cost of operating a fermenter agitator to the cost of the increased air pressure required. However, when the total capital and operating costs of fermentation plants (utilities included) for the two methods of fermentation are compared, the non-mechanically agitated fermenter design is cheaper. The questions are, how much mixing horsepower is available from aeration, versus how much turbine horsepower is effective for aeration and mixing? D. N. Miller [11] of DuPont, describing his results of scale-up of an agitated fermenter states, "both K_La and gas hold-up increase with an increasing gas rate and agitator speed. Gas sparging is the 'stronger' effect and tends to be increasingly dominant as gas rate increases. At superficial gas velocities, 0.49 ft/sec and higher, very little additional mass transfer improvement can be gained with increased mechanical energy input." Otto Nagel and associates [12] found in gas-liquid reactors that the mass transfer area of the gas in the liquid is proportional to the 0.4 power in the energy dissipation. Thus for a 50 hp agitator, 12 hp directly affects the mass transfer area of oxygen. The upper impellers mainly circulate the fluid and contribute very little to bubble dispersion and oxygen transfer. Most of the agitator's power is spent mixing the fluid. (To understand mixing theories see Brodkey, Danckwerts, Oldshue or other texts [13].) The primary function of mixing for aerobic fermentations is to increase the surface area of air bubbles (the interfacial surface area) to minimize the bubble diameter. The fermenter is not the same as a chemical reactor where first and second order reactions occur between soluble reactants. The dissolution rate of oxygen into fermentation broth is controlled by diffusion. The consumption of soluble oxygen by the organism is an irreversible reaction and unless sufficient oxygen diffuses across the air-liquid surface area, the fermentation will cease aerobic metabolism. Methods of forcing more air into solution are: more interfacial surface area, more air/oxygen, higher air pressure, reduced cell volume, or controlling metabolism by reduced carbohydrate feed rates. Not all of these options are practical because of shear, foaming and control devices.

6.2 Fermenter Height

The height-to-diameter (H/D) ratio of a fermenter is very important for oxygen transfer efficiency. Tall, narrow tanks have three major advantages compared to short, squat fermenters. Bubble residence time is longer in taller vessels than shorter ones. The air pressure is greater at the sparger resulting in higher dissolved oxygen in taller vessels. The third advantage is shown in Table 6.4, namely that for a vessel of constant volume, as the H/D ratio increases, the volume of air required is reduced even though the superficial linear velocity remains constant. At the same time, bubble residence time and sparger air pressure increase. For larger volume fermenters, even greater vertical heights are used. The conclusion is that fermenter height is the most important geometrical factor in fermenter design. Conversely, shorter vessels need more air and/or more mechanical agitation to effect the same mass transfer rate of oxygen. The majority of industrial fermenters are in the H/D range of 2–3. The largest sizes are about 10^6 liters.

TABLE 6.4 Effect of Air Requirements on Geometric Fermenter Design

H/D	F	D	scfm	Bubble Residence Time	Sparger Pressure
2	27.3	13.7	3,522	1	12.3
3	35.8	11.9	2,683	1.3	16.0
4	43.4	10.8	2,219	1.6	19.4

Constant: 30,000 gal tank; 24,000 gal run vol; 0.4 ft/sec SLV.

TABLE 6.5 Air Compressor Horsepower per Fermenter

H/D	scfin	30 psig Compressor (hp)	50 psig Compressor (hp)
2	3,522	429	609
3	2,693	327	464
4	2,219	270	384

Constant: 30,000 gal tank; 24,000 gal run vol; 0.4 ft/sec SLV.
Note: Basis of hp is (8 hp)/(0.7)

It is thought that the cost of compressing air sufficient for air agitation alone is prohibitive. However, as seen in Table 6.5, if the fermenters are tall, the power consumption is less than for short squat tanks. Careful selection of compressors with high efficiencies will keep power costs at a minimum.

6.3 Mixing Horsepower by Aeration

The theoretical agitation effect of aeration alone can be easily calculated. There are two separate forces, the first caused by the free rise of bubbles. The bubbles rise from the sparger at a pressure equal to the hydrostatic pressure of the liquid and as they rise to the surface, the gas bubble pressure remains in constant equilibrium with the hydrostatic pressure above it until it escapes from the liquid surface. The temperature of the air in the bubble is equal to the fermentation temperature and remains constant due to heat transfer from the fermentation broth. These conditions describe an isothermal expansion of gas; gas pressure and gas volume change at constant temperature. Using the formula from Perry and Chilton, [14] the theoretical horsepower for the isothermal expansion of air can be calculated.

$$\frac{\text{hp}}{1,000\ \text{scfm}} = 4.36 P_2 \ln\frac{P_2}{P_1}$$

where:

P_1 is the hydrostatic pressure (absolute)
P_2 is the (absolute) pressure above the liquid

Figure 6.8 shows the curves at different superficial linear velocities and the relationship of horsepower to height of liquid in a fermenter. These curves are the mixing energy (power per unit volume) released by rising bubbles to the liquid.

Thus in a fermenter:

1. The horsepower per 1000 gallons (*P/V*) can be increased by adding more air.
2. The effect of aeration scale-up by superficial linear velocity (SLV) is not proportional to (*P/V*). However, by using these curves scale-up at constant (*P/V*) can be used to determine the required SLV.
3. Experience indicates that the *P/V* relationship is not affected by non-Newtonian fluids below 6000 cps apparent viscosity.
4. If the air temperature at the bottom of the fermenter is less than the liquid temperature, there is a gain in *P/V*. This is due to the fact that at a lower temperature, the air density is greater, and heat is transferred from the broth to the bubbles (isothermal expansion) resulting in more work (*P/V*) or kinetic energy imparted to the broth by turbulence.
5. If the fermenter vent valve is restricted to increase the pressure above the broth, it has the effect of reducing *P/V*, but oxygen transfer increases due to the greater partial pressure of oxygen.

There have been reports of air dispersion with improved oxygen transfer using static mixers attached to the air ring. Two papers on static mixers were given by Smith and Koch at the Mixing (Engineering Foundation) Conference in Rindge, NH (1977). Additional papers can be found in the waste treatment field.

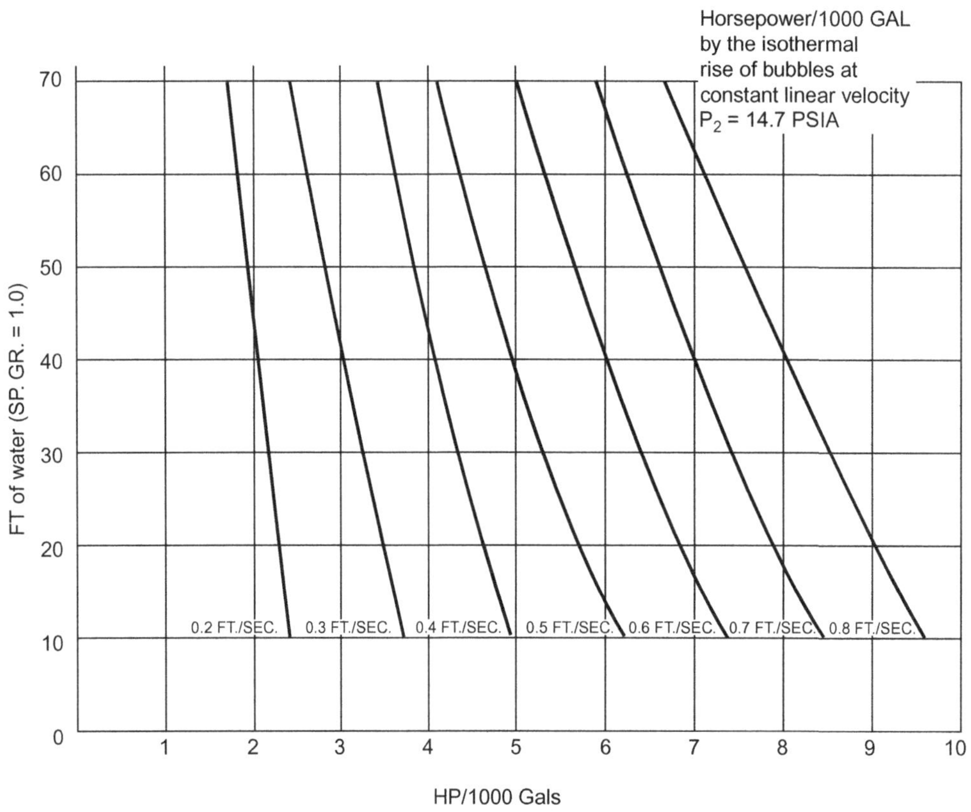

FIGURE 6.8 Isothermal bubble rise curve horsepower/1000 gal.

There is additional energy to be gained from aeration. In order for the air to enter a tank below the liquid surface, the pressure in the sparging device must exceed the static head pressure. Thus the mass of air has a determinable velocity through the orifices of the sparger. The force exerted against the liquid is $F = MV^2/2g$. That is, for a fixed mass flow rate of air, the force varies as the velocity squared. The velocity of air through a nozzle is a function of the (absolute) pressure ratio on each side of the orifice, and it can be increased to sonic velocity. The time of flow through an orifice is so short there is no heat transferred from the broth to the air and the air temperature drops. The expansion of air at sonic velocity is isentropic (adiabatic). The horsepower obtained by the isentropic expansion of air (at any pressure ratio) is (see Perry and Chilton.) [14]:

$$\text{hp} = \frac{144}{33,000}\left(\frac{k}{k-1}\right)P_1V_1\left[\left(\frac{P_2}{P_1}\right)^{\frac{k-1}{k}} - 1\right]$$

where:

V_1 = initial volume before orifice (ft^3)
P_1 = initial pressure (absolute)
P_2 = (absolute) pressure after expansion
$k = C_p/C_v$

Assuming that $k = 1.50$, then:

$$\text{hp} = 0.0131P_1V_1\left[\left(\frac{P_2}{P_1}\right)^{0.333} - 1\right]$$

Figure 6.9 illustrates the adiabatic hp/1000 scfm in fermenters. Important features of using high velocity aeration are the following:

1. Increasing the liquid volume in the fermenter, such as feeding, reduces the horsepower. Conversely, removal of portions of broth will increase the horsepower.
2. The curves show the horsepower range at the air orifice from zero to sonic velocity which can be obtained by knowing the ungassed liquid height (differential pressure cell), the air pressure upstream of the orifice, and the scfm of air used.
3. Increasing the air pressure above the liquid reduces the horsepower (see Fig. 6.10).

The total theoretical horsepower of mixing by aeration alone is the sum of the isothermal and isentropic horsepower. At normal operating conditions, it is possible to double the agitation (P/V) by increasing the velocity of the air without increasing the scfm. It is easy to calculate the size of the orifices to give any desired velocity (up to sonic velocity), and the mixing horsepower. Conversely, one can scale up aeration by horsepower per unit volume and determine the air required, i.e., it is possible to scale

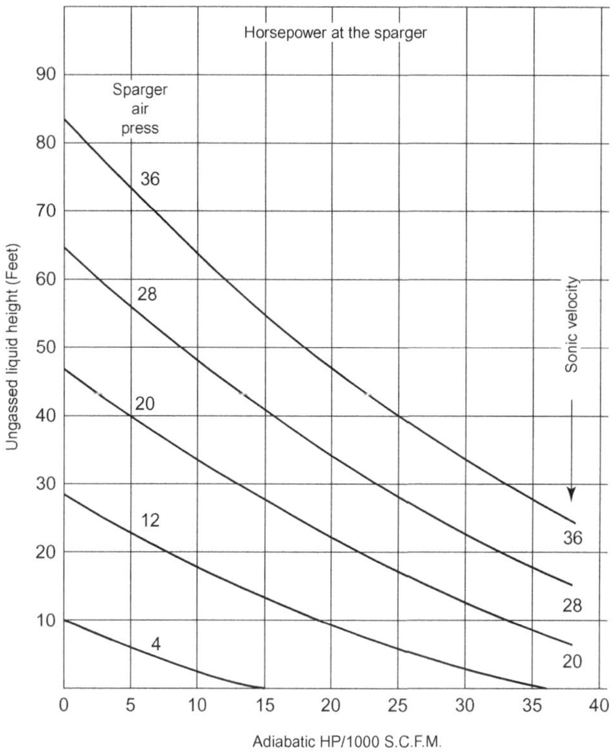

FIGURE 6.9 Isentropic horsepower/1000; scfm.

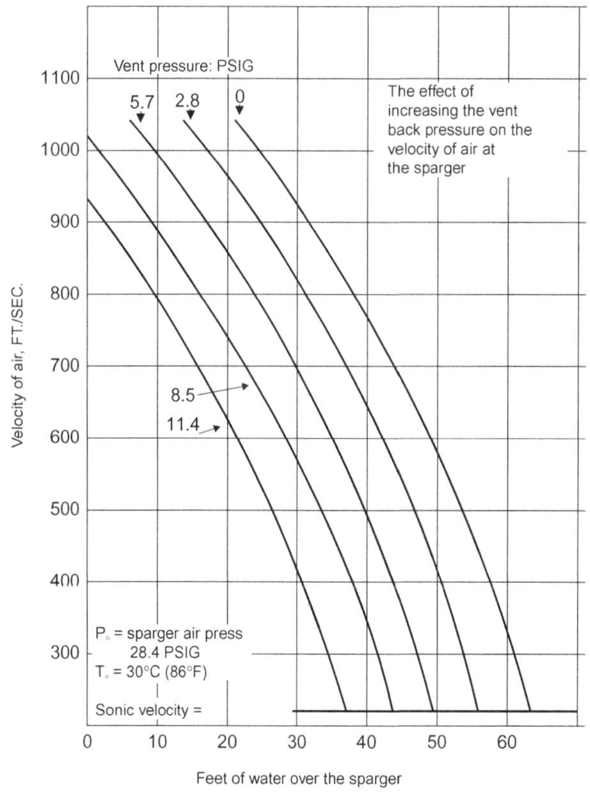

FIGURE 6.10 Isentropic horsepower reduction by back pressure.

up mechanical horsepower used in a pilot scale fermenter to a production vessel which is not mechanically agitated.

6.4 Air Sparger Design

Air sparger design of large fermenters is one of the least discussed topics of the fermentation industry. Most companies design their own air spargers. Some companies have designed and tried a wide variety of ideas. Agitator manufacturers insist that the air ring emit the air bubbles at the optimum radius of the first turbine. However, in air-agitated fermenters, the engineer must be creative to consider both the best mixing and oxygen transfer effects to be obtained from an air sparger. Not much is gained in mixing or mass transfer by having more than one orifice per 10 ft^2 of cross-sectional area.

6.5 Comparison of Shear of Air Bubbles by Agitators and Jets

Visualize a filled fermenter with a variable speed agitator at rest with a Rushton turbine. When aeration is started, large bubbles will rise and impinge on the underside of the turbine disc and escape around the perimeter. When the agitator is rotated at very slow speed, the large bubbles will accumulate behind the turbine blades and small bubbles will trail off into the liquid. If aeration and the agitator speed are increased, the volume of air behind each blade becomes larger, and smaller bubbles trail off into the liquid from the mass of air behind each blade. Due to the design of a Rushton turbine, the liquid and the small air bubbles move horizontally (or radially) into the liquid. As the speed of the turbine increases, the fluid velocity caused by the pumping action of the turbine produces a profile of shear stresses. Theoretically, the turbine speed could be increased to obtain any fluid velocity up to sonic velocity, but the power cost would be high since the energy must move the mass of liquid. Similarly, a limit is reached by increasing the aeration under a rotating turbine. The limit results when the air rotating behind each turbine blade fills all the space in the arc back to the front face of the next turbine blade. All the blades then are spinning in an envelope of air, or the impeller is flooded. For details see Klaas van't Riet [15].

Now visualize the action of a submerged jet of air in liquid. At very low air flow velocities, the bubbles are large. They rise as independent bubbles at the orifice. When the velocity of air through the orifice increases, the air projects as a cone into the liquid and small bubbles shear off. The maximum velocity is reached when the ratio of absolute hydrostatic pressure outside the orifice divided by the absolute air pressure in the orifice is 0.528. This determines sonic velocity. Four regions of the air cone or jet are conceptually drawn in Fig. 6.11. Region I is called the potential core of air with a uniform velocity. The outermost annular cone, Region II, is an *intermittency zone* in which flow is both turbulent and non-turbulent. Region III lies between the potential core and the liquid and is characterized by a

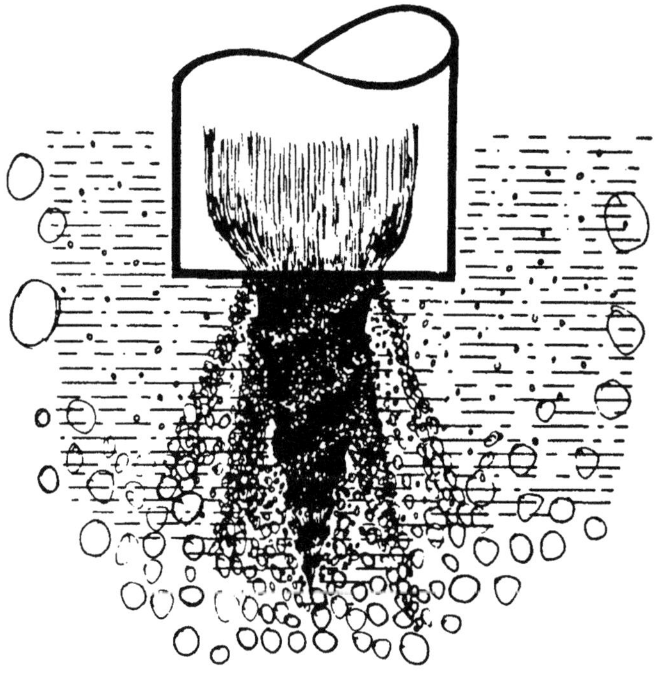

FIGURE 6.11 Region of mixing and shear for a submerged jet.

high velocity (shear) gradient and high intensity of turbulence. Region IV is the mixing zone where the fluid and air merge beyond the potential core. It is a totally turbulent pattern (see Brodkey) [16]. A cloud of very fine bubbles is produced. Oxygen mass transfer is enhanced because of the increased surface area of the very small bubbles. In low viscosity fluids, much more coalescence of bubbles will occur than in high viscosity fluids.

Comparing now the bubble dispersion produced by agitators and jets, it should be clear that agitators shear large air bubbles by moving fluid, and jets produce fine bubbles by forcing high velocity air past relatively stationary fluid. This explains the high energy cost and low efficiency of mechanical agitation. Another benefit of air agitation is that it is a variable horsepower device depending upon the quantity of air and the velocity of the jet.

6.6 The Effect of Shear on Microorganisms

The effect of shear on microorganisms must be determined experimentally. Most bacteria and yeast can withstand very high shear rates. On the other hand, filamentous organisms are less predictable; some are stable and others are ruptured very easily. In the latter case, the air velocity must be reduced below the "critical" shear rate the organism can tolerate. Lowering the speed and/or moving the vertical blades nearer to the shaft can reduce the shear rate of Rushton turbines. The average bubble diameter will become larger and more air (SCFM) may have to be added for adequate dissolved oxygen.

Unfortunately, laboratory and/or pilot plant experiments cannot easily test air agitation because it requires long bubble residence time (tall vessels) with minimal wall drag. The short vessels of the laboratory and pilot plant are precisely where mechanical agitation is absolutely required to achieve good fermentation performance compared to tall air-agitated fermenters. If anyone with large (tall) mechanically agitated fermenters wants to experiment with air agitation, it will be necessary to remove the turbines from the agitator shaft. If the vessel has cooling coils, the mixing section is somewhat like a "draft tube" giving good top to bottom mixing. The air sparger will have to be replaced. Normally, air velocities of 0.75 Mach provide sufficient levels of increased agitation, mass transfer and shear. There is a reasonable body of literature, although scattered, about air agitation, frequently under gas-liquid reactor design and occasionally in bioengineering journals. D. G. Mercer read a paper at the Second World Congress of Chemical Engineering. Dr. M. Charles (Lehigh University) [17] published his experimental work on this subject. Also, see Shapiro, A. H. [18]; Townsend, A. A. [19]; Forstall and Shapiro [20].

6.7 Other Examples of Jet Air/Liquid Mixing

The Buss loop reactor is a system to increase the dissolving of gas into a liquid which contains a dissolved chemical and a catalyst. Normally the reaction is first order and the reaction rate is dependent upon gas diffusion rate. An example is the hydrogenation of glucose to sorbitol. The rate of reaction and yield is increased as follows. The liquid is pumped from the bottom of the reactor, externally up to and through an eductor and discharged subsurface into the agitated vessel contents. The reason for its success is that a high velocity eductor mixes and shears a gas into very fine bubbles of very large surface area better than sparging beneath an agitator. The Buss company, of Bern, Switzerland, designs these systems.

A variation of this design is a production fermenter which has several eductors welded to discharge horizontally and tangentially just above the bottom dish. Broth is continuously pumped from the bottom of the fermenter to all eductors simultaneously. Sterile air is provided to the suction side of the eductor (exterior to the tank). The fermenter has an agitator of very minimal horsepower to keep the solids suspended during sterilization. The pressure of the air supply is low (about 8 psig). A low horsepower blower is used to overcome the pressure drop over the air filter and line losses. Not too surprising, however, the total horsepower of the blower, the circulation pump, and tank agitator have the same power consumption per unit liquid volume as fermenters with conventional agitators, turbines, i.e., about 15 hp/1000 gallons of liquid is minimal for the average commercial aerobic fermentation. Therefore, pumping non-compressible water through eductors at high velocity to shear air into small bubbles is no more efficient than agitator mixing. The reverse case of compressing a gas to expand isentropically at high velocities into a liquid does have significant advantages for large volume fermenters. (See Bailey and Ollis [21].)

6.8 Mechanical Versus Non-mechanical Agitation

In summary, the need for mechanical agitation beyond oxygen mass transfer is not clearly understood in fermentation broths which have a viscosity before inoculation of a Newtonian fluid, but change to pseudoplastic (non-Newtonian) after growth starts. Air-agitated fermenters exist in industry today for a wide range of products. It is a viable alternative to mechanically agitated systems. The advantages are the following:

1. Improved sterility because of no top- or bottom-entering agitator shaft.

2. Construction of very large fermenters is possible because the design is not limited by motor size, shaft length and its weight.
3. Refrigeration requirements are reduced 20 to 35% because of no mechanical agitation—see Table 6.2.
4. Since no agitator, gear box or crane rail is needed, less structural steel is used and cheaper fermenter design results.
5. No maintenance of motors, gear boxes, bearings or seals.
6. The air-agitated fermenter is a variable mixing power unit, like a variable speed drive with no motor and drive noise.
7. Air compressors can be steam driven to reduce power cost and continue to operate during power outages in large plants that have minimal power generation for controls.

7.0 TROUBLESHOOTING IN A FERMENTATION PLANT

The art of troubleshooting is acquired by an alert individual after much experience actually operating a fermentation department. There are electrical, instrumental, mechanical and other physical problems arising from time to time which must be solved. However, the usual source of trouble in a fermentation department is a foreign, or contaminating, microorganism or bacteriophage in the fermentation media. This results in one or more of the following problems:

1. Inoculum cannot be used for the fermenters.
2. Complete or varying degrees of inhibition of product formation in the fermenters.
3. Possibly the product is produced, but contaminating products cannot be separated from the final product.
4. The fermentation broth cannot be filtered or otherwise processed due to physical broth characteristics.

A fermentation department attempts to operate its culture lab, seed and main fermenters at 100% aseptic conditions. Good performance is when contamination occurs in 1% or less of the batches. Naturally, any occurrence of such problems raises the cost of production, and the worst situation is when production is brought to a stop due to contamination. First, it is necessary to have standard operating procedures which are rigidly followed. Second, it is important to have well-trained operators who report back all problems, errors and irregularities, and third, the supervisors and engineer must listen to the operators, and react or explain the information, as the case may be, to close the communication loop and motivate the operators.

The basic sources of contamination in a fermentation department usually fall into the following categories:

1. Contamination of the plant stock culture due to poor techniques.
2. Contamination by the raw materials used in the fermentation.
3. Contamination attributed to inadequate sterilization of the equipment, air or media involved in a fermentation plant.
4. Contamination attributed to inadequate procedures, or procedures not being followed, or insufficient operator training.
5. Contamination caused by lack of a definite routine maintenance schedule on all the fermentation equipment.
6. Contamination by bacteriophage.

The following is a list of procedures, steps, or operations that need to be followed closely and checked routinely to prevent contamination in the six categories stated above.

1. *Contamination in the Culture Laboratory*

   a. Check the stock culture used for the presence of foreign microorganisms and bacteriophages.
   b. Check the flask, etc., used to inoculate the seed fermenter for contamination.
   c. Check the sterilization procedures used on flasks, media, etc.
   d. Check the techniques of the operator who inoculated the flask, etc.
   e. Check the sterilizer, including the temperature and pressure gauges, and operation. Be sure all air is removed from the sterilizing chamber. Use controls.
   f. Check sterility of the sterile area where the culture was transferred by using exposure plates.
   g. Carry out all practical tests for contamination by the best possible techniques using agar plates, several different media and microscopic examination. Use both static tubes and shake flasks. Incubate samples at different temperatures.
   h. Keep the laboratories clean, especially work benches or hoods where transfers are made. Check the germicidal solution used to wash the benches for growth.
   i. Presterilize the cotton used for plugs.
   j. When using grains or other materials which do not dissolve, use a finely ground type to be sure there is no lumping, and that all the material is wetted and suspended. Dry lumps can present sterilizing problems.
   k. If raw materials are heavily contaminated with spores, prepare the media and incubate at 30°C for several hours. When the spores have germinated, sterilize the media.

2. *Contamination in Raw Materials*

a. Purchase all dry materials in a finely ground form.
b. For those materials which do not dissolve, suspending first in cold water usually prevents lumping. Use adequate agitation in the mix tank.
c. Pump the blend over a vibrating or rotating screen to separate all lumps larger than the size that can be sterilized by the time-temperature limitation of the sterilization cycle. Discard the oversized lumps. (Calculate the maximum lump size by the Gurney, Laurie charts. See Perry and Chilton [14].)
d. Batch sterilizing in fermenters is frequently a problem because the slow agitation speed is not adequate to keep all the suspended material equally dispersed. Some may settle out in poor circulation areas of the vessel; some materials may enter the air ring while filling the vessels. Solutions to these problems can be: prehydrolyze the starch or protein with enzymes; use a longer sterilizing time or higher temperatures; use a continuous sterilizer (abandon batch sterilization).

3. *Contamination from Equipment*

Contaminated inoculum tanks:

a. Check inoculum used, and the procedures of inoculating the tank with the operator.
b. Thoroughly inspect the tank inside for cleanliness, including the head.
c. Check tank and accessories for leaks in gaskets, coils, jacket, hatches, air lines from the sterile side of the air filters, etc., by pressurizing the equipment at 20 psig air pressure and going over all joints with a soap solution on a brush. Leakage of air produces foam and large bubbles.
d. Calibrate the temperature and pressure gauges on the tank for accuracy.
e. Inspect the bottom valve and gasket, sample line valve, and the inoculation fitting.
f. Inspect the inside tank fittings such as instrument bulbs, gauges, spargers, brackets, ladders, nuts, bolts, hangers, etc., for dead spots, holes, pits, corrosion, or for sources of debris (old dried up mycelia, media, etc.) which has not been washed or boiled out of the tank.
g. Inspect the sterile air system, the packing and condition of the filters, gaskets, lines, etc. If when steaming out an air filter dark brown media runs out, there has been a media blowback. The filter should be completely cleaned out, restored and sterilized.
h. Talk to the operators about the actual process used for cleaning, sterilization, cooling, inoculation, tank operation and sampling procedures.
i. Check the operators' skills and techniques in taking aseptic samples from the tank for sterility tests. Unless the tank is grossly contaminated on a slide, you may have had a bad sample, due to poor technique on their part.
j. Check the anti-foam and the complete system for addition of the anti-foam, or other substances, such as for pH control, etc.

Contaminated fermenters:

a. Check thoroughly the inoculum tank used to inoculate the fermenter.
b. Inspect the inoculum line or hoses used, and verify that the operator has used the correct technique.
c. Check tank cleanliness and method of cleaning; personally inspect the inside of the tank for sources of debris in cracks, crevices, flanges, bolts, hangers, coils, ladders, top of the tank, etc. The tank should be spotless.
d. Check all the accessories for leaks, while the tank is under 20 lb. pressure, with a soap solution.
e. If coils or jackets are used, test hydrostatically for pressure drop.
f. Check all sensors for temperature, pH, DO, differential pressure cells, etc. Sometimes a crack in a thermometer bulb will continually contaminate a tank due to internal pressure changes.
g. Check out the air system, filters, lines, packing, packing density, and lower than normal pressure drop.
h. Check the anti-foam system and determine if anti-foam is sterile.
i. Check the operators' techniques for running tanks, additions, nutrient recharges, partial broth harvests, etc.
j. Inspect bottom valve, gaskets, sample line and valve, vent line valves, vacuum breakers, etc.
k. If the fermenter is fed continuously (or intermittently), with a sterile nutrient, check the nutrient feed tank as if it were a fermenter.
l. Inspect closely and critically the inoculum and sterile feed transfer lines and procedures. Make sure when they are sterilized that no steam condensate accumulates in the line, but is bled off and all parts of the lines are up to sterilizing temperature.
m. Be certain the relative humidity of the non-sterile fermenter air before the "sterile" filter is and has been less than 85% at all times.

n. Check that pH and DO probes have been removed and cleaned after every run, that the probe holder has been brushed and cleaned with a hypochlorite or a formaldehyde solution.

o. When repeated runs in a vessel become contaminated while others remain sterile, it is customary to remove valves for cleaning or replacement. Replace gaskets. Remove and clean all instrument sensors; "high boiling" of the vessel with Na_2CO_3 or Na_3PO_4 and possibly a germicide may be required.

4. *Operating Procedures*

Training of personnel

a. A training schedule and program should be standard for all new operators in fermentation departments. How and why every operation is important, and similarly, all procedures should be explained to each person.

b. All operators should perform each task exactly alike and in the same sequence. No variation in procedures by individuals or shifts can be permitted.

c. There should be a basic operating manual for each process. That is, if a plant makes antibiotics A, B, and C, there should be three manuals detailing all the steps necessary from seed fermenter media preparation to fermenter harvesting, with what to do if certain variables get out of control, and when a supervisor has to be notified for a decision. No deviations from standard procedures should be allowed except those in writing from the supervisor.

d. Like regularly scheduled safety meetings, regular technical meetings with operators are necessary to instruct and update technical and procedural changes.

e. Operators must be encouraged to feed back their observations, ideas and errors without fear (such as wrong material or quantity being used, a foam over, a blowback into the air filters occurring, pipe leaks, etc.).

5. *Lack of Maintenance as a Source of Contamination*

a. Braided packing on agitator shafts of sterile vessels is always a problem. Germicidal solutions are helpful.

b. Mechanical seals on agitator shafts of sterile vessels usually have a sterilizing liquid circulating which acts as a lubricant as well.

c. Calibration of temperature and pressure gauges for accurate sterilizing temperatures are critical.

d. Check all flanged sterile piping for leaks. Repeated sterilization cycles of lines will lengthen bolts and loosen flanges which can result in material getting between the gasket and flange. This might result in a leak and a source of contamination. The bolts on flanges should be tightened when the lines are hot, after an hour or so at 120°C.

e. Check list for packed-bed type air filters:

Filter should be properly packed to the correct density (12–16 lb/ft^3). Preweigh the exact amount of material for the volume of the filter.

The filter bed should be inspected periodically, on a routine basis and new fiber added to the top if the rest of the fiber is all right, and no channeling is noted.

Any time a filter is subjected to a blowback of media from the tank due to loss of pressure, or a vacuum, it should be cleaned out and repacked.

Occasionally, check from the sterile side of the filter for leaks at the gaskets, valves, piping, threaded connection, etc. Also, check the spargers in the tank for debris or pockets. Make sure filters are dried out before being placed in actual service.

f. Repairs on the internals of the sterile vessels, such as bottom bearings, the air ring, coils, baffles, etc., must be done competently so they do not come apart during the vibration of sterilizing or operating. Work tools, gloves, etc., must not be left inside the vessel.

g. Pressure relief valves and vacuum breakers must be tested for accuracy and routinely cleaned every run.

h. If hoses are used to make sterile transfers, frequent inspection and replacements must be made.

i. Each sterile vessel should have an inspection record that shows all maintenance and the date completed.

6. *Bacteriophage Contamination*

How a bacteriophage enters the fermentation process is usually a matter of speculation, although it is known in some instances that it can accompany the culture itself. Bacteriophages affecting bacterial and fungal fermentation can be found in soils, water, air, and even in the raw materials. They are widely distributed in nature. Their presence usually results in lysis of the cells which results in reduction of cell mass, no further product formation, no oxygen uptake, no heat production, etc. Confirmation of the presence of a bacteriophage can be made with phage plaque plates or by using ultra-filters and isolating the specific virus itself. The filtrate can be tested in the laboratory with the culture.

The only immediate recourse to a bacteriophage attack is to substitute at once an immune strain if one is available. If one is not available, it is better to produce another product while the microbiologists in the culture laboratory isolate a new strain of culture which is resistant to the phage. This usually requires several weeks, which must include yield testing as well as new master lots of reserve culture.

From a plant operation standpoint, the appearance of phage is treated as a bacterial contamination. Equipment checks and cleanup are no more or less rigid.

8.0 GENERAL COMMENTS

a. Usually if a tank is contaminated in the first 24 hours, it is due to contaminated inoculum, poor sterilization of tank accessories and contents, or unsterile air. If contamination comes in after 24 hours, one should check the air supply, nutrient recharges, anti-foam feeds, loss of pressure during the run, lumps in the media, media blowbacks, humid air (wet air filters), etc.

b. The use of air samplers or bubblers on the sterile air system for detection of contamination is merely an exercise in futility. The result means absolutely nothing, as the sample tested is such a minute part of the total air supplied to the tank.

c. Be sure sterile samples from the plant have the correct tank number, time and date.

d. Use Tempilstiks™ to check sterilizing temperatures.

e. During a siege of contamination in the plant, it may also be advisable to make some blank tank runs.

f. Permanently installed feed systems can create problems. Some chemicals (such as urea, sugar, NaOH, $CaCl_2$, etc.) tend to weep from valves and flanges. It is particularly important after new gaskets have been installed in flanges to tighten up all the bolts after sterilization and to check for leaks occasionally. Where possible, the feed should be kept hot (60°C). Reliable valves on feed lines and removable stainless steel strainers with bypass lines at all critical points will improve operations.

g. Vacuum breakers, anti-foam meters, etc., need special attention during the turn around time of every tank. A germicidal rinse prior to placing the tank on the line is considered good practice.

h. If a plant has a continuous sterilizer, keeping the cooling water pressure less than the media pumping pressure will result in media being forced into the cooling water section if leaks develop. The installation of a conductivity probe and a suitable alarm system on the exit cooling water line will detect an immediate leak.

i. Checking for stress corrosion cracking of stainless steel should be carried out by the supervisor at suitable intervals. This occurs in coils and jackets where water is cooled by cooling towers. This can be prevented or minimized by regular water analysis checks for chlorides which are the usual causative agents. The use of chlorides in the media makeup can also cause this problem in the equipment itself. Cracking will occur from the side of metal which is in tensile stress.

j. It is important to have good housekeeping throughout all fermentation areas. Good sanitation practices involve daily wash-downs of the area and equipment, otherwise the work area will become infested with organisms which can outgrow the commercial cultures.

k. Tanks which are used for batch sterilization of media and have internal cooling coils are subject to a rapid wearing of the coils on the hangers. This is because during heating, the coils expand relative to the tank wall. However, when water is added to the coils to cool the batch, the coils contract while the tank shell is still expanded; the resulting friction will wear away the coil and eventually cause a leak.

l. Thermowells should be welded on the inside of the tank so media cannot get into the threads.

m. Sometimes media can get between the foam probe and its covering. After a contaminated run, check the foam probe and lining. Clean with a germicide before replacing.

REFERENCES

[1] H.J. Peppler, D. Perlman (Eds.), Microbial Technology, second ed., vol. 1, Academic Press, New York, 1979.

[2] S. Aiba, A.F. Humphrey, N.F. Millis, Biochemical Engineering, Academic Press, New York, 1965.

[3] C.F. Bruno, L.A. Szabo, Fermentation Air Filtration Upgrading by Use of Membrane Cartridge Filters, American Chemical Society Meeting, New York, 1981.

[4] C.A. Perkowski, Fermentation Process Air Filtration via Cartridge Filters, American Chemical Society Meeting, New York, 1981.

[5] A. Wiseman (Ed.), Topics in Enzyme and Fermentation Biotechnology, John Wiley, New York, 1979.

[6] N.J. Peppler, Microbial Technology, Reinhold Pub. Corp., New York, 1967.

[7] S.H. Lin, A. Theoretical, Analysis of thermal sterilization in continuous sterilizer, J. Ferment. Technol. 53 (2) (1975) 92.

[8] S.H. Lin, Residence time distribution of flow in continuous sterilization process, Process Biochem. 14 (7) (1979).

[9] M.H.J. Ashley, J. Mooyman, Continuous Sterilization of Media, American Chemical Society Meeting, New York, 1981.

[10] D.I.C. Wang, C.L. Cooney, A.L. Demain, P. Dunnill, A.F. Humphrey, M. Lilly, Fermentation and Enzyme Technology, John Wiley and Sons, New York, 1979.

[11] D.N. Miller, Scale-up of agitated vessels gas-liquid mass transfer, Am. Inst. Chem. Engrs. J. 20 (3) (1974) 445.

[12] O. Nagel, Energy Engineering Systems Seminars, 2:835–876 (1979).

[13] General Texts:, P.V. Danckwerts, Gas-Liquid Reactions, McGraw Hill, 1970. C.S. Ho, J.Y. Oldshue, Biotechnology Processes Scale-Up and Mixing, American Institute of Chemical Engineers, 1987.

[14] R.H. Perry, C.H. Chilton (Eds.), Chemica Engineers' Handbook, fifth ed., McGraw Hill, New York, 1973.

[15] K. van't Reit, Turbine Agitator Hydrodynamics and Dispersion Performance, Ph.D. Thesis, University of Delft, Netherlands (1975).

[16] R.S. Brodkey, Turbulence in Mixing Operations, Theory and Applications to Mixing and Reaction, Academic Press, New York, 1975.

[17] M. Charles, Oxygen Transfer and Mixing Characteristics of the Deep-Jet-Aeration Fermenter (personal correspondence).

[18] A.H. Shapiro, Dynamics and Thermodynamics of Compressible Fluid Flow, McGraw Hill, New York, 1953.

[19] A.A. Townsend, The Structure of Turbulent Shear Flow, Cambridge University Press, 1956.

[20] W. Forstall, A.H. Shapiro, Momentum and mass transfer in coaxial jets, J. Appl. Mech. 29 (1967).

[21] J.E. Bailey, D.F. Ollis, Biochemical Engineering Fundamentals, Ch. 9, McGraw-Hill Book Co., 1977.

Chapter 7

Agitation

James Y. Oldshue

1.0 THEORY AND CONCEPTS

Fluid mixing is essential in fermentation processes. Usually the most critical steps in which mixers are used are in the aerobic fermentation process. However, mixers are also used in many auxiliary places in the fermentation process and there are places also for agitation in anaerobic fermentation steps.

This chapter will emphasize the aerobic fermentation step, but the principles discussed can be used to apply to other areas of fermentation as well.

Table 7.1 divides the field of agitation into five basic classifications, liquid-solid, liquid-gas, liquid-liquid, miscible liquids and fluid motion. This can be further divided into two parts—on the left are shown those applications which depend upon some type of uniformity as a criterion, while the processes on the right are typical of those that require some type of mass transfer or chemical reaction as a criterion.

On the left-hand side, visual descriptions of flow patterns and other types of descriptions of the flow patterns are helpful and important in establishing the effect of mixing variables on these criteria. In general, they are characterized by a requirement for high pumping capacity rather than fluid shear rate, and studies to optimize the pumping capacity of the impellers relative to power consumption are fruitful.

The other types of processes involve more complicated extensions of fluid shear rates and the determination of which mixing variables are most important. This normally involves experimental measurements to find out exactly the process response to these variables that are not easy to visualize and characterize in terms of fluid mechanics.

In order to discuss the various levels of complexity and analysis of these mixing systems, some of the fluid mechanics of mixing impellers are examined and then examples of how these are used in actual cases are shown.

2.0 PUMPING CAPACITY AND FLUID SHEAR RATES

All the power, P, applied to the systems produces a pumping capacity, Q, and impeller head, H, shown by the equation:

$$P \propto QH$$

Q has the units of kilograms per second and H has the units of Newton meters per second. Power then would be in watts.

The power, P, drawn by mixing impellers in the low and medium viscosity range is proportional to:

$$P \propto N^3 D^5$$

where D is impeller diameter and N is impeller speed. The pumping capacity of mixing impellers is proportional to ND^3.

$$Q \propto ND^3$$

These three equations can be combined to yield the relationship that

$$(Q/H)_p \propto D^{8/3}$$

where $(Q/H)_p$ is the flow to head ratio at constant power.

This indicates that large impellers running at slow speeds give a high pumping capacity and low shear rates since the impeller head or velocity work term is related to the shear rates around the impeller.

High pumping capacity is obtained by using large diameter impellers at slow speeds compared to higher shear rates obtained by using smaller impellers and higher speeds.

3.0 MIXERS AND IMPELLERS

There is a complete range of flow and fluid shear relationships from any given impeller type.

Fermentation and Biochemical Engineering Handbook.

TABLE 7.1 Classification of Mixing Processes

Physical Processing	Application Classes	Chemical Processing
Suspension	Liquid-Solid	Dissolving
Dispersions	Liquid-Gas	Absorption
Emulsions	Immiscible Liquids	Extraction
Blending	Miscible Liquids	Reactions
Pumping	Fluid Motion	Heat Transfer

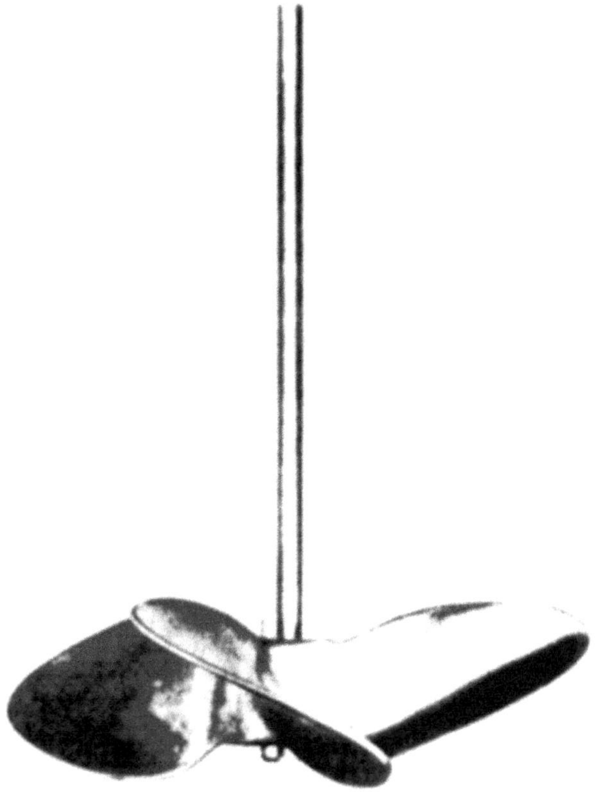

FIGURE 7.1 Photograph of square-pitch marine type impeller.

FIGURE 7.2 Photograph of radial flow, flat blade, disc turbine.

Three types of impellers are commonly used in the low viscosity region, propellers, Fig. 7.1; turbines, Fig. 7.2; and axial flow turbines, Fig. 7.3. Impellers used on small portable mixers shown in Fig. 7.4, are often inclined at an angle as well as being off-center to give a good top-to-bottom flow pattern in the system, Fig. 7.5. Large top-entering drives usually use either the axial flow turbine or the radial flow flat blade turbine. For aerobic fermentation, the radial flow disc turbine is most common and is illustrated in Fig. 7.6.

To complete the picture, there are also bottom-entering drives, shown in Fig. 7.7, which have the advantage of keeping the mixer off the top of all tanks and required superstructure, but have the disadvantage that if the sealing mechanism fails, the mixer is in a vulnerable location for damage and loss of product by leakage.

Figure 7.8 illustrates side-entering mixers that are used for many types of blending and storage applications.

3.1 Fluidfoil Impellers

The introduction of fluidfoil impellers, as shown in Fig. 7.9a through 7.9f, give a wide variety of mixing conditions suitable for high flow and low fluid shear rates. Fluidfoil impellers use the principles developed in airfoil work in wind tunnels for aircraft. Figure 7.10a shows what is desirable, which is no form separation of the fluid, and maximum lift and drag coefficients, which is what one is trying to achieve with the fluidfoil impellers. Figure 7.10b shows what happens when the angle and the shape is such that there is a separation of the fluid from the airfoil body. The A310 impeller (Fig. 7.9a) was introduced for primarily low viscosity fluids and, as can be seen, has a very low ratio of total blade surface area

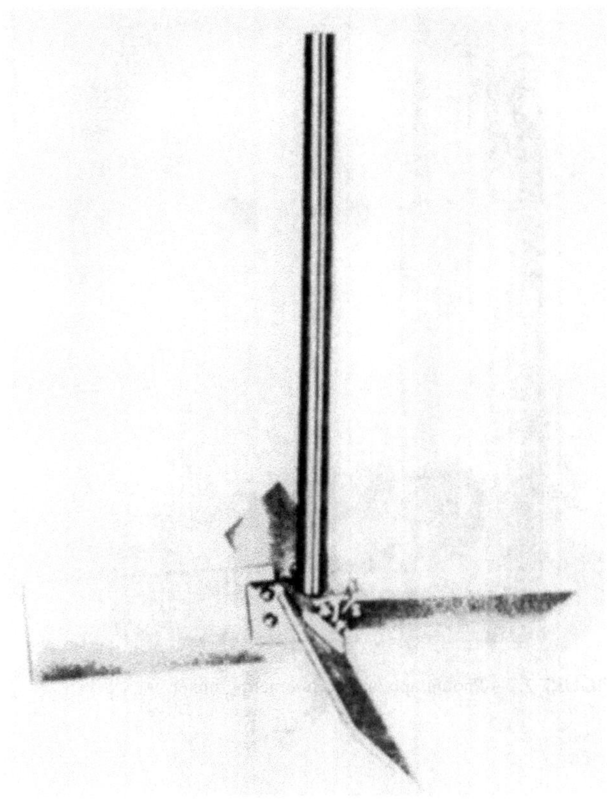

FIGURE 7.3 Photograph of typical 45° axial flow turbine.

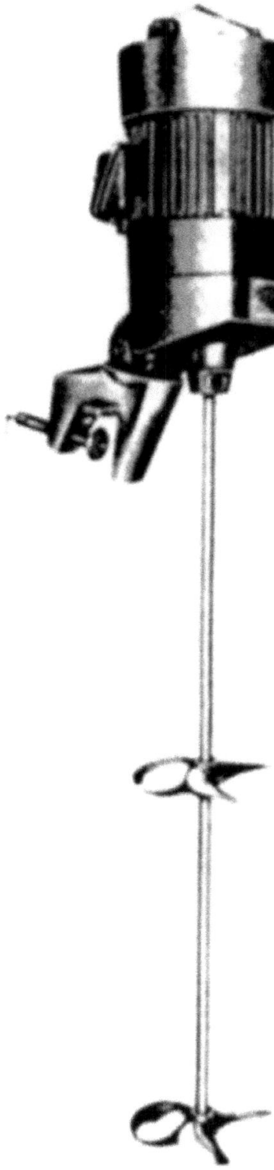

FIGURE 7.4 Photograph of portable propeller mixer.

compared to an inscribing circle that is shown in Fig. 7.11. When the fluid viscosities are higher, the A312 impeller is used (shown in Fig. 7.9b) which is particularly useful in fibrous materials.

To give a more responsive action in higher viscosities, the A320 is available which works well in the transition area of Reynolds numbers. When gas-liquid processes are used, the A315 (Figure 7.9d) has a still higher solidity ratio. It is particularly useful in aerobic fermentation processes. Impellers in Figs. 7.9(a–d) are formed from flat metal stock.

To complete the current picture, when composite materials are used, the airfoil can be shaped in any way that is desirable. The A6000 (Fig. 7.9e) illustrates that particular impeller type. The use of proplets on the end of the blades increases flow about 10% over not having them. An impeller which is able to operate effectively in both the turbulent and transitional Reynolds numbers is the A410 (Fig. 7.9f) which has a very marked increase in twist angle of the blade. This gives it a more effective performance in the higher viscosity fluids encountered in mixers up to about 3 kW.

One characteristic of these fluidfoil impellers is that they discharge a stream that is almost completely axial flow and they have a very uniform velocity across the discharge plane of the impeller. However, there is a tendency for these impellers to short-circuit the fluid to a relatively low distance above the impeller. Very careful consideration of the coverage over the impeller is important. If the impeller can be placed one to two impeller diameters off bottom, which means that mixing is not provided at low levels during draw off, these impellers offer an excellent flow pattern as well as considerable economies in shaft design.

To look at these impellers in a different way, three impellers have been compared at equal total-pumping capacity. Figure 7.12 gives the output velocity as a function of time on a strip chart. As can be seen in Fig. 7.12

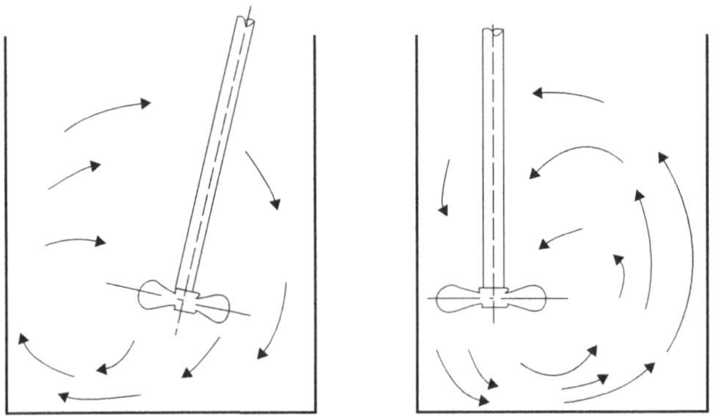

FIGURE 7.5 Flow pattern, propeller, top-entering, off-center position for counterclockwise rotation.

FIGURE 7.6 Series 800 top-entering mixer.

the fluidfoil impeller type (A310) has a very low velocity fluctuation and uses much less power than the other two impellers. For the same flow, the A200 impeller has a higher turbulent fluctuation value. The R100 impeller has still higher power consumption at the same diameter than the other two impellers, and has a much more intense level of microscale turbulence.

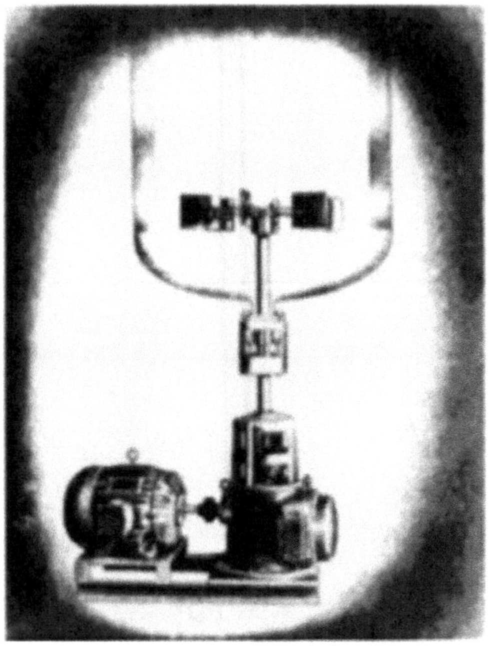

FIGURE 7.7 Photograph of bottom-entering mixer.

FIGURE 7.8 Photograph of side-entering propeller mixer.

The fluidfoil impellers are often called "high efficiency impellers", but that is true only in terms of flow, and makes the assumption that flow is the main measure of mixing results. Flow is one measure, and in at least half of the mixing applications is a good measure of the performance that could be expected in a process. These impellers are low in efficiency in providing shear rates—either of the macro scale or the micro scale.

The use of computer generated solutions to problems and computational fluid dynamics is also another approach of comparing impellers and process results. There are

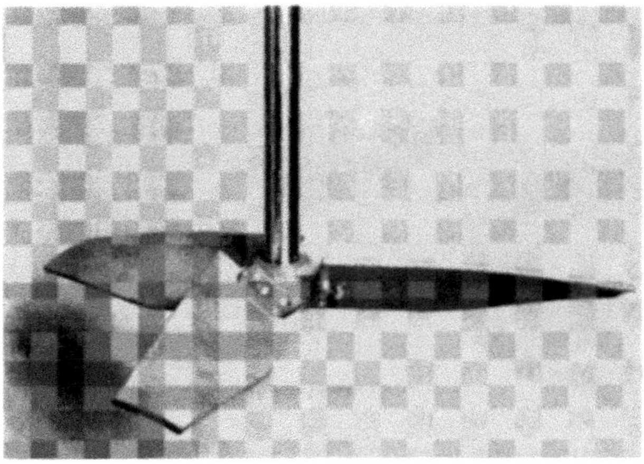

FIGURE 7.9a A310 fluidfoil impeller.

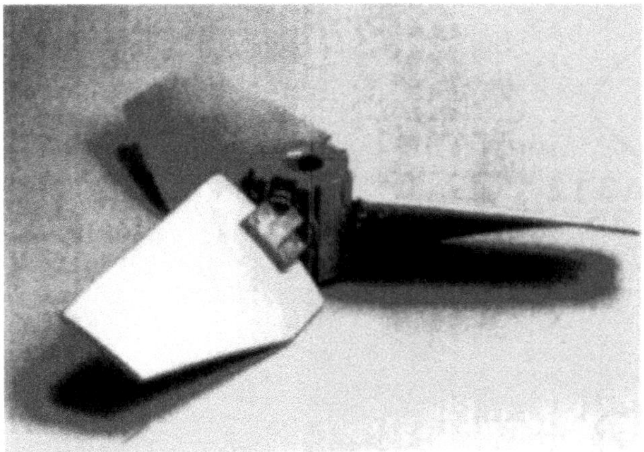

FIGURE 7.9b A312 fluidfoil impeller.

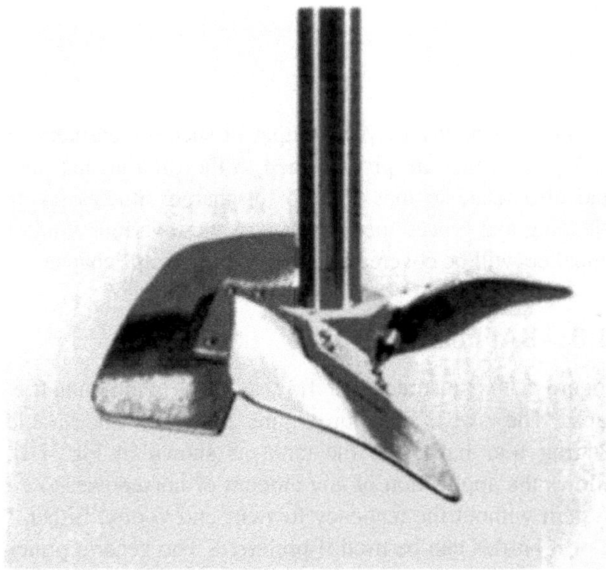

FIGURE 7.9c A320 fluidfoil impeller.

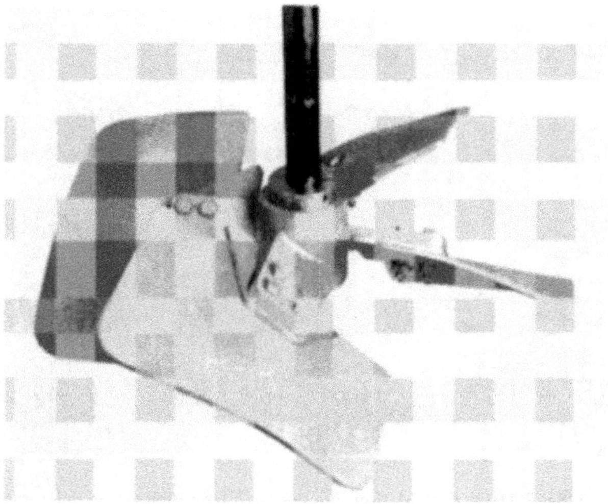

FIGURE 7.9d A315 fluidfoil impeller.

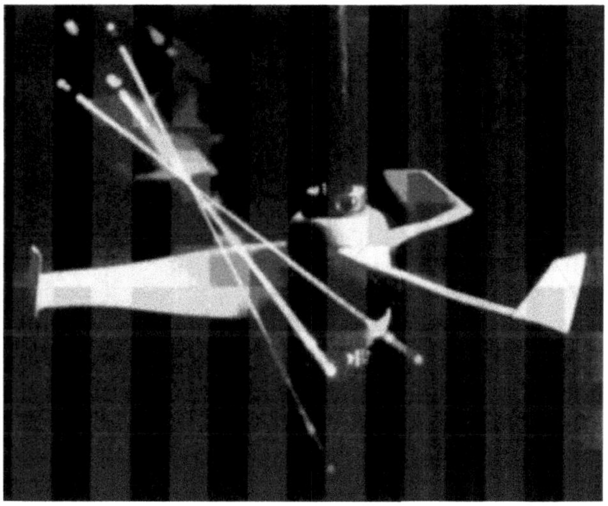

FIGURE 7.9e A6000 fluidfoil impeller made of composite materials.

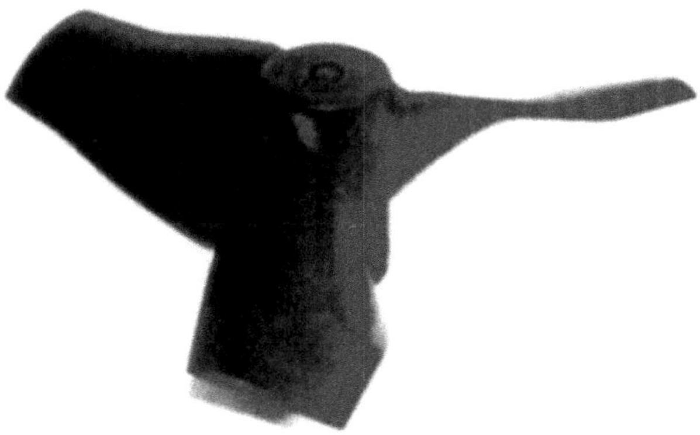

FIGURE 7.9f A410 fluidfoil impeller made from composite materials with high twist angle ratio between tip and hub.

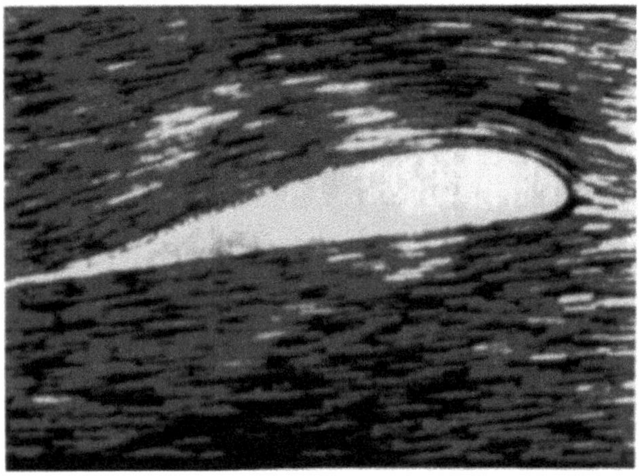

FIGURE 7.10a Typical flow around airfoil positioned for maximum lift; minimum drag.

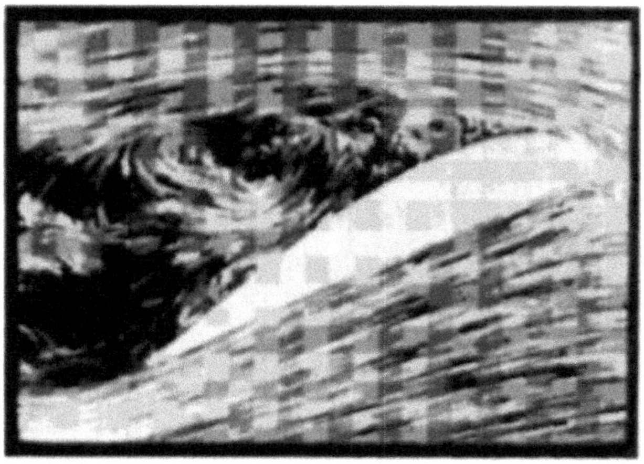

FIGURE 7.10b Typical profile of airfoil at an angle of attack that gives fluids separation from the airfoil surface.

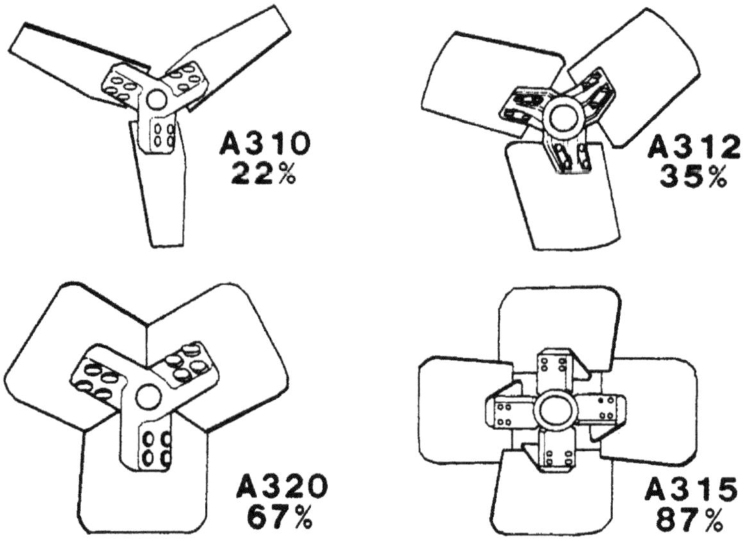

FIGURE 7.11 Solidity ratio defined as a ratio of blade area to circle area. The solidity ratio for four different fluidfoil impellers is shown.

software packages available. It is very helpful to have data obtained from a laser velocity meter on the fluid mechanics of the impeller flow and other characteristics to put in the boundary conditions for these computer programs.

As an example of other types of programs that can be worked on, Fig. 7.13 shows a velocity profile from an A410 impeller, Fig. 7.14, a map of the kinetic energy dissipation in the fluid stream and in the third one (Fig. 7.15) model of heavier than the liquid particles in a random tracking pattern.

Additional models can be made up using mass transfer, heat transfer and some reaction kinetics to simulate a process that can be defined in one or more of those types of relationships.

The laser velocity meter has made it possible to obtain much data from experiments conducted in transparent fluids. Figure 7.16 shows a typical output of such a measurement, giving lines that are proportioned to the fluid at that point and also relate to that angle of discharge. Studies on the blending and process performance of these various fluidfoil impellers will be covered in later sections of this chapter.

4.0 BAFFLES

Figure 7.17 illustrates the flow pattern in an unbaffled tank. The swirl and vortex are normally undesirable. Putting four baffles in the tank, as shown in Fig. 7.18, allows the application of any amount of horsepower to the system without the tendency to swirl and vortex. Either 3, 6 or 8 baffles can be used if preferred. The general principle is to use the same total projected area as exists with four baffles, each 1/12 the tank diameter in width.

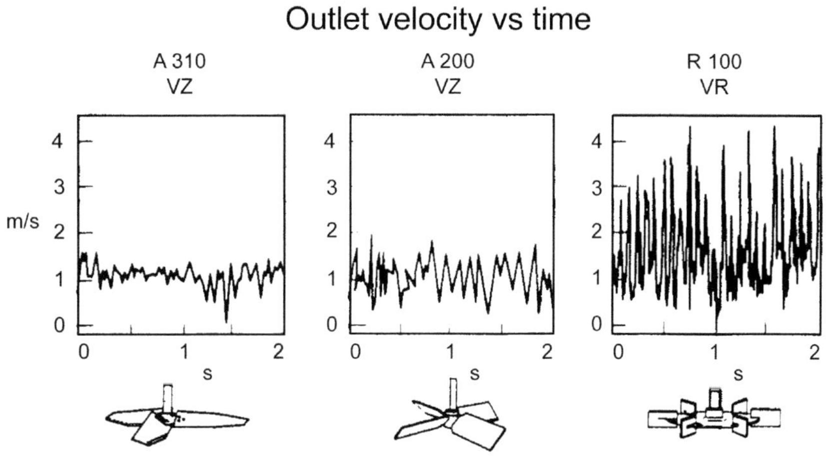

FIGURE 7.12 Velocity trace with time for three different impeller types, A310 fluidfoil, A200 axial flow turbine and R100 radial flow turbine. The impellers are compared at equal discharge pumping capacity, equal diameter and at whatever speed is required to achieve this flow. The power required increases from left to right.

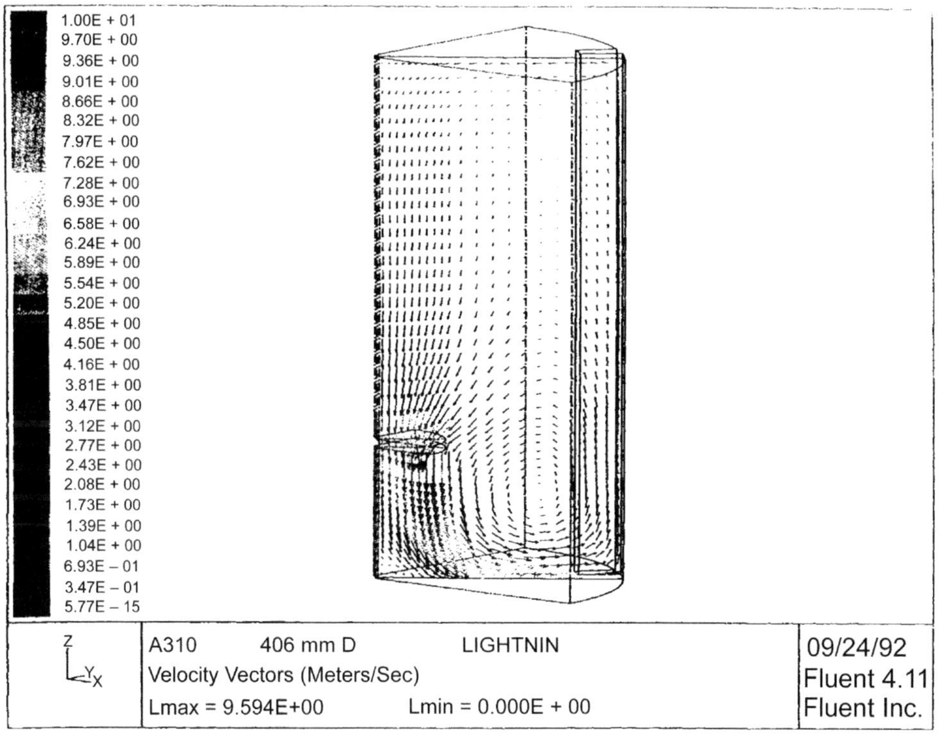

FIGURE 7.13 Typical trace of velocity from a fluidfoil A410 impeller.

For square or rectangular tanks the baffles shown in Fig. 7.19 are typical. At power levels below 1 hp/1000 gal., the corners in square or rectangular tanks often give sufficient baffling that additional wall baffles are not needed.

5.0 FLUID SHEAR RATES

Figure 7.20 illustrates flow pattern in the laminar flow region from a radial flat blade turbine. By using a velocity probe, the parabolic velocity distribution coming off the blades of the impeller is shown in Fig. 7.21. By taking the slope of the curve at any point, the shear rate may be calculated at that point. The maximum shear rate around the impeller periphery as well as the average shear rate around the impeller may also be calculated.

An important concept is that one must multiply the fluid shear rate from the impeller by the viscosity of the fluid to get the fluid shear stress that actually carries out the process of mixing and dispersion.

$$\text{Fluid shear stress} = \mu(\text{fluid shear rate})$$

Even in low viscosity fluids, by going from 1 cp to 10 cp there will be 10 times the shear stress of the process operating from the fluid shear rate of the impeller.

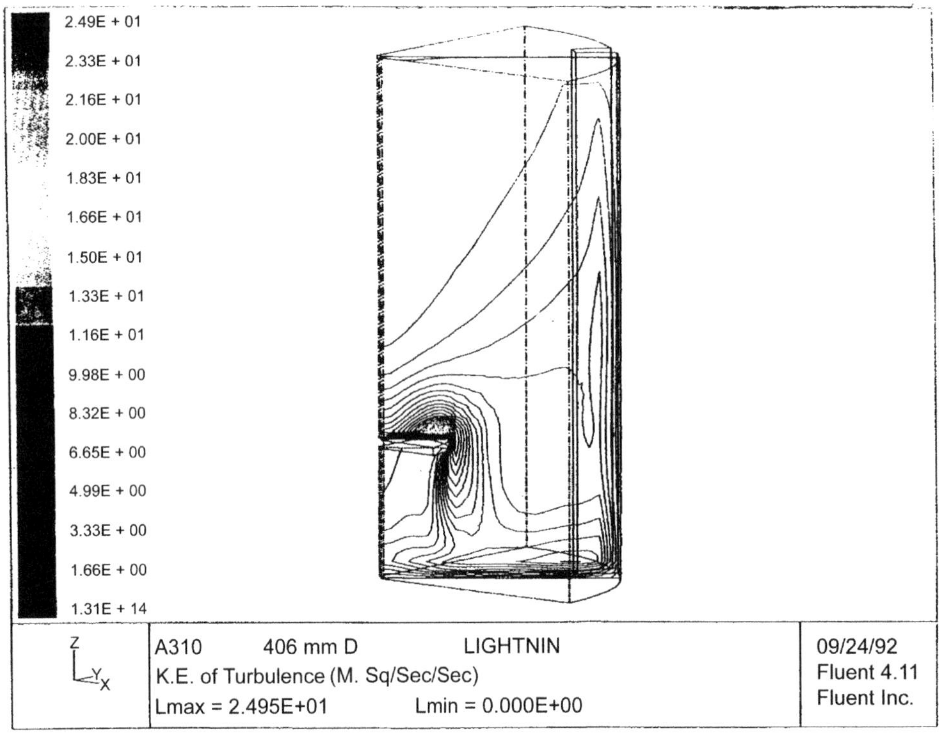

FIGURE 7.14 A typical map of the kinetic energy dissipation in the fluid stream in a mixing vessel.

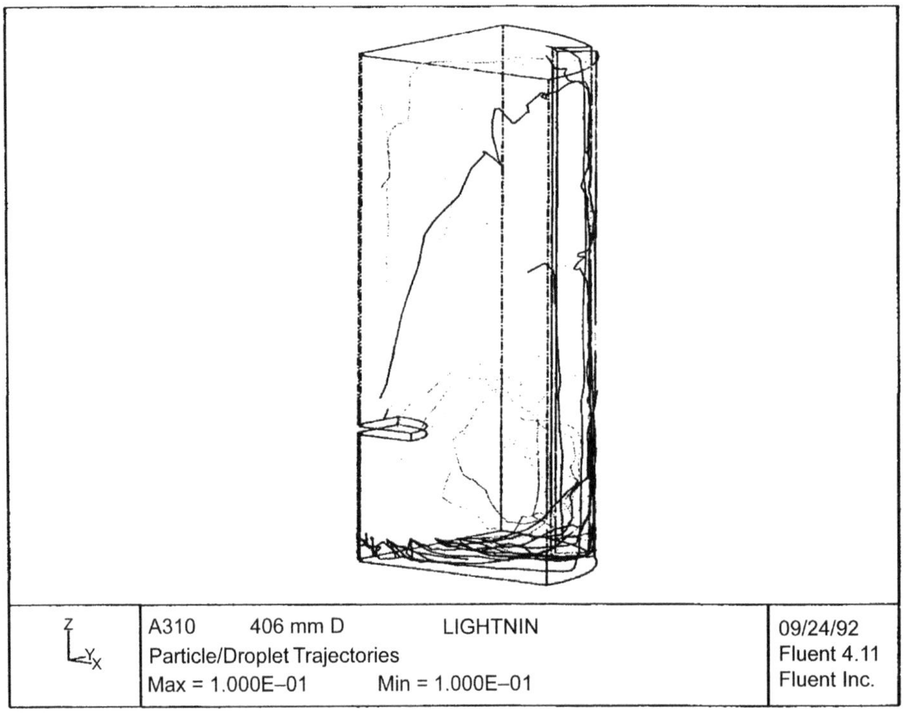

FIGURE 7.15 A model of a particle tracking pattern when the particle is heavier than the liquid in the tank.

Figure 7.22 shows the flow pattern when there is sufficient power and low enough viscosity for turbulence to form. Now a velocity probe must be used that can pick up the high frequency response of these turbulent flow patterns, and a chart as shown in Fig. 7.23 is typical. The shear rate between the small-scale velocity fluctuations is called microscale shear rate, while the shear rates between the average velocities at this point are called the

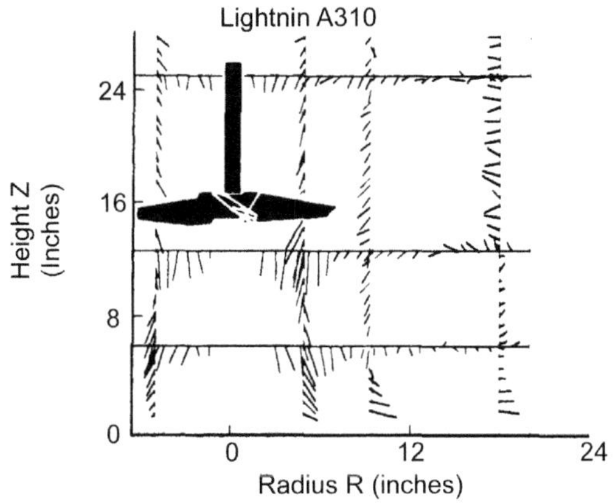

FIGURE 7.16 Typical laser output from the measurement of velocities by means of a Doppler velocity meter.

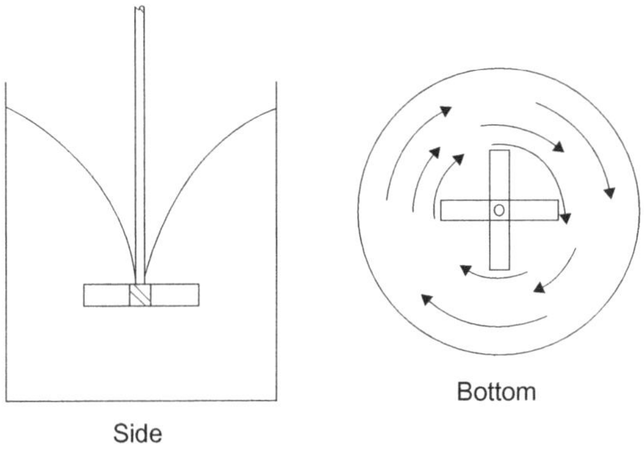

FIGURE 7.17 Vortexing flow pattern obtained with any type impeller which is unbaffled.

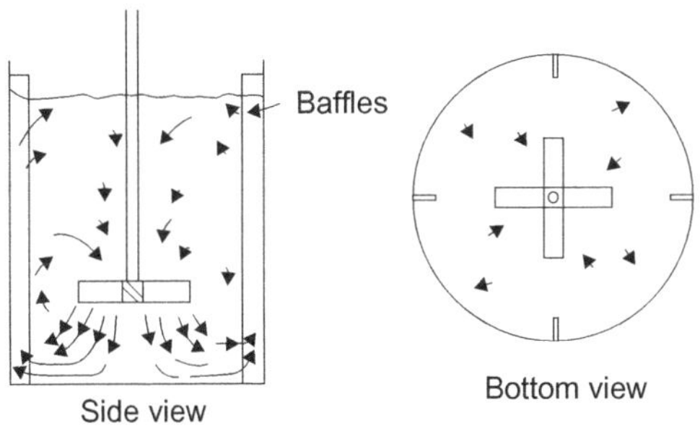

FIGURE 7.18 Flow pattern obtained with any type of impeller.

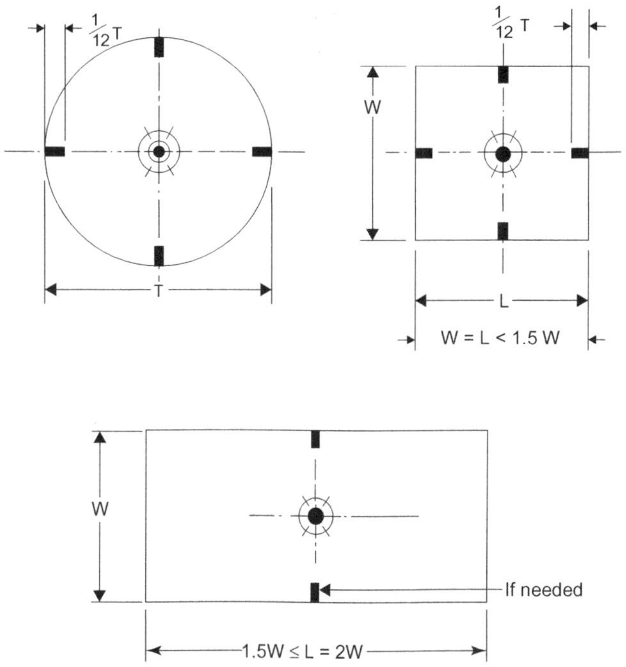

FIGURE 7.19 Suggested baffles for square and rectangular tanks.

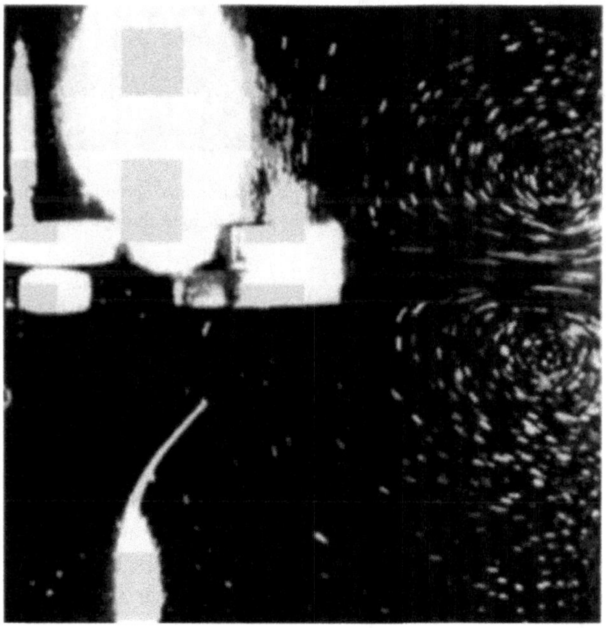

FIGURE 7.20 Photograph of radial flow impeller in a baffled tank in the laminar region, made by passing a thin plane of light through the center of the tank.

macroscale rates. These macro scale shear rates still have the same general form and are determined the same way as shown in Fig. 7.21.

Table 7.2 describes four different macro scale shear rates of importance in a mixing tank. The parameter for the micro scale shear rate at a point is the root mean square velocity fluctuation at that point, RMS.

5.1 Particles

The consideration of the macro- and micro scale relationships in a mixing vessel leads to several helpful concepts. Macroscopic particles that are greater than 1,000 microns in size are affected primarily by the shear rate between the

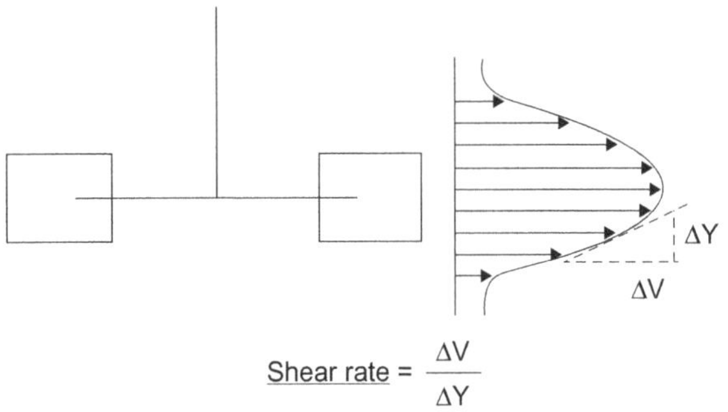

FIGURE 7.21 Typical velocity pattern coming from the blades of a radial flow turbine showing calculation of the shear rate $\Delta V/\Delta Y$.

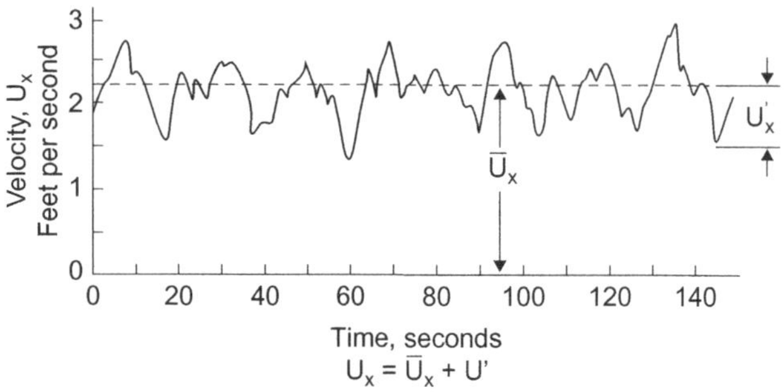

FIGURE 7.23 Schematic, typical velocity fluctuation pattern obtained from high frequency velocity probe placed at a point in the mixing vessel.

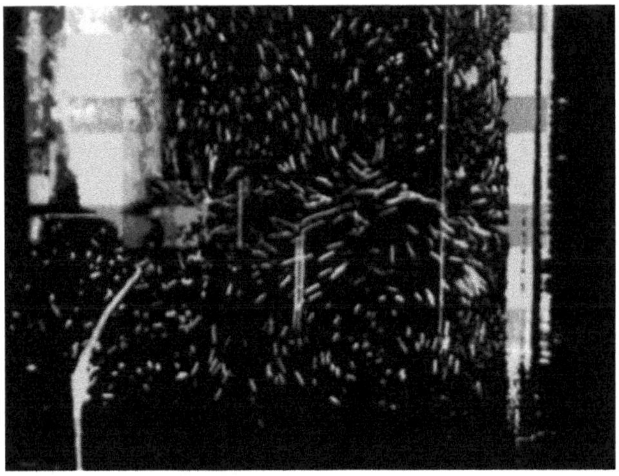

FIGURE 7.22 Photograph of flow patterns in a mixing tank in the turbulent region, made by passing a thin plane of light through the center of the tank.

TABLE 7.2 Average Point Velocity
Max. imp. zone shear rate
Ave. imp. zone shear rate
Ave. tank zone shear rate
Min. tank zone shear rate
RMS velocity fluctuations

average velocities in the process. They are an essential part of the overall flow throughout the tank, and determine the rate at which flow and velocity distribute throughout the tank. The shear rate experienced by macroscopic particles is a predictor of the visual appearance of the tank in terms of surface action, blending or particle suspensions.

The other situation is for the micro scale particles. They are particles less than 100 microns and they see largely the energy dissipation which occurs through the mechanism of viscous shear rates and shear stresses and ultimately the scale at which all energy is transformed into heat.

The macro scale environment is effected by every geometric variable and dimension and is a key parameter for successful scale up of any process, whether micro scale mixing is involved or not. This has some unfortunate consequences on scale up since geometric similarity causes many other parameters to change in unusual ways, which may be either beneficial or detrimental, but are quite different than exist in a smaller pilot plant unit. On the other hand, the micro scale mixing condition is primarily a function of power per unit volume and the result is dissipation of that energy down through the micro scale and onto the level of the smallest eddies that can be identified as belonging to the mixing flow pattern. An analysis of the energy dissipation can be made in obtaining the kinetic energy of turbulence by putting the resultant velocity from the laser velocimeter through a spectrum analyzer. Figure 7.24a shows the breakdown of the energy as a function of frequency for the velocities themselves. Figure 7.24b shows a similar spectrum analysis of the energy dissipation based on velocity squared and Fig. 7.24c shows a spectrum analyzer result from the product of two orthogonal velocities, V_R and V_z, which is called the *Reynolds stress* (a function of momentum).

An estimation method of solving complex equations for turbulent flow uses a method called the K-ε technique which allows the solution of the Navier-Stokes equation in the turbulent region.

5.2 Impeller Power Consumption

Figure 7.25 shows a typical Reynolds number-Power number curve for different impellers. The important thing about this curve is that it holds true whether the desired process job is being done or not. Power equations have three independent variables along with fluid properties: power, speed

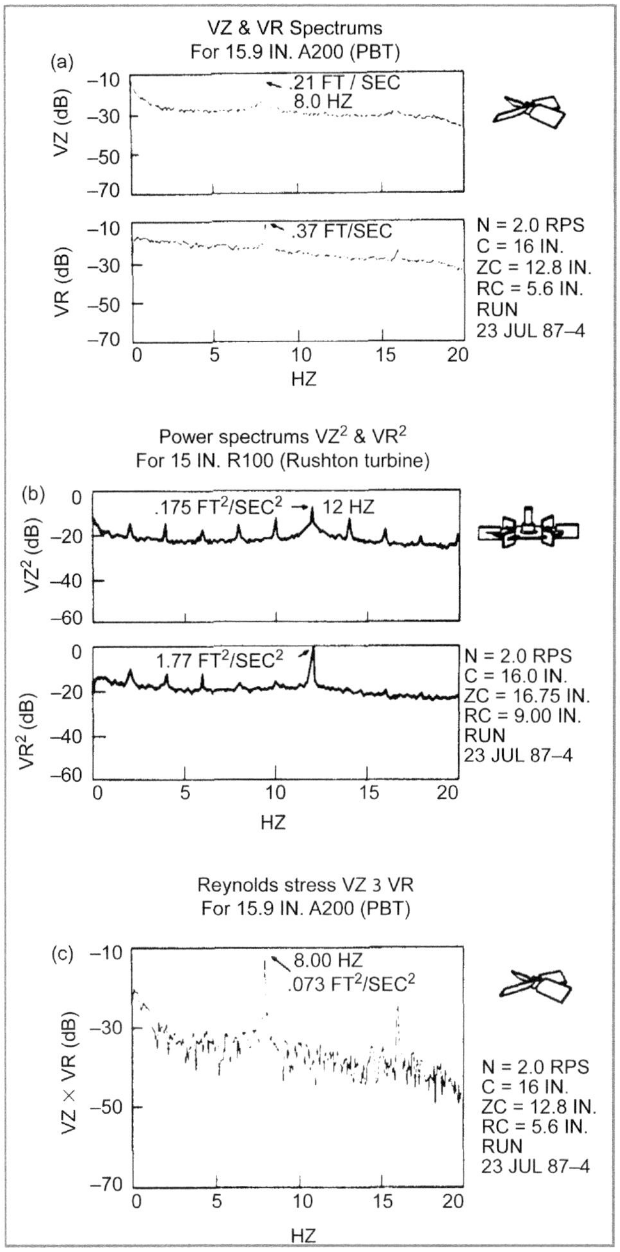

FIGURE 7.24 Typical spectrum analysis of the velocity as a function of (a) velocity frequency fluctuation, (b) the frequency of the fluctuations using the square of the velocity to give the energy dissipation, and (c) the product of two orthogonal velocities versus the frequency of the fluctuations. The product of two orthogonal velocities is related to the momentum in the fluid stream.

and diameter. There are only two independent choices for process considerations.

For gas-liquid operations there is another relationship called the *K factor* which relates the effect of gas rate on power level. Figure 7.26 illustrates a typical *K* factor plot that can be used for estimation. Actual calculation of *K* factor in a particular case involves very specific

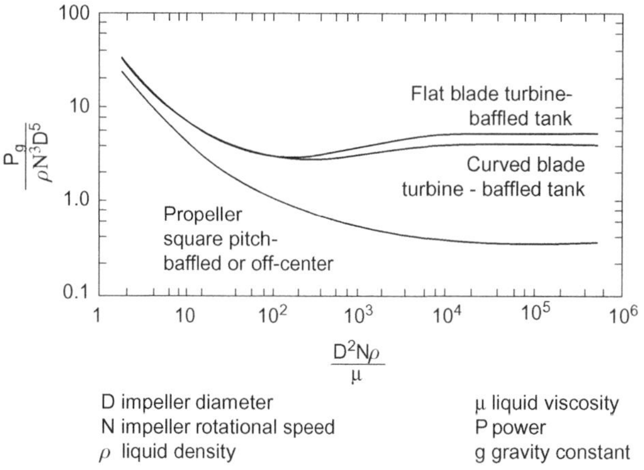

FIGURE 7.25 Power number/Reynolds number curve for the power consumption of impellers.

combinations of mixer variables, tank variables, and fluid properties, as well as the gas rate being used.

Commonly, a physical picture of gas dispersion is used to describe the degree of mixing required in an aerobic fermenter. This can be helpful on occasion, but often gives a different perspective on the effect of power, speed and diameter on mass transfer steps. To illustrate the difference between physical dispersion and mass transfer, Fig. 7.27 illustrates a measurement made in one experiment where the height of a geyser coming off the top of the tank was measured as a function of power for various impellers. Reducing the geyser height to zero gives a uniform visual dispersion of gas across the surface of the tank. Figure 7.28 shows the actual data and indicates that the 8-inch impeller was more effective than the 6-inch impeller in this particular tank.

Also, the 8-inch impeller with standard blades was more effective than the 8-inch impeller with narrow blades. These results all indicate that in this range of impeller-size-to-tank-size ratio, pumping capacity is more important than fluid shear rate for this particular criterion of physical dispersion.

Looking now at some actual published mass transfer rates, Fig. 7.29 shows the results of some experiments reported previously and Figs. 7.30 through 7.33 show some additional experiments reported which give further clarification to Fig. 7.29.

In Fig. 7.29, the ratio of mixer horsepower to gas expansion horsepower is shown with the optimum *D/T* range from a mass transfer standpoint in air-water systems. At the left of Fig. 7.29, it can be seen that large *D/T* ratios are more effective than small *D/T* ratios. This is in an area where the mixer power level is equal to or perhaps less than the gas expansion power level. Moving to the right, in the center range it is seen that the optimum *D/T* ratios are on the order of 0.1 to 0.2. This corresponds to

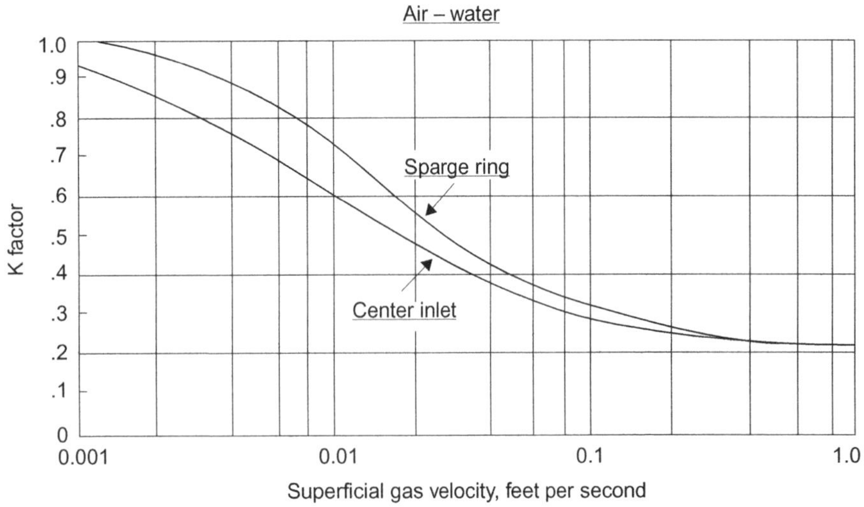

FIGURE 7.26 Typical curve of *K* factor, power drawn with gas on versus power drawn with gas off, for various superficial gas velocities.

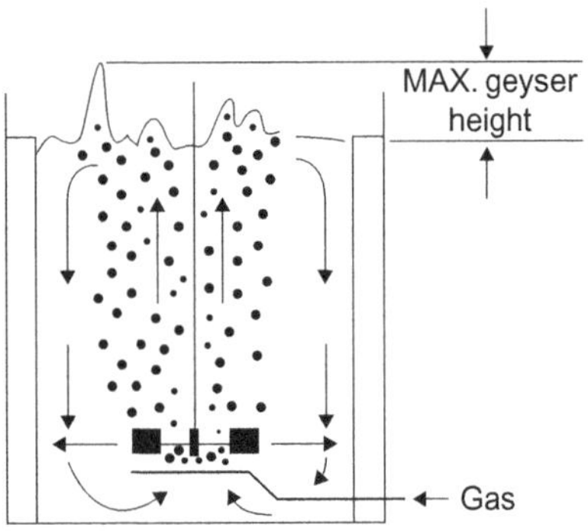

FIGURE 7.27 Schematic of geyser height.

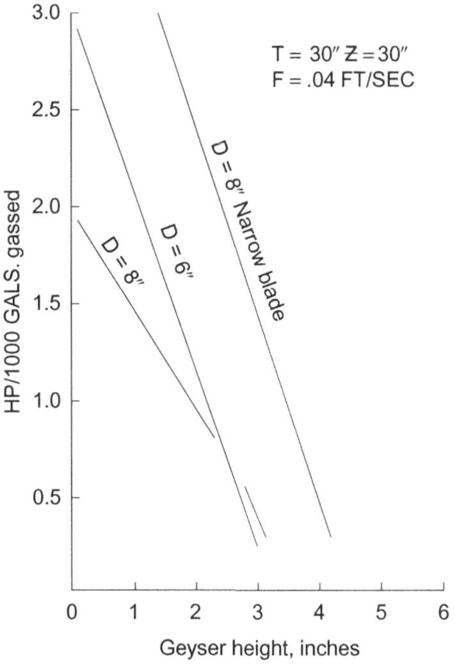

FIGURE 7.28 Plot illustrating measurement of geyser height.

an area where the mixer power level is two to ten times higher than the expansion power in the gas stream. Thus shear rate is more important than pumping capacity in this range, which is a very practical range for many types of gas-liquid contacting operations, including aerobic mass transfer in fermentation.

At the far right of Fig. 7.29 is shown high mixer power levels relative to the gas rate, and it can be seen that D/T makes no difference to the mass transfer. This occurs in some types of hydrogenation, carbonation, and chlorination. In those cases, the power level is so high relative to the amount of gas added to the tank that flow to shear ratio is of no importance.

In Figs. 7.30 through 7.33, the gas rate is successively increased in each of the four figures. At the low gas rate, the 4-inch impeller is more effective than the 6 or 8-inch impeller under all power levels. At the higher gas rates, the larger impellers become more effective at the lower gas rates, while the smaller impellers are more effective at the higher power levels, fitting generally into the scheme shown in Fig. 7.29.

A sparge ring about 80% of the impeller diameter is more effective than an open pipe beneath the impeller or sparge rings larger than the impeller. Figure 7.34 shows this effect and indicates that the desired entry point for the gas is where it can pass initially through the high shear zone around the impeller.

This has led to the common practice today of using the distribution of power in a three-impeller system, for example, 40% to the lower impeller and 30% to each of the two upper impellers, Fig. 7.35.

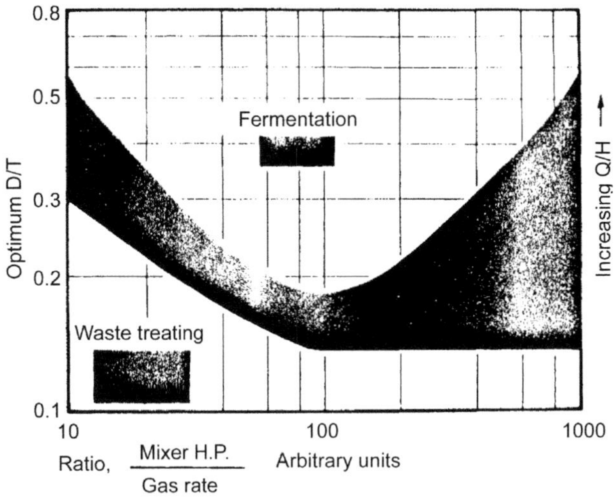

FIGURE 7.29 Effect of horsepower-to-gas rate ratio at optimum *D/T*.

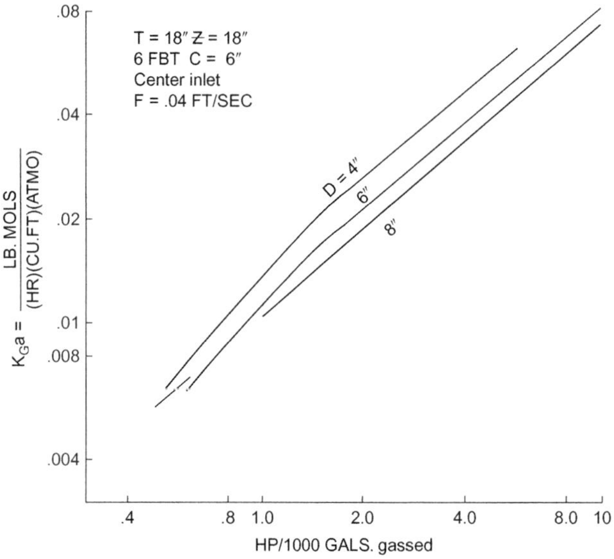

FIGURE 7.31 Effect of horsepower and impeller diameter on mass transfer coefficient at 0.04 ft/sec gas velocity.

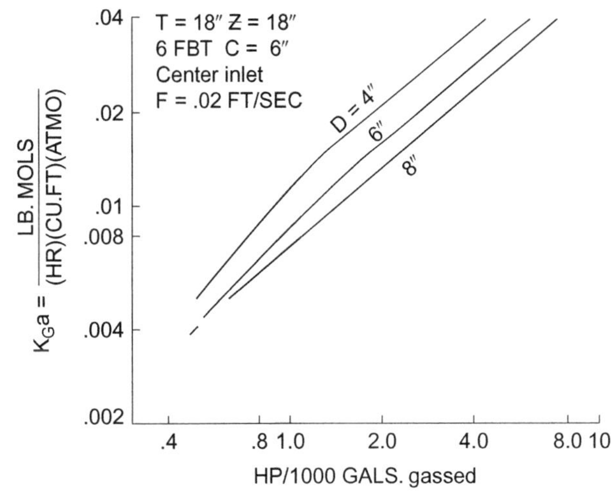

FIGURE 7.30 Effect of sparge ring diameter on mass transfer performance of a flat blade turbine, based on gassed horsepower at gas velocity F = 0.02 ft/sec.

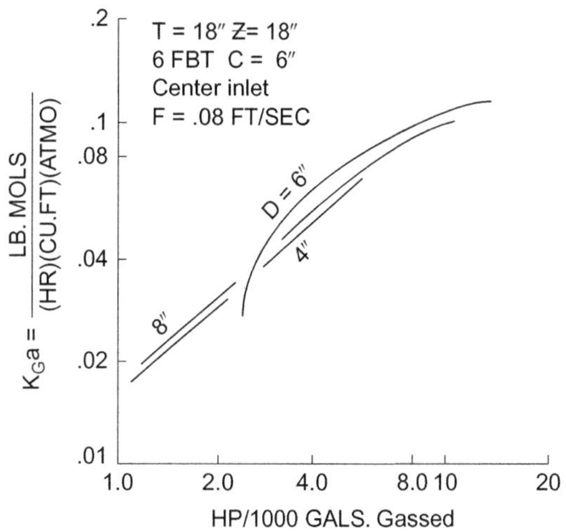

FIGURE 7.32 Effect of horsepower and impeller diameter on mass transfer coefficient at 0.08 ft/sec gas velocity.

In regard to tank shape, it has turned out over the years that about the biggest tank that can be shop-fabricated and shipped to the plant site over the highways is about 14 ft (4.3 m) in diameter. As fermentation volumes have gone from 10,000 gallons (38 m^3) to 50 or 60 thousand gallons, tank shapes have tended to get very tall and narrow, resulting in *Z/T* ratios of 2:1, 3:1, 4:1, or even higher on occasion. This tall tank shape has some advantages and disadvantages, but tank shape is normally a design variable to be looked at in terms of optimizing the overall plant process design.

This leads to the concept of mass transfer calculation techniques in scale up. Figure 7.36 shows the concept of mass transfer from the gas–liquid step as well as the mass transfers step to liquid–solid and/or a chemical reaction. Inherent in all these mass transfer calculations is the concept of dissolved oxygen level and the driving force between the phases. In aerobic fermentation, it is normally the case that the gas–liquid mass transfer step from gas to liquid is the most important. Usually the gas–liquid mass transfer rate is measured, a driving force between the gas and the liquid calculated, and the mass transfer coefficient, K_Ga or K_La obtained. Correlation techniques use the data shown in Fig. 7.37 as typical in which K_Ga is correlated versus power level and gas rate for the particular system studied.

If the data are obtained on small scale, then translation to larger scale equipment means an increase in superficial gas velocity, *F*, because of the change in liquid level in the

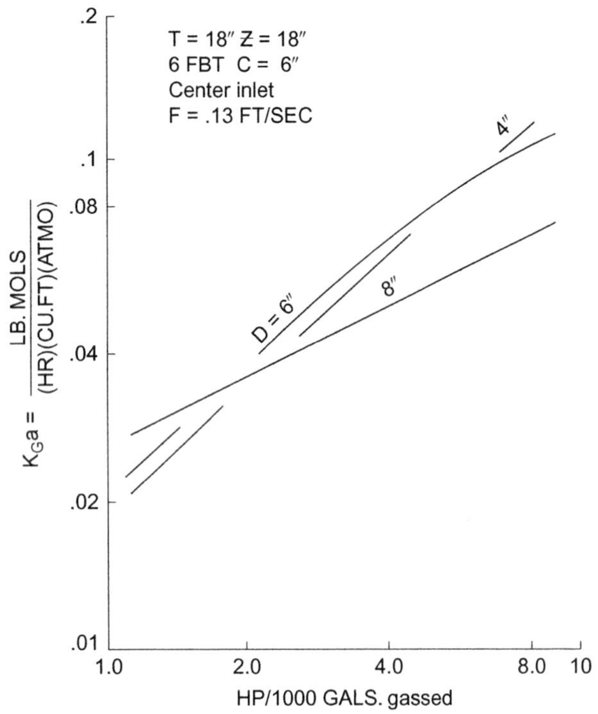

FIGURE 7.33 Effect of horsepower and impeller diameter on mass transfer coefficient for gas velocity of 0.13 ft/sec.

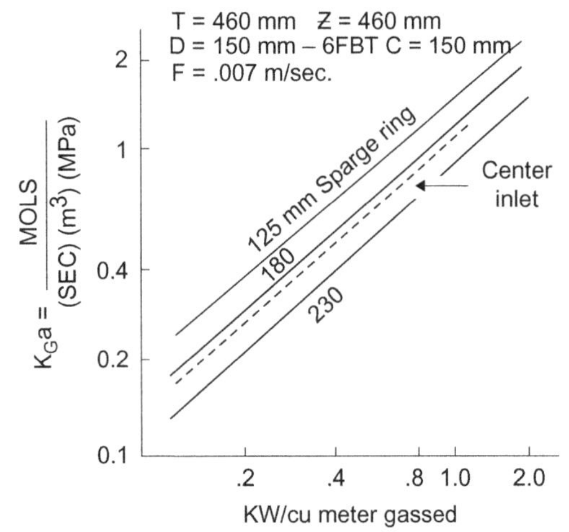

FIGURE 7.34 Gas-liquid mass transfer data for 150 mm turbine in 460 mm tank at 0.07 m/sec superficial gas velocity.

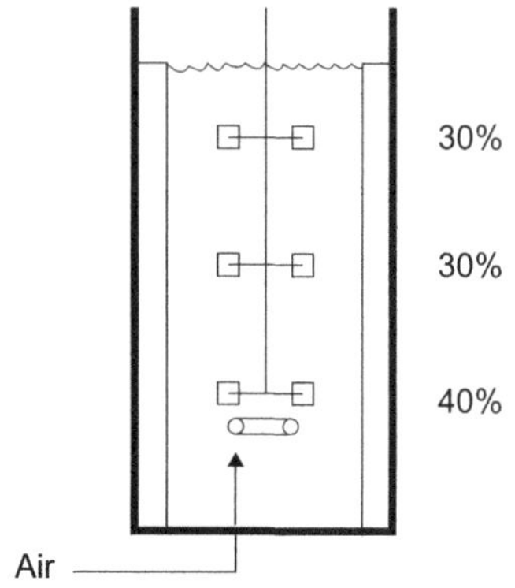

FIGURE 7.35 Typical power consumption relations for triple impeller installation, giving higher horsepower in proportion to the lower impeller.

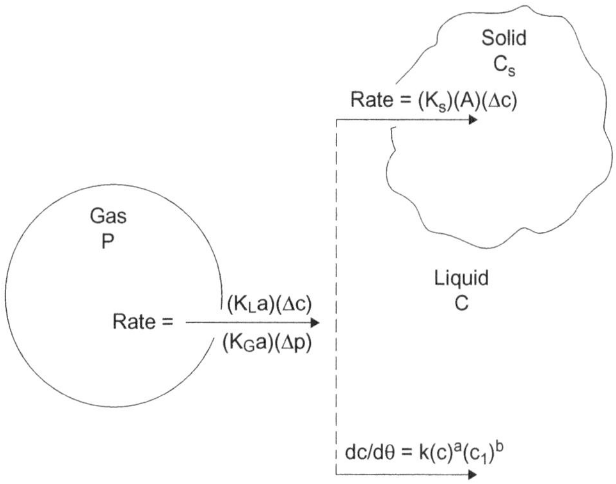

FIGURE 7.36 Schematic showing gas-liquid mass transfer step related to the other steps of liquid-solid mass transfer and chemical reaction.

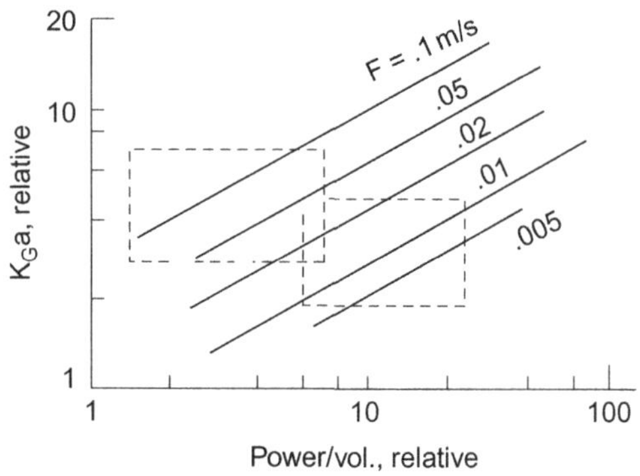

FIGURE 7.37 Typical $K_{G}a$ versus horsepower and gas velocity correlation; box on right indicates typical pilot plant experiments; box on left indicates typical full-scale range.

large-scale system. This normally pushes the data toward the left and would possibly result in a lower power level being required full-scale for the same mass transfer rate.

This has to be examined with great care, because any change in the power level will change the liquid-solid mass transfer rate; change the blend time, the shear rate and, therefore, the viscosity of non-Newtonian broth, and could necessitate many other process considerations.

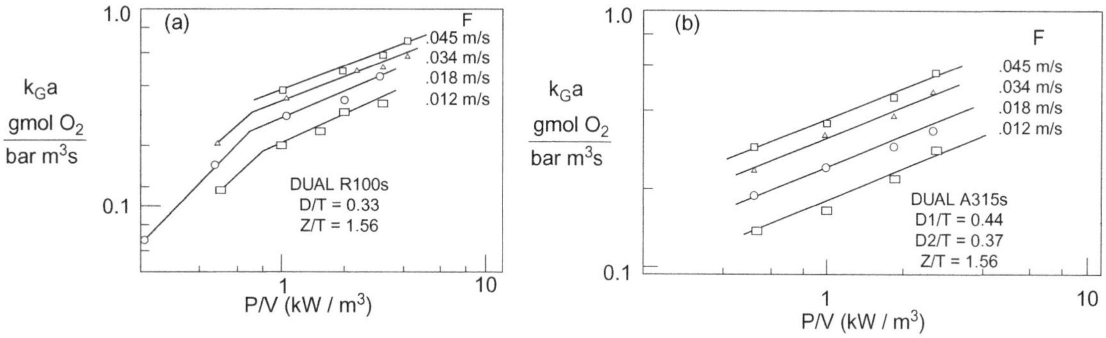

FIGURE 7.38 Typical curves of $K_G a$ versus power and various gas rates for radial flow turbines, (a) R100; (b) A-315.

5.3 Mass Transfer Characteristics of Fluidfoil Impellers

Experiments made with the sulfite oxidation technique evaluate the overall $K_G a$ relationship for radial flow turbines and a very typical curve, shown in Fig. 7.38a gives the value of $K_G a$ versus power and various gas rates. One will notice that there is a break in the curve which occurs about at the point where the power of the mixer is approximately two or three times higher than the power in the gas stream. A curve taken on similar conditions for the A315 impeller does not show the break point (Fig. 7.38b), and matching of the two curves shows that at the low end of the power levels the A315 results in a higher mass transfer relationship, while at higher power levels, the R100 is somewhat better. However, with ±20%, which is with reasonable accuracy for these kind of measurement comparisons, the mass transfer performance is quite similar. The difference of performance on a given fermentation application should thus be higher or lower in terms of the mass transfer coefficient and needs to be studied in detail when a retrofit is desired.

One large difference between A315 and R100 impellers in fermentation is the blend time. Every R100 impeller sets up two flow pattern zones. Thus, in a large fermentation tank with three or four R100 impellers there are six to eight separate mixing zones/cells in the vessel. If the A315 impeller is used for the one or more different impeller positions, it sets up one overall flow pattern which gives one complete mixing zone and results in a blend time on a batch basis of approximately one-half to one-third the time it takes on the R100 configuration. It is quite typical in current practice to use a radial flow turbine at the bottom while using a series of A315 impellers (either one, two or three, on the top positions). This has the overall tendency to reduce the macro- and micro scale shear rates and also can either increase productivity at the same power level or retain the original productivity at a reduced power level. This, in a fermentation process, is of very great importance economically.

However, before retrofitting a large fermentation tank it should be realized unless there is some process data arising from understanding the relationship between the mass transfer and the biological oxidation requirement, retrofitting existing radial flow turbine installations with A315 impeller types does not always give an improvement in process result. The average is normally about two or three times as frequent for a plus result as for neutral or negative results.

It is very difficult to study the effect of fluidfoil impellers in the pilot plant since the pilot plant in general has much shorter blend time and a much more uniform blending composition than appears in the full-scale tank. Thus, putting fluidfoil impellers in the pilot plant improves the blending under a situation where the blending is already much improved over full-scale performance.

6.0 FULL-SCALE PLANT DESIGN

There are four ways in which mixers are often specified when considering installation of more productive units in a fermentation plant. This can involve either a larger tank with a suitable mixer or improvement of the productivity of a given tank by a different combination of mixer horsepower and gas rate. These are listed below:

a. Change in productivity requirements based on production data with a particular size fermenter in the plant.
b. New production capacity based on pilot plant studies.
c. Specification of agitator based on the sulfite absorption rate in aqueous sodium sulfite solution.
d. Specification of the oxidation uptake rate in the actual broth for the new system.

6.1 Some General Relationships in Large Scale Mixers Compared to Small Scale Mixers

In general, a large scale-mixing tank will have a lower pumping capacity per unit volume than a small tank.

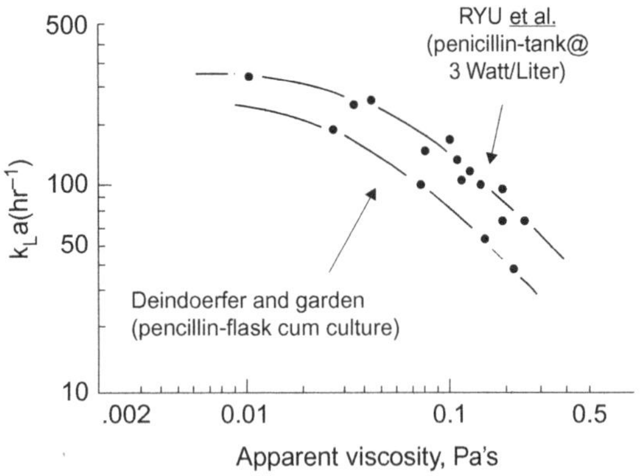

FIGURE 7.39 Illustration of the decrease in K_La with increase in apparent viscosity.

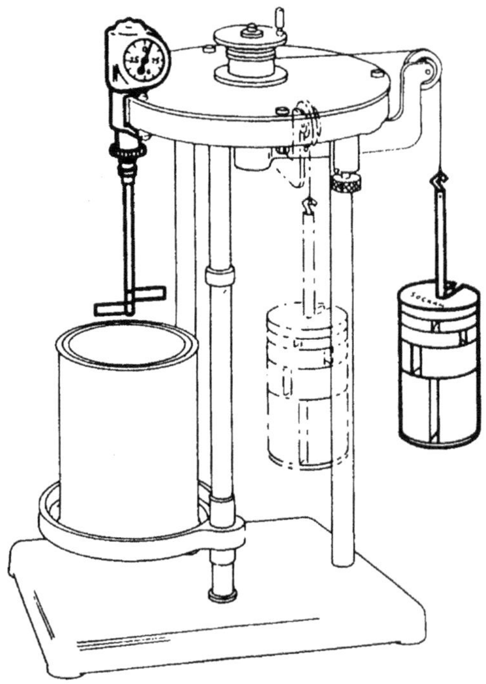

FIGURE 7.40 Schematic illustrating Stormer viscosimeter.

This means that its blend time and circulation time will be much larger than in a pilot tank.

There is also a tendency for the maximum impeller shear rate to go up while the average impeller zone shear rate will go down on scale up. In addition, the average tank zone shear rate will go down as will the minimum tank zone shear rate.

This means that there is a much greater variety of shear rates in the larger tank, and in dealing with pseudo plastic slurry it will have a quite different viscosity relationship around the tank in the big system compared to the smaller system.

Micro scale shear rates operate in the range of 300 microns or less, and are governed largely by the power input.

The power input from the gas per unit volume will increase on scale up. This is because there is a greater head pressure on the system, and there is also an increasing gas velocity.

It may be that the power level for the mixer may be reduced since the energy from the gas going through the tank is higher in order to maintain a particular mass transfer coefficient, K_Ga; however, this changes the relative power level compared to the gas and other mass transfer rates, such as the liquid-solid mass transfer rate. The capacity for the blending type flow pattern is not affected in the same way with changes in the mixer power level as is the gas-liquid mass transfer coefficient.

6.2 Scale up Based on Data from Existing Production Plant

If data are available on a fermentation in a production-size tank, increasing, in a relative proportion, the various mass transfer, blending and shear rate requirements for the full-scale system, may make scale up. For example, it may be determined that the new production system is to have a new mass transfer rate of $x\%$ of the existing mass transfer rates, and there may be specifications put on maximum or average shear rates, and there may be a desire to look at changes in blend time and circulation time. In addition, there may be a desire to look at the relative change in CO_2 stripping efficiency in the revised system.

At this point, there is no reason not to consider any size or shape of tank. Past tradition for tall, thin tanks, or short, squat tanks, or elongated horizontal, cylindrical tanks does not mean that those traditions must be followed in the future. To illustrate the principle involved in the gas-liquid mass transfer, look at Fig. 7.36 which gives the three different mass transfer steps commonly present in fermentation. The mass transfer rate must be divided by a suitable driving force, which gives us the mass transfer coefficient required. The mass transfer coefficient is then scaled to the larger tank size and is normally related to superficial gas velocity to an exponent, power per unit volume to an exponent, and to other geometric variables such as the D/T ratio of the impeller.

A thorough analysis takes a look at every proposed tank shape, looks at the gas rate range required, calculates the gas phase mass transfer driving force, and then calculates the required K_Ga to meet that. Reference is made to data on the mixer under the condition specified and to various D/T ratios to obtain the right mixer horsepower level for each gas rate.

At this point, the role of viscosity must be considered. Figure 7.39 shows the effect of viscosity on mass transfer coefficient. It is necessary to measure viscosity with a viscosimeter which mixes while it measures viscosity. Figure 7.40 illustrates the Stormer viscosimeter that is one device that can be used to establish viscosity under mixing conditions with shear rates that can be established.

In looking at the new size tank, estimates should be made of the shear rate profile around the system, and then using the relationship that viscosity is a function of shear rate, and the fact that it is shear stress

$$\text{Shear stress} = \mu(\text{shear rate})$$

that actually carries out the process, one can then estimate the viscosity throughout the tank, and the product of viscosity and shear rate to give the shear stress. Estimates can then be made of how different the proposed new tank may be compared to the existing known performance of the production tank.

One relationship that cannot be changed simply going from a small to large scale is the fact that the Reynolds number normally increases in the large tank over what it was in the small tank. The Reynolds number is typically anywhere from 10 to 50 times higher in the large vessel than in the small. This means that the fluid in the pilot scale will appear much more viscous in terms of flow pattern and many other parameters than it will in the full-scale tank. It is usually not practical where conducting a process to change viscosity between pilot plant and full scale, but if one is interested in getting an idea of the flow pattern and some of the macro scale effects, then a synthetic fluid of a lower viscosity than the actual could be substituted in the full scale work to give a better picture of the expected flow pattern.

At this point discussion of the quantitative and qualitative nature of available data is desirable. The user, production, research and engineering, and purchasing department should have discussions with the suppliers and technical personnel to arrive at satisfactory combinations of proposals.

6.3 Data Based on Pilot Plant Work

To keep ratios of impellers, gas bubbles and solid clumps in the fermentation related to full scale, the impeller size and blade width in the small scale must always have a physical dimension two or three times bigger than the particle size of concern.

It is possible to model the fermentation biological process from a fluid mechanics standpoint, even though the impeller is not related properly geometrically to the gas–liquid mass transfer step. Thus, one scale of pilot plant might be usable for one or two of the fermentation mass transfer steps, and/or chemical reaction steps, but might not be suitable for analysis of other mass transfer steps. The decision, then, is based on how suitable existing data are for any steps that are not modeled properly in the pilot plant.

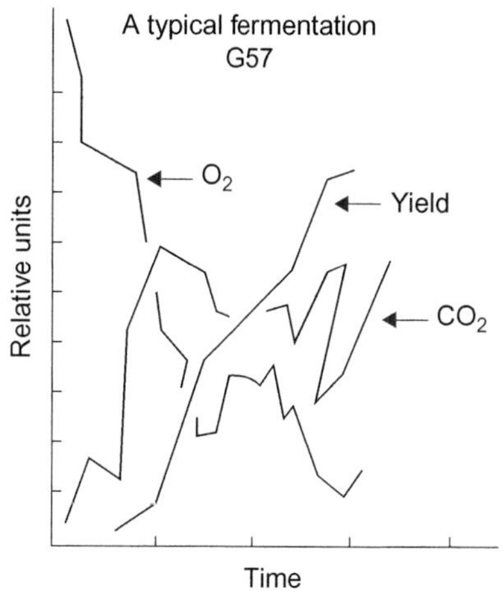

FIGURE 7.41 Schematic of typical data from fermentation showing the change in oxygen content of gas, CO_2 content in liquid and fermentation yield.

Ideally, data should be taken during the course of the fermentation about gas rate, gas absorption, dissolved oxygen level, dissolved carbon dioxide level, yield of desired product, and other parameters which might influence the decision on the overall process. Figure 7.41 shows a typical set of data for this situation.

If the pilot plant is to duplicate certain properties of fluid mixing, then it may be necessary to use non-geometric impellers and tank geometries to duplicate mixing performance and not geometric similarity. As a general rule, geometric similarity does not control any mixing scale-up property whatsoever.

It may also not be possible to duplicate all of the desired variables in each run, so a series of runs may be required changing various relationships systematically and then a synthesis made of the overall results.

One variable in particular is important. The linear superficial gas velocity should be run in a few cases at the levels expected in the full-scale plant. This means that foaming conditions are more typical of what is going to happen in the plant and the fermenter should always be provided with enough head space to make sure the foam levels can be adequately controlled in the pilot plant. As a general rule, foam level is related to the square root of the tank diameter on scale up or scale down.

In duplicating maximum impeller zone shear rates on a small scale, there may be a very severe design problem in the mechanics of the mixer, or the shaft speed, mechanical seals and other things. This means that careful consideration must be given to the type of runs to be made and whether the pilot plant or the semi-work-scale equipment must be available at all times to duplicate the maximum impeller zone shear rates in the plant or whether this sort of data will be obtained on a different type of unit dedicated to that particular variable.

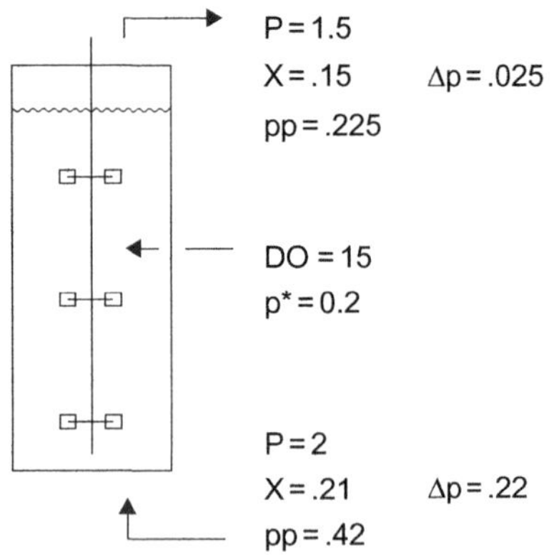

FIGURE 7.42 Typical driving force for larger fermenter.

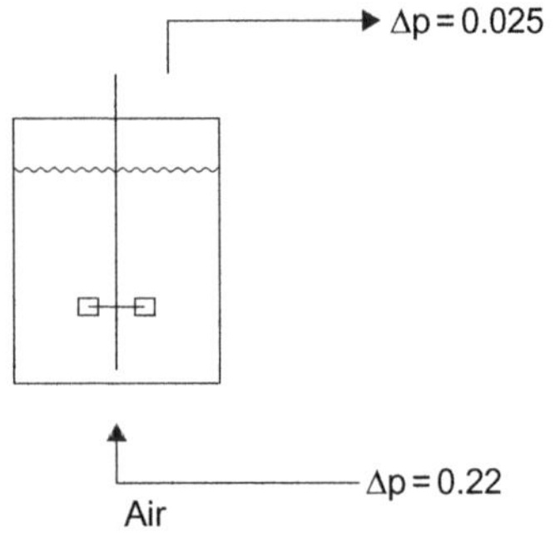

FIGURE 7.43 Driving force for small laboratory fermenter.

Figure 7.37 shows what often happens in the pilot plant in terms of correlating mass transfer coefficient, K_Ga, with power and gas rate in the pilot plant. This curve is then translated to a suitable relationship for full scale. It is possible to consider that with the higher superficial gas velocity, the power level may be reduced in the full scale to keep the same mass transfer coefficient. The box on the right in Fig. 7.37 shifts to the box on the left. This should be considered, but it should be borne in mind that this changes the ratio of the mixer power to gas power level in the system; changes the blend time; changes the flow pattern in the system; the foaming characteristics and also can markedly affect the liquid-solid mass transfer rate if that is important in the process.

In all cases, a suitable mass transfer driving force must be used. Figure 7.42 illustrates a typical case for fermentation processes and illustrates that there is a marked difference between the average driving force, the log-mean driving force, and the exit gas driving force. In a large fermenter, it is this author's experience that gas concentrations are essentially step-wise stage functions and a log-mean average driving force has been the most fruitful. Figure 7.43 illustrates a small laboratory fermenter with a *Z/T* ratio of 1, and in this case, depending on the power level, an estimate must be made of the gas mixing

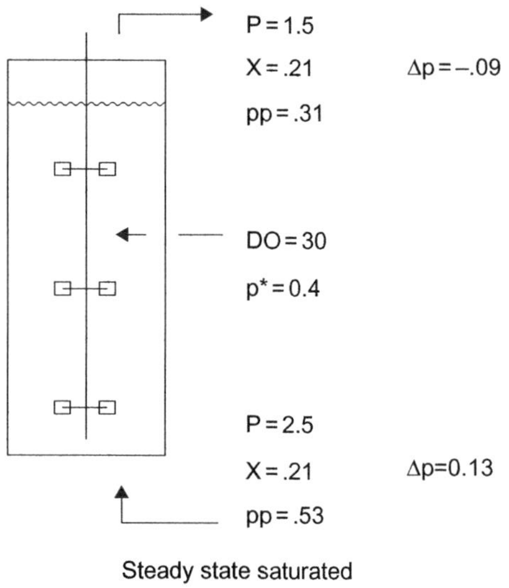

FIGURE 7.44 Driving force for unsteady state saturation run carried to equilibrium.

characteristics and an evaluation made of the suitability of the exit gas concentration for the driving force compared to the log-mean driving force. This is one area that needs to be explored in the pilot program and the calculation procedures.

Just to indicate another peculiarity in the waste-treating industry, it is quite common to run an unsteady state reaeration test in which the tank is stripped of oxygen; air is started with the mixer running and the dissolved oxygen level increase is monitored until the tank is saturated and no further mass transfer occurs. At that point, the DO level is usually between the saturation value at the top and the saturation value at the bottom condition, Fig. 7.44. This means that for steady state there must be enough absorption in the bottom and enough stripping at the top; a very peculiar mass transfer situation results compared to what is happening on any waste treatment production or fermentation mixer. Running experimental tests and basing a lot of calculations on that particular driving force would give markedly different results from those obtained with the mixer operating normally.

6.4 Sulfite Oxidation Data

There are data using excess sodium sulfite with suitable catalysts which keep the dissolved oxygen level at zero, and the data have been obtained on small and large size fermentation tanks on this basis. One caution is that the data should have been taken when the tank was completely clean of antifoams that may be residual from the fermentation process. This antifoam can cause marked differences in the mass transfer coefficient.

TABLE 7.3 Some Gas–Liquid Applications

Agitator Power (hp)	Fermenter Size (gal)	Power Levels	
		(hp/gal)	(kW/m^3)
100	2,000	50/1,000	9.9
900	40,000	23/1,000	4.5
3,500	100,000	35/1000	6.9
100	10,000,000	0.01/1,000	0.002

If someone has a relationship between the sulfite oxidation number and the performance required in the fermenter, this is a perfectly valid way to specify equipment and tests can be run to give an indication of the overall mass transfer rate ensuing.

6.5 Oxygen Uptake Rate in the Broth

If it is desired to relate fermenter performance to oxygen uptake rate in the broth, this number can be specified along with suitable desired gas rates, and the mixer estimated, based on this performance. Again, someone must have the link between this particular mass transfer specification and the actual performance of the fermenter.

If this number is based on pilot plant data, then the effect of the different shear rates and different blend times on the mass transfer relationship, viscosity and the resulting fermentation must be considered.

6.6 Some General Concepts

Table 7.3 gives some typical power levels used in various gas-liquid mixing operations, including waste treating and fermentation, to give some idea of the range of variables. Obviously, the tremendous variety of units used precludes an attempt to guess at a mixer based on general overall approximations.

The lower impeller does the major part of the work on dispersing gas in the system and it is typical practice today to put a high proportion of the power into the lower impeller, somewhat similar to what is shown in Fig. 7.35. Multiple impellers do have zoning action in terms of blending, which is not a great factor in a fermentation which takes seven days, but there are instantaneous differences in the mass transfer, blending and concentration profiles in a tank with multiple impellers.

In addition, the role of the lower impeller in both mass transfer and mixing must be considered and the desirability of having multiple impellers in the tank can be considered.

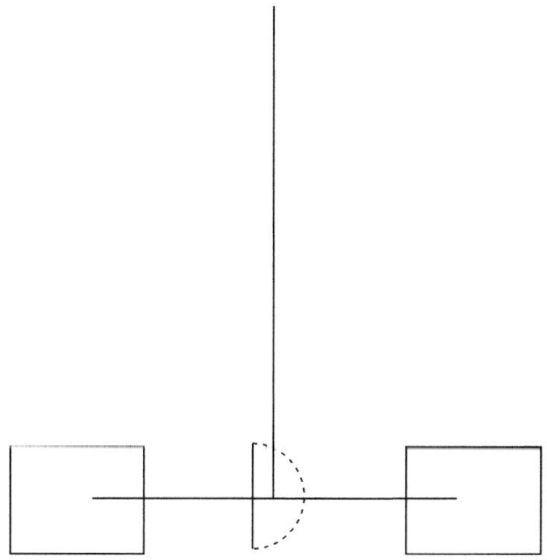

FIGURE 7.45 Typical dual power number impeller with streamlined back of blade.

If it is desired to consider axial flow impellers in a gas-liquid system for any reason, it should be remembered that the upward flow of gas tends to negate the downward action of the pumping capacity of the axial flow turbine. A radial flow turbine must have three times more power than the power in the gas stream for the mixer power level to be fully effective. On the other hand, the axial flow impeller must have eight to ten times more power than in the gas stream for it to establish the axial flow pattern.

6.7 Reverse Rotation Dual Power Impellers

In gas-liquid systems, one of the reasons that the power of the impeller is lower with the gas on than with the gas off is that the gas bubbles collect behind the impeller blade. This streamlines the blade, reducing power. Looking at Fig. 7.45, speculation can be made on what would happen if one were to fill up and streamline the back of the impeller with solid material. What happens is that with no gas rate, Fig. 7.46, the impeller draws less horsepower with the back of the blade streamlined than with the back of the blade flat.

When the gas is turned on, the flat impeller blade has a K factor, which is the ratio of impeller power with gas on to power with gas off, and changes markedly with gas rate, typical of impellers of that type, while the impeller with the streamlined back of the blade has much less change in power with gas rate.

The schematic relationship shown in Fig. 7.46 gives a wide variety of power consumption availabilities without gas and with gas by having the mixer and motor capable of being reversed electrically, and opens up a wide variety of process options.

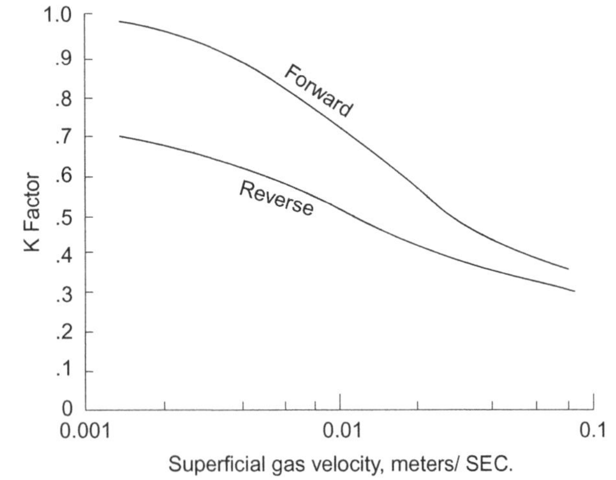

FIGURE 7.46 Power characteristics of dual power number impellers.

7.0 FULL SCALE PROCESS EXAMPLE

There is no way that a mixer can transfer oxygen from gas to liquid any faster than the solid microorganisms can utilize the oxygen in their growth process. If the mixer is capable of supplying the oxygen faster than the organisms can use it, the main effect will be to increase the dissolved oxygen level, C, to balance out the mass transfer equation

$$\text{O.U.R} = K_L a(C^* - C) = K_s(C - C_s{}^*)$$

and the dissolved oxygen level may or may not have an effect on the growth process.

On the other hand, if the organism can utilize oxygen faster than the aerator can provide it, the dissolved oxygen will tend toward zero, although this may affect the resulting oxygen demand of the organism and bring the two demands even more closely into balance.

It is normally helpful to break the fermentation process down into several distinct steps and examine the role that mixing plays in these various steps. Then, the total effect on the process result from the combination of these different steps can be examined.

One of the first requirements is to get a measure of the effectiveness of the existing mixer in the process. This section takes the perspective that there is an existing full-scale fermenter that is carrying out a certain process. The basic questions covered here are: *(i)* what is the role of mixing in this particular process? *(ii)* what are the possible advantages and disadvantages of increasing the mass transfer ability of the agitator to take advantage of the maximum potential of the present strain of microorganism in the process? and *(iii)* what is the potential advantage of providing a mixer that will provide adequate mass transfer for both an increased productive strain at the same cell concentration, or will provide proper oxygen mass transfer at an increased cell concentration?

In looking at the performance of a mixer in a tank with a particular starting concentration of microorganisms, it is possible to determine the kinetics of the antibiotic production that produces the growth of the microorganisms throughout the process. Typical data is shown in Fig. 7.41 previously.

One factor that can add considerable confusion to the analysis is the observation that understimulation or overstimulation of the growth rates of microorganisms in their initial and log growth phase can change their ability to produce antibiotic at the maximum yield point in the cycle and affect the ultimate total yield obtained at the end. It is entirely possible that increasing the mass transfer rate available to the fermentation can have a detrimental effect on total yield because it changes the metabolic situation in the organisms during the first few hours of fermentation, which affects their ultimate potential for total yield.

This effect must be carefully distinguished in analyzing the use of a higher mass transfer ability agitator, which can take advantage of increased respiration requirements of new improved strains or higher cell concentrations during the total cycle.

It is also common that fermentations made in different parts of the world, although supposedly somewhat similar, because of inherently different conditions of processing can give different results in equipment that is quite similar.

The use of higher mixer mass transfer abilities can be examined in two ways:

a. The effect it has on a given type and concentration of starting seed, which includes biomass growth rates and total yield.
b. The effect it has on production from a new, more productive strain or an increased initial seed concentration.

8.0 THE ROLE OF CELL CONCENTRATION ON MASS TRANSFER RATE

Within a given batch run, the cell concentration changes as a function of time. In addition, the viscosity goes up with cell concentration at a given point on the time curve of a fermenter. Figures 7.47 and 7.48 give typical data showing the change in viscosity as a function of the number of days of fermentation for different kinds of systems.

On the other hand, Fig. 7.49 shows the change in mass transfer rate with viscosity, which is caused largely by a change in cell concentration of the total process. It is true that the rate of oxygen transferred per MJ goes down as the cell concentration goes up. However, this cost must

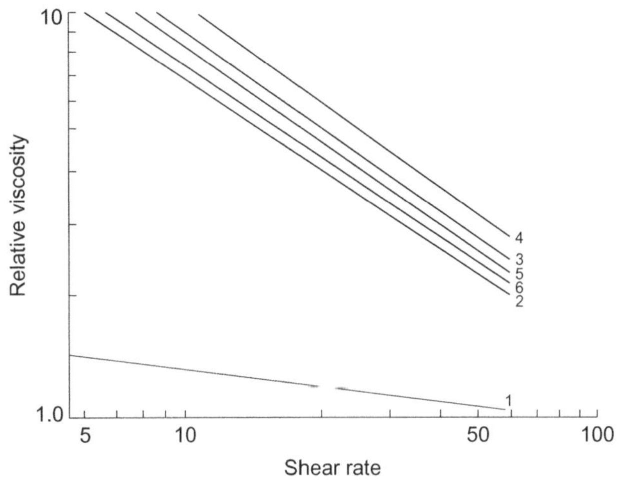

FIGURE 7.47 Viscosity at various shear rates for various days in the fermentation cycle.

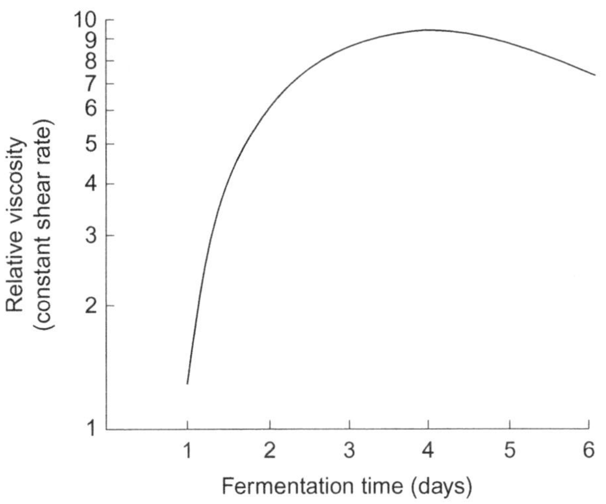

FIGURE 7.48 Viscosity at constant shear rates for various days in the fermentation cycle.

be balanced against the increasing productivity of a given dollar investment in fermentation tank, piping and total plant cost. Analysis needs to be made of the role that mixer cost, including both capital and power, plays in the total productivity cost in order to evaluate desirability of going in this direction.

A previous paper by Ryu and Oldshue treated an example where the final cell concentration was changed from 10 to 12 to 20 g/l, and the oxygen mass transfer dropped from 10 to 8.3 to 6.4 mols oxygen/MJ.

Looking at Table 7.4, the cost of electrical power and other essentials listed a capital cost of $900/kW (1982 cost about $2000/kW) if installed mixer capacity is used, including the associated blower and air supply, and including the installation of the equipment, with the electrical hookup. This is for a *D/T* ratio of 0.37.

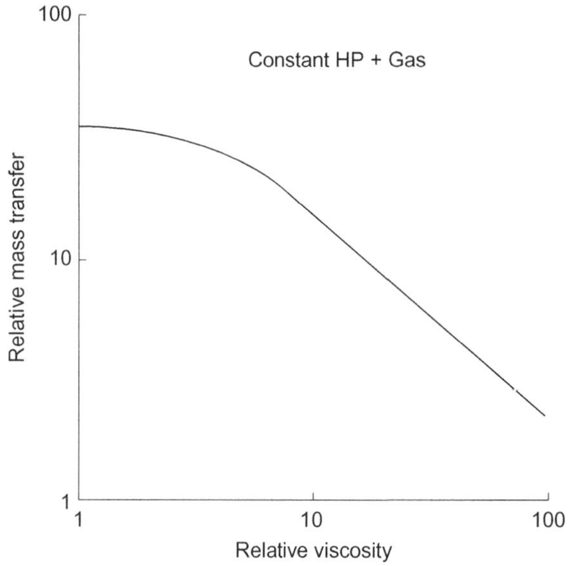

FIGURE 7.49 Decrease in mass transfer rate with viscosity for a given mixer and air rate.

TABLE 7.4 Cost of Mixing for Production of Antibiotics (Based on Oxygen Transfer Rate)

Cost of electrical power	**0.70¢/MJ**
Equipment Cost (expressed as power cost)	0.8¢/MJ
Efficiency of oxygen mass transfer	
Dilute system, 10 g/l*	10 mols O_2/MJ
More concentrated, 20 g/l*	6.4 mols O_2/MJ
Power and Equipment cost	1.5¢/MJ
Cost of dissolved oxygen	
10 g/l	0.15¢/mol O_2
20 g/l	0.23¢/mol O_2
Production cost of antibiotics	46¢/kg
Fractional cost for mixing (antibiotics production)	0.6–1.6%**
Production Yield	1 kg/200 mols O_2

**Cell concentration.*
***Of production cost.*

Electrical power is assumed at 0.7¢/MJ (to obtain ¢/kW-hour multiply by 3.6). The equipment is amortized, using present worth, over a 5-year period, which results in a figure of 0.8¢/MJ. Total cost of the equipment and operation is therefore 1.5¢/MJ. The cost of dissolving oxygen is 0.15¢/mol O_2 dissolved at 10 g/l. At 20 g/l, it is 0.23¢/mol O_2.

TABLE 7.5 Comparison of New Mixer to Original Mixer

	Original Low Power Mixer (A)	New High Power Mixer			
		(D)	(E)	(F)	(G)
Agitation Power, kW	150	380	380	380	380
Aeration Power, kW	37	95	95	95	95
Relative product yield (arbitrary units)	100	200	180	200	200
Cell concentration, g/l	10	20	20	20	20
Oxygen uptake rate, g O_2/hr	0.5	1.0	0.9	1.0	1.0
Maximum available oxygen transfer rate, g O_2/l/hr	0.7	1.1	1.1	1.1	1.1
Fixed fermentation cost, $/kg	30	30	30	30	30
Variable ferm. cost, $/kg	30	15.2	16.67	15.16	15.34
Total production cost, $/kg of product	60	45.2	46.7	45.16	45.34
Cost of oxygen transfer operation (mixing equipment, power), $/kg	0.42	0.52	0.57	0.58	0.72
Present cost of oxygen transfer operation	0.7	1.1	1.2	1.3	1.6
Present cost savings	–	25	22	25	24.5
Maximum impeller zone shear rate (relative)	1.00	1.30	1.30	1.15	1.00

Assuming that it takes 200 mols of oxygen to produce 1 kg of product and that there is a production cost of $60 total per kg of product, the percent cost of oxygen in the dilute system is approximately 0.7% of the total production cost. There is also assumed in this example that there is a fixed cost of $30/kg which does not change with the agitator, and that the variable fermentation cost goes down as the productivity of the particular tank in the process increased.

This is listed in Column A of Table 7.5. Column D gives the results from the paper by Ryu and Oldshue, which described the use of a 500 hp mixer operating at 20 g/l. While the percent of cost due to the mixer has increased, the total production cost per kg of product has gone down 25% to a value of approximately $45.2/kg.

Assume that this higher power mixer, having a maximum impeller zone shear rate 30% higher than Case A, decreased the growth ability of the microorganisms due to increased shear on these particles, changing the floc structure, etc. Further assume that this cut the production of penicillin to 90% of the value it could have had based on cell concentration only. This means that the mixer is producing less product than Column D would indicate, and the production costs have gone up to $46.7/kg (Case E), because all the additional capacity of the larger aerator cannot be used. It can be seen that the aerator has given the ability to transfer 1.1 g O_2/l/hr, in contrast to the 200 hp unit value of 0.7 g O_2/l/hr.

Assume that studies in the laboratory indicate that if the shear rate is cut down to where it is only 15% higher than Case A, then the organism retains its growth potential. This mixer in Case F has a *D/T* ratio 40% higher and therefore, instead of $900/kW, costs $1200/kW, including the associated blower. Putting this into the cost example, even though it changes drastically the initial cost of the equipment, the productivity is improved to the point that the actual production cost is approximately $45.2/kg as it was in Case D.

If studies indicate that the shear rate has to be cut back to the same as it was in Case A, this means the mixer cost is now $1575/kW because of the increased torque and *D/T*, and it does raise the production cost up to $45.3 (Case G), but is still a very small percentage of the total production cost, and is a very small percentage in terms of mixer cost of the total production.

The main point here is that in this particular example, mixer horsepower and capital cost can effect tremendous changes in productivity because of their low cost in terms of the total cost.

9.0 SOME OTHER MASS TRANSFER CONSIDERATIONS

Desorption of CO_2 is an essential part of effective fermentation. The pressure and liquid depth that enhances absorption of oxygen discourages desorption of CO_2. Tall, thin tanks with the same volume of air, yielding a higher superficial velocity, normally give more pounds of oxygen transfer per total horsepower of mixer in air than do short, squat tanks. There also is less absorption of CO_2 under the same conditions. Therefore, some idea of the role of CO_2 desorption rates, backpressure of CO_2 and other things must be obtained in order to evaluate this particular phenomenon. In addition, the fluid mixing pattern in the fermenter must be considered. As broth becomes more viscous, and tanks become taller, more impellers are used and there is a possibility of much longer top to bottom blending times being involved which do affect the dissolved CO_2−oxygen level throughout the system. In general, the dissolved CO_2−oxygen level will assume some value intermediate between the values that would be predicted based on concentration driving forces at the bottom and the top due to the gas stream.

10.0 DESIGN PROBLEMS IN BIOCHEMICAL ENGINEERING

1. A mixer applying 150 kW to the mixer shaft is operating in a batch fermentation at a cell concentration of 20 g/l. Associated with the mixer is a blower, which is providing air at a total expansion horsepower of 37 kW leaving the sparge ring. The cost of power is 0.5¢/MJ. Use an overall energy efficiency for the equipment of 0.9.

   The cost of the mixer plus the associated blower required, plus installation of both is $900/kW with an impeller diameter to tank diameter ratio of 0.35.

   By using large diameter impellers at slower speeds, the maximum impeller zone fluid shear rate can be changed, and the cost of the mixer/kW must be changed accordingly. The cost of the mixer can be approximated to change inversely proportional to the maximum impeller zone fluid shear rate to the 2.3 exponent

   $$\text{Cost} \propto (\text{MIZSR})^{-2.3}$$

   The particular antibiotic requires 200 mols of oxygen for each kg of product produced. The total production cost of the antibiotic is estimated as $60/kg, of which $30 is a fixed cost, independent of productivity of the fermenter, and $30 is the cost associated with the actual fermentation tank itself.

   At 20 g/l cell concentration, the mixer is capable of transferring 6.4 mols of oxygen per MJ.

   It is proposed to increase the solids concentration in the system to 40 g/l, which will effectively double the productivity of the fermentation tank itself. The oxygen transfer activity of the mixer is lowered, due to the increased viscosity, to 4 mol of oxygen per MJ.

   Assuming that the mixer is operated for 250 days per year, and using a 5-year evaluation period, calculate the cost of mixing, capital and operating, in this process, and the percentage cost of mixing under the present operation.

   At 40 g/l of cells, the process horsepower required may be estimated as being proportional to the cell concentration to the 1.4 exponent

   $$(\text{Process mixer hp}) \propto (\text{cell concentration})^{1.4}$$

   This takes into account the transfer rate due to the change in viscosity and the additional transfer rate needed because of the increase in total biomass in the system.

   Calculate the cost of mixing in the new revised system at 40 g/l, the total mixer horsepower (air horsepower is in the same proportion as at 20 g/l), and reduction in antibiotic production cost.
2. The new larger mixer has a higher maximum impeller zone shear rate, which is estimated at 1.4 times as high as a small unit at the same D/T ratio. Assume that this higher shear rate has cut the productivity of the increased cell concentration to 90% of its normal value. In this case, assuming that the mixer is not changed, calculate the cost of mixing and the percent mixing cost/total product cost, and the savings compared to the original 150 kW mixer.
3. A new mixer has been designed at the same total horsepower, but with a shear rate 1.2 times as high as the previous 150 kW unit.

   Calculate the new capital cost of mixing and calculate the total mixing cost and increase in productivity over the original smaller unit.
4. A large diameter impeller at a slower speed would require a larger mixer drive, to reduce the shear rate to the same as it was in the original 150 kW unit. Again calculate the cost of the mixer, mixing cost per MJ and calculate the percent mixing cost/total product cost, as well as the percent savings over the original 150 kW mixer.

11.0 SOLUTION—FERMENTATION PROBLEMS

Problem 1

$$0.5\,¢\,\text{MJ} \times \frac{187}{150} \times \frac{1}{0.9} = 0.7\,¢/\text{MJ}$$

$$\frac{900 \times 100}{3.6 \times 5 \times 250 \times 24} = 0.8\,¢/\text{MJ}$$

$$\text{Total} = 1.5\,¢/\text{MJ}$$

$\frac{200}{6.4} = 31.3$ MJ/kg of product

This yields 47¢/kg

$\frac{0.47}{60} = 0.8\%$ cost of mixing

$\left(\frac{40}{20}\right)^{1.4} = 2.6$

$150 \times 2.6 = 390$ kW mixer

$37.0 \times 2.6 = 96.2$ kW air

$\frac{200}{4} = 50$ MJ/kg product

$= 75$¢/*kg*

$15.00 + (0.75 − 0.47) = $15.28 kg − Operating cost

+30.00 − Fixed cost

$45.28 − Total cost

Mixing cost $= \frac{0.75}{45.28} = 1.6\%$

Cost savings $= \frac{60 - 45.28}{60} = 25\%$

Problem 2

$\frac{75¢}{0.9} = 83¢/\text{kg}$

$\frac{\$30/\text{kg}}{1.8} + (0.83 - 0.47) = \$17.02$ − Operating cost

+30.00 − Fixed cost

$47.02 kg − Total cost

$\frac{0.83}{47.02} = 1.8\%$ Cost of mixing

$\frac{\$60 - \$47.02}{\$60} = 22\%$ Cost saving

Problem 3

Mixer cost $= \left(\frac{1.2}{1.4}\right)^{2.3} = (1.17)^{2.3}$

$= 1.4$

0.8¢/MJ × 1.4 = 1.12

+0.7

1.82¢/MJ

$\frac{200}{4} = 50$ MJ/kg

$50 \times 1.82 = 91$¢/kg

$15.00 + (0.91 − 0.47) = $15.44

+30.00

$45.44

$\frac{0.91}{45.44} = 2\%$ Cost of mixing

$\frac{60 - 45.44}{60} = 24\%$ cost saving

Problem 4

Cost $= (1.4)^{+2.3} = 2.17$

2.17 × 0.8 = 1.74¢/MJ

+0.7

2.44¢/MJ

2.44 × 50 = $1.22/kg

$15 + (1.22 − 0.47) = $15.75

+30.00

$45.75

$\frac{1.22}{45.75} = 2.7\%$ Cost of mixing

$\frac{60 - 45.75}{60} = 24\%$ cost saving

LIST OF ABBREVIATIONS

D	Impeller diameter
K_La	Gas-liquid mass transfer coefficient based on partial pressures
K_Ga	Gas-liquid mass transfer coefficient based on liquid concentrations
MJ	Megajoule
N	Impeller
P	Power
p	Pressure
p^*	Equilibrium partial pressure
Q	Impeller pumping capacity
S.R.	Shear Rate
S.S.	Shear Stress
μ	Viscosity
W	Width of square or rectangular tank
F	Superficial gas velocity, $\frac{\text{total air}}{\text{cross-section area of tank}}$
K_s	Liquid solid coefficient

C*	Equilibrium oxygen concentration corresponding to partial pressure in air stream
C	Liquid oxygen concentration
DO	Dissolved oxygen concentration
H	Impeller head
K	Factor ratio of horsepower with gas to power with gas off at constant speed
Z	Liquid level
T	Tank diameter
O.U.R.	Oxygen Uptake Rate
MIZSR	Maximum impeller zone shear rate
g/l	Grams per liter
kWh	Kilowatt-hour
kW	Kilowatt
CI	Center Inlet gas introduction
kg	Kilogram

FURTHER READING

[1] F.H. Deindoerfer, E.L. Gaden, Appl. Microbiol. 3 (1955) 253.

[2] J.Y. Oldshue, Fermentation mixing scale-up techniques, Biotech. Bioeng. VIII (1966) 3–24.

[3] J.Y. Oldshue, Suspending solids and dispersing gases in mixing vessels, Ind. Eng. Chem. 61 (1969) 79–89.

[4] J.Y. Oldshue, Spectrum of fluid shear rates in mixing vessel, Chemeca '70 Australia, Butterworth, 1970.

[5] J.Y. Oldshue, C.K. Coyle, G.L. Connelly, Gas-liquid contacting with impeller mixers, Chem. Eng. Prog. (1977) 85–89.

[6] J.Y. Oldshue, C.K. Coyle, et al., Fluid mixing in the optimization of fermentation production, Process Biochem. 13 (11) (1978) England.

[7] D.Y. Ryu, J.Y. Oldshue, A reassessment of mixing costs in fermentation processes, Biotech. Bioeng. XIX (1977) 621–629.

Part III

Recovery

Chapter 8

Filtration

Celeste M. Todaro

1.0 INTRODUCTION

The theoretical concepts underlying filtration can be applied towards practical solutions in the field. Comprehension of the basic principles is necessary to select the proper equipment for an application.

Theory alone, however, can never be the basis for selection of a filter. Filtration belongs to the physical sciences, and thus conclusions must be based on experimental assay. It is, however, helpful in understanding why a slurry is more suitable for one design of filtration equipment than another. Methods of optimization in the field can also be predicted by having a background in the theory.

Slurries vary significantly in filtration characteristics. Even batch to batch variation in product particle size distribution and slurry concentration will greatly influence filterability and capacity of a given filter. It is, therefore, essential to evaluate a slurry in laboratory tests at a vendor's facility or at one's plant with rental equipment to prove the application.

There are three (3) types of pharmaceutical filtrations: depth, cake, and membrane. Cake and depth are coarse filtrations, and membrane is a fine, final filtration. Membrane filtration and cross-flow filtration are discussed in Ch. 7.

1.1 Depth Filtration

Examples of depth filtration are sand and cartridge filtration. Solids are trapped in the interstices of the medium. As solids accumulate, flow approaches zero and the pressure drop across the bed increases. The bed must then be regenerated or the cartridge changed. For this reason, this method is not viable for high solids concentration streams as it becomes cost prohibitive. Cartridge filtration may be used as a secondary filtration in conjunction with a primary, such as cake filtration.

2.0 CAKE FILTRATION

Rates of filtration are dependent upon the driving force of the piece of equipment chosen and the resistance of the cake that is continually forming. Liquid flowing through a cake passes through channels formed by particles of irregular shapes.

3.0 THEORY

3.1 Flow Theory

Flow rate through a cake is described by Poiseuilles' equation:

$$\frac{dV}{Ad\theta} = \frac{P}{\mu\left[\alpha\left(\frac{W}{A}\right) + r\right]} \quad (8.1)$$

V = volume of filtrate
A = filter area surface
θ = time
P = pressure across filter medium
α = average specific cake resistance
w = weight of cake
r = resistance of the filter medium
u = viscosity

In other words,

$$\frac{\text{Flow Rate}}{\text{Unit Area}} = \frac{\text{Force}}{\text{Viscosity[Cake Resistance + Filter Medium Resistance]}}$$

3.2 Cake Compressibility

The specific cake resistance is a function of the compressibility of the cake.

$$\alpha' = \alpha' P^s \quad (8.2)$$

where

$$\alpha' = \text{constant}$$

As s goes to 0 for incompressible materials with definite rigid crystalline structures, α' becomes a constant.

Fermentation and Biochemical Engineering Handbook.

For the majority of products, resistance of the filter medium is negligible in comparison to resistance of the cake, thus Eq. (8.1) becomes

$$\frac{dV}{d\theta} = \frac{AP}{\mu\alpha(W/A)} \quad (8.3)$$

Incompressible cakes have flow rates that are dependent upon the pressure or driving force on the cake. In comparison, compressible cakes, i.e., where s approaches 1.0, exhibit filtration rates that are independent of pressure as shown below.

$$\frac{dV}{d\theta} = \frac{A}{\mu\alpha(W/A)} \quad (8.4)$$

The above equations are detailed in *Perry's Chemical Engineer's Handbook* [1].

Compressible cakes are composed of amorphous particles that are easily deformed with poor filtration characteristics. There are no defined channels to facilitate liquid flow as in incompressible cakes.

Fermentation mashes are typical applications of compressible materials, usually having poor filterability in contrast to purified end products that are postcrystallization. These products precipitate from solutions as defined crystals.

4.0 PARTICLE SIZE DISTRIBUTION

Modification and optimization of a slurry, whether amorphous or crystalline, in the laboratory can yield significant improvements in filtration rates. By modeling the process in the laboratory, one can model what is occurring in the plant.

It is evident that attention paid in the laboratory to the factors affecting particle size distribution will save on capital investments made for separation equipment and downstream process equipment. Specific cake resistance (α) can be determined in the laboratory over the life of a batch, to evaluate if time in the vessel and surrounding piping system is degrading the product's particle size to the point it impedes filtration, washing and subsequent drying.

Factors such as agitator design, agitation rates, pumps, slurry lines and other equipment, which can unnecessarily reduce the particle size, should be taken into consideration. Increasing the particle size in the slurry, and narrowing the particle size distribution will result in increased flow rates. Large variations in particle size will increase the compressibility of a cake per unit volume. Since small particles have greater total cumulative surface areas, they will have higher moisture contents. For example, flour and water, when filtered with the same pressure or driving force as sand and water, will have a higher residual moisture level, thereby increasing the downstream dryer size.

In the plant, the type of pump and piping system used to feed the filter are often of great importance, as time spent on crystallization and improving crystal size and particle size distributions can be quickly undone through particle damage. Recirculation loops and pumps for slurry uniformity may not always be necessary.

A review of the most commonly used process pumps are discussed below:

Diaphragm pumps. These offer very gentle handling of slurries and are inexpensive and mobile. However, the pulsating flow can cause feeding and distribution problems in some types of filtration systems, e.g., conventional basket centrifuges. They can also interfere with process instrumentation e.g., flowmeters and loadcells.

Centrifugal pumps. Probably the most common source of particle attrition problems is the centrifugal pump. The high shear forces inherent to these pumps, particularly in the eye of the impeller, make some crystal damage inevitable in all but the toughest crystals. This damage is exacerbated on recirculation loops, which involve multiple passes through the pump. Recessed impellers will reduce this damage, but will often still degrade particles to the point where filtration becomes very difficult.

Positive displacement pumps. The minimal shear operation of progressing cavity or lobe pumps make them ideal for slurries. The non-pulsating flow is beneficial in most processes, but they are significantly more expensive and less portable than diaphragm pumps.

Additionally, a significant amount of attrition can be caused by the particles "rubbing" against each other. Therefore, long lengths of pipe, 90° elbows, throttling valves, control valves, and restrictions of any kind, should be avoided where possible. However, the type of pump employed is usually more significant.

If the feed vessel can be mounted directly above the filter (to reduce the possibility of blockages), then gravity feeding with some pressure in the vessel is normally the best and least expensive arrangement. Minimal shear agitators should be used at speeds sufficient to enhance the solids in the slurry and provide uniformity. Unnecessarily high speeds here can degrade the particles.

The "harder" the crystal, the more brittle and easier to break. Particle shape will also play a part, i.e., spherical crystals don't break easily, needles do, etc.

In general, this will lessen the problem of particle size deterioration and the fewer lines and shorter runs will reduce pluggage.

5.0 OPTIMAL CAKE THICKNESS

As the cake thickness of a product varies, filtration rates and capacity will also change. Equation (8.4) shows that rates increase as the cake (W/A) mass decreases; thus, thin cakes yield higher filtration rates. This is particularly the case with amorphous materials or materials with high specific cake resistance. As α' increases, maximizing $dV/d\theta$ requires W/A to decrease.

In continuous operations this can be done easily. In batch operations however, often filtration equipment cannot efficiently operate with extremely thin cakes. The long discharge times required to remove residual product in preparation for the next cycle, etc., make operation at a product's optimal cake thickness inefficient. Thus, if it requires a significant portion of the cycle time to unload the solids and only a 1/4″–1/2″ of cake is in the equipment, the effective throughput will be reduced, compared to operating with a cake thickness of 3–4 inches or greater.

6.0 FILTER AID

For amorphous materials, sludges or other poor filtering products, improved filtration characteristics and/or filtrate clarity are enhanced with the use of filter aids. Slurry additives such as diatomaceous silica or perlite (pulverized rock), are employed to aid filtration. Diatomite is a sedimentary rock containing skeletons of unicellular plant organisms (diatoms) [2]. These can also be used to increase porosity of a filter cake that has a high specific cake resistance.

$$\text{Porosity} = \frac{\text{Volume of Voids}}{\text{Volume of Filter Cake}}$$

Addition of filter aid to the slurry, in the range of 1–2% of the overall slurry weight, can improve the filtration rates. Another rule of thumb is to add filter aid equal to twice the volume of solids in the slurry. By matching the particle size distribution of the filter aid to the solids to be filtered, optimum flow rates are achieved. One should also use 3% of the particles, above 150 mesh in size, to aid in filtration [3].

Precoating the filter medium prevents blinding of the medium with the product and will increase clarity. Filter aid must be an inert material, however, there are only a few cases where it cannot be used. For example, waste cells removed with filter aid cannot be reused as animal feed. Filter aid can be a significant cost, and therefore, optimization of the filtration process is necessary to minimize the addition of filter aid or precoat. Another possible detriment is that filter aid may also specifically absorb enzymes.

A typical application for these filter aids is the filtration of solids from antibiotic fermentation broths, where the average particle size is 1–2 microns and solids concentration are 5–10%. Being hard to filter and often slimy, fermentation broths can also be charged with polymeric bridging agents to agglomerate the solids, thereby reducing the quantities of filter aid required.

7.0 FILTER MEDIA

Filter media are required in both cake filtration and depth filtration. Essential to selection of a filter medium is the solvent composition of the slurry and washes, and the particle size retention required of the solids.

Choice of the fabric, i.e., polypropylene, polyester, nylon, etc., is dependent upon the resistance of the cloth to the solvent and wash liquor used. Chemical resistance charts should be referenced to choose the most suitable fabric. The temperature of the filtration must also be considered.

Fabrics are divided into three different types of yarns: monofilament, multifilament, and spun. They can be composed of more than one of these types of fabric. Monofilaments are composed of single strands woven together to form a translucent or opaque fabric. Very smooth in appearance, its weave is conducive to eliminating blinding problems.

Multifilament cloths are constructed of a bundle of fibers twisted together. Only synthetic materials are available in this form, since long continuously extruded fibers must be used. Spun fabrics are composed of short sections of bound fibers of varying length. Retention of small particles is increased as the number of fibers or filaments in a bundle increases. The greater the amount of twist in the yarn, the more tightly packed the fabric, which contributes to retention. This twist will also increase the weight of the fabric and frequently extends filter cloth lifetime.

Polyester, nylon and polypropylene are common materials found in monofilament, multifilament and spun materials. Natural fibers such as cotton and wool are found only as spun material. This results in a fuzzy appearance. The effect of the type of yarn on cloth performance is shown in Table 8.1.

There are also combination materials, mono/multifilment for example. These fabrics offer higher permeabilities while maintaining particle retention. They can often resist blinding in some types of equipment.

Three fabric types are available, i.e., woven, nonwoven and knit. Woven fabrics are primarily what is used industrially. Yarns are laid into the length and width at a predetermined alignment. The width is called the *fill* direction and the length, the *warp* direction. They are at 90° angles and usually the yarn count in the warp direction is the higher figure [4].

Different weaving patterns of these materials will also vary cloth performance. Plain, twill and satin weaves are three of the most common. Their effect on cloth performance is shown in Table 8.2.

A nonwoven material, for example, would be a felt. They are pads of short nonrandom fibers, made of rigid construction suitable for many types of filtration equipment.

TABLE 8.1 Effect* of Type of Yarn on Cloth Performance [4]

Maximum Filtrate Clarity	Minimum Resistance To Flow	Minimum Moisture In Cake	Easiest Cake Discharge	Maximum Cloth Life	Least Tendency To Blind
Spun	Monofil	Monofil	Monofil	Spun	Monofil
Multifil	Multifil	Multifil	Multifil	Multifil	Multifil
Monofil	Spun	Spun	Spun	Monofil	Spun

*In decreasing order of preference.
(Courtesy of Clark, J. G., *Select The Right Fabric,* Chemical Engineering Progress, *November 1990.*)

TABLE 8.2 Effect* of Weave Pattern on Cloth Performance [4]

Maximum Filtrate Clarity	Minimum Resistance To Flow	Minimum Moisture In Cake	Easiest Cake Discharge	Maximum Cloth Life	Least Tendency To Blind
Plain	Satin	Satin	Satin	Twill	Satin
Twill	Twill	Twill	Twill	Plain	Twill
Satin	Plain	Plain	Plain	Satin	Plain

*In decreasing order of preference.
(Courtesy of Clark, J. G., *Select the Right Fabric,* Chemical Engineering Progress, *November 1990.*)

The particle size distribution of the material and the clarity required will dictate the micron retention of the medium. Fabrics tend to have a nominal micron retention range as opposed to an absolute micron retention rating. When using precoat on a machine that leaves a residual heel of solids, a more open cloth can be used.

As discussed in the theory section of this chapter, the filter medium is an insignificant resistance to flow, in comparison to the cake. However, if the filter medium retains a high amount of fines, the subsequent cake that builds up becomes more resistant to filtration, thus the degree of clarity required in the filtrate can be a trade-off to capacity.

Air permeability is a standard physical characteristic of the medium's porosity and is defined as the volume of air that can pass through one square foot filter medium at 1/2 inch water column pressure drop of water pressure [4]. Increasing air permeability often decreases micron retention, but doesn't necessarily have to. Two materials with the same air permeability can have different micron retentions. Weave pattern, yarn count (threads/inch), yarn size, etc., all contribute to retention. Heat treating or calendaring a material will also influence the permeability as well as the micron retention. Filter cloth manufacturers can provide assistance in fabric selection as well as information on fabric permeability and micron retention.

8.0 EQUIPMENT SELECTION

More than one equipment design may be suitable for a particular application. Often the initial approach is to replace it in kind. However, it is wise to evaluate the features of the present unit's operation in light of the process requirements and priorities. For example, is it labor intensive? Are copious volumes of wash required?

Ever-increasing environmental concerns may make it necessary to evaluate the existing process to reduce emissions, operator exposure, limit waste disposal of filter aid, or reduce wash quantities requiring solvent recovery or wash treatment. Breakdown of an old piece of equipment often provides the opportunity and justification to improve plant conditions. New "grass roots" designs may have the tendency to revert to industry standards. This is also the opportunity to improve conditions or substantiate the current equipment of choice.

8.1 Pilot Testing

Various small scale test units and procedures are available to determine slurry characteristics and suitability for a particular application. Buchner funnel, and vacuum leaf test units can be purchased or rented from vendors to perform in-house tests, or one can have tests conducted at the vendor's facility. Pilot testing on the actual equipment would be the optimum with a rental unit in the plant. In either case, slurry integrity must be maintained to ensure accurate filtration data.

Slurry taken fresh from the process in-house will yield the best results as product degradation over time, process temperature, effects of process agitators, pumps, etc., must be taken into consideration when shipping product

TABLE 8.3 Guide to Filter Selection

Slurry Characteristics	Fast Filtering	Medium Filtering	Slow Filtering	Very Dilute	Dilute
Cake Formation	in/sec.	in/sec	0.05 to 0.25 in/min	0.05 in/min	no cake
Normal concentration	20%	10 to 20%	1 to 10%	5%	0.1%
Settling rate	rapid, difficult to suspend	fast	slow	slow	
Leaf test rate, lb/hr/sq ft	500	50 to 500	5 to 50	5	–
Filtrate Rate, gal/min/sq ft	5	0.2 to 5	0.01 to 0.02	0.01 to 2	0.01 to 2
Filter Application					
Continuous Vacuum					
Multicompartment drum					
Single-compartment drum					
Dorrco					
Hopper dewaterer					
Top feed					
Scroll-discharge Tilting-pan					
Belt-discharge					
Continuous vacuum disk					
Continuous vacuum Precoat					
Continuous pressure Precoat					
Batch vacuum leaf					
Batch nutsche					
Batch pressure filters					
plate-and-frame					
vertical leaf					
tubular					
horizontal plate					
cartridge edge					

to vendors for conducting tests. Should the particles suddenly be smaller, slower than usual filtrations will be seen and vice versa.

Of course, if equipment is presently in operation at the plant on the particular product, invaluable data can be obtained. Optimization of the filter should be done, perhaps with the vendor's help, to be sure that over-sizing of the next piece of equipment does not occur. Variance of precoat, cake thickness, wash, etc., if not already done on the process, will enable fine-tuning of the process as well as confirm the data for the next system's design.

9.0 CONTINUOUS VS. BATCH FILTRATION

Continuous and batch equipment can be used in the same process by incorporating holdup tanks, vessels or hoppers between them. However, the overriding factor is often one of economics. High volume throughputs in the order of magnitude of a several hundred gallons per hour or greater usually require continuous separation. The size of batch equipment escalates in these cases, resulting in tremendous capital outlay. It is for this reason the rotary vacuum filter has been historically used in the fermentation industry.

10.0 ROTARY VACUUM DRUM FILTER

10.1 Operation and Applications

Raw fermentation broth is an example of a large volume production. Rotary drum vacuum filters (RVF's) have traditionally been found in this service. Slow-settling materials or more difficult filtrations with large scale production requirements are typical applications for this type of equipment. For an overview of filter selection versus filtering rates, see Table 8.3, which is excerpted by special permission from *Chemical Engineering/Deskbook Issue*, February 15, 1971, by McGraw Hill, Inc., New York, NY 10020 (Figure 8.1).

The basic principle on an RVF is a hollow rotating cylindrical drum driven by a variable speed drive at 0.1–10 revolutions per minute. One-third of the drum is submerged in a slurry trough. As it rotates, the mycelia suspension is drawn to the surface of the drum by an internal vacuum. The surface is the filter medium mounted on top of a grid support structure. Mother liquor and wash are pulled through the vacuum line to a large chamber and evacuated by a pump.

Applicable to a broad range of processes, e.g., pharmaceutical, starch, ceramics, metallurgical, salt, etc., many variations of the RVF have been developed, however, the fundamental cylinder design remains the same.

FIGURE 8.1 Rotary drum vacuum filter. *(Courtesy of Komline Sanderson, Inc.)*

The cylinder is divided into compartments like pieces of a pie (see Figure 8.2), and drainage pipes carry fluid from the cylinder surface to an internal manifold.

Filter diameters range from three to twelve feet, with face lengths of one to twenty-four feet, and up to 1000 ft^2 of filtration area [5]. Filtration rates range from 5 GPH per square foot to 150 GPH per square foot. Moisture levels are, of course, dependent upon particle size distribution and tend to range from 25% to 75% by weight and cake thickness tends to be in the 1/8–1/2 inch range, as most applications are for slow-filtering materials.

With the exception of the precoat applications, RVF's do not usually yield absolutely clear filtrate. Although still widely used, rotary vacuum filters are, in some cases, being replaced by membrane separation technology as the method of choice for clarification of fermentation broths and concentrating cell mass. Membranes can yield more complete filtration clarification, but often a wetter cell paste.

The drum is positioned in a trough containing the agitated slurry, whose submergence level can be controlled. As the drum rotates, a panel is submerged in the slurry. The applied vacuum draws the suspension to the cloth, retaining solids as the filtrate passes through the cloth to the inner piping and, subsequently, exiting the system to a vapor-liquid separator with high/low level control by a pump. Cake formation occurs during submergence. Once formed, the cake dewaters above the submergence level and is then washed, dewatered and discharged.

Discharge mechanisms will vary depending upon cake characteristics. Friable, dry materials can use a "doctor" blade as in Figure 8.3. Difficult filtrations requiring thinner cakes incorporate a string discharge mechanism. This is the primary method for starch and mycelia applications. A series of 1/2 inch spaced strings rest on the filter medium at the two o'clock position. The cake is lifted from the drum as shown in Figure 8.4. Fermentation broths containing grains, soybean hulls, etc., are applications for this type of discharge mechanism. The solids may be used for animal feed stock, or incinerated. String or belt discharge mechanisms facilitate cake removal and, therefore, can eliminate the need for filter aid.

Continuous belt discharge (Figure 8.5) is employed for products that have a propensity for blinding the filter medium. A series of rollers facilitate cake removal in this case.

Precoated rotary vacuum drum filters (Figure 8.6) are used by filtering a slurry of filter aid and water first, then subsequent product filtration. Difficult filtering materials, which have a tendency to blind, are removed with a doctor blade. Precoat is removed along with the slurry to expose a new filtration surface each cycle.

The progressively advancing blade moves 0.05 to 0.2 mm per revolution. Vacuum is maintained throughout the cycle, instead of just during submergence, so that the precoat is retained. Once the precoat is expended, the RVF must be thoroughly cleaned, and a fresh coat reapplied.

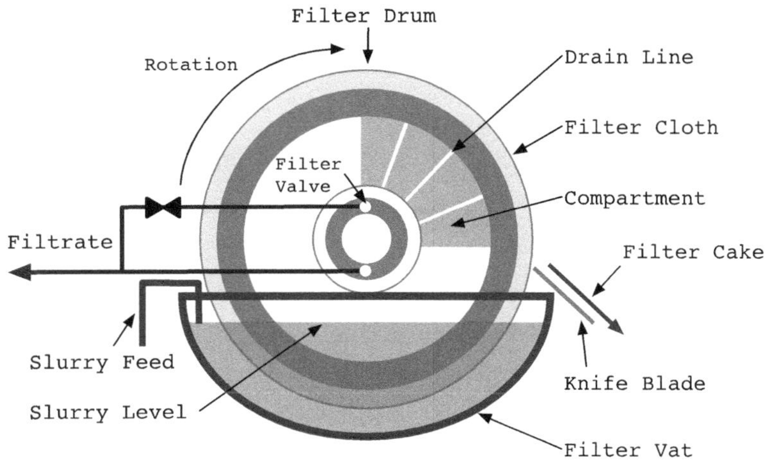

FIGURE 8.2 Rotary vacuum filter schematic.

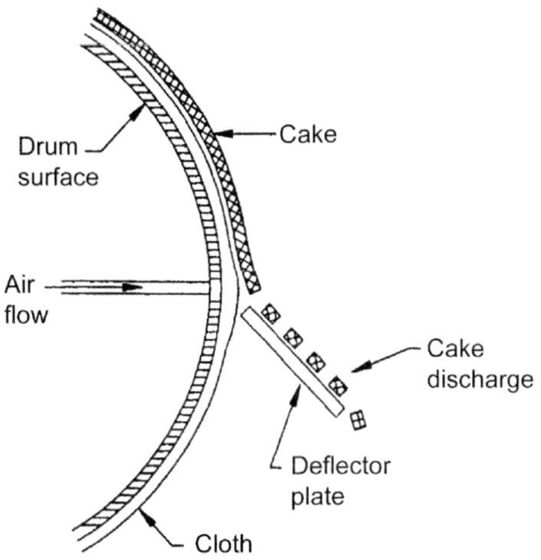

FIGURE 8.3 Cake discharge mechanism.

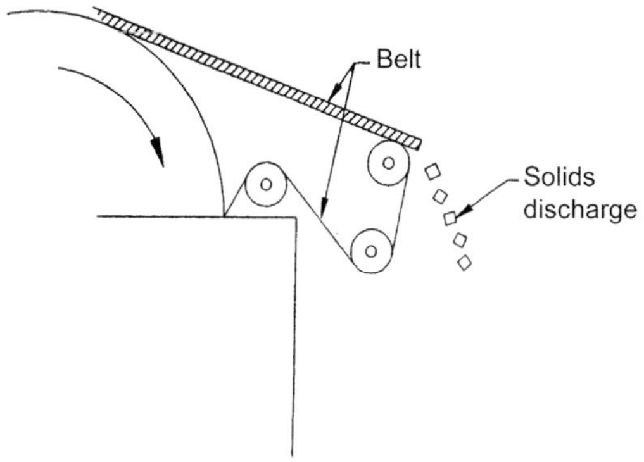

FIGURE 8.5 Belt discharge.

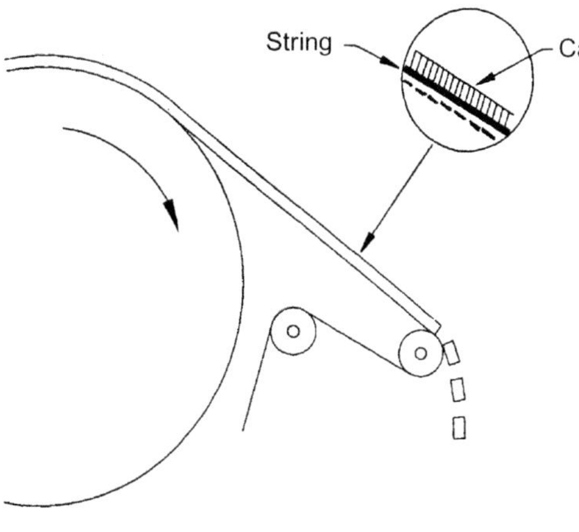

FIGURE 8.4 String discharge.

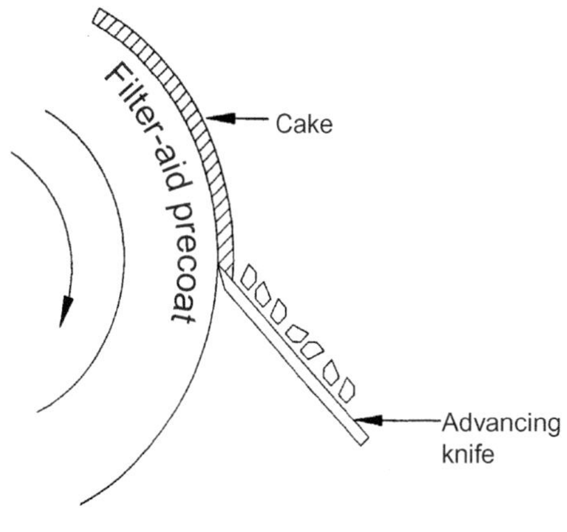

FIGURE 8.6 Precoat rotary vacuum filter.

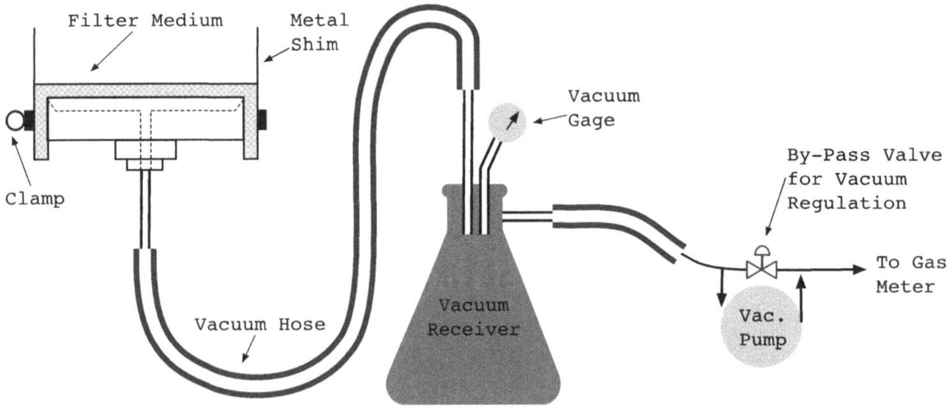

FIGURE 8.7 Pressure leaf test.

10.2 Optimization

Pressure leaf tests are used to model the operating cycle of a RVF. The cycle, consisting of cake formation, dewatering, washing, dewatering and discharge, is simulated by the apparatus shown in Figure 8.7.

The test leaf is immersed in the agitated slurry for cake formation, then removed for drying. If washing is required, the leaf is placed in the wash liquor and then dried again. Discharge from the leaf will indicate type of discharge mechanism required. By varying the time of the portions of the cycle, rotational speeds can be simulated.

It is recommended that optimization be carried out by developing three different cake thicknesses. From this, a capacity versus cake thickness curve can be developed. Additional parameters that have to be evaluated are vacuum level, wash requirements, slurry concentration, and slurry temperature. If cake cracking occurs, the wash should be introduced earlier to avoid channeling.

Several leaf tests should be performed for repeatability. Data collected will permit scaleup to plant scale operations. Significant data will be pounds of dry cake per square foot per hour, gallons of filtrate per square foot per hour, filtrate clarity, wash ratios, (pounds of solids/gallon of wash), residual moistures, filter media selection, knife advance time, precoat thickness, solids penetration into precoat, and submergence level should also be evaluated. For the optimization equation, refer to Peters and Timmerhaus, and Tiller and Crump.

11.0 NUTSCHES

11.1 Applications

The nutsche filter is increasingly prevalent in postcrystallization filtrations. It would not be used directly from the fermenter. Relatively fast filtrations with predictable crystal structures, often found in the

FIGURE 8.8 Agitated nutsche type pressure filter. *(Courtesy of COGEIM SpA).*

intermediate and final step purifications of antibiotic drugs, work well on this batch filter. Batch sizes range from 100 to 7500 gallons.

11.2 Operation

The term *nutsche* is derived from the German word for sucking. Vacuum is applied at the bottom of a vessel that contains a perforated plate. A filter cloth, screen, perforated plate, or porous ceramic plate may be the direct filtration medium (see Figure 8.8). Subsequently, products

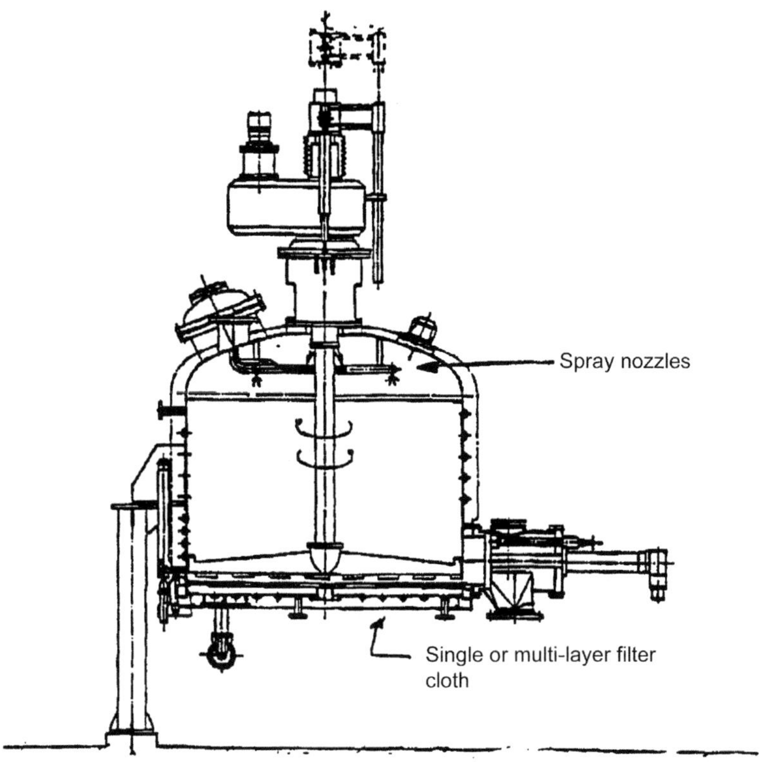

FIGURE 8.9 Agitated nutsche type pressure filter. *(Courtesy of COGEIM SpA).*

should have lower cake resistances and well-defined crystal structures to facilitate separation. The driving force for the separation is vacuum and/or pressure.

With an agitated vessel, the blade can be used to smooth or squeeze the cake, eliminating cracks, when rotated in one direction or for reslurrying and/or discharging the cake when rotated in the opposite direction. The rotation of the agitator can be by electric motor with variable speed drive; however, the translational movement is achieved by a separate hydraulic system. The agitator requires a stuffing box or mechanical seal for pressure or vacuum operation of the unit. Filling is accomplished by gravity feed or pump. Large cakes, in the 10–12 inch range, are developed. When plug flow displacement washing is not effective, and as diffusion of impurities through the cake becomes difficult, reslurrying is the required method. Displacement washing is more efficient and minimizes wash quantities, however, may not always be possible. Filtering, reslurrying and refiltering can all be accomplished in the same unit, thus achieving total containment. See Figure 8.9.

The vessel can also be jacketed for heating and/or cooling and the agitator blade heated. This design can now be a reactor in combination with a filter-dryer or alone as a filter-dryer (Figure 8.10) (see also Chapter 17). This is particularly advantageous for dedicated production of toxic materials requiring an enclosed system. Operator exposure and product handling are minimized.

The nutsche can have limitations for difficult filtrations, as the thick filter cakes can impede filtration. A two-stage system for filtration and drying can offer greater flexibility in plant operations, especially if either the filtration or drying step is rate-limited. Predictable crystals that filter and dry well are the best applications for this all-encompassing system.

Mechanical discharge incorporating the agitator facilitates solids removal centrally or a side discharge is possible. A residual heel of product will be left as the agitator is limited on how close to the screen or filter medium it can go. Residual heels can be reworked by reslurrying or remain until the campaign changes. For frequent product changes, the nutsche can be provided with a split-vessel design. Upon lowering the bottom portion, free access to the inside of the vessel and the filter bottom itself for cleaning purposes is possible. Some manufacturers have air-knife designs that remove the residual heel. Heels as low as one-quarter inch can be obtained.

Materials of construction can vary widely depending upon the application. Typically, 304 or 316 stainless steel, and Hastelloy are supplied, although many other types of material of construction are available. Metal finishes, in keeping with good manufacturing practices (GMP), particularly for areas in contact with final products, require

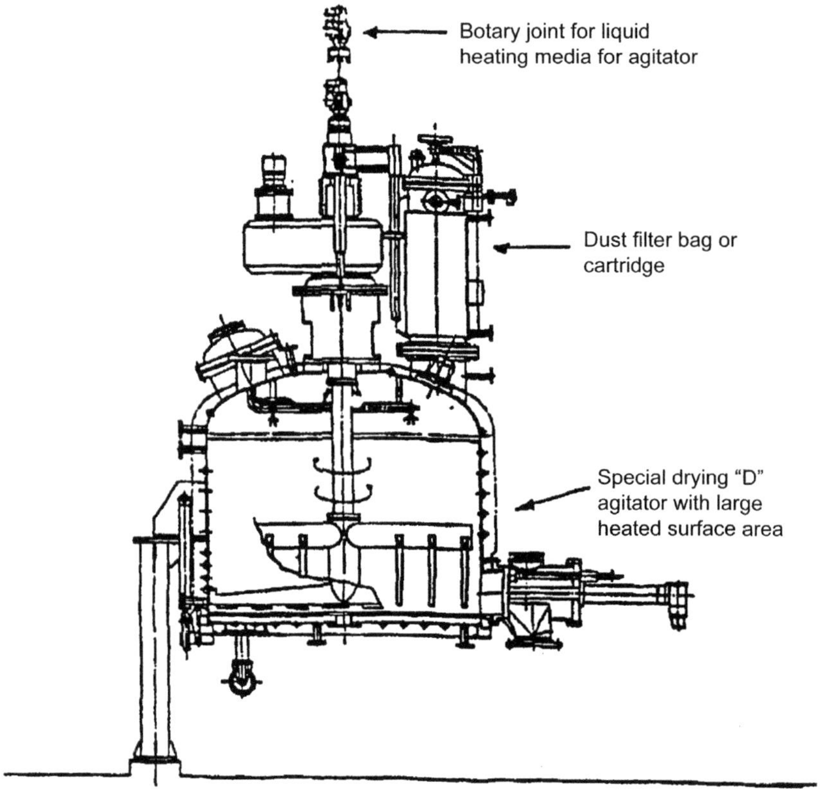

FIGURE 8.10 Agitated nutsche type filter/dryer. *(Courtesy of COGEIM SpA).*

welds to be ground smooth. Finishes can be specified in microns, Ra, or grit. The unit, Ra, is the arithmetical average of the surface roughness in microinches. The rms is the root mean square of the surface roughness in microinches; rms = 1.1 Ra.

A mechanical finish of 400 grit is an acceptable pharmaceutical finish, however, mechanical polishing folds the surface material over itself. When viewed under a microscope, jagged peaks and crevices are visible. Product on the micron level can be accumulated in these areas. Electropolishing of the surface is often used to eliminate these peaks and valleys to provide a more cleanable surface. A layer of the surface material is removed in this case. A mechanical finish of 400 grit is achieved by progressively increasing the grit spec from 60 up to 400. If a 400 grit surface was to be electropolished, the amount of material removed would result in an equivalent 180–220 grit surface roughness. Therefore, a mechanical finish of approximately 180–220 grit need only be specified when electropolishing. A considerable cost savings is realized. It is always advisable to specify the Ra value of the surface whether electropolishing is specified or not.

Filter areas will range from 0.5 to 16 m^2. For large-scale processing, significant floor area is occupied per unit area of filtration [1]. Those products that tend to blind filter media, i.e., colloidal slurries, gelatinous and protein compounds, will require alternate equipment, filtration or centrifugation.

11.3 Maintenance

When used for dedicated production, maintenance is minimal. The agitator sealing system (usually a stuffing box or mechanical seal), however, must be maintained.

Filter cloth change and O-ring changes would be the primary maintenance required. This depends on the filter design. The split vessel design allows for easy access. A removable bottom which can be fixed to the vessel through a bayonet closure system is completely hydraulically controlled and can be lowered in 1–2 minutes. By first using the spray nozzles and flushing the system with a solvent that the product is soluble in, operator exposure will be minimized. Cleaning between final products for 99% validation, can take 1–2 twelve-hour shifts.

Screen lifetime will depend upon the type of screen used. Various types of filter cloths or monolayer metal screens can be used. A multilayer sinterized filter screen is also available. Installation of filter cloths and screens is usually by the use of clamping rings and hold-down bars screwed on the bottom.

12.0 BHS AUTOPRESS

12.1 Applications

Products are processed in an enclosed atmosphere without operator or environmental exposure. FDA requirements, and conformation to good manufacturing practices (GMP) requiring containment of product, are continual issues. Applications are as alternative to centrifugation, nutsche filters or replacement of plate and frame filter presses, and products that are compressible and amorphous in nature with high specific cake resistances.

Fermented products in the post-broth stage, can be applications as well as disrupted cell fermentation. Protein hydrolysate has also been an application. Antibiotics are applications in pre or post-crystallization can also be an application if solids are found to be compressible. Replacement of cartridge filter systems where high filter replacement costs occur as well as low volume waste treatment streams can also be processed.

FIGURE 8.11 Agitated nutsche type pressure filter. *(Courtesy of COGEIM SpA).*

12.2 Operation

This unit is a fully automated, totally enclosed filter press. The core of the system is a pressure chamber. It can be connected to peripheral equipment, such as a dryer or bin, for a totally contained system or it can also be a single stage filter-dryer for some applications. It has a pressurized filter housing that encloses vertical circular filter plates with synthetic or sintered metal filter media. The housing pressure ranges from full vacuum to 150 psig.

FIGURE 8.12 Fully Automated Autopress for gas-tight and safe filtration *(Courtesy of BHS-Sonthofen GmbH).*

A gas-inflated membrane seals the annular space between the housing and the filter plates. The plates can be squeezed bringing pressures up to 600 psig. This helps prevent cake cracking prior to washing. Washing can be done in the forward or reverse direction of rotation. Drying can be accomplished by vacuum or blowing heated gas through the cake. Discharge is done by elastomeric knives. Cake thicknesses are usually 20 mm but range from 5–25 mm.

13.0 MANUFACTURERS

13.1 Rotary Drum Vacuum Filters

Andritz
Lexington, KY
Denver Equipment Company
Colorado Springs, CO
Eimco
Division of Envirotech
Salt Lake City, UT
Komline-Sanderson, Inc.
Peapack, NJ
Peterson Fuller, Inc.
Salt Lake City, UT

13.2 Nutsches

Cogeim	Cogeim
Charlotte, NC	Dalmine, Italy
Rosenmund, Inc.	Rosenmund
Charlotte, NC	Switzerland

13.3 Autopress

BHS Filtration
Charlotte, NC

REFERENCES

[1] R.H. Perry, D.W. Green, I.O. Maloney (Eds.), Perry's Chemical Engineer's Handbook, sixth ed., McGraw Hill Book Co., New York, 1984 (pp. 19–65, 19–89).
[2] R.H. Rees, C.W. Cain, Let diatomite enhance your filtration, Chem. Eng. 8 (1990) 76–79.
[3] C. Orr (Ed.), Filtration Principles and Practices, Part I, Marcel Dekker, Inc., New York, 1977.
[4] J.G. Clark, Select the right fabric, Chem. Eng. Prog. (1990).
[5] H.C. Vogel (Ed.), Fermentation and Biochemical Engineering Handbook, first ed., Noyes Publications, New Jersey, 1983.

FURTHER READING

[1] D.A. Dahlstrom, in: W.J. Mead (Ed.), Encyclopedia of Chemical Process Equipment, Reinhold, New York, 1964, pp. 417–438.
[2] C.E. Silverblatt, H. Risbud, F.M. Tiller, Batch, continuous processes for cake filtrations, Chem. Eng. 4 (1974) 127–136.
[3] F.M. Tiller, et al., Theory and Practice of Solid-Liquid Separation, second ed., University of Houston, 1975.
[4] G.R.S. Smith, How to use rotary, vacuum, pre-coat filters, Chem. Eng. 83 (1976) 84–94.
[5] C.W. Cain, Putting the principals to work filter-cake filtration, Chem. Eng. (1990) 72–75.
[6] P.J. Lloyd, Particle characterization, Chem. Eng. 4 (1974) 120–122.
[7] D.W. Cheape Jr., Leaf tests can establish optimum rotary-vacuum-filter operation, Chem. Eng. 5 (1982) 141–148.
[8] M.S. Peters, K.D. Timmerhaus, Plant Design and Economics for Chemical Engineers, second ed., McGraw Hill, 1968.
[9] F.M. Tiller, J.R. Crump, How to increase filtration rates in continuous filters, Chem. Eng. 5 (1977) 183–187.

Chapter 9

Cross-Flow Filtration

Ramesh R. Bhave

1.0 INTRODUCTION

Cross-flow filtration (CFF) also known as tangential flow filtration is not of recent origin. It began with the development of reverse osmosis (RO) more than three decades ago. Industrial RO processes include desalting of sea water and brackish water, and recovery and purification of some fermentation products. The cross-flow membrane filtration technique was next applied to the concentration and fractionation of macromolecules commonly recognized as ultrafiltration (UF) in the late 1960's. Major UF applications include electrocoat paint recovery, enzyme and protein recovery and pyrogen removal [1–3].

In the past ten to fifteen years or so, the applications sphere of cross-flow filtration has been extended to include microfiltration (MF) which primarily deals with the filtration of colloidal or particulate suspensions with size ranging from 0.02 to about 10 microns. Microfiltration applications are rapidly developing and range from sterile water production to clarification of beverages and fermentation products and concentration of cell mass, yeast, *E. coli* and other media in biotechnology related applications [1–4].

Table 9.1 shows the types of separations achievable with MF, UF and RO membranes when operated in cross-flow configuration. For MF or UF application, the choice of membrane materials includes ceramics, metals or polymers, whereas for RO at the present only polymer membranes are predominantly used. Although cross-flow filtration is practiced in all the above three types of membrane applications, the description of membrane characteristics, operational aspects and applications will be limited to MF and UF, where the cross-flow mode shows the greatest impact on filtration performance compared with dead end filtration. Figure 9.1 shows the schematic of cross-flow filtration including the critical issues and operational modes for clarification or concentration using a semipermeable polymeric or inorganic membrane.

Despite the growing use in a broad range of applications, cross-flow filtration still largely remains a semi-empirical science. Mathematical models and correlations are generally unavailable or applicable under very specific and well-defined conditions, owing to the complex combination of hydrodynamic, electrostatic and thermodynamic forces that affect flux and/or retention. Membrane *fouling* is not yet fully understood and is perhaps the biggest obstacle to more widespread use of CFF in solid-liquid separations. Membrane cleaning is also not well understood. The success of a membrane-based filtration process depends on its ability to obtain a reproducible performance in conformance with the design specifications over a long period of time with periodic (typically once a day) membrane cleaning.

2.0 CROSS-FLOW vs. DEAD END FILTRATION

The distinction between cross-flow and dead end (also known as through-flow) filtration can be better understood if we first analyze the mechanism of retention. The efficiency of cross-flow filtration is largely dependent on the ability of the membrane to perform an effective surface filtration, especially where suspended or colloidal particles are involved. Table 9.2 shows the advantages and versatility of cross-flow filtration in meeting a broad range of filtration objectives [1–3,5]. Figure 9.2 illustrates the differences in separation mechanisms of CFF versus dead end filtration.

High recirculation rates ensure higher cross-flow velocities (and hence Reynold's number) past the membrane surface which promotes turbulence and increases the rate of redispersion of retained solids in the bulk feed. This is helpful in controlling the concentration polarization layer. It may be of interest to note that polarization is controlled essentially by cross-flow velocity and not very much by the average transmembrane pressure (ATP). It should also be noted that higher particle or molecular diffusivity under the influence of high shear can enhance the filtration rates. Since diffusivity values of rigid particles (MF) under turbulent conditions are typically much higher than those for colloidal particles or dissolved macromolecules (UF) microfiltration rates tend to be

Fermentation and Biochemical Engineering Handbook.

TABLE 9.1 Separation Spectrum

Nominal Size of Species	Examples of Species Separated	Process	Remarks
100–500 Dalton	Organic acid, acetic acid, citric acid, amino acids	UF	Product recovered in permeate
		RO	Product in concentrate
200–2,000 Dalton	Antibiotics, penicillin, cephalosporin	MF/UF	Product recovered in permeate
10,000–2,00,000 Dalton	Proteins/polysaccharides	UF	Species retained by the membrane is concentrated in retentate. Some losses may occur in permeate.
0.01–0.3 μm	Viruses, Interferon colloidal	UF	Species is concentrated in retentate
0.1–1 μm	Silica	MF/UF	Product in concentrate phase
0.1–10 μm	*E. coli*, Pseudomonas diminuta, mammalian cells,	MF	Species retained by the membrane Is concentrated in the retentate
	Microorganisms from air	MF	Permeate sterile air
	Oily emulsions	MF/UF	Oils retained by the membrane are concentrated in retentate
1–100 μm	Bacteria cells, yeasts, molds	MF	Species retained by the membrane Is concentrated in retentate

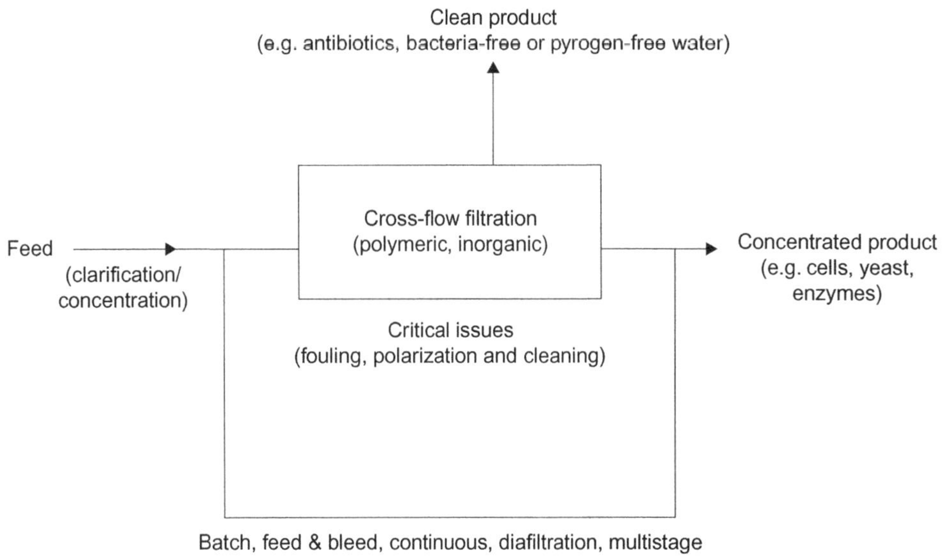

FIGURE 9.1 Schematic of cross-flow filtration.

much higher than ultrafiltration rates under otherwise similar conditions [6].

On the other hand, in dead end filtration the retention is achieved by particle or gel layer buildup on the membrane and in the pores of the medium such as when a depth type filter is used. This condition is analogous to that encountered in packed-bed geometries.

In dead end filtration, the applied pressure drives the entire feed through the membrane filter producing a filtrate which is typically particle-free while the separated particles form a filter cake. The feed and filtrate travel concurrently along the length of the filter generating one product stream for every feed. In CFF, one feed generates two product streams, retentate and permeate. Per pass recovery in through-flow mode is almost 100% (since only the solids are removed) whereas in the cross-flow mode the per pass recovery typically does not exceed 20% and is often in the 1 to 5% range. Recirculation of retentate is thus necessary to increase the total recovery at the expense of higher energy costs.

As the filtration progresses, the filter cake becomes increasingly thicker which results in a reduced filtration rate (at a constant transmembrane pressure). When the flow or transmembrane pressure (depending on the control strategy) approaches a limiting value, the filtration must be interrupted in order to clean or replace the

TABLE 9.2 Cross-Flow Filtration: Key Advantages

Process Goal	Cross-flow Filtration	Dead End Filtration
Ability to handle wide variations in particle size	Excellent	Generally poor
Ability to handle wide variations in solids concentration	Excellent	Poor or unacceptable
Continuous concentration with recycle	Excellent	Poor or unacceptable
Waste minimization	Superior	Can minimize waste if handling low solids feed where cartridge disposal is infrequent
High product purity or yield	Excellent; but may require diafiltration to overcome excessive flux loss at higher recovery	Performance is generally acceptable except in situations involving high solids or adsorptive fouling

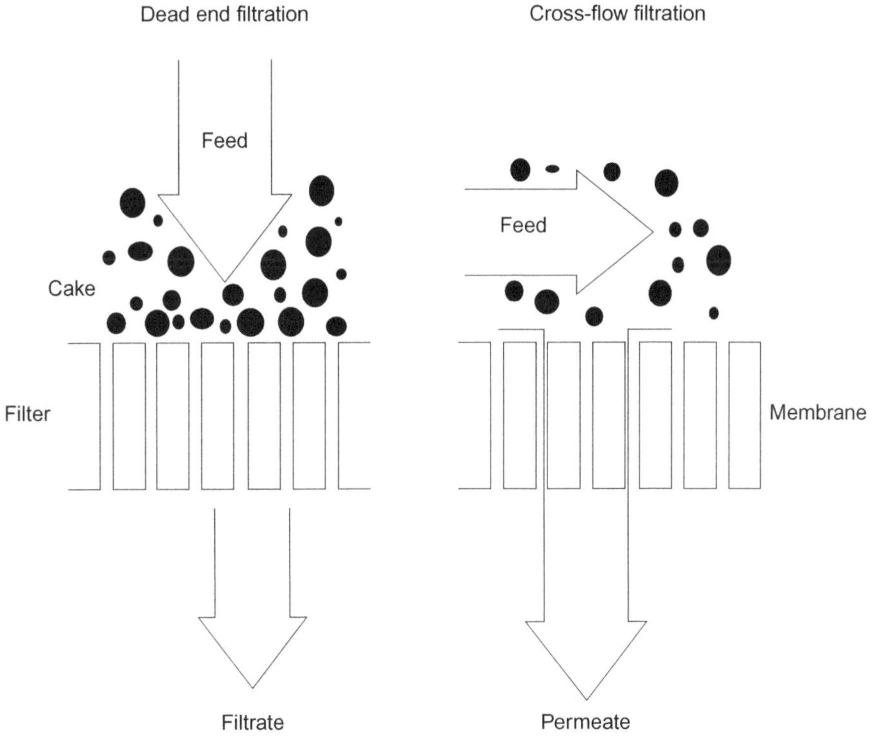

FIGURE 9.2 Cross-flow versus dead end filtration.

membrane filter. This discontinuous mode of operation can be a major disadvantage when handling process streams with a relatively high solid content.

Cross-flow filtration can overcome this handicap by efficient fluid management to control the thickness of the concentration-polarization layer. Thus, feed streams with solid loading higher than 1 wt.% may be better suited for CFF whereas feed streams containing less than 0.5 wt.% solids may be adequately served by dead end filtration. However, if the retained solids constitute the product to be recovered or when the nature of solids is the cause of increased fouling, cross-flow filtration should be considered. CFF is also the preferred mode when particle size or molecular weight distribution is an important consideration, such as in the separation of enzymes, antibiotics, proteins and polysaccharides from microbial cell mass, colloidal matter and oily emulsions. Tubular cross-flow filters are being used to continuously concentrate relatively rigid solids up to 70 wt.% and up to 20 wt.% with gelatinous materials.

3.0 COMPARISON OF CROSS-FLOW WITH OTHER COMPETING TECHNOLOGIES

Cross-flow filtration as a processing alternative for separation and concentration of soluble or dissolved components competes with traditional equipment such as dead end cartridge filtration, pre-coat filtration and centrifugation. The specific merits and weaknesses of each of these filtration alternatives are summarized in Table 9.3. In addition to the ability to handle wide variations in processing conditions, other considerations may need to be addressed for economical viability of cross-flow filtration. These are briefly discussed below. A more detailed discussion on process design aspects, capital and operating cost considerations is presented in Section 6.7.

1. Energy Requirements. Centrifugal devices typically require high maintenance. In contrast, cross-flow filtration requires minimal maintenance with low operating costs in most situations except for large bore (>6 mm) tubular membrane products operating under high recirculation rates. The energy requirements in dead end filtration are typically low.
2. Waste Minimization and Disposal. CFF systems minimize disposal costs (e.g., when ceramic filters are used) whereas in diatomaceous (DE) pre-coat filtration substantial waste disposal costs may be incurred, particularly if the DE is contaminated with toxic organics. Currently, in many applications, DE is disposed of in landfills. In future, however, this option may become less available forcing the industry to use cross-flow microfiltration technology or adopt other waste minimization measures.
3. Capital Cost. Many dead end and DE based filtration systems can have a relatively low capital cost basis [2]. On the other hand, CFF systems may require relatively higher capital cost. Centrifuges can also be capital intensive especially where large-scale continuous filtration is required.

4.0 GENERAL CHARACTERISTICS OF CROSS-FLOW FILTERS

The performance of a cross-flow filter is primarily defined by its efficiency in permeating or retaining desired species and the rate of transport of desired species across the membrane barrier. Microscopic features of the membranes greatly influence the filtration and separation performance [1,3].

TABLE 9.3 Comparison of Cross-Flow Filtration vs. Competitive Technologies

Process Conditions	Cross-flow Filtration	Dead End Filtration	Precoat Filtration	Centrifugation
Low solids 0–1% by volume	Can handle efficiently but needs high flux to be cost effective	Can handle effectively; low cost	Can handle effectively; low cost	Can handle but may be expensive
Medium solids (1–10%) by volume	Can adequately handle: economics depends on flux	Not well suited	Can handle adequately and economically	Can handle adequately and economically
High solids (10 to 70% by volume)	May not be economical at > 25% solids (with few exceptions) for continuous process	Not well suited	May handle the solids; high operating cost	Can handle high solids; high capital and maintenance
Emulsified liquids	Can handle efficiently	Not well suited	Not well suited	Not well suited
Small density differences or fine particles	Well suited due to wide range of pore diameters UF/MF	Can handle adequately	Not well suited	Cannot handle efficiently
Separation of macromolecular solutes	Can handle very efficiently: cost effective altenative	Not well suited	Not well suited; low throughput	Not feasible
Solvents and/or high temperature	Can handle adequately using chemically/thermally resistant membranes	Not well suited	Not well suited in open system	May be difficult to handle
Continuous fractionation of solids	Not well suited	Can handle but performance sensitive to operating conditions	Not feasible	Can handle adequately

- The nature of the membrane material
- Pore dimensions
- Pore size distributions
- Porosity
- Surface properties such as zeta potential
- Hydrophobic/hydrophilic character
- Membrane thickness

From an operational standpoint, the mechanical, thermal and chemical stability of the membrane structure is important to ensure long service life and reliability. Table 9.4 summarizes the influence and significance of these features on the overall performance of a cross-flow filter.

The discussion on the general characteristics of polymeric and inorganic membranes is treated separately partly due to their differences in production methods and also due to important differences in their operating characteristics.

4.1 Polymeric Microfilters and Ultrafilters

Symmetric polymeric membranes possess a uniform pore structure over the entire thickness. These membranes can be porous or dense with a constant permeability from one surface to the other. Asymmetric (also sometimes referred to as anisotropic) membranes, on the other hand, typically show a dense (nonporous) structure with a thin (0.1–0.5 μm) surface layer supported on a porous substrate. The thin surface layer maximizes the flux and performs the separation. The microporous support structure provides the mechanical strength.

Polymeric membranes are prepared from a variety of materials using several different production techniques. Table 9.5 summarizes a partial list of the various polymer materials used in the manufacture of cross-flow filters for both MF and UF applications. For microfiltration applications, typically symmetric membranes are used. Examples include polyethylene, polyvinylidene fluoride (PVDF) and polytetrafluoroethylene (PTFE) membrane. These can be produced by stretching, molding and sintering fine-grained and partially crystalline polymers. Polyester and polycarbonate membranes are made using irradiation and etching processes and polymers such as polypropylene, polyamide, cellulose acetate and polysulfone membranes are produced by the phase inversion process [1,7,8].

Ultrafiltration membranes are usually asymmetric and are also made from a variety of materials but are primarily made by the phase inversion process. In the phase inversion process, a homogeneous liquid phase consisting of a polymer and a solvent is converted into a two-phase system. The polymer is precipitated as a solid phase (through a change in temperature, solvent evaporation or addition of a precipitant) and the liquid phase forms the pore system. UF membranes currently on the market are also made from a variety of materials, including polyvinylidene fluoride, polyacrylonitrile, polyethersulfone and polysulfone.

Microfiltration membranes are characterized by bubble point and pore size distribution whereas the UF membranes are typically described by their molecular weight cutoff (MWCO) value. The bubble point pressure relates to the largest pore opening in the membrane layer. This is

TABLE 9.4 Influence of Membrane Characteristics on Filtration or Separation Performance

Property	Flux	Rejection	Influence or Significance
Asymmetric	high	marginal	Flux is higher compared with symmetric membranes.
Symmetric	high	marginal	Particle retention in porous structure provides higher surface area per unit volume. This also makes them susceptible to irreversible fouling.
Bubble point	marginal	high	Critical factor for membrane Integrity.
Pore dimensions	high	marginal to substantial	Must be carefully optimized to provide high flux combined with high retention.
Pore size distribution	marginal	substantial	Narrow pore size distribution often provides better separation efficiency.
Porosity	high	marginal or none	Higher porosity typically results in higher permeability and thus can Improve flux.
Zeta potential	marginal or none	marginal to substantial	Relates to charge effects and can influence fouling due to adsorption/precipitation
Hydrophobic	marginal	can be significant	Rejection of water may be important for sterility purposes and in non-aqueous separations
Hydrophilic	marginal	can be significant	Provides good wetting of membranes: can increase transport of aqueous solutions; can minimize fouling due to organic substances

TABLE 9.5 Polymeric Microfilters and Ultrafilters

Material	Microfilter	Ultrafilter	Configuration
Acrylic polymer	X	X	HFF
Cellulosic polymer	X	X	FS, PS, SW, HFF
Nylon based polyester	X	X	HFF, PS, FS
Polyamide		X	HFF, FS
Polybenzamidazole		X	FS, SW
Polycarbonate	X		FS
Polyethersulfone		X	SW, T
Polyethylene	X		FS
Polypropylene	X		HFF, FS, T
Polysulfone		X	HFF, SW, T, FS
Polytetrafluoroethylene	X	X	FS, T
Polyvinylidene fluoride	X	X	SW, T, FS, PF
PF – Plate and Frame		SW - Spiral Wound	
PS – Pleated Sheet		T - Tubular (including wide channel)	
FS – Flat Sheet		HFF - Hollow Fine Fiber	

measured with the help of a bubble point apparatus [1,9]. The average pore diameter of a MF membrane is determined by measuring the pressure at which a steady stream of bubbles is observed. For MF membranes, bubble point pressures vary depending on the pore diameter and nature of membrane material (e.g., hydrophobic or hydrophilic). For example, bubble point values for 0.1 to 0.8 μm pore diameter membranes are reported to vary from 1 bar (equals about 14.5 psi) to 15 bar [1]. However, due to the limited mechanical resistance of some membrane geometries (e.g., tubular and to some extent hollow fiber) such measurements cannot be performed for smaller pore diameter MF and UF membranes. The bubble point apparatus can also be used to determine the pore size distribution of the membrane.

Since the majority of UF membranes have dense surface layers, it is difficult to characterize them with a true pore size distribution. Therefore, polymeric UF membranes are described by their ability to retain or allow passage of certain solutes. The MWCO values for UF membranes can range from as low as 1000 dalton (tight UF) to as high as 200,000 Dalton (loose UF). This roughly corresponds to an "equivalent" pore diameter range from about 1 nanometer (nm) to 100 nm (0.1 μm) as described in Ref. 10.

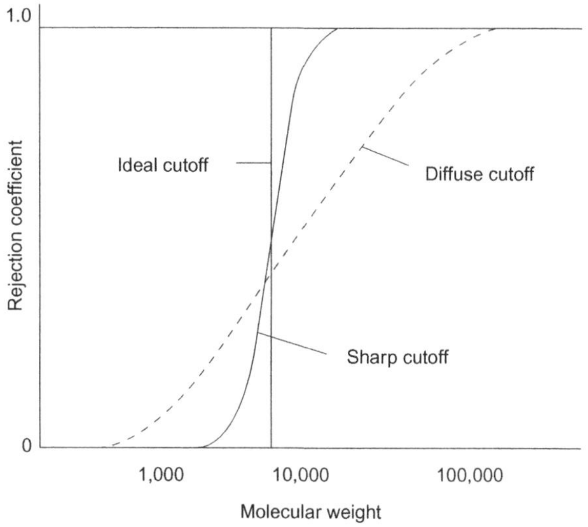

FIGURE 9.3 Rejection coefficient as a function of molecular weight cutoff of an ultrafiltration membrane.

Different membrane materials with similar or identical MWCO value may show different solute retention properties under otherwise similar operating conditions. If adsorption effects are negligible, such a result can be attributed primarily to the differences in their pore size distributions. This is illustrated in Figure 9.3. It can be seen that, although the two membranes are rated by the same MWCO value, their retention characteristics are distinctly different (sharp versus diffuse).

Polymeric cross-flow filters are available in many geometries. These are listed in Table 9.6. It is obvious that no single geometry can provide the versatility to meet the broad range of operating conditions and wide variations in properties. Some cross-flow filters such as cartridge filters have low initial capital cost but high replacement costs and tubular filters may show longer service life but higher operating costs. The optimization of CFF for a specific application may depend on economic and/or environmental factors and is almost impossible to generalize.

4.2 Inorganic Microfilters and Ultrafilters

Cross-flow membrane filters made from inorganic materials, primarily ceramics and metals, utilize entirely different manufacturing processes compared with their polymeric counterparts [3]. Although carbon membranes do not qualify under the inorganic definition, they will be included here due to the similarities with inorganic membranes with regard to their material properties such as thermal, mechanical and chemical resistance as well as similarity in production techniques. Table 9.7 lists the various commonly used materials and membrane geometries in MF and UF modules.

TABLE 9.6 Polymeric Cross-Flow Filters: Module Geometries

Module Geometry	Special Features/Remarks
Flat Sheet	Typical spacing between sheets is 0.25 to 2.5 mm and are used for laboratory evaluations (small surface area modules).
Hollow Fine Fiber	The internal diameter generally ranges form 0.25 to 1 mm. This type of module geometry cannot handle large amounts of suspended solids or fibrous materials.
Plate and Frame	Flat sheet membrane elements are assembled in plate and frame devices to handle larger processing volumes.
Pleated Sheet	Typical spacing between sheets is 0.25 to 2.5 mm. The sheets are enclosed in cylindrical cartridge. Not suitable to handle high solids.
Spiral Wound	Typical spacing between the membrane sheets is 0.25 mm. Not suitable to handle high solids.
Tubular	The internal diameter can range from 2 to 6 mm. Suitable for handling higher solids loading.
Tubular (Wide Channel)	The internal diameter is typically greater than 6 mm and can be as high as 25 mm. The advantage is lower pressure drop and ability to handle high solids/fibrous materials at the expense of higher energy cost.

TABLE 9.7 Inorganic Cross-Flow Filters: Membrane Materials and Module Geometries

Membrane Material	Manufacturer/ Trade Name	Module Geometry	Remarks
α-A lumina	USF/Membralox® Ceraflo®		MF; 0.2 to 5 μm
Zirconia	USF/Membralox®	Tubular/multichannel monolith	UF with pore diameter 20 nm to 100 nm
Titania	USF/Membralox®	Multichannel	MWCO 1000 and 5000 Dalton
Zirconia	TechSep/Kerasep	Multichannel	MWCO 10,000
Ceramic oxides/ Cordierite	Ceramem	Honeycomb monolith	MF/UF with internal channel diameter of 1.5 to 2 mm
Zirconium hydroxide (dynamic)	DuPont/Carre	Tubular	Dynamically formed zirconium hydroxide or $Zr(OH)_4$ – polyacrylic acid membranes for MF/UF
Glass	Asahi Glass Schott Glass/Bioran	Tubular	Mostly for UF, although some MF membranes are available
Stainless Steel	Mott. Pall	Tubular	Mostly for MF pore diameters 0.5 μm and higher

The slip-casting process makes commercial ceramic membranes. This consists of two steps and begins with the preparation of a dispersion of fine particles (referred to as *slip*) followed by the deposition of the particles on a porous support [11].

A majority of commonly used inorganic membranes are composites consisting of a thin separation barrier on porous support (e.g., Membralox® zirconia and alumina membrane products). Inorganic MF and UF membranes are characterized by their narrow pore size distributions. This allows the description of their separative performance in terms of their true pore diameter rather than MWCO value, which can vary with operating conditions. This can be advantageous in comparing the relative separation performance of two different membranes independent of the operating conditions. Their bubble point pressures, in addition, can characterize MF membranes. Due to their superior mechanical resistance bubble point measurements can be extended to smaller diameter MF membranes (0.1 or 0.2 μm), which may have bubble point pressure in excess of 10 bars with water [9].

Typical pore size distributions of inorganic MF and UF membranes are shown in Figure 9.4. The narrow pore size distribution of these membrane layers is evident and

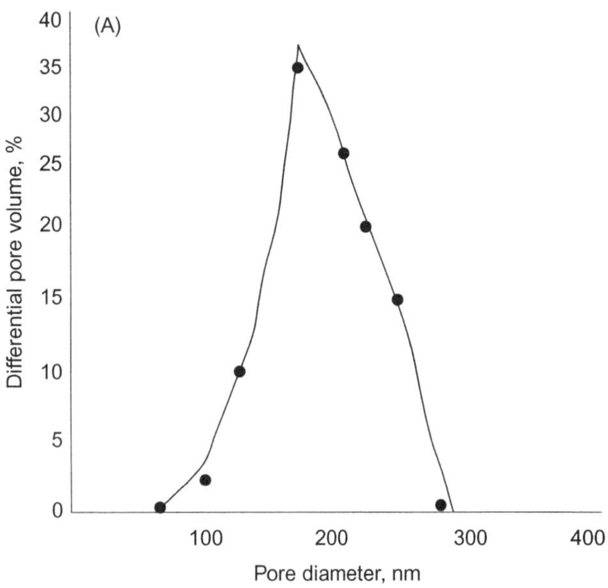

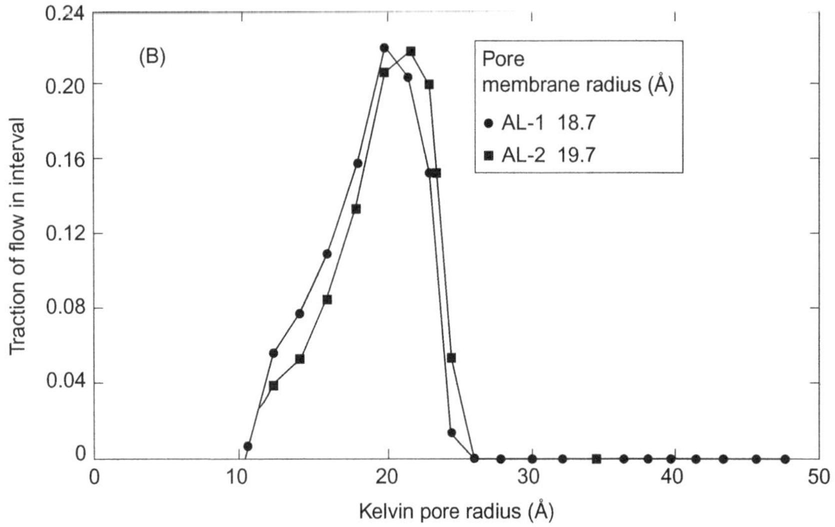

FIGURE 9.4 (A) Typical pore size (diameter) and (B) typical pore size (radius) distributions of inorganic MF and UF membranes.

is primarily responsible for their superior separation capabilities. The manufacturing processes for inorganic membranes have advanced to the point of delivering consistent high quality filters, which are essentially defect free. Inorganic MF and UF membranes also display high flux values (see Table 9.8), which they owe to their composite/asymmetric nature combined with the ability to operate at high temperatures, pressures and shear rates.

Two kinds of membrane geometries are predominantly used, the tubular multi-lumen and the multichannel monoliths with circular, hexagonal or honeycomb structures. The number of channels can vary from 1 to 60.

5.0 OPERATING CONFIGURATIONS

There are several operating configurations that are used in industrial practice depending on flow rate of the product, product characteristics and desired final concentrations of the product which is either to be retained by the membrane or recovered in the permeate.

5.1 Batch System

Figure 9.5 shows a simple batch system consisting of a feed tank, a membrane module and a feed pump, which also serves as a recirculation pump. The recirculation

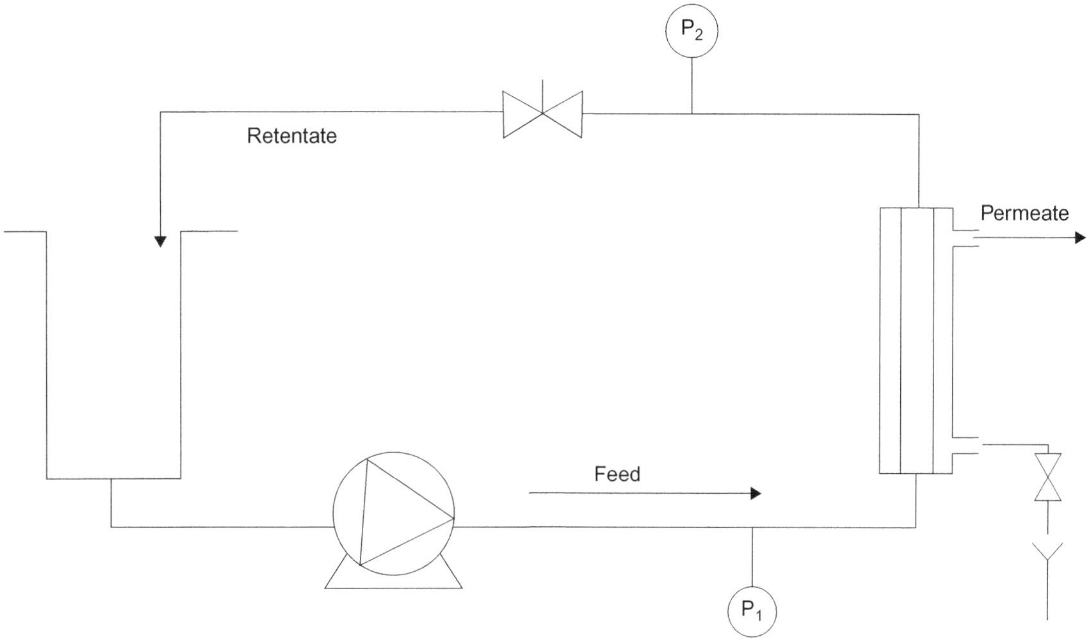

FIGURE 9.5 Schematic of a batch (open-loop) system.

pump maintains the desired cross-flow velocity over a certain range of trans membrane pressures depending on the type of pump and its characteristics (centrifugal or positive displacement). The filtration continues until the final concentration or desired permeate recovery is achieved, unless the flux drops to an unacceptable level. For the retention of suspended solids (e.g., bacteria, yeast cells, etc.) the final concentration factor can be anywhere from 2 to 40 (and higher in some applications where a recovery of >98% is required). In order to minimize the concentration effects, a ratio of concentrate flow rate to permeate flow rate of about 10 to 1 is maintained (assuming the density differences are not significant). This ensures that at any given time the concentration of solids in the recirculation loop is only about 10% higher than that in the feed loop. Depending on the operating cross-flow velocity and viscosity of retentate, the pressure drop along the length of the module can vary from 0.5 bars to more than 2 bars. This often necessitates the use of more than one parallel loop and limits the number of modules in series depending on pump characteristics.

The open loop configuration has some advantages in terms of its simplicity, but also has some disadvantages especially when the product is sensitive to heat or shear effects (e.g., intracellular products, some beverages, and enzymes). Furthermore, when higher cross-flow velocities are required (which is typically the case in many applications) the recirculation rates necessary to sustain them may not be achievable in the open loop configuration, especially if it is also desirable to maintain a concentrate to permeate ratio of at least 10.

This problem can be overcome by placing a feed pump between the feed tank and the recirculation pump, as shown in Figure 9.6. The discharge pressure of the recirculation pump must be at least greater than the pressure loss along the flow channels in the module or several modules connected in series while maintaining the desired recirculation rate.

The pipe sizes for feed and return lines for the closed loop operation are much smaller than that for the open loop system, which can also reduce the capital and operating cost. The feed tank size can also be much smaller for the closed loop, which then allows shorter residence times for heat, or shear sensitive products.

The average flux (J_{av}) in the batch configuration may be estimated using

$$J_{av} = J_f + 0.33(J_f - J_f) \qquad (9.1)$$

Where

J_f = flux at the final concentration
J_i = initial flux

5.2 Feed and Bleed

Batch Mode. The closed loop operation shown in Figure 9.6 may not be suitable in many situations such as when processing large volumes of product and where high product recoveries (>95%) are required. It is well known that flux decreases with an increase in the concentration of retained solids, which may be suspended particles or macro solutes. When high recoveries are

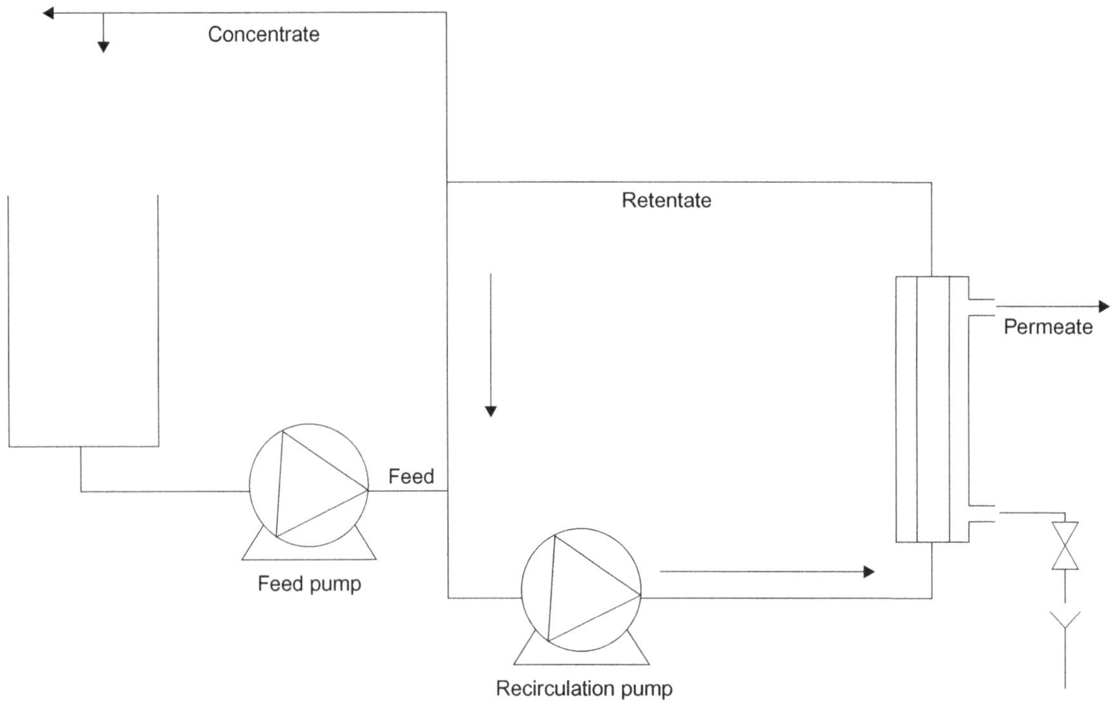

FIGURE 9.6 Schematic of a batch (closed-loop) system.

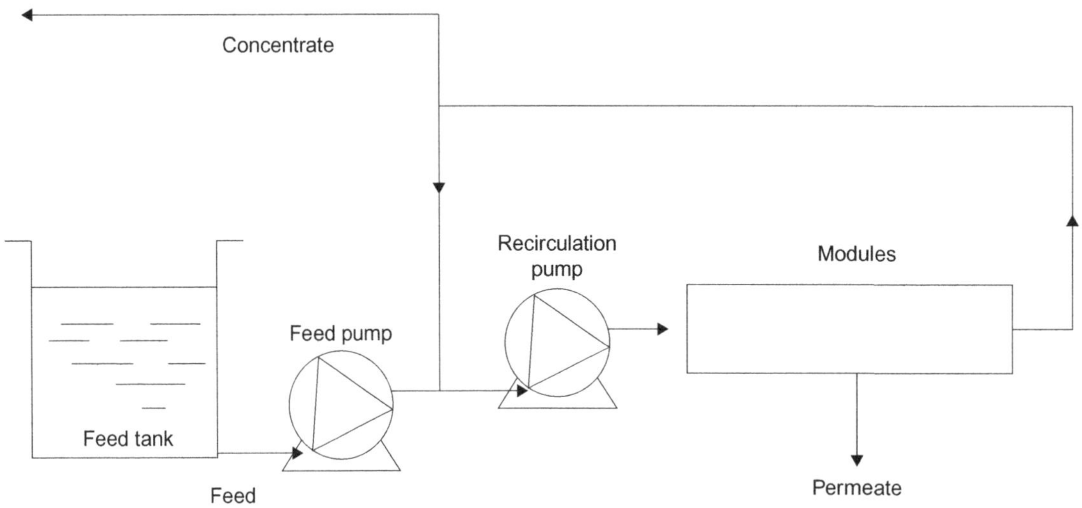

FIGURE 9.7 Schematic of a closed loop feed and bleed system.

required, high retentate solids must be handled by the cross-flow filtration system. For instance, when a 95% recovery is desired, the concentration of solids in the loop must be 20 times higher than the initial feed concentration (assuming almost quantitative retention by the membrane). If the filtration proceeds beyond the 95% recovery, much higher solids concentration in the retentate loop will result which could adversely affect the flux. Figure 9.7 shows the schematic of a batch feed and bleed system.

A constant final concentration in the retentate loop can be maintained by bleeding out a small fraction, either out of the system or to some other location in the process. This operation is described as a batch feed and bleed and is commonly used in the processing of many high value biotechnology products such as batch fermentations to recover vitamins, enzymes and common antibiotics [12]. The CFF system will require larger surface area since the system must be designed at the flux obtained at the final concentration factor (e.g., 20 for 95% recovery).

Continuous Mode. When large volumes are processed the batch feed and bleed system is replaced with a continuous system shown in Figure 9.8. The size of the feed tank is much smaller compared to that for the batch system. However, since the concentration of solids changes with time, the permeation rate decreases with time. This requires the adjustment of feed flow to the recirculation loop. This value is obtained by adding total permeation rate to the bleed rate. The concentration buildup in the continuous feed and bleed mode of operation is somewhat faster than the batch mode, which translates into a higher surface area requirement due to lower flux at higher solids concentration. Such an operating configuration, however, serves very well in many large-scale fermentation broth clarifications (e.g., common antibiotics such as penicillin and cephalosporin) and is used when long holding times are not a concern.

For continuous processes, the lowest possible system dead volume will enable the operation with low average holding times. This may be important in some applications, especially those involving bacteria-laden liquids. Low system dead volume is also desirable for batch or continuous processes to minimize the volumes of cleaning solutions required during a cleaning cycle.

Diafiltration. The product purification or recovery objective in most UF operations is achievable by concentrating the suspended particles or micro solutes retained by the membrane while allowing almost quantitative permeation of soluble products (such as sugars, salts, low molecular weight antibiotics) into the permeate. This approach to concentration of solids obviously has limitations since concentration polarization effects limit recoveries. This limitation can be overcome by the use of diafiltration [1,13,14]. The process involves the selective removal of a low molecular weight species through the membrane by the addition and removal of water. For example, in many antibiotics recovery processes, the broth is concentrated two- to fivefold (depending on the extent of flux reduction with concentration). This corresponds to a recovery of 50 to 80%. Higher recoveries are obtainable by adding diafiltration water or solvent in nonaqueous medium. The permeate leaving the system is replaced by adding fresh water, usually through a level controller, at the rate which permeate is removed. Diafiltration efficiency can be varied by the mode of water addition. Figure 9.9 shows the schematic for a batch and continuous diafiltration process. Diafiltration can be performed at higher temperatures to facilitate higher permeation rates. A possible disadvantage would be the dilution of the product requiring further concentration (e.g., by evaporation).

5.3 Single vs. Multistage Continuous System

Single stage continuous configuration may not be economical for many applications since it operates at the highest concentration factor or lowest flux over most of the process duration. Multistage continuous systems on the other hand, can approximate the flux obtained in the true batch mode, depending on the number of stages. The concentrate from each stage becomes the feed to the next stage. The number of stages required will depend on the final recovery or retentate concentration. Figure 9.10 shows the schematic of a three-stage continuous system.

The optimum number of stages will depend on the application, but typically lie between 2 and 4, with the

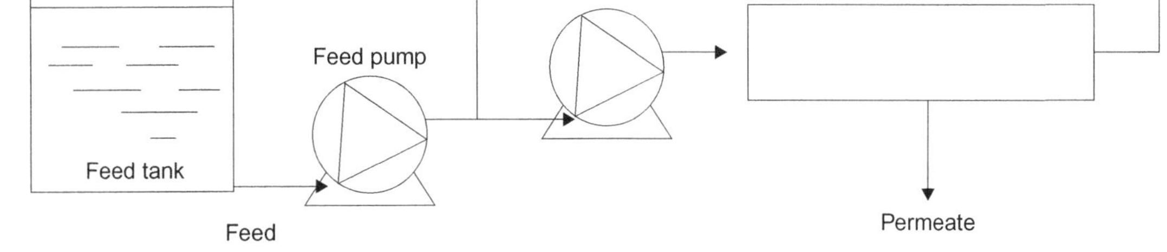

FIGURE 9.8 Continuously fed closed-loop batch feed and bleed system.

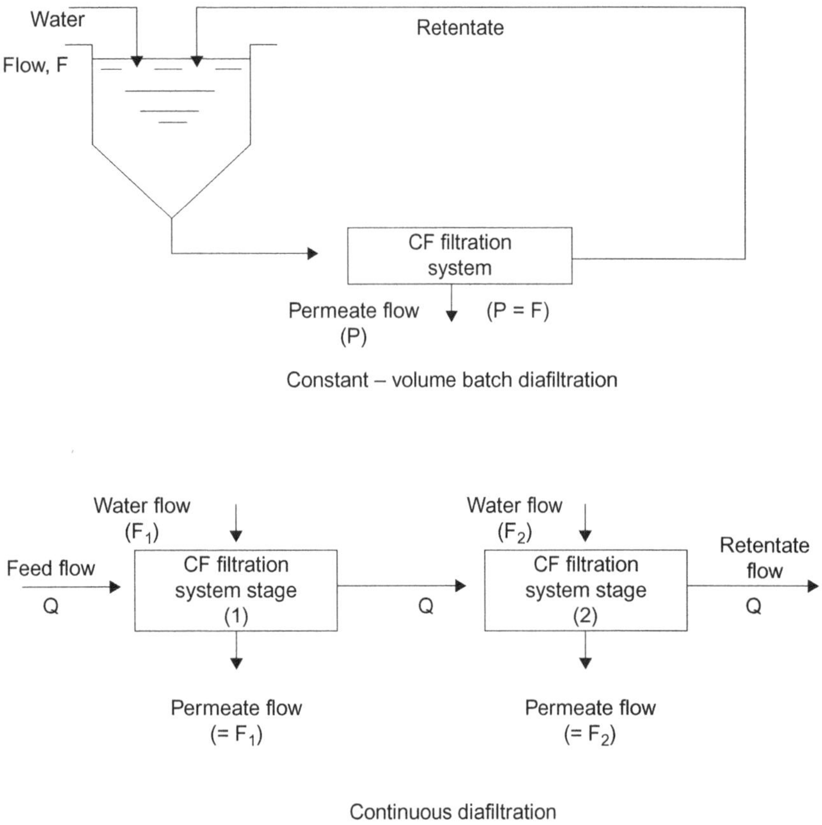

FIGURE 9.9 Batch and continuous diafiltration process modes.

greatest benefit resulting from a single stage to a two stage continuous system. The biggest advantage in using multistage continuous configuration, especially in fermentation and biotechnology applications, is the minimization of residence time, which may be crucial in preventing excessive bacterial growth or to handle heat labile materials [15]. The other advantage of a continuous system is the use of a single concentrate flow control valve. As membrane fouling and/or concentration polarization effect begins to increase over the batch time, flux decreases. This requires the continuous or periodic adjustment of concentrate flow, which may be accomplished with the ratio controller.

One disadvantage with a multistage system is the high capital cost. It is necessary to have one recirculation pump per loop, which drives the power requirements, and operating costs much higher compared with the batch feed and bleed configuration.

6.0 PROCESS DESIGN ASPECTS

6.1 Minimization of Flux Decline with Backpulse or Backwash

Almost all cross-flow filtration processes are inherently susceptible to flux decline due to membrane fouling (a time-dependent phenomenon) and concentration polarization effects, which reflect concentration buildup on the membrane surface. This means lower flux (i.e., product output), which could drive the capital costs higher due to the requirement of a larger surface area to realize the desired production rate. In some situations, the lower flux could also result in lower selectivity, which means reduced recoveries and/or incomplete removal of impurities from the filtrate. For example, removal of inhibitory metabolites such as lactic acid bacteria [16] or separation of cells from broth while maximizing recovery of soluble products [2].

Backwashing or back pulsing with permeate can help remove excessive membrane deposits and hence minimize flux decline [3]. Cross-flow micro- and ultra filters typically operate as surface filtration devices with insignificant pore plugging. If severe pore plugging occurs, back pulse will most likely be ineffective in preventing precipitous flux decline. This type of irreversible fouling may only be corrected by cleaning by chemical and/or thermal heat treatment.

An essential difference between a back pulse and a backwash is the speed and force utilized to dislodge accumulated matter on the membrane surface. In back pulsing, periodic counter pressure is applied, typically in a fraction of a second (0.1–0.5 seconds), while generating high permeate backpressure (up to 10 bar). Backwash on the other hand is relatively gentle where permeate backpressure values may increase up to 3 bar over a few seconds

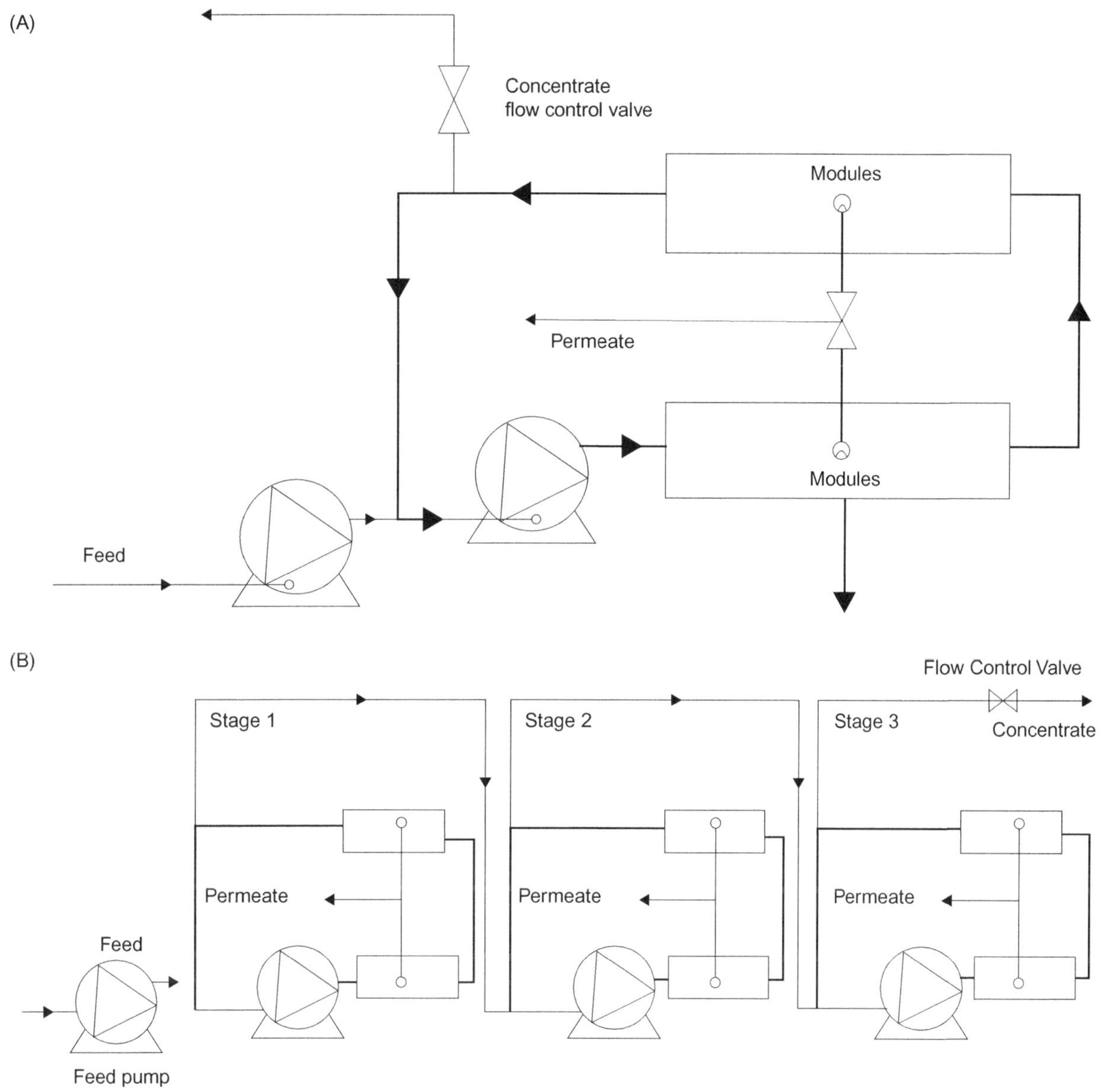

FIGURE 9.10 (A) Single-stage continuous cross-flow filtration system. (B) Continuous multistage cross-flow filtration system.

duration. Backwash is commonly used with polymer MF/UF filters due to their lower pressure limitations [4,17] compared with inorganic MF/UF filters where back pulsing is used [3]. The maximum benefit of back pulse or backwash is obtained when the retentate pressure during instantaneous reverse filtration is lowest and the applied permeate backpressure is highest.

Depending on the operating configuration, a periodic back pulse may be applied on the entire filtration system or when several modules are operating in series, subsequent application will produce more effective results. In the latter case, the retentate pressure may be higher as a result of pressure loss through the interconnected feed channels. It is recommended that when a back pulse or backwash is used, it is applied for the shortest duration possible (to minimize the loss of productivity), it uses minimum permeate volume, and begins simultaneously with the filtration process [3].

Back pulsing is less effective for some smaller pore diameter UF membranes (MWCO <30,000 or pore diameter less than 0.02 μm) and where dense layers are formed or gelatinous products are filtered. It is important to bear in mind that, although back pulsing has the ability to minimize the concentration polarization effects and produce a higher average flux, a certain portion of the permeate is consumed (1 to 3% by volume). If permeate is the product of interest, then the net realized flux will be average flux minus permeate volume used during back pulsing.

6.2 Uniform Trans Membrane Pressure Filtration

In the conventional cross-flow filtration described in the previous section, the trans membrane pressure (TMP) along the feed flow channels varies substantially from the feed end of the module to the exit or retentate end. This occurs due to the pressure loss in the feed channels to maintain the desired flow rate (and hence cross-flow velocity). The shell side or the permeate side is held at a constant pressure. There may be several important consequences which can contribute to a relatively lower flux or loss in separation efficiency. A major consequence is the formation of a nonuniform layer of suspended solids, colloidal matter, and/or gel-forming micro solutes retained on the membrane. It is not uncommon to experience a TMP value up to 50% higher at the module inlet compared to that at the outlet, especially at high shear rate or cross-flow velocity. This could result in a substantially lower average flux. In some applications (e.g., milk or cheese concentration, whey concentration and fermentation broth clarification for product recovery) significant differences in the retention characteristics have also been observed [18]. In many biotechnology related applications, where MF or UF membranes are used, the primary objective is to retain particles (e.g., whole cells or lysed cells, yeast, colloidal matter, and/or macro solutes such as enzymes, pyrogens, proteins, and in some situations oily emulsions). In order to accommodate the wide variations in particle size distributions, a pore diameter is selected that is small enough to retain all the particles or macro solutes, but large enough to allow the permeation of smaller molecular weight soluble products such as common antibiotics, mono- and disaccharides, organic acids and soluble inorganic salts.

Nonuniform TMP values over the filtration surface area may cause substantial (up to 50%) reduction in the product recovery in the permeate. A novel approach to improving the flux and/or product recovery utilizes the concept of a uniform trans membrane pressure [3,19]. This is achieved by varying the permeate side pressure with an independent recirculation pump to adjust the TMP to a constant value. A schematic of the UTP and conventional cross-flow configuration is shown in Figures 9.11 and 9.12, respectively. The TMP profiles for

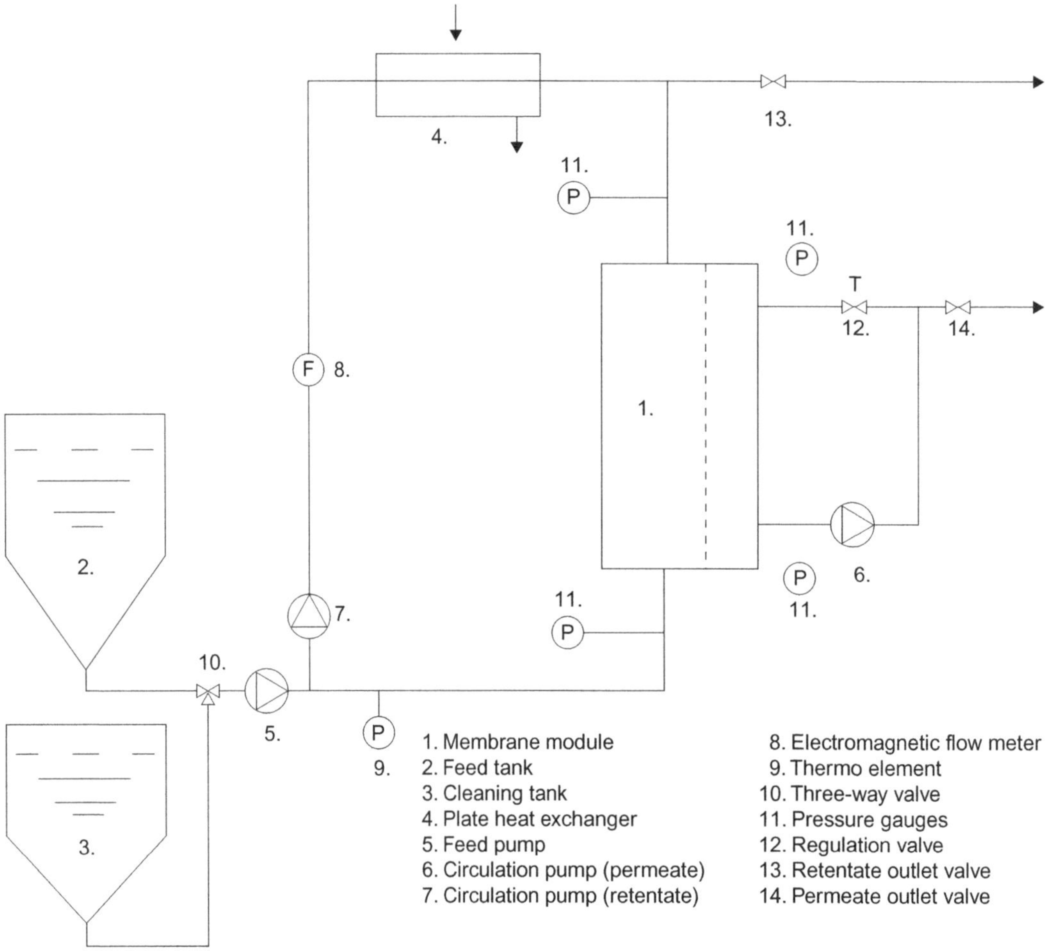

FIGURE 9.11 Schematic of microfiltration with uniform transmembrane pressure.

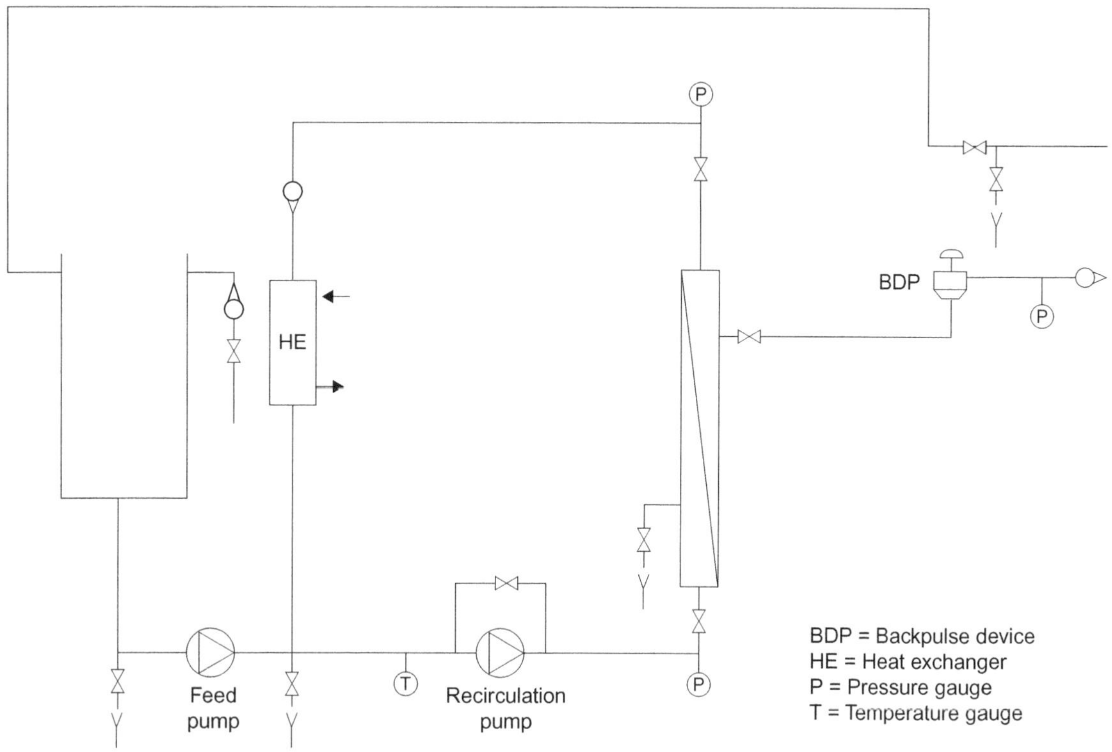

FIGURE 9.12 Schematic of cross-flow filtration feed and bleed system equipped with backpulsing capability.

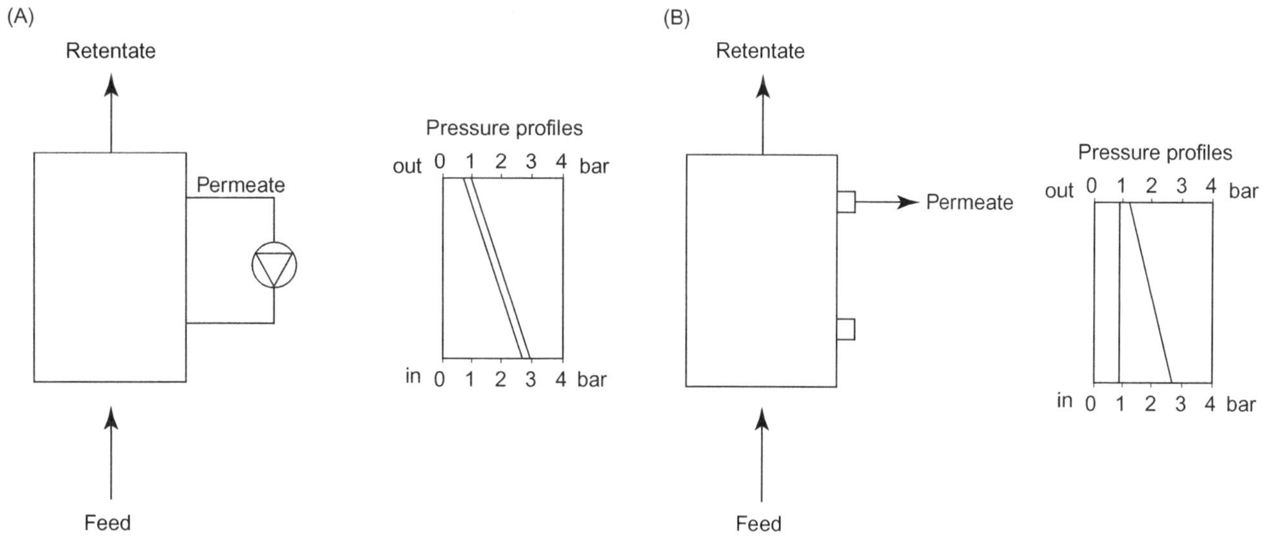

FIGURE 9.13 Transmembrane pressure profiles: (A) uniform transmembrane pressure operation, and (B) conventional cross-flow operation.

the two operational modes are shown in Figure 9.13. Flux improvements up to 500% have been achieved compared with the conventional cross-flow mode in many important food, beverage and biotechnology applications.

An additional benefit is reduced fouling which means longer duration of operation for batch processes and easier cleaning of membrane modules for repeated usage. The only major requirement is the ability of the membrane structure to withstand backpressures up to 5 bar on shell side when filtering high viscosity products such as gelatins, or feed streams with high dissolved solids (20 to 70 wt.%).

6.3 Effect of Operating Parameters on Filter Performance

A number of operating parameters need to be studied to optimize the overall filtration performance. Critical among these are the cross-flow velocity, trans membrane pressure, pore diameter, or MWCO and concentration of the retained species at the end of a batch operation or steady state concentration in continuous filtration. This latter parameter can be related to the recovery of product in the permeate or retentate. Other important operating variables are temperature (and hence viscosity), pH, back pulse or backwash, and pretreatment.

Membrane Pore Diameter or Molecular Weight Cutoff. The value of membrane pore diameter will have a major influence on the permeation and separation characteristics for most process filtration applications. The intrinsic membrane permeability is related to the pore diameter for many microfiltration membranes whereas, for ultrafiltration membranes, it is typically indicative of the solute retention properties. Tables 9.8 and 9.9 provide typical permeability and retention data for many common MF and UF membranes, respectively [1–3,7].

The permeability and retention characteristics listed in the tables, however, should only be used as a guide since the actual filter performance may be dependent on a number of other variables and operating conditions. In addition, for many MF/UF membranes, especially those made of polymeric materials; initial flux and retention properties may significantly alter with repeated use in aggressive conditions or over a longer (6 months to 1 year) period of operation.

It is evident that the smaller the pore diameter, the lower the pure solvent (in most cases water) flux, and the higher the ability to retain macro solutes, colloidal and particulate matter. Users should also be aware that pure solvent flux values are seldom realized in practice and are often at least about an order of magnitude lower in most industrial applications due to effects of fouling and concentration polarization. As a rough rule of thumb, for maximum retention, the pore diameter should be at least about 40 to 50% lower than the smallest particle diameter under the operating conditions. This includes consideration of shear or particle agglomeration/deagglomeration effects. The nominal MWCO on the other hand should be at least 20 to 30% of the smallest molecular weight of the species to be retained. This is due to the fact that for most membranes, particularly polymeric UF, the MWCO characteristics may be diffused [8] rather than sharp (see Section 4.1). Further, secondary layer formation on the membrane surface due to adsorption, fouling and gel polarization will also influence the retention of UF membranes [1,3].

Practical considerations, however, require a compromise between the ideal goals and process economics. One major factor is the lack of reliable information and/or

TABLE 9.8 Typical Permeability Values of Cross-Flow Microfilters*

Pore Diameter μm	Membrane Material	Manufacturer/ Trade Name	Permeability® L/hr-m^2
0.03	Polycarbonate	Nuclepore, Pall	20–300
0.05	Polyethylene/polypropylene	Memtek, Celgard	50–250
	Cellulosic polymers	Koch, Millipore	400–600
	Polysulfone	Amicon, Millipore	
0.1	Nylon	Enka	1800
	Zirconia	Membralox®	1000–2000
	Polysulfone	Amicon	
		Koch, Millipore	
0.14-0.2	Fluoropolymer	W.L. Gore, Millipore	600
	Zirconia	Kerasep	
0.2	PVDF	Durapore	7000
	Polyolefins	Celgard, Memtek,	1500
	Carbon	Carbone Lorraine	2000–3000
	α-Alumina	Membralox®, Ceraflo®	
0.45	PVDF	Durapore	17,100
0.5	Stainless steel	Pall	1500
0.5–1	α-Alumina	Membralox®, Cerano®	5,000–10,000
2–5	Polycarbonate	Nuclepore	50,000–100,000
	α-Alumina	Membralox®	15,000–45,000

with water at 1 bar and 20°C unless otherwise noted.

TABLE 9.9 Retention Characteristics of Cross-Flow Ultrafilters*

Molecular Weight Out Of (MWCO)	Membrane Material	Manufacturer	Remarks®
1000	Cellulosic polymers Polyethersulfone	Amicon Koch	
5000	Polysulfone Polyacrylonitrile	Koch Asahi	
10,000	Zirconia Cellulosic polymers Polysulfone γ-Alumina	Sartorius Fluid Systems/UOP USF	Carbosep M5 Membralox® 5 nm
20,000	Polyamide Zirconia Cellulosic polymers Polysulfone Polyamide	Dorr-Oliver Tech Sep Sartorius Koch, Millipore Hoechst	Carbosep M4
50,000–75,000	Cellulosic polymers Fluoropolymer Polysulfone Zirconia	Tech Sep USF	Carbosep M1 Membralox® 0.02 μm
100,000	Fluoropolymers Polyolefins	Koch Hoechst-Celanese Memtek	Celgard 2400
200,000	Fluoropolymer	Koch	Carbosep M9
300,000	Polysulfone Zirconia	Koch. Millipore Tech Scp USF	Membralox® 0.05 μm
500,000	Polyamide Fluoropolymer	Dorr-Oliver Koch	

May show equivalent or lower MWCO depending on solute and may be influenced by operating conditions.

molecular weight distribution of macro solutes. As a result, application specialists or process engineers typically recommend a pore diameter which is about 75% of the smallest particle size or a MWCO value of about 50–60% lower than the smallest macro solute. The objective is to maximize flux without sacrificing solute retention below the set minimum requirements.

Cross-Flow Velocity. The cross-flow velocity, which is also a measure of the shear or turbulence in the flow channels, may have a strong influence on flux. The actual shear or turbulence will depend on several factors such as channel diameter; viscosity and density of retentate and can vary over the duration of the filtration (especially for batch operations). This can be characterized by the calculation of Reynolds's number on the retentate stream. High Reynolds's numbers (>4000) indicate turbulent flow whereas those below 2000 show laminar flow. The objective is to use a high cross-flow velocity to maximize flux by minimizing the gel polarization layer within the constraints imposed by the allowable pressure-drop or system limitations. It should also be noted that for many applications flux increases with cross-flow velocity. This is illustrated in Figure 9.14 [20]. The extent of flux improvement will depend on process stream, flow regime (laminar or turbulent) and characteristics of the gel polarization layer formed due to concentration buildup at the membrane/feed interface [3].

Blatt et al. [21] have shown that the mass transfer coefficient can be related to the cross-flow velocity by

$$k \propto R\,e^{a}\,Sc^{0.33} \tag{9.2}$$

The value of k can be approximated by

$$k \propto D/\delta \tag{9.3}$$

The value of a can vary from 0.3 to 0.8 in laminar flow and 0.8 to 1.3 in turbulent flow. In the absence of particles (e.g., cells):

$$k \propto v^{0.3-0.5} \quad \text{(low particle loading)} \tag{9.4a}$$

$$k \propto v^{0.6-08} \quad \text{(high particle loading)} \tag{9.4b}$$

For turbulent flow:

$$k \propto v^{0.8-10} \quad \text{(low particle loading)} \tag{9.4c}$$

$$k \propto v^{1.1-1.3} \quad \text{(high particle loading)} \tag{9.4d}$$

This behavior has been explained by the so-called "tubular pinch effect," which enhances movement of

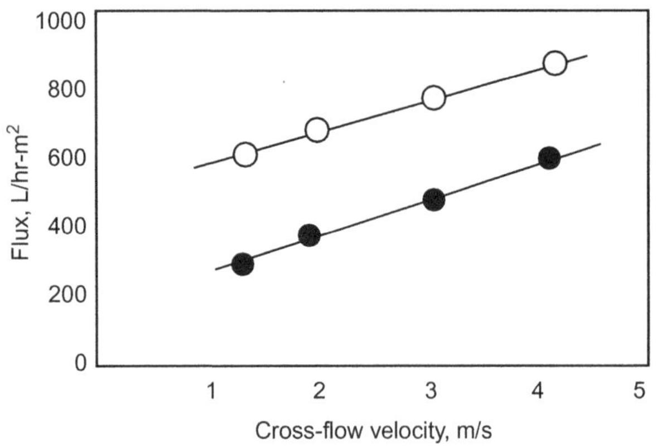

FIGURE 9.14 Effect of cross-flow velocity on flux. Yeast concentration, dry-g/L: (○) 8.5; (●) 30.

particles away from the boundary layer thus reducing concentration polarization effect (see Sec. 3.3).

For turbulent flow, the pressure drop along the flow channel may be estimated by using the following empirical approximation:

$$\Delta P \propto v^{1.75-1.85} \tag{9.5}$$

Under laminar flow conditions,

$$\Delta P \propto v \tag{9.6}$$

This indicates that a higher cross-flow velocity under turbulent conditions can result in more than proportional increase in the pressure drop requiring larger pump discharge pressure to maintain a specified recirculation rate. This limits the number of modules that may be placed in series to minimize capital costs. Typical range of cross-flow velocity values is 2 to 7 m/s. The choice of pump is critical to obtain efficient fluid recirculation. It is critical to understand the shear sensitivity of the fluid/particle to be processed to determine the optimal cross-flow velocity in situations where shear-sensitive materials are involved.

Concentration of Solute or Particle Loading. It is essential to distinguish or separate the effects of membrane fouling from concentration polarization effects.

Membrane Fouling. Pretreatment of the membrane or feed solution prior to filtration may be desirable within allowable limits. The various treatment options are discussed in Section 6.3. At the start of a filtration run, the solute or solids concentration is relatively small and progressively builds as the permeate is removed from the system. If a substantial flux decline is observed at low solids concentration, membrane-fouling aspects are believed to be important. A flux decrease with an increase in solids concentration is largely due to concentration polarization and can be minimized through efficient fluid hydrodynamics and/or back pulsing [3,22,23].

Several approaches have been developed to control membrane fouling. They can be grouped into four categories: (*a*) boundary layer control [21,24–26]; (*b*) turbulence inducers/generators [27]; (*c*) membrane modifications [28–30]; and (*d*) use of external fields [31–34]. In CFF membrane, fouling can be controlled utilizing a combination of the first three approaches (*a*, *b* and *c*). The external field approach has the advantage of being independent of the hydrodynamic factors and type of membrane material [35].

Membrane fouling is primarily a result of membrane-solute interaction [36]. These effects can be accentuated or minimized by proper selection of membrane material properties such as hydrophobicity/hydrophilicity or surface charge, adjustment of pH, ionic strength and temperature leading to solubilization or precipitation of solutes. Increased solubilization of a foulant will allow its free passage into the permeate. If this is undesirable, precipitation techniques may be used which will enhance the retention of foulants by the membranes. Membrane fouling is generally irreversible and requires chemical cleaning to restore flux.

It is important to recognize that fouling in bioprocessing differs from that occurring with chemical foulants. Bio fouling originates from microorganisms. Microbes are alive and they actively adhere to surfaces to form biofilms. Thus, in addition to flux decrease there may be significant differences in solute rejection, product purity, and irreversible membrane fouling resulting in reduced membrane life. For economic viability of CFF it is imperative that a good and acceptable cleaning procedure is developed to regenerate fouled membranes without sacrificing membrane life.

Concentration Polarization. The concentration of the species retained on the membrane surface or within its porous structure is one of the most important operating variables limiting flux. Concentration effects in MF/UF can be estimated by using the following most commonly used correlation [12,37].

$$J = k \ln[C_g/C_b] \tag{9.7}$$

where

J = flux
k = mass transfer coefficient
C_g = gel concentration at the membrane surface
C_b = bulk concentration of solute retained by the membrane

In membrane filtration, some components (dissolved or particulate) of the feed solution are rejected by the membrane and these components are transported back into the bulk by means of diffusion. The rate of diffusion will depend on the hydrodynamics (laminar or turbulent) and on the concentration of solutes. If the concentration

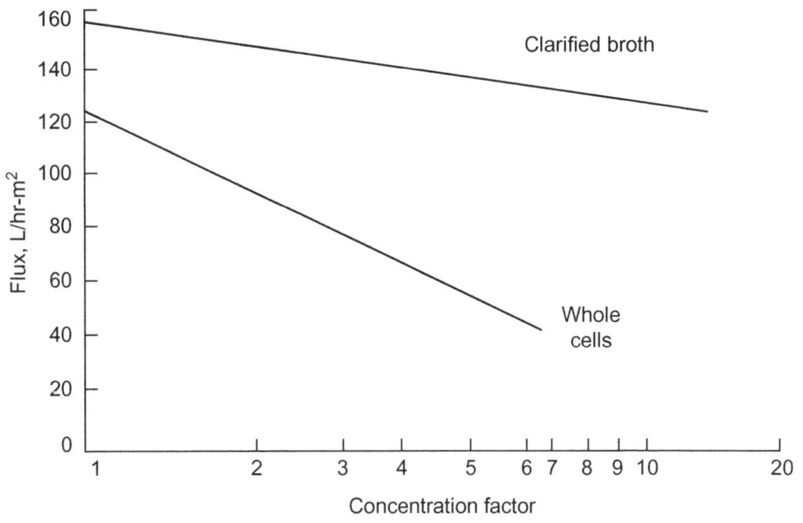

FIGURE 9.15 Typical dependence of flux on concentration.

of solute at the surface is above saturation (i.e., the solubility limit) a "gel" is formed. This increases the flow resistance with consequential flux decrease. This type of behavior, for example, is typical of UF with protein solutions.

In practice, however there could be differences between the observed and estimated flux. The mass transfer coefficient is strongly dependent on diffusion coefficient and boundary layer thickness. Under turbulent flow conditions particle shear effects induce hydrodynamic diffusion of particles. Thus, for microfiltration, shear-induced diffusivity values correlate better with the observed filtration rates compared to Brownian diffusivity calculations [6]. Further, concentration polarization effects are more reliably predicted for MF than UF due to the fact that macro solutes diffusivities in gels are much lower than the Brownian diffusivity of micron-sized particles. As a result, the predicted flux for ultrafiltration is much lower than observed, whereas observed flux for micro filters may be closer to the predicted value.

Typically MF fluxes are higher than those for UF due to their higher pore diameter values, which contribute to higher initial fluxes. However, polarization effects dominate and flux declines with increase in concentration (or % recovery) more sharply in MF than in UF, in general accordance with Eq. (9.4a–d) under otherwise similar conditions. Figure 9.15 shows the typical dependence of flux on concentration [14]. The higher the concentration of the retained species on the membrane compared with its initial value, the higher will be % recovery. However, if the desired product is in the permeate, then % recovery will be dependent on the ratio of the batch volume to its final value for batch filtration or the ratio of concentrate in permeate to that in the feed for continuous filtrations [38,39].

Trans Membrane Pressure. The effect of trans membrane pressure on flux is often dependent on the influence of concentration polarization at a specified cross-flow velocity and solids loading. For MF or UF at low solids concentration and high cross-flow velocity, flux may increase linearly with TMP up to a certain threshold value (1 to 3 bar), and then remain constant or even decrease at high TMP values. This is illustrated in Figure 9.16 [20]. At high solids loading, the threshold value may be lower (0.5 to 1.5 bar) and may also require higher cross-flow velocity to offset gel polarization effects. For each application the optimum value may be considerably different and must be empirically determined.

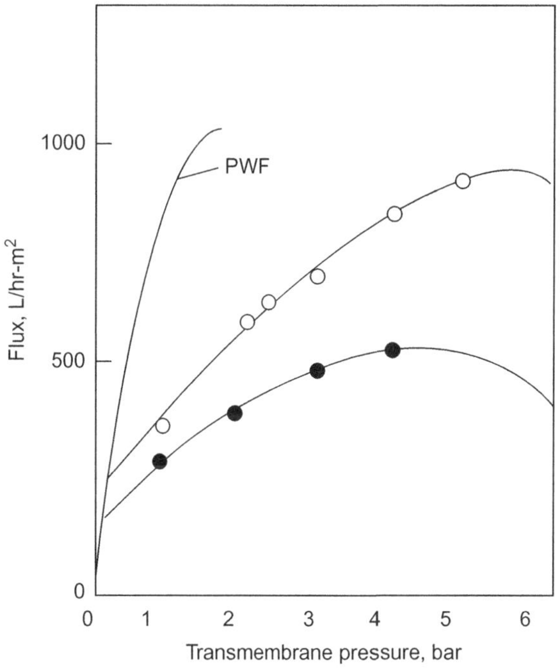

FIGURE 9.16 Effect of transmembrane pressure on flux. Yeast concentration, dry-g/L: (○) 8.5; (●) 30.

An optimum TMP value is one, which maximizes flux without additional energy costs and helps minimize the effects of membrane fouling. In general, the higher the solids concentration, the higher the cross-flow velocity and hence TMP to balance the effects of concentration polarization. In most practical situations, the cross-flow velocity and TMP may be interrelated. One useful approach involves performing pressure excursion studies to determine the optimal flux by varying the TMP at a fixed cross-flow velocity until the threshold TMP is attained and then repeat the tests by selecting a lower or higher cross-flow, depending on the observed trend [40].

In many biotechnology applications, such as fermentation broth clarifications to produce common antibiotics, optimal values of TMP are in the range of 2 to 3 bar (15 to 30 psi) especially at high cell mass concentrations (>30 wt.%) and cross-flow velocity range of 4 to 7 m/s [2,40]. In the operation of commercial systems, often several modules (2 to 4) are interconnected to minimize pump costs. This results in significantly higher TMP on the feed end compared to that at the exit (or retentate). Thus, the TMP on the exit end may be closer to the optimal value whereas at the inlet it may be substantially higher (6 or 7 bar), unless the permeate backpressure on each module is controlled independently.

Temperature and Viscosity. The operating temperature can have a beneficial effect on flux primarily as a result of a decrease in viscosity [3,41]. There is an additional benefit for shear thinning viscoelastic fluids, where the viscosity reduces with an increase in shear (i.e., cross-flow velocity). Typical examples are clarification of fermentation broths and concentration of protein solutions [3,42]. It must be noted that for most fermentation and biotechnology related applications, temperature control is necessary for microbial survival and/or for product stability (e.g., antibiotics, enzymes, proteins and other colloidal materials).

For mass transfer controlled operations, such as when concentration polarization is dominant, flux enhancement due to temperature increase will depend on the value of mass transfer coefficient. This is related to the cross-flow velocity, diffusion coefficient and viscosity [21]. Thus, for example, even though the viscosity may be reduced by a factor of 5, the increase in flux may only be about 50%, due to the nonlinear dependence of flux on viscosity in these situations. For the permeation of a clean liquid (solvent) across a micro porous membrane, however, flux increase may be predicted by the Stokes-Einstein relation [1] and will be approximately inversely proportional to the viscosity of the permeate.

pH, Isoelectric Point and Adsorption. In the filtration of proteins and colloidal substances, the solution pH can have a measurable effect on flux, especially around the isoelectric point where they tend to destabilize and precipitate. In addition, the surface charge or isoelectric point of the membrane material must also be considered. For example, most inorganic membranes are made out of materials such as silica, zirconia, titania and alumina, which have a charge on the surface with isoelectric pH variation from 2 to 9 [3]. Similarly polymeric membranes such as cellulose acetate, polyamides and polysulfones also carry surface charges [7]. Surface charge effects may alter the fouling resistance due to changes in the zeta potentials and could have a substantial influence on flux and/or separation performance. For example, proteins with isoelectric point of 8 to 9 will have a positive charge in a neutral or acidic solution and negative charge in alkaline solutions (pH > 9). However, if the above proteins in a neutral medium are filtered through an alumina membrane (isoelectric pH 9) there will be a minimal adsorption on the membrane due to similar charge characteristics. At a solution pH of 8 or 9, which is also the isoelectric pH of protein, however, proteins may precipitate out of solution. This may have a beneficial effect on the flux through a MF or UF membrane. This also illustrates the interactive effects of solution pH, isoelectric pH (of solutes and membrane material) and adsorption.

Feed Pretreatment. In most fermentation and biotechnology related applications, feed pretreatment is not a viable option. This is due to the fact that any alterations in the feed properties, especially through the addition of precipitants or flocculants, will likely contaminate the product and or adversely affect its characteristics.

Prefiltration is recommended when applicable to remove larger particles and other insoluble matter. However, minor pretreatment chemistries may be allowable, such as pH adjustment to precipitate or solubilize impurities or foulants to maximize flux or retention. For example, adjusting the pH away from its isoelectric point can reduce protein adsorption and fouling [5]. The selection of a suitable pore diameter or MWCO value is done on the basis of the smallest particle size or smallest macro solute present in the feed.

6.4 Membrane Cleaning

Likewise, to the inevitable phenomena of membrane fouling, all membrane based filtration processes require periodic cleaning. Without a safe, practical, reproducible, cost effective and efficient cleaning procedure, the viability of cross-flow filtration may be highly questionable. Membrane cleaning process must be capable of removing both external and internal deposits. In some special situations, such as strongly adsorbed foulants, recirculation alone may not be adequate and soaking of the membranes in the cleaning solutions for a certain period of time will be necessary. The ultimate success of a membrane process will be largely impacted by the ability of the cleaning procedure to fully regenerate fouled membranes to obtain reproducible initial flux at the start of the next filtration cycle.

The ease of finding an effective cleaning process often depends on the thermal and chemical resistance of the membrane material. In other words, the higher the resistance, the easier it is to develop a suitable cleaning procedure. The choice of a cleaning solution depends on several factors such as the nature of the foulants, and material compatibility of the membrane elements, housing and seals. A few general guidelines are available concerning the removal of foulants or membrane deposits during chemical cleaning [3,41].

Common foulants encountered in biotechnology related applications are inorganic salts, proteins, lipids and polysaccharides. In some food or biochemical applications, fouling due to the presence of citrate, tartrate and gluconates may be encountered. Inorganic foulants (e.g., precipitated salts of Ca, Mg and Fe) can be removed with acidic cleaners, whereas proteinaceous and other biological debris can be removed with alkaline cleaners with or without bleaching agents or enzyme cleaners. Many acidic and alkaline cleaners also contain small quantities of detergents, which act as complexing or wetting agents to solubilize or remove insoluble particles, colloidal matter and/or to break emulsions. Oxidizing agents such as peroxide or ozone are also sometimes used to deal with certain type of organic foulants [43]. In addition, organic solvents may be required to solubilize organic foulants that are insoluble in aqueous cleaning solutions.

For many polymeric MF/UF membrane modules, material compatibility considerations limit the use of higher cleaning temperatures and strongly acidic/alkaline/oxidizing solutions. Further, with time and repeated cleaning, polymeric filters are susceptible to degradation. The service life of a hydrophobic type is typically a period of 1 to 2 years and up to 4 years for fluoropolymer based membranes [5]. On the other hand, inorganic membranes can be cleaned at elevated temperatures in strongly alkaline or acidic solutions and can withstand oxidizing solutions or organic solvents. The typical useful service life of inorganic membranes exceeds 5 years and may be used for 10 years or longer with proper cleaning, and good operating and maintenance procedures [3,5].

A careful choice of cleaning solutions and procedures will extend the service life of the membrane. In many polymer membrane filtration systems, membrane replacement costs constitute a major component of the total operating cost. Extending the service life of the membrane modules will have a major impact on the return on investment and can be a determining factor for the implementation of a membrane-based filtration technology. Table 9.10 summarizes the various key parameters that must be considered in developing a cleaning regimen to regenerate fouled membranes.

Product losses during cleaning may be important especially when high recoveries (>95%) are required and the desired product is located in the retentate phase. Additional product loss will occur in the fouled membrane elements. These combined losses may range from 0.5% to 3% which is significant when recovering high value-added product.

TABLE 9.10 Membrane Cleaning: Key Considerations

Type of Foulant	Example	Cleaning Solution	Filter Material Compatibility
Inorganic	Precipitated Ca, Mg, Fe	Moderate to strongly acidic	Some polymeric (PVDF or PTFE) and most inorganic filters.
Organic	Citrate, tartrate gluconate	Acidic/alkaline solutions	Most polymeric or inorganic filters.
Proteins	Enzymes, yeast	Mild to moderately alkaline	Most polymeric and inorganic filters.
	Pectins	Strongly alkaline preferably with chlorine	Some polymeric (PVDF or PTFE) and most inorganic filters.
Biological debris	*E. coli*, bacteria cell walls	Moderately alkaline	Most polymeric or inorganic filters.
Fats/Oils	Stearic acid, oleic acid	Strongly alkaline with oxidizing agents or chloride	Some polymeric (PVDF or PTFE) and most inorganic fillers.
Polysaccharide	Starch, cellulose	Strongly alkaline/acidic or oxidizing solutions	Some polymeric (PVDF or PTFE) and most inorganic filters.

6.5 Pilot Scale Data and Scaleup

Scaling up membrane filter systems must proceed in a logical and progressive series of steps. It is practically impossible to extrapolate data from a laboratory scale system to design a production scale system [44]. To ensure commercial success, it is often necessary to supplement laboratory data with a pilot system capable of demonstrating the viability of the process. This is typically followed-up with extensive testing using demonstration scale or semi-commercial scale filtration systems to obtain long-term flux information and to establish a cleaning procedure to regenerate fouled membrane modules. This exercise is especially important to determine the useful life of membranes. At least a 3 to 6 month testing is recommended regardless of the scale of operation. Pilot scale studies will also allow production of larger quantities of materials for evaluation purposes to ensure that all the separation and purification requirements are adequately met.

It is necessary to ensure that the feed stream characteristics are representative of all essential characteristics, such as age of feed sample, temperature, concentration of all components (suspended and soluble), and pH. The filtration time needed to perform a desired final concentration of retained solids or percent recovery of product passing across the membrane filter (the permeate) must also be consistent with the actual process requirements.

Effects of sample age, duration of exposure to shear and heat, may be very important and must be considered. In the demonstration scale phase, the operating configuration (e.g., batch, feed and bleed, continuous) specified for the production scale system must be used. Careful consideration must be given to the total pressure drop in the flow channels at the desired cross-flow velocity at the final concentration, to ensure proper design of the feed and recirculation pumps.

It is important to generate the flux data on a continual basis as illustrated in Figure 9.17. This type of information is very vital to identify any inconsistencies in the filtration performance and/or to determine if there is any irreversible membrane fouling. Reproducible performance will also be helpful to validate the membrane cleaning regimen for the application.

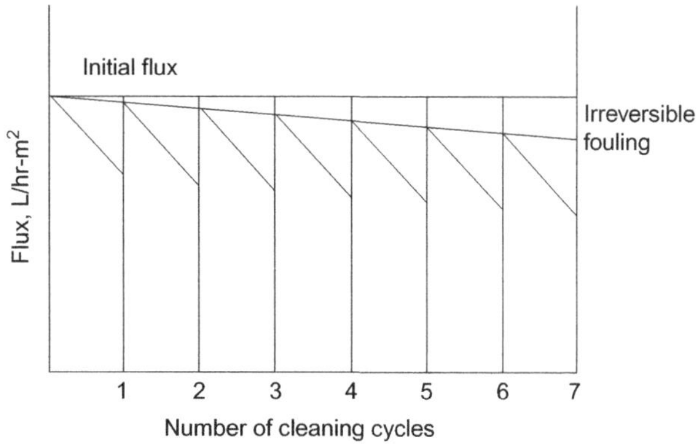

FIGURE 9.17 Filtration performance after several operating cycles and daily cleaning.

6.6 Troubleshooting

Filtration equipment must function in a trouble-free manner and perform in accordance with the design basis. Although most carefully designed, engineered and piloted cross-flow filtration systems will perform to design specifications, occasional failures are not uncommon. For proper troubleshooting of CFF systems, the user must be familiar with the principles of membrane separation, operating and cleaning procedures, influence of operating variables on system performance and equipment limitations.

In this chapter, the principles of membrane separations when operating in the cross-flow configuration are discussed in detail along with the influence of operating variables on flux and separation performance. However, proper start-up and shutdown procedures must be followed to maximize the system performance. For instance, the formation or presence of gas or vapor microbubbles can cause severe pore blockage especially for MF and in some UF applications. Therefore, care must be taken to remove air or gas from the feed and recirculation loop at the start of a filtration run to ensure that no air is drawn or retained in the system. This type of operational problem may not only occur during normal filtration but also during backpulsing.

When troubleshooting the cleaning operation, a good understanding of the foulants and process chemistry is highly desirable. A thorough understanding of materials of construction of the seals/gaskets is required for a proper choice of cleaning regimen. The membrane manufacturer's guidelines must be properly implemented and combined with the process knowledge and feed characteristics. When working with new or dry membranes, it may be necessary to properly wet the membrane elements. For microporous structures, the use of capillary forces to wet the membrane and fill the porosity is recommended.

6.7 Capital and Operating Cost

The manufacture and purification of many biotechnology products derived from fermentation processes involves several separation steps. Up to 90% of the total manufacturing cost may be attributed to various separation processes to raise the concentration of the product in

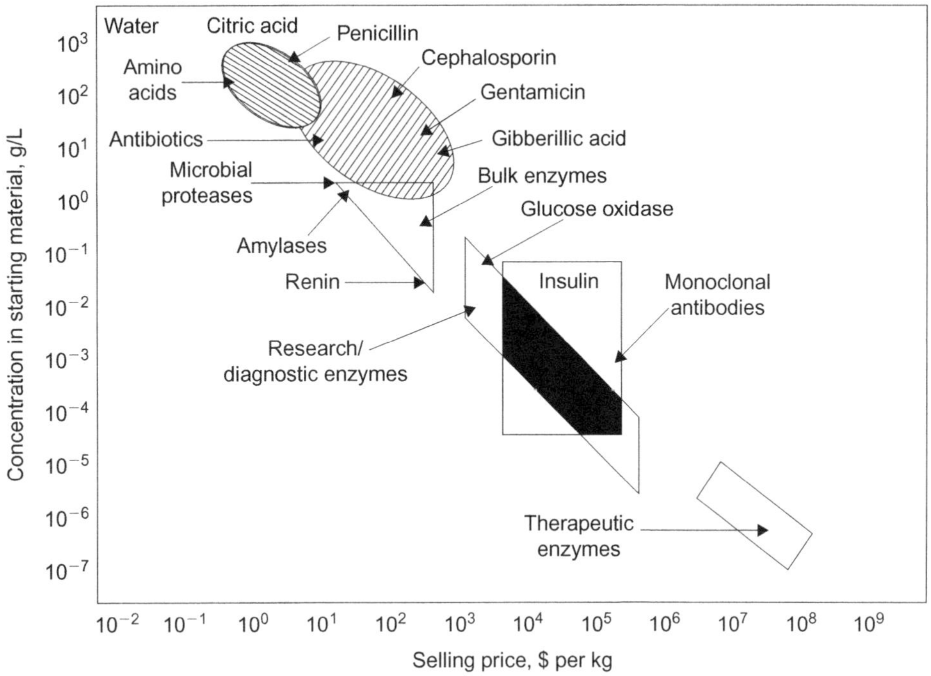

FIGURE 9.18 Separation costs for various biological products as a function of their concentration in starting material.

solution from parts per million or several percent to the final concentrated form. Figure 9.18 illustrates the contribution of separation costs to the selling price of a biological product [45].

The most economical system design is achieved by the consideration of both capital and operating costs. When comparing the overall filter performance against a competing technology, care should be taken to ensure that the total cost or payback is based on the life-cycle rather than solely on the basis of initial capital or operating costs.

It is easier to compare two competing technologies or products on the basis of initial capital cost alone. However, this approach may be erroneous unless operating, maintenance and replacement costs are considered along with differences (or savings) in the value of product recovered or lost. For example, clarification of a fermentation broth with a cross-flow filter may cost up to 4 times higher compared to the capital cost of a pre-coat filter, but the operating cost may be only about 50% of that incurred with pre-coat filter [12]. The disposal cost of the filter aid will also add to the savings of CFF over pre-coat filtration [46]. The higher capital cost can be justified through cost savings yielding a reasonable payback (typically in the range of 3 to 4 years). Cross-flow filtration competes with many traditional separation and filtration technologies such as centrifuges, rotary pre-coat filters, cartridge filters, chemical treatment and settling and a filter press. The advantages and disadvantages of some of these alternatives were briefly discussed in Section 3.0. This section will highlight key items that make up the major portion of the capital and operating costs in cross-flow filtration.

The cross-flow filter accounts for a major portion of the capital cost. The relative percentage contribution to the total capital cost will vary from about 20% for small systems up to 50% for larger systems. Thus, replacement costs, when the CFF has a useful service life of only about a year, can be as much as 50% of the total system cost. Inorganic filters cost more than their polymer counterparts but can last about 5 to 10 years. Capital costs associated with feed pump and recirculation pump(s) represents anywhere from 5 to 15% of the total capital costs. The largest contribution to the operating cost in many cross-flow filtration systems is in the energy consumption for recirculation [1]. For example, in the production of common antibiotics such as penicillin or cephalosporin, high recirculation rates are maintained (corresponding to a cross-flow velocity in the range 5 to 8 m/s) to minimize concentration polarization [2]. Energy requirements under turbulent flow conditions are also significantly higher than under laminar flow situations, under otherwise similar conditions. In addition, total energy costs must also consider heating and/or cooling requirements as well as to deliver the desired head pressure to overcome hydraulic pressure drop.

Operating costs must also consider equipment maintenance, cost of cleaning chemicals and labor costs. CFF systems in general have substantially lower maintenance and labor costs compared with other competing technologies. Cleaning chemical costs are typically low and account for only about 1 to 4% of the total operating costs [3].

6.8 Safety and Environmental Considerations

The proper and efficient operation of a cross-flow filtration system requires a design based on sound engineering principles and must rigorously adhere to safe engineering practices. CFF systems must be equipped with high pressure switches to safely diffuse a high pressure situation and must also use materials and design criteria per American Society for Testing and Materials (ASTM) standards. Proper insulation is required in accordance with Occupational Safety and Health Administration (OSHA) regulations for high surface temperatures or hot-spot when operating at elevated temperatures. For corrosive chemicals, proper handling and disposal procedures must be followed for operator safety.

Containers approved by OSHA and other regulatory agencies must be used when transporting or transferring hazardous chemicals. In addition, proper procedures must be followed when mixing chemicals, either within the manufacturing process or while handling waste solutions.

The majority of CFF processes are operated in a closed configuration which minimizes vapor emissions. Some traditional techniques such as centrifugal processes may generate aerosol foaming in the air (e.g., pathogens) which is highly undesirable.

7.0 APPLICATIONS OVERVIEW

Due to the highly proprietary nature of fermentation of biochemical products, the published descriptions on cross-flow filtration performance are very limited. This section will review some of the more important types of applications where cross-flow filtration is used. The performance descriptions are limited by available published information which is often incomplete. As a result, at best, only qualitative or general comparisons can be made between the various technology alternatives.

7.1 Clarification of Fermentation Broths

Fermentation broths tend to be very dilute and contain complex mixtures of inorganic or organic substances [1,47]. The recovery of a soluble product (MW range 500–2500 dalton) such as an antibiotic, organic acid or animal vaccine from fermentation broth takes several processing steps. The first step is the clarification of broth to separate the low molecular weight soluble product from microorganisms and other particulate matter such as cells, cell debris, husks, colloids and macromolecules from the broth medium [2,48]. In this step, microporous membrane filters (MWCO 10,000 to 500,000 dalton) compete with pre-coat vacuum filter or centrifuge.

When membrane filters are used, the soluble product is recovered in the permeate. This step is followed with diafiltration of the concentrate (continuous or batch) to improve yield. The permeate is then subjected to final concentration.

Many filtration processes operate in a batch configuration at or near ambient conditions (e.g., 20–30°C for penicillin) with some exceptions (e.g., 2–5°C for certain yeast fermentations and 80°C for some higher alcohols).

Batch times can range from 12 to 22 hours depending on the desired final concentration and the required number of diafiltration volumes. At the end of a batch run, membranes are chemically cleaned. Cleaning may take up to 3 hours and involve the use of an alkaline or acidic solution, or both, with a final sanitization step (e.g., 200 ppm NaOCl solution, a dilute solution of sodium bisulphite or a bactericide/fungicide). In some cases, steam sterilization may be performed at the end of each run especially when using inorganic membrane filters.

Today many industrial fermentation broth clarifications are performed using cross-flow MF/UF membrane modules [2,12]. The advantage of CFF over traditional separation processes is not only in superior product flow rates but also in higher yields or lower product losses. Using diafiltration, up to 99% recovery can be obtained [12,49].

7.2 Purification and Concentration of Enzymes

Enzymes are proteins with molecular weights in the range of 20,000 to 200,000 dalton and are predominantly produced in small batch fermenters. UF often combined with diafiltration is widely used in the industry to produce a variety of enzymes such as trypsin, proteases, pectinases, penicillinase and carbohydrates [1,14].

UF offers many advantages over traditional processes such as vacuum evaporation or vacuum evaporation with desalting. These include higher product purity and yields (concentration factor 10 to 50), lower operating costs, ability to fractionate when the molecular sizes of the components differ by a factor of at least 10. The availability of a wide range of MWCO membranes enables the selection of a suitable membrane to maximize flux without substantially compromising retention. UF can also minimize enzyme inactivation or denaturation by maintaining a constant pH and ionic strength. Other techniques such as solvent precipitation, crystallization or solvent extraction may sometimes denature the product owing to phase change [8].

UF performance, however, may be influenced by process variables such as pH, nature of ions and ionic strength, temperature and shear. For example, Melling [50] has reported the effect of pH on the specific enzyme activity of *E. coli* penicillinase in the pH range 5 to 8. Effects of shear inactivation associated with pumping effects are described by O'Sullivan et al. [12] Recessed impeller centrifugal pumps or positive displacement pumps may be used to minimize enzyme inactivation due to shear.

7.3 Microfiltration for Removal of Microorganisms or Cell Debris

In recent years there has been a significant interest in the use of microorganism-based fermentations for the production of many specialty chemicals [51–53].

The product of interest may be produced by either an extracellular or intracellular process relative to the microorganisms. In either of these situations, one of the key steps is the efficient removal of microorganisms or cell debris from the fermentation broth [54,55]. In biotechnology terminology, this step, where cells are separated from the soluble components of the broth, is described as cell harvesting.

Filtration is often preferred over centrifugation due to problems associated with poor separation which results in either reduced product yield or purity. Aerosol generation during centrifugation could be a major problem. This can be alleviated in the CFF mode due to the closed nature of system operation. Additionally, centrifuges may require high energy inputs since there is no appreciable density difference between the bacterial cell walls and the surrounding medium. Pre-coat filtration, when applicable, will suffer from reduced product yield and lower filtration rates (e.g., 0.7 to 16 L/hr-m^2) [52].

The processing steps differ depending on the location of the product relative to the microorganisms. For extracellular products, maximizing broth recovery by clarification is important since the product is in solution. When the product is located within the cell walls, concentration of cell mass is required followed by cell rupture and recovery of products from the cell debris.

Cross-flow microfilters, such as microporous hollow fiber or tubular, are preferred over plate and frame or spiral wound which are prone to plugging due to their thin channel geometry. In addition, CFF can be operated in the continuous mode with backwashing or backpulsing which has a beneficial effect on filtration performance. Process economics dictates the use of high cell concentration to maximize product yields but may hinder the recovery of soluble products (e.g., production of penicillin, cephalosporin). In other situations, high biomass concentrations may hinder the efficient removal (e.g., lactic acid, propionic acid) of inhibitory metabolites [56]. Similar situations exist in the production of acetone-butanol [57] organic acids and amino acids from micro-organism-based fermentations [43].

Concentration of Yeast or *E. coli* Suspensions. The concentration of yeast or *E. coli* is often performed using microporous membrane filters. For example, in the production of ethanol by fermentation, yeast cells are used as biocatalyst [58]. It is necessary to ensure adequate recycling of cell mass to minimize the production of inhibitory products. Typically, a membrane with a pore diameter in the range 0.02 μm to 0.45 μm is used, which represents a good compromise between the requirement to maintain relatively high flux at high cell concentrations while minimizing pore plugging and adsorptive surface fouling. These types of microorganisms are not shear sensitive, which allows the use of high shear rates to reduce concentration polarization effects. Initial cell mass concentration may vary from 5 to 25 gm (dry wt.)/ L. Final concentrations up to 100 gm (dry wt.)/L or cell densities up to 10^{14} cells/L can be achieved by cross-flow filtration with diafiltration [4,20,54].

Table 9.11 shows the performance of polymeric and ceramic filters for the separation and concentration of yeast and *E. coli* suspensions. The ceramic filters, due to their superior mechanical resistance, can be backpulsed to reduce flux decline during concentration. This is illustrated in Figure 9.19 for the filtration of yeast suspension with 0.45 μm microporous cellulose triacetate membrane [4]. Polymeric membranes can be backwashed at pressures up to about 3 bar. The data in Figure 9.20 show the flux improvement with backpulsing using 0.2 μm microporous alumina membrane [20].

7.4 Production of Bacteria-free Water

Bacteria are living organisms composed of a single cell in the form of straight or curved rods (bacilli), spheres (cocci) or spiral structures. Their chemical composition is primarily protein and nucleic acid. Bacteria can be classified by particle sizes in the range of about 0.2 to 2 μm. Some forms of bacteria can be somewhat smaller (~0.1 μm) or somewhat larger, up to 5 μm. Microfiltration can be an effective means of bacteria removal since the pores of a microfilter are small enough to retain most forms of bacteria while maintaining relatively large flow rates for the transport of aqueous solution across the membrane barrier [59].

The relative efficiency of bacteria removal will, however, depend on the level of bacterial contamination and downstream processing requirements. The filter ability to retain bacteria is commonly expressed in terms of the Log Reduction Value (LRV). The LRV is defined as the logarithm of the ratio of total microorganisms in the challenge to the microorganisms in the filtered fluid when a filter is subjected to a specific challenge. A 0.2 μm filter is challenged with *Pseudonomas diminuta* microorganisms and a 0.45 μm filter is challenged with *Serratia*

TABLE 9.11 Concentration of Yeast and *E. coli* with Cross-Flow MF/UF

Feed	Concentration		Pore Diameter μm	Type of Membrane	Average Flux L/h · m²	Reference
	Initial	Final				
Yeast Suspension (reconstituted)	5	80	0.2	Polymeric hollow fiber (polysulfone)	50$^{\Delta}$	[54]
	3	30	0.45	Polymeric flat sheet (cellulose triacetate)	100*	[4]
	8.5	30	0.2	Ceramic multichannel Membralox® alumina	500*	[21]
E. coli (reconstituted)	1.1	36	0.2	Polymeric hollow fiber (polysulfone)	14$^{\Delta}$	[54]

$^{\Delta}$*concentration of organisms expressed in wet wt%.*
**concentration of organisms expressed in dry g/L.*

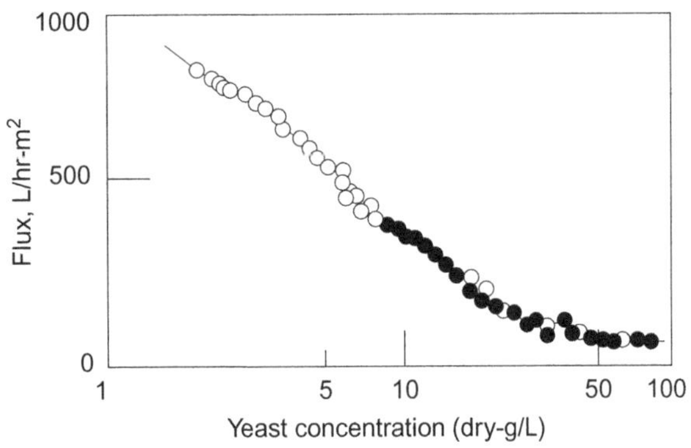

FIGURE 9.19 Filtration of yeast suspension with 0.45 mm microporous cellulose triacetate membrane. Initial yeast concentration, dry-g/L: (○) 3; (●) 8.

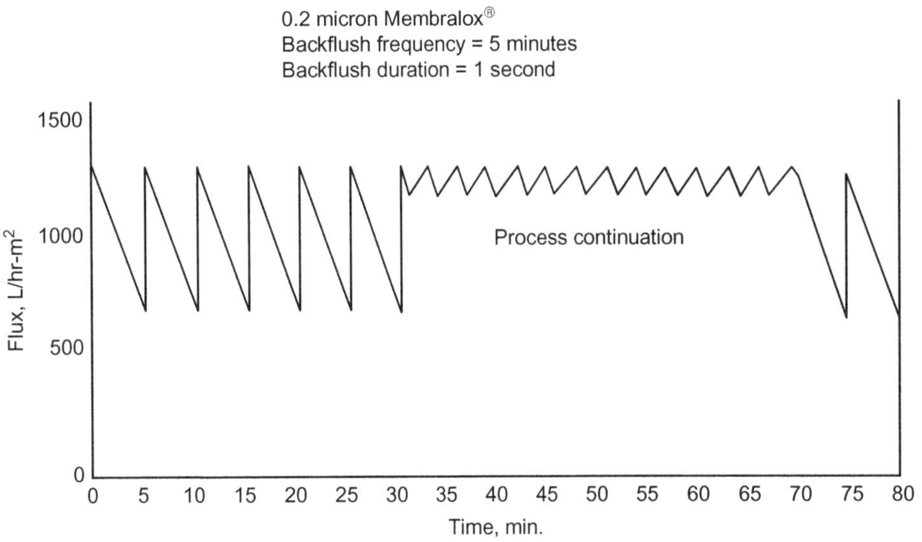

FIGURE 9.20 Effect of backpulsing on flux stability with 0.2 mm microporous alumina membrane.

marcescens using guidelines recommended by the Health Industry Manufacturers Association (HIMA).

Although cross-flow filtration can be effectively used for sterile filtration, dead end filtration can adequately serve these applications when the amount of contaminant is generally small (less than 1000 bacteria/mL). Cross-flow filtration may be more useful when high loads ($>10^7$ microorganisms/mL) of bacteria are involved requiring removal efficiency with a LRV value greater than 7 [60]. At high bacterial loadings, there may be significant membrane fouling and/or concentration polarization which could reduce flux and cause irreversible fouling. At high bacterial loadings, microporous membrane filters operating in the dead end configuration may be limited by low flux and require frequent cartridge replacement due to rapid pore plugging.

Table 9.12 shows the typical LRV values obtained using a polymeric and ceramic microfilter. Sterile filtration requires 100% bacteria retention by the membrane, whereas in many industrial bacteria removal applications the presence of a small quantity of bacteria in the filtrate may be acceptable. For example, drinking water obtained by microfiltration may contain nominal counts of bacteria in the filtrate which is then treated with a disinfectant such as chlorine or ozone. The use of ceramic filters may allow the user to combine the sterile filtration with steam sterilization in a single operation. This process can be repeated many times without changing filters due to their long service life (5 years or longer).

7.5 Production of Pyrogen-free Water

Distillation was used in the past to produce high purity water. Distillation and Reverse Osmosis (RO) filtration are the only acceptable methods listed in the USP for producing Water for Injection. Bulk Pharmaceutical and Biotechnology industries and some do however use Ultra Filtration (UF) to minimize endotoxins in those drug substances that are administered parenterally.

Distilled water is free from inorganic salts but may contain low-boiling organics. Water purity or quality can be measured by several analytical test methods. The most common water quality measure is its electrical resistance. Pure water resistivity is about 18 M-ohms. A triple distilled water typically shows a resistivity of only about 3 M-ohms. Today the combination of UF, RO, ion exchange and activated carbon is capable of producing 18 M-ohms water [8].

Ultrafiltration is used to remove pyrogens and other microorganisms from high purity water. Pyrogens are lipopolysaccharides (also known as endotoxins) with molecular sizes ranging from 20,000 dalton (~ 0.005 μm) up to about 200,000 dalton (~ 0.1 μm) produced from bacterial cell walls. Pyrogens induce fever when injected into animals or humans and cannot be removed by autoclaving or microfiltration [1,62].

The lipopolysaccharide molecule is thermally unstable and destruction requires exposure to temperatures 250°C and higher. Endotoxins can be removed using the principle of molecular size exclusion by reverse osmosis (RO) or ultrafiltration. Reverse osmosis can be used but may cause retention of low molecular weight salts which is highly undesirable in the preparation of certain non-pyrogenic parenteral solutions [63]. Ultrafiltration, on the other hand, with a 10,000 MWCO membrane can effectively remove pyrogens along with other microorganisms (not removed by prior separation techniques) without retaining salts.

Typical UF performance for pyrogen removal with a polymeric and ceramic membrane is shown in Table 9.13. It can be seen that both types of UF membranes can adequately remove pyrogens. The choice of UF membrane (ceramic or polymeric) will depend on operating conditions or other special process requirements. Ceramic membrane ultrafiltration can achieve a 5 log reduction in pyrogen level. These UF membranes have been validated for the production of water meeting the requirements of pyrogen-free water for injection (WFI) standards [64].

8.0 GLOSSARY OF TERMS

Adsorption It relates to the adherence of ions, molecules or particles with the membrane surfaces in contact (internal and external). In CFF this phenomenon could reduce the flux or change the retention characteristics. The tendency to form an adsorbed layer on the membrane surface may depend on the nature of the membrane surface. Typically hydrophilic surfaces adsorb less

TABLE 9.12 Typical LRV with Microporous Filters

Pore Diameter μm	Type of Membrane Filter	No. of Challenge Feed	Organisms* Permeate	LRV	References
0.2	Membralox® alumina	4.2×10^{11}	21	10.3	[59]
0.2	Membralox® alumina	8.4×10^{10}	0	>10.9**	[59]
0.22	Millipore, PVDF	1×10^{10}	0	>10**	[61]

**The number of organisms (Pseudonomas diminuta) in the feed were at least 10^7/mL.*
***indicates sterile permeate (ie. zero bacterial count).*

TABLE 9.13 Pyrogen Removal with Ultrafiltration Membranes

Pore Diameter or Molecular wt. cut-off (MWOO)	Type of Membrane	Pyrogens (ng/mL)		LRV	Reference
		Feed	Permeate		
5 nm	Membralox® Zirconia**	1800	negative*	>5.16	[62]
10.000 MWCO	AmiconH10 P10	1000	negative*	>4.9	[1]

*below the detection limit of 0.0125 ng/mL.
**not a commercial product.

strongly than the hydrophobic surfaces, especially when organics are involved.

Asymmetric/composite membrane This typically consist of a thin (0.5 to 20 microns) fine-pore layer responsible for separation and a support or substrate with single or multiple layers having progressively larger pores which provide the required mechanical strength. This type of structure maximizes the flux by minimizing the overall hydraulic resistance of the permeate (filtrate) flowing across the membrane structure.

Backpressure It is the pressure generated by restricting retentate flow.

Backpulse This is achieved by rapid (typically lasting a fraction of a second) application of periodic counterpressure on the permeate side, typically with the help of an automatic time switch or a microprocessor, to push back a specific (as low as possible) permeate volume in the opposite direction. It is used in many CFF applications (especially with ceramic membranes) as an effective technique to disrupt, reduce or destroy the concentration-polarization boundary layer. Backpulsing also helps to minimize particle/gel infiltration into the microporous structure. Typical backpulse frequencies (cycle times) are in the range of 3 to 10 minute.

Backwash This is similar to the backpulse technique but is less intensive in terms of the pressures applied across the membrane to dislodge particles/gels from the membrane surface. Backpulse is typically carried out at pressures exceeding 4 bar and often in the range 6 to 10 bar, whereas a backwash is carried out at lower pressures (e.g., 2 to 3 bar).

Boundary layer This refers to the layer adjacent to the membrane surface along the periphery of the feed channel. This contributes a major portion of the total resistance to transport and is often the controlling factor in determining the flux.

Bubble point It relates to the largest pore diameter of the membrane. The bubble point is the smallest pressure difference at which the first gas bubbles appear from a liquid-saturated membrane pressurized by an inert gas. The bubble point test is also used for checking the physical integrity of the membrane for the presence of defects such as cracks or pinholes.

Cake layer This refers to the layer resulting from the physical deposition of solids, primarily on the membrane surface.

Channel It is the opening or section of a module through which the fluid enters (feed) and exits (retentate). Most commercial modules have multiple lumens or channels.

Clean water flux It refers to the original flux of filtered deionized water with a virgin membrane. In most applications, after cleaning is performed the clean water flux is restored to approximately within 10% of its original value. It is seldom recovered to its original value after membrane cleaning due to monomolecular irreversible adsorption of foulants.

Concentration factor The ratio of initial feed volume or weight to the volume or weight remaining at the end of filtration. The calculations differ for batch versus other modes such as batch feed-and-bleed or continuous cascade configuration.

Concentration polarization This occurs when solutes or particles rejected by the membrane accumulate on or near the membrane surface. As a direct consequence of this condition, there is an increase in the resistance to solvent transport resulting in a flux decrease and possibly changes in the retention characteristics.

Continuous diafiltration Under this mode of operation, water or other solution is continuously added to the feed or retentate at a rate equal to permeate rate.

Cross-flow velocity This is the average rate (m/s or ft/sec) at which the feed or retentate flows parallel to the membrane surface.

Dynamic membrane This type of membrane structure, also referred to as *formed-in-place membrane*, can be produced in several different ways. A commonly used technique involves introducing solutions of organic or inorganic polyelectrolytes into the feed channel(s) and forming filtering layers on a porous support by applying a pressure difference. Another process uses the components in the feed stream itself to form a filtration layer (or cake).

Feed and bleed A continuous mode of CFF where the feed is pumped into the recirculation loop at a rate which equals the summation of the permeation rate and the flow rate of retentate. It allows for increased control over the effect of concentration on the filtration performance.

Flow excursion It describes the optimization of retentate flow at a constant transmembrane pressure.

Gel layer This refers to the concentration polarization boundary layer with the highest solute (gel forming components of the feed) concentration.

Hydrophilic A property characterized by a strong tendency to bind or adsorb water. Examples of hydrophilic materials are carbohydrates, vegetable gums, pectins and starches, and some complex proteins such as gelatin and collagen.

Hydrophobic It is related to the water repelling property of a membrane material or a substance. This property is characteristic of all oils, fats, waxes, resins, as well as finely divided powders such as carbon black and magnesium carbonate.

Internal pore fouling This is caused by the deposition of material inside the porous structure which often leads to significant flux decline and irreversible fouling. Internal pore fouling can be due to adsorption, precipitation, pore plugging or particle adhesion.

Membrane fouling A phenomenon characteristic of all membrane-based filtration processes in which the membrane adsorbs or interacts with feed components. Membrane fouling causes a flux decrease and may also increase the retention of certain components in the feed. Membrane fouling is typically a time-dependent phenomenon and often independent of concentration. In some situations a partial dependence on concentration may be observed.

Molecular weight cutoff It refers to be smallest molecular weight of a macrosolute for which the membrane shows at least 90% rejection. This value is typically determined under a set of well-defined conditions using model compounds (e.g., polyethylene glycols, dextrans and proteins such as BSA) at low concentration.

Normalized flux Permeation rate per unit filtering surface, normalized to a given set of operating conditions such as constant temperature, pressure and/or concentration.

Permeability This is defined as flux per unit transmembrane pressure for a given solvent at a fixed temperature. It differs from process flux and should only be used with respect to clean liquids.

Pressure excursion This refers to the incremental increases in transmembrane pressure while maintaining a constant retentate flow.

Rejection or retention coefficient This describes the ability of the membrane to retain the desired species from the feed on the membrane surface. Since the rejection is often dependent on membrane characteristics and operating parameters, these must be clearly stated so that a fair comparison can be made between different types of membranes for a given application. It is defined as: $R = 1\text{-}C_P/C_R$, where, C_p is the concentration of the species in permeate and C_R is its concentration in the retentate. If a significant passage of the species occurs, then an average concentration is used.

Reynold's number It describes the nature of hydraulic regime such as laminar flow, transitional flow or turbulent flow. It is defined as the ratio of the product of hydraulic diameter and mass flow velocity to that of fluid viscosity. Mass velocity is the product of cross-flow velocity and fluid density. Laminar flow exists for Reynold's numbers below 2000 whereas turbulent is characterized by Reynold's numbers greater than 4000.

Retentate It refers to that portion of the feed that does not cross the filtering surface in a single pass. It is also described as concentrate since in many situations the depleting particle-free permeate leaves higher solids in the retentate stream.

Transmembrane pressure It is the average driving force for permeation across the membrane. Neglecting osmotic pressure effects for most MF/UF applications, it is defined as the difference between the average pressure on the feed (or retentate) side and that on the permeate (or shell side).

Zeta potential It relates to the electrokinetic potential across the interface of all solids and liquids and specifically to that of the diffuse layer of ions surrounding a charged colloidal particle. Such a diffuse aggregation of positive and negative electric charges surrounding a suspended colloidal particle is largely responsible for colloidal stability.

ACKNOWLEDGMENT

The author wants to thank Dr. H. S. Muralidhara for his careful review of the manuscript. He also made many useful suggestions and some contributions pertaining to the discussions on membrane fouling, cleaning and concentration polarization.

APPENDIX: LIST OF MEMBRANE MANUFACTURERS (MICROFILTRATION AND ULTRAFILTRATION)

Manufacturer	Comments
A/G Technology Corp. 34 Wexford Street Needham, MA 02194 USA	Hollow fiber UF and MF
Amicon Corporation (W Grace). R. Scientific Systems Design 21 Hartwell Avenue Lexington, MA 02173 USA	Primarily laboratory scale UF hollow fiber membranes
Anotec/Alcan Anotec Separations, Ltd. Wildmere Road Banbury, Oxon, UK 0X167 JU	Ceramic disc filters MF and UF
Asahi-Kasei Hibiya Mitsui Building 1-2 Yurakucho, 1-chome, Chiyoda, Tokyo, Japan	Hollow fiber membranes
Asahi Glass 2-1-22 Marunouchi Chiyoda-ku, Tokyo 100, Japan	Tubular SiO_2-based microfilters
Asahi Glass 1185 Avenues of the Americas, 20th Floor New York, N. Y. 10036, USA	
Berghof Forschungsinstitut Berghof GmbH P. O. Box 1523 74 Tubingen 1 W. Germany	Hollow fiber polymide modules
Brunswick Technetics Membrane Filter Products 4116 Sarrento Valley Blvd. San Diego, CA 92121, USA	Asymmetric Polysulfone cartridge filters
Carbone-Lorraine Le Carbone Lorraine Tour Manhattan Cedex 21 F-92095, France	Tubular MF and UF carbon membranes

Carbone-Lorraine 400 Myrtle Avenue Boonton, N. J. 07005 USA	
Carre/Du Pont Du Pont Separation Systems Glasgow Wilmington, DE 19898 USA	Dynamic ceramic and composite membranes on tubular support. Now sold through Graver
Ceramem Ceramem Corporation 12 Clematis Avenue Waltham, MA 02154 USA	Honeycomb multichannel ceramic membranes on microporous coerdierite support
Daicel 3-8-1 Tornaomom Building Kasumigasiki Chiyoda-ku Toyko, Japan	Tubular and spiral wound modules; cellulosic and on-cellulosic membranes. MOLSEP brand
De Danske Sukkerfabriker (Dow Separations) 6 Tietgensvej P. O. Box 149 DK-4900 Nakskov Denmark	Plate and frame UF modules; CA, polysulfone and polyamide membranes. Thin film composite membranes (Filmtec).
Dorr-Oliver, Inc. 77 Havemeyer Lane Stamford, CT 06904 USA	Plate - type UF modules
Fairey Fairey Industrial Ceramics, Ltd. Filleybrooks Stone Staffs ST15 OPU, UK	Porous tubular and disc ceramic microfilters
Gelman Sciences, Inc. 600 South Wagner Road Ann Arbor, MI 48106 USA	Primarily MF
Hoechst Celanese Separation Products Div. 13800 South Lakes Drive Charlotte, NC 28273 USA	Flat sheet and hollow fiber MF/UF
Koch Membrane Systems 850 Main Street Wilmington, MA 01887 USA	Tubular and spiral wound membranes
Kuraray Company, Ltd. Project Development Department 12–39 1-chome, Umeda, Kita-ku Osaka 530 Japan	Polyvinyl alcohol UF and MF
Memtec Memtec Laboratories Oakes Road Old Toongabbie New South Wales Australia	Polyamide membranes
Memtek 22 Cook Street Billerica, MA 01866 USA	PVDF microfiltration membranes
Mott Mott Metallurgical Corporation Farmington Industrial Park Farmington, CT 06032 USA	Disc and tubular metallic microfilters
Millipore Millipore Corporation Ashby Road Bedford, MA 01730 USA	Broad line of polymeric membrane filters
NGK NGK Insulators, Ltd. Shin Maru Building 1-5-1, Marunouchi Chiyodo-ku, Tokyo 100 Japan	Tubular and multichannel ceramic UF and MF membranes
Nitto Denko America, Inc. 5 Dakota Drive Lake Success, N. Y. 11042 USA	Tubular and hollow fiber systems
Nuclepore Corporation 7035 Commerce Circle Pleasanton, CA 94566 USA	Nuclear track-etch membranes (capillary-pore type).
Osmonics, Inc. 5951 Clearwater Drive Minnetonka, MN 55343 USA	UF spiral wound membranes
Pall Pall Porous Metal Filters Cortland, NY 13045 USA	Disc and tubular multichannel ceramic MF and UF membranes
PCI Patterson-Candy International Laverstoke Mill Whitchurch Hampshire RG287NR England	Tubular UF membranes and systems
Rhone-Poulenc Chemie 21 rue Jean Goujan 25360 Paris, France	Plate and frame membrane filters Kerasep ceramic MF and UF membranes
Sartorius Sartorius-membranfilter GmbH Weender Landstrasse 94–108 P. O. Box 142 3400 Gottingen, W. Germany	Mainly for MF
Sartorius Filters, Inc. 26575 Corporate Avenue Hayward, CA 94545 USA	
Schleicher & Schuell, Inc. D-3354 Dassel Kr. Einbeck W. Germany	Laboratory MF and UF membranes
Schott Glaswerke Postfach 2480 D-6500 Mainzx Germany	Bioran® tubular and SiO_2-based ultrafilters
Toto Company, Ltd. 1-1 Nakajima 2-Chome Kokura-ku Kita-kyashu-shi 802, Japan	Tubular composite ceramic filters for MF and UF

U. S. Filter Membralox® Products Group 181 Thorn Hill Road Warrendale, PA 15086 USA	Membralox® tubular and multichannel ceramic MF, UF and NF membranes. Ceraflo® MF membranes.
Societe Ceramiques Techniques, Usine de Bazet B. P. 113 650001, Tarbes, France	

REFERENCES

[1] M. Cheryan, Ultrafiltration Handbook, Technomic Publishing Co, Lancaster, 1986.
[2] L. Mir, S.L. Michaels, V. Goel, Cross-flow microfiltration: applications, design and cost, in: W.S. Ho, K.K. Sirkar (Eds.), Membrane Handbook, Van Nostrand, Reinhold, New York, 1992, pp. 571–594.
[3] R.R. Bhave (Ed.), Inorganic membranes: Synthesis, Characteristics and Applications, Van Nostrand Reinhold, New York, 1991.
[4] K. Matsumoto, S. Katsuyama, H. Ohya, Separation of yeast by cross-flow filtration with backwashing, J. Fermentation Technol. 65 (1987) 77–83.
[5] S.L. Michaels, Cross-flow microfilters, Chem. Eng. (1989) 84–91.
[6] R.H. Davis, Theory of cross-flow filtration, in: W.S. Ho, K.K. Sirkar (Eds.), Membrane Handbook, Van Nostrand, Reinhold, New York, 1992, pp. 480–505.
[7] S.S. Kulkarni, E.W. Funk, N.N. Li, Ultrafiltration, in: W.S. Ho, K.K. Sirkar (Eds.), Membrane Handbook, Van Nostrand, Reinhold, New York, 1992, pp. 391–453.
[8] P.A. Schweitzer (Ed.), Handbook of Separation Techniques for Chemical Engineers, McGraw-Hill, New York, 1979, pp. 2–3 to 2–103.
[9] H.P. Hsieh, General characteristics of inorganic membranes, in: R.R. Bhave (Ed.), Inorganic Membranes: Synthesis, Characteristics and Applications, Van Nostrand, Reinhold, New York, 1991.
[10] D.J. Paulson, R.L. Wilson, D.D. Spatz, Cross-flow membrane technology and its applications, Food Technol. (1984) 77–111.
[11] A.J. Burggraaf, K. Keizer, Synthesis of inorganic membranes, in: R.R. Bhave (Ed.), Inorganic Membranes: Synthesis, Characteristics and Applications, Van Nostrand, Reinhold, New York, 1991.
[12] T.J. O'Sullivan, A.C. Epstein, S.R. Korchin, N.C. Beaten, Applications of ultrafiltration in biotechnology, Chem. Eng. Progress January (1984) 68–75.
[13] P. Ng, J. Lundblad, G. Mitra, Optimization of solute separation by diafiltration, Separ. Sci. 11 (5) (1976) 499–502.
[14] B.R. Breslau, The theory and practice of ultrafiltration, proc. scientific conference corn-refiners association, Food Food Chem. 17 (1982) 36–90.
[15] A.C. Epstein, Batch versus continuous systems for ultrafiltration, Proc. Fifth Annual Membrane Technology Planning Conference, Cambridge, 1987.
[16] M. Taniguchi, N. Kotani, T. Kobayashi, High-Concentration cultivation of lactic acid bacteria in fermenter with cross-flow filtration, J. Fermentation Technol. 65 (2) (1987) 179–184.
[17] S. Ripperger, G. Schulz, Microporous membranes in biotechnical application, Bioprocess Eng. 1 (1986) 43–49.
[18] N. Olesen, F. Jensen, Microfiltration: the influence of operation parameters on the process, Michwissenschaft 44 (8) (1989) 476–479.
[19] R.M. Sandblom, Filtering Process, U. S. Patent 4,105,547, Aug. 8, 1978.
[20] K. Matsumoto, M. Kawahara, H. Ohya, Cross-Flow filtration of yeast by microporous ceramic membrane with backwashing, J. Ferment. Technol. 66 (2) (1988) 199–205.
[21] W.E. Blatt, A. Dravid, A.S. Michaels, L. Nelson, Solute polarization and cake formation in membrane ultrafiltration: causes, consequences and control techniques, in: J.E. Flinn (Ed.), Membrane Science and Technology, Plenum Press, New York, 1970.
[22] M. Bennasar, B. Tarodo de la Fuente, Model of the fouling mechanism and of the working of a mineral membrane in tangential filtration, Sci. Des. Ailments 7 (1987) 647–655.
[23] R. Glimenius, Microfiltration—State of the Art, Desalination 53 (1985) 363–372.
[24] H. Beechold, Ultrafiltration and electro-ultrafiltration, in: J. Alexander (Ed.), Colloid Chemistry, 1, The Chemical Catalog Company, 1926.
[25] G. Belfort, Membrane modules; comparisons of different configurations using fluid mechanics, J. Membrane Sci. 35 (1988) 245–270.
[26] P.L.T. Brian, Concentration polarization in reverse osmosis desalination with variable flux and incomplete salt rejection, Ind. Chem. Fund 4 (1965) 438–445.
[27] G.M. Rios, M. Rakotoarisoa, B. Tarodo de la Fuente, Basic transport mechanisms of ultrafiltration in the presence of fluidized particles, J. Membrane Sci. 34 (1987) 331–343.
[28] P. Langer, R. Schnabel, Porous glass membranes for downstream processing, Chem. Biochem. Eng. Q 2 (4) (1988) 242–244.
[29] P. Langer, S. Breitenbach, R. Schnabel, Ultrafiltration with porous glass membranes, Proc. Int. Techn. Conf. on Membrane Separation Processes, Brighten, U. K, 1989.
[30] P.H. Hodgson, A.G. Fane, Cross-flow microfiltration of biomass with inorganic membranes: the influence of membrane surface and fluid dynamics, Key Eng. Mater. 61 (62) (1992) 167–174.
[31] C.C. Mermann, High frequency excitation and vibration studies on hyperfiltration membranes, Desalination 42 (1982) 329–338.
[32] H.S. Muralidhara, The combined field approach to separations, Chem. Tech. (1988) 224–235.
[33] M. Bier (Ed.), Electrophoresis, vol. 1, Academic Press, New York, 1959.
[34] J.D. Henry, L.F. Lawler, C.H.A. Kuo, A solid/liquid separation process based on cross-flow and electrofiltration, A. I. Ch. E. J. 23 (6) (1977) 851–859.
[35] H.S. Muralidhara, Membrane fouling: techno-economic implications, Proceedings of the Second International Conference on Inorganic Membranes, Trans Tech Publications, Zurich, 1991.
[36] H. Attia, M. Bennasar, B Trodo de la Fuente, Study of the fouling of inorganic membranes by acidified milks using scanning electron microscopy and electrophoresis, J. Dairy Res. 58 (1991) 39–50.
[37] Z. Knez, Ultrafiltration Processes in Biotechnology, Proc. Filtech Conf. (1987) 194–203.
[38] J.N. Johnson, Cross-flow microfiltration using polypropylene hollow fibers, Fifth Annual Membrane Technology Planning Conference, Cambridge, Oct. 1987.

[39] M. Mateus, J.M.S. Cabral, Pressure-Driven membrane processes for steroid bioseparation: a comparison of membrane, hydrodynamic and operating aspects, Bioseparation 2 (1991) 279–287.
[40] E. Raaska, Cell harvesting with tangential flow filtration: optimization of tangential flow, Proc Membrane Technology Group Symposium, Tylosand, Sweden, May 28–30, 1985.
[41] H.P. Hsieh, R.R. Bhave, H.L. Fleming, Microporous Alumina Membranes, J. Membr. Sci. 39 (1988) 221–241.
[42] W. Muller, Microfiltration Process and Apparatus, U. S. Patent, 5,076,931, Dec. 31, 1991.
[43] U. Merin, G. Daufin, Separation processes using inorganic membranes in the food industry, Proc. of the First International Conference on Inorganic Membranes, Montpelier, 1989.
[44] R. Gabler, S. Messinger, Scaling-up membrane filter systems, Chem. Tech. (Oct., 1986) 616–621.
[45] J. Haggin, Membranes play growing role in small-scale industrial processing, Chem. Eng. News July (11) (1988) 23–32.
[46] R.R. Bhave, J. Guibaud, B. Tarodo de la Fuente, V.K. Venkataraman, Inorganic membranes in food and biotechnology applications, in: R.R. Bhave (Ed.), Inorganic Membranes: Synthesis, Characteristics and Applications, Van Nostrand, Reinhold, New York, 1991.
[47] F.N. Inman, Filtration and separation in antibiotic manufacture, Proc. Filtration Soc. Lond. January (24) (1984).
[48] S. Ripperger, G. Schultz, W. Pupa, Membrane separation processes, biotechnology, Proc. Fifth Annual Membrane Technology Planning Conference, Cambridge, Oct., 1987.
[49] R.A. Tutunjan, R. Breslau, Pharm. Tech. Conference, New York, 1982.
[50] I. Melling, Process. Biochem. 9 (7) (1974).
[51] D. Gerster, R. Veyre, Mineral ultrafiltration membranes in industry, in: S. Sourirajan, T. Matswura (Eds.), Reverse Osmosis and Ultrafiltration, Amer. Chem. Soc., Washington, D.C., 1985, pp. 225–230.
[52] P.A. Belter, E.L. Cussler, W.S. Hu, Bioseparations: Downstream Processing for Biotechnology, John Wiley, New York, 1988.
[53] M.L. Leiva, G. Tragardh, Chem. Tech. 35 (8) (1983) 381.
[54] M. Lasky, D. Grant, Use of microporous hollow fiber membranes in cell harvesting, Anal. Lab. Nov./Dec. (1985) 16–21.
[55] E. Bjurstrom, Chem. Eng. February (18) (1985).
[56] P. Boyaval, C. Corre, Biotechnol. Lett. 9 (11) (1987) 801.
[57] M. Minnier, E. Ferras, G. Goma, P. Soucaille, presented at the VII International Biotechnology Symposium, New Delhi, 1984.
[58] H. Hoffmann, T. Scheper, K. Schugerl, Use of membranes to improve bioreactor performance, Chem. Eng. J. 34 (1987) B13–B19.
[59] J.R. Morgart, J.L. Filson, J. Peters, R.R. Bhave, Bacteria Removal by Ceramic Microfiltration, U. S. Patent #5,242,595, 1993.
[60] T.J. Leahy, M.J. Sullivan, Validation of bacterial-retention capabilities of membrane filters, Pharm. Tech. 2 (11) (1978) 65–75.
[61] J.L. Filson, R.R. Bhave, J. Morgart, J. Graaskamp, Pyrogen Separations by Ceramic Ultrafiltration, U. S. Patent 5,047,155, 1991.
[62] W. Reinholz, Making water for injection with ceramic membrane ultrafiltration, Pharm. Tech. (Sep., 1995) 84–96.
[63] A.E. Allegra Jr., V. Goel, Virus removal from therapeutic proteins, IMSTEC 1992 Conference, Sydney, Nov., 1992.
[64] V. Goel, M.A. Accomazzo, A.J. DiLeo, P. Meier, A. Pitt, M. Pluskal, Deadend microfiltration; applications, design and cost, in: W.S. Ho, K.K. Sirkar (Eds.), Membrane Handbook, Van Nostrand, Reinhold, New York, 1992, pp. 506–570.

FURTHER READING

[1] http://www.fda.gov/ICECI/Inspections/InspectionGuides/ucm074905.htm

Chapter 10

Distillation for Recovery of Biofuels and Bio-Based Chemicals

Steve Licht

1.0 INTRODUCTION AND THEORY

1.1 Introduction with Historical Background

Distillation originated for the recovery of natural fermentation products. Today, purification of the highest volume fermentation products (fuel grade ethanol and alcoholic beverages) is accomplished by distillation. Thus, the field of fermentation and biochemical engineering has always relied on distillation as one of the primary methods for product purification.

Distillation, as a laboratory process and a manufacturing method, originated in the distant past. The practice of distillation, as it is known today, required the pre-existence of skillful pottery or glass making, and basic metalworking, for construction of the distillation apparatus.

Distillation was conceptualized in the earliest days as a process wherein heating and boiling a liquid mixture caused "spirits" to be stripped from that liquid and move from the liquid into the actively boiling bubbles, as invisible vapor. It was learned by experience that if these spirits passed upward through equipment wetted with liquid flowing downward, the spirits would become more concentrated or "rectified." "To rectify" means "to make correct." Thus the distillation operation was known to accomplish two things, either one of which might have been the main objective of the alchemist or the distillery worker:

- Stripping of the spirits (more volatile substances) from the starting liquid
- Rectification of the spirits to a stronger, more concentrated condition

Nothing about these objectives has changed—they fully apply to the implementation of distillation today. After recent centuries' increased understanding of physics and chemistry, we have a *better understanding* of the distillation process in terms of the way molecules behave according to laws of thermodynamics, and our ability to *predict and model* the performance of distillation processes, using graphical methods and computers, has increased significantly.

The objective of this chapter is to permit the reader to understand how a distillation system is used for the recovery of a fermented organic chemical (or a synthetic organic or inorganic chemical) and how such a system is designed for new applications.

Originating a robust and practical distillation process concept for a new application is a creative challenge.

Distillation process synthesis is accomplished by studying the actual vapor-liquid equilibrium (VLE) relationships between all pairs of volatile chemical compounds present in the multi-component mixture, and it requires recovery and purification of one (or more) of the volatile components. Conceptualization of the process flowsheet is followed by process modeling using tried-and-true graphical methods or computer simulators. Then begins process optimization, which seeks the combination of maximum product yield, minimum energy usage, simplicity instead of complexity, minimal capital investment, and lowest operating cost that will give the greatest economic return.

1.2 How a Distillation System Works

This chapter will employ a modern inquiry-based teaching method to guide the reader to an improved understanding of what a distillation process consists of and how a distillation system works:

- The text will pose a series of questions about individual, reasonably sized features of a distillation system—either the distillation science, the equipment, the controls, the design, or plant operations.

Fermentation and Biochemical Engineering Handbook.

- After each question, the text will provide evidence, in the form of discussion, bullet points, diagrams, or photographs, leading the reader to formulate answers to the most recently posed question.
- The text will then confirm the answer before moving on to other questions.

In this manner, large and complex processes may be understood from one point of view, then another, and so on, until the whole becomes clear, as the sum of its parts.

In this introductory section, we will begin with a focus on *what is flowing*—into, out of, and inside the various parts of the distillation system—without trying to explain, in any depth, *why*. We will address *how* the system works, and *why* it works, later.

- What is flowing inside a distillation system while it is operating?
  - A hot fluid is providing heat.
  - A cold fluid is providing cooling.
  - Feed material (a mixture of several chemical compounds) is flowing in.
  - Purified distilled product (containing the target compound) is flowing out.
  - Waste streams (containing compounds other than the target for purification) are flowing out.
  - Liquid reaching the column bottom is being boiled, over and over again.
  - Column overhead vapors are being condensed.
  - Part of the condensed overhead flows back to the top of the column as liquid reflux.
  - Gases that cannot be condensed are flowing out.
- What is flowing inside of the simple distillation system shown in block diagram form in Fig. 10.1?
  - Figure 10.2 tabulates the quantities, composition, and properties of the one stream flowing into and the three streams flowing out of the ethyl acetate system flow example given in Fig. 10.1:
    - Stream 1: Feed mixture containing volatile compounds and dissolved solids in addition to the target compound, ethyl acetate
    - Stream 4: Purified product ethyl acetate
    - Stream 2: Overhead distilled waste containing volatile non-target compounds from the feed
    - Stream 3: Bottom waste containing dissolved solids and non-volatile compounds
  - Figure 10.3 shows ethyl acetate column internal flows: quantities of liquid flowing downward and vapor flowing upward through each tray, where they mix, reach vapor-liquid equilibrium, separate, and flow to the next stage: up for vapor, down for liquid.
  - Figure 10.4 is a graphical representation of the flows tabulated in Fig. 10.3.
  - Figure 10.5 gives the ethyl acetate distillation column temperature profile during continuous on-stream operation at steady state.
    - Please note that Tray 1 in this table, and in all the computer-generated simulations in this chapter, is the overhead condenser. Tray 2 is the top theoretical stage in the column, which coincides with the highest elevation tray inside the distillation column.
    - The conventional practice for numbering distillation process theoretical trays during the design calculations is to designate Tray 1 as the condenser; Tray 2 as the highest tray inside the top of the column; the maximum tray number representing the reboiler at the column bottom (in this example, Tray 21); while Tray 20 is the lowest tray inside the column. This practice is not very intuitive compared to building floor numbers. It is the reverse of typical numbering schemes for multi-story buildings, where the floor numbers start with 0 or 1 at the bottom and get higher going up. It may be counterintuitive to number the floors in a tall structure from the top down, but it is how computer process simulators universally number the trays, so we must conform to this practice.
    - This numbering convention will be used throughout this chapter because it is used in most (if not all) modern engineering and scientific articles about distillation theory or practice.
  - Figure 10.6 presents the column temperature profile as a graph.

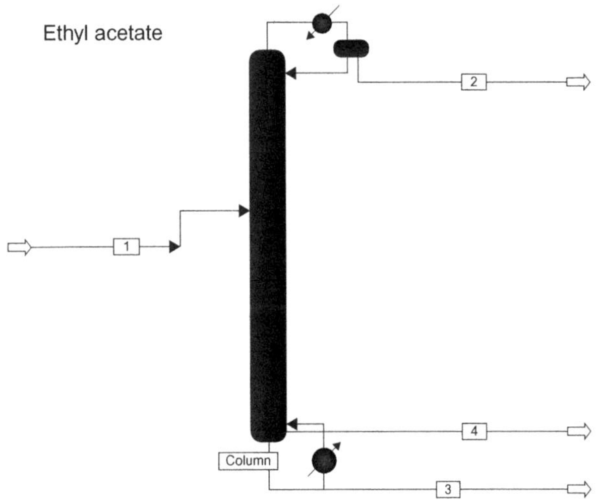

FIGURE 10.1 Ethyl acetate distillation system block flow diagram.

Now, to continue progressing toward understanding of *what* is happening in the distillation system (to be followed by probing into *how and why*), we will move on to other major questions of

Stream name Stream description Phase		1 Feed liquid	2 OVHDS Waste liquid	4 Main product vapor	3 Crud liquid
Temperature	KG/HR	2310.000	1751.617	548.383	10.000
Pressure	C	10.000	64.721	82.587	82.910
Molecular weight	BAR	4.000	1.000	1.209	1.220
		69.989	65.710	87.950	88.086
Weight comp. percents					
H_2O		4.3290	5.7033	0.0182	0.0007
Ethanol		6.2771	8.2564	0.0689	0.0093
EOAC		72.6407	63.9912	99.7706	99.9513
IPA		0.5195	0.6659	0.0611	0.0146
MEAC		16.2338	21.3832	0.0811	0.0241
Weight comp. rates	KG/HR				
H_2O		100.0000	99.9000	0.1000	0.0001
Ethanol		145.0000	144.6210	0.3780	0.0009
EOAC		1677.9999	1120.8801	547.1246	9.9951
IPA		12.0000	11.6632	0.3353	0.0015
MEAC		374.9999	374.5526	0.4450	0.0024
Enthalpy	M*KCAL/HR	0.031	0.077	0.069	0.000

FIGURE 10.2 Ethyl Acetate Distillation Stream Flows with Conditions and Compositions.

Tray	Flow rate	
	Liquid KG-MOL/HR	Vapor KG-MOL/HR
1	179.2	
2	181.0	205.8
3	181.1	207.6
4	181.3	207.7
5	181.5	207.9
6	181.8	208.2
7	182.1	208.5
8	182.4	208.8
9	225.2	209.1
10	225.8	218.9
11	226.4	219.4
12	227.0	220.0
13	227.7	220.7
14	227.8	221.3
15	225.9	221.5
16	222.3	219.5
17	222.8	216.0
18	224.2	216.4
19	224.9	217.9
20	225.3	218.6
21		225.1

FIGURE 10.3 Ethyl Acetate Distillation Column Internal Flow Rates.

inquiry. The first question was "What is flowing?" The second question, starting just below this paragraph, is "What is heating?" The third question will appear about one page later: "What is boiling?" Subsequent questions, the evidence presented, and discussions and conclusions, will follow without further editorial interruptions (like this one) except for discussion of unexpected conclusions (like the previous discussion of tray numbering) or transitions to new subjects.

- What is heating?
  - Heat is flowing energy.
  - Heat is the flow of energy from hot to cold material.
  - Objects don't contain heat.
  - Heat flow = heat transfer.
  - "Specific heat" is the same as "heat capacity": Both mean how much heat is needed to raise the temperature of one gram of something by one degree Celsius in the metric system, or one pound by one degree Fahrenheit in the traditional English system, now used primarily in the United States.
- Heat transfer
  - Means that heat goes from one material (hot) to another material (cold)
- Heat can be transferred in three ways:
  - Conduction
  - Convection
  - Radiation
- Heat transfer by conduction:
  - Is the way heat flows through solids
  - Is the way heat flows through motionless liquids
  - Is very, very slow through gases and through air pockets inside of bulk solids or bulk liquids
- Heat always flows from higher to lower temperature places.
  - [200°C →*Heat flows this way*→ 196°C]
  - Conductors (metals) are good at conducting heat.
  - Insulators (insulation) conduct heat poorly.

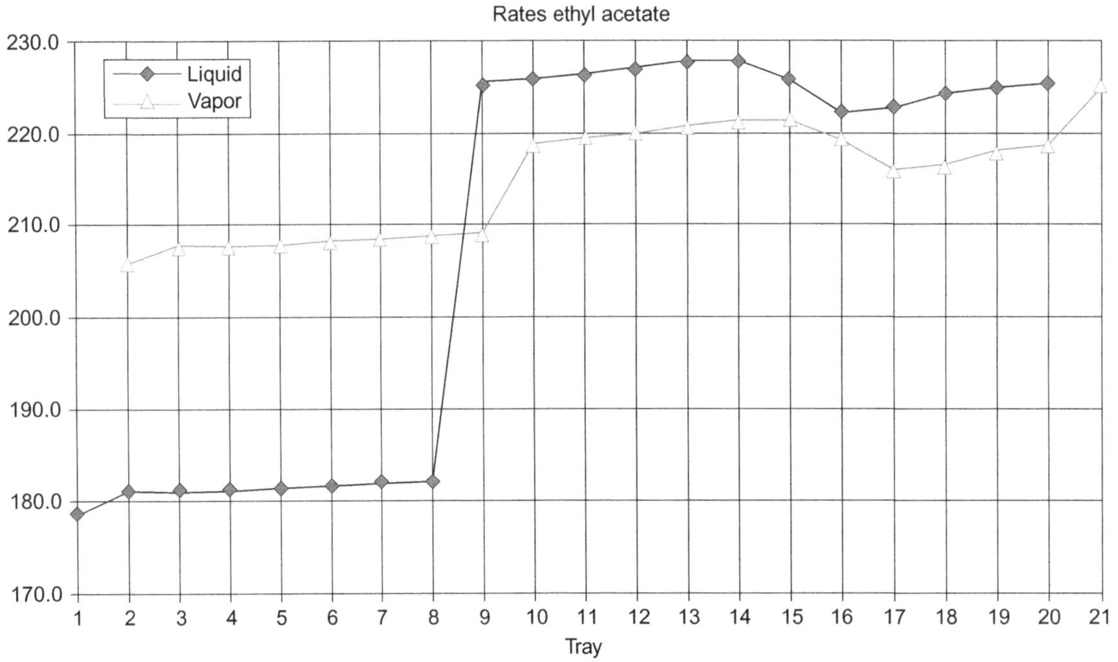

FIGURE 10.4 Ethyl Acetate Distillation Column Internal Flow Graph.

Tray	Temperature C
1	64.7
2	68.2
3	69.0
4	69.6
5	70.1
6	70.5
7	70.9
8	71.2
9	71.5
10	72.0
11	72.4
12	72.8
13	73.1
14	73.5
15	74.1
16	75.9
17	79.2
18	81.3
19	83.1
20	82.6
21	82.9

FIGURE 10.5 Ethyl Acetate Distillation Column Temperature Profile.

- Heat transfer by convection:
  - Means that warmed or cooled fluid moves from one place to another
  - Natural convection means the fluid is not being stirred or pumped.
    - Examples: Teapot on stove, tank with cooling coil
  - Forced Convection means the fluid moves by stirring or pumping.
    - Examples: A stirred pan, jacketed reactor, shell and tube heat exchangers
- Heat transfer by radiation:
  - Is what sunshine is
  - Is heat radiating from a fire
  - Is what is meant by "red-hot"
  - Is what goes on in the fireboxes in boilers, furnaces, and fired heaters
  - Radiation can go through empty space.
  - Radiated heat can either be reflected (mirror), transmitted (clear glass), or absorbed.
- What is boiling?
  - Boiling means adding heat to a liquid and making it vaporize.
  - The liquid must be in contact with a hot surface that is hotter than the boiling point of the liquid at whatever pressure there is in the "boiler" or "reboiler."
  - In distillation systems, the same material goes back to be boiled again and again; hence the traditional term "reboiler".
- Types of boiling: Nucleate boiling and film boiling
  - Nucleate boiling:
    - Liquid touches all of the hot surface.
    - Considered more safe
    - Provides good heat transfer rate
  - Film boiling:
    - Surface is so hot it is smothered in vapor.
    - Considered unsafe due to surface hot spots
    - Provides poor heat transfer rate

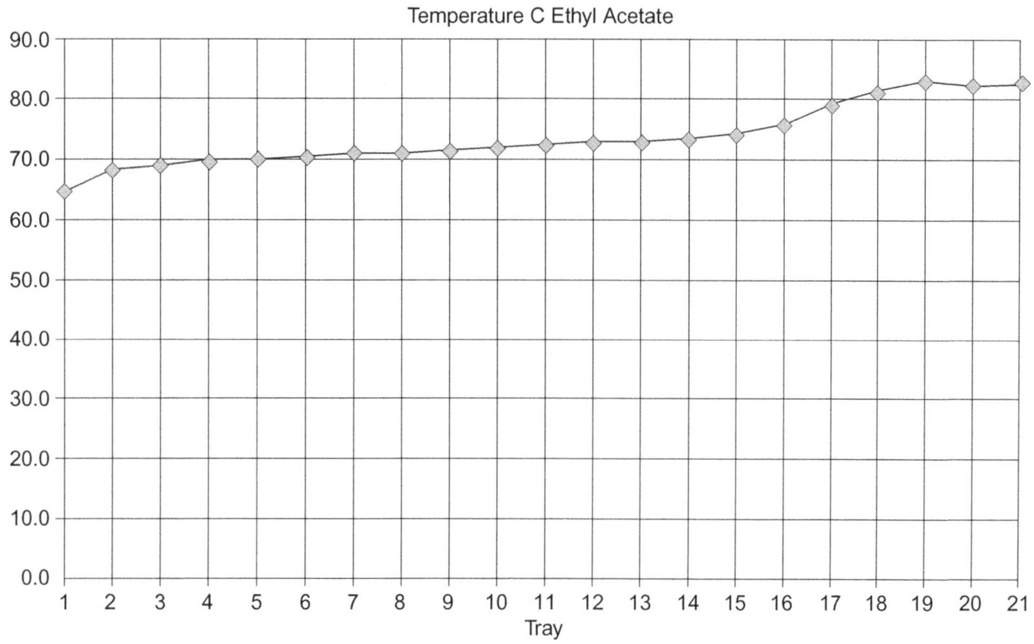

FIGURE 10.6 Ethyl Acetate Distillation Column Temperature Profile Graph.

- What is condensing?
  - Condensation is the opposite of boiling.
  - Condensation means removing heat from a vapor and making it turn back into a liquid.
  - The vapor must be in contact with a cold surface that is colder than the boiling point (also called the dew point) of the vapor at whatever pressure there is in the "condenser."
  - Condensers usually are shell and tube heat exchangers.
- What is cooling?
  - Cooling is merely heat transfer (already discussed) with the goal of making something colder rather than hotter.
  - In a distillation system, cooling is used to remove heat from vapors and condense them to liquids, and also to cool liquids to lower temperatures, often for safe storage or further processing.
- What is contacting?
  - The purity or composition of the liquid and the vapor streams change stage by stage as vapor *contacts* liquid flowing in the opposite direction in the column, mixes, and reaches vapor-liquid and thermodynamic equilibrium before being separated into liquid and vapor streams once again, and sent onward to the next stage (up for the vapor, down by gravity for the liquid).
  - Trays are mounted horizontally inside distillation columns.
  - Figure 10.7 is a photograph taken during manufacture of a copper float valve tray section. When in service, vapor will come up through slots (from back to front in this photograph) capped by floating check valves, and liquid will flow across the tray (from top to bottom in this photograph).
  - Figure 10.8 is a shop photograph of the deck of a valve tray with round valve caps (rather than the rectangular valve caps shown in Fig. 10.7) and with a two-pass liquid path. On this tray, the liquid will

FIGURE 10.7 Copper Valve Tray Deck Section, Normally Horizontal during Operation.

come in from above, along the centerline (defined in this photograph by five hexagonal shapes without punched trays). Half of the liquid will pass sideways across the tray toward the left, and half toward the right, falling into the downcomers (which look like rain gutter downspouts) seen in the lower and middle left and upper right edges of the photo.

- Figure 10.9 is a shop photograph of the underside of the valve tray that will be mounted directly above the tray shown in Fig. 10.8. The undersides of the round valve caps (loosely mounted in punched round holes) are clearly visible. Also visible is the line of central circular copper downcomers, whose locations correspond to hexagonal flat patterns on the deck top in Fig. 10.8. Liquid will enter the tray shown in Fig. 10.9 at both outside edges, then flow toward the centerline. Thus, with two-pass trays, the liquid flows from center to edge or from edge to center on alternating odd- and even-numbered trays, throughout the whole column. Vapor flows upward through the valve caps, in the same direction on every tray.

- What is pumping?
  - There are different types of pumps for different duties.
  - Positive displacement pumps:
    - Cannot have flow restricted while they are running
    - Can maintain a fixed flow rate despite variations in backpressure

FIGURE 10.8 Deck of Float Valve Tray with Round Valve Caps and Two-Pass Liquid Flow through Alternating Center and Side Downcomers.

FIGURE 10.9 Underside of Float Valve Tray, Two-Pass, Showing Copper Valve Caps and Row of Center Downcomers.

- Can be much easier to prime than centrifugal pumps
- Useful for transfer pumps that must be frequently drained
- Centrifugal pumps:
  - Can vary in flow rate across a wide range
  - Any change in backpressure changes the flow rate.
  - Can be cut back to zero flow for approximately one minute without damage
  - One hour at zero flow will overheat and damage a centrifugal pump.
- What requires control?
  - A control strategy (manual or automatic) is required for the following:
  - Feed input
  - Heat input
  - Column product purity control
  - Column bottom effluent
  - Reflux
  - Column top distillate
  - Other feed, product, or internal streams
  - System pressure
- Manually or automatically set control valves carry out the selected control strategy.

1.3 Theory of Multi-Component Vapor-Liquid Equilibrium (VLE) Relationships that Determine Distillation Process Feasibility and Capability

1.3.1 Actual VLE Relationships in Multi-Component Mixtures

- Actual multi-component mixture distillation behavior is determined by examining all binary pairs of components.
- The McCabe-Theile diagram in Fig. 10.10 depicts a methylene chloride–methanol binary distillation process
- The scale of the X-axis is mole fraction in the liquid of the more volatile compound in the binary pair.
- The scale of the Y-axis is mole fraction in the vapor of the more volatile compound in the binary pair.
- In this example, the more volatile compound is methylene chloride, and the less volatile compound is methanol. It is noteworthy that there is an azeotrope (a point where X = Y on the equilibrium curve) near the upper right corner of the diagram.
- Blue curve represents equilibrium concentration, y = vapor, x = liquid.
- The red lines have slopes equal to liquid/vapor molar flow ratio.
- Green steps show the change in concentration per distillation stage.

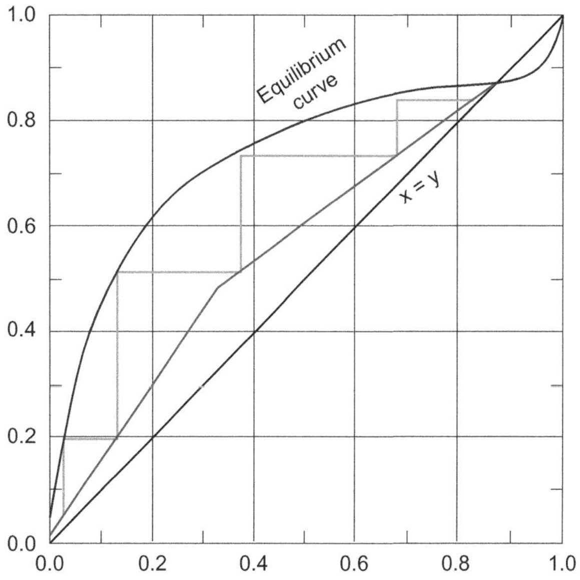

FIGURE 10.10 Vapor-Liquid Equilibrium Diagram for Methylene Chloride–Methanol Mixtures, with McCabe-Thiele Graphical Constructions Representing Theoretical Stages of Separation.

- The equilibrium curve crosses the x = y line at the azeotrope value, 87% molar methylene chloride.
- It is not possible to distill pure methanol from a mixture more than 87% methylene chloride.
- It is not possible to distill pure methylene chloride from a mixture with more than 13% molar methanol.
- The higher the red lines are above the x = y line, reflux is lower, less energy is used, but more stages are required.
- Equilibrium curves closer to the x = y axis require more reflux and more separation stages than broadly spaced curves.
- The minimum reflux amount occurs where the red line rises up to almost touch the blue line.
- Below minimum reflux, your column disappears.
- If the reflux ratio is less than the minimum reflux ratio for this system, the separation of a whole column equals only that of a one-stage process.

- Azeotropes are barriers that can be crossed by adding a third component that raises or lowers the volatility of component #1 or #2.
- Extractive distillation
  - Uses a less volatile third component to carry component #1 down the column, leaving pure #2 in the distillate
- Heterogeneous azeotropic distillation
  - Uses a more volatile third component (entrainer) to carry component #1 up the column, leaving pure #2 as the column bottom product.

- Liquid-liquid extraction
  - Can be used to reduce the amount of one component, moving the feed concentration to the other side of the azeotropic point
- Combinations of all binary pair VLE relationships fully define the distillation behavior of all non-reacting, non-ionic multi-component mixtures, no matter how many components are present!
  - In a four-component system, there are six binary pairs, as follows:
    - Component #1 with #2
    - Component #1 with #3
    - Component #1 with #4
    - Component #2 with #3
    - Component #2 with #4
    - Component #3 with #4
- Binary pair interactions do not fully define VLE in reactive or ionic systems. Examples include the following:
  - Ammonia-water solutions
  - Hydrochloric acid solutions
  - Amines with water and acid gases
  - Acetic acid mixed with amides
  - Compounds that form dimers or polymers
- In the absence of azeotropes, there is no theoretical limit to the purity that can be achieved by distillation with a sufficiently large number of stages
- Categories of actual binary pair VLE behavior
  - Both pure components may be liquid at the temperature and pressure of the system, or one pure component may be a fixed gas at these conditions.
  - Component pairs may be azeotropic or non-azeotropic.
  - Azeotropic pairs may form either a high or a low boiling azeotrope.
  - The two components may be fully miscible, or when mixed they may form partially miscible liquid phases or practically immiscible liquid phases.
  - Either one or both components may exhibit ionic or reactive behavior

1.3.2 TXY and XY Diagrams Graphically Illustrate Distillation Processes [Temperature is (T), Liquid Mole Fraction is (X), Vapor Mole fraction is (Y)]

- TXY and XY diagrams graphically illustrate VLE behavior of most binary systems (Fig. 10.11).
- TXY and XY diagrams are not well suited to systems in which one component is a fixed gas with limited solubility, because the corresponding shape is not a visible smooth curve in the central part of the diagram but instead a steeply rising curve very close to the outer edges of the composition diagram. The design principles employed with these diagrams are successfully used by computer programs for process simulation, but such systems' diagrams are not very informative to the human eye (Fig. 10.12).

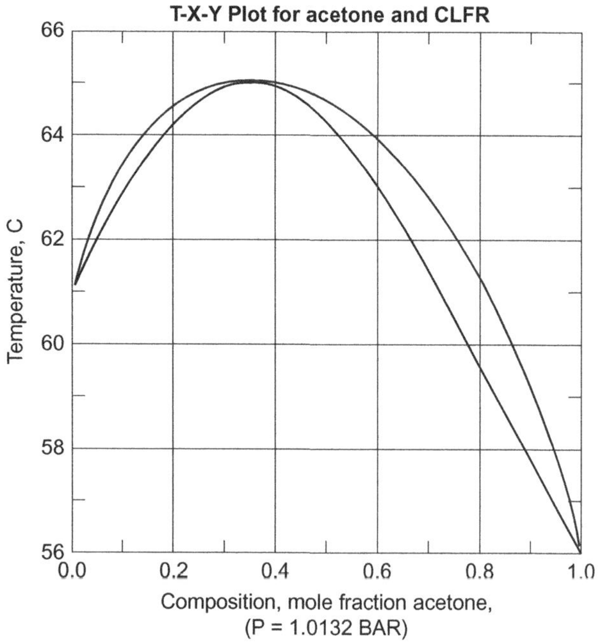

FIGURE 10.11 TXY Acetone and Chloroform Showing a High-Boiling Azeotrope (Top Center).

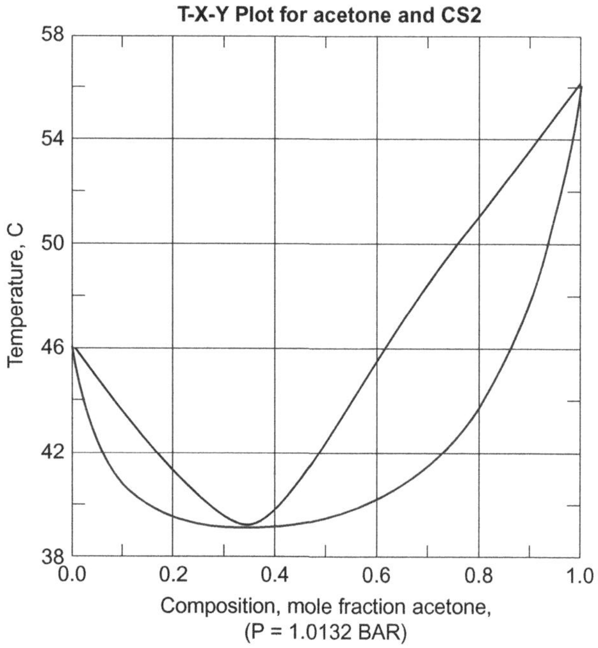

FIGURE 10.12 TXY Plot for Acetone and Carbon Disulfide Showing a Low-Boiling Azeotrope (Bottom Center).

- With practically immiscible binary liquid systems, these diagrams are not visually informative, for the same reason: the equilibrium curve can be very steep (though not as steep as for a fixed gas) and lie very close to the edges of the diagram—too close to be informative.
- TXY and XY diagrams do not provide sufficient information for determining the behavior of ionic, associative, or reactive components. The diagrams *are* accurate for any two pairs of compounds reaching phase equilibrium, including non-reactive or *unreacted but potentially reactive* molecules, but the relationship is time dependent.
- With ionic, associative, or reactive compounds, some molecules will change from one form to another; for example, a portion of a monomeric compound changes to dimeric binary form, removing some of the monomer from the mixture and thus changing the mass balance. This can be computer modeled, but the reactions (which are time dependent) must be taken into account, as well as thermodynamic equilibrium.
- XY diagrams determine how to un-mix any two components of a liquid mixture. We will next discuss how to evaluate this possibility; we will use the X-Y diagram for methanol and ethanol as shown in Fig. 10.13:
  - The equilibrium curve will tell you, at any point along it, whether a component will tend to go UP the column (y > x) or DOWN the column (y < x).

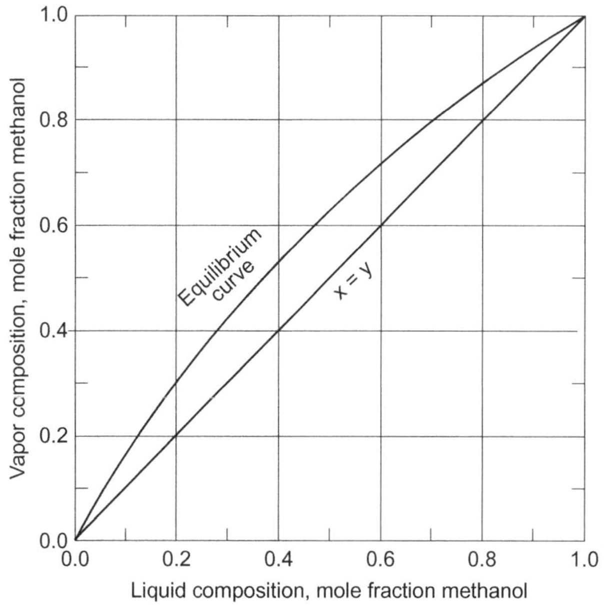

FIGURE 10.13 XY Plot for Methanol and Ethanol with No Azeotrope.

  - As illustrated by Fig. 10.14, which shows a close-up of the lower left-hand corner of the X-Y diagram for methylene chloride and water, in the absence of an azeotrope, there is no theoretical limit to the degree of purity that can be achieved by distillation.
  - As illustrated in Figs. 10.15, 10.16, and 10.17, if the equilibrium curve crosses the x = y line at any point, there is an azeotrope at the cross-over concentration.
  - Simple distillation with one feed and two components cannot produce product streams whose compositions cross over to opposite sides of azeotrope barriers.
- XY diagrams provide a road map to separation scheme synthesis.
  - Two high-purity products can be recovered from non-azeotropic binary mixtures.
  - One (but not two) high-purity products can be recovered from an azeotropic binary mixture.
  - Which component can be purified depends on the feed concentration's location on the XY diagram relative to the azeotrope concentration.
  - The broader the area between the equilibrium curve and the x = y line, the easier the separation is:
  - Lower energy input is required; less reflux flow is needed, and fewer stages of separation are sufficient.
  - If the equilibrium curve approaches the x = y line asymptotically, a pinch point occurs, where very little concentration improvement can be made by distillation

1.3.3 Suitability, Advantages, and Limitations of VLE Thermodynamic Models

- Types of VLE models (Summary)
  - VLE model based on a linear gas solubility model and pure liquid component properties: Henry's law, Raolt's law, steam distillation
  - VLE models based on reduced pressure and temperature (T/T critical) generalized compressibility correlations:
    - RK, SRK
  - VLE models based on activity coefficient concept, with actual binary VLE measurements regressed to produce "best-fit" binary interaction parameters (BIPs):
    - Wilson, Van Laar, NRTL, UNIQUAC
  - VLE model based on activity coefficient concept, with molecule functional group contribution sums (regressed from many real binary VLE pairs) used to calculate predicted (not directly measured or regressed) BIPS:
    - UNIFAC

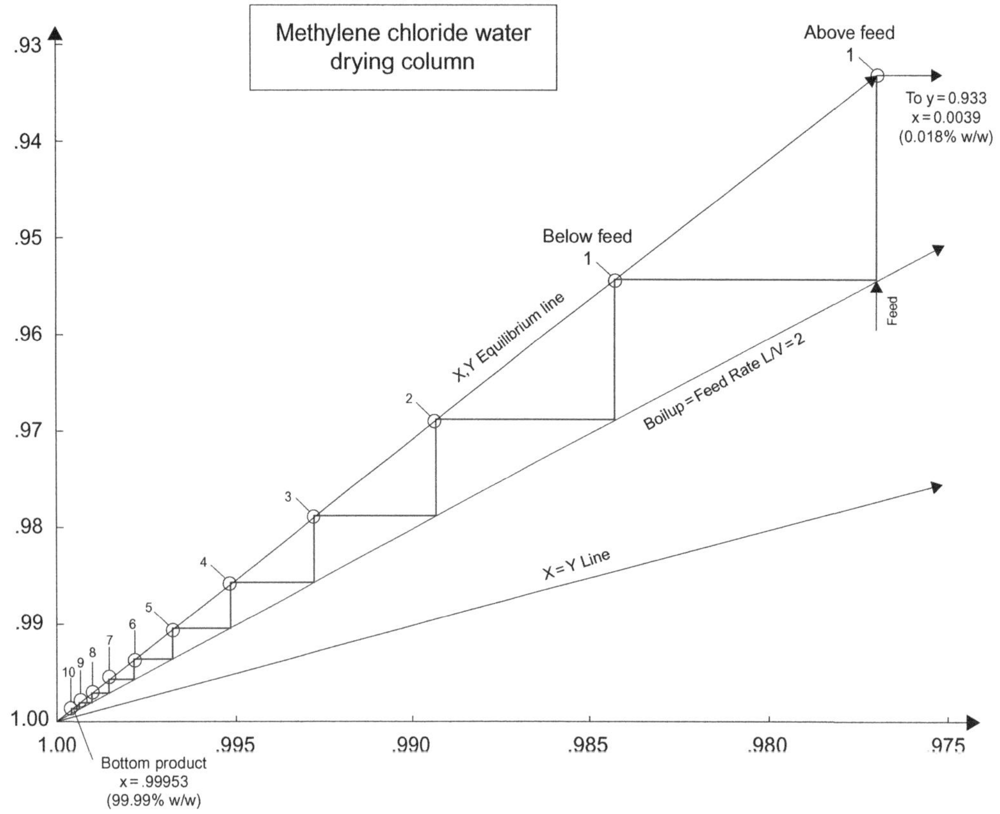

FIGURE 10.14 XY Plot for Methylene Chloride–Water with McCabe-Thiele Graphical Construction Showing Steps toward Increasingly High Purity.

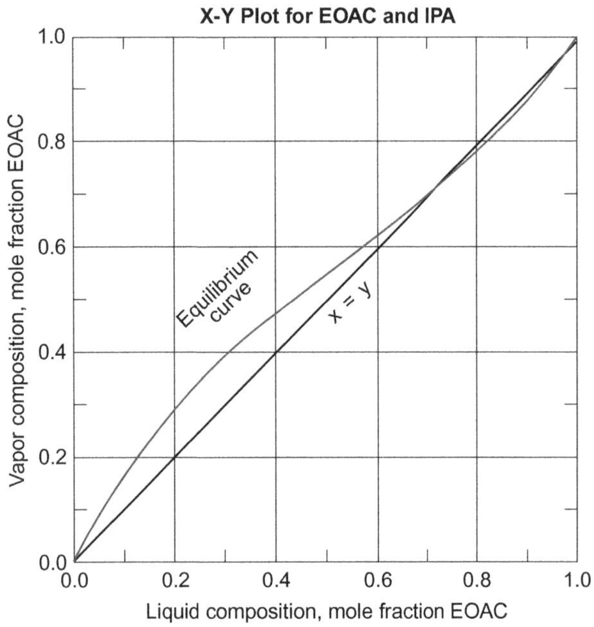

FIGURE 10.15 XY Plot for Ethyl Acetate and Isopropanol Showing Low-Boiling Azeotrope at X = Y = 0.72.

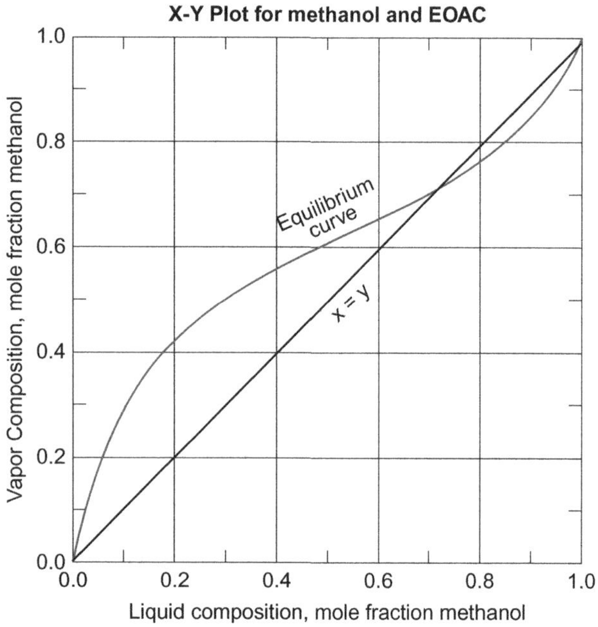

FIGURE 10.16 XY Plot for Methanol and Ethyl Acetate Showing Low-Boiling Azeotrope at X = Y = 0.70.

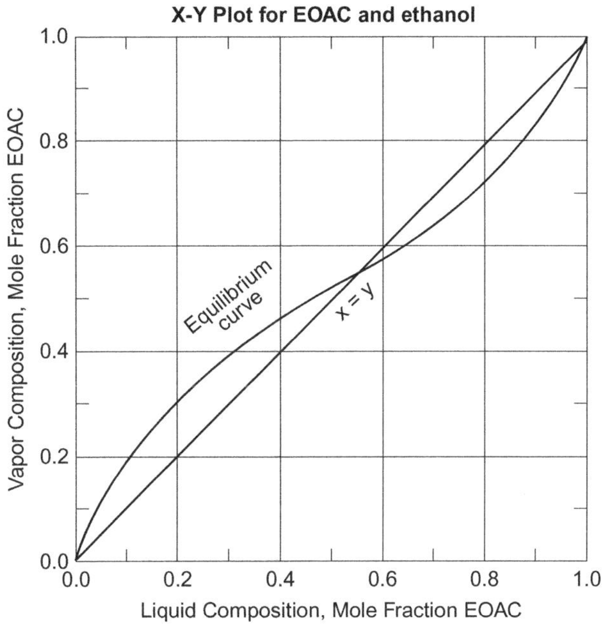

FIGURE 10.17 XY Plot for Ethyl Acetate and Ethanol Showing Low-Boiling Azeotrope at X = Y = 0.55.

- VLE model based on a linear gas solubility model and pure liquid component properties: Henry's law
  - Henry's law is useful for modeling fixed gases dissolved in liquids.
  - For the Henry's law VLE model to be a good representation of actual chemical behavior, the fixed gas molecules must not dissociate or react in the liquid phase.
  - Henry's law relates the gas partial pressure to its concentration in the liquid by a linear proportion.
  - Unfortunately, the Henry's law proportionality constant is very temperature dependent.
- VLE model based on simple interpolation of pure component properties: Raoult's law
  - Raoult's law states that the vapor pressure of each component in a liquid solution equals its pure component vapor pressure at system temperature, multiplied by that component's mole fraction in the liquid phase.
  - Raoult's law is similar to the ideal liquid solution density mixing rule.
  - Systems in which there is very little gain or loss of volume when pure component liquids are mixed are the same systems for which the mixture VLE is likely to be well modeled by Raoult's law.
  - Pure hydrocarbons of many types, at a temperature less than 70% of critical temperature, or more than 130% of critical temperature, and liquefied inert gases can be accurately modeled by Raoult's law.
- VLE model based on summation of two liquid phase vapor pressures: "steam distillation"
  - Classical steam distillation accurately models an ancient process by which fragrances and essential oils are distilled from plant materials.
  - Steam distillation processes involve boiling water and another poorly water miscible compound together at the same time.
  - In the steam distillation VLE model, the water and the other compound exhibit their own vapor pressure at the system temperature.
  - Each of the two act as if the other's vapor was not even there: each acts as if it were being distilled, alone, at reduced system pressure.
  - This model is useful in many petroleum refining and hydrocarbon applications where water is present.
- Non-ideal VLE models based on pure component compressibility
  - VLE models based on reduced pressure and temperature (T/T critical) generalized compressibility correlations were developed primarily for petroleum and petrochemical uses.
  - Most lab measurements and development work date from 1930 through 1970.
  - Redlich-Kwong (RK) and Soave-Redlich-Kwong (SRK) correlations, in original and various "extended" versions, continue to be used today and work well for hydrocarbons.
- VLE models based on the "activity coefficient" concept
  - The activity coefficient concept applies correction factors to Raoult's law partial pressures.
  - Activity coefficients are useful for calculation purposes for systems that do not closely follow one of the VLE models that are based on pure component properties.
  - Activity coefficients are multiplied times the partial pressure as computed by Raoult's law to yield a corrected partial pressure.
  - Correction factors or activity coefficients frequently range between 0.1 and 100, with activity coefficients over 1000 common in partially miscible liquid systems.
  - Attempts have been made to relate activity coefficient models to molecular interaction forces, but no unifying theory seems to explain non-ideal VLE across a broad range of system types.
  - As a consequence, there are activity coefficient models that reproduce the VLE behavior of certain systems very well and other systems very poorly.

- Evaluation of and selection of VLE models based on activity coefficient concept
  - Usually the best result in modeling binary or multi-component VLE will occur when actual binary VLE measurements have been regressed to produce "best-fit" binary interaction parameters (BIPs) for one of the common activity coefficient models.
  - Models of this type in common use in 1975 included Wilson and Van Laar, whereas the NRTL and UNIQUAC models are most often used today, making their library BIPs easier to find, but not necessarily making their accuracy better.
  - Some of these models follow the actual VLE curves of certain binary pairs better than others, while some fail to model XY behavior well.
  - Always generate XY VLE from your specific BIPs and VLE model, and compare to the best measured data you can find, before judging the accuracy of your model.
  - Do not ignore binary pairings for which there are no library BIPs available.
  - Some method must be used to fill in missing BIPs or very misleading, false, fictional results can arise during computer simulation.
- Another type of VLE model based on the activity coefficient concept: UNIFAC
  - The UNIFAC model does not use library BIPs for each binary pair of components, as do the activity coefficient models discussed previously.
  - Instead, the UNIFAC model examines the structure of each molecule, with molecular functional group contributions (regressed based on many real binary VLE pairs) summed for each molecule's structure.
  - UNIFAC uses the group contribution sums to calculate predicted (not directly measured or regressed) BIPs that can be used in other activity coefficient–based VLE models.
  - UNIFAC is commonly used by process simulators to generate predicted BIPs where library BIPs are lacking for some component pairs.
  - UNIFAC, as a predictor, can lead to false results, so always generate XY diagrams or do lab work to evaluate them.
  - Try not to start relying on UNIFAC so much that you overlook using actual binary data in binary activity coefficient models.
- Vapor-Liquid-Liquid Equilibrium (VLLE) models
  - One vapor–two liquid phase equilibrium (VLLE) models exist for NRTL, UNIQUAC, and other activity coefficient correlations whose form permits two liquid phases to exist.
  - These models can work very well when their interaction parameters have been regressed from carefully measured VLLE data for the chemical components in question.
  - VLLE models applied using BIPs derived from VLE data alone can give results that are wrong yet comfortably misleading.
  - Never overlook making actual liquid-liquid partitioning measurements on which to base VLLE process development.
- Thermodynamic restrictions to binary VLE models
  - Thermodynamic theory leads to many restrictions on the possible VLE behavior of mixtures.
  - People have developed many thermodynamic consistency tests that can be applied to determine whether VLE data or VLE correlation outputs are thermodynamically consistent.
    - **Warning No. 1:** Activity coefficient data or XY models produced by computations using any of the VLE models will nearly always pass the thermodynamic consistency test, no matter how far the computed results deviate from actual system behavior
    - **Warning No. 2:** 100% genuine and highly accurate VLE measurements can, and frequently do, fail thermodynamic consistency tests.
    - We must not reject carefully measured data that does not conform to our theories, and instead we must examine the theories and compare to experiments for validity.
- VLE models for binary pair behavior, plus material and energy balancing, determine the compositions calculated by a computer process simulator.
- How to tell if a VLE model is a good description of actual VLE
  - Prepare TXY and XY diagrams using the selected VLE model and the values of the BIPs that you hope to use in the computer process simulator.
  - Compare to measured actual binary system TXY and XY.
  - Shapes should be very similar.
  - Temperatures should be very similar.
  - The model must be checked to be certain that any azeotrope shown in the model exists in the real system.
  - The azeotrope concentration must be correct.
- Activity coefficient-based VLE models often fail at pinch points.
  - Often the "best fit" BIPs for modeling the VLE of a binary system with a pinch point will show an azeotrope cross-over instead!
  - The acetone-rich end of the acetone-water binary system often shows an azeotrope above 99% acetone, when in fact this is not an azeotropic system.
  - Process simulation computer models used in such situations will fail unless the binary VLE is adjusted to get rid of fictional azeotropes.

2.0 DEVELOPMENT OF A DISTILLATION APPLICATION

2.1 Using VLE Information for Conceptual Distillation Process Synthesis

- How to determine what variety of distillation process will "un-mix" and purify the desired compound
  - It is time to turn off the computer!
    - Even if you are going to use a computer to simulate the distillation process and let it do all the hard work of performing material and energy balances, first you must decide what process you will ask the computer to model.
  - Determine what goes up and what goes down.
    - List the boiling temperatures of all components and all azeotropes existing among those components, in ascending order, with the coolest temperature at the top.
    - The azeotropes can be thought of as behaving like pure components in this evaluation.
    - This list will tell you what goes up the most easily.
  - Look at the list to find a cut point, where you can let everything above that point go up and the rest go down.
    - If you need to collect the second most volatile material on the list, consider a partial condenser, and consider a liquid side draw near the top of the column.
    - If you need to collect or remove a material mid-way through the list, consider a liquid side draw and possibly a second distillation column.
    - If you need to collect the second least volatile material on the list, consider a vapor or liquid side draw near the bottom of the column.
    - If the least volatile material is your product, collect it from the column bottom.
  - Determine the overhead condensate condition.
    - The overhead condensate may form one liquid phase.
    - The overhead condensate may form two liquid phases, one usually being water.
    - There can be two organic liquid phases or two organic liquids plus water—this does happen!
    - Determine the specific gravity of the phases.
    - If the overhead condensate forms two or three liquid phases, decide which phase the reflux must come from, heavy or light.
    - Determine which phases represent product or waste streams.
  - Determine if side draws are required—liquid or vapor.
    - Determine if side draws are required to remove mid-range components from the column.
    - Determine whether each side draw is better taken as a liquid or a vapor.
    - Determine relatively how far up on the column each side draw should be.
    - Determine whether the main distillation column should be split into two columns to permit mid-range component recovery rather than side draw.
  - Determine if pinch points will make purification impractical.
    - Look at the VLE diagrams closely to determine if pinch points appear.
    - If pinch points are found, investigate thoroughly to be certain that they are pinches rather than azeotropes.
    - Evaluate the L/V ratio required to make progress purifying the key components in the pinch zones.
    - Look at the boil-up heat and column size required to purify in the pinch zones.
    - Evaluate the economics of recovering the pinched product to decide whether it is worthwhile.
- Consider thermal decomposition.
  - Look at the thermal stability of all of the compounds present in the column.
  - If any compound is suspected of undergoing thermal decomposition at the column's operating temperature, investigate what temperature would be safe, and consider lowering the column operating pressure to achieve a safe maximum exposure temperature.
  - Investigate thermal stability improvement through the removal of destabilizers or the addition of stabilizers.
- Beware of ionic compounds and reactive component pairings.
  - While ionic compounds can be distilled in their neutral non-ionic forms, care is required to make certain that the ionic equilibrium drives the majority of the compound into its neutral form.
  - In the case of mixtures containing ammonia or amines, the pH must be high.
  - In the case of HCl and other acid gases, the pH must be low.
  - Heat of solution greatly affects the energy requirement for distillation of ionic compounds.
- Distillation of reactive compounds
  - Consideration of reaction equilibrium and reaction rates is required to correctly design a distillation system that includes reactive compounds.

    - If reactive compounds are present but it is not desired that they be permitted to react within the distillation system, then the operating pressure and temperature may be lowered until the reaction potential is small, or reaction inhibitors may be added.
- Accommodate fixed gases.
    - If any fixed gases are present in the feed material, their distillation equilibria and reactivity potential must be assessed.
    - Inert gases are usually present as dissolved substances in most distillation feed materials. Examples of inert gases include nitrogen and methane.
    - The system pressure control and venting system must be prepared to release gases that are dissolved in the feed but do not significantly re-dissolve in the liquid distillation products.
- Determine if pump-around loops are desired.
    - Pump-around loops may be included in a distillation system design.
    - Usually these loops provide cooling at a point different than the column condenser, either to condense vapor to permit liquid product withdrawal in large quantities, or to balance the vapor flow rates in different sections of the column relative to the physical capability of the installed distillation trays or packing.
- Determine if extractive distillation or entrainer distillation would be useful.
    - If your examination of the component VLE does not point toward a likely candidate distillation process, consider extractive or entrainer distillation.
    - Extractive distillation pumps an additional, less volatile, liquid feed stream into a distillation column above the primary feed inlet.
    - Its effect is to reduce the relative volatility of one or more components that otherwise would have gone overhead with desirable product.
    - Entrainer distillation traps a large physical inventory of an additional light component in the decanter, the reflux system, and the upper part of the column, where it enhances the volatility of one or more compounds that otherwise would go down the column into the bottoms product.
    - Both extractive and entrainer distillation typically require that a second column be provided for regeneration.
- Make initial educated guesses about reflux ratio, heat input, and number of distillation stages.
    - Look at one of the key component pair distillation XY diagrams.
    - Construct a tried-and-true McCabe-Theile diagram on that XY diagram.
    - Use the McCabe-Theile diagram to determine minimum L/V above and below the feed point.
    - On a McCabe-Theile diagram constructed with a reasonable reflux ratio, 1.5 times the minimum reflux, use graphical construction to determine the approximate number of theoretical stages.

2.2 Using a Computer Process Simulator to Model a Candidate Distillation Process

- Set up a topographical flow sheet model.
- Select components from the computer process simulator's embedded component library or create new components or pseudo components.
- Select VLE correlation options.
- Retrieve VLE BIPs from the library.
- Evaluate TPXY relationships calculated from the selected VLE model.
- Adjust BIPs or in-fill using UNIFAC until all binary pair TPXY are close to the truth.
- Input temperature, pressure, state, and quantity data for feed streams, cut streams, and side draws.
- Set up model constraints and input equipment and convergence specifications.
- Define equipment:
    - Flash blocks for setting feed stream conditions, mixers where multiple feeds go to one column, feed preheaters, product coolers, interchangers
- Define distillation column fixed parameters:
    - Number of stages, feed and product locations, condenser and reboiler type, column pressure profile
- Define distillation column calculation constraints:
    - One constraint will determine the column bottom condition (use fixed reboiler heat input for the first trial run) and another constraint will determine the column top condition (use fixed reflux ratio for the first trial run).
- Set convergence criteria and damping factors.
- Provide estimated distillation column internal flow & temperature profiles
    - Column flow and temperature profiles must be provided so the computer simulator will have a starting point for its calculations.
    - Usually it is sufficient to input a reasonable top and bottom temperature for the column, and reasonable top and bottom molar flows, and tell the simulator program to interpolate and fill in intermediate stage data.

- After one converged run has been made on the system you are studying, successive runs usually will converge more readily if the column profile computed during a previous run is used as the starting point, rather than the coarsely interpolated initial estimated profile.

- Run the initial simulation, adjusting input parameters until one run converges with reasonable results.
- If your first set of input parameters runs to convergence with reasonable results (rather than zero overhead vapor flow or other correct but trivial converged solutions), you are very lucky:
  - You input a good guess for the initial tower profile.
  - If your first run does not converge, heed the warnings given in the simulator output.
  - It is very easy to underestimate the heat input required, or to broadly miss the correct column flows, on the high or low side, so change the input and try again.
  - It is also very easy to mistake the engineering units used by the simulator model for other similar units, so when the process simulator results seem to be off by a large or familiar factor (like 1.8, 2.2, 454, or 1000), check the units, re-input, and try again.
- After one simulation has converged: Evaluate, adjust, rerun.
  - After one simulation has converged, evaluate the output product stream's quantity and quality, and the energy used.
  - Adjust the primary column input parameters to improve deficiencies in quantity or quality:
  - Move the feed point up or down; increase the number of theoretical stages.
  - Increase the reboiler heat input and increase the reflux ratio to improve product purity, but reduce the heat input and the reflux ratio to decrease hot and cold utility usage.
  - Do not be afraid to make many variations and run the simulator over and over again to see what happens if this or that increases or decreases.
  - Limit the size of step changes made to input parameters, to increase the chance of new runs converging successfully.
- The opportunity window for optimization opens now and closes when the final simulation run becomes the design basis for column, internals, and overall distillation system design.
- Make use of the process optimization opportunity.
  - Now that the simulation gives close to the desired result, it is time to run more simulations in a parametric block.
    - One parameter to reduce is the number of theoretical stages—try 10%, 20%, 50%, or 80% fewer!
  - Adjust the feed point because the result may surprise you:
    - Try moving the feed up and down because either may yield improved purity.
  - Step all side draws up and down in position and in flow rate.
  - In the optimization runs, permit the more advanced column top and bottom convergence specification criteria their chance to decrease energy input or reflux quantity:
    - Try using purity specifications or flow quantity specifications instead of heat input and reflux ratio as in the initial runs.
  - Run column tray or packing sizing software on these optimization run outputs to provide physical size input for cost estimation.
    - The optimal design will balance fixed and variable life cycle costs.

2.3 Selection of Column Internal Contacting Equipment

- What trays, or which packing?
  - Unless low pressure drop (<0.1 psi per stage) is necessary, for column diameter >2 feet (500 mm), consider valve trays first.
  - If pressure drop of near 0.01 psi per stage is required in non-aqueous, low-pressure systems, consider structured packing first.
  - For low-pressure drop in aqueous systems, consider random packing first at or above atmospheric pressure, and consider structured packing first under vacuum.
  - For high vacuum conditions (1 to 20 mmHg absolute), consider woven wire mesh packing first and knit wire packing second.
  - Valves for valve trays are available round or rectangular, with directional patterns such as checkered, basket weave:
    - All will work, with variations in cost and maximum and minimum rate at a given diameter and tray spacing key among selection criteria.
    - Floating valves act as check valves on the tray deck, reducing liquid weeping at low vapor rates and improving performance during turndown.
    - Punched raised louvers, also called fixed valves, resist fouling and are mechanically very strong, which is good when turndown is not important.

2.4 Rate the Selected Physical Distillation Column for Alternative Operations

- Often a distillation system will be used for more than one process
- After simulating the selected primary process and optimizing the physical distillation column for that process, it is time to simulate alternative processes.

- If the required alternative processes can be performed in the physical system selected for the primary process, no design modifications will be needed, but their process information should appear in the design basis because it is not certain which system will demand the highest duty or dictate maximum size of the reboiler, condenser, coolers, pumps, instruments, or auxiliaries.
- If the alternative process requires more separation stages, a different feed point, different decanter or reflux provisions, or much higher or lower energy input, incorporate this information into the design basis specification, in addition to the complete primary design.

3.0 DESIGN OF A DISTILLATION SYSTEM FOR A NEW APPLICATION

3.1 Laboratory Testing for Design of a Commercial Scale Distillation System

- Laboratory distillation apparatus mimics commercial systems (Fig. 10.18).
- Laboratory column trays or packing are commonly designed to give one theoretical tray per 3 to 10 centimeters (1 to 4 inches) of column height. This permits the apparatus to fit within a reasonable ceiling height (Fig. 10.19).
- Lab requires tiny distillation trays or tiny packing.
- Packing of the type shown in Fig. 10.20 will operate with a Height Equivalent to a Theoretical Plate

FIGURE 10.18 Lab Distillation Apparatus for Batch Distillation of Ethanol.

FIGURE 10.19 Lab Distillation Column Side View, Showing Packing Inside.

FIGURE 10.20 Lab Distillation Packing, Loose–Scale Shows Size in Inches.

(HETP) of 3 to 10 centimeters. Thus, 10 to 30 theoretical stages of distillation can be accomplished in a 1-meter-tall packed column section. Such sections can be stacked to give 100 or more theoretical stages in one column. Distillation with more than 200 stages in one column is not done in practice due to construction limitations for tall equipment.

- Distillation lab analytical requirements include Gas Chromatograph (GC) for resolution of highly volatile compounds, High Performance Liquid Chromatography (HPLC) for identification of non-volatile compounds, moisture balance with infrared light to measure residual solvent content of solids, and a Karl Fischer titration apparatus for the measurement of the water content of solid or liquid samples.
- Lab and computer simulation flowsheet modeling of IPA recovery system options
  - Figure 10.21 is an outline diagram of a lab IPA evaporator diagram that used no reflux, thereby allowing a packed column to operate as one theoretical stage.
    - Figure 10.22 shows the lab IPA evaporator block flow diagram, and Fig. 10.23 shows the computer simulation stream data corresponding to an operation performed in the apparatus from Fig. 10.21.
      - Figure 10.24 illustrates a lab IPA–water binary distillation system diagram, with two packing sections, for the recovery of purified IPA instead of simple evaporation and condensation (as in the evaporator of Fig. 10.21).
    - Figure 10.25 shows the lab IPA–water distillation system block flow diagram, and Fig. 10.26 shows the computer simulation stream data corresponding to an operation performed in the apparatus from Fig. 10.24.
      - Lab IPA distillation computer simulation stream data

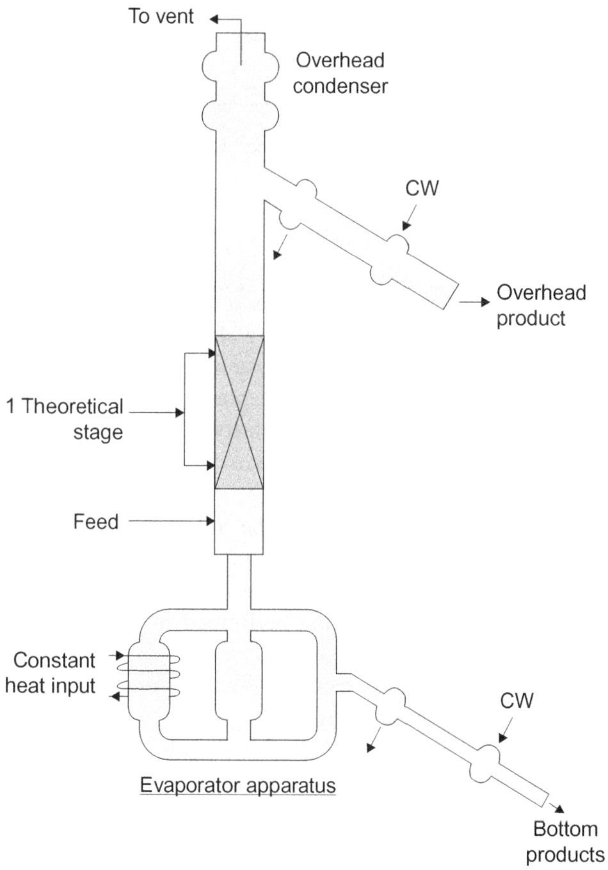

FIGURE 10.21 Lab Evaporator Apparatus (Distillation Column without Reflux).

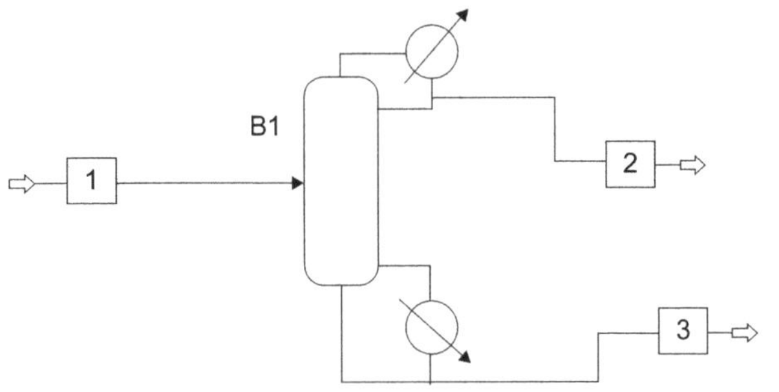

FIGURE 10.22 Process Simulation Computer Model Flowsheet for Lab Evaporator.

IPA/DBP/water				
Stream ID		1	2	3
Temperature	C	15.0	82.2	83.7
Mass flow	KG/HR	800.000	710.000	90.000
Mass fraction				
Water		0.002	0.002	81 PPB
ISOPR-01		0.983	0.998	0.867
METHA-01				
DIBUT-01		0.015	trace	0.133

FIGURE 10.23 Simulation of Flows for Lab Evaporation of IPA, Scaled Up to Commercial Size.

- Figure 10.27 shows a further development from Fig. 10.24: a lab IPA–methanol–water distillation system with three packing sections and two vapor draw streams. This system is used to purify the volatile product IPA in the presence of another volatile compound, methanol.
- Figure 10.28 shows the lab IPA–methanol–water block flow diagram, and Fig. 10.29 shows the computer simulation stream data corresponding to an operation performed in the apparatus from Fig. 10.29.
  - Lab IPA–methanol–water computer simulation stream data

3.2 How to Design a Completely New Distillation System

- Distillation system design overview
  - In this chapter, we will not attempt to provide detailed instructions about designing a distillation system and its associated equipment; such instructions are readily available elsewhere—see, for example, *Distillation Design* by Henry Z. Kister, Reference [2]. However, we will briefly addresses how to specify the following:
  - Process duties
  - Materials
  - Equipment
  - Instruments
  - Utilities

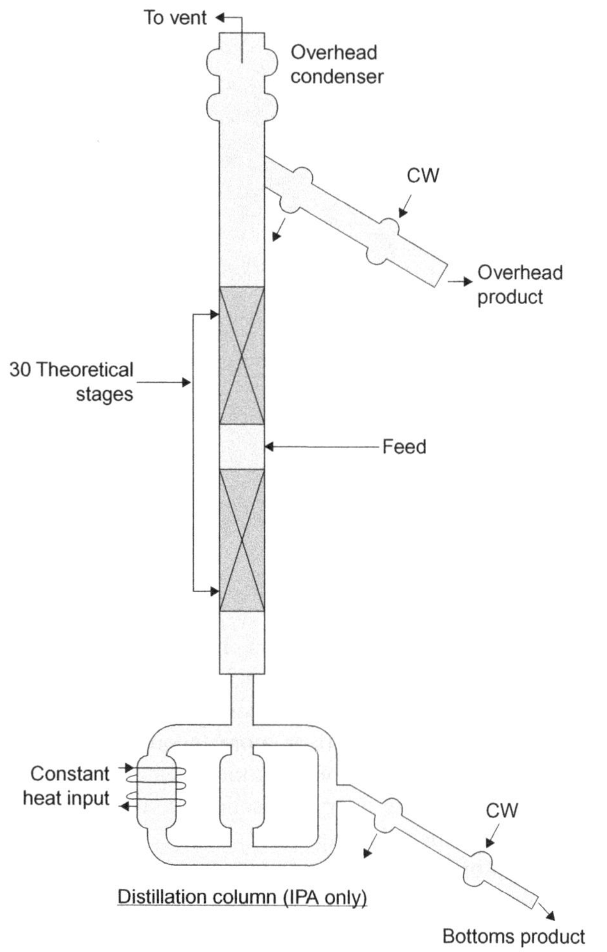

FIGURE 10.24 Lab Distillation Apparatus Diagram for Binary Distillation of IPA and Water.

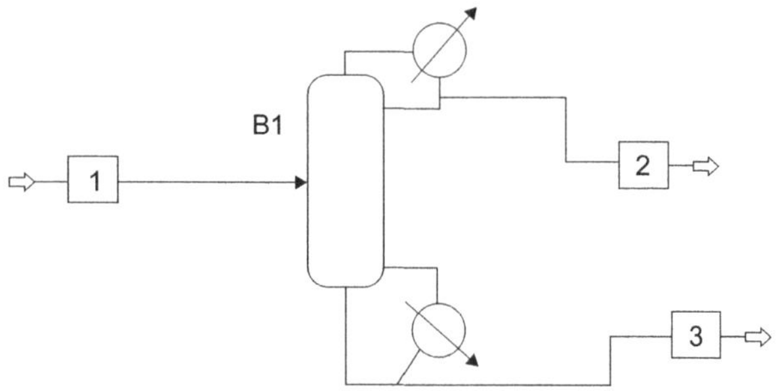

FIGURE 10.25 Process Simulation Flowsheet for Lab Binary Distillation of IPA and Water.

IPA/Water				
Stream iD		1	2	3
Temperature	C	15.0	81.9	82.8
Mass flow	KG/HR	800.000	150.924	649.079
Mass fraction				
Water		0.002	0.010	100 PPM
ISOPR-01		0.998	0.990	1.000
METHA-01				
DIBUT-01				

FIGURE 10.26 Simulation of Flows for Lab Binary Distillation for Recover of IPA, Scaled Up to Commercial Size.

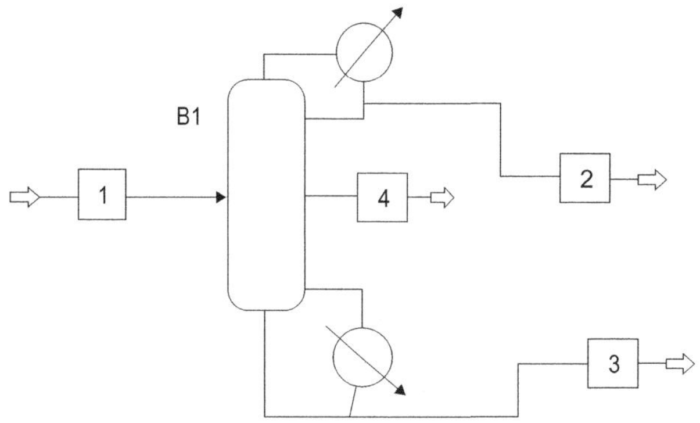

FIGURE 10.28 Process Simulation Flowsheet for Lab Ternary Distillation of IPA, Methanol, and Water.

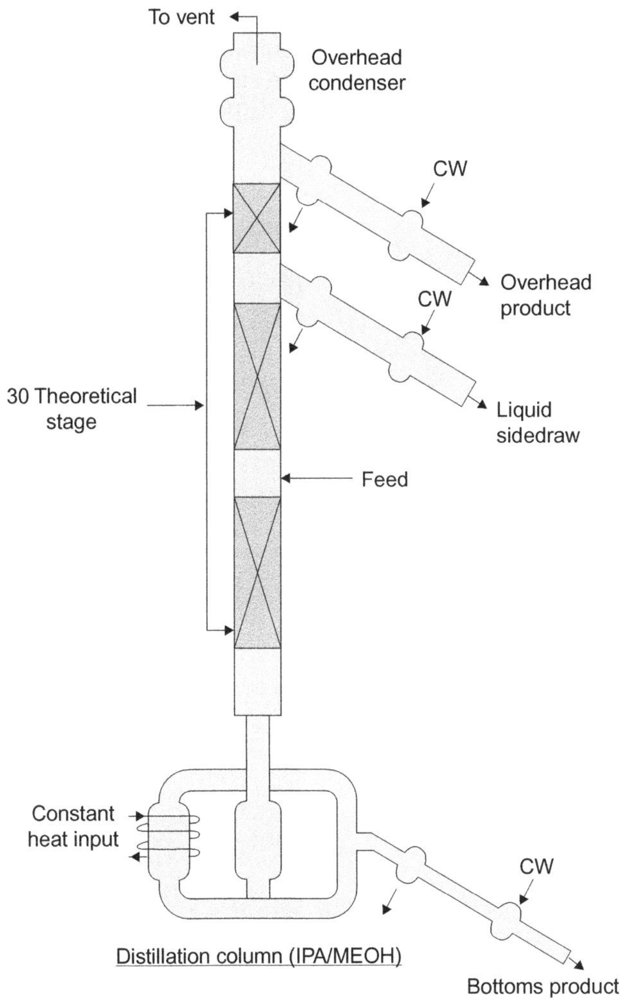

FIGURE 10.27 Lab Distillation Apparatus Diagram for Ternary Distillation of IPA, Methanol, and Water.

IPA/methanol/water					
Stream ID		1	2	3	4
Temperature	C	15.0	66.3	82.8	74.6
Mass flow	KG/HR	500.000	27.000	446.000	27.000
Mass fraction					
Water		0.002	0.008	102 PPM	0.028
ISOPR-01		0.948	0.249	1.000	0.790
METHA-01		0.050	0.744	918 PPB	0.182
DIBUT-01					

FIGURE 10.29 Simulation of Flows for Lab Ternary Distillation for Recover of IPA from Mixture with Methanol and Water, Scaled Up to Commercial Size.

- This section also discusses how to conduct the following:
  - Safety reviews
  - Environmental reviews
  - Operability reviews

- Design basis information
  - Determine what is right or wrong about earlier preliminary designs and any referenced prior designs.
  - Collect materials and information that describe the new intended system and previous reference systems.
    - Block flow diagrams, material balances, preliminary Piping and Instrument Diagrams (P&IDs), reference system P&IDs, design memos, specification sheets, equipment quotes, catalog cuts, utility conditions
  - Find the most current versions of design information.
  - Determine what may be true or false in these documents so that you can avoid repeating past mistakes.
- Define major equipment list.
  - This list contains all of the equipment required to make the actions of the block flow diagram happen.
  - Equipment that will heat, boil, condense, cool, pump, contact, decant, or contain materials
- Specify equipment for the major equipment list.
  - Determine duties, materials, types, models.
  - What needs to be done by equipment to accomplish what the block flow diagram describes?
  - What equipment characteristics, capabilities, and capacities will be required?

- Design equipment in order:
- Distillation columns and internals
- Reboilers and condensers
- Other heat exchangers
- Pumps
- Decanters
- Batch pot and product receivers
- Other tanks and agitators

- Equipment specification sheets
  - Heat exchangers (condenser, reboiler, coolers, others)
  - Shell and tube exchangers are most common, but do not ignore other types.
    - Shell and tube heat exchanger baffles, rods, tubesheet, tubes
- Figure 10.30 shows a typical shell and tube heat exchanger, partially assembled in a fabrication shop. All the important components are visible: tubes, tubesheets, baffles, baffle support rods, and large diameter pipes and pipe caps, which will be welded to form the shell and the heads.
- Figure 10.31 shows a small shell and tube heat exchanger tube bundle lifted from its shell during a field investigation.
- Figure 10.32 is a close-up of the tubesheet surface of the bundle lifted in Fig. 10.31, nicely illustrating the appearance of a shell and tube heat exchanger tubesheet with rolled-in tubes. Note that the tube ends are not precisely flush with the tubesheet surface. This is intentional.
- Figure 10.33 shows a typical small shell tube heat exchanger two-pass top head bonnet, showing the gasket surface. This bonnet was taken off the top of the same heat exchanger bundle seen in Figs. 10.31 and 10.32.
  - Plate and frame heat exchangers
  - Plate and frame heat exchangers are used to reduce cost and space requirements when fluids are compatible with gasket materials and are not prone to plugging with solids. They have high heat transfer coefficients due to their small passage size. Figure 10.34 is a photograph of a typical small plate and frame heat exchanger, installed and in operation. The heat exchanger is the blue rectangular block in the center of the photograph, with one pipe attached to each of the four corners. This is a typical piping arrangement for plate and frame heat exchangers.
- Shell and tube heat exchanger specification sheets must state the following:
  - Shell side/tube side performance
  - Shell side/tube side construction
  - Specifications for tubing, baffles, gaskets, etc.
  - Sketches may be useful.

FIGURE 10.31 Small Shell and Tube Heat Exchanger Tube Bundle Lifted from Shell.

FIGURE 10.30 Shell and Tube Heat Exchanger During Assembly. Baffles and Rods Are Attached to the Nearside Tubesheet. Tubes Fit Through the Holes in the Baffles.

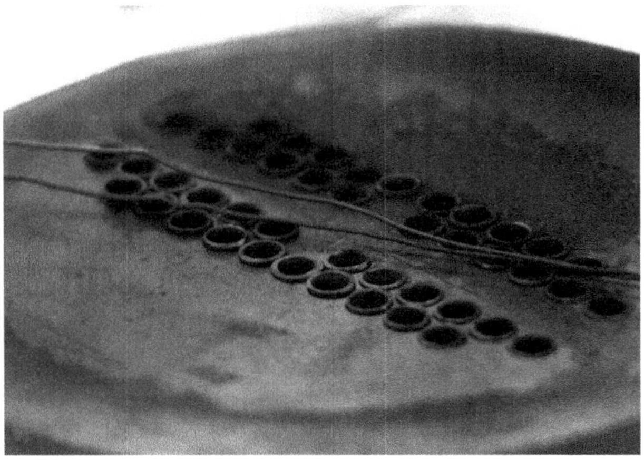

FIGURE 10.32 Small Shell and Tube Heat Exchanger Tubesheet with Rolled-in Tubes.

FIGURE 10.33 Top Head Bonnet for Two-Pass Small Shell and Tube Heat Exchanger.

FIGURE 10.34 Plate and Frame Heat Exchanger Installed and in Service.

- Vessel specification sheets are required for all vessels not purchased as catalog items:
  - Distillation columns
  - Decanters
  - Batch pots
  - Product receivers
  - Other tanks
- Pump design calculations
  - Be certain of appropriate pump types for each specific duty.
    - Positive displacement pumps:
      - Cannot have flow restricted while they are running
      - Can maintain a fixed flow rate despite variations in backpressure
      - Can be much easier to prime than centrifugal pumps
      - Useful for transfer pumps that must be frequently drained
    - Centrifugal pumps:
      - Can vary in flow rate across a wide range
      - Any change in backpressure changes the flow rate
      - Can be cut back to zero flow for approximately one minute without damage
      - One hour at zero flow will overheat and damage a centrifugal pump.
    - Centrifugal pump calculations:
      - First determine:
        - Discharge head
        - Flow rate
        - NPSHA
      - Use a pump rating computer program or printed catlog information for selection of models.
    - Centrifugal pump design specification completion
      - Materials
      - Impellor size
      - Seals
      - Motors
      - Couplings
      - Baseplates
      - Seal-less options
    - Positive displacement pump calculations
      - First determine:
        - Discharge head (maximum)
        - Flow rate
        - NPSHA
      - Use catalog or equal program for selection of models
    - Positive displacement pump design specification completion
      - Materials
      - Seals
      - Motors
      - Speed reducers
      - Variable speed drives
      - Couplings
      - Baseplates
- Start drawing the new distillation system P&ID
  - Draw by hand or with a computer drafting or graphics design program
  - Equipment duties and connection sizes must be known.
  - Consider how to fill the system, start it, shut it down, and empty it.
  - What valves and instrumentation are required to operate this plant safely?

- How to create a new P&ID
  - Start with a logical arrangement of equipment and main process pipe lines from the block flow diagram.
  - Add main process pipe lines from the block flow diagram.
  - Add more detailed process path piping.
  - Add utility supply and return and vent piping.
  - Add instruments to measure and control the process and utilities.
  - Add valves for control, protection, and isolation.
  - Add pressure safety equipment for any section containing a vessel that can be blocked in.
  - Every section of pipe must be tagged with size, fluid description, material code, insulation, and heat tracing
- Develop a control scheme and add the control hardware to the P&ID.
  - Create and design control loops and interlocks.
  - Verify the presence and select the type of measurement instruments for flow, pressure, temperature, and level.
  - Review the pipe line map to make certain that the equipment can be filled, emptied, and recycled.
  - Check and verify the presence of safety equipment:
    - Pressure relief PSV or PSE for any section containing a vessel that can be blocked in
    - High-level switches and low-level switches
    - High-pressure switches if appropriate
  - Control strategy must be determined for:
  - Feed input
  - Heat input
  - Column product purity control
  - Column bottom effluent
  - Reflux
  - Column top distillate
  - Other feed, product, or internal streams
  - Control loops must be created and designed to carry out selected control strategy.
  - Add process indicators and safety interlocks.
  - Add indicators for information or data recording purposes.
  - Interlock devices are included for protection of pumps and overall system safety.
    - Continue development of P&ID after creation of all pipe lines, instruments, and controls until all instruments are selected, tagged, and numbered for inclusion in lists and specifications.
- Conduct P&ID operability review
  - Make certain that the system allows for:
  - Filling
  - Starting up
  - Sampling
  - Shutting down
  - Emptying
- Conduct P&ID environmental review:
  - No uncontrolled vent emissions
  - Seals on pumps and instruments up to good practices
  - Means included to stop accidental spills or vent releases
  - Plugs or blind flanges to prevent accidental releases
  - Containment requirements for spills in the area
- Conduct preliminary P&ID safety review
  - Use Hazard and Operability Study (HAZOP) procedure or other safety review procedure.
  - Identify possible abnormal performance.
  - Include mitigating measures:
    - Check valves, interlocks, level switches, pressure safety valves, burst disks, extra valves or caps on drains and vent connections
- Revise P&ID in accordance with recommendations of reviews:
  - Operability
  - Environmental
  - Safety
- Back-check the revised P&ID to make certain that:
  - Review comments have been incorporated correctly
  - No new problems have been created
- Complete the P&ID development process:
  - By attaching summary information to the P&ID, about equipment on the P&ID:
    - Pumps
    - Distillation columns
    - Decanters
    - Vessels
    - Heat exchangers
- Develop an instrument list
  - The instrument list will be extracted from the P&ID.
  - Can be created automatically or by hand
  - Automation can be better if working from scratch, but needs non-interfering humans to accept its automatically generated numbering.
  - Instrument list contents (for each instrument)
    - Quantity
    - Tag #
    - Type
    - Service
    - Range
    - Connections
    - Material
  - Example of instrument list categories:
    - Field instruments, electric and mechanical
    - Control panel devices, electrical and pneumatic
    - Control valves and actuated valves

- Develop pipe line list
  - Extracted from P&ID
  - All lines listed by type
    - Process vapor
    - Process liquid
    - Steam
    - Nitrogen
    - Cooling water
- Create a workable system layout.
  - Create plan, elevation, and 3D layout views (Fig. 10.35):
    - Where will everything sit?
    - How much space will it take?
    - How will people access the area?
    - How will equipment be accessed? (Fig. 10.36)
    - How will the system be assembled? (Fig. 10.37)
    - How will the system be maintained?
- Proceed to detail engineering, fabrication, and assembly.
  - Detail engineering includes everything:
    - Control panel
    - Instruments
    - Vessels
    - Gaskets
    - Etc.
  - Follow the P&ID to the letter during detail design, or propose changes using a change control system.
    - Mark up redline versions to correct any mistakes or record changes to the P&ID.
    - Issue updated versions for construction.
    - Field verify as-built version of P&ID.
  - Pay attention to how the system will look. Distillation systems are tall and will be seen from a distance.

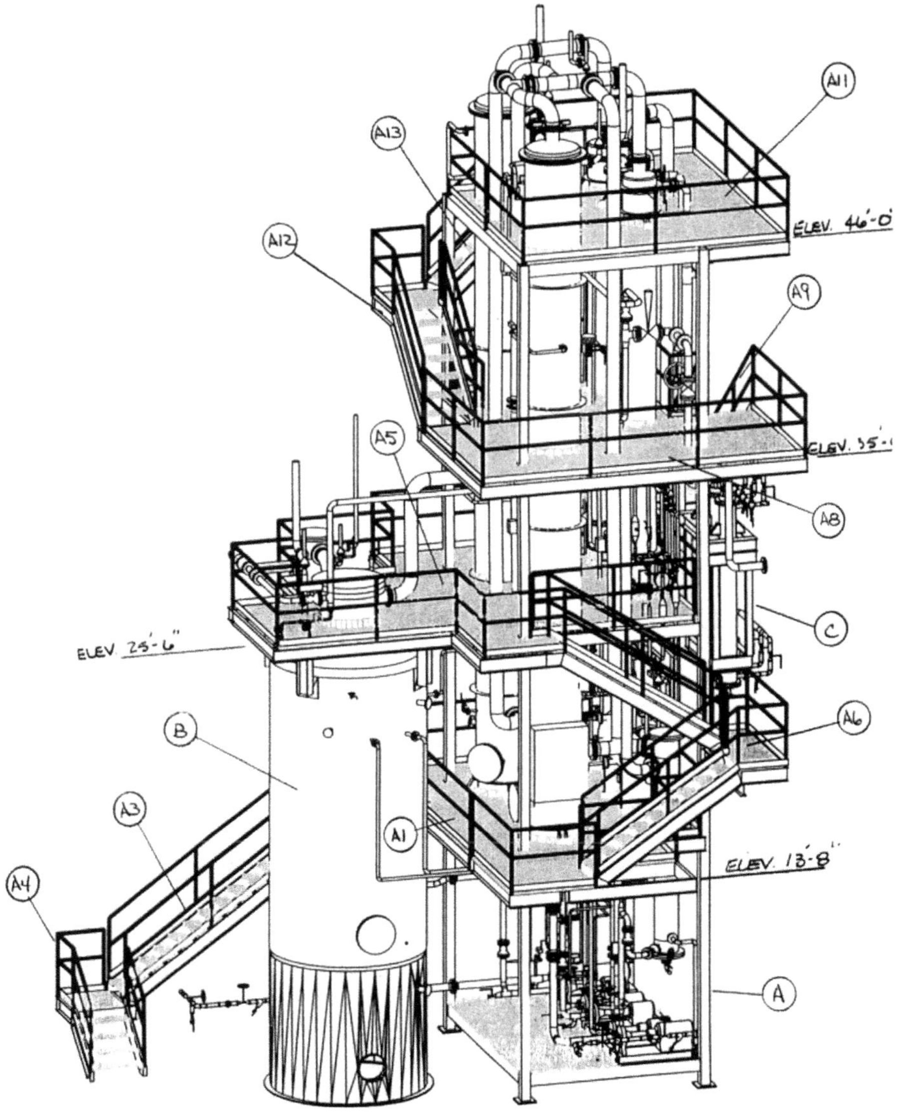

FIGURE 10.35 3 Dimensional Computer Aided Design (3D CAD) Drawing of Distillation System Showing All Major Equipment.

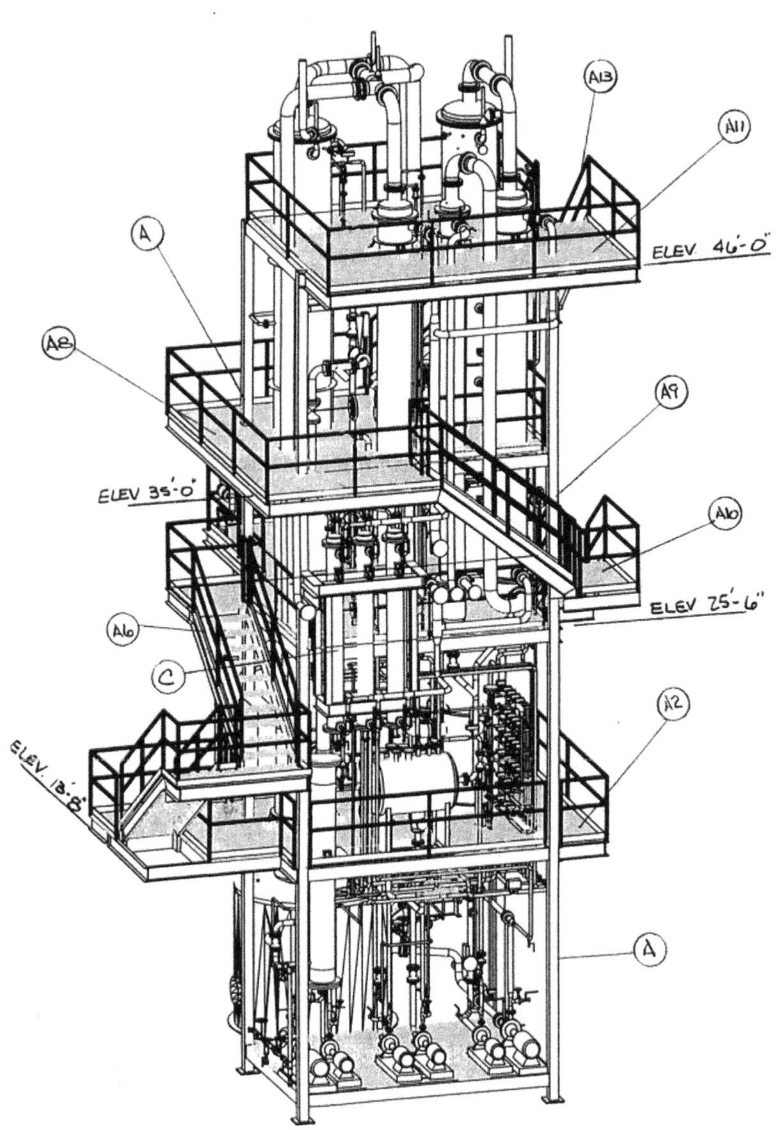

FIGURE 10.36 3D CAD Drawing of Distillation System Showing Access to Pumps, Heat Exchangers, Decanter, and Major Piping Tie-Ins, on the "Back Side".

4.0 CONTROL AND AUTOMATION OF DISTILLATION SYSTEMS

4.1 Practical and Theoretical Aspects of Controlling a Commercial Distillation System

- Distillation process control means *controlling the purity and quantity of product distilled from mixed feed material* by controlling the equipment elements that make up the physical distillation system.
  - Many levels of control simplicity or complexity are possible.
    - Each has its merits and is appropriate someplace.
  - First, understand what goes on inside a distillation system that requires control before determining how simple or complex your system control scheme and control hardware should be.
    - Plan how to start up, line out, shut down, and handle emergencies. Such planning helps to provide a robust control scheme.
  - Consider the diverse types of field instrumentation available, the types of process controllers available, and their merits and limitations before selecting control strategies and specifying them on P&ID's and in instrument lists.
    - Computerized Programmable Logic Controller (PLC) and Distributed Control System (DCS) hardware and software design presents additional challenges.
    - Full automation is possible, following ISA S88 model for batch plant control or simpler sequential function chart logic.

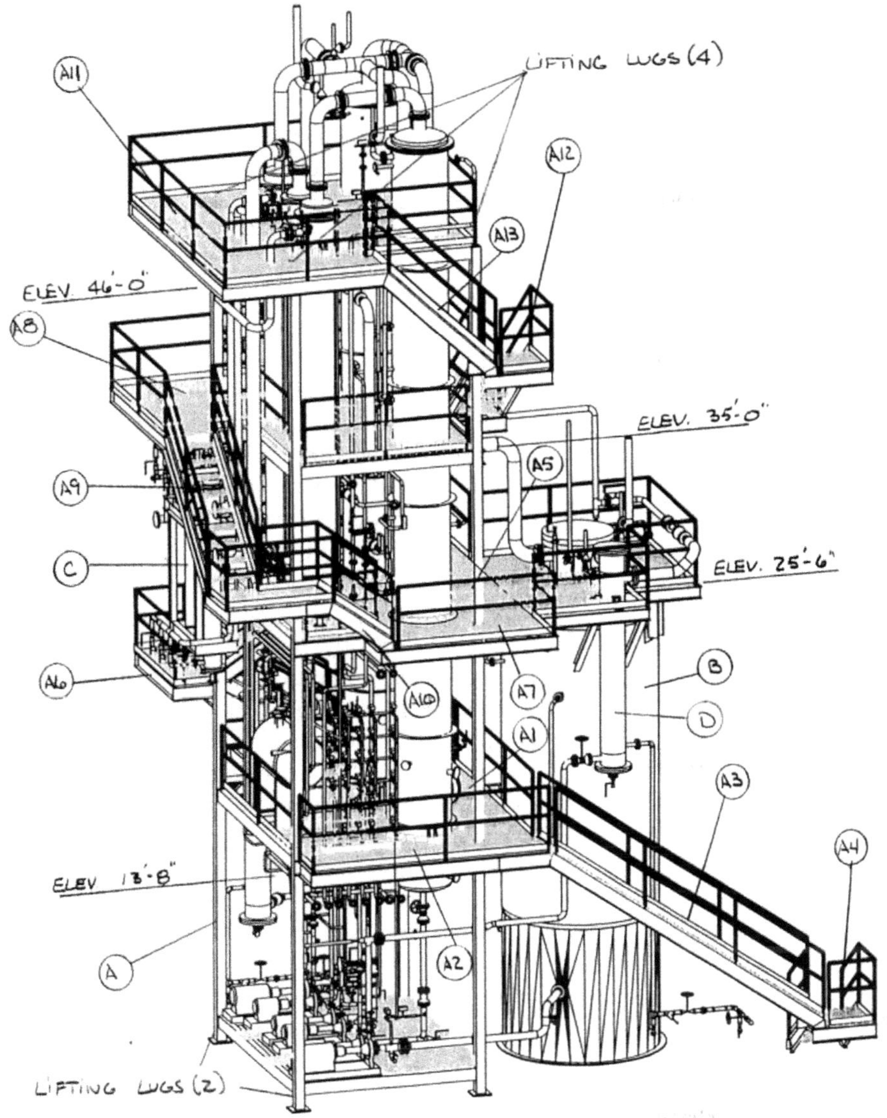

FIGURE 10.37 Distillation System 3D CAD Showing Stairway Access to Grade.

- What requires control?
  - A control strategy (manual or automatic) is required for the following:
    - Feed input
    - Heat input
    - Column product purity control
    - Column bottom effluent
    - Reflux
    - Column top distillate
    - Other feed, product, or internal streams
    - System pressure
  - Manually or automatically set control valves carry out the selected control strategy.
- What needs monitoring and recording?
  - In addition to control loops, process indicators and safety interlocks require monitoring and recording.
  - Indicators are for operator information or for data recording purposes.
  - Interlock devices monitor their inputs and respond if required for protection of pumps and overall system safety.
- What other things are standing by just in case they are needed?
  - Safety equipment is standing by:
  - Pressure safety valve (PSV) or burst disk pressure safety element (PSE) relieves pressure in each

section of the system containing a vessel that can be blocked in.
  - High-level switches and low-level switches
  - High-pressure switches
  - Interlocks triggered by switches or alarms
- Understand what is going on inside this distillation system before planning how to stop or start it or how to handle problems.
- What electrical devices, utilities, and equipment are turned on?
  - Control panel power
  - Pump motor power
  - Agitator motor power
  - Instrument air
  - Cooling source
  - Heating source
  - Fire protection
  - Phone, radio, security
- Control panel devices requiring power supply turned on:
  - Panel AC and DC
  - Uninterruptable Power Supply (UPS) input and output
  - PLC or DCS
  - Human - Machine Interface (HMI) or supervisory control and data acquisition (SCADA)
  - PC operator stations
  - Alarm panels
  - Analyzers
  - Printers
- Motor control center (MCC) devices turned on:
  - MCC 3-phase supply
  - Motor starters for all motors in use
  - Breakers to instrument power transformers
  - Breakers to lighting panels
  - Breakers to UPS feeds
  - Pumps and agitator motors switched on—possible ways:
    - On/off switch at MCC
    - Local on/off switch by motor or on panel
    - Local hand/off/auto switch by motor or in panel
    - Automatic run from control panel relay logic
    - Manual run from HMI
    - Automatic run from PLC or DCS system
- Compressed air supply for instruments active, dry, and up to pressure:
  - Instrument air compressors running
  - Instrument air dryer working
  - Air supply pressure OK
  - Instrument air filters OK
  - Air valves on headers and branch lines to instruments are open.
  - Air bleed and drain valves are closed.
- Prerequisites to preparing a new distillation system control strategy
  - Steady-state material balance must exist.
  - Equipment process duties and connection sizes must be known.
  - Consider how to fill the system, start it, shut it down, and empty it.
  - What valves and instrumentation are required to operate this plant safely?
  - What control scheme will keep this system going and safely making good product day after day?
- Instruments in a distillation system may be categorized as follows:
  - Sensors touch the process and "feel" or sense conditions.
  - Transducers change this "sensation" into a standard signal that can be transmitted elsewhere, either by pneumatic air pressure, milliamp current, voltage, HART (Highway Addressable Remote Transducer) digital protocol added on top of milliamp signals, or directly by digital field buses.
  - Indicators give a readout for you to see and understand.
  - Controllers actively respond to process and setpoint changes, trying to maintain the process measured variable equal to its setpoint value.
  - Control valves move as directed by the control system to change flow characteristics as required to modify the process measured variable.
- Field instrumentation
  - How field instrumentation works can be difficult to understand because there are so many different types and designs of field instruments.
  - However, most of the action and most of the problems occur in field instruments, not at the control panel, so it is worthwhile to understand how the field instruments work and what they are sensing.
  - Field control instruments perform control actions locally.
  - Control panel instruments perform indication or control actions remotely.
- Distillation process control: general introduction
  - Control of processes can be accomplished many different ways.
  - Some of these ways are safer, more accurate, or more automatic than others.
  - Manually controlled distillation systems are still best for simplicity and long-term reliability, especially with unsophisticated operations and maintenance staff.
  - The more advanced automation that can be implemented on PLC and DCS platforms is only justified in very high production environments.
- Types of process control
  - Uncontrolled or "wild," like city water temperature
  - Manual control, like a room heated by wood stove
  - On-off control, like a room heated by a furnace with a thermostat

- Feedback control, like a car cruise control
- Feed-forward control, like a multi-stream product blending system
- Advanced control modes, like multi-fuel boiler combustion controls
- Most common types of process control are manual, on-off, and feedback control.

- Typical feedback control loop
  - Applies to control of temperature, pressure, liquid level, or flow.
  - Can be proportional (P), proportional integral (PI), or proportional integral derivative (PID)
  - Loop dynamics will depend on the settings employed for gain, reset time, and derivative time.
  - Loop stability will depend on how well these settings suit the natural process value ranges (as compared to calibrated instrument ranges), the speed of changes, the length of time lags, and the amplitude of noise interfering with measurement.
- Switching between manual and automatic process control modes in a feedback flow control loop
  - In manual mode, you control the output (that is, open or close control valve) directly from the control panel or computer interface screen.
  - In automatic mode, you adjust the setpoint only, and the controller controls the output (that is, opens or closes the control valve) to keep the process steady.
  - Some DCS systems use different terms for these same concepts:
    - What is called "automatic" here is called "cascade" there, and what is "manual" here is "automatic" there!
- Stable processes
  - Stability of process under control is essential to safe operation of the equipment being controlled and the whole plant facility.
    - Stable processes stay within safe bounds of temperature, pressure, flow, and liquid level.
    - Unstable processes can get too hot or too cold, be under- or overpressurized, flow too fast or too little, run tanks empty or overflow them.
    - All processes are open loop stable without controllers—they will steady out by themselves.
    - All processes can be made closed loop unstable if poorly designed or poorly tuned controllers are added.
  - Cascaded feedback control loops add stability to second-order or long time lag processes.
    - Cascaded feedback loops use the output of one loop (usually level or temperature control) as the setpoint for another loop (usually flow control), thereby isolating the primary control loop from disturbances occurring in the second control loop.
- The same control loops can be implemented on many different hardware platforms:
  - Pneumatic controllers with 3–15 psig I/O air signals to and from the field
  - Electronic controllers, single loop or multi-loop, 4–20 mA or other AC or DC I/O signals to and from the field
  - Microprocessor computer based, PLC or DCS, electronic signals to and from the field
  - PLC or DCS with digital signals added to analog signals to and from the field (HART), or digital bus communication (field bus, AS-I bus (Actuator-Sensor-Interface), etc.)
- More complicated control concepts also fit onto pneumatic and electronic platforms:
  - Ratio control
  - Mathematical operation blocks such as square root extraction, addition, multiplication, Boolean logic
  - Cascade: output of one controller used as the setpoint for another controller
  - Fuzzy logic means that:
    - Multiple inputs or outputs used for one process control application
- Sequencing of actions is possible when microprocessor-based PLC or DCS systems are used.
  - The same control loops that can be implemented on pneumatic or electronic stand-alone controllers can be implemented on PLC or DCS platforms.
  - Sequential function charts can be provided to cause actions such as startup, shutdown, emergency actions, and operating sequences to occur automatically.
  - Batch plant processing concepts can be implemented in PLC systems and DCS systems, usually following the ISA S88 standard.
- Data collection and storage is available with PLC and DCS systems, or separate data historians.
  - DCS systems have built-in data historians.
  - PLC systems usually provide data historian capabilities within the SCADA system or via other linked software.
  - Data collection and storage from electronic or pneumatic controllers require transducers to digital protocols, similar to PLC and DCS I/O cards.
  - Stand-alone pneumatic and electronic controllers probably should be replaced with PLC or DCS systems when large electronic data collection and storage requirements are desired.
- Relative merits and drawbacks of alternative distillation process control schemes
  - A control strategy (manual or automatic) is required for the following:
    - Feed input
    - Heat input
    - Column product purity control
    - Column bottom effluent

- Reflux
- Column top distillate
- Other feed, product, or internal streams
- System pressure

- Some general truths about what you can and cannot control in a distillation process
  - There is one parameter that always must be controlled:
    - System pressure as measured at one point, usually in the column head space or condenser vent line
  - The number of other different things that you can independently control in a distillation process usually is equal to the number of streams entering the process, plus the number of streams leaving; reflux does not count.
  - Typically, you try to control one thing about each stream:
    - Either control its purity, control its flow rate, or alter the liquid level in a vessel where this material accumulates, thereby controlling inventory in the system.
  - If you try to control *two things* about any *one stream*, then you must accept that *you will control nothing about one other stream* in exchange for this privilege.
    - For example, if you control reflux flow independently, you surrender something else.
  - It is most common to control the feed flow rate, the heat input rate, and the purity or flow rate of all except one of the column effluent streams—top, bottom, or side draw.
- Simplicity is better than complexity; that successes may be repeated and that failures not be repeated—hence, these guidelines.
  - No control scheme works all of the time, but these guidelines will try to make suggestions that will work most of the time.
  - First think about how you would manually control any distillation system you design if the only instruments available were dial thermometers, visual sight level indicators, rotameters or visual "look box" flow indicators, specific gravity measurement floats, and manual globe valves.
    - If there is little automated technology in use in the region where this system is installed, that is probably how the system should be controlled.
  - The technology-enabled, more sophisticated control systems only function by doing things that people could do manually if they were constantly vigilant, well trained, and did not mind cold or heat or high places.
    - "Everything should be made as simple as possible, but not simpler." *Albert Einstein*

- Control of reflux flow to the distillation column
  - During startup, total reflux is usually required, so the controls must allow for return of 100% of the condensed overhead vapor as reflux.
  - Either the reflux must return to the column naturally by gravity, or reflux accumulator level control is required to pump all the liquid back as reflux.
  - Usually it is best from the control point of view to keep reflux under some sort of natural level control, manipulating the distillate product stream flow based on temperature control for purity, with the balance of the condensate returning as reflux.
- Control of reflux flow to the distillation column
  - Remember that condensed vapor rate minus reflux equals distillate product rate, so if you choose to control the reflux rate independently, you lose the opportunity to control the distillate flow.
  - In some systems where distillate purity is sensitive to top temperature, controlling the top purity by temperature control cascaded to reflux flow control is a good scheme—but this might not get you through startup.
  - Trying to control reflux by its ratio to the distillate flow rarely yields stable control, except during very smooth and steady operation periods.
  - A system with zero reflux loses one of its other independent control opportunities.
- Control of partial condenser vapor product rate or composition
  - You may not think that your system condenser is running as a partial condenser, but it nearly always is.
    - Fixed gases dissolved in the feed materials and low boilers under the column top conditions largely pass through any condenser provided.
  - Some distillation systems vent a significant amount of gas through their partial condensers, some less, but rarely is the vent flow equal to zero.
    - It is difficult or impossible to control the vent vapor stream from a partial condenser by measurement either of its temperature or its flow rate.
  - What is most practical to control is the temperature of the coolant inside the condenser.
    - The vent vapor temperature nearly always is very close to the coolant temperature.
  - If utilities are not available at the desired temperature, or if the coolant temperature must be adjustable by the process, use a tempered coolant recirculation loop.

- Control of an overhead waste stream as a decanted second phase
  - When the overhead condensate stream forms two or three liquid phases, one being aqueous and the other(s) organic, it is almost always desirable to let the phases settle by gravity, and let the aqueous phase go a different way than the organic.
  - Letting the phases enter a decanter vessel under very calm internal conditions usually is sufficient for practical success in settling.
    - The lower the difference in specific gravity between the phases, the lower the settling velocity of the mixed phase materials, so low, wide decanters work more quickly than tall, thin ones.
  - External or internal manometer principle or DP control loops will succeed at keeping the interface within a controlled range if both phases are of known unchanging specific gravity.
    - If different specific gravities must be accommodated, a device that detects the physical location of the interface, such as a capacitance probe or weighted magnetic float, will work better than DP or manometer loops.
  - Part of the reason why organic phases do not separate from one another so easily as an aqueous phase separates from an organic phase is that their surface tensions are nearly equal; therefore, the organic phases may form continuous intermingled sheets (like wood grain) rather than discrete drops that readily rise or fall and merge when they touch.
- Distillate product purity and flow control
  - Whether the distillate product is a desirable product or instead is a waste stream, it is still most common to control its composition by control of temperature somewhere in the upper part of the column rather than rely solely on a column bottoms product composition control scheme.
  - It is more common to control the top product purity than the bottom because the top of the column is usually much quicker to respond to a flow change with a measurable temperature change (because of fewer time lag opportunities).
  - It is usually best to control distillate purity by temperature control from a point inside the distillation column, near the top, but in a sufficiently impure zone that the normal tray compositions at this point display a pleasantly large and predictable change in temperature between the stages.
  - The column temperature controller output can be used to control the distillate valve directly, but it is far better to cascade the temperature controller output to the setpoint of a distillate flow control loop, thereby providing a degree of damping that greatly helps stabilize the reflux flow and thus the composition and temperature profile in the column.
  - When the reflux flow is fixed by flow control without reset from decanter or reflux drum level or is cascaded from column temperature control, then the distillate flow cannot be independently controlled and must be driven by reflux drum level.
- Temperature control setpoint pressure correction method
  - Usually when distillation column temperature is being controlled, it is for the purpose of controlling the composition of an effluent stream.
  - The temperature setpoint value at the control point usually can be expressed as a difference between the temperature measured at this point and the boiling point temperature of the pure component that dominates the column inventory at this point.
    - The actual temperature experienced inside the column shifts with column pressure and composition.
    - The desired temperature difference between the control point temperature and the pure component boiling temperature is a function of composition only and does not shift with pressure.
  - The effects of varying pressure on measured temperature can be filtered out of the purity control scheme by taking the system pressure at the temperature measurement point, possibly calculated by lever arm rule from other pressure measurement points, and then using Antione's equation for vapor pressure to calculate the pure component boiling temperature at system pressure.
    - Control is then based on the temperature difference from the calculated pure boiling point rather than on absolute pressure measurement.
    - This scheme has worked in practice many times, in diverse applications.
- Control of feed flow into a distillation system
  - A fixed feed flow rate is usually best for continuous distillation systems, as it is for most production line operations.
  - The feed flow rate may be changed either continuously or in step changes at discrete points in time, to match the distillation system processing rate with the production rate of upstream processes.
  - A distillation system will usually keep working well through step changes in feed rate or feed composition as long as the mass flow of the most difficult impurity to remove is kept below a fixed maximum rate, and the mass flow of the primary

product to be purified is also kept below a fixed maximum rate.
    - Temperature control at some point in the distillation column may be cascaded to control of the feed flow rate, but this loop would have a large inherent time lag and thus would be apt to cause cycling.
    - If the bottoms level is used to control feed flow, a large process time lag is likely to combine with a very insensitive process variable (bottoms level) to spawn a bounded cycle in the value of most distillation system flows and compositions.
  - Zero feed flow is normal for a batch distillation system, but be aware that with zero feed, all of the parameters of the distillation system operation (inventory, level, temperature, purity) will vary over time.
- Control of a second feed stream going into a distillation system
  - A second feed stream may mean that there are multiple feeds into the column containing similar components, or the second feed may be very different in composition, as in extractive distillation.
  - If the second feed stream is just an alternate feed containing the same primary product as the first stream, then either or both feed streams may be varied independently, so long as the rules about maximum impurity mass flow rate and maximum primary product mass flow rate are respected.
  - If the second feed stream is an extractive distillation agent, it should be flow controlled, with its setpoint value determined by a ratio flow controller that watches the first feed stream flow rate.
- Control of distillation process heat input
  - Distillation systems operate with the best overall stability when their heat input rate is steady.
    - When the heat input comes from steam heating, then it is best to control the flow rate of steam going to the reboiler
    - When hot oil or other sensible hot fluid is used for heating, then the flow of this liquid must be controlled, the temperature of the heating media in and out must be measured, the temperature difference must be multiplied times the flow rate to determine the heat input, and the flow rate setpoint must be adjusted to maintain constant heat input.
    - Heat input control in mechanical vapor recompression (MVR) systems usually is achieved by controlling the motor speed or inlet vane position of the MVR compressor.
  - Heat input control for multiple effect column trains, where one column heats another column, can only be done at the first effect column, where the fresh heat source is—and it is crucial that the heat input be steady and stable, because there are high-order time lags inherent in such a design.
  - Occasionally it is appropriate to use column lower zone temperature controller output or bottoms level controller output, cascaded to the heat input flow controller to trim the heat input rate, but such a loop must be low gain, primarily integral, with an experience-tuned derivative feature to ensure the stability of this loop when the feed flow or feed composition changes.
- Control of column differential pressure
  - Column differential pressure is often used as a control parameter by design engineers, but the writer has yet to see any column differential pressure control working in automatic mode without it causing continual cycling of the column flows, temperatures, and purity.
  - The reason for this cycling is that the column pressure drop is a function of multiple independent variables.
    - Column differential pressure naturally varies slowly over time in a well-run distillation system, in response to changes in liquid traffic and vapor traffic up and down the column, which themselves may be responses to changes in feed composition or other uncontrolled parameters.
  - Control engineers sometimes choose to set heat input rate by control of column differential pressure, but there are drawbacks that may exceed the benefits of doing so.
    - An empty distillation column never can be started up using such a control feature, because the relationship between heat input rate and the column pressure drop is constantly changing before the column is fully loaded with liquid and vapor traffic.
    - Load changes made while lining the system out tend to start cycles that never seem to go away if the column differential pressure is used to control heat input or feed rate.
    - After a large column is well lined out, differential pressure control may be used in a slow and insensitive PI loop to trim either the heat input or, in some cases, the liquid feed rate.
    - In very low pressure operations with packing, where the measured differential pressure is indeed well related to vapor flow rate, differential pressure control in a low-gain PI loop may be used to adjust the heat input in order to keep the packing vapor rate below the flooding point, where fractionation efficiency would suffer.

- Control of column bottom level and effluent flow
  - The column bottoms stream may be permitted to flow out of the column by gravity. There is no need to control this stream unless it must be pumped to get to its destination or unless it must be held back to prevent pressurized vapor from blowing through, as in a steam trap.
    - Most commonly, the bottoms feed stream is indeed pumped away with a simple level control loop that responds to load changes in the same manner as natural gravity outflow.
    - Frequently, the column bottoms stream is controlled by level cascaded to flow control. Doing so achieves a more steady flow rate, which is important when the bottoms stream will pass through heat recovery and cooling equipment.
  - The bottoms level controller can also be cascaded to the reboiler steam controller, but this is generally only useful in columns where the bottoms product is a vapor side draw, or in any other column with a very small or intermittent need for bottoms flow—such as columns in which most of the feed will be vaporized, leaving only a small solids-containing bottom purge stream.
- Control of side draw streams, liquid and vapor
  - Most often, liquid side draw streams are simply flow controlled, with the purity as measured by analytical techniques or by process temperature used by the human operator to adjust the flow setpoint.
    - Sometimes the liquid side draw flow responds well to cascade control from one of the column temperature controllers, by the same principle as distillate purity control.
  - Vapor side draws rarely are suitable for direct flow control in the vapor state, but if they are to be injected as a vapor into other equipment, this must be done using a Differential Pressure (DP) type, turbine, or vortex shedding flow meter.
  - If vapor side draws are being condensed, it is usually easier and more reliable to measure the average flow rate of the condensed liquid leaving the vapor side draw condenser.
    - If the vapor side draw condenser liquid effluent is flow controlled and the condenser is a type that will tolerate partial flooding with liquid, then controlling the flow of liquid out will also achieve control of the vapor going into the condenser.
  - The vapor side draw flow rate setpoint may be taken from cascaded output from the column bottoms level or one of the column temperature controllers.
- Column pressure control—positive pressure systems
  - In distillation systems that are meant to operate at atmospheric pressure, the system usually can have a vent that is open to the air (often through a flame arrestor) or to a vapor collection header whose pressure is controlled near atmospheric by the end-of-pipe emission control device (oxidizer, condenser, scrubber, or flare).
  - If the gases in the vapor collection header should not be permitted to enter the distillation system, as they would at shutdown, then a vent breathing control is required and usually uses inert gas for breathing in.
  - In distillation systems that must operate at positive pressures, a pressure control loop using the column head pressure or condenser vent pressure is usually connected to a split range output, with one end of the range opening a vent valve and the other opening a pressurized gas source valve.
  - Control of condenser coolant temperature can also be used on one side of the split range pressure control, instead of a pressurizing gas source, but the dynamics of this control scheme make it slow to respond to load changes, so it is better for use in relatively steady operations.
- Column pressure control—vacuum systems
  - In distillation systems meant to operate at negative gauge pressure, usually an absolute pressure transmitter in the column head or in the condenser vent line is used in a simple pressure control loop, with output going to restrict a valve located between the distillation system and the vacuum source.
  - If the vacuum source is dedicated to this distillation system, the pressure control loop output alternatively may be used to open a discharge to feed recycle valve for a vacuum pump that operates as a fixed volume device, or an air bleed valve, or a steam supply valve for a steam ejector.
  - Rarely is gas makeup required for control of vacuum operations, because most real equipment experiences some air in-leakage, and most distillation feed materials contain some dissolved gas.
- Control of column pressure in multiple effect columns
  - When one column's vapor is used to heat another column's reboiler, the first column's pressure depends on the condensing capability provided by the second column.
    - Generally, multiple effect trains of distillation columns experience pressure float, with increasing load increasing all the columns' pressures, and none except the lowest pressure column experiencing independently controllable fixed pressure.
  - The operating pressure of the upstream partner in a pair of multiple effect columns can be increased

from its natural value dictated by heat transfer, but only by restricting vapor flow (rarely practical) or by limiting condensing heat transfer surface available (usually practical) by permitting the level of condensate to rise in the downstream column's reboiler.

– This control scheme causes inventory changes in the upstream column's condensate and thus is only advisable for distillation systems on long-term steady operating campaigns.

4.2 PLC and DCS System Software Development for Control of a Distillation System

- Overview of PLC and DCS control system similarities and differences
  - Software development for PLC and DCS systems typically begins with user requirement specification (URS) development, review, and approval.
  - Well-designed PLC and DCS systems from many suppliers look and feel the same to the plant operator and can have the same functionality. Selection between PLC and DCS systems is an equipment purchase decision usually guided by criteria other than the systems' technical capability.
  - Good automated manufacturing practice (GAMP) guidelines were developed for the production of validated control and batch management software. They apply to FDA-regulated applications and are often used as a guideline for the development of software in other industries as well. GAMP prescribes a "waterfall" sequence of project document development, starting with the most general and proceeding downward to the most detailed. Coding and testing follows the reverse order, starting at the bottom with the most detailed, and moving upward to the broader system operation scope. The GAMP model is summarized in Fig. 10.38.
  - PLC and DCS software design diverges at the functional design specification (FDS) and detailed design specification (DDS) levels.
    – PLC control software typically is based on ladder logic, with human interface software developed separately using an interface program (SCADA) that employs different tag numbers and names than the PLC operating ladder logic.
    – DCS control software usually is based on an object linking and embedding graphical control module concept and sequential function charts, with only one database of tag numbers and names used both in control logic and in operator interface screens.
  - Good automated manufacturing protocols (GAMP) provide guidelines for management of the control hardware and software life cycle.

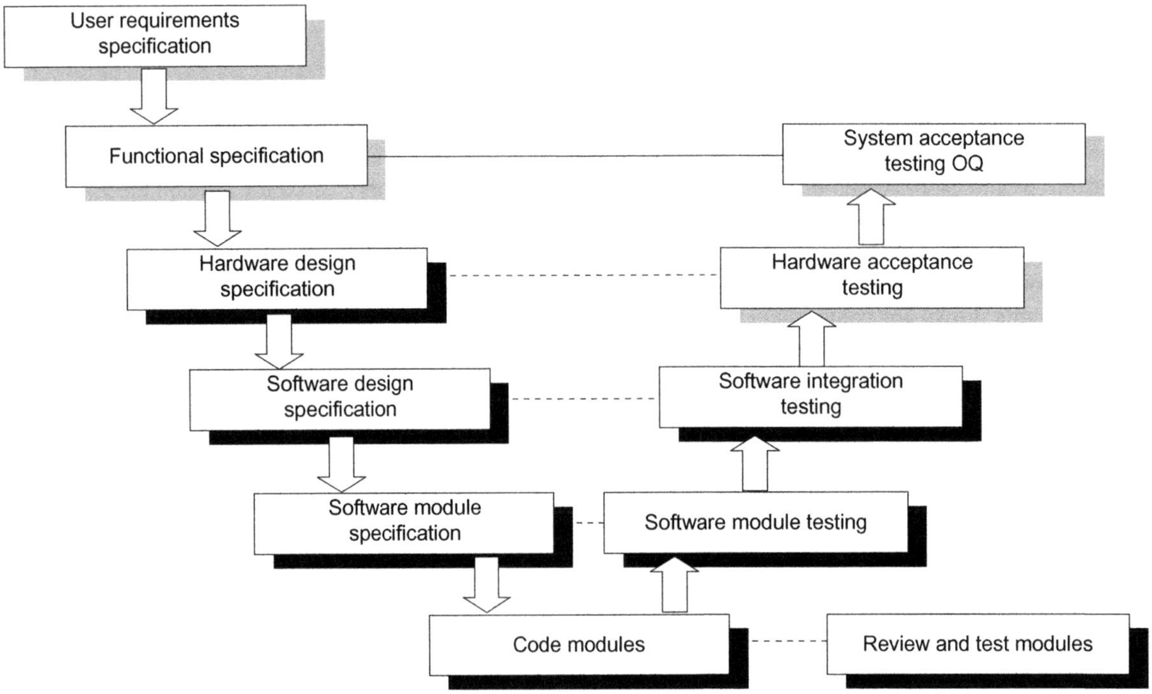

FIGURE 10.38 GAMP Control Software Development and Testing Model.

- User requirements specification (URS)
  - URS is the key document for control hardware specification and control software development.
  - The URS should be prepared jointly by a process engineer, a control engineer, and someone who knows the production methods, data collection requirements, and site safety standards.
- Functional design specification (FDS)
  - The FDS is usually prepared by control engineers who will perform the control system programming.
  - The purpose of the FDS for a relatively small and simple system is to completely specify how the required control strategy (depicted on the P&IDs and in the URS) will be implemented on the selected control platform.
  - The purpose of the FDS for a relatively large or complex control system is to specify in broad terms what the approach will be to implementing the required control strategy.
  - The FDS should be read carefully by the team that prepared the URS, and discrepancies should be challenged now—not left to be challenged later!
- Detailed design specification (DDS)
  - The DDS will contain all the detailed information required to select the correct control modules, define all the I/O, and implement the required control strategy.
    - The physical model must be completely defined in the DDS.
    - The procedural model must also be completely defined in the DDS.
    - Safety features and interlocks must be completely defined and tabulated.
    - The contents and functionality of the graphical operator interface must also be defined.
  - The DDS also must be carefully checked by technically astute people who understand the process and the user's requirements.
  - Discrepancies among the DDS, FDS, and URS should be resolved before control programming begins, or they may cause major trouble later.
- Control software and hardware factory acceptance testing (FAT)
  - Control software should be thoroughly tested, usually with simulated inputs (that is, in an office, not connected to a real live process), before customer acceptance.
  - FAT testing should begin with the lowest level of control module integrations (I/O blocks) and proceed through to the highest level (sequential function charts or recipe execution).
  - Hardware FAT testing should be performed using the completed and FAT tested software configuration, preferably loaded on the hardware that will be included in the final installation—not substitute or loaner hardware.
- Control software and hardware site acceptance testing (SAT)
  - Control software should be thoroughly tested after installation in the completed site hardware installation, within minimal simulated inputs and most I/O communicating with the field hardware of the actual process, before customer final acceptance of the site control installation.
    - SAT testing should begin with the lowest level of control module integrations (I/O blocks) and proceed through to the highest level (sequential function charts or recipe execution).
    - Software-to-software interactions that have been thoroughly tested in the office FAT environment do not have to be completely tested again at the SAT, but some randomly sampled FAT tests should be repeated during the SAT to weed out possible problems.
  - Hardware FAT testing should be performed using the completed hardware installation, with all wiring in place and power supplied by the intended normal sources—preferably, no temporary wiring is allowed.
- Simplicity of design pays off in simplicity of testing.
  - Software that is designed with more complexity than is required will also need much more testing effort than is required.
  - Software testing is supposed to follow all possible software branches, so building too many branch points in and out of a software path (compared to programming several long, non-branching, parallel paths that perform the same function) multiplies the testing time and paperwork.
  - Think of simplicity of design while preparing the URS and especially while reviewing the FDS.

4.3 Implementing Fully Automated Operation Using ISA S88.01 Model for Batch Control

- ISA-S88.01-1995 Batch control models and terminology
  - Standard for batch manufacturing control
  - Defines models, concepts, and terminology
  - Promotes modularity and flexibility
  - Emphasizes good practices for design and operation
  - Applicable to batch distillation processes
  - Can be used to manage a continuous distillation process operation from initializing parameters through startup, operation, and shutdown
- S88 standard phases of control project execution
  - Functional specification or FDS phase defines the automation requirements.

- Design specification or DDS phase defines how the application software meets the functional specification requirements.
- Implementation or configuration phase applies to the configuration of the application software.
- Testing phase verifies that the application software meets the functional specification and design specification requirements.
- The GAMP control hardware and software life cycle guidelines apply to all of these phases.

- S88 model types: physical, procedural, process models
  - Physical model defines the hierarchy of equipment used in the batch process and the equipment's physical points of interaction with the control system, I/O, discrete logic, and analog control loops.
    - The S88 physical model diagram is depicted in Fig. 10.39.
  - Procedural model defines the control sequences and actions that enable equipment to perform the process steps.
    - The S88 procedural model diagram is shown in Fig. 10.40.
  - Process model defines the process functionality from a high level.
    - The S88 process model diagram is shown in Fig. 10.41.
- The S88 procedural state matrix diagram is illustrated in Fig. 10.42.
  - S88 Procedural state matrix definitions
    - **Idle**: No action. Waiting for a start command from the operator or a batch recipe instruction.
    - **Running**: Normal operation of equipment. Begins when the start command is received.
    - **Complete**: The running state has completed. Waits for a reset command to transition to idle.
    - **Pausing** (not pictured): The operator has issued a pause command during the running state. The running logic progresses to the next defined pause point.
    - **Paused** (not pictured): The running state is paused and no action is taken. At this point, the operator or a batch recipe can issue a hold, stop, or abort command. When the operator issues the resume command, the running logic continues from the pause point and the state returns to running.
    - **Holding**: Equipment is placed in a safe state. The running state is disrupted and placed in

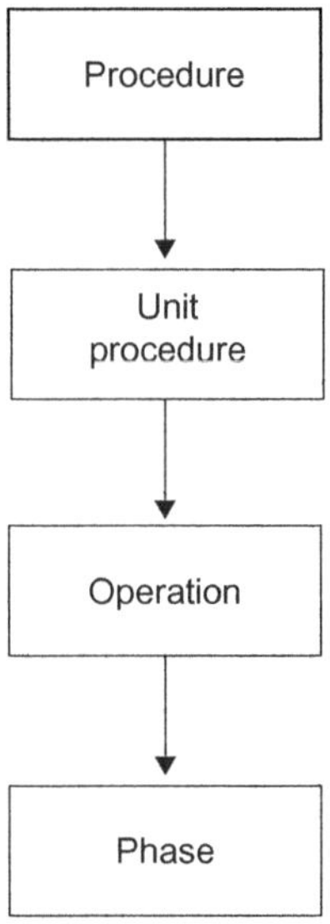

FIGURE 10.40 Diagram Defining S88 Procedural Model Hierarchy.

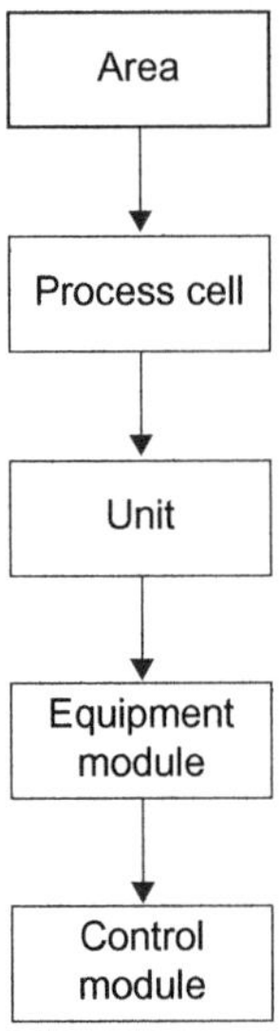

FIGURE 10.39 Diagram Defining S88 Physical Model Hierarchy.

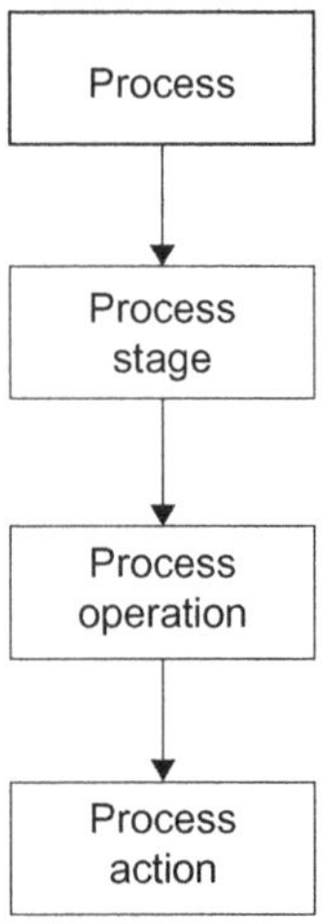

FIGURE 10.41 Diagram Defining S88 Process Model Hierarchy.

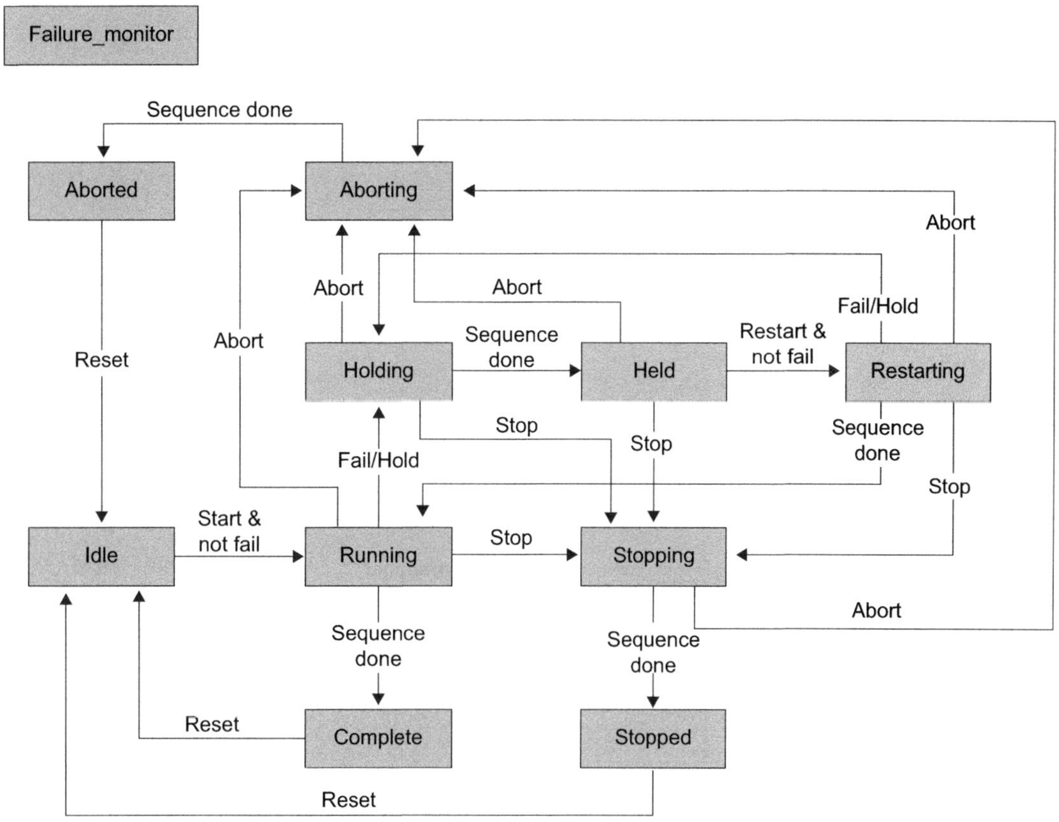

FIGURE 10.42 S88 Procedural State Matrix.

holding when an exception to normal operation is detected or the operator issues the hold command.

- **Held**: Holding state has completed. No actions are taken. At this point, the operator or a batch recipe can issue a restart, stop, or abort command.
- **Restarting**: The restart command has been issued by the operator while the state is held. Takes action to return equipment to normal operation. Once restarting finishes, the active state transitions to the running state.
- **Stopping**: Equipment is placed in a safe state. The current state is disrupted when the operator issues the stop command.
- **Stopped**: Stopping state has completed. At this point, the recipe cannot be restarted.
- **Aborting**: Equipment is placed in a safe state. The current state is disrupted when the operator issues the abort command.
- **Aborted**: Aborting state has completed. At this point, the recipe cannot be restarted.

- Figure 10.43 illustrates how a general pressure control equipment module includes the physical model and one or more (in this example, three) procedural models.

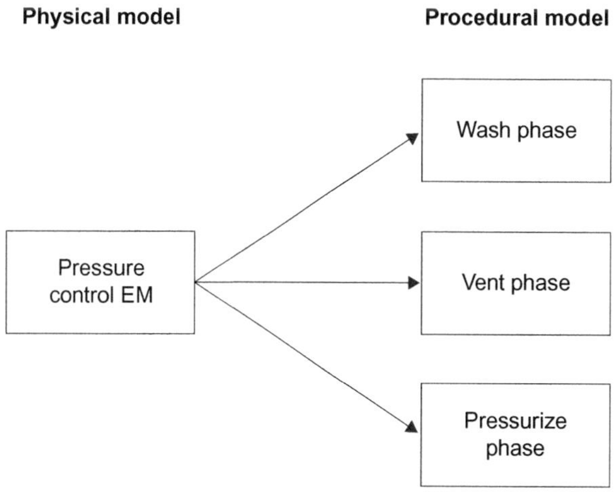

FIGURE 10.43 General Pressure Control Equipment Module Phase Diagram.

- As seen in Fig. 10.44, physical elements such as flow control loops are components in the physical model.

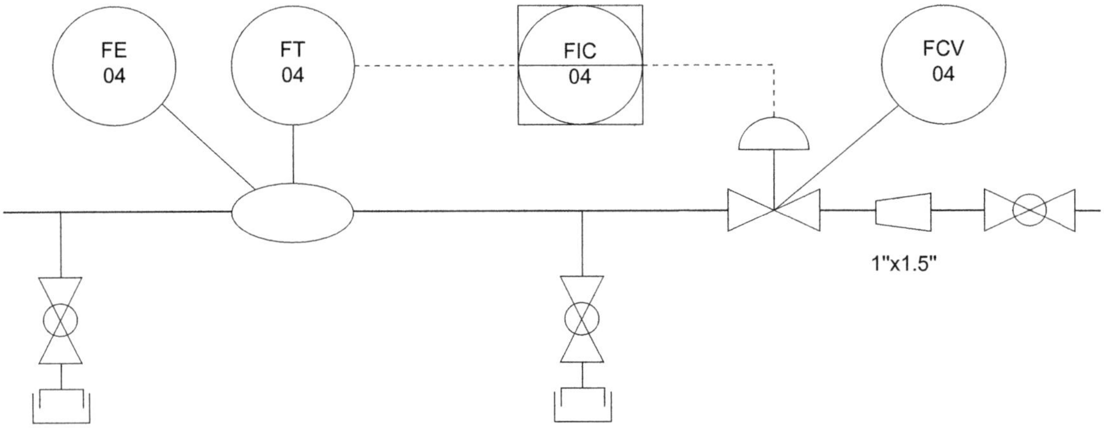

FIGURE 10.44 General Flow Control Physical Model.

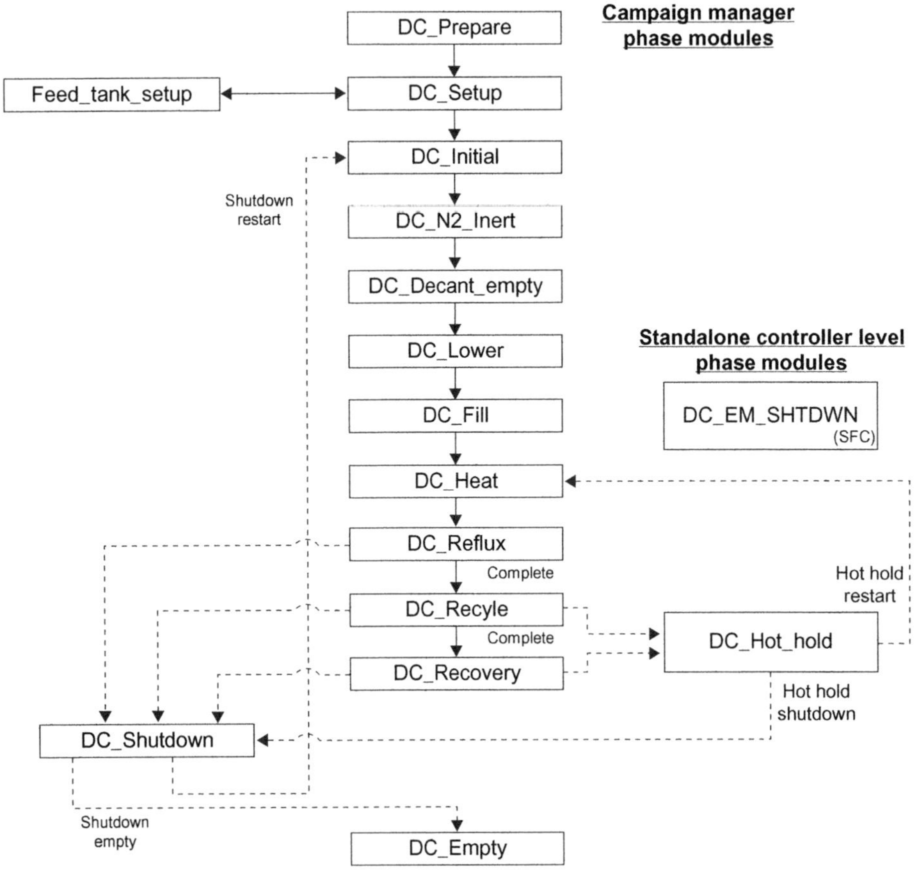

FIGURE 10.45 Operational Procedural Model, Top-Level Sequential Flow Chart.

- Sequential function charts, subdivided into phases and steps, make up the procedural model, as shown in Figure 10.45 for one fully automated, multi-purpose, multi-product, GAMP-compliant distillation system:
- S88 batch recipe procedural model example, as used to control a distillation system
- The following presents a summary description of implementing a distillation production operation recipe and a system cleaning operation recipe on a DCS control platform running a batch executive following the S88 procedural model.
- This information is intended to define the flow of the procedural model during these operations and

explain how the phases are implemented within each recipe.
- Recipe descriptions are presented in the next sections, with a phase map for each operation and a brief summary of the phase actions, followed by safety interlock charts.

- S88 procedural model example—for distillation system operation
  - The S88 procedural model includes all procedures for initializing, starting, running, and shutting down a distillation column and its associated process systems.
  - Figure 10.45 provides an overview of the phase structure.
  - The recipe itself does not include three independent stand-alone phases: DC_HOT_HOLD, DC_SHUTDOWN, and DC_EM_SHTDWN
    - These phases run independently at the controller's module level.
    - The logic in these phases executes at all times and waits for process or operator conditions to initiate its logic.
  - Availability of S88 hold states versus external hold phases
    - Once the column achieves the reflux state, the operator will be prohibited from holding the batch campaign—instead, either the hot hold or shutdown phases will need to be started to bring the system to a safe holding state.
    - The hot hold and shutdown phases interface with the batch phases via discrete flags.
    - During the reflux, recycling, and recovery phases, a hot hold condition executes the hot hold logic to place the system process equipment into a holding state, in which the column continues distillation with a lower input, thus allowing it to return to full recovery with minimal effort.
- Phase map for the automated distillation recipe operation
  - The diagram in Fig. 10.45 displays the phase map for the distillation recipe operation.
  - The following are brief descriptions of the phases shown in Fig, 10.45 for the distillation system operation:
    - DC_PREPARE
      - Checks to see that no control modules or equipment modules associated with the unit have simulation mode selected or interlocks bypassed.
      - If any modules are in simulation mode or have interlocks bypassed, the operator will be prompted and the phase will not continue until the problem has been addressed.
    - DC_SETUP
      - Configures the column control strategies for the correct setup based on recipe.
      - This phase sets unit parameters, configures master PID controllers for the cascaded loops, and, where multiple process variables are used, configures the recipe selected PV.
    - DC_INITIAL
      - Runs the DC_SET_CASCADE composite, which sets equipment to cascade mode and drives it to its fail-safe state.
      - All automated solenoid valves are set to the fail-safe condition (usually closed), controllers are set to manual with 0% output, and all pumps are stopped.
      - This phase also runs the DC_LOCAL_DISABLE composite, which disables operator control of automated equipment.
    - DC_N2_INERT
      - Lowers the concentration of oxygen in the system by pressurizing the complete distillation system with nitrogen to a predetermined pressure and releasing that pressure multiple times.
    - DC_DECANT_EMPTY
      - Pumps out any material in the decanter's main and overflow compartments prior to filling the column with liquid.
      - For each compartment, if the level is above a given setpoint, the material is pumped out to either a recovery tank or a waste location based on the response to an operator prompt.
      - When the level in each compartment is below a low setpoint value, then the phase completes.
    - DC_LOWER
      - Checks the levels of the distillation column and the batch pot and determines whether or not to lower the liquid levels of the unit.
      - If the level in the unit is above the given setpoint, liquid is removed until it reaches a target value, recirculation is enabled, and the phase completes.
      - If the level in the unit is below the given setpoint, then the phase completes.
    - DC_FILL
      - Checks the levels of the distillation column and the batch pot and determines whether or not to raise the level to a setpoint value within a given dead band.
      - If the level in the unit is below the setpoint, liquid will be fed until the measured level reaches a target value.

        - Once the level in the unit is within the specified setpoint range, recirculation is enabled and the phase completes.
        - If the level in the unit is above the given setpoint at the start of the phase, recirculation is enabled and the phase completes.
    - DC_HEAT
        - Initiates the heating sequence for the distillation unit.
        - The steam valves (distillation column and batch pot as necessary) will be ramped open at a specified rate, to a desired steam flow rate value.
        - The phase continues until the level of material in the decanter increases by 5%, and then the phase completes.
    - DC_REFLUX
        - Configures the column equipment for achieving total reflux.
        - Depending on the recipe setting of decanter mode, miscible, immiscible heavy, or immiscible light, the phase controls the reflux from the correct decanter compartment level controller.
        - The controller output is cascaded to the setpoint of the reflux controller.
    - DC_RECYCLE
        - Transitions the column from a total reflux condition to the recycle to feed configuration in which all distillate, product, and waste streams are recycled back to the feed tank.
        - During this phase all column flow, level, and temperature controls are functioning in automatic.
        - This phase enables the automatic column feed control, the decanter distillate take-off, the decanter waste source take-off, the column and batch pot bottoms waste, and the side draw product flow.
    - DC_HOT_HOLD
        - Transfers the process equipment to a holding state, in which the column continues distillation with a lower input, thus allowing it to return to full recovery with minimal effort.
    - DC_SHUTDOWN
        - Performs a controlled shutdown of the distillation column unit equipment.
    - DC_EMPTY
        - Empties the distillation column and blows nitrogen through the column and associated equipment after the column has been shut down.
    - DC_EM_SHTDWN
        - Performs a rapid emergency shutdown of the distillation column unit equipment.
        - This logic is not executed in a phase.
        - Instead, this logic resides in a sequential function chart embedded in a controller resident control module.
        - The logic drives all equipment to their safe states and can either be operator initiated from a button on the operator interface or initiated from the control module on detection of emergency shutdown conditions.
- Availability of S88 hold states versus use of external hold phases
    - Once the column achieves the reflux state, the operator will be prohibited from holding the batch campaign—instead, either the hot hold or shutdown phase will need to be run to bring the system to a safe holding condition.
    - The hot hold and shutdown phases interface with the batch phases via discrete flags.
    - During the reflux and boilout phases, a hot hold condition executes the hot hold logic to place the system process equipment to a holding state, in which the column continues distillation with a lower energy input, thus allowing it to return to full recovery with minimal effort.
    - Once the hot hold state is achieved, the operator is prompted to either shut down or return to column recovery.
    - Once the shutdown state is achieved, the operator is prompted to either continue to the rinse and purge phase or return and restart the cleaning procedure.
- Figure 10.46 depicts the S88 procedural model for the cleaning/decontamination operation for the same fully automated multi-purpose distillation system whose operational procedural model (shown in Fig. 10.45) was discussed previously.
    - The cleaning/decontamination recipe includes all procedures necessary to rinse, fill, heat, and boil out the distillation column for cleaning purposes.
    - Fig. 10.46 provides an overview of the phase structure.
    - Phase map—distillation system cleaning/decontamination recipe
    - As in the previous section, the recipe itself does not include three independent stand-alone phases, DC_HOT_HOLD, DC_SHUTDOWN, and DC_EM_SHTDWN
        - These phases run independently at the module level of the controller.
        - The logic for these phases executes whenever the measured process conditions (pressure, level, etc.), or the operator, triggers initiation.

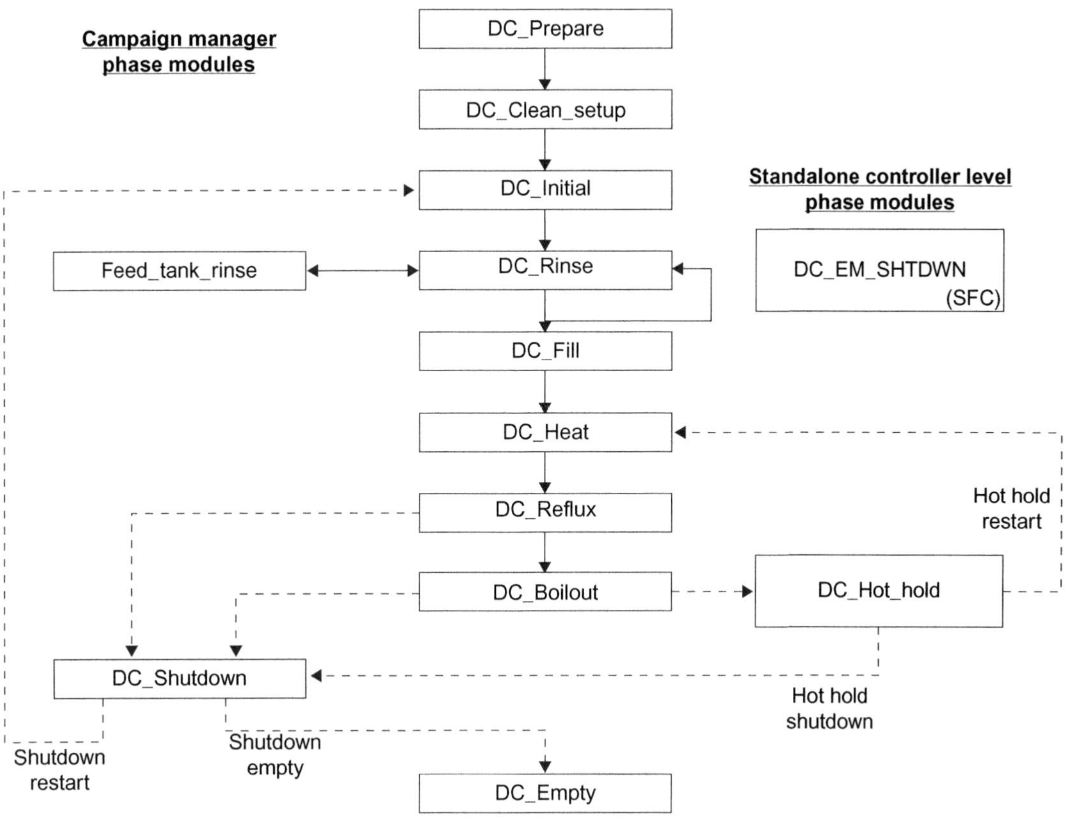

FIGURE 10.46 Cleaning/Decontamination Procedural Model, Top-Level Sequential Flow Chart.

- Phase overview—system cleaning/decontamination recipe
  - The following are brief descriptions of the phases shown in Fig. 10.46 that are unique to the cleaning/decontamination sequence and are not included among the modular phases that cleaning/decontamination shares in common with the distillation operation.
- DC_RINSE
  - Fills the distillation column unit equipment with process water, circulates the water for a specified amount of time, and discharges the contents of the column via the bottom discharge route until the column is empty.
- DC_BOILOUT
  - Performs a boil-out of the distillation column unit equipment, including all receiver vessels.
  - It flushes condensate through all the various outlet lines, to purge residue from past recovery operations.
  - This phase is intended to execute during the cleaning recipe after the column has been filled, heated to boiling, and has achieved total reflux.

- Other procedural model logic is always active and standing by.
  - E-Stop—Module level logic executed as a result of a predetermined triggering condition:
    - Anytime E-Stop is triggered, the currently active phase transitions to the HELD state and then completes.
  - ShutDown—Module level phase executed as a result of a triggering condition:
    - Anytime ShutDown is called for, the currently active phase transitions to the HELD state and then completes in a manner determined by a polling of discrete flags that indicate the system condition existing before ShutDown was triggered.
  - ESTOP_MONITOR interlock actions have no phase dependence.
  - SHUTDOWN_MONITOR actions are phase dependent.
  - PHASE_FAILURES logic module has phase-dependent output.
  - Individual parameter-based interlock actions are always active.

5.0 DISTILLATION PLANT OPERATIONS

- Here we present generic, universal methods and logical rules governing how to safely operate, shut down, or start up a distillation. These rules apply to most non-reactive distillation applications in typical manufacturing operations, in a multitude of industries.
- Site-specific operating procedures must always take precedence over application of the generic methods given here. Never operate a manufacturing distillation system without learning and following its specific standard operating procedures (SOP).
- This section will assist in understanding *why* the actions dictated by the SOP are necessary parts of a safe operation.

5.1 Startup Preparations and Normal Operation

- Understand what is going on inside the distillation system before planning how to stop or start it, or how to handle problems.
  - When the distillation system is operating correctly, what is flowing?
    - A hot fluid is providing heat.
    - A cold fluid is providing cooling.
    - Feed material is flowing in.
    - Purified distilled product is flowing out.
    - Waste streams are flowing out.
    - Column bottoms are being boiled.
    - Column overheads are being condensed.
    - Part of the condensed overhead flows back to the top of the column as liquid reflux.
    - Gases that cannot be condensed are flowing out.
- Preparations to make prior to starting up a distillation system
  - Make certain the required electrical devices, utilities, and equipment are turned on.
    - Control panel power
    - Pump motor power
    - Agitator motor power
    - Instrument air
    - Cooling source
    - Heating source
    - Fire protection
    - Phone, radio, security
    - Control panel devices requiring power supply turned on:
      - Panel AC and DC
      - UPS input and output
      - PLC or DCS
      - HMI or SCADA
      - PC operator stations
      - Alarm panels
      - Analyzers
      - Printers
  - Ensure that motor control center (MCC) devices are turned on:
    - MCC three-phase supply
    - Motor starters enabled for all motors in use
    - Breakers to instrument power transformers
    - Breakers to lighting panels
    - Breakers to UPS feeds
    - Pumps and agitator motors switched on. Possible ways:
      - On/off switch at MCC
      - Local on/off switch by motor or on panel
      - Local hand/off/auto switch by motor or in panel
      - Automatic run from control panel relay logic
      - Manual run from HMI
      - Automatic run from PLC or DCS system
  - Compressed air supply for instruments must be active, dry, and up to pressure:
    - Instrument air compressors running
    - Instrument air dryer working
    - Air supply pressure OK
    - Instrument air filters OK
    - Air valves on headers and branch lines to instruments are open
    - Air bleed and drain valves are closed
- How to fill an empty distillation system to prepare for startup
  - First make certain that no one is still working on or inside of equipment.
  - Make certain that all equipment and piping is in place, blinds are removed, drains are closed, and normally open manual valves are open.
  - Only the distillation column base and reboiler circulation loop need to be filled to normal liquid levels before a startup.
  - Other equipment, such as the decanter and reflux drum, can remain empty at this time.
- How to start up a cold (but not empty) distillation system
  - Make certain that all cooling systems and vent collection systems are working.
  - With a forced circulation reboiler, start the reboiler circulation pump.
  - Slowly ramp open the heat source to the reboiler (steam or hot oil) to near normal heat input.
  - When material begins boiling and the column level decreases, add more feed to top up the column bottom as required, whenever required.
  - When the column is hot all the way to the top, condensate will appear in the reflux equipment.

- How to line out a newly started or restarted distillation system to achieve a total reflux condition
  - When condensate appears in the reflux equipment, start the reflux pump, allowing all of the reflux to return to the column.
  - Keep topping up with feed material as needed.
  - Ramp the heat input rate up to normal.
  - When column temperatures line out and no more material top-ups are required, the system is ready to begin accepting continuous feed.
- How to start feed, recycle, and line out a distillation system to normal compositions and flows
  - Ramp up feed rate quickly to 75% of normal.
  - Make certain that control loops for bottoms flow out and distillate out are functioning.
  - Frequently bottoms and distillate are recycled back to the feed tank until on-spec.
  - When column temperatures stabilize, increase feed to 100% of normal, and allow to re-stabilize.
- When and how to recognize that a distillation system has begun to make good product
  - All the flow rates in the system will be near their usual values.
  - All the column temperatures will be near their usual values.
  - A trend recording of the temperatures and flows will show straight lines.
  - On-line analyzers and density meters will show values within good quality specifications.
- How to continue long-term operation with on-spec quality
  - Primarily, it is the designers' duty to make certain that the distillation system includes robust control configurations that ensure quality.
  - If you must control something manually to maintain on-spec quality, increasing the amount of reflux usually improves the distillate quality, and increasing the heat input usually improves the bottoms stream quality.
  - Decreasing the feed rate improves quality at the top and bottom of the column in most systems.
  - Flow blockages often show up as quality crashes.

5.2 Performance Testing

- Performance testing is done after the system has reached steady-state operation.
  - Guidelines for performance specification preparation for a distillation system
    - Performance specifications must have both maximum and minimum values for every component in the feed and products.
    - Take care not to write impossible performance specification language:
      - It is easy to promise that a given product rate will be produced from a given feed range, when only certain parts of the range provide sufficient incoming material to make the promised output.
  - Performance test requirements are tuned to industries and have legal ramifications in many locations.
    - Have performance test specifications prepared by technical personnel but reviewed by others with contractual and legal expertise.
  - Usually a well-designed distillation system meets its performance test specifications easily on first try if the feed materials are truly as specified, without unexpected components present.

5.3 Troubleshooting

- What to do when quality measurements fail
  - First, resample and reanalyze to be sure that the quality failed, not the measurement.
  - If uncertain about what to do to improve quality, decrease the feed rate by 25% as a first step.
  - If quality improves, gradually bring the feed rate back up toward the normal value.
  - If quality did not improve, increase the reflux.
  - If quality still did not improve, increase the heat input, watching out for the signs of column flooding (high pressure drop across column).
  - If nothing you do works, call someone to help.
- Counterintuitive actions can achieve or recover quality.
  - Sometimes the right thing to do looks at first like you are doing the opposite of the correct thing!
  - Decreasing the heat input may improve both top and bottom product quality if the column is starting to foam or jet-flood.
  - When the overhead product coming out of the decanter in an azeotropic partially miscible distillation system contains too much water, often the correct action is to add lots of fresh water to the decanter, to sweep away co-solvency compounds and permit better partitioning.
- Operational problems not beginning with the process equipment may be caused by utility variations:
  - Utility quality
  - Utility quantity
  - Supply pressure
  - Temperature
  - Unsteadiness
  - Cycling

- Never neglect, always do something about:
  - Unusual noises
  - Leaks
  - Odors
  - Vibration
  - Degrading condition of equipment
  - Degrading condition of work area
- Problems that might come along during normal operation—what to watch for and what to do
  - Problems that might occur while a distillation system is operating can be sorted into these categories:
    - Quantity problems
    - Quality problems
    - Integrity problems
    - Safety problems
- Quantity problems can refer to quantity of feed, product, wastes, utilities, containers, lab services
  - The quantity of available feed or the feed flow rate may increase or decrease.
  - There may be too much or not enough product or waste materials coming out of the system.
  - There may be too little or excessive flow of steam, hot oil, cooling water, glycol, nitrogen, air, etc.
  - There may not be enough tanks or drums to hold products or wastes, or vials to hold samples.
  - The lab might not be able to analyze samples as fast as demanded.
- What to do about quantity problems
  - Most distillation systems continue operating fine with any feed rate within a broad range.
  - Not enough or too much product or waste flow should always be investigated until understood.
  - Insufficient utilities eventually lead to quality problems or safety problems if not restored.
  - There are always more product containers somewhere—slow down production as a last resort.
  - If the lab is overloaded, negotiate priorities or help the lab staff by doing analyses yourself if you can.
- Quality problems can refer to feed, product, wastes, utilities, lab services.
  - Feed quality too high will require changes in the operation, possibly lower feed rate.
  - Feed quality too low will require changes in operation, possibly higher feed rate.
  - Waste product quality deviations require attention to be certain product is not lost.
  - Utility quality deviations require monitoring because they may lead to other problems.
  - Lab service quality problems may require technical or organizational solutions.
  - Integrity problems are serious and will not go away by themselves if ignored.
  - No visible leaks should be tolerated other than minor quantities of steam, water, air.
  - Flange gaskets, threaded connections, and valve stem packing require occasional tightening.
  - Broken supports should always be fixed in an improved manner so they will not break again.
  - Noisy or vibrating equipment or piping should receive maintenance attention as quickly as possible.
  - Chemical leaks or hazardous utility leaks require quick action to stop the leak and repair the source.

5.4 Safety Concerns and Emergency Handling

- Safety problems can be little or big, but little problems left uncorrected tend to get bigger.
  - Steam leaks get bigger and become burn hazards or burst hazards.
  - Chemical leaks at flanges and valve packing become toxic or explosive vapor clouds or slick corrosive or flammable liquid pools.
  - Broken supports lead to more broken supports and eventually falling of equipment parts.
  - Noisy bearings, pump seals and couplings, and electric motors never heal themselves.
- Major safety problems demand action because they threaten to become true emergencies.
  - Any loud or repeating snapping or cracking sound demands emergency shutdown and evacuation of all personnel from the area.
  - Alert supervisors, and do not enter the compromised area alone.
  - Any large and quick pressure increase demands emergency shutdown and venting.
  - Steadily rising temperature measurements require treatment as if a fire already is burning.
  - Flames from solvent fires often are not visible in daylight.
  - Never walk into any area where solvent is spilled or a solvent fire is suspected.
- Emergency handling: intervention and shutdown
  - If safety interlocks or emergency systems are triggered, do not prevent them from acting.
  - When there is an emergency in a system equipped with emergency stop (E-Stop) buttons, use them.
  - If required to intervene manually, first block any energy inputs to the system (steam, hot oil), then shut off all electric motors, then block all material inputs to the system.
  - It is usually better to let pressure safety valves and burst disks relieve system pressure than to open vent valves to atmosphere during an emergency.

5.5 Distillation System Shutdown

- How to make a smooth shutdown
  - Ramp the feed flow rate down to zero.
  - Ramp the heat input flow rate down to zero.
  - Keep all other instruments functioning, control loops in automatic, pumps running, and utilities flowing.
  - As material output flows decrease naturally, close their flow paths and stop pumps.
  - When all pumps have stopped and vessels contain only safe residual liquid heels, vent or inert the system as prescribed for shutdown periods.
- When and why to empty a distillation system
  - A distillation system should be emptied when the next production material is not compatible with the liquid heel left inside after a shutdown.
  - Some or all of the equipment or piping may need to be emptied and blinded off to make it safe for work to be done during the shutdown period.

  Drain all pipelines to vessels or to drain connections, pump all vessels empty, blow out lines with nitrogen or steam, water rinse, air dry.

REFERENCES

Materials in this chapter, including photographs and diagrams, were previously published with copyright registered to the author, Steve Licht, in the following print book and e-book, Reference [1]:

[1] S. Licht, Distillation Without Equations, Author House Publishers, Bloomington, Indiana, 2005.

[2] H.Z. Kister, Distillation Design, McGraw-Hill, New York, 1992.

Chapter 11

Solvent Extraction

David B. Todd

1.0 EXTRACTION CONCEPTS

Liquid-liquid extraction is a unit operation frequently employed in the pharmaceutical industry, as in many others, for recovery and purification of a desired ingredient from the solution in which it was prepared. Extraction may also be used to remove impurities from a feed stream.

Extraction is the removal of a soluble constituent from one liquid into another. By convention, the first liquid is the *feed* (F) which contains the solute at an initial concentration X_f. The second liquid is the *solvent* (S) which is at least partially immiscible with the feed. The solvent may also have some solute present at an initial concentration of Y_s, but usually Y_S is essentially zero.

The solvent does the extraction, so the solvent-rich liquid leaving the extractor is the *extract* (E). With the solute partially or completely removed from the feed, the feed has become *refined* so the feed-rich liquid leaving the extractor is the *raffinate* (R).

When the feed and solvent are brought together, the *solute* (A) will distribute itself between the two liquid phases. At equilibrium, the ratio of this distribution is called the *distribution coefficient* (m):

$$m = \frac{Y_A}{X_A} = \frac{\text{concentration of } A \text{ in extract phase}}{\text{concentration of } A \text{ in raffinate phase}}$$

The distribution coefficient, m, is a measure of the affinity of the solute (A) for one phase (E, S) over the other phase (F, R). The concentration of A may be expressed in various units, but for ease of subsequent calculations, it is preferable to express the concentration on a solute-free basis for both phases. For example, in the extraction of acetone from water with toluene:

$$X = \frac{\text{weight acetone}}{\text{weight acetone} - \text{free water}}$$

$$Y = \frac{\text{weight acetone}}{\text{weight acetone} - \text{free toluene}}$$

$$m = \frac{Y}{X}$$

Although the units of m appear to be dimensionless, they actually are (weight acetone-free water)/(weight acetone-free toluene).

If more than one solute is present, the preference, or *selectivity*, of the solvent for one (A) over the other (B) is the *separation factor* (α).

$$\alpha_{AB} = \frac{m_A}{m_B}$$

The separation factor (α_{AB}) must be greater than unity in order to separate A from B by solvent extraction, just as the relative volatility must be greater than unity to separate A from B by distillation.

The analogy with distillation can be carried a step further. The extract phase is like the vapor distillate, a second phase wherein the equilibrium distribution of A with respect to B is higher than it is in the feed liquid (liquid bottoms).

Extraction requires that the solvent and feed liquor be at least partially immiscible (two liquid phases), just as distillation requires both a vapor and a liquid phase.

Extraction requires that the solvent and feed phases be of different densities.

Even though extraction may successfully remove the solute from the feed, a further separation is required in order to recover the solute from the solvent, and to make the solvent suitable for reuse in the extractor. This recovery may be by any other unit operation, such as distillation, evaporation, crystallization and filtration, or by further extraction.

Extraction is frequently chosen as the desired primary mode of separation or purification for one or more of the following reasons:

1. Where the heat of distillation is undesirable or the temperature would be damaging to the product (for example, in the recovery of penicillin from filtered broth).
2. Where the solute is present in low concentration and the bulk feed liquor would have to be taken overhead (most fermentation products).
3. Where extraction selectivity is favorable because of chemical differences, but where relative volatilities overlap.

Fermentation and Biochemical Engineering Handbook.

4. Where extraction selectivity is favorable in ionic form, but not in the natural state (such as citric acid).
5. Where a lower form or less energy can be used. The latent heat of most organic solvents is less than 20% that of water, so recovery of solute from an organic extract may require far less energy than recovery from an aqueous feed.

1.1 Theoretical Stage

The combination of mixing both feed and solvent until the equilibrium distribution of the solute has occurred, and the subsequent complete separation of the two phases is defined as one theoretical stage (Fig. 11.1). The two functions may be carried out sequentially in the same vessel, simultaneously in two different zones of the same vessel, or in separate vessels (mixers and settlers).

Extraction may also be performed in a continuous differential fashion (Fig. 11.2), or in a sequential contact and separation where the solvent and feed phases flow countercurrently to each other between stages (Fig. 11.3).

2.0 DISTRIBUTION DATA

Although data for many systems are available in the literature [1], in many cases it will be necessary for the engineer to obtain the distribution information for his own specific application.

The simplest method is to mix solvent and feed liquors containing varying quantities of solute in a separatory funnel, and analyze each phase for solute after settling. Where feed and solvent are essentially immiscible, the binary plot, such as shown in Fig. 11.4, is useful. For later ease of calculation, it is desirable to express concentrations on a solute-free basis. If there is extensive miscibility, a ternary plot (Fig. 11.5) would be preferable. Tie lines represent the equilibrium between the coexisting phases.

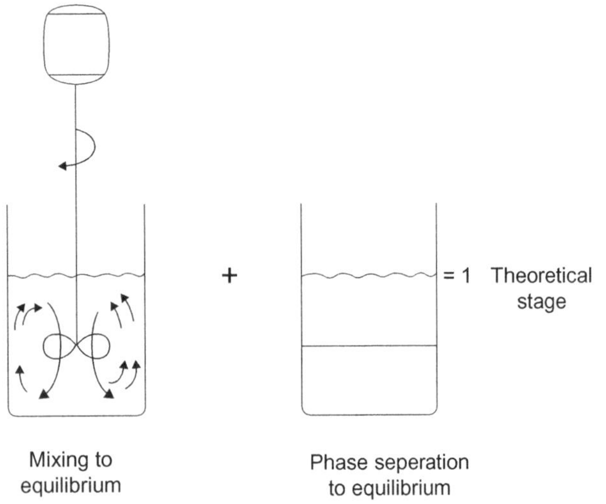

FIGURE 11.1 Theoretical stage.

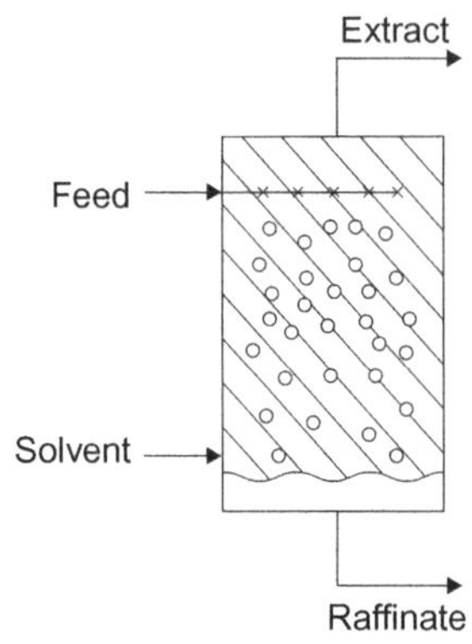

FIGURE 11.2 Differential extraction.

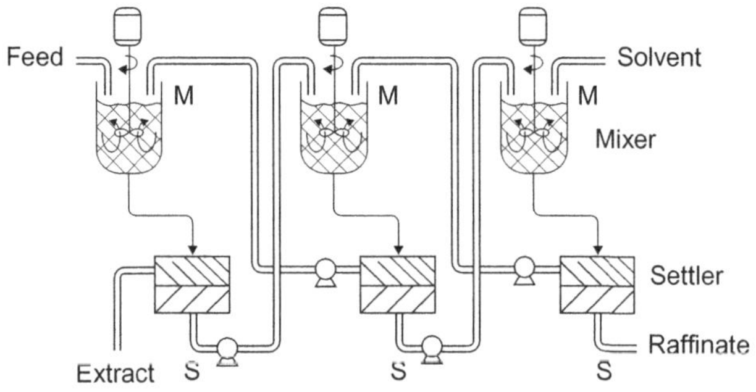

FIGURE 11.3 Sequential contact and separation.

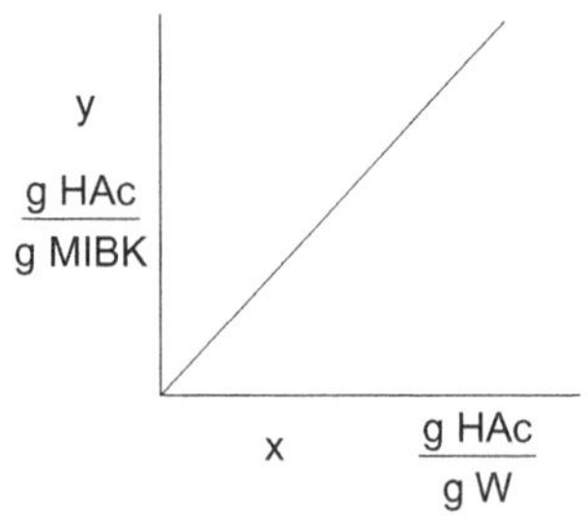

FIGURE 11.4 Binary plot of distribution data.

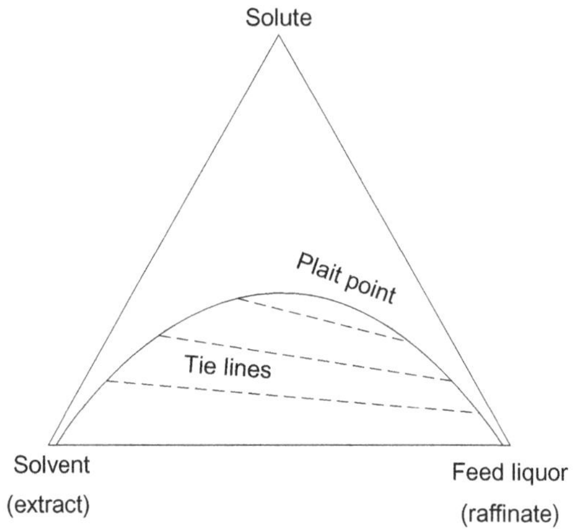

FIGURE 11.5 Ternary plot of distribution data.

Plotting the data on log-log graphs may be helpful in understanding some of the underlying phenomena and interpolating or extrapolating meager data. An example is shown in Fig. 11.6 for the distribution of phenol between water and various chlorinated methanes. In the dilute region, the limiting slope is generally always unity. However, as the solute becomes more concentrated, there may be a tendency for solute molecules to associate with each other in one of the phases. Thus, the equilibrium data in Fig. 11.6 suggest that the phenol molecules form a dimer in the organic phase, probably by hydrogen bonding, leading to a slope of 2 in the distribution plot.

The possibility of complex formation in one of the phases illustrates the concern that many industrial extraction processes involve not only the physical transfer of molecules across an interface, but also that there may be a sequence of chemical steps (which may be rate limiting) that have to occur before the physical transfer can take place.

Whenever the distribution coefficient is greatly different than unity, there is an implication that there exists an *affinity* of the solute for that specific solvent, and this affinity may involve some loose chemical bonding.

Examples of computer programs for predicting and correlating equilibrium data are described by Lo, Baird, and Hanson [2].

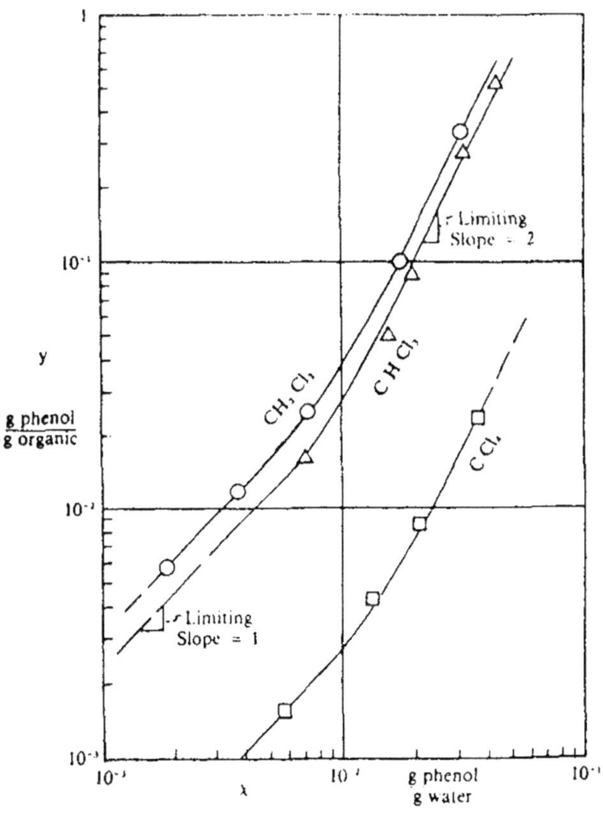

FIGURE 11.6 Distribution of phenol between water and chlorinated methanes.

3.0 SOLVENT SELECTION

The molecular formula of the solute may suggest the type of solvent which may be selective for its extraction, based on probable affinities between related functional groups. Thus, to extract organic acids or alcohols from water, an ester, ether, or ketone (of sufficient molecular weight to have very limited solubility in the aqueous phase) might be chosen as the solvent. The pH of aqueous phase feeds may also be very important. The sodium or potassium salts of an organic acid may well prefer the aqueous media at pH > 10, but in the acidulated form may readily extract into the organic phase if the pH is low.

Specific factors taken into consideration in the selection of a solvent include:

1. *Selectivity*—the ability to remove and concentrate the solute from the other components likely present in the feed liquor.
2. *Availability*—the inventory of solvent in the extraction system can represent a significant capital investment.
3. *Immiscibility with the feed*—otherwise there will need to be recovery of the solvent from the raffinate, or a continual and costly replacement of solvent as make up.
4. *Density differential*—too low a density difference between the phases will result in separation problems, lower capacity, and larger equipment. Too large a density difference may make it difficult to obtain the drop sizes desired for best extraction.
5. *Reasonable physical properties*—too viscous a solvent will impede both mass transfer and capacity. Too low an interfacial tension may lead to emulsion problems. The boiling point should be sufficiently different from that of the solute if recovery of the latter is to be by distillation.
6. *Toxicity*—must be considered for health considerations of the plant employees and for purity of the product.
7. *Corrosiveness*—may require use of more expensive materials of construction for the extraction process equipment.
8. *Ease of recovery*—as transfer of the solute from the feed still entails the further separation of solute from the solvent, solvent recovery will need to be as complete and pure as possible to permit recycle to the extractor as well as minimizing losses and potential pollution problems.

4.0 CALCULATION PROCEDURES

Sizing the equipment required for a given separation will depend upon both the flow rates involved and the number of stages that will be required.

With a binary equilibrium plot, Fig. 11.7, the distribution of extract and raffinate following one stage of contact is readily determined. Representing a mass balance of the solute transferred:

$$(Y_S - Y_E)S = (X_F - X_R)F$$
$$\frac{(Y_S - Y_E)}{(X_F - X_R)} = \frac{F}{S}$$

Thus, a line can be drawn from X_F, with a slope of F/S to the intersection with the equilibrium line, thus establishing Y_E and X_R.

For multiple contact, Fig. 11.8, the operating line can be written around some point in the column between stage "n" and (n + 1):

$$S(Y_{n+1} - Y_S) = F(X_n - X_R)$$
$$(Y_{n+1} - Y_S) = \frac{F}{S}(X_n) - \frac{F}{S}(X_R)$$

Since liquid-liquid extraction frequently involves only a few stages, the above equation can be used for an analytical solution.

The desired concentration of extract Y_E is set equal to Y_1, and the raffinate in equilibrium with the first stage, X_1, is determined from the equilibrium curve. With this value of X_1, Y_2 is calculated from the above operating equation; then X_2 is determined from the equilibrium line and the calculation procedure is continued until $X_n \leq X_r$.

A graphical solution is also readily obtainable. The operating line, with slope F/S, is drawn from the inlet and outlet concentrations. The number of stages is then stepped off in the same fashion as with a McCabe Thiele diagram in distillation, as shown in Fig. 11.8.

With a ternary equilibrium diagram, such as Fig. 11.5, the process result can be determined graphically. In Fig. 11.9, the addition of solvent to a feed containing X_F solute will be along the straight line connecting S with X_F. From an overall mass balance, the composition M of the mixture of feed and solvent is determined. With M in the two-phase zone, the overall mixture M separates along a tie line to end points Y_E and X_R on the equilibrium curve. The relative quantities of each phase can be calculated using the inverse lever-arm rule.

With more than one contact, an operating point Q is located outside the ternary diagram, as shown in Fig. 11.10. With a specified solvent/feed ratio and a desired raffinate purity, X_1, with the given feed, X_F the composition of the final extract, Y_n, is fixed by material balance. Point Q is formed by the intersection of the line drawn from Y_n through X_F, with the line drawn from the fresh solvent Y_S through X_1.

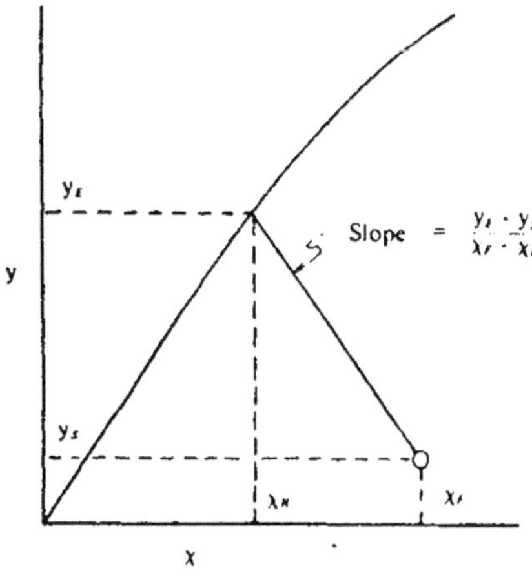

FIGURE 11.7 Graphical solution for single contact.

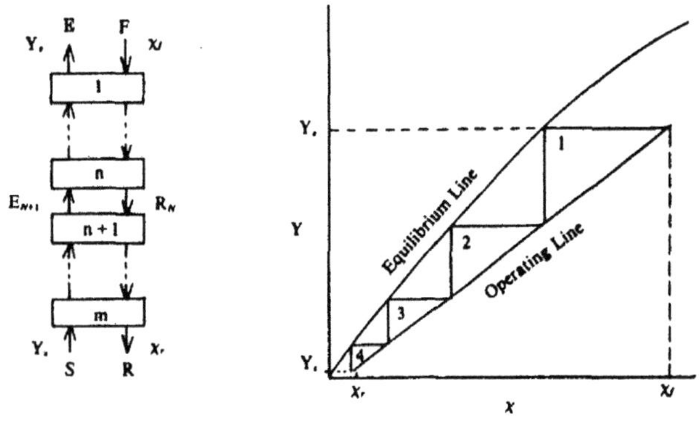

FIGURE 11.8 Graphical solution for multiple contact.

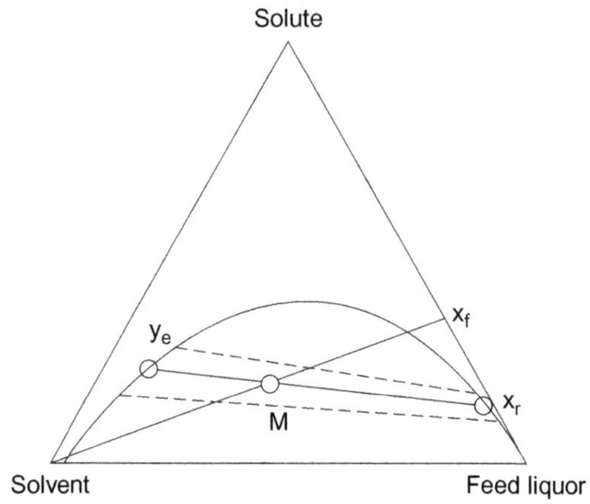

FIGURE 11.9 Graphical solution for single contact with ternary equilibrium data.

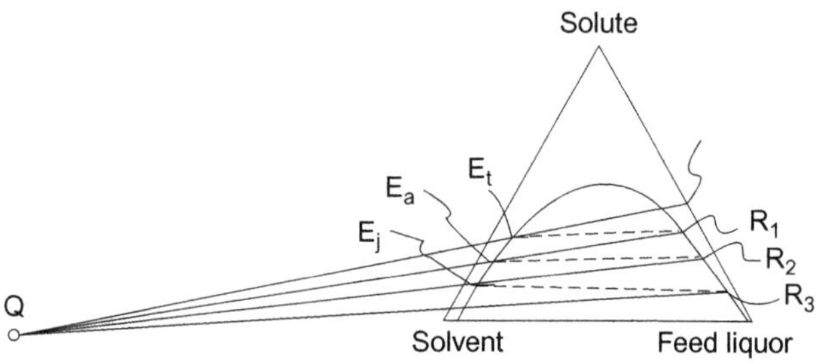

FIGURE 11.10 Graphical solution for multiple contact.

Point M in Fig. 11.9 represented the material balance:

$$F + S = E + R = M$$

Point Q in Fig. 11.10 represents a hypothetical quantity obtained by rearrangement of the above equation:

$$F - E = R - S = Q$$

The material balance for each stage is:

$$F - E_I = R_n - E_{n+1} = Q$$

Thus, a line through Q represents the operating line between stages. The number of stages is obtained by sequentially stepping off first the equilibrium distribution along a tie line, and then to the next stage by a line drawn from point Q through the raffinate to locate the next extract.

4.1 Simplified Solution

If the *distribution coefficient* is constant, and if there is essentially no mutual solubility, the fraction not extracted, Ψ, can be calculated directly as a function of the extraction factor, E, and the number of stages, n.

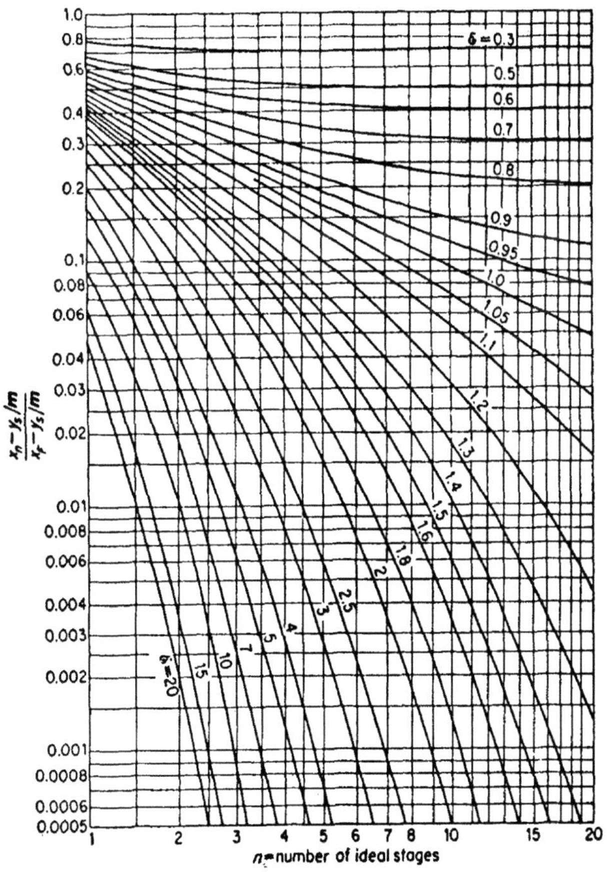

FIGURE 11.11 Countercurrent multistage extraction with immiscible solvents and constant distribution coefficient. *(From:* Liquid Extraction *by R.E. Treybal. Copyright© 1963, McGraw-Hill. Used with the permission.)*

$$\psi = \frac{X_1 - Y_S/m}{X_F - Y_S/m}; \quad E = \frac{mS}{F}$$

$$\psi = \frac{E - 1}{E^{n+1} - 1}; \quad E \neq 1$$

Treybal [3] discusses the derivation of these equations and presents a graphical solution reproduced here as Fig. 11.11.

Even when the two limitations of immiscibility and constant distribution coefficient do not quite hold, Fig. 11.11 does allow a quick estimate of the trade-offs between solvent/feed ratio and number of stages required to obtain a desired degree of extraction (raffinate purity).

The above solutions are all based on *ideal* or *theoretical* stages. Even in discrete stage systems, like mixer-settlers, equilibrium may not be attained because of insufficient time for diffusion of solute across the phase boundary or insufficient time for complete clarification of each stage.

In continuous differential extractors (columns) it has been convenient to think in terms of a height equivalent to a theoretical stage (HETS), and to correlate HETS as a function of system and equipment variables. Alternately, correlations may be obtained on the basis of the height of a transfer unit (HTU), which is more amenable to calculations which separately include the effects of backmixing [2,4].

4.2 Sample Stage Calculation

An aqueous waste stream containing 3.25% by weight phenol is to be extracted with one-third its volume of methylene chloride to produce a raffinate without more than 0.2% phenol. How many stages are required?

Graphical Solution. Figure 11.12 is constructed using the equilibrium data for the distribution of phenol between methylene chloride and water from Fig. 11.6.

The operating line is determined on a solute free basis as follows:

$$X_F = \frac{3.25(100)}{96.75} = \frac{3.36 \text{ g phenol}}{100 \text{ g water}}$$

$$X_R = \frac{0.2(100)}{99.8} = \frac{0.20 \text{ g phenol}}{100 \text{ g water}}$$

Thus, per 100 g of water feed, the amount of phenol removed is:

$$3.36 - 0.20 = 3.16 \text{ g}$$

At a volumetric feed rate of solvent equal to one-third the feed, and a specific gravity of 1.31 for methylene chloride, the weight ratio of solvent to feed is:

$$\frac{W_S}{W_F} = \frac{1}{3}\frac{(1.31)}{(1.0)(0.9675)} = 0.451$$

The phenol removed from the 100 g of water (3.16 g) must be in the extract, which contains 45.1 g of methylene chloride:

$$Y_E = \frac{3.16}{45.1}(100) = 7.01 \frac{\text{g phenol}}{100 \text{ g MeCl}_2}$$

The operating line is drawn from (3.36, 7.01) to (0.20, 0.00) in Fig. 11.12 and the stages stepped off. The stages are counted at the intersections with the equilibrium line. It is seen that the fourth stage produces a raffinate with a value less than required. Thus, the number of theoretical stages is interpolated to be 3.8.

Analytical Solution. The equation for the operating line is determined from the inlet and outlet concentrations. The operating line equation relates the extract concentration of one stage to the raffinate concentrate from the previous stage.

$$Y_{n+1} = \frac{W_F}{W_S}X_n - \frac{W_F}{W_S}X_R$$

$$Y_{n+1} = 2.22X_n - 0.444$$

Starting with Y_E, which is Y_1 for the first extraction stage, the raffinate X_1 in equilibrium is determined from the distribution curve Fig. 11.6:

at

$Y_1 = 7.01,\ X_1 = 1.43$
$Y_2 = 2.22(1.43) - 0.444 = 2.73$
$X_2 = 0.784$ from Fig. 11.6
$Y_3 = 2.22(0.784) - 0.444 = 1.30$
$X_3 = 0.42$ from Fig. 11.6
$Y_4 = 2.22(0.42) - 0.444 = 0.488$
$X_4 = 0.150$ from Fig. 11.6

Since X_4 is less than the observed $X_r = 0.20$, the fractional stage is estimated as follows:

$$\frac{X_3 - X_r}{X_3 - X_4} = \frac{0.42 - 0.20}{0.41 - 0.15} = \frac{0.22}{0.26} = 0.85$$

So the total number of stages is calculated to be 3.85.

Short-Cut Solution. The curved equilibrium relationship means that the Treybal plot, Fig. 11.11, perhaps cannot be used. The required stages can be bracketed by calculating the extraction factor at each end of the extraction. At the dilute end:

$$D = \frac{0.63}{0.20} = 3.15$$

$$E = \frac{W_S}{W_F}D = (0.451)(3.15) = 1.42$$

$$\psi = \frac{0.2}{3.36} = 0.060$$

$n = 4.9$ from Fig. 11.11

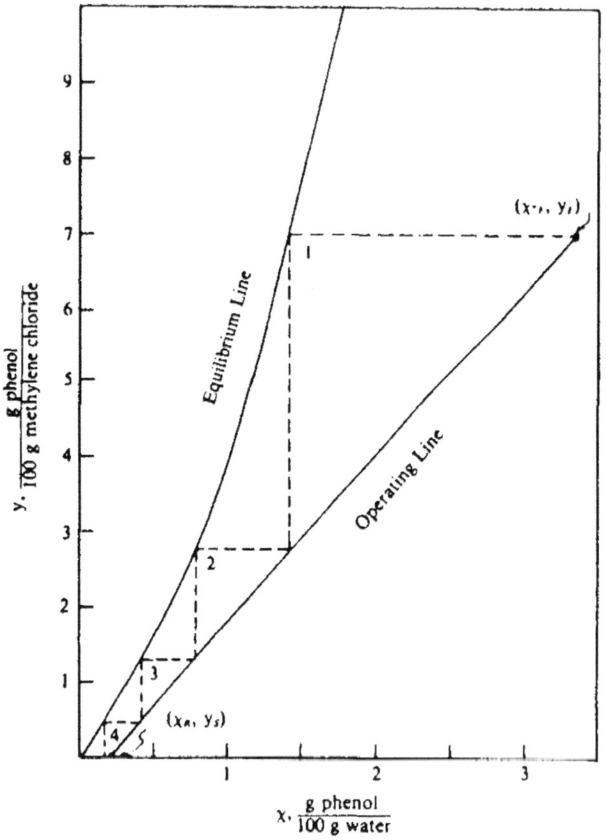

FIGURE 11.12 Stages for sample calculation.

At the concentrated end:

$$D = \frac{7.01}{1.43} = 4.90$$

$$E = (0.451)(4.90) = 2.21$$

$n = 3.5$ from Fig. 11.11

Using an average extraction factor of $E = 1.81$, the number of stages from the Treybal plot is 4.1.

The Treybal plot can be used to provide estimates for other requirements as well. For example, if it were desired to increase the amount of phenol extracted from 94 to 99%, what increase in solvent flow or number of stages would be required?

At $E = 1.81$, $\Psi = 0.01$, $n = 6.3$ stages

At $n = 4.1$, $\Psi = 0.01$, $E = 2.8$ required

Thus, the solvent flow would have to be increased by a factor of:

$$\frac{2.8}{1.81} = 1.55$$

Thus, to increase extraction from 94 to 99% would require 57% more stages or 55% more solvent, or some lesser combination of both.

5.0 DROP MECHANICS

An understanding of the performance of extraction equipment is furthered by an understanding of what may be going on inside individual drops. With the assumption of transfer of a solute *A* from a dispersed feed phase into a continuous solvent, as shown in Fig. 11.13, a concentration profile across the interface would appear to have a discontinuity (Fig. 11.14). The discontinuity is a consequence of the distribution coefficient, and reflects the general practice of choosing a solvent which has a greater preference for the solute than the feed phase has. If activities instead of concentrations were used, there would be no discontinuity at the interface.

Transfer of solute across the interface can be assumed to be controlled by what happens through the immobilized films on both sides of the interface. Handlos and Baron [5] have presented generalized correlations for the calculation of the individual inside and outside coefficients for mass transfer across these films.

Small drops lead to more transfer area and better extraction, but to slower settling and less capacity. Thus, selection of extraction equipment frequently involves a compromise choice balancing efficiency against capacity.

The terminal velocity of liquid drops is the same as solid spheres when the diameter is small. The drag coefficient versus Reynold's number can be recalculated to provide a diameter-free ordinate versus a velocity-free abscissa to facilitate direct solution, as shown in Fig. 11.15. With drops, a maximum velocity is attained, and this maximum has been correlated with a parameter based on physical properties of the system.

The practical sequence of this phenomenon in column extraction is illustrated in Fig. 11.16. Drops larger than d^* won't travel any faster, so there is no capacity gain, and they have less specific area, so there will be an efficiency loss. Drops smaller than d^* will result in more extraction by providing more transfer area and a longer contact time, but at the potential expense of lower capacity.

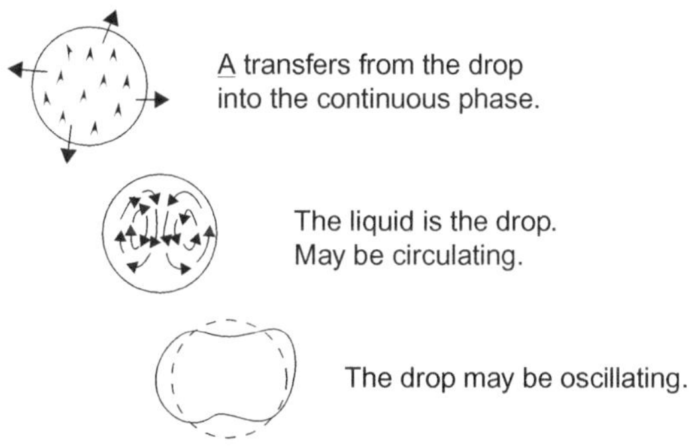

FIGURE 11.13 Drop mechanics.

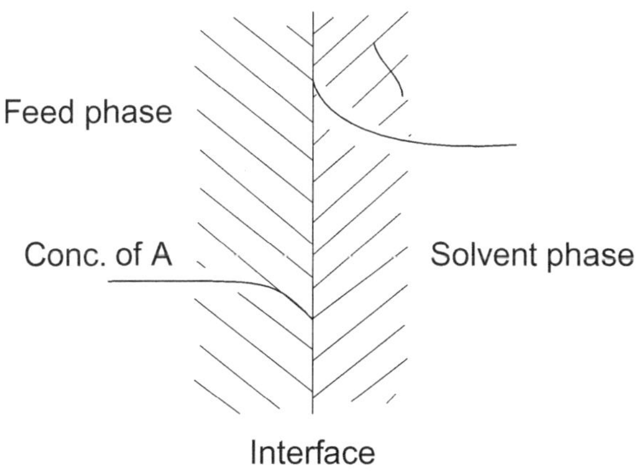

FIGURE 11.14 Solute concentration at the interface.

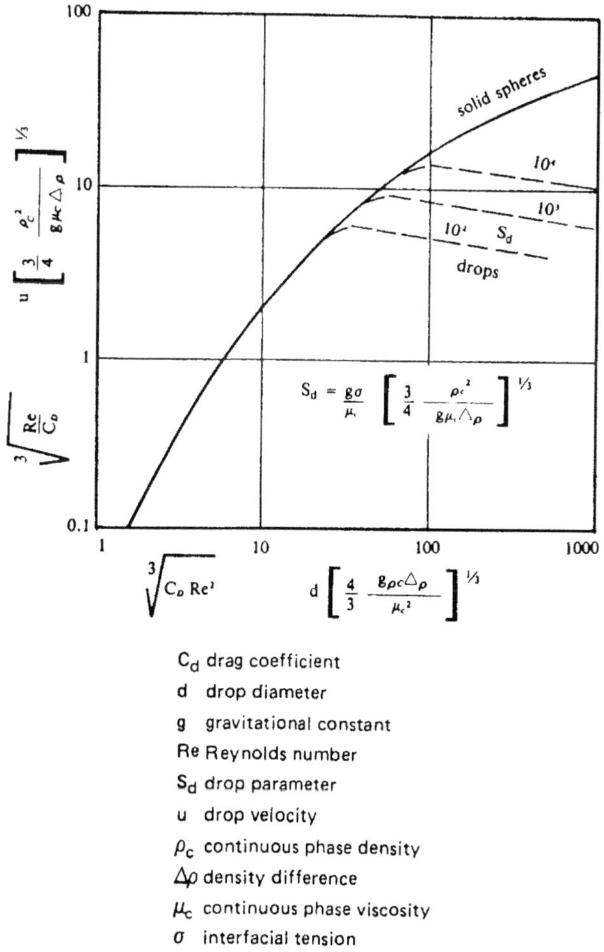

FIGURE 11.15 Dimensionless drop velocity vs. dimensionless drop diameter.

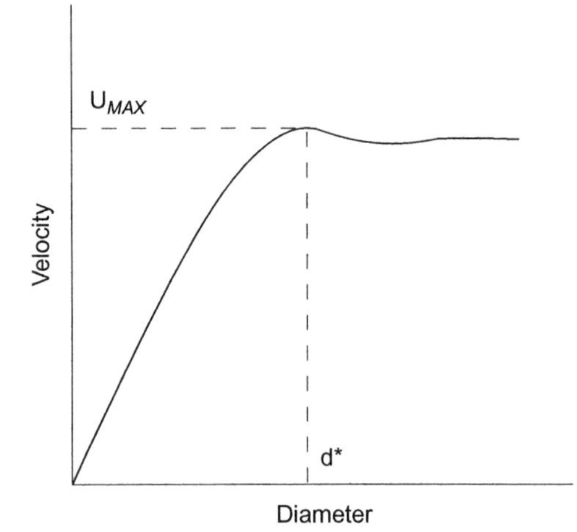

FIGURE 11.16 Drop velocity vs. drop diameter.

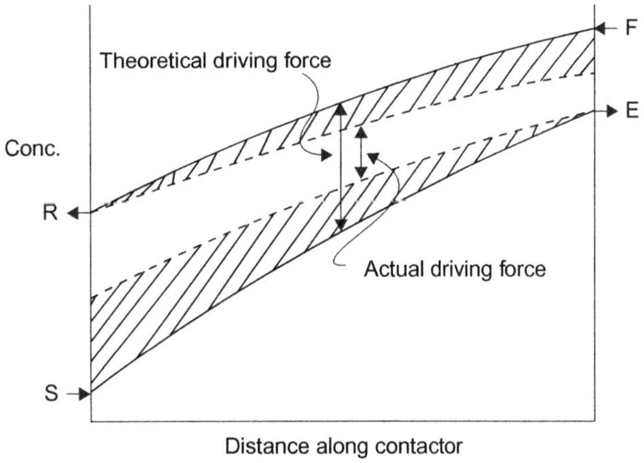

FIGURE 11.17 Effect of backmixing on extraction driving force.

It is generally desirable to provide as uniform a drop size as possible. A wide range in drop sizes may allow the smaller drops to attain equilibrium, but they are en route longer, while the larger drops zip through, not attaining equilibrium.

It is also considered desirable to allow drops to coalesce and be redispersed, as mass transfer from a forming drop is always higher than it is from a stagnant drop.

Backmixing caused by flow patterns induced in the equipment can also deleteriously affect performance by reducing the driving force gradient, as illustrated in Fig. 11.17. Sleicher [6] presents procedures for calculating the consequences of backmixing on overall extraction results.

6.0 TYPES OF EXTRACTION EQUIPMENT

Extraction equipment can be classified by function as providing discrete stages or continuous differential contact.

TABLE 11.1 Classification of Industrial Extraction Equipment

Flow by	Drop Size Control by	Stagewise	Continuous Differential
Gravity alone	Gravity alone	Perforated Place Column	Spray Column Packed Column
	Mechanical rotation	Mixer-Settler	RDC Oldshue-Rushton Column ARD Column Kühni Column Raining Bucket Contactor
	Mechanical reciprocation		Karr Column Pulsed Packed Column Pulsed Perforated Plate Column
Centrifugal Force	Flow through baffles	Westfalia Extractor Robatel Extractor	Podbielniak Extractor Alfa Laval Extractor

Separation may be by gravity alone or by centrifugal force. Additional energy may be applied to control drop size, either by mechanical agitation or pulsation. This classification is shown in Table 11.1, along with major examples of available equipment.

6.1 Non-Agitated Gravity Flow Extractors

Spray Column. The simplest differential extractor is the spray column (Fig. 11.18a), which depends upon the initial dispersion of the dispersed phase to create favorably sized droplets. There is no means provided to redisperse this phase if any coalescence occurs. Although the equipment is simple and inexpensive, it is difficult to obtain more than one stage extraction. The passage of the dispersed phase induces considerable backmixing of the continuous phase, particularly in larger diameter columns.

Packed Column. (Fig. 11.18b) Interphase contact can be improved in the spray column by providing extensive surface for coalescence and redispersion. This surface is provided with packing which provides surface while maintaining a large open area for flow, such as Raschig rings, Berl saddles, and variants thereof. There is some loss in capacity because of the cross section occupied by the packing, but this is more than offset by the gain in improved mass transfer and lessening of continuous phase backmixing.

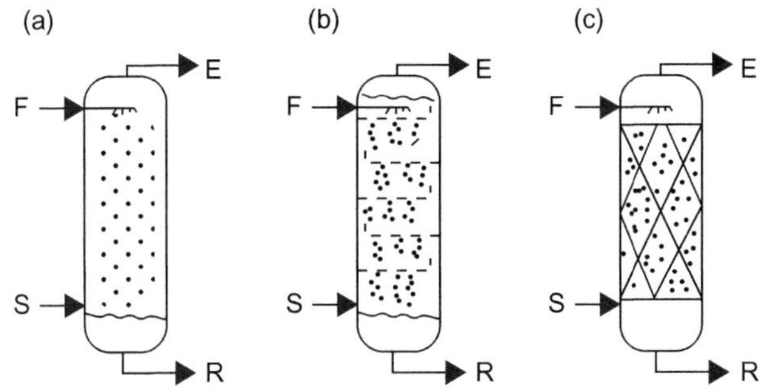

FIGURE 11.18 Non-agitated gravity flow extractors. *(a)* Spray, *(b)* packed, and *(c)* perforated plate.

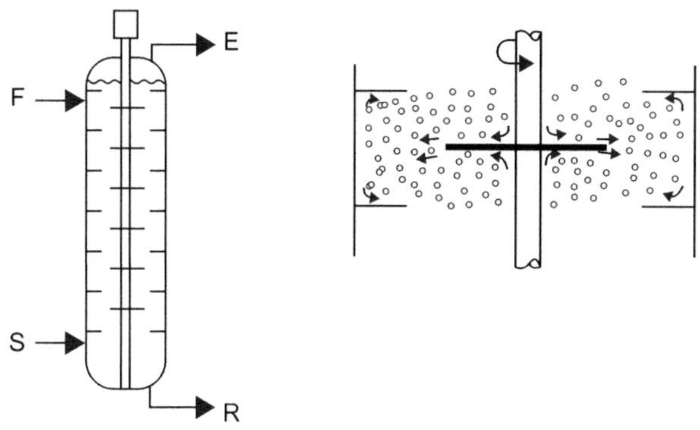

FIGURE 11.19 Rotating disc contactor (RDC).

Packing should be chosen that preferentially is wetted by the continuous phase to discourage formation of rivulets of the dispersed phase bypassing through the column. In large diameter columns, redistribution trays should be installed to overcome potential channeling. Smaller packing size is generally more efficient, but restricts flow more, and is more prone to fouling by trapping solids. Eckert [7] summarizes design criteria for the selection of packing for packed columns.

Perforated Plate Column. (Fig. 11.18c) Sieve trays can be placed in the spray column to cause coalescence and redispersion of the dispersed phase. The trays can be designed to permit flow of both phases through the same perforations, but such trays generally have a quite narrow operating range. Generally, some sort of *downcomer* (or *upcomer*) is provided to allow a separate path for the continuous phase and one-way flow of the dispersed phase through the perforations. The density difference between the two phases and the height of coalesced phase provide the driving force for redispersion through the orifices.

In contrast with vapor-liquid columns, tray efficiencies are very low (5 to 30%) in liquid-liquid systems. The trays do limit continuous phase backmixing as well as provide drop redispersion, but at the expense of reduced capacity.

6.2 Stirred Gravity Flow Extractors

Provision of a shaft through the extraction column allows for repeated redispersion of the drops via various impellers located along the shaft. A variety of industrial equipment is available, with the differences being in the design of the impellers on the shaft for dispersion, and stators in the column for baffling and coalescence. Stirred columns offer the operator increased flexibility in operation by independent control over the dispersion process.

RDC Column. The *rotating disc contactor* (Fig. 11.19) provides for redispersion by a series of discs along the shaft, combined with a series of fixed stators. Vortices are formed in each *compartment*, and the shear of the fluid against the rotor or stator causes the drop breakup. In many instances, performance can be predicted from first principles, relating drop size to the energy input, and calculating slip velocity and mass transfer coefficients based on that diameter and the physical properties of the system (see Strand, Olney & Ackerman [8]).

With increasing rotational speed, efficiency improves as drops become smaller, but maximum capacity is lessened. Increased rotational speed also increases continuous phase backmixing, and causes some segregation of the phases as the lighter phase accumulates around the shaft while the denser phase hugs the wall. At the same energy input, dispersing the light phase leads to smaller drops because all of the light phase must pass over the tips of the spinning discs; whereas dispersion of a denser phase is brought about primarily by fluid motion over the stationary ring baffles.

Oldshue-Rushton Column. This column is similar to the RDC, except that the flat rotor discs have been replaced with turbine type agitators (Fig. 11.20). As with the RDC, the diameter of the agitators can be varied along the shaft to compensate for the progressive change in the physical properties of the system as extraction occurs.

Other variations of stirred columns which are available include the *asymmetric rotating disc* (ARD) contactor, the *Kühni column*, and two types of *Scheibel columns*. The rotor of the ARD is located off center, which permits more elaborate baffling for the necessary transport of flows with less backmixing.

The Kühni column employs radial flow impellers located between perforated plates for compartmentalization. The first Scheibel column used wire mesh zones to promote coalescence and limit backmixing between turbine-agitated mixing zones. A later Scheibel column used a shrouded radial impeller and multiple ring baffles to direct most of the rotor's energy towards dispersion and away from axial mixing.

Raining Bucket Contactor. This contactor consists of a series of scoops located on a slowly rotating, baffled rotor within a horizontal cylindrical vessel (Fig. 11.21).

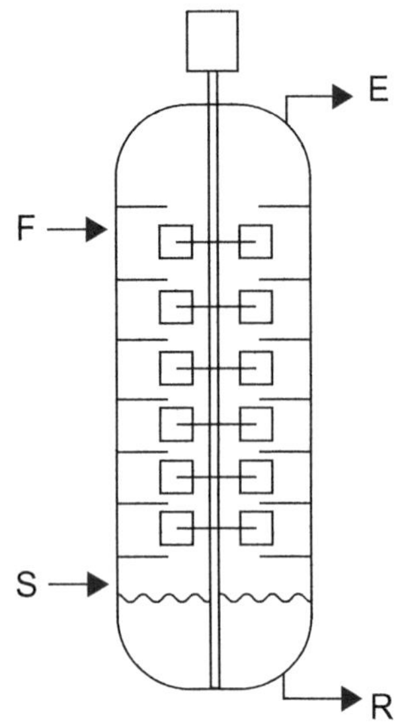

FIGURE 11.20 Oldshue-Rushton column.

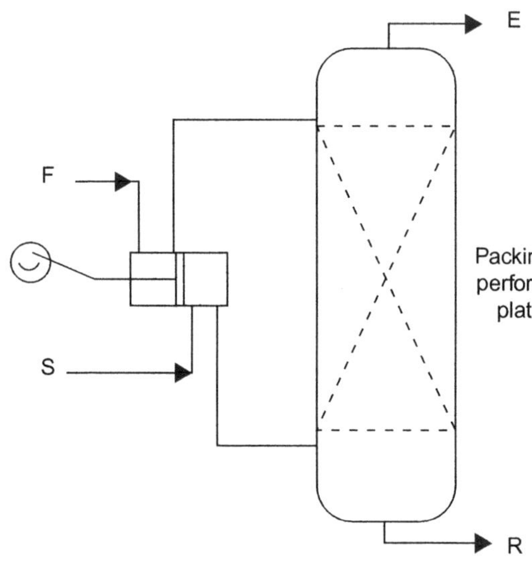

FIGURE 11.22 Liquid pulsed columns.

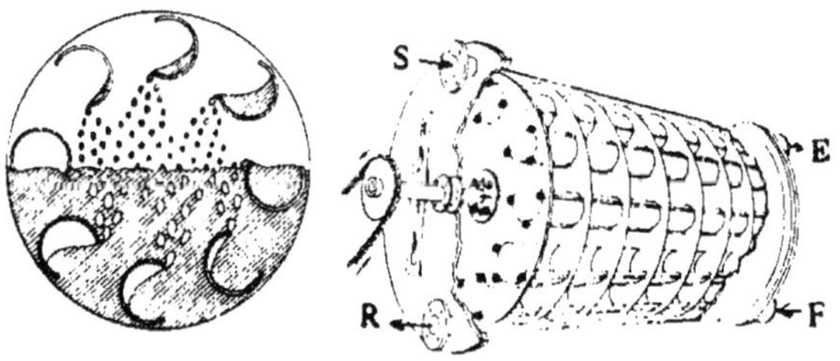

FIGURE 11.21 Raining bucket contactor.

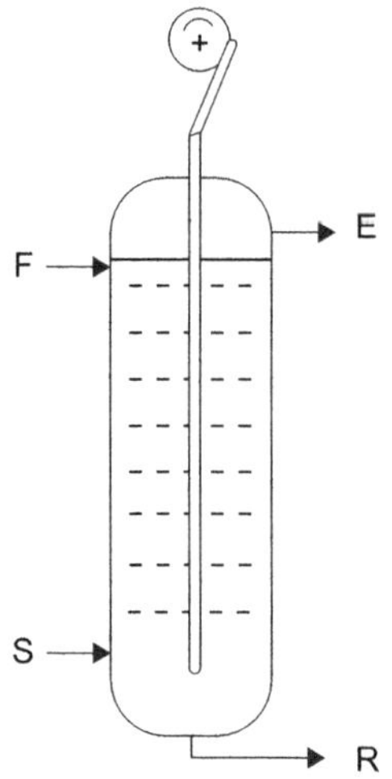

FIGURE 11.23 Karr reciprocating plate column.

An interface is maintained near the middle, and the scoops capture and then allow one phase to rain through the other, and vice versa, once each revolution.

There is little, if any, control of droplet size, but the raining bucket contactor is the only one that disperses each phase in the other. If the flow ratio differs greatly from unity, backmixing of the low flow phase can be serious, and line out with changed operating conditions can take a long time.

6.3 Pulsed Gravity Flow Extractors

Liquid Pulsed Columns. The liquid in a packed or perforated plate column may be pulsed to promote better mass transfer (Fig. 11.22). If a sieve plate column is pulsed, downcomers are no longer required. Pulsing can be caused by a piston pump or by air pulsing external to the column. Frequencies are generally 1 to 3 Hz and amplitude up to 20 mm. Drop size is dependent upon the product of amplitude times frequency. As this product is increased, the smaller diameter drops so produced lead to more holdup and better mass transfer, but to a fall off in capacity. Eventually, at a high enough amplitude x frequency product, backmixing increases to the extent that efficiency also begins to diminish.

Mechanically Pulsed Column. The *Karr column* (Fig. 11.23) consists of perforated plates ganged on a common shaft which is oscillated by an external drive. The perforated area and hole size are much larger than in typical sieve plate operation. At high amplitude x frequency product in larger columns, the tendency for excessive backmixing can be curtailed by installation of some fixed baffles.

Mixer-Settlers. The extraction function of bringing feed and solvent intimately together, and then allowing them to separate is frequently done in mixer-settlers. The functions may be done in separate vessels, or in different portions of the same vessels, or sequentially in the same vessel on a batch basis. As noted earlier in Fig. 11.3, the flows of feed and solvent can be countercurrent to each other through a series of mixer-settlers.

Sizing of the mixer is based upon providing sufficient agitation and sufficient residence time to allow equilibrium to be approached, and thus will depend upon the flows to be processed as well as the physical properties of the two liquids. Since some extractions actually involve a chemical reaction, the time of contact can be very important. If, for reasons of improved mass transfer, it is desired to disperse the high flow phase, it may be necessary to recycle some of the low flow phase to keep an appropriate phase ratio in the mixer different than the feed flow ratio.

The settler must provide a long enough quiescent residence time for the emulsion which is produced in the mixer to break, and a low enough lineal velocity for the two phases to become essentially free of entrainment. In some instances, coalescing material, such as wire mesh, may be installed to lessen entrainment, however, such material should be used with some caution because of the tendency for fouling by accumulation of foreign material.

It is frequently possible to introduce one of the phases into the eye of the impeller, and thus be able to pump one entering fluid while the other flows by gravity from the next upstream and downstream stages, without the need for separate interstage pumps.

6.4 Centrifugal Extractors

Many of the commercial extraction processes encountered in the pharmaceutical industry involve systems which emulsify readily and are exceedingly difficult to separate cleanly. Stability of the solute may also be a factor, and rapid separation may be required to prevent degradation and loss of the product. Centrifugal extractors fill an important niche for just such problems.

The most common centrifugal extractor is the Podbielniak® Contactor, as shown in cutaway view in Fig. 11.24. Essentially it is a sieve plate column that has been wrapped around a shaft and spun to create a multigravitational force to do both the redispersion and the separation. All fluids enter and leave through shaft passageways and mechanical seals.

The performance of centrifugal extractors has been described by Todd and Davies in general detail [9] and specifically for pharmaceutical use [10]. The primary benefits of centrifugal extractors accrue from their compactness and superior clarifying capabilities. Solvent inventory can be held to a minimum. Centrifugal extractors are also particularly appropriate handling high phase ratios, as the low flow phase can be kept continuous without much backmixing, thereby allowing the large flow fluid to be dispersed to provide more mass transfer area.

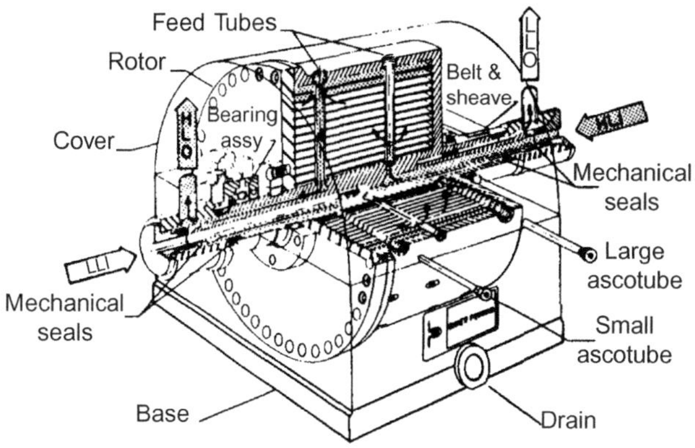

FIGURE 11.24 Podbielniak® centrifugal extractor.

The Podbielniak and Alfa Laval centrifugal extractors are essentially continuous differential contactors. The Westfalia and Robatel centrifugal extractors contain discrete mechanical stages, and flow from one to another is effected by spill over discs and skimmers according to usual centrifugal clarifier practice. As the number of discrete stages is increased, the allowable flow rates are proportionately decreased.

6.5 Equipment Size Calculation

Agitated Columns. The size of an extraction column frequently can be estimated from a knowledge of the flow rates and physical properties, combined with some empirical generalizations.

1. The maximum capacity (at zero stirrer speed or pulsation) is directly related to the terminal velocity of the dispersed phase through the minimum physical constriction in the column.
2. The terminal velocity of the dispersed phase droplets is related to the physical properties of the system by the correlation shown in Fig. 11.15.
3. For many systems, the effect of hindered settling can be approximated by:

$$V_t = \frac{1}{1-h}\left(\frac{V_d}{h} + \frac{V_c}{1-h}\right)$$

   where V_t, V_d, V_c are the superficial lineal velocities of the drop, dispersed phase, and continuous phase, and h is the holdup.
4. Agitated columns are frequently operated so that the capacity is half what it would be at no agitation (zero rpm or pulsation). Agitation is used to reduce droplet diameter to this equivalent point to increase mass transfer rate and mass transfer area.
5. For sizing purposes, the diameter of the column will be chosen so that the column is operating at 75% of the flood point.

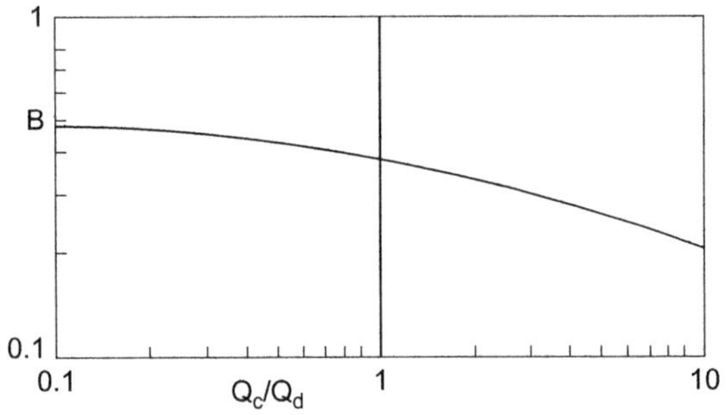

FIGURE 11.25 Empirical constant *B* for determining agitated column diameter.

6. The holdup at flooding can be determined by differentiating the equation in criterion #3. Combining this relationship with all the constants leads to the following equation:

$$D = \frac{0.09}{B} Q_d^{0.5} \left(\frac{\mu_c}{\sigma}\right)^{0.88} \left(+ \frac{\rho_e^2}{\mu_c \Delta\rho}\right)^{0.138}$$

(with D in meters, Q_d in m^3/h, μ_c in poise, σ in dynes/cm, and ρ in g/cc). The factor B is related to holdup and dependent upon phase ratio, as shown in Fig. 11.25.

The countercurrent contact zone height will depend primarily upon the number of stages required (n) and the column characteristics. The effect of backmixing also increases the column diameter. A reasonable first approximation of extraction height (L) required for agitated columns is:

$$L = 0.94n\sqrt{D} \quad (L \text{ and } D \text{ in meters})$$

Most columns also require clarifying zones at each end to provide for coalescence and to minimize entrainment. These zones also are dependent upon column diameter. The combined height required (Z) for the clarifying zones can be approximated by:

$$Z = 3\sqrt{D} \quad (Z \text{ and } D \text{ in meters})$$

Estimates of column size required for three different cases are tabulated in Table 11.2. Case A involves the removal of dioxane from a benzene stream with water as the extracting solvent. Case B involves the recovery of methyl ethyl ketone from a heptane stream with water. Case C is for the removal of phenol from an aqueous stream with methylene chloride.

In addition to the calculated heights and diameters, the total traffic flow (the sum of both flows divided by the column cross-section) is listed. Typical traffic flows for agitated columns are in the 25 to 100 m^3/m^2 hr range.

The manufacturer of the extraction column will likely select the next larger diameter size for which he has standardized components. He may also insist upon some pilot plant test to confirm the capacity and efficiency requirements.

TABLE 11.2 Examples of Column Sizing Calculations

Case			A	B	C
Remove solute from feed with solvent			Dioxane	MEK	Phenol
			Benzene (c)	Heptane (d)	Water (d)
			Water (d)	Water (c)	$MeCl_2$ (c)
Flow	Q_c	m^3/hr	20.6	30.9	7.6
	Q_d	m^3/hr	13.3	16.4	22.7
Ratio	Q_c/Q_d		1.54	1.88	0.33
Constant	B		0.356	0.344	0.447
Viscosity	μ_c	poise	0.0065	0.010	0.007
Int. tens.	σ	dyne/cm	30	45	45
Spec. grav.	ρ_c		0.884	1.00	1.31
	ρ_d		1.00	0.688	1.00
	Δ_ρ		0.116	0.312	0.31
	$Q_d^{0.5}$		3.65	4.06	2.75
	$\left\{\frac{\mu_c}{\sigma}\right\}^{0.088}$		0.476	0.477	0.462
	$\left\{\frac{\rho_c^2}{\mu_c \Delta_\rho}\right\}^{0.138}$		2.61	2.22	2.51
Diameter	D	m	1.146	1.122	0.642
Theo. stages,	n		4	6	4
Ht, contact	L	m	4.03	5.97	3.01
clarif.	Z	m	3.21	3.17	2.40
Total	H		7.24	9.14	5.41
Traffic flow		m/hr	32.9	47.9	93.6

Note: (c) = continuous phase; (d) = dispersed phase.

The manufacturers of other proprietary extraction devices, such as centrifugal extractors, will be able to provide estimates of the probable size equipment required, based on comparisons with similar systems and their own accumulated design experience.

Many pharmaceutical extractions do not lend themselves to simple straightforward analytical solutions. Rarely is there a case of simple extraction of a single solute from a clean feed with pure solvent. There may well be solids present which can stabilize emulsions and cause excessive entrainment. Usually, more than one solute is present, so selectivity as well as extent of extraction becomes important. Also, the solvent may contain residual solute from the solvent recovery section. Again, suppliers of extraction equipment should be contacted for their help in solving real industrial extraction problems.

Packed Columns. Capacity of packed columns is strongly dependent upon the packing being used. As the surface area of the packing is increased to improve efficiency, in general, both the hydraulic radius and the fraction void decrease, thereby increasing resistance to flow and lowering capacity. For a given extraction, the maximum capacity (flooding rate) generally follows the form:

$$V_d^{0.5} + V_c^{0.5} = K$$

where K is a function of packing characteristics and physical properties of the system.

Compared to agitated columns, both diameter and height will have to be larger. Flow redistributer are advisable at periodic intervals to offset the tendency for channeling and bypassing frequently encountered in packed columns. Characteristics of various packings and correlations for capacity and stage height are given by Treybal [3] and Eckert [7].

Mixer-Settlers. The mixing required for adequate dispersion can be determined and scaled-up by the methods outlined by Oldshue [11].

Sizing of settlers poses some uncertainty in that solvent recycle within the process may lead to accumulation of an interfacial *rag*, which tends to stabilize emulsions at the interface. For a first approximation, an arbitrary residence time, like 20 minutes, might be assumed unless bench shake-outs indicate an even longer time required for adequate clarification.

Proprietary Extractors. Manufacturers or proprietary design extraction equipment (such as the Podbielniak Centrifugal Extractor or the RTL (raining bucket) Contactor) provide catalogs listing the relative capacities of the various sizes of equipment which are offered. Pilot equipment is usually available for determining extraction performance, and the manufacturer utilizes both the pilot data and experience with similar systems to provide assured commercial designs.

7.0 SELECTION OF EQUIPMENT

The choice of extraction equipment should be based on the minimum annual cost for the complete package of extractor and accessory equipment, including operating and solvent loss costs.

In addition to the requirements of processing so much feed and solvent with a required number of theoretical stages, there are the practical considerations concerning contamination, entrainment, emulsification, floor space, height requirements, cleanability, and versatility to handle other than design rates. The suitability of various type extractors with respect to each of these considerations is listed in Table 11.3. Not all of the features compared in the table can be equated. The tabulation is provided to show comparisons to aid in the selection of suitable equipment.

TABLE 11.3 Extractor Selection Chart

	Low Cost		High Efficiency	Total Throughput	Flexibility	High Volumetric Efficiency	Lowest Space		Ability to Cope with Systems	
	Capital	Operating					Vertical	Floor	Which Emulsify	with Solids
Mixer settler	3	2	4	4	4	2	5	1	1	3
Spray	4	5	1	3	2	1	1	5	2	4
Perf. plate	4	5	2	2	2	2	1	4	3	2
Packed	4	4	2	2	2	2	1	4	3	1
Pulse	3	3	4	3	4	4	3	4	3	3
Agitated	3	4	4	3	4	4	3	4	3	3
Centrifugal	2	3	4	3	4	5	5	5	5	2
	5 is outstanding				2 is fair					
	4 is good				1 is poor					
	3 is adequate				0 is unsuitable					

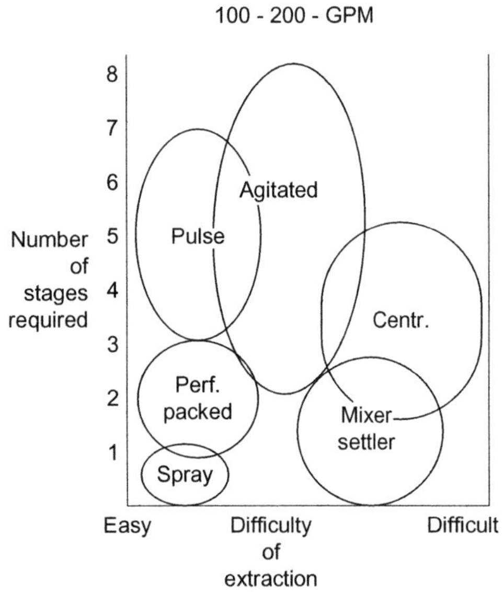

FIGURE 11.26 Extractor selection map.

Other criteria for the selection of an extractor are the ease of separation of the two phases and the difficulty of extraction. For example, if the two phases have a large density difference, or at least one is quite viscous, the energy required to get a good enough dispersion for good extraction may lead to excessive backmixing of the continuous phase.

The extractor selection map depicted in Fig. 11.26 reflects the above considerations plus the number of stages required. Where the degree of extraction exceeds the probable maximum staging achievable in one extractor, the extractors can be used in series.

8.0 PROCEDURE SUMMARY

Liquid-liquid extraction should be considered as a desirable route for product recovery and purification along with fractional crystallization and distillation. The ability to make separations according to chemical type, rather than according to physical properties such as freezing point or vapor pressure, is one of extraction's major attractions. Energy frequently can be saved in the recovery of valuable products from dilute broth solution since a small quantity of a selective solvent can be used, and recovery from the concentrated extract is then facilitated.

Selectivity of potentially attractive solvents can frequently be determined from simple shake-outs over the desired concentration range. From these distribution data, the combinations of amount of solvent and number of theoretical stages can be calculated.

Suppliers of extraction equipment will likely wish to participate in pilot testing to confirm the correlations for capacity and efficiency of the specific equipment being considered.

After installation, the equipment suppliers can also provide technical assistance in bringing the extraction equipment on line and solving problems which may arise from the commercial plant operation with its potential variation in feed and solvent quality and accumulation of impurities.

9.0 ADDITIONAL INFORMATION

With Treybal's book [3] essentially out of print, the *Handbook of Solvent Extraction* by Lo, Baird, and Hanson [2] provides a most comprehensive reference. In addition to the previously cited *Perry's Handbook* chapter on liquid extraction by Robbins [1], *The Essentials of Extraction* by Humphrey, Rocha, and Fair [12], and a three part *A Fresh Look at Liquid-Liquid Extraction* [13], provide briefer, but very useful guidelines. Details of extraction processes specifically involved in pharmaceutical production have been described by King et al. [14], and by Kroner, Hustedt, and Kula [15].

REFERENCES

[1] L.A Robbins, Sec. 15, liquid-liquid extraction, in: R.H. Perry, D.W. Green, J.O. Maloney (Eds.), Perry's Chemical Engineers Handbook, sixth ed., McGraw Hill, New York, 1984.
[2] T. Lo, M.H.I. Baird, C. Hanson, Handbook of Solvent Extraction, John Wiley & Sons, New York, 1983.
[3] R.E. Treybal, Liquid Extraction, second ed., McGraw Hill, New York, 1963.
[4] C. Hanson (Ed.), Recent Advances in Liquid-Liquid Extraction, Pergamon Press, Oxford, 1971.
[5] A.E. Handlos, T. Baron, Mass and heat transfer from drops in liquid-liquid extraction, AIChE J. 3 (1957) 127–135.
[6] C.A. Sleicher, Axial mixing and extraction efficiency, AIChE J. 5 (1959) 145–149.
[7] J.S. Eckert, Extraction variables defined, Hydrocarb. Process. 55 (3) (1976) 117–124.
[8] C.P. Strand, R.B. Olney, G.H. Ackerman, Fundamental aspects of rotating disk contactor performance, AIChE J. 8 (1962) 252–261.
[9] D.B. Todd, G.R. Davies, Performance of centrifugal extractors, Proceedings ISEC, 74, 3:2379–2398.
[10] D.B. Todd, G.R. Davies, Centrifugal pharmaceutical extractions, Filtr. Sep. 10 (6) (1973) 663–666.
[11] J.Y. Oldshue, Mixing, Handbook of Fermentation Engineering.
[12] J.L. Humphrey, J.A Rocha, J.R. Fair, The essentials of extraction, Chem. Eng. 91 (18) (1984) 76–95.
[13] R.W. Cusack, P. Fremeaux, D. Glatz, A Karr, A fresh look at liquid-liquid extraction, Chem. Eng. 98 (2) (1991) 66–76. 98(3):132–138 (Mar. 1991); 98(4):112–120 (Apr. 1991).
[14] M.L. King, A.L. Forman, C. Orella, S.H. Pines, Extractive hydrolysis for pharmaceuticals, CEP 81 (5) (1985) 36–39.
[15] K.H. Kroner, H. Hustedt, M.R. Kula, Extractive enzyme recovery: economic considerations, Process Biochem. 19 (1984) 170–179.

Chapter 12

Evaporation

Howard L. Freese

1.0 INTRODUCTION

As described in the Chemical Engineers' Handbook, "Evaporation is the removal of solvent as a vapor from a solution or slurry. The vapor may or may not be recovered, depending on its value. The end product may be a solid, but the transfer of heat in the evaporator must be to a solution or a suspension of the solid in liquid if the apparatus is not to be classed as a dryer. Evaporators are similar to stills or re-boilers of distillation columns, except that no attempt is made to separate components of the vapor [1]."

The task demanded of an evaporator is to concentrate a feed stream by removing a solvent that is vaporized in the evaporator and, for the greatest number of evaporator systems, the solvent is water. Thus, the "bottoms" product is a concentrated solution, a thick liquor, or possibly a slurry. Since the bottoms stream is most usually the desired and valuable product, the "overhead" vapor is a by-product of the concentration step and may or may not be recovered or recycled according to its value. This determination may be made upon incremental by-product revenues for reusable organic solvents, or upon minimizing incremental processing costs for water vapor that may be slightly contaminated and must be further treated to meet environmental constraints. The solvent vapors generated in an evaporator are nearly always condensed somewhere in the process, with the exception of solar evaporation systems (ponds) which evaporate into the local atmosphere.

All evaporators remove a solvent vapor from a liquid stream by means of an energy input to the process. The energy source is most usually dry and saturated steam, but can be a process heating medium such as: liquid or vapor phase heat transfer fluids (Dowtherm or Therminol), hot water, combustion gases, molten salt, a high temperature process stream, or, in the case of a solar evaporation plant, radiation from the sun.

Evaporation should not be confused with other somewhat similar thermal separation techniques that have more precise technical meanings, for example: distillation, stripping, drying, deodorizing, crystallization, and devolatilization. These operations are principally associated with separating or purifying a multicomponent vapor (distillation), producing a solid bottoms product (drying, crystallization), or "finishing" an already-concentrated fluid material (stripping, devolatilization, deodorizing).

Engineers, scientists, and technicians involved in fermentation processes will usually be concerned with the concentration of aqueous solutions or suspensions, so the evaporation step will be the straightforward removal of water vapor from the process, utilizing steam as a heating medium. The focus will be, then, on the evaporator itself and how it should be designed and operated to achieve a desired separation in the fermentation facility.

2.0 EVAPORATORS AND EVAPORATION SYSTEMS

An evaporator in a chemical plant or a fermentation operation is a highly engineered piece of processing equipment in which evaporation takes place. The process and mechanical computations that are required to properly design an evaporator are many and very sophisticated, but the basic principles of evaporation are relatively simple, and it is these concepts that the engineer or scientist involved in fermentation technology should comprehend.

Often an evaporator is really an evaporation system that incorporates several evaporators of different types installed in series. All evaporators are fundamentally heat exchangers, because thermal energy must be added to the process, usually across a metallic barrier or heat transfer surface, in order for evaporation to take place. Efficient evaporators are designed and operated according to several key criteria:

1. *Heat Transfer*. A large flow of heat across a metallic surface of minimum thickness (in other words, high heat flux) is fairly typical. The requirement of a high heat transfer rate is the major determinate of the evaporator type, size, and cost.
2. *Liquid-Vapor Separation*. Liquid droplets carried through the evaporator system, known as *entrainment*, may contribute to product loss, lower product quality, erosion of metallic surfaces, and other problems including the necessity to recycle the entrainment.

Fermentation and Biochemical Engineering Handbook.

Generally, decreasing the level of entrainment in the evaporator increases both the capital and operating costs, although these incremental costs are usually rather small. All these problems and costs considered, the most cost-effective evaporator is often one with a very low or negligible level of entrainment.

3. *Energy Efficiency*. Evaporators should be designed to make the best use of available energy, which implies using the lowest or the most economical net energy input. Steam-heated evaporators, for example, are rated on steam economy—pounds of solvent evaporated per pound of steam used [2].

The process scheme or flow sheet is a basis for understanding evaporation and what an evaporator does. Since the purpose of an evaporator is to concentrate a dilute feed stream and to recover a relatively pure solvent, this separation step must be defined. Figure 12.1 is a model for any evaporator, whether a simple one-pass unit or a complex multiple-effect evaporation system, which considers only the initial state of the feed system and the terminal conditions of the overhead and bottoms streams. The model assumes: steady-state conditions for all flow rates, compositions, temperatures, pressures, etc.; negligible entrainment of nonvolatile or solid particulates into the overhead, and no chemical reactions or changes in the chemical constituents during the evaporation process.

Example: In the production of Vitamin C, a feed stream containing monoacetone sorbose (MAS), organic salts, and water is to be concentrated. The feed rate is 4,000 lb/hr, and contains 30% by weight water. If the desired bottoms product is 97% solids, how much water is evaporated?

	Feed	Bottoms	Distillate
Water, lb/hr	1,200	87	1,113
MAS and solids, lb/hr	2,800	2,800	None
Total, lb/hr	4,000	2,887	1,113

Usually, a process flow sheet is given which includes much important design information for the complete process. This basic resource document is the key reference for the overall material balance for the process, and includes mass flow rates and complete chemical compositions for every stream in the process network. Other data usually included in the process flow sheet are: temperature and pressure for every process stream, important physical and thermodynamic properties for each stream, identification numbers and abbreviations for each equipment component, and identification and information for every addition and removal of energy or work for the process.

A standard "Heat Exchanger Specification Sheet" is used to specify the evaporator in sufficient detail so that prospective vendors may understand the application and develop a firm quotation. The Tubular Exchanger Manufacturers Association (TEMA) has developed the specification sheet shown in Figure 12.2, which is widely used by engineering and design firms and by heat exchanger and evaporator fabricators [3].

The input data that is needed to complete the heat exchanger specification sheet for an evaporation system can be grouped together in three categories:

Process variables: material balance and flow rates, operating pressure, operating temperature, heating medium temperature, and flow rate.

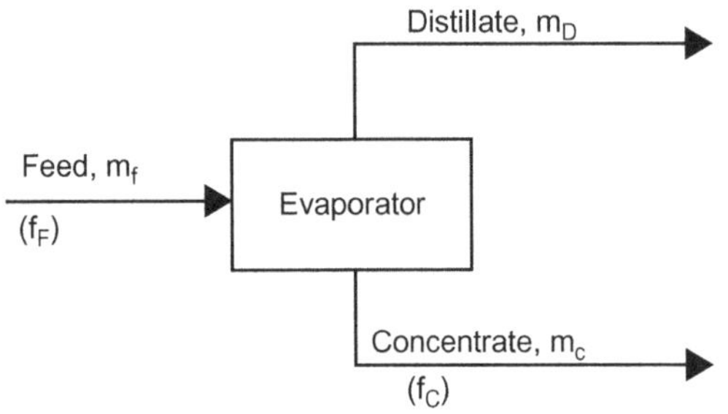

	Feed	Concentrate	Distillate
Water	$(1 - f_F)\, m_F$	$(1 \frac{1}{f_C} - 1)\, f_F m_F$	$(1 - \frac{f_E}{f_C}) m_F$
Solids	$f_F m_F$	$f_F m_F$	–
Total	m_F	$(\frac{f_E}{f_C}) m_F$	$(1 - \frac{f_F}{f_C}) m_F$

Boundary conditions: $0.0 < m_F$
$0.0 < f_C \leq 1.0$

FIGURE 12.1 Model and material balance for evaporators. (*Luwa Corporation.*)

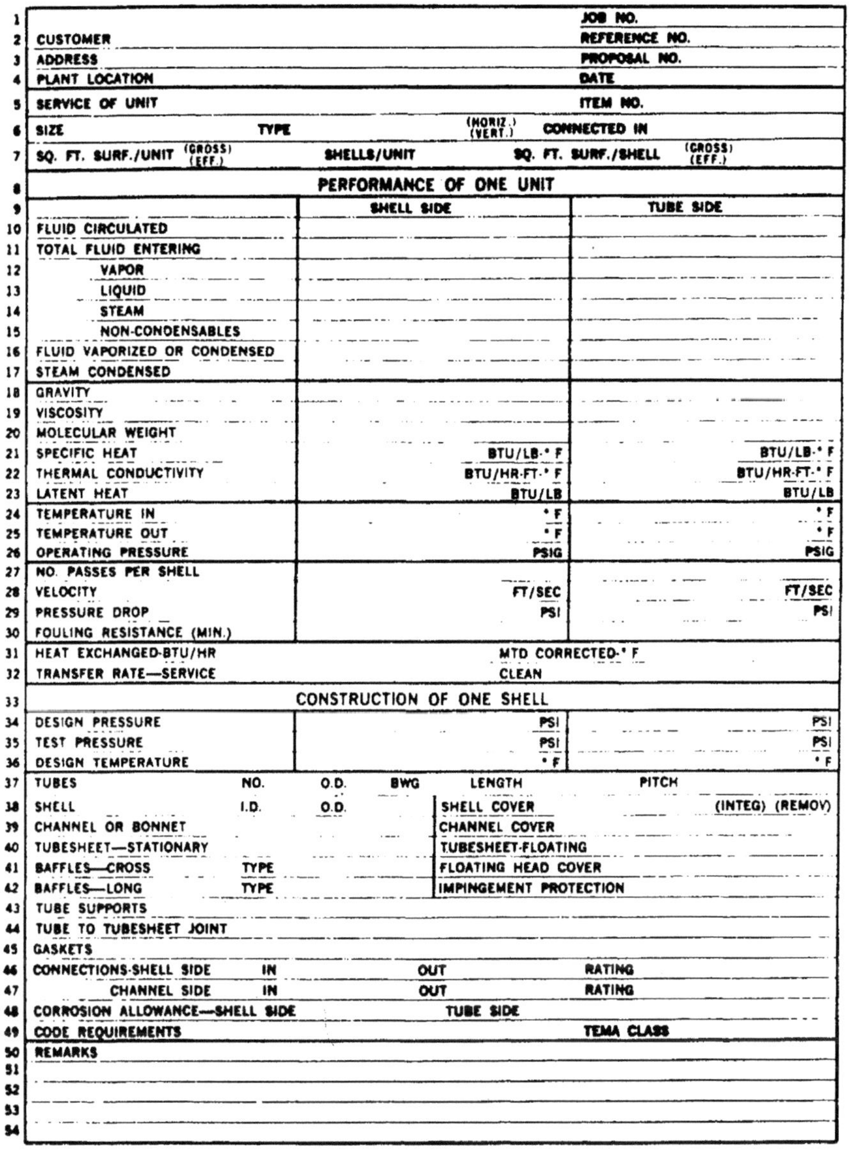

1			JOB NO.	
2	CUSTOMER		REFERENCE NO.	
3	ADDRESS		PROPOSAL NO.	
4	PLANT LOCATION		DATE	
5	SERVICE OF UNIT		ITEM NO.	
6	SIZE	TYPE	(HORIZ.) (VERT.)	CONNECTED IN
7	SQ. FT. SURF./UNIT (GROSS) (EFF.)	SHELLS/UNIT	SQ. FT. SURF./SHELL (GROSS) (EFF.)	
8	PERFORMANCE OF ONE UNIT			
9		SHELL SIDE	TUBE SIDE	
10	FLUID CIRCULATED			
11	TOTAL FLUID ENTERING			
12	VAPOR			
13	LIQUID			
14	STEAM			
15	NON-CONDENSABLES			
16	FLUID VAPORIZED OR CONDENSED			
17	STEAM CONDENSED			
18	GRAVITY			
19	VISCOSITY			
20	MOLECULAR WEIGHT			
21	SPECIFIC HEAT	BTU/LB·°F	BTU/LB·°F	
22	THERMAL CONDUCTIVITY	BTU/HR-FT·°F	BTU/HR-FT·°F	
23	LATENT HEAT	BTU/LB	BTU/LB	
24	TEMPERATURE IN	°F	°F	
25	TEMPERATURE OUT	°F	°F	
26	OPERATING PRESSURE	PSIG	PSIG	
27	NO. PASSES PER SHELL			
28	VELOCITY	FT/SEC	FT/SEC	
29	PRESSURE DROP	PSI	PSI	
30	FOULING RESISTANCE (MIN.)			
31	HEAT EXCHANGED-BTU/HR		MTD CORRECTED-°F	
32	TRANSFER RATE—SERVICE		CLEAN	
33	CONSTRUCTION OF ONE SHELL			
34	DESIGN PRESSURE	PSI	PSI	
35	TEST PRESSURE	PSI	PSI	
36	DESIGN TEMPERATURE	°F	°F	
37	TUBES	NO. O.D. BWG	LENGTH	PITCH
38	SHELL	I.D. O.D.	SHELL COVER	(INTEG) (REMOV)
39	CHANNEL OR BONNET		CHANNEL COVER	
40	TUBESHEET—STATIONARY		TUBESHEET-FLOATING	
41	BAFFLES—CROSS	TYPE	FLOATING HEAD COVER	
42	BAFFLES—LONG	TYPE	IMPINGEMENT PROTECTION	
43	TUBE SUPPORTS			
44	TUBE TO TUBESHEET JOINT			
45	GASKETS			
46	CONNECTIONS-SHELL SIDE	IN	OUT	RATING
47	CHANNEL SIDE	IN	OUT	RATING
48	CORROSION ALLOWANCE—SHELL SIDE		TUBE SIDE	
49	CODE REQUIREMENTS		TEMA CLASS	
50	REMARKS			
51				
52				
53				
54				

FIGURE 12.2 Heat exchanger specification sheet. *(© 1978 by* Tubular Exchange Manufacturers Association, *all rights reserved.)*

Physical property data: specific gravities, viscosity-temperature relationships, molecular weights, and thermodynamic properties.

Mechanical design variables: pressure drop limitations, corrosion allowances, materials of construction, fouling factors, code considerations (ASME, TEMA, etc.).

3.0 LIQUID CHARACTERISTICS

The properties of the liquid feed and the concentrate are important factors to consider in the engineering and design of an evaporation system. The liquid characteristics can greatly influence, for example, the choice of metallurgy, mechanical design, geometry, and type of evaporator [4]. Some of the most important general properties of liquids that can affect evaporator design and performances are:

Concentration—Most dilute aqueous solutions have physical properties that are approximately the same as water. As the concentration increases, the solution properties may change rapidly. Liquid viscosity will increase dramatically as the concentration approaches saturation and crystals begin to form. If the concentration is increased further, the crystals must be removed to prevent plugging or fouling of the heat transfer surface. The boiling point of a solution may rise considerably as the concentration progresses.

Foaming—Some materials, particularly certain organic substances, may foam when vapor is generated. Stable foams may be carried out with the vapor and, thus, cause excessive entrainment. Foaming may be caused by dissolved gases in the liquor, by an air leak below the liquid level, and by the presence of surface-active agents or finely divided particles in the liquor. Antifoaming agents, may suppress foams by operating at low liquid levels, by mechanical methods, or by hydraulic methods.

Temperature Sensitivity—Many fine chemicals, food products, and pharmaceuticals can be degraded when exposed to only moderate temperatures for relatively brief time periods. When processing or handling heat sensitive compounds, special techniques may be needed to regulate the temperature/time relationship in the evaporation system.

Salting—Salting refers to the growth on evaporator surfaces of a material having a solubility that increases with increasing temperature. It can be reduced or eliminated by keeping the evaporating liquid in close or frequent contact with a large surface area of crystallized solid.

Scaling—Scaling is the growth or deposition on heating surfaces of a material which is either insoluble, or has a solubility that decreases with temperature. It may also result from a chemical reaction in the evaporator. Both scaling and salting liquids are usually best handled in an evaporator that does not rely upon boiling for operation.

Fouling—Fouling is the formation of deposits other than salt or scale. Fouling may be due to corrosion, solid matter entering with the feed, or deposits formed on the heating medium side.

Corrosion—Corrosion may influence the selection of the evaporator type, since expensive materials of construction usually dictate that evaporator designs allowing high rates of heat transfer are more cost effective. Corrosion and erosion are frequently more severe in evaporators than in other types of equipment, because of the high liquid and vapor velocities, the frequent presence of suspended solids, and the high concentrations encountered.

Product Quality—Purity and quality of the product may require low holdup and low temperatures, and can also determine that special alloys or other materials be used in the construction of the evaporator. A low holdup or residence time requirement can eliminate certain types of evaporators from consideration.

Other characteristics of the solid and liquid may need to be considered in the design of an evaporation system. Some examples are: specific heat, radioactivity, toxicity, explosion hazards, freezing point, and the ease of cleaning. Salting, scaling, and fouling result in steadily diminishing heat transfer rates, until the evaporator must be shut down and cleaned. While some deposits can be easily cleaned with a chemical agent, it is just as common that deposits are difficult and expensive to remove, and that time-consuming mechanical cleaning methods are required.

4.0 HEAT TRANSFER IN EVAPORATORS

Whenever a temperature gradient exists within a system, or when two systems at different temperatures are brought into contact, energy is transferred. The process by which the energy transport takes place is known as *heat transfer*. Because the heating surface of an evaporator represents the largest portion of the evaporator cost, heat transfer is the most important single factor in the design of an evaporation system. An index for comparing different types of evaporators is the ratio of heat transferred per unit of time per unit of temperature difference per dollar of installed cost. If the operating conditions are the same, the evaporator with the higher ratio is the more "efficient."

Three distinctly different modes of heat transmission are: conduction, radiation, and convection. In evaporator applications, radiation effects can generally be ignored. Most usually, heat (energy) flows as a result of several or all of these mechanisms operating simultaneously. In analyzing and solving heat transfer problems, it is necessary to recognize the modes of heat transfer that play an important role, and to determine whether the process is steady state or unsteady-state. When the rate of heat flow in a system does not vary with time (i.e., is constant), the temperature at any point does not change and steady-state conditions prevail. Under steady-state conditions, the rate of heat input at any point of the system must be exactly equal to the rate of heat output, and no change in internal energy can take place. The majority of engineering heat transfer problems are concerned with steady-state systems.

The heat transferred to a fluid, which is being evaporated, can be considered separately as sensible heat and latent (or "change of phase") heat. Sensible heat operations involve heating or cooling of a fluid in which the heat transfer results only in a temperature change of the fluid. Change-of-phase heat transfer in an evaporation system involves changing a liquid into a vapor or changing a vapor into a liquid, i.e., vaporization or condensation. Boiling or vaporization is a convection process involving a change in phase from liquid to vapor. Condensation is the convection process involving a change in phase from vapor to liquid. Most evaporators include both sensible heat and change-of-phase heat transfer.

Energy is transferred due to a temperature gradient within a fluid by convection; the flow of energy from the heating medium, through the heat surface of an evaporator and to the process fluid occurs by conduction. Fourier observed that the flow or transport of energy was proportional to the driving force and inversely proportional to the resistance [5].

$$\text{Flow} = f(\text{potential} \div \text{resistance})$$

Conductance is the reciprocal of resistance and is a measure of the ease with which heat flows through a homogeneous material of thermal conductivity k.

$$\text{Flow} = f(\text{potential} \times \text{conductance})$$

A potential or driving force in a process heat exchanger or evaporator is a local temperature difference, ΔT. Figure 12.3 illustrates an example of conduction through composite walls or slabs having different thickness and composition. The conductance, also known as the *wall coefficient*, is given by: $h_w = k/x_w$ (e.g. Btu/hr ft^2 °F) [6]. By selecting a conducting material, such as copper or carbon steel, which has a relatively high value of thermal conductivity, and by designing a mechanically rigid but thin wall, the wall coefficient could be large. Fouling problems at surfaces x_0 and x_3 must be understood and accounted for. A stagnant oil film or a deposit of inorganic salts must be treated as a composite wall, too, and can seriously reduce the performance of an evaporator or heat exchanger over time. This phenomenon has been accounted for in good evaporator design practice by assigning a *fouling factor, f*, for the inside surface and the outside surface based upon experience [7]. The *fouling coefficient* is the inverse of the fouling factor:

$$h_{fo} = 1/f_o \text{ outside fouling coefficient}$$
$$h_{fi} = 1/f_i \text{ inside fouling coefficient}$$

Note that the bulk fluid temperatures (designated T_a and T_b in Figure 12.3) are different than the wall or skin temperatures (T_0 and T_3). Minute layers of stagnant fluid adhere to the barrier surfaces and contribute to relatively important resistances that are incorporated into a *film coefficient*.

$$h_o = \text{outside film coefficient}$$
$$h_i = \text{inside film coefficient}$$

The magnitude of these coefficients is determined by physical properties of the fluid and by fluid dynamics, the degree of turbulence known as the Reynolds number or its equivalent. Heat transfer within a fluid, due to its motion, occurs by convection; fluid at the bulk temperature comes in contact with fluid adjacent to the wall. Thus, turbulence and mixing are important factors to be considered, even when a change in phase occurs as in condensing steam or a boiling liquid.

The development of heat transfer equations for the tubular surface in Figure 12.4 is similar to that for the composite walls of Figure 12.3 except for geometry. It is quite important to differentiate between the inner surface area of the tubing and the outer surface area, which could be considerably greater, particularly in the case of a well-insulated pipe or a thick-walled heat exchanger tubing. Unless otherwise specified, the area A, used in determining evaporator sizes or heat transfer coefficients, is the surface through which the heat flows, measured on the process or inside surface of the heat exchanger tubing.

The derivation of specific values for the inside and outside film coefficients, h_i and h_o, is a rather involved procedure requiring a great deal of applied experience and the use of complex mathematical equations and correlations; these computations are best left to the staff heat transfer specialist, equipment vendor, or a consultant. Listed are four references that deal specifically with evaporation and the exposition and use of semi-empirical equations for heat transfer coefficients [8–11].

If steady state conditions exist (flow rates, temperatures, composition, fluid properties, pressures), Fourier's equation applies to macro-systems in which energy is transferred across a heat exchanger or an evaporator surface:

$$Q = UA\Delta T$$

The term U is known as the overall heat transfer coefficient and is defined by the following equation:

$$1/U = 1/h_o + 1/h_{fo} + 1/h_w + 1/h_{fi} + 1/h_i$$

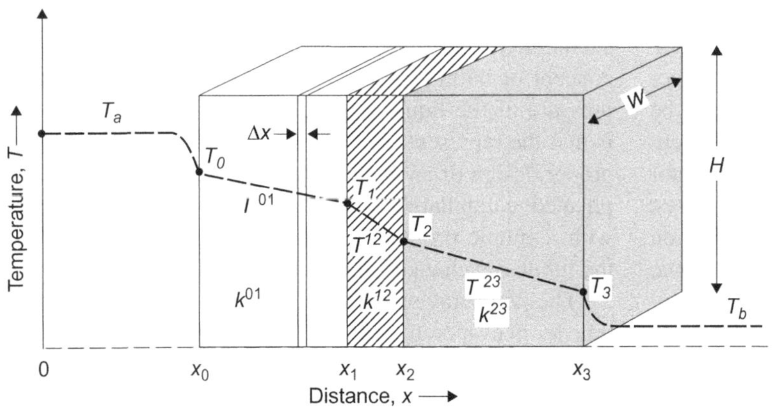

FIGURE 12.3 Heat conduction through a composite wall, placed between two fluid streams T_a and T_b. *(From* Transport Phenomena *by R. B. Bird, W. E. Stewart, and E. N. Lightfoot, 1960, p. 284. Used with permission of John Wiley & Sons, Inc.)*

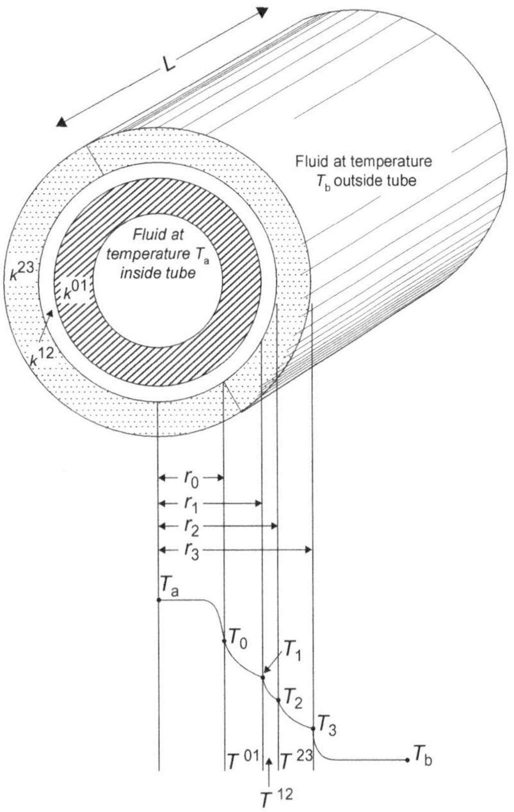

FIGURE 12.4 Head conduction through a laminated tube with fluid at temperature T_a inside and fluid temperature T_b outside. *(From* Transport Phenomena *by R. B. Bird, W. E. Stewart, and E. N. Lightfoot, p. 287, John Wiley & Sons, Inc., 1960. Used with permission.)*

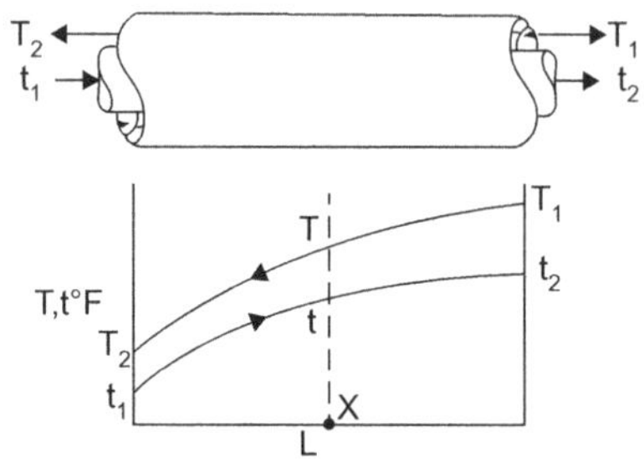

$$\Delta t = \text{LMTD} = \frac{(T_1 - t_2) - (T_2 - t_1)}{\ln (T_1 - t_2)/(T_2 - t_1)} = \frac{\Delta t_2 - \Delta t_1}{\ln \Delta t_2/\Delta t_1}$$

FIGURE 12.5 Logarithmic mean temperature difference in a counterflow heat exchanger with no phase changes. *(Luwa Corporation.)*

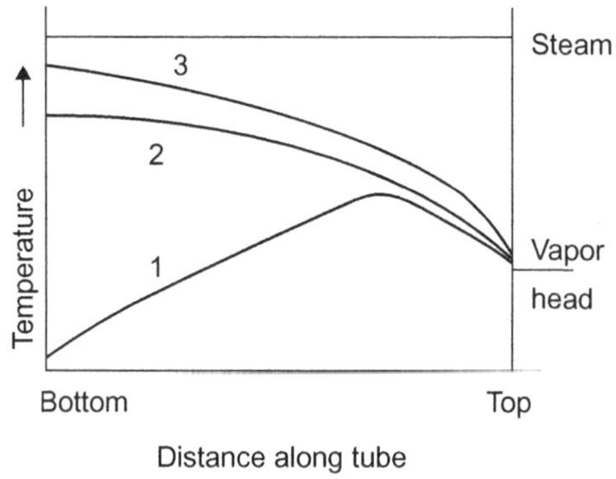

FIGURE 12.6 Temperature variations in a long-tube vertical evaporator. *(1)* Feed not boiling at tube inlet. *(2)* Feed enters at boiling point. *(3)* Same as curve 2, but feed contains 0.01% surface active agent. *(From* Chemical Engineers' Handbook, *edited by R. H. Perry and C. H. Chilton, 5th ed., p. 11–29. ©1973, McGraw-Hill. Used with permission.)*

Example:

$h_o = 1000$ Btu/hr ft^2 °F	Condensing steam
h_{fo} = very large	Clean steam
$h_w = 39{,}000$	1" #16 BWG copper tubing
$h_{fi} = 500$	Inside fouling coefficient
$h_i = 600$	Aqueous solution of inorganic salt

$$U = (0.001 + 0.0 + 0.003 + 0.002 + 0.0016)^{-1}$$
$$= 213 \text{ Btu/hr ft}^2 \text{ °F}$$

The evaporator design engineer determines the heat load, Q, and the driving force, ΔT, from the Heat Exchanger Specification Sheet. If an overall coefficient, U, can be obtained from operating or pilot plant data (or can be calculated, as in the example above), the required evaporator surface area, A, can be obtained. In most types of evaporators, the overall heat transfer coefficient can be a strong function of the temperature difference, ΔT. Because the driving force is not constant at every point along a heat exchanger or evaporator surface, a LMTD (Log Mean Temperature Difference) and LMTD correction factors are used in the Fourier equation to represent ΔT. Figure 12.5 shows how the LMTD can be calculated using terminal temperatures (i.e., inlet and outlet temperatures) for a heat exchanger in the simple case where no change of phase occurs.

In a steam-heated evaporator, both the heating medium and the process fluid undergo a phase change and most of the energy transferred is latent heat. Some sensible heat may be involved if the feed stream is to be preheated and if the condensate undergoes some sub cooling. Further, some types of evaporators (for example, a submerged tube forced-circulation evaporator) involve the concept of boiling point elevation, due to the hydrostatic pressure of the liquid phase. The point to be emphasized is that the representative driving force, ΔT, utilized in the proper design of an evaporator involves some rather complicated computations and correction factors, compared with a simple problem of the transfer of sensible heat in the tubular exchanger illustrated in Figure 12.5.

The temperature difference used in computing heat transfer in evaporators is usually an arbitrary figure, since it is really quite impossible to determine the temperature of the liquid at all positions along the heating surface (for example, see Figure 12.6). The condensing temperature of

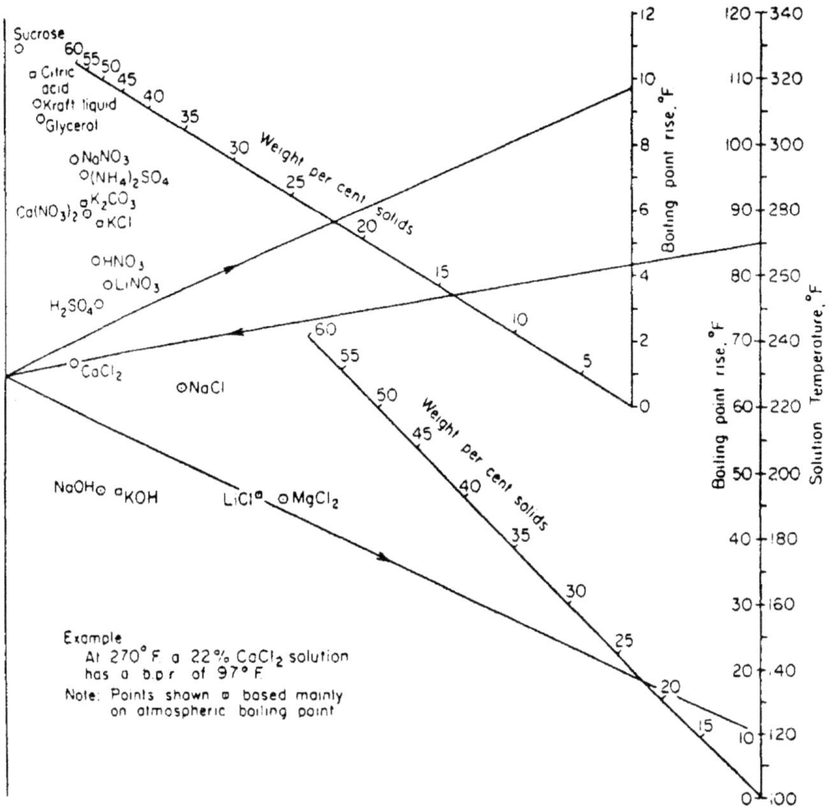

FIGURE 12.7 Boiling-point rise of aqueous solutions. *(From* Chemical Engineers' Handbook, *edited by R.H. Perry and C. H. Chilton, 5th ed., p. 11–31. ©1973, McGraw-Hill. Used with permission.)*

steam, the more common heating medium, can usually be determined simply and accurately from a measurement of pressure in the steam side of the heating element, together with use of the steam tables. In a similar manner, a pressure measurement in the vapor space above the boiling liquid will give the saturated vapor temperature which, assuming a negligible boiling-point rise, would be substantially the same as the boiling liquid temperature. Temperature differences calculated on the basis of this assumption are called *apparent temperature differences* and heat-transfer coefficients are called *apparent coefficients*.

Boiling-point rise is the difference between the boiling point of a solution and the boiling point of water at the same pressure. Figure 12.7 can be used to estimate the boiling-point rise for a number of common aqueous solutions. When the boiling-point rise is deducted from the apparent temperature difference, the terms *temperature difference corrected for boiling-point rise* and *heat-transfer coefficient corrected for boiling-point rise* are used. This is the most common basis of reporting evaporator heat transfer data, and is the basis understood in the absence of any qualifying statement [12, pp. 11–31].

5.0 EVAPORATOR TYPES

Most evaporators consist of three main elements or parts: a heating unit (*calandria*), a region for liquid-vapor separation (sometimes called a *vapor head, flash chamber*, or *settling zone*), and a structural body to house these elements and to separate the process and heating fluids. One simple way to classify evaporators is:

1. Heating medium separated from evaporating liquid by tubular heating surfaces
2. Heating medium confined by coils, jackets, double walls, flat plates, etc.
3. Heating medium brought into direct contact with evaporating liquid (e.g., a submerged combustion evaporator)
4. Heating with solar radiation [12, pp. 11–27]

By far, most evaporators used in the process industries fall into the first category, having tubular heating surfaces. In the natural circulation evaporators, movement of liquid across the heating surface is induced by the boiling process itself, the two-phase mixture of liquid and vapor being less dense than the column of liquid behind it, which pushes it forward and upward. For some thicker fluids, liquids with a high solids content, or liquids which have a tendency to react or foul on a heated surface, a forced circulation evaporator may be a better choice; a centrifugal pump circulates liquid through a loop around the heating unit at a much higher velocity than is possible in a natural circulation evaporator.

Evaporators can be designed to operate batch wise, continuously or in a semi-batch or *campaign* fashion, but once an evaporator system is designed to operate in one of these modes, it is not easy to change from one type of operation to another from the standpoint of available hardware and process instrumentation.

The specialty evaporators make up the second classification of evaporator types. These are generally much smaller and simpler than the tubular evaporation systems, and are often batch or multipurpose evaporators. The third group is a unique classification and the direct-fired, submerged combustion evaporator is the best example of this type.

The last classification includes the solar evaporation system, the oldest evaporation principle employed by man and, in concept, the simplest evaporation technique. Solar evaporators require tremendous land areas and a relatively cheap raw material, since pond leakage may be appreciable. Solar evaporation generally is feasible only for the evaporation of natural brines, and then only when the water vapor is evaporated into the atmosphere and is not recovered.

Evaporators may be operated either as *once-through* units, or the liquid may be recirculated through the heating elements. In once-through operation, all the evaporation is accomplished in a single pass. The ratio of evaporation to feed is limited in single pass operation; single pass evaporators are well adapted to multiple-effect operation, permitting the total concentration of the liquid to be achieved over several effects. Mechanically agitated thin-film evaporators are generally operated once through. Once-through evaporators are also frequently required when handling heat-sensitive materials.

Recirculated systems require that a pool of liquid be held within the equipment. Feed mixes with the pooled liquid and the mixture circulates across the heating element. Only part of the liquid is vaporized in each pass across the heating element; unevaporated liquid is returned to the pool. All the liquor in the pool is therefore at the maximum concentration. Circulatory systems are therefore not well suited for evaporating heat sensitive materials. Circulatory evaporators, however, can operate over a wide range of concentrations and are well adapted to single-effect evaporation.

There is no single type of evaporator that is satisfactory for all conditions. It is for this reason that there are many varied types and designs. Several factors determine the application of a particular type for a specific evaporation result. The following sections will describe the various types of evaporators in use today and will discuss applications for which each design is best adapted.

A number of different evaporator designs are illustrated in Figure 12.8, and the variations based upon these concepts are many. Often, physical properties and materials handling considerations for the feed or the bottom streams (e.g., solids content, viscosity, heat sensitivity) will indicate that one evaporator type will be better suited for the duty than other types [13, pp. B-48].

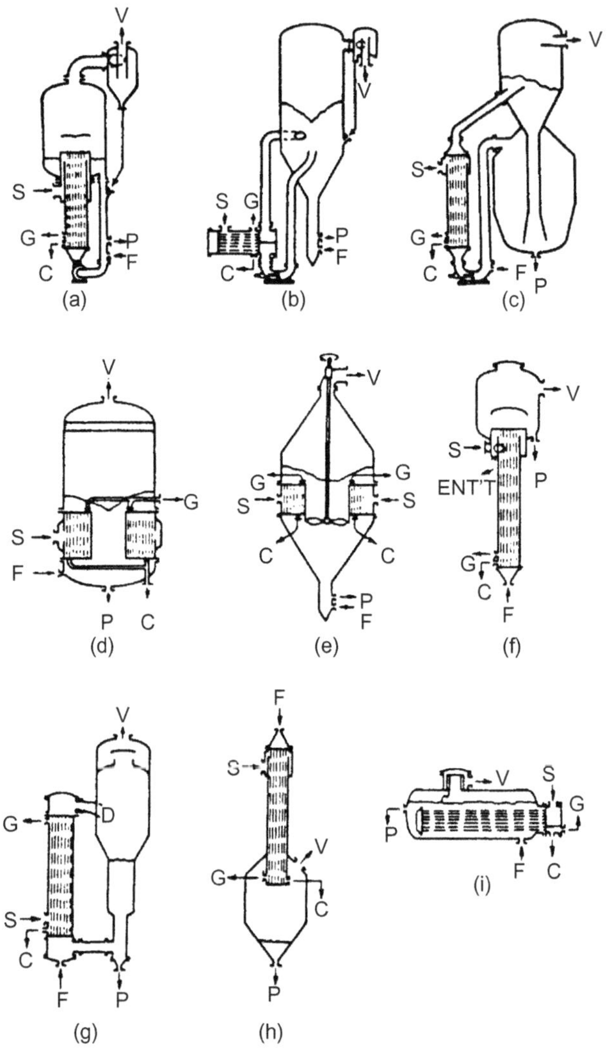

FIGURE 12.8 Evaporator types. *(a)* Forced circulation. *(b)* Submerged-tube forced circulation. *(c)* Oslo-type crystallizer. *(d)* Short-tube vertical. *(e)* Propeller calandria. *(f)* Long-tube vertical. *(g)* Recirculating long-tube vertical. *(h)* Falling film. *(i)* Horizontal-tube evaporator. C, condensate; F, feed; G, vent; P, product; S, steam; V, vapor, ENTT, separated entrainment outlet. *(From* Chemical Engineers' Handbook, *edited by R. H. Perry and C. H. Chilton, 5th ed., p. 11–28, ©1973, McGraw-Hill. Used with permission.)*

5.1 Jacketed Vessels

When liquids are to be evaporated on a small scale, the operation is often accomplished in some form of jacketed tank or kettle. This may be a batch or continuous operation. The rate of heat transfer is generally lower than for other types of evaporators and only a limited heat transfer area is available. The kettles may or may not be agitated.

Jackets may be of several types: dimpled jackets, patterned plate jackets, and half-pipe coil jackets. Jacketed evaporators are used when the product is somewhat viscous, the batches are small, good mixing is required, ease of cleaning is important, or when glass-lined steel equipment is required.

5.2 Horizontal Tube Evaporators

The earliest fabricated evaporator designs incorporated horizontal tubes. A vertical tank-like cylinder housed a horizontal tube bundle in the lower portion of the vessel, and the vapor space above the tubes served to separate the entrained liquid from the rising vapors. A later design based on a horizontal body and a removable U-type bundle is illustrated in Figure 12.8(i). Another modification, the *kettle type* re-boiler, is similar and is more often employed as a bottoms heater for a distillation column than as an evaporator.

Initial investment for horizontal tube evaporators is low, but heat transfer rates may also be relatively low. They are well suited for non-scaling, low viscosity liquids. For several scaling liquids, scale can sometimes be removed from bent-tube designs by cracking it off periodically by shock cooling with cold water; or, removable bundles can be used to confine the scale to that part of the heat transfer surface, which is readily accessible.

Horizontal tube evaporators may be susceptible to vapor binding, and foaming liquids cannot usually be handled. The short tube variety is seldom used today except for preparation of boiler feed water. The kettle-type reboiler is frequently used in chemical plant applications for clean fluids.

The advantages of horizontal tube evaporators include relatively low cost for small-capacity applications, low headroom requirements, large vapor-liquid disengaging area, relatively good heat transfer with proper design, and the potential for easy semiautomatic descaling. Disadvantages include the limitations for use in salting, or scaling applications, generally.

5.3 Short-Tube Vertical Evaporators

The short-tube vertical evaporator, Figure 12.8(d), also known as the calandria or Robert evaporator, was the first evaporator to be widely used. Tubes 4′ and 8′ long, often 2″ to 3″ in diameter, are located vertically inside a steam chest enclosed by a cylindrical shell. The early vertical tube evaporators were built without a *downcomer* but did not perform satisfactorily, so the central downcomer appeared very early. There are many alternatives to the center downcomer; different cross-sections, eccentrically located downcomers, a number of downcomers scattered over the tube layout, downcomers external to the evaporator body.

The short-tube evaporator has several advantages: low headroom, high heat transfer rates at high temperature differences, ease of cleaning, and low initial investment. Disadvantages include large floor space and weight, relatively high liquid holdup, and poor heat transfer with low temperature differences or with high product viscosity. Natural circulation systems are not well suited for operation at high vacuum. Short-tube vertical evaporators are best applied when evaporating clear liquids, mild scaling liquids requiring mechanical cleaning, crystalline product when propellers are used, and for some foaming products when inclined calandrias are used. Once considered "standard," long tube vertical units have largely replaced short tube vertical evaporators.

Circulation of liquid across the heating surface is caused by the action of the boiling liquid (natural circulation). The circulation rate through the evaporator is many times the feed rate. The downcomers are therefore required to permit the liquid to flow freely from the top tube sheet to the bottom tube sheet. The downcomer flow area is, generally, approximately equal to the tubular cross-sectional area. Downcomers should be sized to minimize holdup above the tube sheet in order to improve heat transfer, fluid dynamics, and minimize foaming. For these reasons, several smaller downcomers scattered about the tube nest are often the better design.

5.4 Propeller Calandrias

Natural circulation in the standard short tube evaporator depends upon boiling. Should boiling stop, any solids suspended in the liquid phase will settle out. The earliest type of evaporator that could be called a forced-circulation device is the propeller calandria illustrated in Figure 12.8(e). Basically a standard evaporator with a propeller added in the downcomer, the propeller calandria often achieves higher heat transfer rates. The propeller is usually placed as low as possible to avoid cavitation and is placed in an extension of the downcomer. The propeller can be driven from above or below. Improvements in propeller design have permitted longer tubes to be incorporated in the evaporator.

5.5 Long-Tube Vertical Evaporators

More evaporator systems employ this type of design than any other because it is so versatile and is often the lowest cost per unit of capacity. Long-tube evaporators normally are designed with tubes 1″ to 2″ in diameter and from 12′ to 30′ in length. A typical long-tube evaporator is illustrated in Figure 12.8(f). Long-tube units may be operated as once through or as recirculating evaporation systems.

A once-through unit has no liquid level in the vapor body, tubes are 16′ to 30′ long, and the average residence time is only a few seconds. With recirculation, a level must be maintained, a deflector plate is often provided in the vapor body, and tubes are 12′ to 20′ long. Recirculated systems can be operated either batch wise or continuously.

Circulation of fluid across the heat transfer surface depends upon boiling and the high vapor velocities associated with vaporization of the liquid feed. The temperature of the liquid in the tubes is far from uniform and relatively difficult to predict. These evaporators are less sensitive to changes in operating conditions at high temperature differences than at lower temperature differences. The effects of hydrostatic head upon the boiling point are quite pronounced for long-tube units.

The long-tube evaporator is often called a *rising* or *climbing film evaporator* because vapor travels faster than the liquid upward through the core of the tube, therefore dragging the liquid up the tube in a thin film. This type of flow can occur only in the upper portion of the tube. When it occurs, the liquid film is highly turbulent and high heat transfer rates are realized. Average residence times are low, so long-tube vertical evaporators can be utilized for heat sensitive materials.

The long-tube vertical evaporator offers several advantages: low cost, large units, low holdup, small floor space, good heat transfer over a wide range of applications. Disadvantages include: high headroom is needed, recirculation is frequently required, and they are generally unsuited for salting or severely scaling fluids. They are best applied when handling clear fluids, foaming liquids, corrosive liquids, and large evaporation loads.

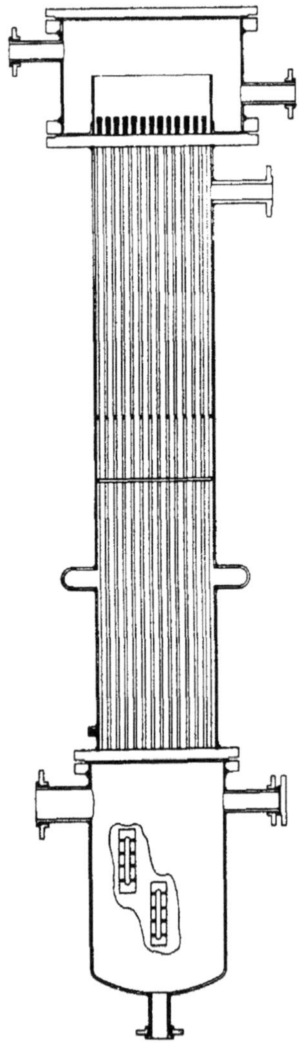

FIGURE 12.9 Falling film evaporator for urea concentration; bottom vapor takeoff. (*Henry Vogt Machine Company.*)

5.6 Falling Film Evaporators

Falling film evaporators, Figure 12.8(h), are long-tube vertical evaporators that rely upon gravity flow of a thin fluid layer from the top of the tubes, where the liquid is introduced, to the bottom of the unit where the concentrate is collected. Evaporation takes place on the surface of the falling liquid film that is highly turbulent. The fluid pressure drop across the process side of a falling film evaporator or re-boiler system is very low and usually negligible, due to the gravity flow [13, pp. B-60]. Separation of entrained liquid from the vapor is usually accomplished in a chamber at the bottom of the tubes, although some units are designed so that the volatiles flow upward against the descending liquid film and are removed at the top of the unit.

Feed to a falling film evaporator is usually introduced under the liquid level maintained at the top of the tubes, so that a reservoir of rather low velocity liquid is available for liquid distribution to the many vertical tubes. In falling film evaporator and re-boiler design, equal fluid distribution among the tubes and film initiation are very important factors. For this reason, a number of sophisticated and very effective hydraulic distributing devices have been developed to handle different types of process fluids [14]. In order to achieve uniform liquid loading and evaporation rates in each tube, and to ensure that sufficient liquid is available in every tube to maintain the liquid film (thus avoiding dry or hot spots), particular attention must be paid to liquid distribution. Figure 12.9 is a cross section of a urea concentrator and Figure 12.10 illustrates some of the many tube distributors or ferrules that can be inserted into the flush upper end of the evaporator tubes.

Heat transfer rates in falling film evaporators are relatively high even at low temperature differences across the liquid film; thus, these evaporators are widely used for heat sensitive products because of uniform temperatures

FIGURE 12.10 Tube distributors for falling film evaporators. (*Henry Vogt Machine Company.*)

and short residence times. Generally, moderately viscous fluids and materials with mildly fouling characteristics can easily be handled in falling film evaporators in series for heavy evaporation loads, and part of the liquid can be pumped and recycled to the top of the unit.

The least expensive of the low residence time evaporators, falling film evaporators, offer many advantages, particularly for large volumes of dilute material. These advantages include: large unit sizes, low liquid holdup, small floor space, and good heat transfer over a wide range of conditions. Falling film units are well suited for heat sensitive materials or for high vacuum application, for viscous materials, and for low temperature differences. Occasionally, rising and falling film evaporators are combined into a single unit.

5.7 Forced Circulation Evaporators

Evaporators in which circulation is maintained, independent of the evaporation rate or heating temperature, through the heating element are known as *forced circulation evaporators*. Forced circulation systems are illustrated in Figures 12.8a,b. Forced circulation systems are more expensive than comparable natural circulation evaporators and are, therefore, used only when necessary.

A choice of a forced circulation evaporator can be made only after balancing the pumping cost, which is usually high, with the increase in heat transfer rates or decrease in maintenance costs. Tube velocity is limited only by pumping costs and by erosion at high velocities. Tube velocities are usually in the range of 5 to 15 feet per second. Sometimes the pumped fluid is allowed to vaporize in the tubes. This often provides high heat transfer rates, but increases the possibility of fouling. Consequently, this type of evaporator is seldom used except where headroom is limited or the liquids do not scale, salt, or foul the surface.

The majority of applications are designed so that vaporization does not occur in the heat exchanger tubes. Instead, the process liquid is recirculated by the pump, is heated under pressure to prevent boiling, and is subsequently flashed to obtain the required vaporization. This type of evaporator is often called the submerged-tube type because the heating element is placed below the liquid level and uses the resulting hydrostatic head to elevate the boiling point and to prevent boiling in the tubes. The heating element may be installed vertically (usually, single pass), or horizontally (often, two-pass as shown in Figure 12.8b).

The recirculation pump is a crucial component of the evaporation system, and the following key factors need to be considered when establishing the recirculation rate and the pump capacity:

1. Maximum fluid temperature permitted
2. Vapor pressure of the fluid
3. Equipment layout
4. Tube geometry
5. Velocity in the tubes
6. Temperature difference between the pumped fluid and the heating medium
7. Pump characteristics for the pumps being evaluated with the system

A recirculating pump should be chosen so that the developed head is dissipated as pressure drops through the circuit of the system. It is important that the pump and system be properly matched. The fluid being pumped is at or near its boiling point and, therefore, the required NPSH (net positive suction head) is usually critical. The pump should operate at this design level. If it develops excessive head, it will handle more volume at a lower head. At the new operating point, the required NPSH may be more than is available, and cavitation will occur in the pump. If insufficient head is provided, the velocities may not be sufficiently high to prevent fouling; lower heat transfer rates may result; or the fluid may boil in the heating element with subsequent fouling or decomposition.

Forced circulation evaporators offer these advantages: high rate of heat transfer; positive circulation; relative freedom from salting, scaling, and fouling; ease of cleaning; and a wide range of application. Disadvantages include: high cost; relatively high residence time; and the necessity for centrifugal or propeller pumps with associated maintenance and operating costs. Forced circulation evaporators are best applied when treating crystalline products, corrosive products, or viscous fluids. They are also

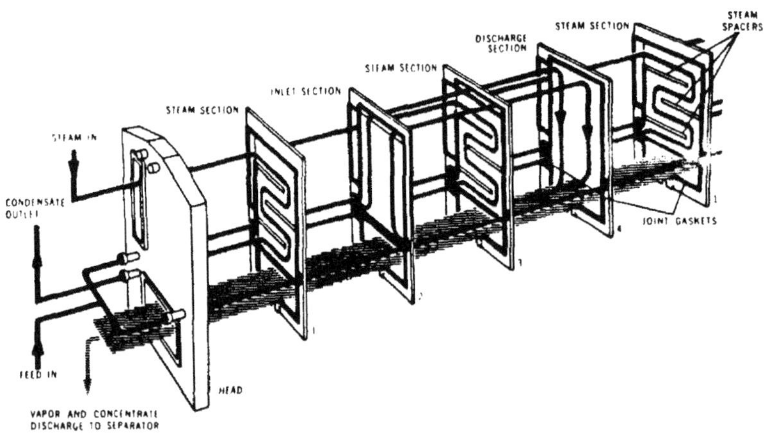

FIGURE 12.11 Plate evaporator, rising/falling film type. (*APV Company, Inc.*)

well suited for vacuum service, and for applications requiring a high degree of concentration and close control of bottoms product concentration.

5.8 Plate Evaporators

Plate evaporators may be constructed of flat plates or corrugated plates, the latter providing an extended heat transfer surface and improved structural rigidity. Two basic types of heat exchangers are used for evaporation systems: plate-and-frame and spiral-plate evaporators. Plate units are sometimes used because of the theory that scale will flake off such surfaces, which can flex more readily than curved tubular surfaces. In some plate evaporators, flat surfaces are used, each side of which can serve alternately as the liquor side and the steam side. Scale deposited while in contact with the liquor can then be dissolved while in contact with the steam condensate. There are still potential scaling problems, however. Scale may form in the valves needed for cycling the fluids and the steam condensate simply does not easily dissolve the scale produced.

A plate-and-frame evaporator, like the one illustrated in Figure 12.11, is so named because the design resembles that of a plate-and-frame filter press. This evaporator is constructed by mounting embossed plates with corner openings between a top carrying bar and a bottom guide bar. The plates are gasketed and arranged so narrow flow passages are formed when a series of plates are clamped together in the frame. Fluids pass through the spaces between the plates, either in series or parallel flow, depending on the gasketing which confines the fluids from the atmosphere.

Spiral-plate evaporators may be used instead of tubular evaporators, and offer a number of advantages over conventional shell-and-tube units: centrifugal forces improve heat transfer; the compact configuration results in a shorter liquid pathway; they are relatively easily cleaned and resistant to fouling; differential thermal expansion is accepted by the spiral arrangement. These curved-flow units are particularly useful for handling viscous material or fluids containing solids.

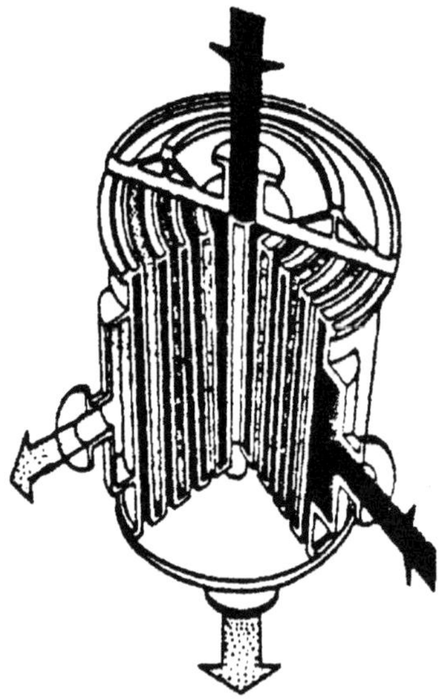

FIGURE 12.12 Spiral plate heat exchanger, both fluids with helical flow pattern. (*Graham Manufacturing Company, Inc.*)

A spiral-plate heat exchanger is constructed by winding two long strips of plate around an open, split center to form a pair of concentric spiral passages. Spacing is maintained along the length of the spiral by spacer studs welded to the plates. In some applications both fluid-flow channels are closed by welding alternate channels at both sides of the spiral plate (Figure 12.12). In other applications, one of the channels is left completely open and the other is closed at both sides of the place, Figure 12.13. These two types of constructions prevent the fluids from mixing.

The spiral heat exchanger can be fitted with covers to provide three flow patterns: *(i)* both fluids in spiral flow,

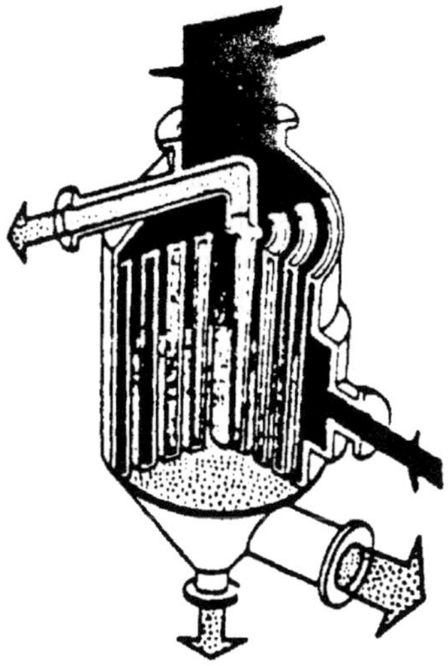

FIGURE 12.13 Spiral plate heat exchanger, one fluid in helical flow and one fluid in axial flow pattern. (*Graham Manufacturing Company, Inc.*)

(*ii*) one fluid in spiral flow and the other in axial flow across the spiral, (*iii*) one fluid in spiral flow and the other in combination of axial and spiral flow.

Plate type heat exchangers (see Figure 12.11) can be designed to operate as rising film, falling film, or rising-falling film evaporators. In some applications the rising and falling films are removed from the plate by the turbulence caused by extremely high vapor velocities. This action reduces the apparent viscosity and tends to minimize scaling.

The volume of product (*holdup*) in the evaporator is very small in relation to the large available heat transfer surface. Plate-and-frame evaporators can generally handle the evaporation of heat sensitive, viscous, and foaming materials. They permit fast start-up and shutdown and are quite compact, so little headroom is required. They are easily cleaned and readily modified.

A major concern is the need for gaskets and the large gasketed area. However, interleakage of fluids cannot occur without rupturing a plate, because all fluids are gasketed independently to seal against the atmosphere. Leakage can be avoided by selecting appropriate gasket materials and following proper assembly procedures [15].

5.9 Mechanically Agitated Thin-Film Evaporators

These evaporators, sometimes called *wiped-film* or *scraped-film evaporators*, rely on mechanical blades that spread the process fluid across the thermal surface of a single large tube (Figure 12.14), not unlike the wiper on the windshield of a car. All thin-film evaporators have essentially three major components: a vapor body assembly, a rotor, and a drive system [16].

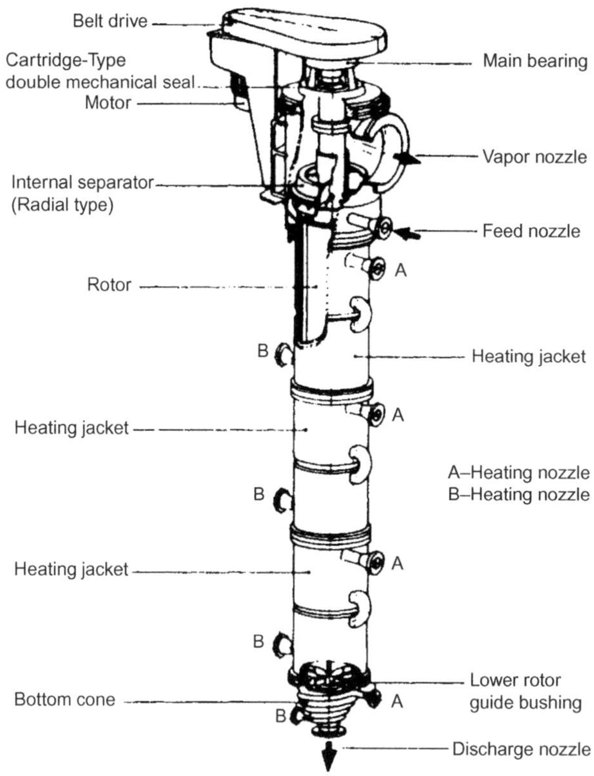

FIGURE 12.14 Mechanically agitated thin-film evaporator, vertical design with cylindrical thermal zone. (*Luwa Corporation.*)

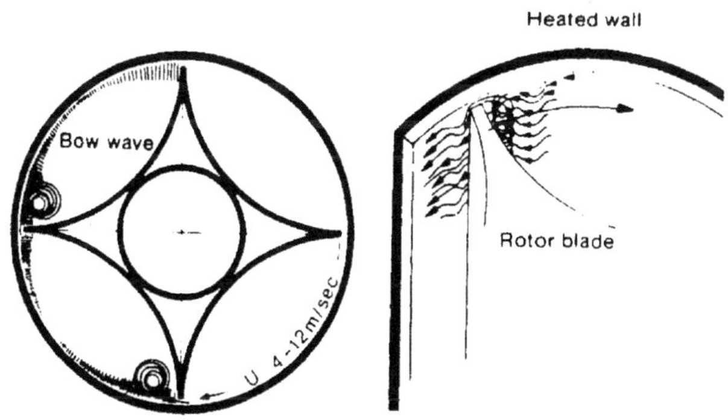

FIGURE 12.15 Distribution of liquid in mechanically agitated thin-film evaporator. (*Luwa Corporation.*)

In this thin-film evaporator design, product enters the feed nozzle above the heated zone and is mechanically transported by the rotor, and gravity, down a helical path on the inner heat transfer surface. The evaporator does not operate full of product; the liquid or slurry forms a thin film or annular ring of product from the feed nozzle to the product outlet nozzle as shown in the cross section of Figure 12.15. Holdup or inventory of product in a thin-film evaporator is very low, typically about a half a

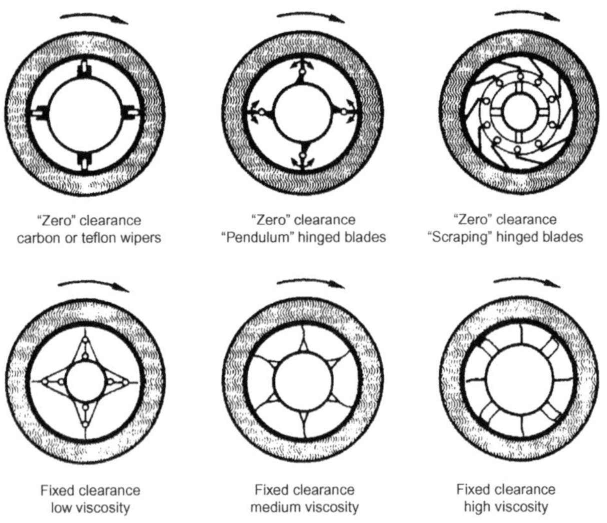

FIGURE 12.16 Six types of rotors for mechanically agitated thin-film evaporators; cross-sectional views. (*Luwa Corporation.*)

pound of material per square foot of heat transfer surface. The high blade frequency, about 8 to 10 blade passes per second, generates a high rate of surface renewal and highly turbulent conditions for even extremely viscous fluids. A variety of basic or standard thin-film evaporator designs are commercially available, including vertical or horizontal designs, and both types can have cylindrical or tapered thermal bodies and rotors.

The rotors may be one of several *zero-clearance* designs, a rigid fixed clearance type, or in the case of tapered rotors, an adjustable clearance construction type (Figure 12.16). One vertical design includes an optional *residence time control ring* at the end of the thermal surface to hold back product and thus build up the film thickness. The majority of thin-film evaporators in operation are the vertical design with a cylindrical fixed-clearance rotor shown in Figure 12.14 [17].

Mechanically agitated thin-film evaporators are used for four general types of applications:

1. Heat sensitive products
2. Fluids with fouling tendencies
3. Viscous materials
4. Liquids containing a large amount of dissolved or suspended solids

The one-pass, plug flow operation of a thin-film evaporator is an advantage for minimizing thermal degradation of a heat sensitive product in an evaporation step. The mean residence time in the evaporator can be just seconds, rather than minutes or hours in a recirculating evaporation system. For this reason, thin-film evaporators are widely used for heat sensitive food, pharmaceutical, and other chemical products. Also, it should be noted that the thin-film evaporator can be operated at a higher temperature to make a better separation, whereas care must usually be taken to keep the product temperature lower in an evaporation system with longer residence times (see Figure 12.21, later in this chapter).

Thin-film evaporators are frequently used for extremely viscous fluids, those in the range of 1,000 to 50,000 centipoise, and for concentrating streams with more than 25% suspended solids. Heat transfer coefficients for these types of materials in a thin-film evaporator are typically much greater than coefficients in any other type of evaporator for the same conditions. Very high temperature difference (e.g., 100 to 200°F) can be maintained to better utilize the heat transfer area by increasing the heat flux, Q/A.

These evaporators are necessarily precision machines and therefore are more expensive than other types, particularly so if compared strictly on equivalent heat transfer area. When the performance for a specific evaporation duty is the basis of comparison, the thin-film evaporator is often the more economical choice because the larger heat transfer coefficient and higher driving force mean much less surface is required than for other evaporators ($A = Q/U\Delta T$). Thin-film evaporator cost per unit area decreases significantly with unit size, and the largest typical unit has 430 square feet of active heat transfer surface [18].

5.10 Flash Pots and Flash Evaporators

The simplest continuous evaporation system is the single stage "flashing" of a heated liquid into an expansion tank or *flash pot* which is maintained at a lower pressure than the feed. The principle is that of an adiabatic (or isoenthalpic) expansion of a saturated liquid from a high pressure to a lower pressure, thus generating a mixture of saturated liquid and vapor with the same total enthalpy at the lower pressure.

First applied for production of distilled water on board ships, flash and multistage flash evaporators have been more recently utilized to evaporate brackish and sea water as well as for process liquids. An aqueous solution is heated and introduced into a chamber that is kept at a pressure lower than the corresponding saturation pressure of the heated feed stream. Upon entering the chamber, a small portion of the heated water will immediately "flash" into vapor, which is then passed through an entrainment separator to remove any entrained liquid and condense the water vapor. A series of these chambers can be maintained at successively lower pressures with vapor flashing at each stage. Such a system is called a *multistage flash evaporator.*

The flashing process can be broken down into three distinct operations: heat input, flashing and recovery, and heat rejection. The heat input section, commonly called a brine heater, normally consists of a tubular exchanger which transfers heat from steam, exhaust gas from a turbine, stack gases from a boiler, or almost any form of heat energy. The flashing and recovery sections consist of adequately sized chambers which allow the heated fluid to partially flash, thereby generating a mixture of vapor and liquid. The vapor produced in this process is passed through moisture separators and directed either to the heat recovery condensers (for multistage units) or to the third section, the reject condensers. Since the evaporator does no work, the heat reject sections receive essentially all of the energy supplied in the heat input section of the evaporator.

Usually, the three sections are combined into one package. In single stage flash evaporators, there are no regenerative stages to recover the energy of the flashed vapor. A multistage system extends the flashing and recovery zone by condensing the flashed vapor in each stage by heating the brine prior to the heat input zone. This reduces the amount of heat required for evaporation. The number of stages or flashes is determined by the economics of each installation. Until recently, flash evaporators were limited to "water poor" areas, where there was an abundance of relatively low cost fuel or energy. The flash evaporator is an extremely flexible system and can be made to operate with almost any form of heat energy. Proper instrumentation must be applied for multistage evaporators which incorporate a large number of stages. The interrelated variables of brine recirculation, makeup and blow down flow rates, brine heater temperature, and final stage liquid level must be properly controlled.

5.11 Multiple Effect Evaporators

The use of multiple effects in series is quite common for evaporation of large amounts of dilute aqueous feed, requiring the evaporation of from thousands to hundreds of thousands of pounds per hour of water. The basic principle is to use heat given up by condensation in one effect to provide the re-boiler heat for another effect. In most multiple effect units, the overhead vapor from one effect is condensed directly in the heating element of the next effect.

Multiple effect evaporators are generally large, complex systems and are normally the most expensive type of evaporator to procure and install, but they can also be the most economical evaporator to operate, thus justifying their high first cost. Perhaps it is best to think conceptually of multiple-effect systems as requiring a higher "up front" investment of total capital in order to significantly reduce the largest variable operating cost, the cost of energy. Simplistically, the addition of a second effect will reduce energy consumption by about 50%; a four-effect evaporator installation will use about 25% of the energy of a single effect evaporator performing the same duty. It is not uncommon to find seven to ten effect evaporators, for example, in the production of pulp and paper, an industry with high-energy costs and one that must evaporate enormous quantities of water. Figure 12.17 is a photograph of a large, outdoor installation of an eight effect, long tube, and vertical evaporator system.

In addition to the reduction in steam usage, there is also a reduction in cooling water required to operate the last effect condenser. Approximately 30 pounds of cooling water must be provided for each pound of steam supplied to the first effect. The increased energy economy of a multiple effect evaporator is gained only as a result of increased capital investment, which tends to increase at about the same rate as the required area increases. A five-effect evaporator will usually require more than five times the area of a single effect because of the staging of the driving force, ΔT, which is less than a single effect evaporator. The only accurate method to predict changes in energy economy and heat transfer surface requirements as a function of the number of effects, is to use detailed heat and material balances together with an analysis of the influence of changes in operating conditions or rates of heat transfer. This, of course, requires a copious amount of engineering effort and computational work, a task performed best by sophisticated computer programs.

The distribution in each effect of the available temperature difference between condensing steam and process liquid can be allocated by the evaporator designer. Once the evaporator is put into operation, the system establishes its own equilibrium. This operating point depends upon the amount of fouling and the actual rates of heat transfer. Usually it is best not to interfere with this operation by attempting to control temperatures of different effects of an evaporator. Such attempts result in a loss of capacity since control usually can be accomplished by throttling a vapor imposing an additional resistance. The pressure loss results in a loss of driving force and a reduction in capacity.

The designer has a number of options to achieve the greatest energy economy with a given number of effects [19, pp. B-118]. These are usually associated with the location of the feed in respect to the introduction of the steam. Figure 12.18 illustrates several methods of operation which are: forward feed, backward feed, mixed feed, and parallel feed.

Usually, heat transfer rates decrease as temperature decreases, so that the last effects have the lowest rates of heat transfer. By leaving the resistance of these effects higher, the designer can increase the temperature difference across them, thus increasing temperature and heat transfer rates in all the earlier effects. It has been shown that the lowest total area is required when the ratio of temperature difference to area is the same for all effects. When the materials of construction or evaporator type vary among effects, lowest total cost is achieved when the ratio of temperature difference to cost is the same for each effect. However, in most cases where evaporator type and materials of construction are the same for all effects, equal heat transfer surfaces are supplied for all effects.

FIGURE 12.17 Multiple-effect long-tube vertical evaporator. (*Ecodyne Division, Unitech Corporation.*)

Often in multiple-effect evaporators the concentration of the liquid being evaporated changes drastically from effect to effect, especially in the latter effects. In such cases, this phenomenon can be used to advantage by *staging* one or more of the latter effects. Staging is the operation of an effect by maintaining two or more sections in which liquids at different concentrations are all being evaporated at the same pressure. The liquid from one stage is fed to the next stage. The heating medium is the same for all stages in a single effect, usually the vapor from the previous effect. Staging can substantially reduce the cost of an evaporator system. The cost is reduced because the wide steps in concentrations from effect to effect permit the stages to operate at intermediate concentrations, which result in both better heat transfer rates and higher temperature differences.

6.0 ENERGY CONSIDERATIONS FOR EVAPORATION SYSTEM DESIGN

The single largest variable cost factor in making a separation by evaporation is the cost of energy. If crude oil is

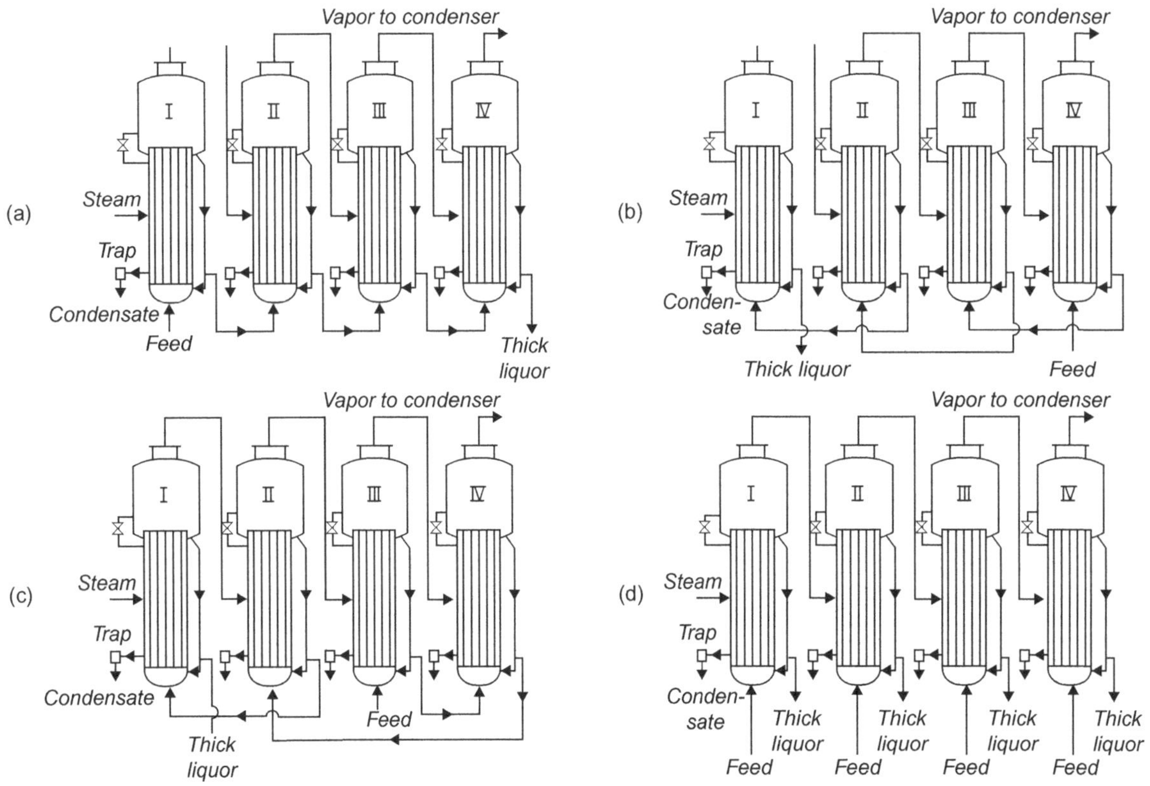

FIGURE 12.18 Patterns of liquid flow in multiple-effect evaporators: *(a)* Forward feed. *(b)* Backward feed. *(c)* Mixed Feed. *(d)* Parallel feed. *(From* Unit Operations of Chemical Engineering *by W. L. McCabe and J. C. Smith, 2nd. ed. p. 464. ©1967 McGraw-Hill. Used with permission.)*

the ultimate source of energy, the typical 2013 cost of about \$630 per m^3 (\$100 per barrel) is equivalent to more than \$16.66 for 1 million kJ. Water has a latent heat of 480 kJ/kg at 760 mm of mercury, absolute, so the energy required to evaporate 1 kg of water exceeds 0.80 cents. Therefore, the efficient utilization of energy is the most important consideration in evaluating which type of evaporation system should be selected.

Energy can never be used up; the first law of thermodynamics guarantees its conservation. When normally speaking of "energy use" what is really meant is the lowering of the level at which energy is available. Energy has a value that falls sharply with level. Accounting systems need to recognize this fact in order to properly allocate the use of energy level.

The best way to conserve energy is not to "use" it in the first place [19, pp. B-91]. Of course, this is the goal of every process engineer when he evaluates a process, but once the best system, from an energy point of view, has been selected, the necessary energy should be used to the best advantage. The most efficient use of heat is by the transfer of heat through a heat exchanger with process-oriented heat utilization, or by the generation of steam at sufficient levels to permit it to be used in the process plant directly as heat. When heat is available only at levels too low to permit recovery in the process directly, thermal engine cycles may be used for energy recovery. Heat pumps may also be used to "pump" energy from a lower to a higher level, enabling "waste" heat to be recovered through process utilization.

Thermal efficiencies of heat exchangers are high, 90–95%. Thermal efficiencies of thermal engine cycles are low, 10–20%. Heat pumps permit external energy input to be reduced by a factor of 4 to 5; however, the energy required in a heat pump is in the work form, the most expensive energy form.

Utility consumption, of course, is one of the major factors which determine operating cost and, hence, the cost of producing the product for which a plant has been designed. In order to select the proper equipment for a specific application it is important to be able to evaluate different alternatives, which may result in a reduction of utility usage or enable the use of a less costly utility. For example, the choice of an air-cooled condenser versus a water-cooled condenser can be made only after evaluating both equipment costs and the costs of cooling water and horsepower.

When heating with steam, a selection of the proper steam pressure level must be made when designing the evaporator. No definite rules for the selection can be established because of changing plant steam balances and availability. However, it is generally more economical to select the lowest available steam pressure level which offers a saturation temperature above the process temperature required. Some evaporator types require relatively low temperature differences. Some products may require low temperature in order to reduce fouling or product degradation.

Maximum outlet temperatures for cooling water usually are dictated by the chemistry of the cooling water. Most cooling water contains chlorides and carbonates; consequently temperatures at the heat transfer surfaces must not exceed certain values in order to minimize formation of deposits or scale, which reduces heat transfer and leads to excessive corrosion. In addition, velocity restrictions must be imposed and observed to prevent corrosion and fouling as a result of sedimentation and poor venting. Stagnant conditions on the water side must always be avoided. In some plants, water consumption is dictated by thermal pollution restrictions.

Unnecessary restraints should not be imposed on the pressure drops permitted across the water side of condensers. All too often, specified design values for pressure drop are too low and much higher values are realized when the unit has been installed and is operating. Not only does this result in more expensive equipment, but also frequently the water flow rate is not monitored and cooling water consumption is excessive, increasing operating costs. Because cooling water consumption is governed by factors other than energy conservation and because cooling water velocities must be maintained above certain values, tempered water systems can be effectively used at locations where cooling water temperatures vary with the season of the year. At some locations a 30°C difference between summer and winter water temperatures is experienced. At such locations a tempered water system may be used in order to reduce both pumping costs and maintenance costs. A tempered water system requires a pump to recycle part of the heated cooling water in order to maintain a constant inlet water temperature.

Evaporative-cooled condensers in many applications give greater heat transfer than air-cooled or water-cooled condensers. The evaporative equipment can do this by offering a lower temperature sink. Evaporative-cooled condensers are frequently called wet-surface air-coolers. Perhaps the best description for this type of equipment is a combination shell-and-tube exchanger and cooling tower built into a single package. The tube surfaces are cooled by evaporation of water into air.

Air-cooled condensers are especially attractive at locations where water is scarce or expensive to treat. Even when water is plentiful, air coolers are frequently the more economical alternative. Elimination of the problems associated with the water side of water-cooled equipment, such as fouling, stress-corrosion cracking, and water leaks into the process, is an important advantage of air-cooled equipment. In many cases, carbon steel tubes can be employed in air-cooled condensers when more expensive alloy tubes would otherwise have been necessary. The use of air-cooled heat exchangers may eliminate the need for additional investment in plant cooling water facilities.

Maintenance costs for air-cooled equipment are about 25% of the maintenance costs for water-cooled equipment. Power requirements for air-coolers can vary throughout the day and the year if the amount of air pumped is controlled. Water rates can be varied to a lesser degree because daily water temperatures are more constant and because water velocities must be kept high to reduce maintenance. The initial investment for an air-cooled condenser is generally higher than that for a water-cooled unit. However, operating costs and maintenance costs are usually considerably less. These factors must be considered when selecting water or air as the cooling medium.

Air-cooled condensers employ axial-flow fans to force or induce a flow of ambient air across a bank of externally finned tubes. Finned tubes are used because air is a poor heat transfer fluid. The extended surface enables air to be used economically. Several types of finned-tube construction are available. The most common types are extruded bimetallic finned tubes and fluted tension-wound finned tubes. The most common fin material is aluminum.

Air-cooled heat exchangers generally require more space than other types. However, they can be located in areas that otherwise would not be used (e.g., on top of pipe racks). A forced draft unit has a fan below the tube bundle that pushes air across the finned tubes. An induced draft unit has a fan above the tube bundle that pulls air across the finned tubes. Using controllable-pitch fans normally controls air-cooled condensers. Good air distribution is achieved if at least 40% of the face area of the bundle is covered with fans. It is most economical to arrange the bundles and select the fan diameters to minimize the number of fans. Controllable-pitch fans permit only the airflow required for heat transfer to be pumped. An important added advantage is the reduction of the power required for operation when ambient air temperature is lower than that used for design. Controllable-pitch fans can result in a 50% reduction in the annual power consumption over fixed-blade fans.

There are many ways to waste energy in pumping systems. As energy costs have continued to climb, it has often been found that a complete pumping unit's initial investment can be less than the equivalent investment value of one electrical horsepower. Calandria circulating pumps require a certain available NPSH. This is usually obtained by elevating the evaporator, often with a skirt. Quite often the designer establishes the skirt height before he selects the calandria-recirculating pump. In the interest of economy he provides a skirt as short as possible, often without realizing that he will be forever paying an energy penalty for a smaller initial capital savings. More efficient pumps often require greater NPSH. Therefore, it is prudent to check the NPSH requirements of pumping applications before establishing skirt heights of evaporator systems.

Heat pumps or refrigeration cycles involves the use of external power to "pump" heat from a lower temperature to a higher temperature. The working fluid may be a refrigerant or a process fluid. Heat pumps use energy that often would otherwise be thrown away in the form of waste heat in effluents or stack gases. The external energy input can be reduced by a factor of 4 to 5, depending on the temperature difference and temperature level of the heat pump system.

There are several ways to increase the steam economy, or to get more evaporation with less steam input, for certain types of evaporation applications. The use of multiple-effect configurations or compression evaporation can be considered for large flow rates of relatively dilute aqueous solutions. Both multiple-effect and compression evaporation systems require a sizable incremental capital investment over single-effect evaporators, and these systems are larger and more complex than the simpler one-stage evaporators. Like the multiple-effect evaporators described above, a reduced level of steam consumption can only justify compression evaporation systems.

In a compression evaporation, a part or all of the evaporated vapor is compressed by a compressor to a higher pressure level and then condensed, usually in the heating element, thus providing a large fraction of the heat required for evaporation [19, pp. B-111]. Energy economy obtained by multiple-effect evaporation can sometimes be equaled in a single-effect compression evaporator. Compression can be achieved with mechanical compressors (commonly known as Mechanical Vapor Recompression or MVR Systems) or with steam jet thermo-compressors (commonly known as Thermal Vapor Recompression or TVR Systems). To achieve reasonable compressor costs and power requirements, compression evaporators must operate with fairly low temperature differences, usually from 5° to 10°C. This results in a large heat transfer surface, partially offsetting the potential energy economy. When a compression evaporator of any type is designed, the designer must provide adequate heat transfer surface and may decide to provide extra area over that required to anticipate reduced heat transfer should fouling occur. If there is inadequate surface to transfer heat available after compression, the design compression ratio will be exceeded causing a thermo-compressor to *break* or *backfire* or a mechanical compressor to exceed the horsepower provided.

Mechanical compression evaporation (Figures 12.19 and 12.20) is generally limited to a single effect. All of the vapor is compressed and condensed, eliminating the

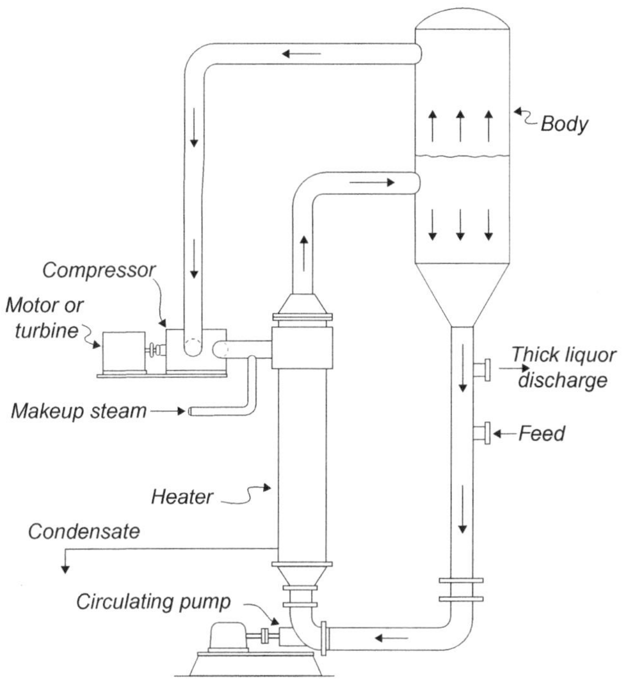

FIGURE 12.19 Mechanical recompression applied to forced-circulation evaporator. *(From* Unit Operations of Chemical Engineering *by W. L. McCabe and J. D. Smith (2nd. ed., 1967), p. 473. ©McGraw-Hill. Used with permission of McGraw-Hill Book Company.)*

cooling water required for conventional or steam jet thermo-compression evaporators; an advantage when cooling water is costly. Mechanical compression is ideally suited for locations where power is relatively inexpensive and fuel is expensive. The greatest advantage of mechanical compression is the high-energy economy. Compressors may be reciprocating, rotary positive displacement, centrifugal, or axial flow. Single stage positive displacement compressors appear to be better suited to compression evaporation because of lower cost and their characteristic fixed capacity, dependent only on speed or discharge pressures. They are, however, limited in developed compression ratios and material of construction. The compressor may be driven with a diesel unit, a steam turbine, a gas turbine or an electric motor. Selecting the compressor drive requires analysis of all factors present at a particular location. One disadvantage of mechanical compression is that most systems require a heat source to initiate evaporation during start-up.

Because the vapor is frequently water, which has a low molecular weight and a high specific volume, compressors are usually quite large and costly. Compressors require high purity of the vapor to avoid buildup on the blades of solids that result from evaporation of liquid as the vapor is superheated by compression. Liquids having high boiling point elevations are not usually adaptable to compression evaporation. Mechanical compression evaporation sometimes requires more heat than is available from the compressed vapor, so the evaporation rate can be controlled by regulating the makeup steam flow to maintain a constant liquor temperature. Usually, mechanical compression results in slightly higher maintenance costs because of the compressor and its drive. Mechanical compression is best suited for atmospheric or pressure operation, for mildly corrosive vapors, for low boiling-point elevation liquids, low temperature differences across the calandria, and where energy economy is important.

Steam jet thermo-compressors can be used with either single or multiple-effect evaporators. As a rule-of-thumb, the addition of a thermo-compressor will provide an improved steam economy equivalent to an additional effect, but at a considerably lower cost. Thermo-compressors have low efficiencies that further diminish when the jet is not operated at its design point. Thermo-compressors in a typical operation can entrain one pound of vapor per pound of motive steam. They are available in a wide range of materials of construction, and can have a wide range of design and operating conditions. They should be considered only when high pressure motive steam is available, and when the evaporator can be operated with low-pressure steam. Motive steam pressures above 60 psig usually are required to justify using thermo-compressors. Steam condensate from thermo-compressors often is contaminated with trace amounts of product and may have to be treated before being returned to the steam generator.

It is relatively easy to design an evaporator using thermo-compression for a given set of operating conditions. However, once the thermo-compressor has been designed and fabricated, its performance characteristics are basically fixed. The design of a thermo-compression evaporation system should include an analysis of the consequences of changing operating points. The characteristics of a thermo-compressor make it difficult to predict performances at conditions different from the design point, so accurate prediction of evaporator performance at other than design conditions becomes impossible.

Because of the unpredictable performance of thermo-compressors, control of evaporators using them is more difficult than for a conventional system where it is necessary to set only steam and feed rates to maintain a constant evaporation rate. One-way to provide flexibility with better operating stability is to use two or more thermo-compressors in parallel. This permits capacity control without loss in energy economy. Thermo-compression evaporators are used for single or double-effect systems where low operating temperatures and improved economy are desired. It costs less to add a thermo-compressor instead of an additional effect, and both have about the same effect on energy economy. The temperature differences across the thermo-compressor should be below 15°C. This evaporator system is not as flexible as multiple-effect systems

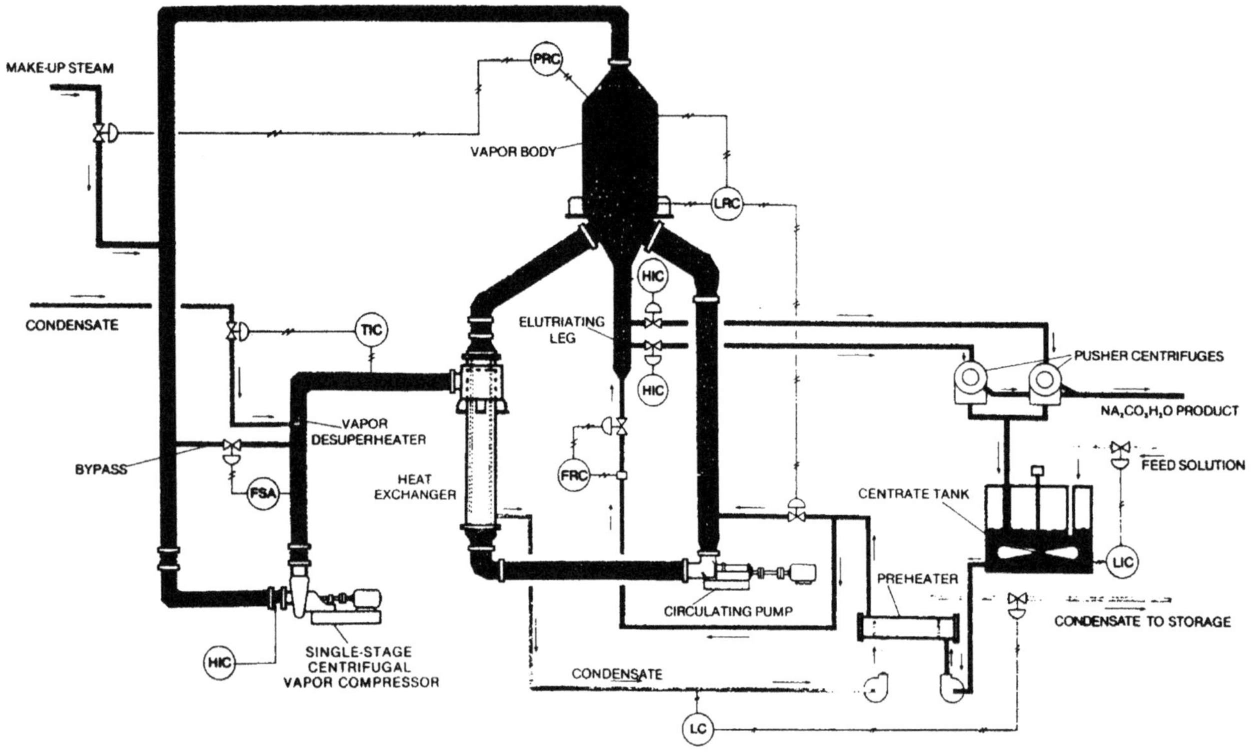

FIGURE 12.20 Single-effect recompression evaporator for soda ash. (*Swenson Division, Whiting Corporation.*)

because of the unpredictable variation of performance characteristics for the thermo-compressor under changing operating conditions.

7.0 PROCESS CONTROL SYSTEMS FOR EVAPORATORS

From the process viewpoint, the two parameters that should be regulated are the concentration and flow rate of the bottoms product. If the composition of the feed stream is constant, good control of the feed rate and the evaporation rate will give the desired concentrated product at the proper production rate (see Figure 12.1). Of course, the method of control can depend upon the evaporator type and method of operation. When evaporation rate is to be maintained at a constant rate, a steam flow controller is generally used. Steam flow control usually is accomplished by throttling the steam, which results in a loss of temperature difference. Steam may, therefore, be uncontrolled to achieve maximum capacity. Steam pressure controllers may be used to protect the equipment or to assure substantially constant temperatures in the front end of a multistage evaporation system. Constant temperatures in the later effects of the evaporator can be controlled with a pressure controller on the last effect.

A control system consists of three parts: a measurement; a control algorithm; and a process actuator. The process actuator (often a control valve) is always a direct user of energy; the measurement may take energy from the process (as in the case of a head-type flow meter); and the control calculation never requires a significant energy supply. However, the correct control calculation is essential for energy-efficient operation of any process.

The well-engineered control system depends on the ability to directly measure the parameter that is to be controlled, or to measure another parameter from which the controlled variable can be inferred. In every case, a measurement of the controlled variable is preferred. A survey of the measurements in a major production unit gave the following distribution of process instrumentation [19, pp. B-149]:

Type of Measurement	Percent
Flow	34
Temperature and analytical	24
Pressure	22
Liquid level	20

Flow rates are the largest single group of process measurements used for control, and flow is the only process variable for which significant energy may be required by the measuring device. Most flows are measured by orifice meters, which are head-type devices that extract head loss from the pumping system. The amount of power required by an orifice, nozzle, or venturi tube meter can be significant. There are many flow sensors available and numerous considerations to be evaluated in the design of a flow metering system. The cost of operating each meter should be evaluated and the type selected should have the best balance between operating, maintenance, and capital costs. Although the energy required to operate a process unit can be reduced if the designer becomes sensitive to the hidden cost in each meter installation, the amount of energy required to operate most process meters is small; and the opportunities for significant reductions in energy usage by modifications of flow meters in lines less than 10 inches in diameter is limited.

A control system requires a mechanism to change the state of the process when a disturbance causes the control variable to move from the desired value. This control mechanism is most often a control valve although it can be a motor, a set of louvers, an electrical power supply, a fan on an air-cooled condenser, etc. Control that is achieved by changing the area of the valve body opening is a direct energy expense to the operating unit.

The control valve is a variable orifice device in which the size of the orifice is adjusted to control a process variable. Consequently, the manufacturer, type, or even the size of a control valve has no effect on the energy dissipated in the control of a selected stream once the process pressure, line size, and pumps have been selected. This energy-independence of the control valve assures that continuous throttling of the flow stream is required to control a process variable. In those cases where a valve is used for shut-off or override control (not a continuous throttling device), energy savings can be realized by selecting a valve with a minimum pressure loss in the full-open position.

Any control system, which is properly designed to control the evaporation process, must maintain both an energy and material balance across the evaporator boundaries. The control system must be able to accommodate some fluctuation in the feed flow rate or composition within a specified range, and still enable the evaporation system to perform the required separation with stable operation. The control system should function to reduce heat input with a reduction in feed rate, or change the evaporation rate as changes in the feed composition occur.

The best control system should be used in the design of evaporator systems. Products, which are off-specification, require additional time, expense, and energy in reprocessing. A properly designed control system can do much to reduce these wastes, and ensure that the evaporation system uses the optimum energy during normal operation.

Product concentration can be controlled by measuring a number of physical properties, most usually specific gravity and boiling point elevation. Control is usually accomplished by controlling the discharge of product from the evaporator. Feed rate and flow rates between effects are then adjusted to maintain constant liquid levels. When this is not possible, throttling the feed may control product concentration. Often there is a considerable time lag before a change in feed rate is reflected by a change in product concentration. Liquid level control in evaporators may be important for product concentration, to prevent scaling, and to maintain heat transfer rates. Level control may also reduce splashing and entrainment.

Several methods are used to control the amount of heat removed in the condenser, including controlling the cooling medium flow rate or temperature, changing the amount of surface available for heat transfer, and introducing inert gases into the condensing vapor.

The condensate from a condenser is sub cooled. Because of the temperature gradient across the condensate liquid film, there is no way to avoid sub cooling even when it is not desired. In some cases, the condensate is purposely sub cooled several degrees in order to reduce product losses through the vent. For total condensers with essentially isothermal conditions, sub cooling results in a pressure reduction, unless something is done to prevent it. This happens because the sub cooled liquid has a vapor pressure lower than the operating pressure. The system pressure for a sub-cooled condensate will be the vapor pressure of a condensate when no inert gases are present. Permitting the system pressure to vary as the degree of subcooling changes is not usually desirable. A constant pressure vent system is normally provided to prevent this pressure kind of variation. Inert gases are introduced or removed as required to ensure that the system pressure drop is reflected only by the friction drop and not by changes in vapor pressure. The inert gases should be introduced downstream of the condenser; introducing inerts upstream of the condenser will reduce the heat transfer rate requiring more heat transfer surface. This approach, however, is sometimes used to control the condenser. When the condenser and condensate tank are closely connected, the condensate tank must be properly sized in order to permit the condensate liquid level to be controlled somewhere in the condensate tank. If the condensate tank is too small, liquid level control can be achieved only by flooding part of the condenser, especially when the condensate is pumped from the tank. Liquid level must be maintained in the condensate tank and not in the condenser.

Control of natural circulation calandrias presents some problems not found in other heating elements. When heating with condensing vapors, changes in condensing pressure affect four variables: heat transfer coefficients, temperature difference, liquid composition, and circulation rate. The same four variables are affected when throttling the flow of a liquid heating medium. The whole mode of operation is changed when one variable is altered, and it is not always possible to predict from experience which direction the change will take. The liquid level on the shell side of kettle-type re-boilers should be sufficient to ensure that all the tubes are covered with the boiling fluid. Controlling by varying process liquid levels may result in fouling of the heat transfer surface as part of that surface is deprived of liquid. In addition, the temperature difference may be affected as the hydrostatic head (which affects boiling temperature) is reduced.

The temperature gradient across the liquid film in falling-film evaporators must be kept relatively low, usually less than 15°C. Excessive temperature difference between the process and utility fluids may result in boiling of the fluid on the heat transfer surface with resulting fouling. Film boiling can also occur with subsequent reduction in the rate of heat transfer. Inert gases are sometimes injected into a falling-film evaporator in order to reduce the partial pressure required to vaporize the volatile component. This technique will often eliminate the need for vacuum operation. Enough inert gas must be injected to achieve the desired results, but too much can produce flooding and entrainment, resulting in poor control.

Steam-heated calandrias with process boiling temperature less than 100°C can present control problems, especially at reduced rates and during start-up. In most such cases, low-pressure steam is used for heating. Control is usually achieved by throttling the entering steam in order to reduce the pressure at which it is condensed. At reduced rates this often results in steam pressures less than atmospheric or less than the steam condensate return system pressure. The steam is usually removed through steam traps which require a positive pressure differential to function. In order for the trap to function, steam condensate floods part of the steam chamber until the steam pressure is sufficient to operate the trap. This leads to poor control and all the problems associated with condensate flooding.

Calandrias heated with sensible heat from a hot liquid are normally controlled by throttling the liquid flow. Usually, good control may be achieved by controlling the temperature of the heating medium. The best utilization of the available heat transfer surface is achieved by maximizing the temperature difference in the calandria, and this is accomplished by designing for high pumping rates for the heat transfer medium and the process fluid in order to achieve nearly isothermal conditions on both sides of the heating surface.

8.0 EVAPORATOR PERFORMANCE

Energy economy and evaporative capacity are the major measures of evaporator performance. When evaporating water with steam, the economy is nearly always less than 1.0 for single-effect evaporators, but in multiple-effect evaporators it is considerably greater than 1.0. Other performance variables to be considered include: product quality, product losses, and decrease in performance as scaling, salting, or fouling occurs.

Designers of evaporation systems strive to achieve high heat transfer rates. This can be justified by a cost/benefit analysis. High rates of heat transfer in theory must often be proved in practice. Evaporators designed for high rates of heat transfer are generally more affected by traces of scale or non-condensable gases.

Product loss requirements may be an important factor in the evaporator design. Provisions to reduce product losses have far less effect on the cost of an evaporator system than does the amount of heat transfer surface. Product losses in evaporator vapor occur as a result of entrainment, splashing, or foaming. Foaming properties of the liquid may at times dictate the selection of evaporator type.

Losses from entrainment result from the presence of droplets in the vapor that cannot separate because of the high vapor velocity. Entrainment is thought to be due mainly to the collapse of the liquid film around vapor bubbles. This collapse projects small droplets of liquid into the vapor. The amount of entrainment is a function of the size distribution of the droplets and the vapor velocity. Bubbles leaving the surface cause droplets of different sizes to be propelled upward. The smaller droplets are caught in the fast moving vapor and are carried upward as entrainment, while the larger ones fall to the surface. The largest size (or mass) drop carried up is dependent on the gas velocity and density, and on other physical properties. At very high gas velocities, large drops produced at the surface are shattered into smaller droplets and all the generated entrainment is carried overhead, flooding the device. This breakup phenomenon occurs when the inertial forces, which cause a pressure or force imbalance, exceed the surface tension forces, which tend to restore a drop to its natural spherical shape. The gas velocity at which this flow crisis develops is the flooding velocity and is given by the following equation:

$$v_{Gf} \cong 0.7 \left[\frac{(\rho_L - \rho_G)\sigma}{\rho_G^2} \right]$$

Equipment containing both gases and liquids in which the gas flows vertically upward will be flooded at velocities exceeding that predicted from the above equation. In practice, most equipment is designed to operate well below the flooding limit. Factors such as disengaging height, convergence effects, and nonuniform gas velocities prevent operation at velocities exceeding roughly half the flooding velocity.

The amount of entrainment from an upward flowing gas can be estimated and is a function of gas velocity, gas and liquid densities, and surface tension. Entrainment can be separated from a gas stream with a variety of mechanisms, including gravity, inertial impaction, interception, centrifugal force, and Brownian motion. Separators can be classified according to mechanism, but it is more useful to categorize them by construction type. Separators in common use include: flash tanks, vane impingement separators, wire mesh separators, Karbate strut separators, centrifugal separators, cyclones, and special separator designs.

Flash tanks are generally used when the liquid entrainment exceeds 20% of the gas flow on a weight basis. Flash tanks may be either vertical or horizontal. Proper sizing of a flash tank should result in a residual entrainment under 3% of the gas rate. Vane impingement, wire mesh, centrifugal, and Karbate strut separators are commercial proprietary design and all compete for similar applications. Performance and cost, however, can vary widely from one type to another. The designer should understand the advantages and disadvantages of each type and the level of separation that each type can achieve.

Except for flash tanks and some special separators, the efficiency of all these separators tends to increase with increasing velocity up to a maximum allowable limit. In this region the efficiency seems to depend primarily on gas velocity and particle size, and to be somewhat insensitive to gas and liquid physical properties. Except for the cyclone and some special separators, there is a predictable maximum allowable velocity. The following equation is commonly used:

$$v_m = F\left[\frac{(\rho_L - \rho_G)}{\rho_G}\right]^{1/2}$$

where v_m is the maximum gas velocity and F is an experimentally derived constant. Both equations indicate that the term $(\rho G)^{-1/2}$ is of primary importance in separator sizing.

9.0 HEAT SENSITIVE PRODUCTS

The world value for end products using biological manufacturing methods was approximately $250 billion in 1980 [20]. This total value can be broken down as follows:

	$ Billion	**$ Billion**
Food		218.4
Baked goods requiring yeasts	41.4	
Butter and cheese	79.2	
Alcoholic beverages	90.6	
Others	7.2	
Drugs		15.2
Biologicals	4.1	
Antibiotics	7.7	
Hormones	3.4	
Others		15.0
Fuel ethanol, amino acids, enzymes	3.2	
Miscellaneous	11.8	
Total	248.6	248.6

Many of these food and pharmaceutical products are heat sensitive; that is, the finished product may be damaged or destroyed if it is exposed to too great a temperature over an extended period of time. Even common products like tomato catsup and penicillin "spoil" or lose their efficacy when exposed to ambient temperatures for long periods of time. The chemical reactions that limit shelf life are strongly temperature dependent. Some biological products may be handled at elevated temperatures in dilute solutions, but may degrade at the same temperature once the concentration exceeds a certain threshold value.

Thus, it is not uncommon for heat sensitive products to be concentrated in two different evaporator types in series—in natural or forced circulation evaporators to perform the bulk of the solvent removal and in a low mean residence time evaporator to finish the concentration step. Because of the recycle and back mixing effects of a natural or forced circulation evaporator, the mean residence time of the average molecule can be several hours. Some evaporators operate without recycle; these are called once-through or single pass evaporators, and have mean residence times measured in minutes or even seconds.

Figure 12.21 represents the "heat history" or the temperature-time relationship as seen by the product, which is to say an average fluid element in the product. After the product reaches the final concentration within

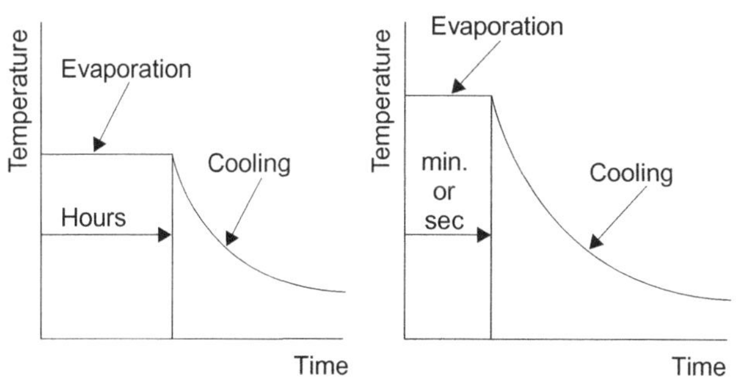

FIGURE 12.21 Heat history of long (*left*) and short (*right*) residence time evaporators. (*Luwa Corporation.*)

the allotted time, it cools to an ambient or final temperature. Because the loss in product quality (degradation, oxidation, polymerization, etc.) is generally due to one or more chemical reactions, this deterioration phenomenon is a rate problem which is determined by chemical reaction kinetics. It is, therefore, possible to evaporate a heat sensitive product at a higher temperature in a short residence time evaporator rather than in a recirculating or other long residence time unit. This could have important positive consequences, including increased energy efficiency and smaller evaporator size. Note that it may make sense to install a low residence time product cooler for the concentrate from a low residence time evaporator, since long residence time cooling could still damage the product.

An average residence time expressed as holding volume divided by discharge rate was frequently used in the past for both single-pass and recirculation evaporators. However, statistical analysis of several types of evaporators has revealed that the actual time of replacement of 97% of the feed in a recirculating evaporator is about 3.2 times the average residence time as defined above. It takes longer to replace a larger percentage of feed.

The actual residence time achieved in any evaporator can be calculated from the equation below:

$$x = 1 - e^{-\theta/r}$$

where

x = fraction of feed removed
θ = time, minutes
r = ratio of holding volume to discharge rate, 1/min

Nearly every supplier of evaporation equipment and systems maintains a pilot plant facility where, for a fee, different evaporation schemes can be set up. Data obtained from several days of testing on small laboratory or pilot plant units can be good predictors of evaporator performance, and these data are very helpful in the scaling-up calculations for production-sized installations. Samplers obtained from the test work can be used to check the mass balances, concentrations, and product quality. Serious operational problems like foaming, plugging, and fouling can occur in even short pilot plant tests and can point to the need for alternative evaporator types or modified designs.

10.0 INSTALLATION OF EVAPORATORS

Many details must be considered when installing heat transfer equipment. Some of these may seem of minor importance but it is precisely these small details that often lead to poor performance, operational difficulties, and increased maintenance.

Vent and drain connections are normally provided, and they should be permanently installed. In some cases, intermittent venting of non-condensables may be acceptable. For vertical exchangers with cooling water on the shell side, it is essential to provide means for venting gases that are released as the water is heated. If these gases are not continuously removed, they collect in pockets; the shell side heat transfer is reduced; and corrosive attack may occur in the gaseous region. In vertical units, the gases tend to collect just beneath the top tube sheet and the tubes in this area corrode rapidly when not surrounded by liquid. Corrosion and stress cracking may occur in this area and solids may also build up on the tubes. Vent connections should be located at positions which enable these gases to be vented and to insure that the tube sheet is swept with water.

Vertical steam-heated exchangers must also be vented to remove carbon dioxide and other gases that can accumulate under the top tube sheet. Corrosion of the shell, tubes, and tube sheet, especially in the area just opposite the steam inlet nozzle, may result if adequate venting is not provided. Sometimes a continuous steam purge or intermittent venting is recommended.

The cooling water side of condensers should operate under a positive pressure whenever possible. Frequently, water-cooled units are mounted high in a structure and the available pressure is barely adequate to deliver water to the user. Often a vacuum exists and boiling of the water may occur. This causes corrosion and hydraulic problems, both in the exchanger and the outlet water piping. Any control valve should be placed in the outlet piping so that the maximum available pressure is realized at the exchanger. No design should be finalized without considering what the pressure at the outlet of the exchanger will be when operating, and the consequence of a siphon effect in the cooling water line.

Generally, U-bend exchangers should not be installed in a vertical position. Vertical U-bends are difficult to vent or drain on the tube side because connections cannot be provided at the U-bend end. Multi-pass exchangers are relatively high and provisions are made to ensure that the tubes operate completely flooded.

Sufficient space should be allowed in the equipment layout to remove the tube bundle from the removable bundle units. Consideration should be given to units expected to require periodic maintenance or cleaning. Early recognition should be given to space requirements of air-cooled exchangers. Equipment layouts should be made recognizing that longer tube lengths result in more economical heat exchangers. Care should be exercised to avoid forcing the equipment to fit the layout, rather than providing a layout to match the equipment.

The manner in which a heat exchanger is piped up can influence its performance. Horizontal units should have inlet and outlet nozzles on the top and bottom of the shell or channel. Nozzles should not be on the horizontal centerline of the unit. In general, fluids should enter the bottom of the exchanger and exit at the top, except when condensation occurs. Units are almost always designed to be counterflow and the piping must reflect this. When concurrent flow has been specified, it is equally important that it be piped to suit.

If the equipment will be used in corrosive service and if the components will have a short life expectancy, the design engineer should select units that are easy to repair. Removable bundle units may be required. In addition, the plant equipment layout should be arranged to facilitate removal and repair. Maintenance costs and production losses can often be reduced by specifying equipment with standard components or designing equipment to be interchangeable among several different services. By anticipating potential maintenance problems, the designer can avoid high maintenance or cleaning expenses and costly shutdowns. To anticipate maintenance problems, the designer needs to be familiar with the plant location, process flow sheet, and anticipated plant operation. The designer will need to know, for example, whether the equipment will have to be periodically cleaned and if chemical or mechanical cleaning methods can be used.

11.0 TROUBLESHOOTING EVAPORATION SYSTEMS

Occasionally, it becomes necessary to investigate the performance of an evaporator in order to evaluate its performance at other operating conditions or to determine why the system is not performing as expected. Fortunately, most conditions that result in an evaporator not meeting expected performance are easily corrected. Troubleshooting, therefore, often means checking for small details, which have a great effect on the performance of the evaporator system. Of course, it is possible that a type of evaporator has been misapplied, the heat transfer surface that has been provided is not adequate for the intended service, or fouling is occurring.

Discrepancies in performance may be caused by deviations in physical properties of fluids, flow rates, inlet temperatures, mechanical construction of the equipment, or by problems caused during the installation of the equipment. The troubleshooter should first check to see that compositions, flows, temperatures, and physical properties agree with those specified for design. He should then examine the equipment drawings to determine if the problem could lie in the manner in which the equipment was constructed, or in the manner in which the equipment has been installed. After these basic items have been reviewed, the checklist below outlines some questions that should be raised.

Calandrias:

1. Has the steam side been vented to remove air or other entrapped gases?
2. Has the steam control valve been adequately sized? What is the actual steam pressure in the steam chamber?
3. Has the steam trap been properly selected and sized?
4. Are the control valve and steam trap functioning correctly?
5. Is steam condensate flooding part of the surface? What is the temperature of the steam condensate? Is the condensate nozzle large enough? Is steam trap piping adequately sized?
6. Is the process liquid level maintained at the proper place? Are liquid level instruments calibrated? Are instrument leads plugged?
7. Is the liquid holdup adequate to prevent surging?
8. Are process compositions and temperatures equal to those used for design? Does the process material contain enough volatiles to provide adequate boiling?
9. What is the temperature of the top head for natural circulation calandrias? A temperature higher than the liquid temperature may indicate inadequate circulation for some reason?
10. Is the available steam pressure equal to that used for design?
11. Are process nozzles adequate?
12. Is the process side adequately blown down or purged?
13. Were debris and other foreign objects removed from the equipment and piping prior to start-up? How often is the unit cleaned? What is the appearance of the equipment before cleaning? Is the cleaning adequate?
14. If a pump is provided, do the pump and the system match? Is the pump cavitating?
15. Has enough backpressure been provided to prevent boiling of the process fluid when the evaporator operation requires this (submerged tube type)?
16. Is entrainment occurring? Are entrainment separators properly sized and installed? Are they plugged?
17. Is dilution important?
18. Are flows adequate to maintain flow regimes used in design? Is the pressure drop out of line?
19. For falling film units, is the unit plumb? Is there equal liquid distribution to each tube? Is the inlet channel vented to remove any flashed vapors? Are flows adequate to ensure that a film is formed? Is the outlet flow rate sufficient to prevent the film from breaking?

Condensers:

1. Has a constant pressure vent system been provided? Has it been properly installed? Inerts should be injected downstream of the condenser, not upstream?

2. Is the vent system adequate?
3. Are condensate connections properly sized? Is liquid being entrained into the condenser? If horizontal, are the tubes level (or sloped toward the outlet)?
4. Is all piping adequate?
5. Is the water side operating under a vacuum?
6. Are the temperatures and composition equal to those used for design?
7. Is the water flow adequate? Properly vented?
8. Were debris or other foreign objects removed from the equipment and piping prior to start-up?
9. If air-cooled, is the inlet piping adequate to effect good distribution? Are fan blades properly pitched? Are motors delivering rated power? Are fan belts slipping? Is there noticeable recirculation of hot exhaust air?

Performance testing is an experimental procedure to help understand the performance of an evaporation system. Tests can be performed to identify and characterize unsatisfactory performance, and often indicate methods to improve operation. Performance tests may also be required to establish that a new evaporator system has met performance guaranteed by the supplier. Tests may be used to determine evaporator capacity under different operating conditions or to obtain data for designing a new evaporator system. The American Institute of Chemical Engineers has published a procedure entitled *Testing Procedure for Evaporators* [21]. This procedure covers methods for conducting performance tests and discusses several factors influencing performance and accuracy of test results.

Tests are conducted to determine capacity, heat transfer rates, steam economy, product losses, and cleaning cycles. Practically all the criteria of evaporator performance are obtained from differences between test measurements. Errors can result when measuring flow rates, temperatures and pressures, concentrations, and steam quality. Factors that can have a great effect on performance include dilution, vent losses, heat losses, and physical properties of fluids.

Frequent causes for poor performance of an evaporator system include the following:

Low Steam Economy:

Steam economy with a fixed feed arrangement can be calculated from heat and material balances. Steam economies lower than those calculated during the design of the unit may be the result of one or more of the following:

1. Leakage of pump gland seal water
2. Excessive rinsing
3. Excessive venting
4. Flooded barometric condensers
5. Dilution from condensate leakage

Low Rates of Heat Transfer:

1. Salted, scaled, or fouled surfaces
2. Inadequate venting
3. Condensate flooding
4. Inadequate circulation

Excessive Entrainment:

1. Air leakage
2. Excessive flashing
3. Sudden pressure changes
4. Inadequate liquid levels
5. Inadequate pressure levels
6. Operation at increased capacity

Short Time Between Cleaning Cycles:

Downtime required for cleaning may not agree with the expected frequency of cleaning. Short cycles may be caused by:

1. Sudden changes in operating conditions (such as pressure or liquid level)
2. Low liquid levels
3. Introduction of hard water or other contaminants during cleaning, rinsing, or from seal leaks
4. High temperature differences
5. Improper cleaning procedures

REFERENCES AND SELECTED READING MATERIAL

[1] R.H. Perry, C.H. Chilton, S.D. Kirkpatrick (Eds.), Chemical Engineers' Handbook, fourth ed., McGraw-Hill, New York, 1963.

[2] P.E. Minton, Course Director Lecture Outline and Notes from Evaporation Technology, The Center for Professional Advancement, East Brunswick, New Jersey, 1978, p. B-2.

[3] Standards of Tubular Exchanger Manufacturer's Association, fifth ed., New York, 1968, pp. 9.

[4] L. McCabe, J.C. Smith, Unit Operations of Chemical Engineering, second ed., McGraw-Hill, New York, 1967, pp. 439–440.

[5] D.Q. Kern, Process Heat Transfer, McGraw-Hill, New York, 1954, pp. 6–7.

[6] R.B. Bird, W.E. Stewart, E.N. Lightfoot, Transport Phenomena, John Wiley & Sons, New York, 1960, pp. 284.

[7] D.Q. Kern, op. cit., pp. 845.

[8] C.H. Gilmour, A Resumé of Expressions for Heat Transfer Coefficients, Union Carbide Corporation, South Charleston, West Virginia, 1959.

[9] D.Q. Kern, op. cit.

[10] W.H. McAdams, Heat Transmission, McGraw-Hill, New York, 1954.

[11] P.E. Minton, op. cit.

[12] R.H. Perry, C.H. Chilton (Eds.), Chemical Engineers' Handbook, fifth ed., McGraw-Hill, New York, 1973, pp. 11–31, 11–27.

[13] P.E. Minton, op. cit., pp. B-48, B-60.

[14] M. Sack, Falling Film Shell-and-Tube Heat Exchangers, 61st National Meeting of American Institute of Chemical Engineers, Houston, Texas, 1967.

[15] P.E. Minton, op. cit., pp. B-65–B-68.

[16] H.L. Freese, Mechanically Agitated Thin-Film Evaporators, from the short course Evaporation Technology, The Center for Professional Advancement, East Brunswick, New Jersey, 1978, pp. D-2.
[17] H.L. Freese, W.B. Glover, Mechanically agitated thin-film evaporators, Chem. Eng. Prog. (1979) 56–58.
[18] R. Fischer, Agitated thin-film evaporators: part 3—process applications, Chem. Eng. (1965) 186.
[19] P.E. Minton, op. cit., pp. B-118, B-91, B-111, B-149.
[20] Opportunities in Biotechnology, 1980 to 1988–1990, T. A. Sheets Company—Management Consultants, Cleveland, 1980.
[21] Testing Procedure for Evaporators, American Institute of Chemical Engineers, New York, 1961.

FURTHER READING

[1] P.E. Minton, op. cit., pp. B-181–B-184.
[2] R.H. Perry, C.H. Chilton, op. cit., pp. 11–38.
[3] R.B. Bird, W.E. Stewart, E.N. Lightfoot, Transport Phenomena, John Wiley & Sons, New York.
[4] Evaporation: A Prime Target for Industrial Energy Conservation, Energy Research and Development Administration, COO/2870-1 UC-95f, Oak Ridge, Tennessee, 1977.
[5] P.E. Minton (Course Director), Evaporation Technology, The Center for Professional Advancement, East Brunswick, New Jersey, 1978.
[6] H.L. Freese, W.B. Glover, Mechanically agitated thin-film evaporators, Chem. Eng. Prog. (1979).
[7] C.H. Gilmour, A Resume of Expressions for Heat Transfer Coefficients, Union Carbide Corporation, South Charleston, West Virginia, 1959.
[8] J.A. Kent (Ed.), Riegel's Handbook of Industrial Chemistry, seventh ed., Van Nostrand Reinhold Company, New York, 1974.
[9] D.Q. Kern, Process Heat Transfer, McGraw-Hill, New York.
[10] W.H. McAdams, Heat Transmission, McGraw-Hill, New York, 1954.
[11] J.M. McKelvey, G.V. Sharps, Fluid transport in thin-film polymer processor, Polym. Eng. Sci. (1979).
[12] P.E. Minton, R.C. Lord, R.P. Slusser, Design of heat exchangers, Chem. Eng. (1970).
[13] A.B. Mutzenberg, N. Parker, R. Fischer, Agitated thin-film evaporators, Chem. Eng. (1965).
[14] D. Perlman (Ed.), Fermentation Advances, Academic Press, New York, 1969.
[15] R.H. Perry, C.H. Chilton (Eds.), Chemical Engineers Handbook, sixth ed., McGraw-Hill, New York.
[16] G.L. Solomons, Materials and Methods of Fermentation, Academic Press, New York, 1969.
[17] Standards of Tubular Exchanger Manufacturers Association, fifth ed., New York, 1968.
[18] Upgrading Existing Evaporators to Reduce Energy Consumption, Energy Research and Development Administration, COO/2870-2, Oak Ridge, Tennessee, 1977.
[19] A. Underkofler, R.J. Hickey (Eds.), Industrial Fermentation, vols. I and II, Chemical Publishing Company, New York, 1954.
[20] F. Widmer, A. Giger, Residence time control in thin-film evaporators, Chem. Process Eng., Lond. (1970).

Chapter 13

Centrifugation

Celeste M. Todaro

1.0 INTRODUCTION

The solids-liquid separation process can be accomplished by filtration or centrifugation. Centrifuges magnify the force of gravity to separate phases, solids from liquids or one liquid from another. There are two general types of centrifuges:

Sedimentation Centrifuges—where a heavy phase settles out from a lighter phase, therefore requiring a density difference and

Filtering Centrifuges—where the solid phase is retained by a medium like a filter cloth, for example, that allows the liquid phase to pass through.

2.0 THEORY

Centrifuges operate on the principle that a mass spinning about a central axis at a fixed distance is acted upon by a force. The force exerted on any mass is equivalent to the weight of the mass times its acceleration rate in the direction of the force.

$$F = ma \tag{13.1}$$

Where

m = mass
a = acceleration rate
F = force

This acceleration rate is zero without a force acting upon it, however, it will retain a certain velocity, v. If forced to move in a circular path, a vector velocity v/r exists as its direction is continually changing.

$$a_c = v\,v/r \tag{13.2}$$

Where

a_c = centrifugal acceleration

$$a_c = w^2 r \tag{13.3}$$

v = velocity
r = radius
w = angular velocity

Should a mass be rotated within a cylinder, the resulting force at the cylinder wall is called a centrifugal force, F_c.

$$F_c = mw^2 r \tag{13.4}$$

This is away from the center of rotation. The equal and opposite force:

$$F_{cp} = -mw^2 r \tag{13.5}$$

is the centripetal force. This is the force required to keep the mass on its circular path.

If a cylindrical bowl holding slurry is left to stand, the solids will settle out under the force of 1 g or gravity. By spinning the bowl the solids will settle under the influence of the centrifugal force generated as well as the force of gravity that is now negligible. Solids will collect at the wall with a liquid layer on top. This is an example of sedimentation in a solid bowl system. By perforating the bowl or basket and placing a filter cloth on the inside wall, one has now modeled a filtering centrifuge similar in principle to an ordinary household washing machine.

This amplification of the force of gravity is commonly referred to as the number of g's. The centrifugal acceleration (a_c) referenced to g is $w^2 r/g$, which is given by the equation:

$$\text{Relative Centrifugal Force } (G) = 1.42 \times 10^{-5} n^2 D_i \tag{13.6}$$

where

n = speed in revolutions/minutes
D_i = diameter of the bowl in inches

The driving force for separation is a function of the square of the rotational speed and the diameter of the bowl; however, there are restrictions in the design of centrifuges that will limit these variables.

An empty rotating centrifuge will exhibit a stress in the bowl called a *self-stress*, S_s.

$$S_s = w^2 r_i^2 \rho_m \tag{13.7}$$

where

w = angular velocity
r_i = radius of the bowl
ρ_m = density of the bowl material

The contents of the bowl also generate a stress or pressure on the inner wall of the bowl. Assuming the radius of the bowl (r_i) is equal to the outer radius of the bowl contents (r_2), we have

$$S_c = w^2 r_2 (r_2^2 - r_1^2) c / 4t \quad (13.8)$$

where

t = thickness of the bowl
ρ_c = density of contents of the bowl
r_1 = inner radius of the bowl contents (solids and liquid)
r_2 = outer radius of the bowl contents (solids and liquid)

The total stress in the bowl wall is:

$$S_T = S_s + S_c$$

$$S_T = w^2 r_2 \left[r_2 \rho_m + \frac{(r_2^2 - r_1^2)\rho_c}{4t} \right]$$

with $D_i = 2r_i$ and in common units:

$$S_T = 4.11 \times 10^{-9} n^2 D_i \left[D_i \rho_m + \frac{(D_i^2 - D_1^2)\rho_c}{4t} \right] \quad (13.9)$$

Centrifuges are designed such that S_s is 45 to 65% of S_t.

D_i, D_1, t	(inches)
n	(rpm)
S_T	(lb/in^2)
ρ_c	(lb/ft^3)

Increasing the bowl speed and its diameter increases the g force, but also increases the self-stress and the stress induced by the process bowl. The design is, therefore, really limited by the material of construction available, however, for a given bowl stress, the centrifugal acceleration is an inverse function of bowl diameter. For example, doubling the rotational speed, and halving the bowl diameter, doubles the acceleration while keeping the total stress relatively constant. It is for this reason that the smallest diameter centrifuges operate at the highest g forces. Tubular centrifuges operate at 2–5 inches diameter with g forces over 60,000. Disk centrifuges operate at 7–24 inches at 14,000 to 5500 g's, while continuous decanter centrifuges with helical conveyors are designed with bowl diameters of 6–54 inches and g forces of 5,500–770 g's. Filtering centrifuges with diameters of 12 to 108 inches have corresponding g forces of 2000 to 260.

3.0 EQUIPMENT SELECTION

Upon review of Table 13.1, it is evident that there are several types of equipment that can be used for the same application. There are also many equipment vendors that can be consulted. In consulting with vendors to narrow the choices, proprietary information may be divulged regarding the nature of one's process. Be sure to sign a secrecy agreement to protect all confidential information.

3.1 Pilot Testing

Preliminary data taken in the laboratory regarding the separation characteristics of a product can be beneficial when beginning the equipment selection process. If one is retrofitting an existing unit with an identical system, it will not be as complex and time consuming as designing a "grass roots" facility. Planned data can then be helpful. However, should the process be altered, one should evaluate the effect on centrifugation.

Careful attention should be taken to ensure that the existing tank design, peripheral pumps, piping and agitators do not provide shear that will cause particle degradation. Capital dollars spent in this area on crystallization studies, the selection of the correct pumps, etc., will directly impact the capacity of your equipment as particle degradation will significantly affect throughput and final residual moisture adversely, as it does filtration. (See Ch. 8 for a more detailed discussion.)

If a process exists and in-plant expertise is available to optimize the equipment, it would be advisable to do so before a purchase. Upon review of the existing design, sufficient improvements can often be made in an older piece of equipment thereby avoiding a more costly investment. Vendors usually offer this type of assistance at no charge from their office or at a daily rate in the field.

3.2 Data Collection

The first step is to collect pertinent information to the process, including a process flow sheet, product information and completion of a typical questionnaire, as shown in Table 13.2.

Knowledge of the most critical aspect of the process can guide the sometimes difficult selection process. For example, the requirement of a very dry product with strict impurity levels suggests a filtering centrifuge. A product with a feed rate of 150 gpm, without wash requirements, would lead us to a continuous sedimentation centrifuge.

A simple Büchner funnel test will indicate fast, medium or slow filtration. Slow filtering materials that have inordinate quantities of particles passing the filter paper will be submicron and difficult to capture in a filtering system. Therefore, a sedimentation centrifuge should be considered.

TABLE 13.1 Product Recovery Fermentation

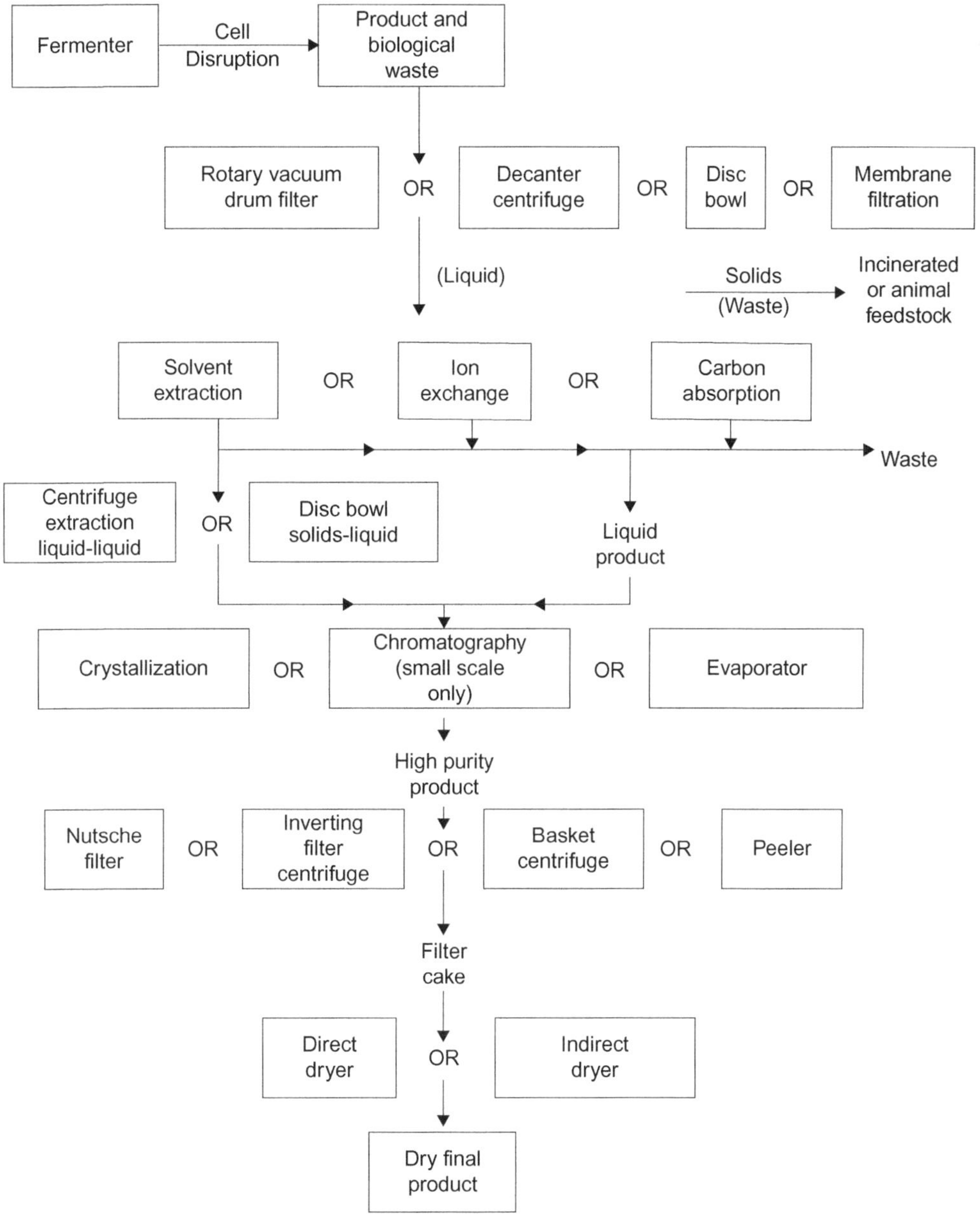

A phenomenon called "cake cracking" can occur and will be evident in this simple test. Not all materials exhibit this. It depends upon the surface tension of the product and its tendency to shrink as dewatering occurs. Amorphous, thixotropic materials will exhibit this more than rigid solids.

If the cake cracks, a liquid level must be maintained on top of the cake before washing to prevent channeling of the wash liquors. A crude estimate of the wash ratio, gallons of wash per pound of dry cake, can also be made in the laboratory. The filter cake developed on a Büchner funnel will have certain characteristics; the product may appear to have a defined crystal structure or be more amorphous. Microscope studies will indicate the shape of particles, which will be helpful in trial runs. Needle crystals, for example, may break easily at high filling speeds, and during discharge on a basket centrifuge with plough platelets tend to pack in compressible beds and may be better suited to sedimentation than filtration. A particle size distribution will help in this analysis and is also required for cloth selection.

Cake compressibility is the ability of a cake to reduce its volume, i.e., porosity, when stress is applied. The resulting cake will display an increase in hydraulic resistance. This is not necessarily caused by an average change in porosity, as a porosity gradient can occur by the redistribution of the solid material. Rigid granular particles tend to be incompressible and filter well even with thick cakes. Materials that are easily deformed such as amorphous or thixotropic materials will respond well to mechanical pressure or operation with thin cakes.

Laboratory test tube centrifuges can determine if there is a sufficient density difference between the two phases to consider sedimentation as an alternative. If there is a sharp separation, one can anticipate the same in the field. One can also answer the following questions. Do the solids settle or float? Is the solid phase granular or amorphous? What is the moisture content? The characteristics of the solids indicate the solids discharge design required, i.e., scroll in decanters, or in disk centrifuges, flow-through nozzles or wall valves.

A laboratory centrifuge can also simulate the operation of a filtering centrifuge and verify product characteristics, filtration rates and wash requirements. Various filter cloths can also be tried using only one-liter samples.

With these data summarized, one can now discuss the application with vendors or consultants with expertise in centrifuge operations to help simplify the selection process. Pilot plant testing can be done with 10–25 gallons at the vendor's facility or with a rental unit for an in-plant trial. If sufficient material is available, semi works tests are recommended as more data can be taken for scale up. Equipment manufacturers should be questioned about how they are scaling up, whether it is based upon volume, filter cloth area, etc., and what accuracy can be expected. Critical to a trial's success is how representative a slurry sample is. This is especially important with fermentation processes that change over time and in-plant trial of reasonable scale may be mandatory.

There can be clear advantages to using a centrifuge over a filter, such as a drier product or a more effective separation. This will be dependent upon the application. However, there can be applications where a nutsche or even several types of centrifuges appear, from small scale testing, to yield similar product quality results. Ultimately, the decision will depend upon "operability," that is installation, maintenance, and day-to-day operation. For these purposes, it is strongly recommended that the plant invest the time and relatively small capital costs for a rental unit to run an in-plant test on the product.

Production scale rental units can be operated by taking a side stream from an existing process. This facilitates comparisons of the new equipment to existing plant operations. Rental cost for one month will be approximately three percent of the purchase price of the test equipment; credit for part of the rental is usually offered against the purchase of a production unit. Rental periods may be limited, however lease options are available.

3.3 Materials of Construction

Various materials are available as with all process equipment, ranging from carbon steel coated with rubber, Halar or Kynar, to stainless steel 304, 316 and higher grades, or more expensive alloys such as titanium, Hastelloy C-22, C276, or C4. These grades of Hastelloy will, however, double the capital outlays.

Coatings should be avoided if possible as product "A" can diffuse into the surface and potentially reverse its path. It therefore has the potential to contaminate product "B." Being permeable, they are also subject to peeling. Coatings can be used most effectively on stationary parts dedicated to liquid use that are, therefore, not exposed to maintenance tools, etc.

Working with an existing process will provide the most reliable data for choosing the most economical material for the service required. A new process may require a corrosion study by an in-plant metallurgist, if questionable. Test coupons are a relatively inexpensive way to conduct testing on a small scale and can be obtained from equipment vendors or from the materials manufacturers themselves. Companies such as Haynes, Allegheny, etc., will also have descriptions of the suitability of their different alloys for various processes. An overview of corrosion testing and materials is presented in *Perry's Chemical Engineer's Handbook*, Sixth Edition, Section 13.23.

4.0 COMPONENTS OF THE CENTRIFUGE

Centrifuges consist of the following components:

- Rotor (bowl or basket) that rotates and contains the product
- Solids discharge unloading system, plough, scroll, inverting basket, nozzle system, etc.

- Drive system to rotate the bowl including main bearing shaft with seals, etc., and motor for electric or hydraulic operation
- Frame to support unit
- Enclosure to contain rotor

5.0 SEDIMENTATION CENTRIFUGES

Sedimentation centrifuges are commonly known as *solid bowl systems*, i.e., perforated bowls that are used to separate materials such as cream from milk, sludges from water in waste water treatment plants, and, of course, the biotechnology materials.

The basic principle of sedimentation is that a fluid consisting of two or more phases is subjected to a centrifugal-force field. As the heavier phase travels away from the axis of rotation, there is an ever-increasing centrifugal force. The centrifuge increases the settling rate to clarify one phase, while simultaneously concentrating the other (usually solids). There is no flow of liquid through a cake; hence difficult filtrations are typical applications. How quickly phases separate will depend upon many factors. Capacities and performance are dictated by the particle size, distribution, solids concentration, and particle shape. Adjustments for these changing factors can be achieved only through experiment, testing the particular application with its deviations.

6.0 TUBULAR-BOWL CENTRIFUGES

Used often in the laboratory, this unit is limited to 4.5 kgs of solids loading with an estimated 10–15 gallons/hour liquid feed rate. Applications include stripping small bacteria or viruses from a culture medium.

6.1 Operation

The simplest sedimentation centrifuge design is the tube type, constructed of a tube 2 to 5 inches in diameter and spun at 62,000 g's. Slurry enters at the bottom of the tube through a feed nozzle and the effluent discharges over a dam at the top. Solids are deposited along the walls as particles intersect the bowl wall and are removed from the fluid. The bowl is suspended from an upper bearing and drive assembly. There is a loose damping assembly at the bottom. By installing two different liquid discharge ports at different radii and elevations, it is feasible to separate two different liquid phases and a solid phase. Solids are unloaded manually when clarity diminishes.

7.0 CONTINUOUS DECANTER CENTRIFUGES (WITH CONVEYOR)

Typical applications in fermentation are thick fermentation broths with high solids concentrations where a relatively drier cake is required. However, protein precipitate cannot be sedimented and animal cell debris, due to their slimy nature, can render scrolling ineffective.

Solids and liquids are discharged continuously in this type of design which can process coarse particles that would blind the discharge system and disks of disk bowl machines. The principle of operation is shown in Figure 13.1.

This unit, often referred to as a decanter, is constructed with a conical bowl and an internal rotating scroll conveyor to propel solids or *beach* them along the inclined wall bowl to then be discharged.

The scroll rotates slower than the bowl at a differential speed of 1/20 to 1/160 of the bowl speed; this differential speed causes translation of the solids along the bowl. Particularly, soft solids can be conveyed with low conveyor differential speeds should higher differential speeds cause resuspensions. Units will have *g* forces up to 6000 and range in diameter from 6" to 48".(1) The solids discharge outlet is usually smaller than the liquid discharge outlet at the opposite end.

By varying the liquid discharge outlet size the pool level or depth of the pond can be controlled. The lower the level, the greater the length of the dry beach section. These units operate below their critical speeds between fixed bearings attached to a rigid frame. Mechanical seals are available for pressure operation up to 150 psig. Operating temperatures are from -87°C to +260°C

7.1 Maintenance

Bearings are the primary concern in this design, however, lifetime will depend upon the service and hours of operation. Abrasive materials can cause excessive wear along the feed zone, the conveyor leading to the beach, and the solids-discharge ports. Refacing with replaceable hard surfacing materials such as Hastelloy or tungsten may be required [1].

A variation on this design is a *screening bowl* machine. After the solids have been pulled from the pool of liquid they will pass under a section of a wedge-bar screen to allow for additional dewatering as well as washing the solids more effectively. This design can of course only be used with particles of 80–100 microns or greater as smaller solids will pass into the effluent.

7.2 Typical Problem for Continuous Decanter Centrifuge with Conveyor

The process has a feed rate of 150 gpm of a fermentation broth at 1.5% solids. We would like to concentrate the feed to 8% solids with no more than 0.2% loss of solids in the effluent. What must the solids phase (underflow) and liquid phase (outflow) throughputs be?

Two mass balance equations must be solved simultaneously.

$$U + Q = 150$$
$$U(.080) + Q(.002) = 150(.015)$$
$$U = 150 - Q$$
$$(150 - 0)(.080) + Q(.002) = 150(.015)$$

Overflow = 125 gpm
Underflow = 25 gpm

$$\% \text{ solids Recovery} = \frac{(25)(.08)}{(150)(0.015)} = 89\%$$

U = Underflow
Q = Overflow

8.0 DISK CENTRIFUGES

Applications for the disk centrifuge (Figure 13.2) can overlap the continuous decanter, but will typically be lower in solids concentration and often finer in particles. Examples are:

1. Cell harvesting, broth clarification for recovery of antibiotics and hormones from the culture medium, for example, mycelia
2. Fractionation of human blood plasma
3. Separation of microorganisms and their fragments when processing fermentation products such as: bakers yeast, single cell proteins, vaccines, amino acids and enzymes
4. Isolation and purification of cell proteins
5. Bacterial cells (*E. coli*) for enzymatic deacylation of penicillin G
6. Mammalian cells

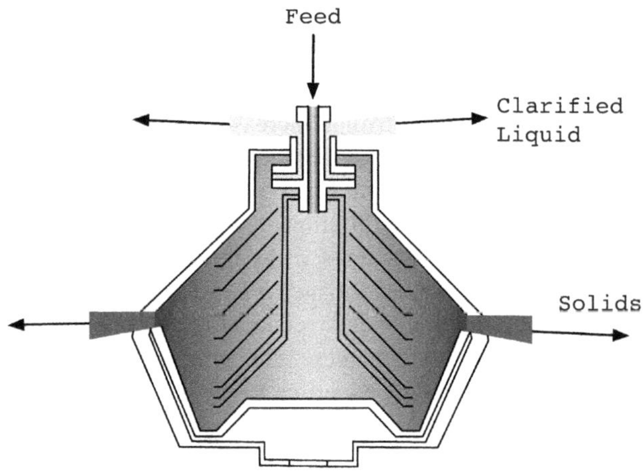

FIGURE 13.2 Disk-bowl centrifuge.

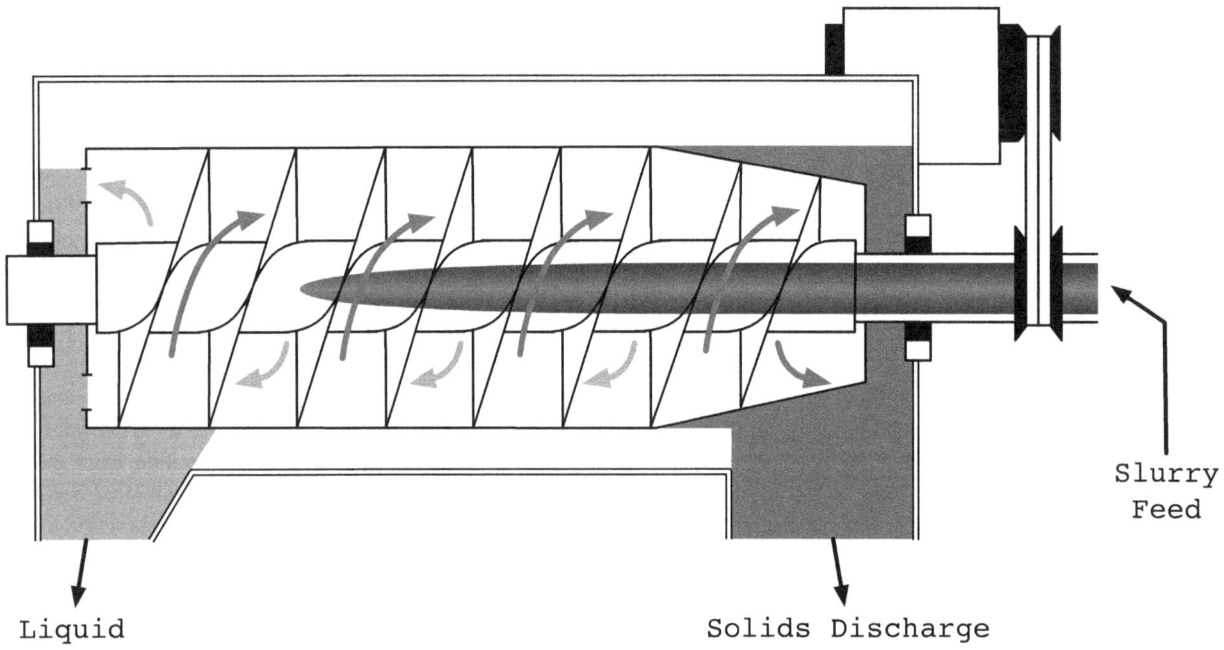

FIGURE 13.1 Solid Bowl Decanter.

8.1 Operation

Solid wall disk centrifuges were designed initially as cream separators. It is a solid bowl design containing a set of stacked disks.

Bowl diameters in a standard disk centrifuge range from 7 to 24 inches and centrifuge *g*-forces of 14,200 to 5,500. A continuous nozzle discharge centrifuge has diameters from 10 to 30 inches and *g*-forces of 14,200 to 4,600. The unit rotates on a vertical shaft as slurry is introduced and pumped down a central pipe beneath the disk stack in close proximity to the bowl wall. The slurry then flows into the disks as particles settle on the underside of the inclined disks and slide to collect along the bowl wall. Liquid continues to move upwards until it overflows a weir and exits the unit.

In a manual design, the bowl is one piece and the system must be stopped and opened up to discharge the collected solids. In a continuous operation, such as a wall-valve-discharge centrifuge, the bowl is made of two cones, a top and a bottom, which periodically separate to release the solids at full rotational speed.

In a nozzle discharge centrifuge, the bowl is a solid, two-cone design. Orifices are located along the maximum diameter to allow solids to flow continuously. Liquid loss must be minimized and orifice sizes are therefore closely matched to the solids capacities. Thickened solids can be recycled to satisfy nozzle flow to maintain a dry effluent. This also circumvents plugging when using larger than usual orifices.

Solids concentrations can range from 15 to 50%. For smaller machines where solids content varies, the intermediate solids discharge design (wall-valve), is preferred. Solids must be of wet toothpaste consistency to flow from these types of disk machines. With the intermittent discharge, however, the solids can be wetter as it is mechanically feasible to open and close the bowl quickly enough to avoid liquid passage.

8.2 Maintenance

To cool the rotor, air is pulled into the casing of the unit and leaves through a frame drain. This air can be a biohazard depending upon the nature of the product. It is therefore recommended to seal the solids discharge frame drain to the solids collection vessel. This air can then be exhausted and handled as waste. Bearings of course, as with any centrifuge, will need to be changed, preferably on a preventive-maintenance basis.

9.0 FILTERING CENTRIFUGES VS. SEDIMENTATION CENTRIFUGES

Cakes will be more compacted in a filtering centrifuge as compared to a sedimentation centrifuge. As solids build at the filter medium surface, a pressure drop is induced by this cake resistance, similar to that in pressure filtration. Sedimentation therefore has an inherent advantage, particularly with difficult filtering, compactible materials.

Applications for filtering centrifuges are usually products granular in nature and relatively incompressible. Centrifugal filters, through the use of a filter medium such as a cloth, are capable of retaining particles down to 1–10 microns. Those using screens as the filter surface will be able to retain 80 micron material as the smallest, without recycling. As many filtering centrifuges leave a residual *heel* of product after solids discharge, finer retention than 70 microns is possible.

Sedimentation centrifuges can even separate to submicron levels as long as there is a density difference. Commercial centrifuges successfully separate one micron particles when specific gravity differences of 0.1 exists. High-speed disk and tubular centrifuges can separate with specific gravity differences as low as 0.02 between phases [1].

Solids from centrifugal filters are drier and powder-like compared to wetter, viscous consistencies from sedimentation units. Of course this is also related to the nature of the products, usually due to particle size. However, if operating with the same material, a centrifugal filter will provide a firmer, drier cake and the wash will also be far more effective due to residence times and a more effective separation.

Capacities or economics are usually the controlling factors in deciding which piece of equipment to use and, even though a filtering centrifuge may yield a superior product quality, the number of machines required may not be justifiable. Production capabilities of sedimentation centrifuges are staggering, over 24,000 gal/h in some cases. Screening centrifuges, pushers, etc., not covered here, process large volumes of materials but of particles greater than 100 microns.

10.0 FILTERING CENTRIFUGES

The driving force for separation in a filtering centrifuge is centrifugal force, unlike filtration where pressure or vacuum is used. The basic principals of Poiseuilles' cake filtration equation can be applied by substituting *P* (pressure) with the stress or pressure induced by the bowl contents, which is a function of centrifugal acceleration. (See Eq. 13.8.)

10.1 Cake Washing

A wash can be introduced to fulfill different requirements:

1. Remove original slurry liquid (mother liquid)
2. Dissolve impurities
3. Alter pH

4. Displace mother liquor with another liquid (often to facilitate drying of a solvent with a lower vapor pressure or eliminate toxic solvent)

Distinctly different types of wash are:

1. *Displacement*—Removal of one liquid in favor of another
2. *Diffusion*—Dissolved materials retained in the capillary liquid and in the surface liquor are transported by the wash medium
3. *Dissolution*—Components of the solid which is composed of different materials of varying solubility are dissolved in the wash medium

One or all of the steps can be used on a product or occur simultaneously. Several steps can be used often, first a displacement wash to remove mother liquors and any associated impurities, followed by a dissolution or diffusion wash.

Two different methods of introducing the wash are:

1. *Flood Washing*—With this method, the wash medium is fed at a faster rate than the case is dewatering, thus a liquid level forms on the top of the cake. This ensures distribution of the wash fluid over the entire cake. Positive displacement is the most effective form of washing. Carried out in a plug flow manner, clean wash fluid contacts the solids without backmixing. Except where retention time is required to allow for mass diffusion of the impurities through the solids, positive displacement washes are more efficient then reslurrying. Redilution of the impurities occurs as reslurrying back mixes impurities into the fresh medium. The cake should be even to achieve this displacement wash, as the wash fluid will seek the path of least resistance on nonuniform cakes. Vertical basket centrifuges in particular have uneven cakes due to the feeding method and can require copious quantities of wash to compensate for the uneven cake.
2. *Spray Washing*—Liquid is supplied via spray nozzles. It is the only effective way of working cakes, which are uneven from the top to the bottom for basket centrifuge. Peeler centrifuges, being unaffected by the force of gravity in the distribution of the cake, tend to have more even cakes, although nonuniformity can still occur due to the feeding mechanism.

11.0 VERTICAL BASKET CENTRIFUGES

11.1 Applications

Vertical basket centrifuges have been the "work horse" for the pharmaceutical industry for many years for intermediate and final filtration steps. Chemical and specialty chemical productions use this type of equipment in a wide range of applications.

In fermentation, typically, post-crystallization steps are processed on this unit, crystalline products that are free-draining. Different designs are available, the simplest being a manual design or under-driven top-discharge. A perforated basket with cloth or screen is the filter media. Filtrate passes through to a filtrate chamber as shown in Figure 13.3. Solids must be dug out or the entire filter bag hoisted out with the solids. Labor intensive with operation exposure a problem, this design is more often employed in pilot than production plants. *G*-forces up to 800 *g*'s are attainable. Introduction of a traversing plough mechanism to this design enables automatic solids discharge through the basket bottom. Speeds are variable in this design through 800 *g*'s.

The cycle is similar to that of a household washing machine, filling, spinning or dewatering, washing, a final spin and unloading. The feed is fed off to one side. This, coupled with the force of gravity in the vertical basket design, can cause an uneven slurry distribution, i.e., a thicker cake at the bottom of the basket or in the middle where the feed pipe is located. Hence, reduced filling speeds may be required depending upon the product. Cones for 360° feed distribution are available for more even loading.

Operating batch-wise, basket centrifuges in general are optimized best when operated almost continuously by feeding the machine at the same rate as the slurry is dewatering, thus maximizing solids concentration. Loading the basket infinitely fast will only produce a basket with the same solids concentration as the feed tank.

The wash can be quantified by mass or volumetric flow to ensure product quality. The intermediate and final spins are usually timed if the basket is automated, however, if out of balance conditions exist during the feed

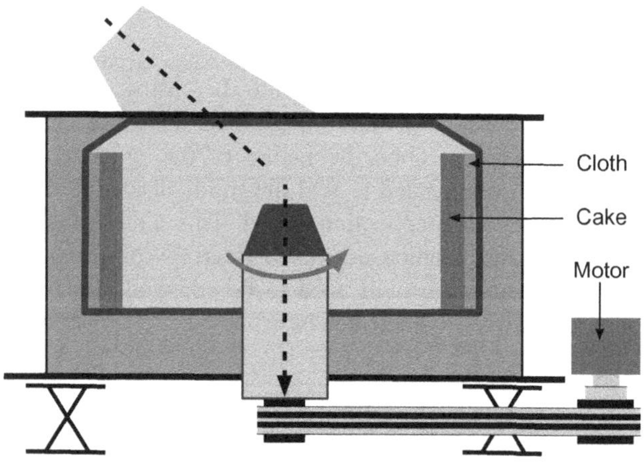

FIGURE 13.3 Vertical basket centrifuge (manual unloading).

cycle, an operator is often required throughout the operation. Operator judgement is often required to determine when the liquid level on the cake disappears. This may be required, for example, to remove all mother liquor, introduce the wash, particularly for difficult filtrations, or to be sure the liquid level remains on the cake to prevent cracking and preferential channeling of the wash liquid. Variable solids concentrations or particle sizes distributions will make it more difficult to fully automatic a standard basket centrifuge as operator judgment may be required at several points in the cycle. Consistent, uniform batches with every filtration can, however, be automated on a time basis.

11.2 Solids Discharge

For solids unloading, a plough cuts out the solids at reduced speeds of 40–70 rpm, and traversing action must be slowed down to prevent the basket from stalling. For certain products, the cake can be sufficiently difficult to remove that the plough cannot remove all of the solids due to tolerances and possibility of damaging the filter cloth, thus a residual heel of solids is left for some products. For some products, this is not a problem and the next cycle can begin. For others, the heel can glaze over and reduce filtration rates on subsequent batches. It must be scraped out manually, dissolved, or an air knife can be used, depending upon the hardness of the heel. Depending upon how problematic the *residual heel*, even the automated vertical basket can be labor intensive.

11.3 Operational Speeds

An average cycle would be filling at 600 rpm, washing at 800 rpm and dewatering at 1000 rpm. With a 48" basket, these are *g* forces of 240, 426 and 667, respectively. Discharge by a plough occurs at less than 100 rpm, or, if manually unloaded, at zero speed.

11.4 Maintenance

If there are significant out-of-balance operating conditions, mechanical parts such as the plough or cake detection can vibrate loose. The bearings and shaft seal components will also have limited lifetimes, depending on the operation.

12.0 HORIZONTAL PEELER CENTRIFUGE

12.1 Applications

The *horizontal peeler centrifuge* (Figure 13.4) is a variation of the vertical basket. Up to 80 inches in diameter and producing as much as 100 tons per day of product applications, this machine has been prevalent in the isolation of beet sugar and starch. The design, characterized as *Ter Meer*, after the inventor, is sometimes used in bulk pharmaceutical productions. Dedicated productions of relatively easy filtrations being processed are applications for this type of equipment.

12.2 Operation

Solids Discharge. This is carried out by an automatic plough or knife. Since the knife cannot contact the filter medium, a heel of product remains in the basket after each discharge. This can prevent fines from passing, but, like in a vertical basket, may become glazed and impervious to filtration. Backwashing the heel or redissolving may be possible. Even changing the depth at which the blade cuts the cake may help. In the Ferrum design, high-pressure air forces the cake off the screen during discharge. This will work in some applications. Solids exit a chute, but can also discharge by screw conveyor (Figure 13.5).

Feed Mechanism. A cake detection device, pneumatic in some designs, activates the feed valve closure when the desired cake depth is reached. A proximity switch monitors this cake depth. This device can also act as a cake distributor to level the load during feeding. Alternative feed designs are available depending upon the vendor.

Wash. The entire cycle is operated automatically on a pre-programmed basis, all by time. The wash can, of course, be by time or volumetric basis, monitored by air in-line flow meter and totalized.

Operational Speeds. During the entire sequence of loading, deliquoring, and unloading, a constant bowl speed is maintained. "Dead time," associated with acceleration and deceleration, is minimized. Maximum

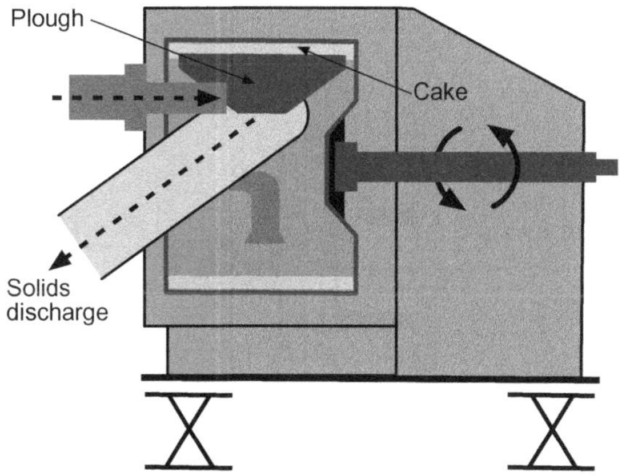

FIGURE 13.4 Peeler centrifuge.

operating speeds depend upon bowl diameter, the larger the diameter the lower the speeds. Sizes range from pilot scale 450 mm to 1430 mm diameter with *g*-forces from 3200 to 1200. Basket speeds range from 3000 rpm to 1200 rpm. Specially designed vibration-damping systems will minimize plant structural supports required. Each manufacturer's design must be evaluated as to what is required. Special cement foundations are often a necessity.

13.0 INVERTING FILTER CENTRIFUGE

Originally designed by the firm Heinkel, inverting filter technology has revolutionized the concept of filtering centrifuges since their introduction into the pharmaceutical market in the early 1980s. The design eliminated the inherent problems in the conventional centrifuge design of the solids discharge process and balance problems long associated with centrifuges.

FIGURE 13.5 Inverting filtering centrifuge. (Courtesy of CelesTech Inc.)

Affecting a fully automatic solids discharge, an inverting filter removes all product from the cloth, thereby eliminating any residual heel. This permits the separation of a wider range of materials than conventional basket centrifuges. Amorphous through crystalline products can be separated in this type of centrifuge as there are no residual solids left on the cloth that can blind or glaze over. Extremely difficult filtrations are therefore possible. Small volumes of fermentation broth through post-crystallization steps are found on this type of unit.

13.1 Operation

Solids Discharge. The centrifuge is horizontally mounted and the cycle is similar to a vertical or peeler centrifuge, i.e., feeding, washing, dewatering and solids discharge. The basket, however, is in two parts, a bowl and a bowl insert. By fixing the end of the filter cloth under a clamping ring on the bowl insert, the filter cloth can be inverted by axially moving the bowl insert. This is shown in Figure 13.6. Rotation of the bowl and bowl insert in unison at reduced speed ensures a complete solids discharge.
Feed Mechanism. An open ended pipe centered in the bowl allows feeding of the slurry 360° around the cloth. The Inverting Filter Centrifuge is horizontally mounted like the peeler, so *g*-force does not effect the distribution of the cake. In addition, bars connecting the front plate to the back plate of the bowl insert serve as a distribution mechanism. Slurry passing the bars is evenly dispersed providing for a uniform cake. As a result, out-of-balance conditions are minimized. A special cement foundation or vibration isolator normally required for centrifuges is not necessary. Without this vibration, a load cell can be used in lieu of cake detectors or "feelers" to monitor the cycle and prevent overfilling the bowl. A typical cycle is shown in Figure 13.7.

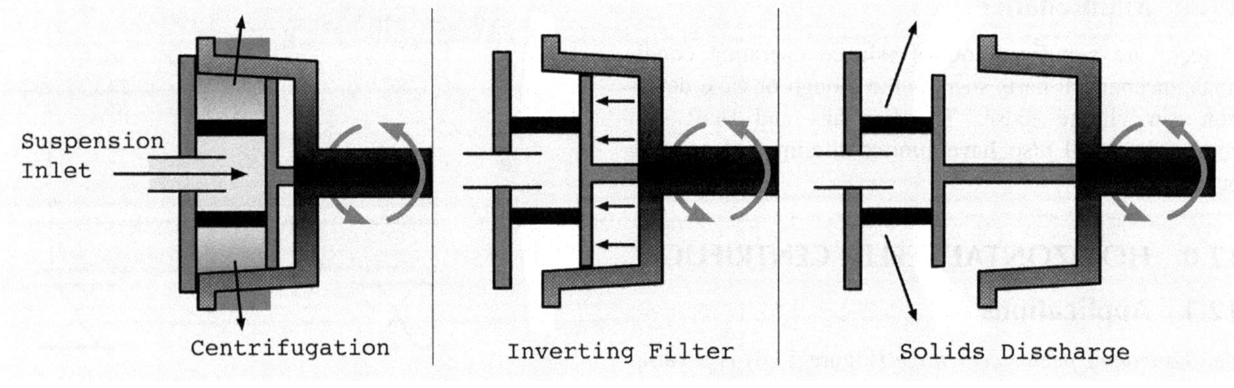

FIGURE 13.6 Inverting filter centrifuge. (Courtesy of Perry Videx.)

Multipurpose Applications. Cake thickness is varied, dependent upon the application. Thin cakes from finer particles or more amorphous, compressible materials versus thick cakes for hard, easy filling crystals.

Discharge time is less than one minute, and even cake distributions allow for higher filling, washing and dewatering speeds, thus the overall cycle is shorter. One can therefore "efficiently" operate with a thin cake as low as 1/4", if necessary, as opposed to 3–6 inches on a conventional basket. If operating with a thin cake on a basket, the residual heel still exists and, as it requires sufficiently longer times for processing, it would be inefficient to operate with such a thin cake.

As a result of relatively thinner cakes and higher *g*-forces, filtration rates per unit filter cloth area can be as high as 20–30 times that of typical basket centrifuges. For that reason, a smaller volume Inverting Filter Centrifuge can replace a larger basket centrifuge. By optimizing based upon cake thickness (see Figure 13.8.), higher productivities will be reached.

Fermentation Applications

Fermentation broth applications with cell debris or micellia removal have been successful in the inverting filter technology. However, the large volumes required could preclude this technology.

Stacked disk centrifuges and solid bowl decanters would be be more likely applications.

Fermentation processes that produce proteins using E. Coli bacteria in the drug productionprocess are also separated in the Inverting Filter Centrifuge.

Post-crystallization fermentation can be found in filtering centrifuges.

These include modern processes such as human insulin and classic processes such as ampicillin, pen-V, erythromycin, penicillin and 7-ACA.

They have all been processed on filtering centrifuges such as normal baskets, and Inverting Filter

Centrifuges. They are precipitated in crystalline form then isolated in the centrifuge, washed and then dried.

Crystallization of fermentation supernatant or intermediates can be used for purification in place of chromatography. Crystallization of the final drug is an alternative to lyophilization (freeze drying)

13.2 Maintenance

By operating with minimal vibration, wear of mechanical parts is reduced. Bearings and shaft seals are changed on a preventive maintenance basis every three to four years.

Regular maintenance is required for the filter cloths and product contacted O-rings. These must be chosen in materials of construction compatible with the product. They are usually changed at the end of a campaign before switching to a new product. For dedicated processes, lifetime will depend upon the product.

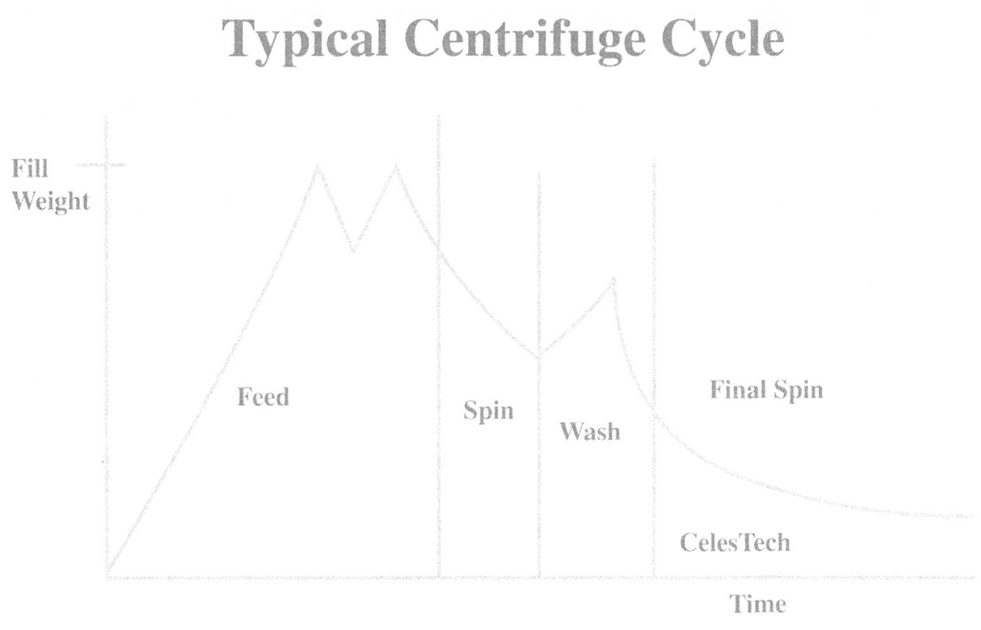

FIGURE 13.7 Centrifuge cycle

14.0 MAINTENANCE: CENTRIFUGE

All rotating equipment will exhibit certain harmonic frequencies upon acceleration and deceleration of the unit. It is the speed at which the frequency of rotation equals the natural frequency of the rotating part.[1] They can be calculated from the moment of inertia, but are best found by experiment, running the unit from zero to maximum speed and noting any increase in vibration or noise of the unit. Operating conditions should pass through these speeds, however, never maintain them for any period of time as, at this speed, any vibration induced by the imbalance in the rotor is compounded resulting in abnormally high stresses. True critical speeds are well above the allowable operating speeds.

14.1 Bearings

As a rotating unit, bearings changes should be planned on a preventive maintenance basis every few years. Bearing noise-monitoring systems can prevent emergency shutdowns. Ninety percent of bearing failures can be predicted months in advance. Ten percent are still unforeseen. Bearing factories produce the highest caliber of any manufactured goods. Defects are not the primary cause of failures. Failures usually stem from:

1. Contamination, including moisture.
2. Overstress
3. Overuse of lubrication including mixing incompatible greases.
4. Defects created on installation or transportation and to a much less extent, insufficient lubrication.

Bearing temperature probes are also available at each bearing point although they will not provide the advance notice that a noise monitoring system will. Scheduled shutdown and changing of shaft seals is recommended on at least a tri-annual basis. Over-greasing of bearings and mixing of incompatible bearing greases can cause more problems than under-greasing. High temperatures of over 100°C will also be exhibited when over-greasing or under-greasing.

The trend towards elimination of hydraulics in pharmaceutical process areas has turned maintenance over to the instrumentation and electrical specialists, as variable frequency drives become the standard. The Inverting Filter Centrifuge in some models eliminates all hydraulics from the centrifuge design to satisfy increasingly stringent cleanroom requirements by pharmaceutical companies. The risk of hydraulic oil in the process area

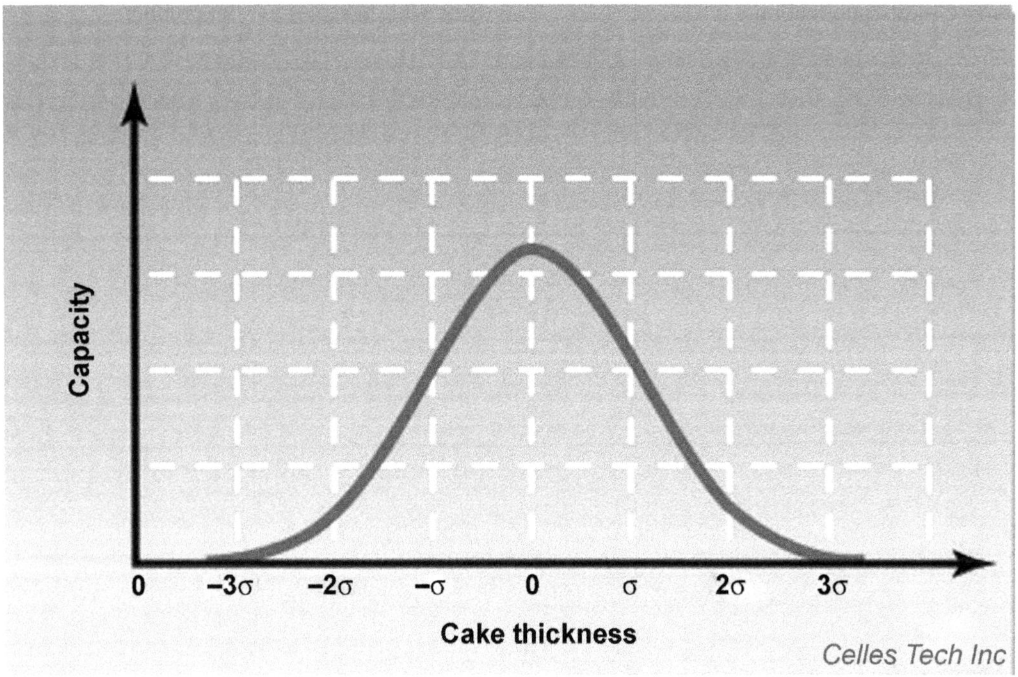

FIGURE 13.8 Optimizing.

has been a concern with respect to contamination with the product.

15.0 SAFETY

Vibration Detection System. To monitor vibration levels, every centrifuge should be equipped with a vibration detection device. Usually mounted on the filtration housing itself, a local transducer will send a signal back to a vibration monitor. After exceeding a certain vibration setpoint (2–3 inches per second for a standard basket centrifuge, 0.75 inches/sec. for an inverting filter) the controller will close all process valves and decelerate the machine to a stable operating condition, therefore, any mother liquors can be spun off. Should the vibration levels still exceed the setpoint, the machine will be given an emergency stop signal.

Out of balance conditions usually occur due to feeding of an unbalanced load or uneven cake. This is more likely in vertical basket centrifuges than horizontally mounted systems. It is less common in an Inverting Filter Centrifuge due to the central feed distribution.

Inerting System with Oxygen Analyzers. Operations with solvents require an inert atmosphere, usually nitrogen, or in some cases, carbon dioxide. A purge of the bearing housing, shaft sealing system and process areas are required in critical blanketing of the system, based upon time to allow for a certain number of volume changes to be performed or the preferred method of blanketing, until a low oxygen setpoint is needed. Oxygen analyzers continuously monitoring a sample gas stream from a vent on the unit confirms the safe oxygen operating level. This level must be below the lower explosive limit for solvent. Three conditions must occur for a fire, a spark, an oxygen rich atmosphere, and the fuel, i.e., solvents. Although a spark is not expected, static electricity can occur in lines, etc.

One should choose an oxygen analyzer with a wet sampling system, i.e., precondition of the sample gas with a prefilter. Entrained solvents in the sample gas can affect readings. A scrubber may be required to remove noxious gases to prevent corrosion to the sampling system. Oxygen analyzers should be evaluated as some systems fail to the unsafe condition of 0% O_2.

16.0 PRESSURE AND CENTRIFUGATION

It is more efficient to mechanically dewater solids than thermally, due to costly energy requirements. Filters such as pressure or vacuum units are used for solids/liquid separation, providing high forces to drive the liquid through the cake. Equipment designed to combine both centrifugation and pressurization has lead to increased dewatering of solids beyond what either process would do alone. (See Figure 13.9.)

A centrifugal field achieves mechanical separation of slurries by emptying the liquid in the capillaries between

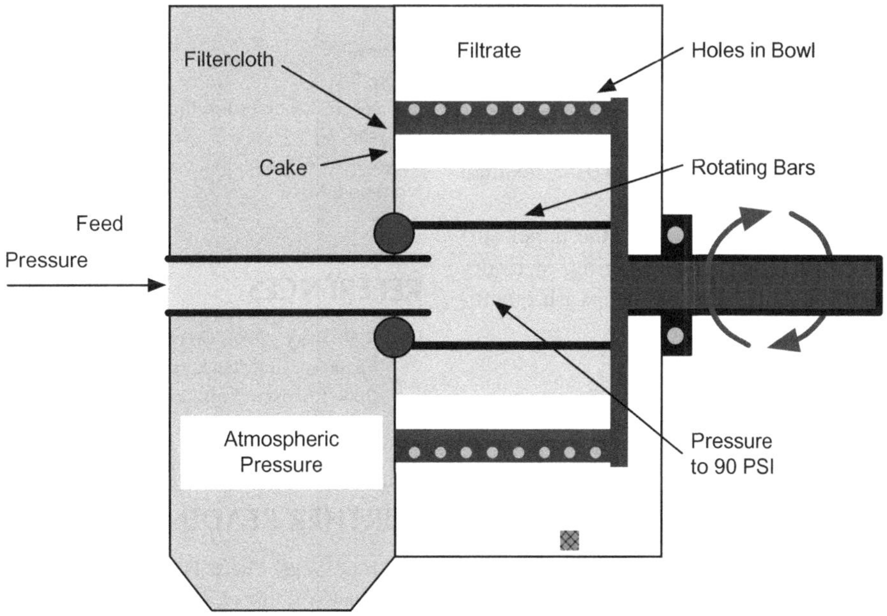

FIGURE 13.9 Pressure in inverting filter centrifuge.

the solids. Larger particles will exhibit faster drainage of these capillaries.

Liquid in the interstices of the solids is retained due to high capillary forces in the micron pores and cracks in the particle. These capillary forces are so high that they can only be removed thermally. This contributes to a certain capillary height of liquid that is independent of the packed bed weight. After dewatering for an extended time, an equilibrium point is reached. Only changing the driving force by increasing centrifugal force will overcome and reduce this equilibrium saturation point. Product with a smaller particle size distribution will have higher capillary forces and thus a higher equilibrium saturation point or residual moisture.

By addition of the driving force pressure, or pressure differential across the packed bed, additional liquid is forced through the capillaries below the equilibrium saturation point, thus reducing the residual moisture.

Initial pressurization of the basket alone, thus avoiding pressurizing the entire centrifuge, can decrease final dewatering steps by as much as 80%. By blowing through the cake at a certain temperature, volume of gas, and pressure, drying will be achieved. Products that are crystalline and easy to filter can be dried in a relatively short period of time, not adding significantly to the overall cycle. Difficult filtering, amorphous materials may see overall cycle times reduced or products previously wet and sticky now easily handled at lower moisture levels going into a dryer. (See Table 13.3.) Downstream drying equipment can then be reduced in size, or possible eliminated. Although the more difficult and time intensive the drying cycle, the less likely that the one stage centrifuge dryer will be efficient as a single unit. In this case a decoupled system with a separate dryer suitable for the product would be more efficient.

The Inverting Filter with pressure added during the process is able to dry products as low as 0.01% residual moisture using hot gas.

With heated gas, it is possible to break the upper surface of the moisture film, aiding in dewatering, or to dry or strip solvents. Steam washing can reduce wash quantities required significantly.

17.0 MANUFACTURERS

17.1 Filtering Centrifuges

Inverting Filter Centrifuges

Comi-Condur SpA
Italy
Heinkel
Germany

Perforated Basket Centrifuges/Vertical or Horizontal

Andritz
Germany
Ketema, Bird, Dorr-Oliver (Acquired by Andritz(formerly Krauss-Maffei))
Broadbent
England
Robatel Pittsfield, MA
Robatel France
Western States Machine Company, Inc.
Hamilton, OH

17.2 Sedimentation Centrifuges

Baker Perkins, Inc.
Saginaw, MI
Andritz
Florence, KY
Centrico, Inc. (Westfalia)
Northvale, NJ
DeLaval (Sharples/Pennwalt)
Warminster, PA

17.3 Oxygen Analyzers

Neutronics
Exton, PA
Orbisphere Laboratories, Inc.
Emerson, NJ
Servomex Company, Inc.
Norwood, MA

REFERENCES

[1] R.H. Perry, D.W. Green, J.O. Maloney (Eds.), Perry's Chemical Engineers Handbook, sixth ed., Sections 21, 27, McGraw-Hill Book Co., New York, 1984.

FURTHER READING

[1] H.C. Vogel (Ed.), Fermentation and Biochemical Engineering Handbook, first ed., Noyes Publications, New Jersey, 1983.
[2] Centrifuges, Filtration and Separation, Nov 1993, p. 6.
[3] C. Wang, D. Demain, L. Humphrey, Fermentation and Enzyme Technology, Wiley-Interscience, John Wiley and Sons, New York, 1979, pp. 262–267.

[4] Equipment Testing Procedure Committee (AIChE), Centrifuges: A Guide to Performance Evaluations, AIChE, New York, 1980.

[5] M.G. Wirtsch, W. Stahl, Model for mechanical separation of liquid in a field of centrifugal force, Min. Process. 11 (1988) 619–627.

[6] B.C. Bershad, K.M. Chaffiotte, W.F. Leung, Making centrifugation work for you, Chem. Eng. 8 (1990) 84–89.

[7] R.J. Wakeman, Modelling slurry dewatering and cake growth in filtering centrifuge, Filtr. Separ. (1994) 75–81.

[8] G. Mayer, W. Stahl, Model for mechanical separation of liquid in a field of centrifugal force, Min. Process. 11 (1988) 619–627.

[9] J. Wright, Practical guide to the selection and operation of batch-type filtering basket centrifuges, Filtr. Separ. (1993) 647–652.

Chapter 14

Drying

Barry Fox, Giovanni Bellini and Laura Pellegrini

Section I

Indirect Drying

Giovanni Bellini and Laura Pellegrini

1.0 INTRODUCTION

The drying operation is often the final step of a manufacturing process. *Indirect drying* will be discussed in this section; it is the process of removing liquid by conductive heat transfer.

Sometimes drying is a part of the manufacturing process itself, as in the case of seasoning of timber or in papermaking, but generally, the reasons for carrying out a drying operation are:

- to reduce the cost of transport
- to ensure a prolonged storage life
- to make a material more suitable for handling
- to avoid presence of moisture that may lead to corrosion
- to provide the product with definite properties

The type of raw material is of extreme importance in the drying process; for instance, to retain the viability and the activity of biological materials such as blood plasma and fermentation products, the operation is carried out at very low temperatures, while more severe conditions can be applied to foodstuffs.

If it is possible to remove moisture mechanically, this will always be more economical than removing it by evaporation. However, it will be assumed in the following that, for the type of raw material and its final use, the removal of volatile substances is carried out by heat.

2.0 THEORY

Drying Definition. Drying is a unit operation in which a solvent, generally water, is separated from a solution, semisolid material or cake/solid pastes by evaporation.

In the drying process, the heat is transferred simultaneously with the mass, but in the opposite direction.

Drying Process Description. The moisture content of a material is usually expressed as a weight percentage on a dry basis. The moisture may be present as:

- *Free moisture.* This is the liquid in excess of the equilibrium moisture content for the specific temperature and humidity condition of the dryer. Practically, it is the liquid content removable at a given temperature and humidity.
- *Bound moisture.* This is the amount of liquid in the solids that exhibits a vapor pressure less than normal for the pure liquid.

In the drying of materials it is necessary to remove free moisture from the surface as well as bound moisture from the interior. The drying characteristics of wet solids can be described by plotting the rate of drying against the corresponding moisture content. A typical drying curve is shown in Figure 14.1 and it can easily be seen that this is subdivided into four distinct sections:

The curved portion, *AB*, is representative of the unsteady state period during which the solid temperature reaches its steady state value, *ts*. *AB* may occur at decreasing rate as well as at the increasing rate shown.

The critical moisture content is thus identified as the average moisture content of the solid at the instant the first increment of dry area appears on the surface of solid.

The critical moisture content depends upon the ease of moisture movement through the solid, and hence, upon the pore structure of the solid, sample thickness and drying rate. Segment *BC* is the constant-rate period. During this period, the drying is controlled simultaneously by heat and mass transfer applied to a liquid-gas interface in dynamic equilibrium with a bulk gas phase.

Moisture flow from within the material to the surface is fast enough to maintain a completely wet surface. The surface temperature reaches the wet-bulb temperature. The rate of drying can be expressed as:

$$\frac{dW}{d\phi} = Kp(ps - pa) \qquad (14.1)$$

Fermentation and Biochemical Engineering Handbook.

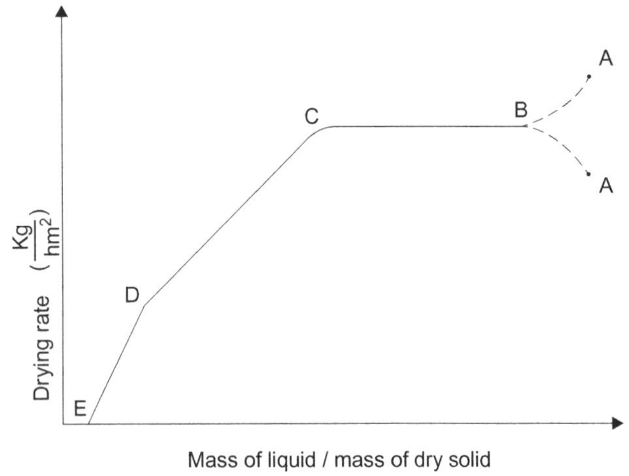

FIGURE 14.1 Drying rate curve.

where $dW/d\phi$ is the rate of drying, i.e., change in moisture with time; *Kp* is the mass transfer coefficient, *ps* is the saturation vapor pressure of the liquid at the surface temperature, *ts*; and *pa* is the partial pressure of water vapor. In addition, the following equation also applies:

$$\frac{dW}{d\phi} = \frac{ha}{\lambda}(ta - ts) = Kp(ps - pa) \qquad (14.2)$$

where λ is the latent heat of vaporization, *ha* is the heat transfer coefficient, *ta* is the dry bulb temperature of the air and *ts* is the temperature of the product surface.

By integrating Eq. (14.2), it is possible to derive the drying time in the constant rate period. Equation (14.2) is derived for heat transfer to the material being dried by circulating air. When large metal sheets or trays are close to the product, it is not possible to ignore the conduction and radiation contribution to heat transfer. In this case, the solid temperature is raised above the air wet-bulb temperature and Eq. (14.2) becomes:

$$\frac{dW}{d\phi} = ha\frac{A1}{\lambda}(ta - ts) + \frac{hcA2}{\lambda}(tc - ts) + \frac{FA3E\delta}{\lambda}(Tr^4 - Ts^4) \qquad (14.3)$$

where *A*1, *A*2, *A*3 are the solid surfaces, respectively, for convection, conduction and radiation heat-transfer, *tc* is the temperature of the heat surface for conductive transfer, Fis a view factor, depending on the geometry, *E* is the emissivity of the surface, δ is the Stefan-Boltzmann constant, *Tr* is the absolute temperature of the radiating surface and *Ts* is the absolute temperature of the product surface. The increase in *Ts* allows the drying at an increased rate, both during the constant rate and the first falling rate period. At the end of the constant-rate period, the movement of the liquid to the solid surface becomes insufficient to replace the liquid being evaporated. The *critical moisture content* is thus identified as the average moisture content of the solid at the instant the first increment of dry area appears on the surface of the solid. The critical moisture content depends upon the ease of moisture movement through the solid and, hence, upon the port structure of the solid, sample thickness and drying rate.

Segment *CD* is the first falling-rate drying period. It is the period between the appearance of the first dry area on the material surface and the disappearance of the last liquid-wet area; drying occurs at a gradually reduced rate. At point *D*, there is no significant area of liquid saturated surface.

During the phase *CD*, Eq. (14.2) is still applicable to the moisture removal rate, provided that *ts* and *ps* are suitably modified and account is taken of the partial dryness of the surface.

Segment *DE* is the second falling-rate. The moisture content continues to fall until it reaches the *equilibrium moisture content, E.* The equilibrium moisture content is reached when the vapor pressure over the solid is equal to the partial pressure of vapor in the atmosphere. This equilibrium condition is independent of drying rate. It is a material property. Only hygroscopic materials have an equilibrium moisture content.

For non-hygroscopic materials, the equilibrium moisture content is essentially zero at all temperatures and humidities. Equilibrium moisture content is particularly important in drying because it represents the limiting moisture content for given conditions of humidity and temperature. The mechanisms of drying during this phase are not completely understood, but two ideas can be considered to explain the physical nature of this process—one is the diffusion theory and the other the capillary theory.

Diffusion Mechanism. In relatively homogeneous solids, such as wood, starch, textiles, paper, glue, soap, gelatin and clay, the movement of moisture towards the surface is mainly governed by molecular diffusion and, therefore, follows Ficks' Law.

Sherwood and Newman gave the solution of this equation in the hypothesis of an initial uniform moisture distribution and that the surface is dry; the following expression is derived (for long drying times):

$$\frac{dW}{d\phi} = \frac{\pi^2 D}{4L^2}(W - We) \qquad (14.4)$$

where $dW/d\phi$ *is* the rate of drying during the falling rate period, *D* is the liquid diffusivity of the solid material, *L* is the total thickness of the solid layer thickness through which the liquid is diffusing, *W is* the moisture content of the material at time, ϕ, and *We* is the equilibrium moisture content under the prevailing drying conditions. Equation (14.4) neglects capillary and gravitational forces.

Capillary Model. In substances with a large open-pore structure and in beds of particulate material, the liquid flows from regions of low concentration to those of high concentration by capillary action. Based on this mechanism, the instantaneous drying rate is given by:

$$\frac{dW}{D\phi} = \frac{h(ta - ts)(W - We)}{2\phi\, L\, l(Wo - We)} \quad (14.5)$$

where ϕ is the density of the dry solid and Wo is the moisture content when diffusion begins to control.

Most biological materials obey Eq. (14.4), while coarse granular solids such as sand, minerals, pigments, paint, etc., obey Eq. (14.5).

Shrinkage and Case Hardening. When bound moisture is removed from rigid, porous or nonporous solids they do not shrink appreciably, but colloidal nonporous solids often undergo severe shrinkage during drying. This may lead to serious product difficulties; when the surface shrinks against a constant volume core, it causes the material to warp, check, crack or otherwise change its structure. Moreover, the reduced moisture content in the hardened outer layer increases the resistance to diffusion. In the end, the superficial hardening, combined with the decrease in diffusive movement, make the layer on the surface practically impervious to the flow of moisture, either as liquid or vapor. This is called *case hardening*.

Reducing the drying rate, thereby flattening the moisture gradient into the solid, can minimize all these problems. Since the drying behavior presents different characteristics in the two periods—constant-rate and falling-rate—the design of the dryer should recognize these differences, i.e., substances that exhibit predominantly a constant-rate drying are subject to different design criteria than substances that exhibit a long falling-rate period.

Since it is more expensive to remove moisture during the falling-rate period than during the constant-rate one, it is desirable to extend as long as possible the latter with respect to the former. Particle size reduction is a practical way to accomplish this because more drying area is created.

An analysis of the laws governing drying is essential for a good dryer design, therefore, it is important to note that, due to the complex nature of solid phase transport properties, only in a few simple cases can the drying rate (and drying time) be predicted with confidence by the mathematical expressions reported above. In these cases, one usually deals with substances that exhibit only, or primarily, constant-rate drying.

For materials that present a non-negligible falling-rate period, the use of specific mathematical equations is subject to a high number of uncertainties and simplifying assumptions are generally required.

It is clear that the purely mathematical approach for designing a drying plant is not possible, given the present state of knowledge.

3.0 EQUIPMENT SELECTION

Several methods of heat transfer are used in the dryers. Where all the heat for vaporizing the solvent is supplied by direct contact with hot gases and heat transfer by conduction from contact with hot boundaries or by radiation from solid walls is negligible, the process is called *adiabatic, or direct drying.*

In *indirect or nonadiabatic drying*, the heat is transferred by conduction from a hot surface, first to the material surface and then into the bulk. This chapter discusses only indirect drying.

The problem of equipment selection can be very complex; different factors must be taken into consideration, for example, working capacity, ease of cleaning, hazardous material, dryer location and capital cost (see Figure 14.2).

The first step refers to the choice of continuous versus batch drying and depends on the nature of the equipment preceding and following the dryer as well as on the production capacity required. In general, only batch dryers will be considered in the following.

Batch dryers include:

- Shelf dryers. They are usually employed for small capacities and when the solvent doesn't present particular problems.
- Vacuum dryers. These are the most-used batch dryers.
- Vacuum dryers are usually considered when:
- Low solids temperature (<40°C) must be maintained to prevent heat causing damage to the product or changing its nature
- When toxic or valuable solvent recovery is required
- When air combines with the product, during heating, causing oxidation or an explosive condition

Before starting work on selecting a dryer, it is good practice to collect all the data outlined in Table 14.1.

In vacuum drying, the objective is to create a temperature difference or "driving force" between the heated jacket and the material to be dried. To accomplish this with a low jacket temperature, it becomes necessary to reduce the internal pressure of the dryer to remove the liquid/solvent at a lower vapor pressure. Decreasing the pressure creates large vapor volumes. Economic considerations arising from concerns of leakage, ability to condense the solvent, size of vapor line and vacuum pump, affect the selection of the operating pressure. Materials handled in vacuum dryers may range from slurries to solid shapes and from granular, crystalline product to fibrous solids. The characteristics of each type of vacuum dryer are discussed below to help make a proper choice Table 14.2.

Vertical Vacuum Pan Dryers. The agitated vertical dryer (Figure 14.3.) has been designed for drying many different products that may come from centrifuges or filters.

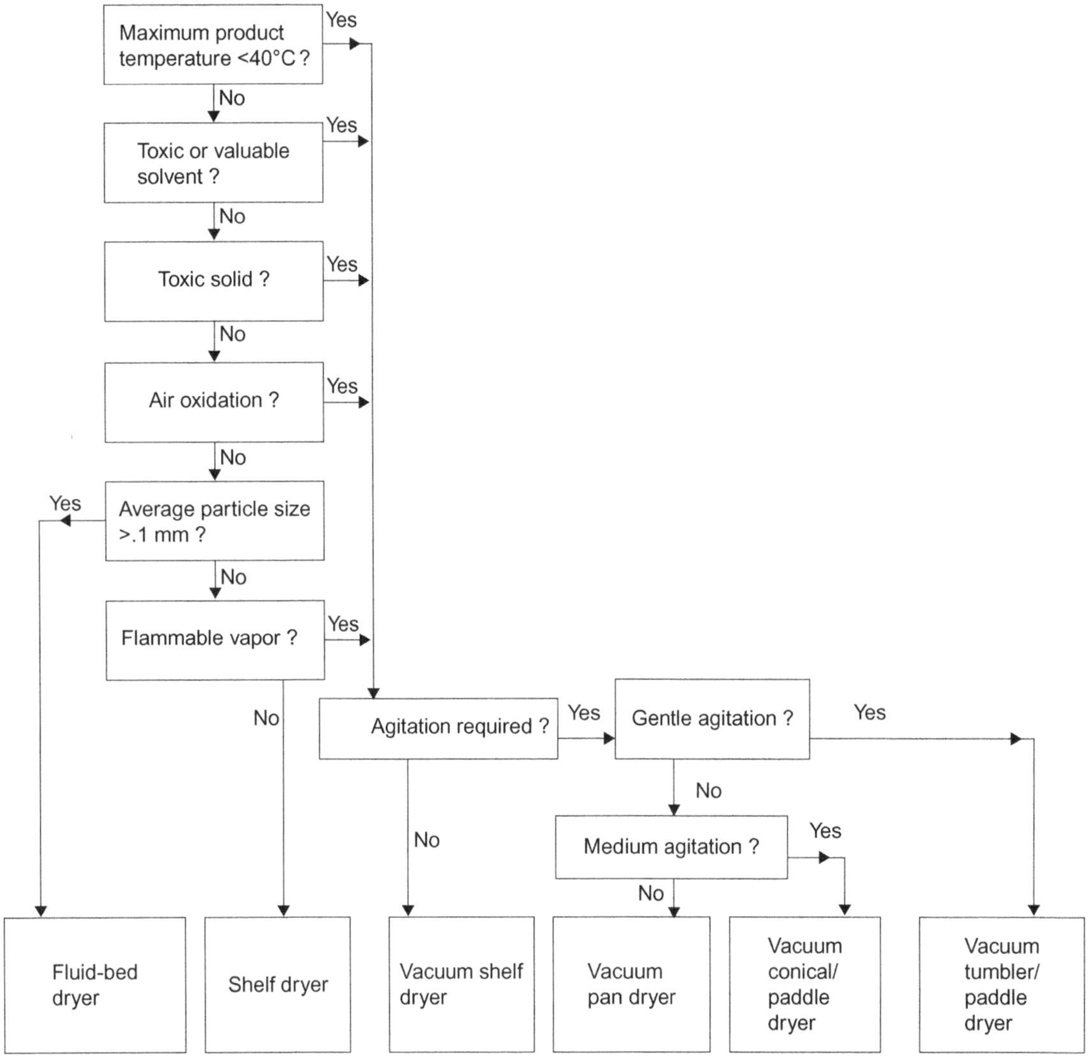

FIGURE 14.2 Flowchart for selection of a batch dryer.

TABLE 14.1 Data to be Assessed Before Attempting Drying Selection

- Production capacity (kg/h)
- Initial moisture content
- Particle size distribution
- Drying curve
- Maximum allowable product temperature
- Explosion characteristics (vapor/air and dust/air)
- Toxicological properties
- Experience already gained
- Moisture isotherms
- Contamination by the drying gas
- Corrosion aspects
- Physical data of the relevant materials

TABLE 14.2 Standard Pan Dryers – Multidry-EV*

Diam. mm	Cylindrical Height, mm	Geometrical Volume, m^3	Agitator Speed rpm	Installed Power kW
700	500	0.3	10–80	11
900	600	0.6	5–55	15
1200	700	1.2	5–40	18
1400	950	2.0	3–35	30
1600	1100	3.0	2–30	45
1800	1400	5.0	2–28	75

**Courtesy of COGEIM SpA.*

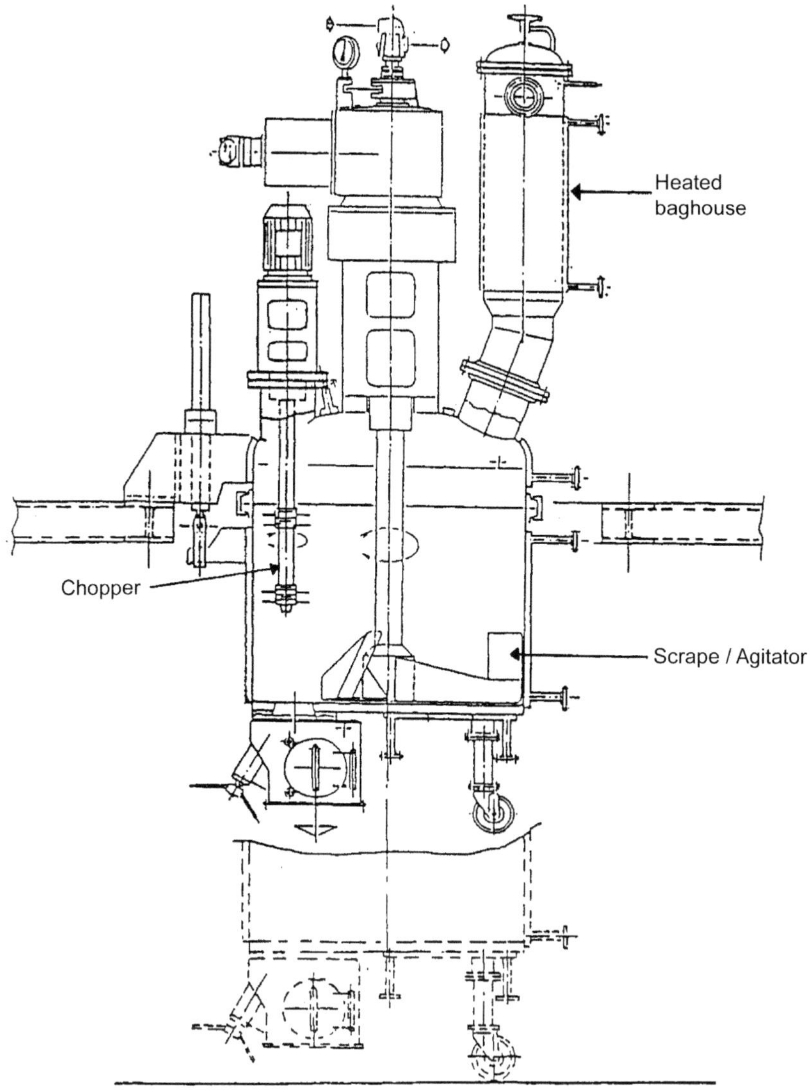

FIGURE 14.3 Multidry-EV Pan Dryer. *(Courtesy of COGEIM SpA).*

Generally, a vertical cylindrical casing forms the body with a flat bottom flanged to the top cover head. The unit is fully heated by an outside half-pipe jacket welded on the cylindrical wall, the bottom and the top head.

The dished head is provided with the appropriate nozzles for feed inlet, instrumentation, heating or cooling medium, vapor outlet, lamp and rupture disk. The dished head and the cylindrical body are separated by means of a hydraulic system to provide easy access to the vessel for inspection or cleaning. A high-powered agitator having two crossed arms located at different heights, is designed for processing products that go through a viscous transition phase (high viscosity). The same dryer can be provided with a different agitator, high speed, which is applicable for low to medium viscous products. The agitator can be totally heated. To eliminate possible agglomerates or lumps formed during the drying process, and discharge problems, a chopper device is supplied. The shaft sealing can be either a stuffing-box or a mechanical seal. A bottom discharge valve for the dry product is hydraulically driven and located in a closed hatch. The geometrical volume of these vertical dryers ranges from a few liters to approximately 500 liters (see Table 14.3).

Materials having average–low density (100–500 kg/m^3) and low–medium viscosities, which require perfect mixing of the dried product, could require another type of vertical dryer.

Here, a dryer having a truncated-cone casing is used (Figure 14.4). The agitator is supplied combining:

- A screw feeder that propels the product upwards.
- An orbital rotation, of the same screw feeder following the geometry of the cone, providing efficient circumferential mixing. The screw is totally supported by the rotating shaft. The shaft and the screw design guarantee

TABLE 14.3 Standard Conical Dryers – Mixodry-EMV*

Diam. mm	Useful Volume, m^3	Installed Power Screw Rotation, kW	Installed Power Screw Revolu., kW
1050	0.3	2.2	0.37
1500	1.0	2.2	0.37
2000	2.0	5.5	0.75
2700	5.0	11.0	1.1
3150	8.0	22.0	1.5
3600	12.0	22.0	1.5

Courtesy of COGEIM SpA.

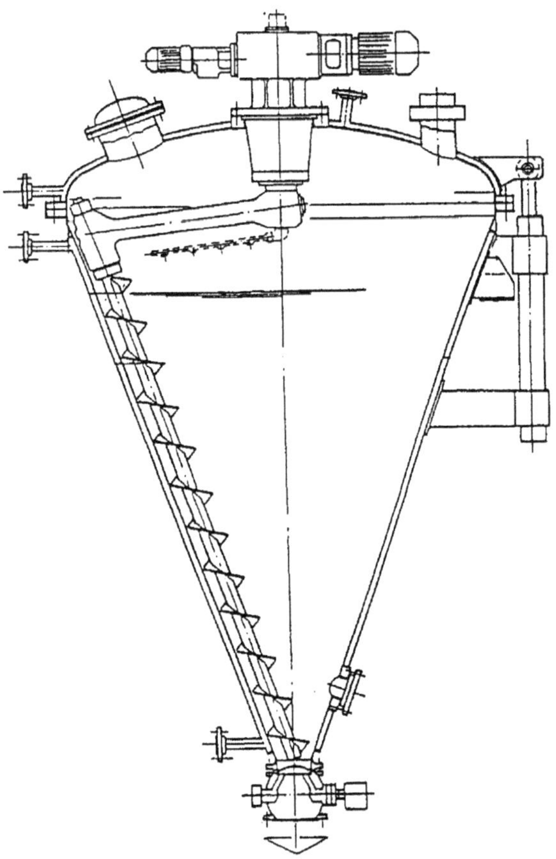

FIGURE 14.4 MIXODRY-EMV Conical Pan Dryer. *(Courtesy of COGEIM SpA).*

containment of the lubricants. A special detector indicates any possible leak before it contacts the product. The geometrical volume of these dryers ranges from a few liters to approximately 12,000–15,000 liters (Table 14.3).

Horizontal Vacuum Paddle Dryers. For products having high specific weight (1000–1500 kg/m^3), middle-high viscosity, and requiring high process temperature and a very large working capacity, the use of the horizontal dryer (Figure 14.5) is recommended. Generally the horizontal dryer is constructed as a jacketed horizontal cylinder with two heads provided with an outside jacket. Nozzles, for product feeding and discharge, for the bag filter, nitrogen inlet, bursting disc is provided. A heated horizontal shaft with radial paddles which scrape the wall, performs the mixing operation and supplies the majority of the total heat flux when compared to the heated outside vessel walls. This is due to the design of the shaft and the agitator blades, which allow the forced circulation of the heating fluid; the design and working conditions are the same as for the outside jacket system.

The rotating agitator blades prevent deposits, which would reduce the total heat-exchange in a short period of time. The scraping blades are designed and installed to facilitate the discharge of the dried product through the product discharge nozzle. This is achieved by reversing rotation. The shaft sealing can be either by a stuffing-box or a mechanical seal. The agitator drive consists of a gearbox coupled to a reversible/variable speed motor. The heating fluid inside the agitator both enters and exits the system through a rotating joint. A discharge plug valve is normally installed in the central part of the vessel and is activated by a pneumatic piston. A jacketed dust filter can be installed with a reverse jet cleaning system. The dryer capacity normally ranges from a few liters to approximately 20,000 liters (Table 14.4).

For products, which need a very high standard quality level, a different version of the horizontal paddle dryer (Figure 14.6) has been designed. The construction differences can be summarized as follows:

- The shaft is supported on only one side to allow the opening of the opposite head for cleaning and inspection
- The design of the paddles provides for self-cleaning of the cylindrical body and heads
- All surfaces are consistent with GMP norms
- The sealing system is a special double mechanical seal with flushing system
- All surfaces in contact with the product are mirror polished.

The geometrical volume ranges from a few liters to approximately 5000 liters (Table 14.5).

Vacuum Shelf Dryer. It is the simplest and oldest vacuum dryer known. It can be used for drying a wide range of materials like solids, free flowing powders, fibrous solids having special forms and shape, practically any material that can be contained in a tray. It finds application where the material is sensitive to heat and so valuable that labor

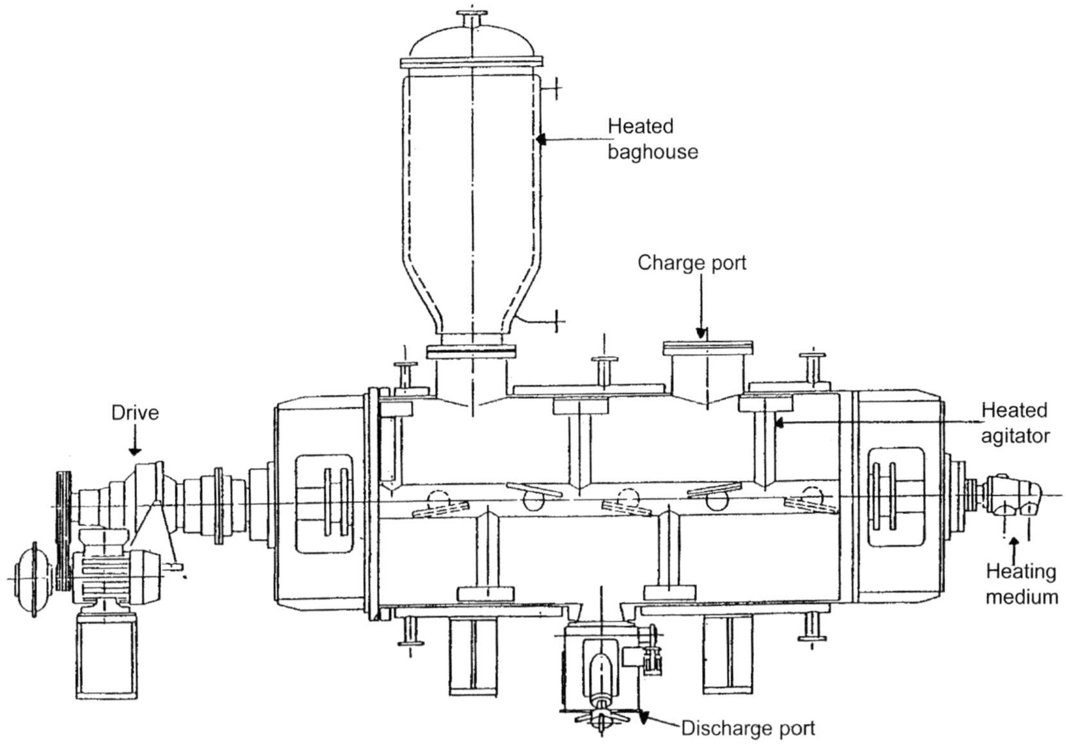

FIGURE 14.5 Chemidry-EO Horizontal Paddle Dryer. *(Courtesy of COGEIM SpA).*

TABLE 14.4 Standard Horizontal Dryers – Chemidry-EO*

Diam. mm	Length, mm	Geometrical Volume, m^3	Agitator Speed rpm	Installed Power kW
500	1000	0.2	6–80	4–5.5
1000	2000	1.6	6–12	11–15
1200	3000	3.4	6–12	18–22
1400	4200	6.4	6–12	30–37
1800	5000	13.0	5–10	55–75
2200	6000	22.0	3–6	75–90

*Courtesy of COGEIM SpA.

costs are insignificant. At the same time, it is normally used when the powder production is very low.

The dryer consists of a vacuum tight cylindrical or rectangular chamber containing a number of heated shelves on which trays are heated. A quick opening permits easy loading, unloading and during maintenance, easy cleaning. Different heating mediums are used, i.e., steam, hot water or hot oil. There are no moving parts inside the vessel, that means no sealing problem and consequently, a good vacuum can easily be maintained. The disadvantages are mainly due to its lower heat transfer rate (long drying time) and impossibility of safely handling toxic products because of the hazards involved in charging and discharging trays.

Tumbler Vacuum Dryer. This dryer is designed for drying of chemical, pharmaceutical products, which are not sticky. Its double cone rotating shape ensures direct contact between the material and the heated surface, resulting in uniform heat transfer.

For optimum drying results approximately 50% to 60% of the total volume is required. Any greater percent fill would greatly restrict the product movement and retard the evaporation rate. A frequent condition that occurs with some sticky materials is the formations of balls, which can be broken by addition of an intensifier bar with the rotating vessel or by intermittent rotation of the vessel. The unit is completely jacketed and designed for circulation of a heating medium. The tumbler is normally gentle in action and the absence of internal moving parts assures against disintegration of crystals or abrasion.

3.1 Testing and Scale-Up

Generally, it is possible to carry-out some tests in the manufacturer's small scale units, but it is necessary to remember that during shipping the material may have changed its

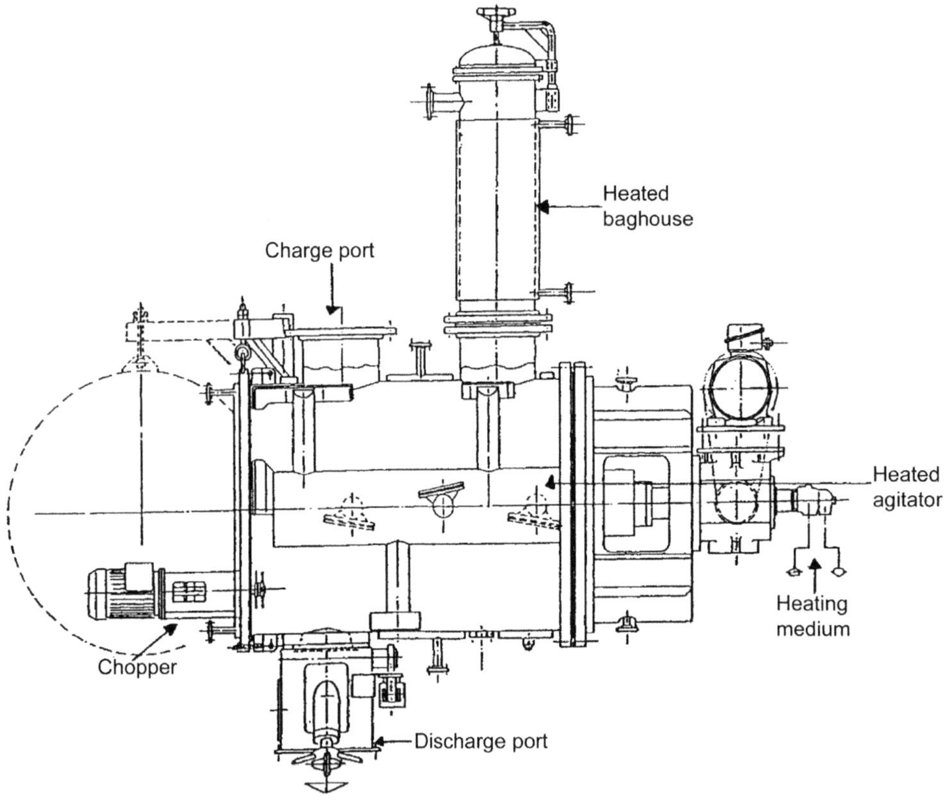

FIGURE 14.6 Steridry-EO Horizontal Paddle Dryer of overhung design. *(Courtesy of COGEIM SpA).*

TABLE 14.5 Standard Horizontal Dryers – Steridry-EO*

Diam. mm	Length, mm	Geometrical Volume, m^3	Agitator Speed rpm	Installed Power kW
500	700	0.14	5–40	4–5.5
1000	1500	1.18	5–25	5.5–11
1150	1650	1.71	5–25	11–15
1300	1800	2.39	5–20	15–18.5
1500	1900	3.36	5–20	18.5–22
1700	2000	4.53	5–15	22–30
1800	2250	5.72	5–15	22–37

**Courtesy of COGEIM SpA.*

property due to chemical or physical modifications, because the quantity of sample is limited, it is not possible to check long-run performance. It is also impossible to evaluate the behavior of the dried material in the plant's solids-handling equipment. If the pilot test is positive, it is good practice, before designing a production unit, to install a small-scale dryer in the plant and investigate what is possible under actual process conditions. It is essential in this test that a representative sample of the wet feed is used and the test conditions simulate as closely as possible the conditions characteristic of the commercial size dryer. The experimental method for measuring product moisture content should be clearly defined and consistent with that used in the industrial plant. It should be noted that a heat transfer coefficient is the main product of the test and, based on this, a scale-up to the final heating surface can be done.

The heat transfer coefficient combines the surface coefficient for the condensing steam, the resistance of the metal wall and the surface coefficient on the working side. Because conditions vary with the type of material involved, the amount of moisture it contains, the thickness of the layer in contact with the surface, the structure of this layer, and many other factors, it is impossible to construct an overall heat transfer coefficient without experimental data.

Scale up of laboratory data is a critical step and requires considerable experience. Since scale up is subject to many factors that are not quantifiable, it is based primarily on experience and is a function of the specific dryer. When the heating surface is known, it is easy to calculate the working volume and the dryer's geometrical volume (Figure 14.7).

For a pan dryer, the percent of the total volume occupied by the batch is called the working volume, which is another critical consideration. As the working volume approaches 100% of the total volume, there is less void space available for material movement and contacting of the heated surface.

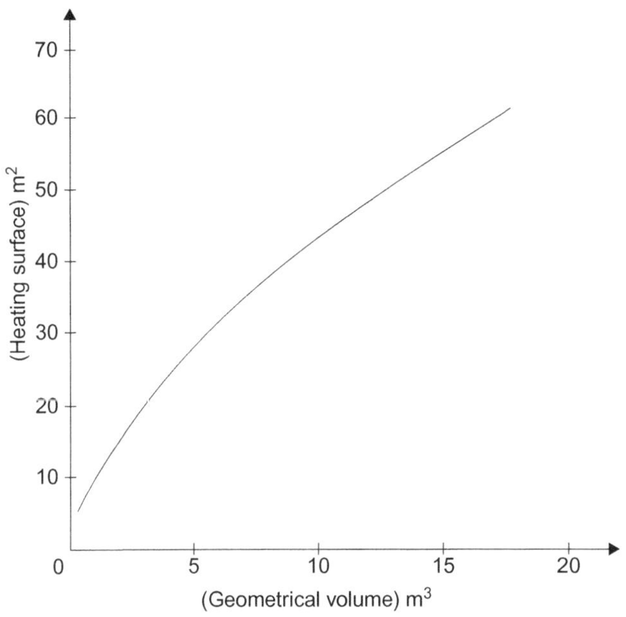

FIGURE 14.7 Total dryer heating surface versus geometrical volume. *(Chemidry-EO).*

In the vacuum batch dryer, approximately 60% total volume is required for optimum drying result. A very simple and approximate equation can be used for the scale up of vertical pan and horizontal paddle dryers:

$$\frac{tb}{ta} = \frac{(A/V)a}{(A/V)b}$$

where

t = Drying time
A = Heat transfer surface, m^2
V = Vessel working volume, m^3
a = Pilot plant
b = Industrial size plant

3.2 Cost Estimation

Capital investment is the total amount of money needed to supply the plant and manufacturing facilities plus the amount of money required as working capital for operation of the facilities. To estimate a fixed-capital investment it is necessary to consider the following costs:

- Purchased equipment
- Instrumentation
- Electrical
- Piping
- Service facilities
- Building

The cost of purchased equipment is the basis for estimating the capital investment. The various types of equipment can be divided conveniently into:

- Process equipment
- Raw materials handling and storage equipment
- Finished products handling and storage equipment

Of course, the most accurate method of determining process equipment cost is to obtain bids from the supplier. When a dryer unit must be evaluated, the following have to be considered.

Dryer Type and Size. If a vacuum pan dryer is selected, it is then necessary to choose its configuration and size. The size is dependent on the capacity needed and this is based on pilot tests. The configuration is dependent on the property of the material to be dried and the pollution specifications. For instance, the system blades-agitator can be heated or not, and the sealing between agitator and dryer body can be accomplished either by a stuffing box or a mechanical seal. The latter can either be double pressurized or simple. The chopper can be installed or not, the agitator rotation can be electrical/frequency converter or hydraulic. All of the hydraulic system has to be considered including all piping connections, etc.

Construction Materials. Normally, the dryers are made of stainless steel. The most common stainless steels used are the type 304 and 316 generally having low carbon content. They contain chromium and nickel at different percents. The addition of molybdenum to the alloy, as in type 316, increases the corrosion resistance at high temperature strength. The presence of chromium increases its resistance to oxidizing agents. The price for the type 304 and 316 is quite similar. If very highly corrosion resistant materials are required then Hastelloy C276 or C22 can be used.

Hastelloy is used where structural strength and good corrosion resistance are necessary under conditions of high temperatures. Compared to stainless steel, the price of a Hastelloy dryer is approximately double. Other less expensive alloys can be used, such as Inconel, 77 percent nickel and 15 percent chromium. Nickel exhibits high corrosion resistance to most alkalis.

Internal Finishing (GMP). For pharmaceutical purpose the internal finishing must be at least 220 grit and the dryer manufactured according to GMP standards. This makes the price of the dryer some 20 to 30 percent (%) higher than the standard design.

Installation. The installation involves costs for labor, foundations, platforms, construction expenses, etc. The installation cost may be taken as a percentage of the dryer cost, approximately 20 to 50 percent, depending upon its sophistication.

If no cost data are available for the specific dryer selected, a good estimate can be obtained by using the logarithmic relationship known as the *six-tenths-factors* rule. A price for a similar one, but having different capacity, is the sole requirement.

$$\text{Cost dryer } A = \text{Cost dryer } B \left[\frac{(\text{Capacity dryer } A)}{(\text{Capacity dryer } B)}\right]^{0.6}$$

This relation should be only used in the absence of any other information.

3.3 Installation Concerns

The auxiliary equipment effects the dryer performance.

Heating System. Depending on the maximum temperature allowed inside the dryer, water ≤98°C or steam low/medium pressure 3–6 bar can be used as heating medium in the jacket. Due to the relatively low temperature required to dry fermentation products, the heating medium is generally circulating pressurized hot water. Either an electric immersion heater or steam in a shell and tube heat exchanger can heat the water. The recirculating pump should always be pumping into the heater so that its suction is from the outlet of the dryer. In addition, the suction side of the pump should always have an air separator to prevent cavitation. The entrapment of air is inevitable in a hot water heating system.

Cooling System. Where cooling of the product is absolutely necessary, a cooling exchanger can be mounted in parallel with the heating exchanger and used at the end of the drying cycle. By turning a couple of valves to direct the flow through the cooling exchanger, the recirculated water will then remove the heat from the product and transfer it to a cooling medium in the cooling exchanger.

Vacuum system. A well-designed system should include:

1. *Dust collector*—which is installed on the top of the dryer and made of a vertical cylindrical casing, complete with an outside jacket. Generally the filter elements are bags fixed on the upper side to a plate and closed on the lower part. The filter bags (Figure 14.8) are supported through an internal metal cage. Cleaning of the bags is obtained by a mechanical shaking device or by nitrogen pressure. Design and working conditions are the same as for the vacuum dryer.
2. *Condenser*—designed according to the scaled-up pilot test evaporation rate. Normally it is a shell and tube unit. It should also be self-draining into a vacuum-receiver, which collects the solvent as well as maintains the vacuum integrity of the entire system. The condensate receiver should have a sight glass so that visual inspection will indicate when it needs emptying. Obviously the receiver should be large enough to contain all of the condensate from one batch of product.

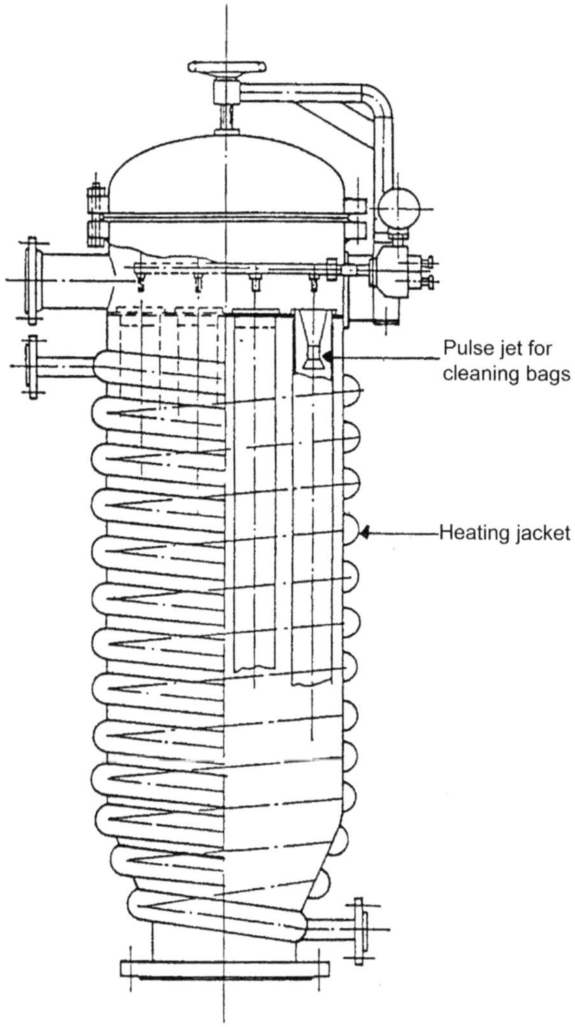

FIGURE 14.8 Standard bag filter. *(Courtesy of COGEIM SpA).*

3. *Vacuum pump*—whose flow-rate depends largely upon the intake of air at the various fittings, connections, etc. Different kinds of vacuum pumps can be used; e.g., rotary—water/oil sealed, reciprocating dry vacuum pump.

If a water sealed-vacuum pump is used, the liquid ring may permit scrubbing the effluents (non-condensed vapors) and removal of the pollution load by controlling the vapor emission. Obviously, when the liquid ring becomes saturated, it must be discharged. This type of device should always be considered where low boiling solvents and hazardous or toxic vapors are involved; better if a closed circuit is considered. This type of pump is simple to operate and requires little maintenance. Depending on service liquid temperature, a single-stage pump will allow a vacuum of 100 to 150 torr. However, it is more usual to employ a two-stage liquid ring pump which will attain 25 torr, and below 10 torr when used in combination with an ejector. The ejectors consist essentially of a steam nozzle, which discharges a high-velocity jet across a suction chamber connected to the equipment to be evacuated. The gas is entrained by the steam and carried into a venturi-shaped diffuser which converts the velocity energy of the steam into pressure energy.

Where it is necessary to operate at the end of the drying cycle below 5 torr, different types of oil sealed rotary pumps can be supplied. Such applications might occur where there is a need to operate at very low drying temperatures.

Hydraulic System. Generally, for vertical pan dryers an hydraulic system is provided for the agitator rotation, the opening closing of the dryer by a rapid device as bayonet and/or TRI-CLAMP and the lowering of the vessel for maintenance or cleaning purposes. The hydraulic components positioned on the dryer are normally:

1. One hydraulic motor for agitator rotation
2. Three hydraulic cylinders for lowering, raising the vessel
3. Hydraulic cylinder for a rapid opening (the number is dependent upon the dimensions of the vessel)

The hydraulic system consists of: oil reservoir, electric motor, hydraulic pump, heat exchange for oil cooling, oil filters, oil level indicator, electric valves and flow distributors. An hydraulic plant, which has been properly installed, and care has been taken during the start-up phase, should enjoy long life and not need much maintenance.

A cardinal principle in the operation of a trouble-free hydraulic system, on which all manufacturers agree, is that the operator continuously monitors the quality as well as the condition of the hydraulic fluid to make certain there are no impurities. The reliability of the hydraulic system is directly related to the integrity of the fluid.

The following periodic checks are recommended:

1. Monthly external cleaning and inspection. This will uncover any leaks, which can then be repaired.
2. Monthly air filter checking and replacement of the dirty cartridge.
3. Weekly oil filter checking.
4. Weekly oil level check, each time the level falls to the minimum, oil must be added
5. Oil replacement on the average every 2,000–3,000 hours.
6. Heat exchanger must be cleaned semiannually.

3.4 Safety Considerations

Where the handling of materials containing highly flammable solvents is concerned, the dryer must be located in a classified area and the electrical parts designed according to the standards specified for this level.

The mechanical, electrical and instrument specification should also include requirements for:

1. *Explosion protection*—a vent should be considered for a safe relief of a positive pressure.
2. *Avoidance of ignition*—potential ignition sources may be electrical equipment, discharge of static electricity or mechanical friction (associated with the agitator). The dryer must be grounded.
3. *Facilitating safe operation*—ventilation should be provided during loading; a supply of inert gas is required for breaking the vacuum.

Most hazards are listed below:

1. Ignition of dust cloud can occur during unloading of a dusty flammable product from the dryer
2. Ignition of bulk powder can occur if a dryer is opened to atmosphere while still hot
3. Ignition of flammable vapor can occur when loading solvent-wet material into the dryer, and also when unloading the product if the dryer has not previously been purged with nitrogen
4. Exothermic decomposition—some heat-sensitive materials may decompose with evolution of large volumes of gas if they are overheated during drying

The danger from an explosion can be reduced in two different ways:

1. The dryer can be designed according to pressure vessel code and consequently be able to contain any possible explosion
2. The process/operating conditions should be altered to insure a higher level of safety

At the same time, the following start-up and shutdown procedures are recommended:

Start-up:

1. Inspect the plant and remove any deposits, check position of valves and settings of temperature and vacuum regulators
2. Purge the dryer with nitrogen
3. Load the wet material in the dryer
4. Start cooling water to the condenser
5. Start the vacuum pump
6. Start the agitator
7. Apply heat to the jacket

Shutdown

1. Switch off the heating medium
2. Wait till the product has cooled for sale discharge
3. Close the vacuum line
4. Stop the agitator
5. Fill the vessel with nitrogen to atmospheric pressure
6. Open the dryer and remove the product
7. Clean the dryer

4.0 EQUIPMENT MANUFACTURERS

1. Vertical vacuum dryers:
   COGEIM SpA
   GLATT GmbH
   Hosokawa Micron Europe
   Moritz
   Patterson Kelley Co.
2. Paddle dryers:
   Buss
   COGEIM SpA
   List
3. Filters or Filter-Dryer Products
   COGEIM, Charlotte, NC
   Jaygo, Mahwah, NJ
   Krauss-Maffei, Florence, KY
   Micro Powder Systems, Summit, NJ
   Rosenmund, Charlotte, NC (De Dietrich)
   Sparkler Filter, Conroe, TX
   Steri-Technologies (Zwag), Bohemia, NY
4. Dryers, Spray
   APV/Crepaco, Tonawanda, NY
   Niro Atomizer, Columbia, MD
5. Dryer/Blenders
   GEMCO, Middlesex, NJ
   J.H. Day, Cincinnati, Ohio
   Micron Powder Systems, Summit, NJ
   Niro-Fielder, Columbia, MD
   Patterson-Kelly, East Stroudsburg, PA
   Processall, Cincinnati, OH
6. Dryers, Freeze
   Edward High Vacuum, Grand Island, NY
   Finn-Aqua, Windsor Locks, CT
   Hull, Hatboro, PA
   Stokes, Warminster, PA
   Virtis, Gardner, NY

5.0 DIRECTORY OF MANUFACTURERS

Buss AG
4133 Pratteln, 1
Basel
Switzerland

COGEIM SpA
Compagnia Generale Impianti
Via Friuli, 19
24044 Dalmine (Bergamo)
Italy

GLATT GmbH
Process Technology
P. 0. Box 42
7852 Binzen/Lorrach

Federal Republic of Germany
GUEDU (De Dietrich)
21140 Semur-En-Auxois
France

Hosokawa Micron Europe
P. O. Box 773
2003 Rt Haarlem
The Netherlands

List AG
4133 Pratteln
Switzerland

Moritz
7, Avenue de Pommerots
B. P. 37
78400 Chatov
France

Pattrerson Kelley Co.
Division of Harsco Corporation
101 Burson St.
P. O. Box 458
E. Stroudsburg, PA 18301
United States

FURTHER READING (FOR SECTION I: INDIRECT DRYING)

[1] W.L. Badger, J.T. Banchero, Introduction to Chemical Engineering, McGraw-Hill Kogakusha.

[2] J.M. Coulson, J.F. Richardson, J.R. Backhurst, J.H. Harker, third ed., Chemical Engineering, vol. 2, Pergamon Press, 1978.

[3] J.C. Forrest, Drying processes, in: N. Blakebrough (Ed.), Biochemical and Biological Engineering Science, Academic Press, London and New York, 1967.

[4] A.S. Foust, L.A. Wenzel, Principles of Unit Operations, second ed., Wiley.

[5] W.L. McCabe, J.C. Smith, Unit Operations of Chemical Engineering, third ed., McGraw-Hill.

[6] P.Y. McCormick, The Key to Drying Solids, Chemical Engineering, August 15, 1988.

[7] M.S. Peters, K.D. Timmerhaus, Plant Design and Economics for Chemical Engineers, McGraw-Hill Kogakusha, 1980.

[8] M.R. Spotts, P.F. Waltrich, Vacuum Dryers, Chem. Eng. January 17 (1977).

[9] C.M. Van't Land, Selection of Industrial Dryers, Chem. Eng. March 5 (1984).

[10] T.H. Wentz, J.R. Thygeson Jr., Drying of wet solids, in: P.A. Schweitzer (Ed.), Handbook of Separation Techniques for Chemical Engineers, McGraw-Hill, New York, 1979.

Section II

Direct Drying

Barry Fox

1.0 INTRODUCTION

The purpose of this chapter is to review various forms of solids dryers and auxiliary components. It is intended to be a practical guide to dryer selection (as opposed to the theory of drying, which is addressed in various technical manuals referenced in the bibliography). From a microscopic viewpoint, the process is simple: water or solvent basically evaporates leaving the solid behind. When viewed macroscopically, it is apparent that the drying process is extremely complicated with many interdependent forces that combine in various dryers to achieve the end result. The information in this article can also help the reader become more familiar with the drying process from beginning to end.

Drying is the process of removing a liquid from a solid. The liquid to be dried can be water or a hydrocarbon based solvent. The solids are usually classified as organic or inorganic, either of which can be completely or partially soluble in the liquid medium. The inorganic materials are generally called salts because they are usually soluble in water. Organic materials are more difficult to dry due to temperature sensitivity. When drying, organic materials can stick to walls and cling to themselves resulting in a tacky consistency.

Direct drying is the process of removing this liquid via the mechanism of *convective* heat transfer. The heat input usually takes the form of preheating a carrier medium (such as air, evaporated solvent or an inert gas) that transfers the sensible heat and acts as an absorbent to take away the liquid in the vapor form. The carrier medium can hold a fixed amount of liquid (saturation) at its defined temperature. The solids release the liquid to the carrier medium as a function of saturation and equilibrium. In essence, the heated gas has a higher saturation affinity for the liquid in the vapor form than does the solid at the gas temperature.

Typical examples of conventional direct dryers are spray, fluid bed, flash, rotary, belt and continuous tray type. In the former three types, the wet solids are suspended in the carrier medium. In the latter three types, the carrier medium passes slowly across the bed of solids. Additionally, there exists some minor tumbling of the solids through the gas stream (carrier medium).

There is a nonconventional form of direct drying that is often overlooked or possibly unknown to the designers of the process. It is applicable to almost any of the forms of dryers mentioned in this chapter. The method is to use the solvent or liquid that is being dried as the carrier medium for the heat transfer. In essence, the moisture that is evaporated from the product is recycled and reheated. It replaces air or inert gas and this hot vapor is used to strip off additional liquid from the wet product. The excess vapors are removed via a vent condenser outside of the closed vapor loop. This procedure can be applied in any of the drying processes mentioned here. One advantage to this method of drying is that the product sees only the vapor with which it is already in contact in the liquid state. A possible reason for using this method is product oxidation when air drying. This method may reduce oxidation if the solvent is used. This method is also more energy efficient when solvents are present since the inert gas that is recycled in the former method needs to be reheated after it has been cooled down to condense the solvents.

2.0 DEFINITIONS

Absolute Humidity—the ratio of mass of vapor (moisture) to mass present in the carrier gas stream. Example: 0.02 pounds of water per pound of air. This number can be used to find the relative humidity on the psychometric charts. It is also useful for cumulative quantities in a stream due to such items as products of combustion (when a gas fired heater is used), and evaporation and ambient quantities. This is necessary for calculating condenser or venting amounts.

Bound Moisture—liquid that is bound to a solid by chemical bonds or physical adsorption in the molecular interstices of the solids.

Capillary flow—the flow of liquid through the pores of a solid.

Critical moisture content—the average moisture in the solids when the constant rate drying period ends.

Diffusion—the process of mass transfer of the liquid from the interstices of the solid to the surface of the solid.

Dry basis—means of measuring moisture content in terms of moisture content per quantity of dry product, for example, pounds of water per pound of dry product. (Also see *Wet Basis.*)

Equilibrium moisture content—the limiting moisture content to which a product can be dried under fixed conditions such as temperature, humidity and pressure.

Evaporative cooling—when drying a solid with free or bound moisture, the effect of a phase change from the liquid state to the vapor state removes energy from the liquid-solid mass. This results in a reduction of temperature in a nonadiabatic operation, whereas in an adiabatic operation of constant heat input, the temperature may drop or more likely it will maintain a level (pseudo-wet bulb) temperature.

Falling rate period—this is the period of drying where the instantaneous drying rate is constantly decreasing.

Feed material—this is the description of the material being dried before it enters the dryer.

Final moisture content—the desired product moisture level required after completion of the drying process.

Free flowing—refers to the feed and product characteristics, as in a free flowing powder. This is the state in which the material being dried would not cling to itself, forming large chunks or possibly bridging in a hopper.

Free moisture—liquid which is promptly removable due to its availability at the interface between the surface of the particles (solids) and the gas stream.

Hygroscopic material—solids having an affinity for liquids due to a chemical or physical attraction between the solids and the liquid.

Initial moisture content—the average moisture contained in the wet material before the start of the drying process. If given in percent, specification of wet or dry basis is necessary.

Plug flow—a term used to describe the breakup of a continuous process into small batch segments. The term may originate from a reactor tube being filled or plugged with small quantities of material using a piston pump. The reactor would process the volume of material in each piston cavity like a small batch, yet when the material is viewed as a large quantity, it appears homogeneous. The term is used in conjunction with semi-continuous operations.

Product—this is the description of the solid material after it has been dried.

Relative humidity—the percentage of water vapor in a gas stream relative to its saturation level. Example: 100% relative humidity is the complete saturation of a carrier gas stream, whereby any further vapor cannot be absorbed by the gas and will condense or precipitate out in the liquid phase. There is an equilibrium between the liquid-solid mass and the gas stream (carrier medium). This equilibrium is a result of a combination of saturation capability of the medium at a given temperature. At higher temperatures, the carrier medium has a higher saturation limit and, therefore, a lower relative humidity, given the same absolute humidity.

Wet basis—means of measuring moisture content in terms of quantity of moisture per quantity of wet material. For example, if we have 1000 lbs. of wet cake with 200 lbs. of water, our moisture content is 20% on a wet basis and 25% on a dry basis. (See *Dry basis.*)

Wet bulb temperature—the dynamic equilibrium temperature attained by a water surface when the rate of heat transfer by convection equals the rate of mass transfer away from the surface.

3.0 PSYCHROMETRIC CHARTS

There are many forms of psychometric charts available from various technical sources as well as many manufacturers of process equipment who have tailored the chart for use with their equipment. These charts are useful for determining moisture content in the air at a given temperature and relative humidity or wet bulb temperature. The type of information obtainable from these charts depends upon which chart one uses, because each is designed differently. Usually, accompanying the chart is a set of instructions for use. Please see the references for examples of these charts and the various forms in which they exist.

4.0 DRYING THEORY

The process of drying solids is usually quantified into three phases:

- *Initial adjustment period—tins* is the stage at which the wet feed material heats up or cools down to the starting drying temperature which is basically referred to as the *wet cake temperature.* For example, the wet feed is introduced to the heated dryer at ambient temperature. During this period the material temperature will start to rise to the wet bulb temperature that may be different from the initial feed temperature. The reason the temperature of the wet cake remains low relative to the gas temperature is a phenomenon known as *evaporative cooling.*

  Constant rate period—this is the stage at which the free moisture is evaporating from the solids at a constant rate. If one were to measure the temperature of the bed or individual particles of wet solids at this point, the temperature would be the wet cake temperature. After the free moisture has evaporated, the cake temperature rises, an indication of the end of the constant rate period. Several stages of this period can occur due to the existence of bound moisture. If bound moisture exists, the energy required to break the bonds is absorbed from the gas stream as heat input. As the bonds break, the bound moisture is released and is removed as surface moisture described above. The quantity of molecules of hydration and the temperature, which the product must reach in order to break these bonds, affect the overall constant drying rate period. One can generally observe a rise in the wet cake temperature, after the free moisture has evaporated, to the temperature that is required to break the bonds. This temperature then becomes the next wet bulb level or isotherm. Several levels of bound moisture may exist during the drying process. (Note: In general, this bound moisture phenomenon occurs mostly with inorganic salts and therefore may not be a major concern of the pharmaceutical or biochemical industry.)

Diffusion, or falling rate period—this is the stage where the rate at which the liquid leaves the solid decreases. The liquid that is trapped inside the particles diffuses to the outside surface of the particle through capillary action. The random path that the liquid must take slows down the drying process at this stage. (See Figure 14.9 for typical graph.)

5.0 FUNDAMENTAL ASPECTS OF DRYER SELECTION

The starting point in determining how to dry certain products is first to ascertain whether the process will be a batch or continuous operation.

If the product is manufactured in relatively small quantities and identification of particular size lots is required, than batch mode is usually the route taken. Full accountability may be achieved when batch processing with the proper controls and procedures in place. The full batch of material to be dried must be enclosed in the dryer. A necessity for the equipment should be that the product dries uniformly.

When manufacturing large quantities of a product that does not require tight batch controls, a more efficient operation (usually less expensive) results by

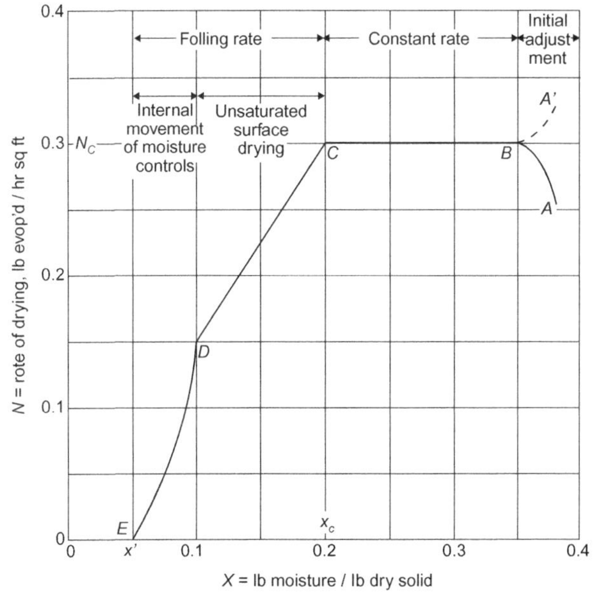

FIGURE 14.9 Typical rate-of-drying curve, constant drying conditions.

drying the product in a continuous or semi-continuous fashion. The product, in one case, can be batch-stored in large vessels and fed at a continuous rate to the dryer. The product is usually dried in small quantities thus requiring a long time to process the entire amount in a smaller, more efficient piece of equipment. In another situation, ideal for continuous operation, the product would be manufactured upstream of the dryer in a true plug flow manner and transferred to the dryer at a constant rate. In other words, the dryer's capacity matches that of the upstream equipment. This is the most efficient manner.

5.1 Batch Direct Dryers

Most direct batch dryers are fluid bed types such as those that retain the batch on a screen while pneumatically fluidizing the product. Mechanically agitated or tumble rotary dryers also exist. If the product is temperature sensitive, the user should consider a vacuum dryer as an alternative. Vacuum or lower pressure can be utilized to assist in drying the product. However, since most of the mass transfer occurs as a result of the heat input transferred via conduction through the walls of the dryer's jacket, which is considered to be an indirect dryer. For more information on indirect dryers please refer to the first section of this chapter.

5.2 Batch Fluid Bed Dryers

In the category of fluid bed dryers, there are two types of processes commonly used to suspend the material—pneumatic and mechanical fluidization.

1. *Pneumatic Fluid Bed Dryers.* In the pneumatic fluidization process, the wet cake is placed in the dryer and dry heated gas is introduced at a very high velocity (under the bed of product) through a fine screen or a porous plate in order to fluidize the product. There is a visible layer of material that is sustained as the gas passes through the bed. The wet gas leaves the chamber through a sock or bag type dust collector that removes the fines and returns them to the batch. [More recently, stainless steel cartridge filters are becoming very popular because they can be *cleaned-in-place* (CIP). The Aeromatic-Fielder Division of Niro has developed this.] If the carrier medium is air containing only clean water vapor, the gas can then be exhausted to the atmosphere if it contains clean water vapor. If the medium is an inert gas, it can be recycled back to the dryer while removing contaminants or solvents via a condenser and filter. However, this inert gas must then be reheated to the proper inlet temperature.

2. *Mechanically Agitated Fluid Bed Dryers.* In the fluidization process, the wet cake is gently lifted by rotating paddle type agitators thus blending the product into the gas stream creating an intimate mixing of the wet solids with the dry gas stream. This results in a very efficient exposure of the wet product's surface area. The advantage of such a dryer is a faster drying time and a lower total energy input due to lower overall energy requirements (see Figure 14.10).

5.3 Batch Rotary Dryers

In a batch rotary dryer, a horizontal cylinder is used to contain the batch while heated air is passed across the length of the cylinder. A jacket can be placed on the outside of the cylinder where steam or hot water is introduced to aid in heat transfer via conduction through the walls. Sometimes an agitator shaft with paddle arms, either heated or unheated, is included in the design to assist in heat transfer and product discharge.

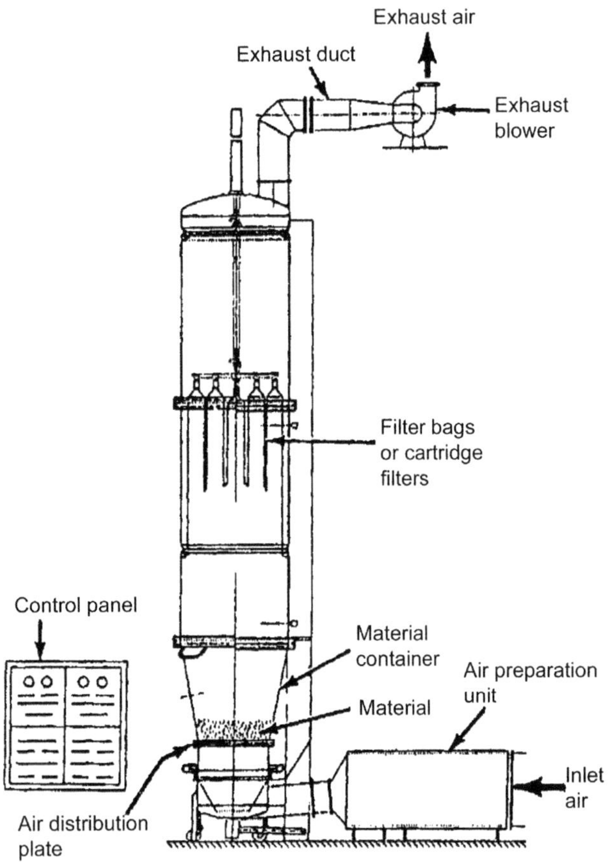

FIGURE 14.10 Typical batch fluid bed dryer.

5.4 Ribbon Dryers

This type consists of a long, jacketed, horizontal cylinder, or a "U" shaped trough, which contains an agitator shaft positioned down the length of the bowl. The purpose of the cylindrical-shaped vessel may be for operation under pressure or vacuum. The agitator spokes are intermittently mounted on the shaft that support inner and outer rows of ribbon flights pitched so as to move the product. The outer ribbon flights usually move the product towards one side of the vessel and the inner ribbons move the product towards the other side. This design would have the discharge port at one end of the dryer. An alternative to this design is to have a center discharge, where the ribbons on one half of the dryer are pitched at 90° to the ribbons on the opposite end.

The drying here is achieved by means of exposing the product to the surface area of the jacketed vessel. The jacket is a shell of metal (usually carbon steel) welded onto a stainless steel vessel body. This design can include a heated shaft for increased surface area exposure. The heat transfer medium used here is generally steam, hot oil, or hot water. Ports must be provided so as to vent the evaporated vapors being removed from the product.

5.5 Paddle Dryers

Whereas a gaseous medium can be used to transfer heat to the product, in most cases the paddle type is considered to be an indirect dryer. It is similar in design to the ribbon dryer. The differences exist when heated (hollow) paddles are used as opposed to flat blades. (See the previous section on indirect drying.) Also, shoe-like paddles or plows can be used which tend to disperse or smear the product against a heated, horizontal, cylindrical wall. The advantage of a heated paddle design is that the surface area exposure to the product being dried has been expanded thus increasing the overall heat transfer rate. Most paddle dryers are designed for use under vacuum that can supplement the indirect drying process.

5.6 Agitated Pan Dryers

An agitated open pan dryer is somewhat more complicated mechanically. This is a short cylinder whose axis and agitator are vertical. The agitator can enter from either the top or the bottom.

As with the paddle dryers, these are mostly considered to be indirect dryers since heat transfer is from the jacket. If the product is a sticky, pasty material one may wish to use this design. The advantage of the pan dryer is the availability of several heated agitator designs that improve the overall heat transfer rate appreciably over a simple heated jacket; the reason is the same as mentioned in the previous section on paddle dryers. As mentioned earlier, venting of the dryer is necessary to remove the evaporated vapors.

5.7 Continuous Dryers

Continuous types of direct dryers are spray, flash or rotary designs, where the product enters in a form suitable to be handled properly by that dryer. Spray dryers can accommodate a feed stream in a slurry or solution form, whereas a flash dryer is intended to take a feed cake that can be broken up into individual pieces without coalescing. The feed characteristics required for flash dryers are that the product must break up if introduced as a cake. If introduced as a paste, it is necessary for the feed to be back mixed with previously dried material so as to firm up the cakes' consistency. Rotary dryers are more flexible in that they can handle a wide variety of feed consistencies.

5.8 Spray Dryers

Spray dryers are large cylindrical chambers with a cone or flat bottom. They also appear in the form of a large cube or box referred to as a *box dryer*. Small nozzles are located in the chamber walls through which the feed material, in the form of a slurry or solution, is atomized to a fine droplet. The droplet comes into contact with a hot gas stream and dries to a powder in the time that it takes for it to fall to the bottom of the chamber. Typical residence times in a chamber are 12 to 30 seconds. Due to the evaporative cooling effect, the inlet temperature on this type of dryer is normally quite high relative to the dry products' temperature limitations. Typical inlet temperatures for spray dryers range from 400°F to 1000°F depending on the application. The higher temperatures are normally for inorganic salt drying and the lower temperatures are normally for organic temperature sensitive products. The resultant individual product is always spherical in shape due to the initial droplet and it will tend to be extremely porous and fracture easily.

Further processing of spray dried product has been done to achieve *instantizing* by creating agglomerates of these spheres, which retain a relatively high surface area compared to the individual particles. The action of adding this powder to water results in a release of energy from the agglomerated bonding forces and the capillary effect of the water traveling into the porous spheres. The net effect is one of quick dissolution of the agglomerated powder. This is highly desirable when searching for a means of *instantizing* a product.

In order to consider spray drying, certain criteria about the material must be met. The feed slurry viscosity must be low enough whereby it can be pumped through either a rotary atomizer, a two fluid nozzle or a high pressure nozzle which have narrow paths. The product must also be able to withstand high inlet temperatures for very short times. Usually the product will reach the outlet temperature.

5.9 Flash Dryers

Flash dryers have many variations in design. The most basic is a long pipe with a fan, using a high velocity air stream to fluidize and move the wet product in the pipe while concurrently drying the product. Wet feed consistency is extremely important here because if the feed is too sticky or tacky, it will tend to lump together in the feed inlet. The use of paddle type back mixers is common to combine dry material with wet cake to provide a consistent friable feed to the dryer. Backmixing 50% of the dry material with the wet cake is not uncommon. The inlet of the flash dryer is sometimes referred to as the throat—it is of the venturi design so that product is whisked into the chamber and exposed to a high velocity and high temperature as quickly as possible. Most products being flash dried are inorganic salts because the process lends itself to materials that are not temperature sensitive and their final particle shape is not critical. The larger chunks of material will abrade quickly in the long pipe as the material traverses the entire distance of the dryer.

5.10 Ring Dryers

A second flash dryer design is a *ring dryer*. The product may be introduced into this type of dryer in a similar fashion as the standard flash dryer. However, instead of the long pipe as the drying chamber, the product is brought into a ring of pipe where the high velocity air stream and centrifugal forces keep the larger particles in the loop, while smaller particles leave via a contoured inner discharge port. The inlet and outlet port are tangential to the ring and two fans are utilized; one as an exhaust fan for providing the motive air and the second for providing the closed loop recirculation and introducing the product into the ring.

5.11 Mechanically Agitated Flash Dryers

A third design is a flash dryer design (combined fluid bed and flash) with a small drying chamber using high speed agitators and complicated swirling arrangements. In this design the cake can be in the pasty or even runny form. Material enters a feed vat where pitched turbine-type blades rotate at a slow speed forcing the cake or paste down towards the bottom of the vat. At the bottom of the vat is a port, which opens to a screw conveyor that sends the material into the flash chamber. The flash chamber has a relatively narrow diameter, is a vertical vessel with an agitator on the bottom, which breaks up the cake. This occurs while hot air (or inert gas) is introduced into the chamber bottom to dry the cake in a co-current fashion. The larger pieces of the wet cake tend to ride on top of the agitator, thus the *fluid bed* part of the design. The fine particles swirl around the drying chamber and leave via a cone shaped weir at the top of the vessel. The weir acts as a barrier to the oversized materials and retains them in the chamber until they break down in particle size and essentially is dry.

5.12 Rotary Tray or Plate Dryers

In a continuous rotary dryer, rotating tray or plate type dryer design, the feed enters and is turned over many times exposing fresh wet surfaces of material to the hot gas stream. In the former, lifters or an agitator may be used to assist the product in moving through the unit and towards the discharge port. In the latter (the Wyssmont Turbo-Dryer), wiper arms are used to displace the product as it falls through slots to successive trays below. This style is ideal for delicate, crystalline pharmaceuticals where particle breakage is a major concern for the producer.

In the plate type design (Krauss Maffei's Plate Dryer), a cylindrical housing with a vertical axis holds a stack of heated plates or discs on a central support. There are rotating wiper arms which plow through and push material towards the center or outside of the heated plates. The plates are sized so that the plate below is under the above plate at the point where the material falls. The plates are designed as a closed conductive surface to transmit the heat up from the bottom of the plate into the bed of material. There are also some radiation effects from the upper plate to the material below. This design can be operated under pressure or vacuum conditions since the heat transfer is by conduction. The limitation of this design is that it can only handle relatively free flowing materials.

5.13 Fluid Bed Dryers

Continuous fluid bed dryers can be used for drying materials which have a consistency whereby the initial material is friable. This is an important consideration, especially in continuous processing, since the cake must continue to move along the path of the bed at a fairly constant rate. Otherwise the result will be a blockage of the unit and shutdown of the processing line until the blockage can be cleared.

The "constant rate" part of the drying process in a fluid bed occurs at the porous plate or screen where the heaviest (and usually the wettest) material rests. This part of the process is the most critical as the material needs to become fluidized here to continue through the bed in stages, over weirs or simply down the deck of the bed. The air flow here needs to be at the highest velocity to compensate for the denser (wetter) material.

The lighter material fluidizes rather easily and is generally dry by the time it "floats" to the top. One manufacturer's design uses weirs to retain the heavy (wet) material in the first zone. In the second zone, the bulk of the moisture has been removed and the lighter material can be dried through the diffusion or *falling rate* period. In most cases there is a desire to cool the product and this can be achieved in a third zone. Such a design is similar in shape to a submarine, with the weirs having a very short height on top of the porous plate, which is located horizontally at the lower middle section of the cylinder. Since the weirs are low relative to the height of the chamber, cool air will mix with the drying air in the open chamber and therefore, it is more efficient to perform the cooling operation in a separate unit.

Fines removal is achieved by placing a recycle fan/duct loop at a peak above the front-center of the top of the chamber. By lifting the fines up into this area they can be sent off to a bag house or cyclone where they can be collected for removal from the batch or for recycling.

When agglomeration is desired, a spray bar can be incorporated to introduce a mist to a point near the inlet throat of the fluid bed unit where the wet material can be mixed with fines, which are recycled as described above. This process is used mainly after a product has been spray dried or if the initial material to the fluid bed is not too wet.

6.0 DATA REQUIREMENTS

In order to assist the purchaser of the dryer, the equipment manufacturer will require certain data about the product to be dried and the operation around the dryer.

One needs to have the answers to questions relating to the physical properties, characteristics and end use of the material to be dried as well as the liquid being removed. The following example is used to illustrate why such data is needed as a guide to select a dryer; notice the terms, which are italicized:

Example 1—Dry 100 lbs. of wet cake with an *initial volatiles content* of 25% *(wet basis) down to a final volatiles content* of less than 1%. This will be achieved in a *batch* operation for lot identification, GMP and high quality standards. The *loose bulk density* of the *wet cake* is 45 lbs./ft.^{3} and the loose bulk density of the dry powder is 12 lbs./ft.^{3} the *temperature limit* of the product is 145°F and the *feed material* is wet with ethanol. The product is a *pharmaceutical, which* will be a *finished product.* The *solvent* is to be recovered for reuse, and its physical properties can be found in a handbook of hydrocarbons. The product is *free flowing* in the dry state, but very *tacky and pasty* in the wet state.

We can see in the above example that there is a significant amount of information about the product to be dried. With more information available about the product, we can select the dryer such that it performs the necessary functions with a better and more efficient operation. Also, a possible benefit may be improved product quality.

Using the above example, let's select a dryer for the operation specified—first, we can calculate the amount of solvent to be dried by taking the total weight of the wet cake and multiplying it by the initial volatiles content. This is the total amount of ethanol in the product as it will be introduced to the dryer. In this case the result is 25 lbs. Now we must calculate the amount of volatiles left in the final dry product. One way of calculating this amount is to first calculate the total solids in the batch. Since we have 100 lbs. of wet cake with 25 lbs. of ethanol, we must have 75 lbs. of dry solids or total solids. Using the total solids in the batch as a basis, we add back the non-dried moisture by the following formula:

$$\text{Final Volatiles in dried product} = (\text{Total solids}) \times (1/[1 - FMC \text{ wet basis}]) - 1)$$

Or, if we carry the unit analysis through:

$$\text{Pounds} = (\text{pounds}) \times (1/[1 - (\%/100)] - 1)$$

Substituting the above numbers we see:

$$75 \text{ lbs.} \times (1/[1 - (1\%/100)]) \text{ or } 75 \times ([1/.99] - 1)$$

or:

$$75 \times (1.0101 - 1) = .7575 \text{ pounds of ethanol.}$$

Therefore, the final product weight is 75.76 pounds.

When the final volatiles content is less than 5% a simplified version of this formula reduces to:

$$\text{Total Final Volatiles} = \text{Total solids} \times \text{FMC}/100$$

or, in this case:

$$75 \text{ lbs.} \times 1\%/100 = .75 \text{ lbs. of ethanol}$$

Depending upon the application, one can choose whichever method is more practical. In any case, the final weight of the product will be about 76 lbs. The next calculation is for the volume that the cake will occupy in the batch. By dividing the wet cake weight by the wet loose bulk density we attain the wet volume of the cake as it will be introduced into the dryer. This is important for several reasons. It is important to know the volume that will be occupied by the cake as it relates to the geometry of the proposed dryer. Some dryers may not be suitable beyond a given working volume due to dryer design characteristics. A more important reason is to actually choose the correct size of the dryer. A comparison must be made between the dry volume and the wet volume. The dryer should be chosen on the basis of the larger of the two operating volumes. Note that there is a difference between operating volume and a manufacturer's stated "Total" volume. The total volume given may include vapor space but it is not meant to contain just product. For instance, if we use the above case, the dry volume is about 75/12 ft^3 or 6.25 ft^3. The wet volume is 100/45 ft^3 or 2.22 ft^3.

Due to the tacky nature of the product, one should begin the dryer selection process with a mechanically agitated style and proceed to test various types from there. If the material were free-flowing, a batch fluid bed dryer with an operating volume of 6.25 ft^3 would be ideal.

7.0 SIZING DRYERS

In order to practically evaluate a design, you need to conduct test work on either a specific manufacturer's laboratory/pilot plant unit or design and build a test unit of your own. The former case is recommended because of the obvious advantages involved in making use of the manufacturer's experience and the relatively low cost. Typically, these costs are only charged by the manufacturers in order to cover their expenses for the laboratory and to avoid becoming a substitute for a company's research facility.

The option of building your own pilot unit may be desirable if there is much test work to be performed in research and development, but the drawback is that it may be difficult to obtain a full-scale model of your own lab design without manufacturing it through a metal fabrication shop. Usually, the best option is to select a reputable manufacturer through references and rent or purchase one of their pilot or laboratory models to conduct serious test work, which can be used to scale up to a production size model.

After conducting test work, most manufacturers are willing to explain the internal features of the design of their unit. This may require sufficient mechanical design details to remove some of the mystique surrounding the manufacturer's design.

7.1 Spray Dryers

Sizing a continuous spray dryer begins with defining the hourly quantity which is broken down into solids and liquids. The quantity of the liquid present is used to calculate the energy input necessary to evaporate that amount of liquid. For example:

Example 2—Dry 1000 lbs./hour of a slurry/solution containing 90% water and 10% solids. The temperature limit of the product is 170°F. The feed temperature is 65°F and the viscosity is less than 100 centipoise.

The above information is all that is required to perform a preliminary sizing on a spray dryer. The temperature limit of the product is given at 170°F. Using 160°F as the outlet temperature of the dryer should allow us the safety necessary so as not to exceed the temperature limit of the dry powder (product). The fact that we get evaporative cooling while drying the spherically shaped droplets to a similarly shaped powder allows us to use an inlet temperature of about 320°F. Thus the temperature difference (ΔT) is 160°F. This is the primary driving force in motivating the water to leave the product due to the enormous difference in saturation equilibrium between the wet droplet/dry powder and the very hot dry air.

To illustrate further, the relative humidity of air at 320°F is about 1% and the absolute humidity of the air is about 0.027 lbs. of water per pound of dry air. This already includes moisture from the products of combustion of the natural gas used for heating in an open system. This is a long way from saturation. Our objective is to calculate how much air will be required to "carry" the water out of the spray drying chamber. In order to avoid condensation in a bag house, where the temperature may drop to 110°F depending upon indoor vs. outdoor installation factors, we will use 100° as our saturation limit. This means that at 100°F our relative humidity may equal 100% and our absolute humidity is equal to 0.042lbs. water per pound of dry air. Working backwards on the air-water saturation charts,[5] at 160°F we will have a relative humidity of 20% in the air. The difference between the absolute humidities is our factor for calculating the volume of air required to carry the water out of the chamber: 0.042−0.027 = .015 lbs. of water/lb. of dry air. We have 900 lbs./hr of water to evaporate, therefore, we need at least 60,000 lbs./hr of air. At a density of 0.075 lbs./cubic foot at 70°F, we need an inlet volume of 13,333 cubic feet per minute (came).

In spray drying, typical residence times are based on certain manufacturers configurations for the material to dry as it free falls and is swept through the chamber. These range from 12 to 25 seconds. Choosing a residence time here of 20 seconds yields a chamber volume of about 4,444 cubic feet. Using a cone bottom configuration and a 1:1 diameter to straight side ratio, the dimensions required are for the chamber to be about 16 feet in diameter and a height for the cone bottom and cylinder of about 32 feet.

Next we will calculate the energy consumption. We have 900 lbs./hr of water to evaporate. Practically, we use 1000 BTU/lb. of water as the energy consumption for evaporation. This means we need 900,000 BTU/hr. As a coincidence, the amount of heat generated from burning natural gas is also about 1000 BTU/cubic foot. Therefore, if we now divide the 900,000 by 1000 we have 900 cubic feet of natural gas per hour as our energy consumption for evaporation. As a general rule, we must add some amount of energy consumption due to radiation heat transfer losses through the shell of the dryer. We will add 10% here since the temperature is not very high. For higher inlet/outlet temperatures such as 1000°F−400°F, a number such as 20% would be acceptable as an estimate.

7.2 Flash Dryers

These generally follow the same rule as used in the example above, however, one major difference is that one needs to know the fluidizing velocity of the wet cake or back mixed material to dry. The factors involved in determining the fluidizing velocity are particle size, particle density, particle shape, bulk density, and medium for fluidization. Since there are too many factors to place into a reliable equation, the most expedient method of determining this quantity is for tests to be conducted. Once this velocity is known, the volume of air required to dry (which can be calculated as in the previous example) is divided by the velocity, thus resulting in the diameter of the flash dryer pipe. The residence times are generally the same as those in spray dryers.

7.3 Tray Dryers

The basis for sizing convection type tray dryers usually requires the testing of three parameters on a fixed tray area—residence time, air velocity across the bed, and bed depth. Basically, this form of dryer requires a very long time to dry material relative to spray, flash and fluidized bed types. Much of the performance of this dryer has to do with the turning over of the cake/bed so that new surfaces are exposed to the air stream which is at a constant temperature. The time requirement can range from 10 minutes for some products to upwards of 24 hours or longer. The length of the residence time is so important that a difference of 30 minutes may require the selection of a larger or smaller size dryer with a significant impact on the price.

Example 3—Dry 100 lbs./hr (wet cake) of a pharmaceutical cake with 35% moisture coming from a plate and frame filter press in 1" thick pieces. The temperature limit of the dry material is about 190°F. The final moisture content is desired at 2% or less. It is known that on a continuous tray dryer with a bed depth of 2", it takes 3 hours to dry the material on a tray area of 1 ft^2. The loose bulk density of the wet cake is 60 lbs./ft^3. Calculating the initial amount on the tray as 1/6 of a ft^3 (2″/12″), or 5 lbs.

and dividing by the 3 hour drying time, we have a drying factor of 1.67 lbs./hour-ft^2. Dividing this factor into the 100 lbs./hr required we see that the area required is about 60 ft^2. This is effective area and not total area, thus any places where material is not present in a 2" layer must not be accounted for as area. This refers to slots in between trays, as in a Wyssmont type Turbo-Dryer, or, if the material is very free flowing, it may be 2" high in the middle of the bed, but it's angle of repose gives it a 1" layer depth on the edges of the tray.

Manually loaded shelf dryers are not continuous so we would have to calculate a batch-surge arrangement to accommodate the continuous operation.

7.4 Fluid Bed Dryers

For the continuous vibrating deck type units, we would perform the sizing on a similar basis to that for flash drying, although we would probably not need to back mix here since the fluid bed can handle a denser, pastier material.

Simply stated, the units are sized based on the amount of area required to dry the quantity desired in a finite time period. Unfortunately, each manufacturer uses different designs for their bed screen or porous plate. The way to size the dryer would be based on an effective drying yield defined in units of pounds of solvent per hour-ft^2. Knowing the amount of solvent or moisture to be removed per hour, one can easily calculate the area of the fluid bed. Getting the manufacturer to divulge the yield may prove difficult since they may not have the data for your particular solvent or may not want to divulge it due to the competitive nature of the business.

7.5 Belt or Band Dryers

These dryers are very similar to continuous tray dryers, from a process viewpoint, yet different in layout and conveying design. The volumetric throughput of these units can be calculated with the usable surface area of the belt, the layer depth of the material and the speed of the belt. Process considerations also include the temperature of the air above the material and the temperature of the material bed itself. The product is generally dry when the bed temperature rises or approaches the air temperature. This is an indication that the moisture which had been evaporating and cooling down the bed of product no longer exists, thus indicating that the product is dry.

Process designs vary among manufacturers, however, there is generally an air flow from the top or sides blowing down or across the bed of material in a zoned area. This zone may have its own batch dryer with a fan, a heater, instrumentation and duct work, or it may be manifold such that the air flow is regulated to maintain a certain temperature in that section using the evaporative cooling effect to control the outlet temperature from that zone.

In the case of the downward air flow pattern, the belt is porous, whereas in the case of the cross flow pattern, the belt may be solid. The selection of air flow/belt design would depend on the particle size and shape of the product being dried. One would choose a porous belt when drying a large granular material, which would not fall through the pores of the belt (which can be a wide metal or nylon wire mesh or, with a smaller particle size, could be a tight braided type screen). For very small particles, or those, which can become dusty, one should choose the solid belt design with the cross flow air pattern. This will probably result in a longer drying time, but the alternative of dry powder falling through the cracks of a porous belt will result in a loss in productivity or yield. Another factor for choosing the solid belt may be if the product being dried is time-temperature sensitive in which case material which falls through the cracks may decompose and may also pose a contamination problem for the good product passing through the dryer.

8.0 SAFETY ISSUES

Most drying applications require a review of the safety issues by responsible personnel within the user's company. Some of the matters to be discussed may affect the decision as to which type of dryer to use. Relevant questions to ask (and the concerns they raise) may include the following:

1. Is the product solvent wet? Inclusion of a solvent recovery system should be required for emissions and personnel protection. The emissions requirements are mandated by federal, state and/or local regulations and pertain to both atmospheric (air) and sewer (liquid) pollution by the hydrocarbon based or other solvents. For personnel protection, appropriate OSHA regulations should be followed pertaining to the proper breathing respirators, eye and skin protection.
2. Is the product dusty or hazardous? Even if the product is water wet, consideration should be given to the fact that the product may be toxic, flammable or hazardous in other ways. This would entail a *hazard analysis* review of what if situations. For example, What if the product escapes from the confinement of the dryer, or what if air gets into the dryer from the surrounding environment? Some drying processes may require the addition of a fire or explosion suppression system. One such system uses an infra-red detector to sense a cinder combined with a sonic detection device to sense the shock wave of a deflagration. Halon gas is immediately released into the drying chamber to suppress and smother the deflagration before it ruptures

the chamber. For more information the reader should contact the suggested manufacturers in the reference list at the end of the chapter. Other safety devices to investigate are rupture discs, relief valves, emergency vents and conservation breather vents. Also explosion containment is a valid path to follow with a manufacturer.

3. Is the dryer in a safe area? The dryer's environment may be another consideration to look at from the standpoint of personnel access to emergency stop buttons on the equipment. The location should be where an employee can reach the button when in a panic state. Proper dust collection equipment should be specified for a situation where the system is exhausting gases. This can be in the possible form of a bag house, a cyclone, or a scrubber. It is also possible to have a situation where two or three of the above are required.
4. Is the dryer built to safe standards? During the course of drying the material, the vessel may be subjected to positive or negative pressure. As such, it may need to be regulated by standards such as the ASME (American Society for Testing and Materials) Code, which defines mechanical specifications, such as wall thicknesses, for various materials of construction. This must be defined by the manufacturer, but the purchaser should be aware of the process needs so as to inform the manufacturer to insure the proper design.

8.1 Specific Features

If there is a part of the design, which can result in someone getting hurt due to temperature of a hot metal surface, OSHA regulations apply. An electrical panel being washed down with water would be regulated by NEMA classifications. The steam pressure required to heat a vessel may need to be 100 psig for the drying operation. As such, the jacketed heat transfer surface area on the vessel will need to be built according to the ASME Code. If there is a moving part in the dryer, which operators may be exposed to, or where possible injury may result, this requires serious, consideration of limit switches on access doors, etc. These are used quite often to protect the operator from opening a unit with moving parts. Attention should be paid to any possible moving parts, which can coast after the door is opened and the limit switch has shut off the electrical circuit. The above-mentioned regulations are only examples of possible situations, which may be encountered. It is the responsibility of the user of the equipment to provide the manufacturer with enough information as to the intended use of the equipment so as to allow the manufacturer input for safety.

9.0 DECISIONS

When choosing a particular dryer design one should consider the following factors:

1. Has this design been used for this product/process before?
2. Is the equipment design produced by several manufacturers? This allows the purchaser to choose several potential vendors. Each may have similar designs, but the purchaser should consider individual features that offer advantages to the process, production or maintenance departments.
3. What are the cost factors involved between continuous and batch designs? Has continuous processing been considered more valuable than the quality, which can be defined by batch integrity?
4. Is the design capable of meeting your stringent quality standards with regards to the overall cleanability of the design? This may have to do with the internal quality of the welds and polishing of the machine. Another factor may be the outside support structure itself, which may need to be redesigned by the manufacturer in order to make the unit easier to clean or inspect. What are the implications of cleanup between batches? How clean does the equipment have to be when changing products?
5. Will the unit selected fit into an existing area? Does the area need to be enlarged? Will permits be required? In selecting a spray dryer, generally, a good rule of thumb is to select the largest dryer, which can fit into a given space based on the height available.
6. Have the auxiliary systems (materials handling, heating, solvent recovery, dust collection, etc.) been given enough consideration with safety factors for product changes? For example, the dryer may be large enough to handle the intended capacity, but the heating system may be too small.

10.0 TROUBLE SHOOTING GUIDE

Some of the problems encountered in drying are a result of the following actions by the user:

1. Changing the product formulation that the dryer was originally intended to process.
2. Changing upstream process equipment. Example—The dryer was designed to process material from a vacuum belt filter at 30% W.B. which is now coming from a filter press with a moisture content of 40% W.B. This will probably lengthen the drying time and may affect the product quality due to possible changes in feed characteristics. If a flash dryer is being used, the finer material may overheat because it is not wet enough and the length of the tube fixes the processing time.

3. Changing the solvent used to process the material to be dried. This will affect the performance of the solvent recovery system. It may also affect the materials of construction if a particularly nasty solvent is used such as methylene chloride. For resistance to chloride attack, a dryer of stainless steel construction may now need to be made of a higher nickel alloy such as Hastelloy or Monel.
4. Changing the process parameters. Temperature, pressure and work (agitators, mechanical action) all affect the end product in some way. Temperature is the most visible parameter. Agitation can be critical in both breaking up a product in a lump form or it may cause the undesirable effect of churning the product into a very difficult paste. Products, which are thixotropic or dilatant, are most affected by agitation.

Additional suggestions:

1. Product melting or liquefying in dryer. Check if there is bound water present and reduce product temperature exposure.
2. Case Hardening. Break up feed material and increase air flow or agitation to speed up drying process.
3. Occasional discoloration or charring. Check for exposed hot surfaces or residual product holdober.

11.0 RECOMMENDED VENDORS LIST

Aeromatic/Fielder Div. Niro, Inc. 9165 Rumsey Road Columbia, MD 21045	Batch and Continuous PneumaticFluid Bed and Spray Dryers
Aljet Equipment Co. 1015 York Road Willow Grove, PA 19090	Continuous Flash Ring Dryers
Barr & Murphy Ltd. Victoria Ave. Westmount, Quebec H3Z 2M8	Continuous Flash Ring and Spray Dryers
Fenwall Safety Systems 700 Nickerson Road Marlborough, MA 01752	Fire and Explosion Suppression Systems
Fike Metal Products 704S 10 Street Blue Springs, MO 64105	Fire and Explosion Suppression Systems, Rupture Discs
Compline San Komline-Sanderson Corp. 100 Holland Ave. Peapack, NJ 07977	Continuous Paddle Dryers
Krauss Maffei Corp.(Andritz) Process Technology Div. PO Box 6270 Florence, KY 41042	Continuous Plate Dryers
Paul 0. Abbe, Inc. 139 Center Ave. Little Falls, NJ 07424	Batch Dryers, Mechanical Fluid Bed Dryers
Protectoseal Company 225 W Foster Ave. Bensenville, IL 60106	Emergency, Conservation and Breather Vents (Safety and Explosion Venting)
Wyssmont Co. 1470 Bergen Blvd. Ft. Lee, NJ 07024	Continuous Rotary Tray Dryers

FURTHER READING (FOR SECTION II: DIRECT DRYING)

[1] R.H. Perry, C.H. Chilton, Chemical Engineer's Handbook, fifth ed., McGraw-Hill, 1973.

[2] Babcock and Wilcox, Steam/Its Generation and Use, thirty-eight ed., The Babcock and Wilcox Company, 1972.

[3] J.M. Smith, H.C. Van Ness, Introduction to Chemical Engineering Thermodynamics, third ed., McGraw-Hill, 1975.

[4] Randall N. Robinson, Chemical Engineering Reference Manual, fourth ed., Professional Publications, 1987.

[5] R.E. Treybal, Mass Transfer Operations, second ed., McGraw-Hill, 1968. p. 187, Fig. 7.5, p. 582, Fig. 12.10.

Part IV

Purification

Chapter 15

Crystallization

Stephen M. Glasgow

1.0 INTRODUCTION

Crystallization is one of the oldest methods known for recovering pure solids from a solution. The Chinese, for example, were using crystallization to recover common salt from water some 5000 years ago.

The perfection and beauty of the crystal that fascinated the early tribes now leads to a product of high purity and attractive appearance. By producing crystals of a uniform size, a product that has good flow, handling, packaging, and storage characteristics is obtained.

Crystallization is still often thought of as an art rather than a science. While some of the aspects of art are required for control of an operating crystallizer, the discovery by Miers of the metastable region of the supersaturated state has made it possible to approach the growth of crystals to a uniform size in a scientific manner.

To produce pure crystalline solids in an efficient manner, the designer of crystallization equipment takes steps to ensure the control of:

1. The formation of a supersaturated solution
2. The appearance of crystal nuclei
3. The growth of the nuclei to the desired size

2.0 THEORY

The first consideration of the equipment designer is the control of the formation of a saturated solution. In order to do this, it is necessary to understand the field of supersaturation.

2.1 Field of Supersaturation

The solubility chart divides the field of the solution into two regions: the subsaturated region where the solution will dissolve more of the solute at the existing conditions, and the supersaturated region.

Before Miers identified the metastable field, it was thought that a solution with a concentration of solute greater than the equilibrium amount would immediately form nuclei. Miers' research and the findings of subsequent researchers determined that the field of supersaturation actually consists of at least three loosely identified regions (Figure 15.1):

Metastable region—where solute in excess of the equilibrium concentration will deposit on existing crystals, but no new nuclei are formed.

Intermediate region—where solute in excess of the equilibrium concentration will deposit on existing crystals and new nuclei are formed.

Labile region—where nuclei are formed spontaneously from a clear solution.

The equipment designer wishes to control the degree of supersaturation of the solution in the metastable region when designing a batch crystallizer. In this region, where growth takes place only on existing crystals, all crystals have the same growth time and a very uniform crystal size is obtained.

When designing a continuous crystallizer, the designer wishes to control the degree of supersaturation in the lower limits of the intermediate region. In continuous crystallization, it is necessary to replace each crystal

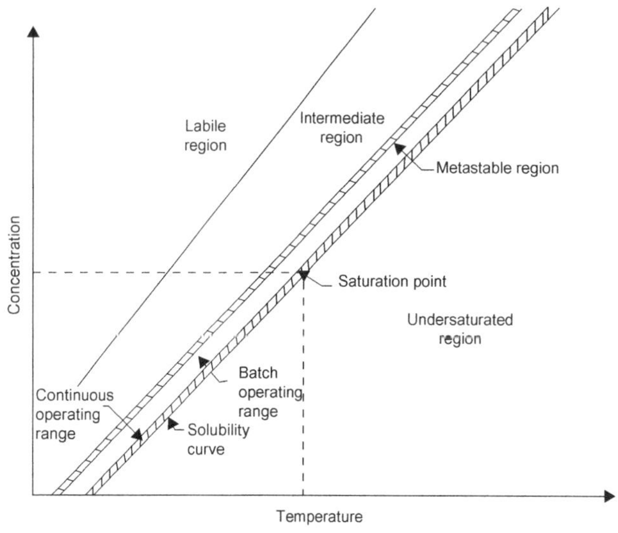

FIGURE 15.1 Solubility curve.

Fermentation and Biochemical Engineering Handbook.

removed from the process with a new nuclei. It is also necessary to provide some degree of crystal size classification if a uniform crystal size is to be obtained.

Solutions of most organic chemicals can, as a general rule, attain a considerably higher degree of supersaturation than inorganic chemicals. The formation of crystalline nuclei requires a definite orientation of the molecules in the solution. This requires the proper orientation of several molecules at the moment of a random collision. Since the number of possible orientations increases with increasing complexity of the molecule, considerably higher degrees of supersaturation can be obtained for solutions of chemicals with complex molecules.

2.2 Formation of a Supersaturated Solution

If a solution is to have only a slight degree of supersaturation, then a cyclic system in which large quantities of liquor are supersaturated uniformly is required. The solution must then be brought back to saturation before feed liquor is allowed to enter the system and the mixture is again supersaturated in the next cycle.

The removal of the metastable supersaturation is a slow process. A large amount of crystal surface is required to allow for the large number of random collisions necessary to remove the supersaturation generated during the cycle. The proper orientation of both the molecules in solution and the molecule on the crystal surface is required for deposition, and the increased complexity of the molecule increases the number of collisions required for proper orientation.

If the supersaturation generated during the cycle is not completely removed, the level of supersaturation attained during the following cycle is increased. This increase from cycle to cycle will continue until the supersaturation level of the solution exceeds the metastable region and enters the labile region, where spontaneous nucleation occurs. The occurrence of spontaneous nucleation means loss of control of crystal size.

Supersaturation is clearly the most important single consideration for any crystallization process. By giving proper attention to the degree of supersaturation generated during each cycle and its proper release during the design stage, half the battle will be won. Supersaturation should be controlled by making certain only small changes in temperature and composition occur in the mass of mother liquor.

2.3 Appearance of Crystalline Nuclei

Usually the crystallization equipment is charged with a clear feed solution. As this solution is saturated, it is important to control the increase in supersaturation as the labile region is approached. This is important since the formation of an excessive number of nuclei will cause a continuous crystallizer system to have an extremely long period before desired crystal size can be achieved and prevent a batch system from ever producing desired crystal size during that particular run.

Once initial nucleation has been achieved successfully, the control of secondary nucleation becomes important. Since crystal growth is a surface phenomenon, each nuclei formed is available to absorb the supersaturation generated by the cycle. This means that only one nuclei is to be formed for each single crystal removed if a constant crystal size is to be maintained.

When an excessive number of nuclei are formed during operation of the crystallizer, the average size of the final product is reduced. As an example of this effect, one can assume the formation of 1 lb. of 200 mesh nuclei. Assuming that no further new nuclei are formed, this 1 lb. would weigh 8 lbs. if grown to 100 mesh crystals. Following this trend further, it is found that growth to 60 mesh crystals will result in 38 lbs.; 14 mesh crystals would yield 7000 lbs. (see Figure 15.2).

Secondary nucleation is constantly occurring. It occurs when a crystal collides with the vessel wall or with another crystal. To control this collision-induced nucleation the number of crystals in the system must be controlled.

Increasing the local supersaturation into the labile region will also cause secondary nucleation. This occurs when there are local cold spots caused by radiation from the vessel wall, subcooling caused by subsurface boiling and build up of residual supersaturation in solutions with high viscosity and insufficient agitation. This calls attention to the need for insulation of the vessel, for control to ensure that boiling occurs at the liquid-vapor interface, and for provision for sufficient agitation of the solution in the vessel.

Mechanically induced nucleation can result from excessive agitation caused by an impeller sweeping through a solution in the metastable region of supersaturation or turbulence caused by violent boiling. Limiting the tip speed of a pump or agitator and limiting the escape velocity at the vapor-liquid interface can minimize this type of secondary nucleation.

After the control of supersaturation, control of nuclei formation is the most important consideration in the design of crystallization equipment. If a constant number of crystals are maintained in the crystallizer, then a constant surface area for crystal growth will be available. This will result in good control of product size.

2.4 Growth of Nuclei to Size

As noted above, crystal growth is a surface phenomenon. Given sufficient agitation, the depositing of solute on the surface is controlled by proper orientation of the molecules, rather than by film diffusion to the surface; the

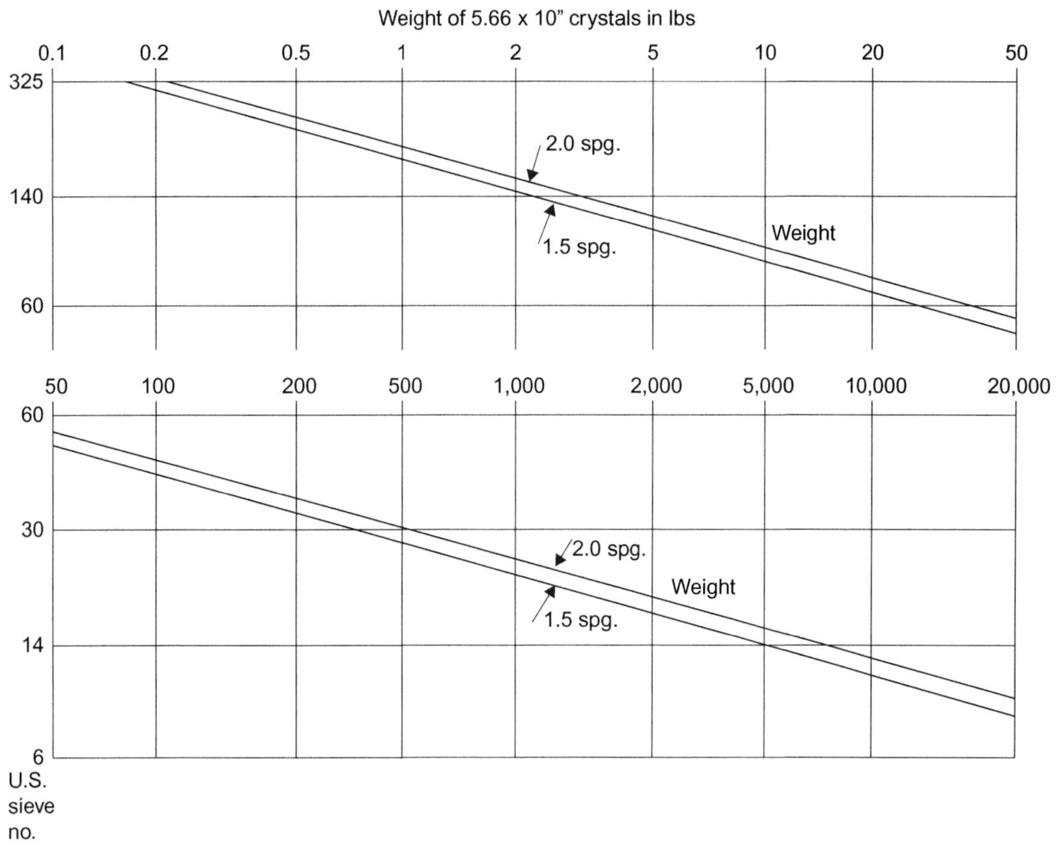

FIGURE 15.2 Increase of weight with crystal size.

crystal growth rate approaches zero order with increasing driving force. Since growth becomes a function of time only, the crystal must be retained in the crystallizer for a sufficient amount of time to allow it to grow to the desired size.

The growth type crystallizers maintain the crystals in a fluidized bed (thereby providing both agitation and size classification of the crystals). The supersaturated solution flows through the fluidized bed and releases the supersaturation to the crystal surface.

Not all crystals will remain in the crystallizer the calculated retention time. This is only a statistical average. Since there will be a range of growth times, there will be a distribution of crystal sizes. The more narrow the range of actual retention times, the more narrow the crystal size distribution.

3.0 CRYSTALLIZATION EQUIPMENT

The type of equipment to be used in a crystallization process depends primarily upon the solubility characteristic of the solute. Solutions from fermentation processes can be classified as follows:

1. Chemicals where a change in solution temperature has little effect on the solubility. An example is hexamethylene tetramine as shown in Figure 15.3. The supersaturated solution is produced by evaporation of the solvent. The equipment needed here is called an evaporative crystallizer (see Figure 15.4).
2. Chemicals, e.g., fumaric acid, which show only a moderate increase in solubility with increasing temperature. A combination of evaporation and cooling may be used to produce the supersaturated solution. Depending upon the yield required, this operation might be carried out in either a vacuum cooling crystallizer or an evaporative crystallizer (see Figure 15.5).
3. Chemicals, e.g., adipic acid, which show a large increase in solubility with increasing temperature. Cooling the solution can be an effective way to produce the supersaturated solution, although a combination of evaporation and cooling can also be employed. In addition to the two types of crystallizers mentioned above, a cooling crystallizer may be used (see Figure 15.6).

One of the more important features of the Oslo type crystallizer is that the container for crystal growth has certain elements of design similar to all modes of operation (evaporative, vacuum cooling, cooling). In the crystal growth container a supersaturated solution of uniform temperature and concentration is conducted upward through a dense fluidized bed of crystals. The crystals are kept fluidized by this upward flow of liquor. This results in a classifying action in the crystal growth container, which keeps the large crystals suspended in the bottom layer of the suspension and the smallest crystals in the top layer, with the intermediate sizes suspended between. If the process dictates the need for crystals being present throughout the system, the fluidized bed may be expanded to allow a portion of the crystals to overflow the crystal growth container into the circulation loop.

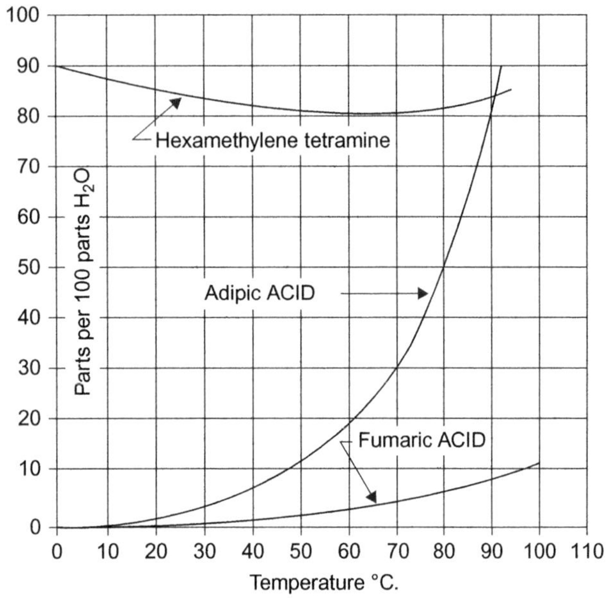

FIGURE 15.3 Effect of temperature rise on solubility in water.

3.1 Evaporative Crystallizer

A properly designed crystallizer should result in reasonably long periods between clean outs, uniform crystal growth, and minimal flashing in the vaporization container to reduce entrainment. These objectives are attained by keeping supersaturation well below the upper limit of the metastable region in all parts of the crystallizer, and by maintaining a large fluidized suspension of crystals in the crystal growth container to provide sufficient surface for desupersaturation.

In the Oslo design this is accomplished by continuously mixing the feed liquor with a large amount of circulating mother liquor. The mixture is passed through a heat exchanger, where the heat required by the process is added by raising the temperature of the circulating mixture to a few degrees (3–6°F) above the operating temperature of the crystallizer.

The heated solution is passed into the vaporization container where the temperature is lowered to the operating temperature by vaporization of an equivalent amount of the solvent. The supersaturated solution thus produced, flows down a central pipe and upward through the crystal growth container. As the supersaturated liquor passes the fluidized crystals, the supersaturation is released to the surface of the crystals, allowing for uniform growth.

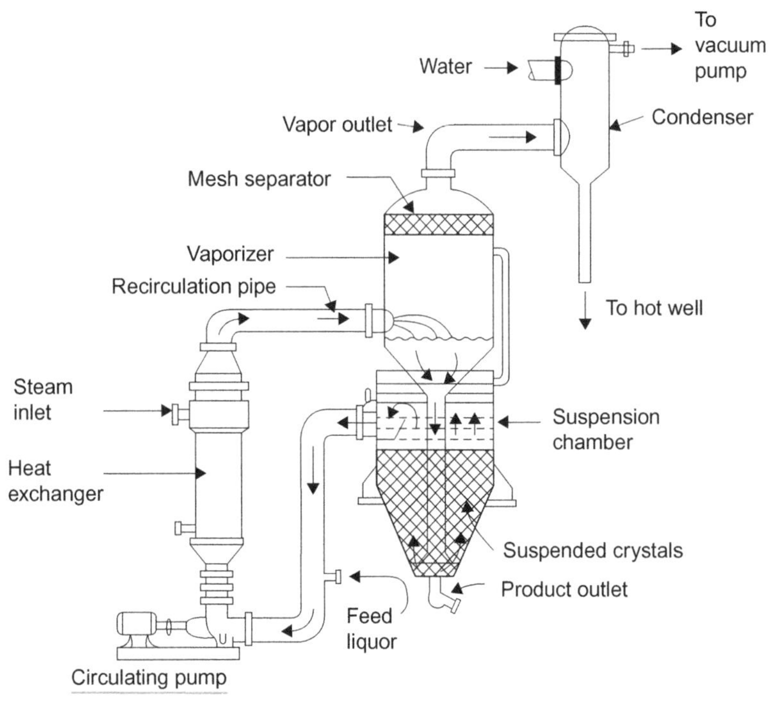

FIGURE 15.4 Oslo evaporative crystallizer.

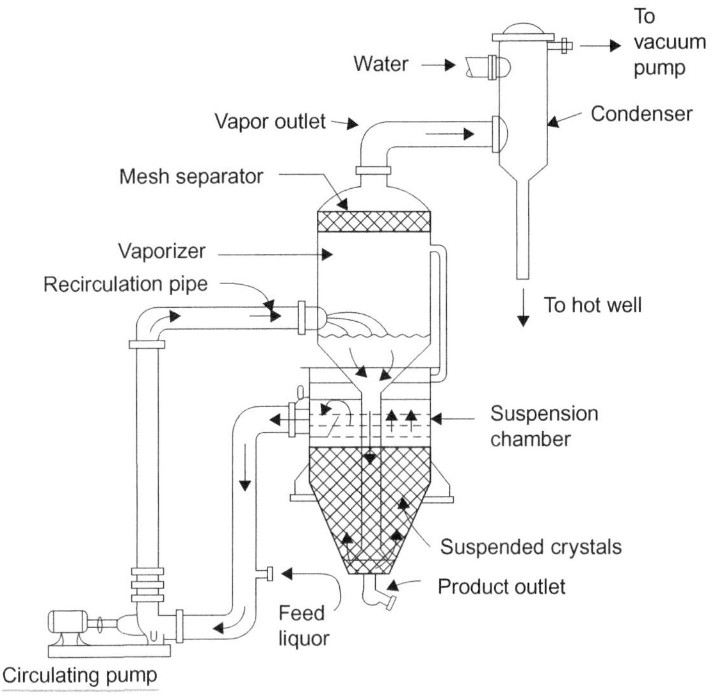

FIGURE 15.5 Oslo vacuum cooling crystallizer.

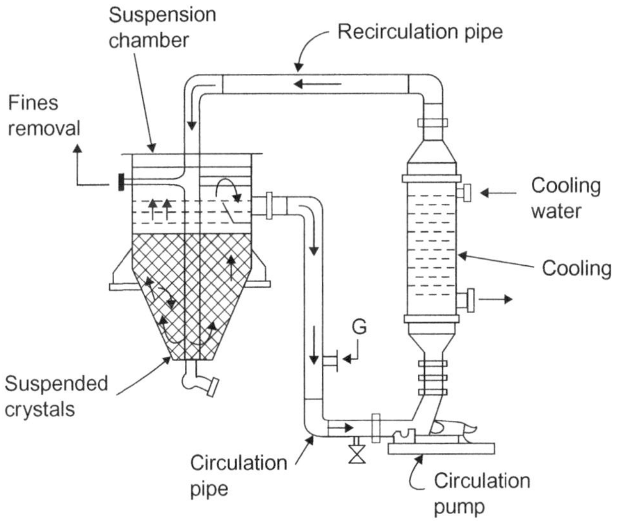

FIGURE 15.6 Oslo cooling crystallizer.

The now saturated mother liquor is passed out of the crystal growth container into the circulation loop where it is again mixed with fresh feed liquor and the cycle repeated.

In the crystal growth container a sufficient quantity of crystals is maintained in a fluidized bed to achieve almost complete release of supersaturation. The individual crystals must be kept in constant motion, as they are by the fluidization, to prevent their growing together, but the motion must not be so violent as to cause excessive secondary nucleation. The amount of crystals required is a function of the crystal species, the solution and its impurities, the operating parameters, and the desired crystal size.

The heat added to the system must be done in such a manner that no boiling occurs in the heat exchanger tubes. Boiling in the tubes would cause scaling, and hence, result in frequent shutdowns for clean out.

3.2 Vacuum Cooling Crystallizer

The elements of design for a vacuum cooling crystallizer are the same as for the evaporative crystallizer except a heat exchanger is not required. The operating features are also similar. In this case, the heat for evaporation is supplied by the sensible heat of the feed and the heat of crystallization.

If it is desired to operate at a temperature which results in the solution having a vapor pressure below the vapor pressure of the available coolant, a steam-jet booster may be used in the vacuum system.

3.3 Cooling Crystallizer

The crystal growth container is similar to the other type crystallizers outlined above, but the supersaturated solution is produced differently. A vertically arranged shell-and-tube heat exchanger is used to remove the sensible heat of the feed and the heat of crystallization.

By eliminating the evaporation, the vaporization chamber is eliminated and the vessel is now designed to operate at atmospheric pressure.

To keep the supersaturation of the solution in the metastable region, the temperature drop through the heat exchanger must be comparatively small. To prevent scaling of the heat exchanger surface, the temperature difference between the mother liquor and the coolant must be kept small.

3.4 Batch Crystallization

Both batch and continuous operation are used in industry. The final choice between a batch and continuous process will be made in favor of the one which gives the most favorable evaluated cost.

In some cases, where the final solution has a very low concentration of the product, or the final solution has a high viscosity, or there is a large quantity of impurities, the batch crystallizer will be chosen because it can produce a crystal quality not achievable by a continuous crystallizer.

The basic design criteria used for a continuous crystallizer also apply to a batch crystallizer. These criteria are to:

1. Maintain the solution in the metastable region of supersaturation
2. Provide a large fluidized bed of crystal to allow effective, efficient release of supersaturation
3. Minimize secondary nucleation

The batch crystallizer is filled with hot feed solution and then cooled, either by evaporation of solvent by lowering of the operating pressure (vacuum cooling) or by using a heat exchanger and a coolant fluid. As the feed is cooled, a supersaturated solution is produced. From this supersaturated solution, the crystalline nuclei are formed. The crystals are grown to their final size as further cooling continues to produce supersaturation as the driving force. At the end of the batch cycle, the magma is removed from the crystallizer and sent to the dewatering equipment to recover the crystals.

4.0 DATA NEEDED FOR DESIGN

The first and most important piece of information required is a solubility curve. If solubility data for the specific solution is not available, information which is at least representative must be supplied.

The next set of information required is the physical properties of the solutions. These are viscosity, specific heat, specific gravity, boiling point elevation and thermal conductivity. While all these data may not be available, those available will give the experienced designer the information required to make an intelligent "guesstimate" of the missing physical property values.

The third set of data includes those variables set by the plant. These are quality and quantity of utilities available; composition, temperature, and quantity of feed solution; and finally, desired production rate and quality (size distribution) of final product.

The final data the designer hopes for are pilot plant data from tests he has conducted. It is here that the designer determines what level of supersaturation the solution can support, the crystal surface area required for desupersaturation, the effect of secondary nucleation, and the residence time required for growth to desired size. Some of these values are measured directly while others are implied by indirect measurements.

Although the major suppliers of crystallization equipment have extensive experience in crystallization and can often design equipment which will operate satisfactorily from the solubility curve and the values for the physical properties, it is still advisable to conduct pilot plant studies on typical solutions from an operating commercial plant or process pilot plant. The presence of impurities, pH of the solution, and solubility of the product at the operating temperature all have an effect on crystal growth rate, shape and purity. By running commercial solutions in a pilot plant, the designer can detect problems which may arise during the crystallization process and possible overall process problems may be anticipated.

Due to the importance of pilot plant test data, all of the major crystallization equipment suppliers maintain a well-equipped pilot plant and have an experienced and knowledgeable staff. These operators, engineers, and designers have defined the parameters for scale up very well, so well that scalps of over 2,000 to 1 have been made successfully.

5.0 SPECIAL CONSIDERATIONS FOR FERMENTATION PROCESSES

The preceding sections dealt with the design considerations for crystallization in general terms. Now emphasis is directed to areas of special concern for the processing of organic chemicals.

5.1 Temperature Limitation

Because the properties of an organic chemical can be altered by prolonged exposure to high temperature, an upper temperature limit of 70°C is set for solutions produced by fermentation processes. The value can be raised or lowered if test work so indicates.

5.2 High Viscosity

Most aqueous solutions of high molecular weight organic chemicals have a high viscosity. Since the upper temperature of the solution is limited, this problem cannot be overcome by raising the operating temperature. The high viscosity dampens the turbulence of the solution in the fluidized bed, making effective and efficient contact of the crystals and the supersaturated solution very difficult. Inefficient contact leads to a buildup in residual supersaturation and hence, excessive nucleation. Often equipment to remove fine (very small crystals) must be supplied as part of a crystallization system for aqueous solutions of organic compounds.

5.3 Long Desupersaturation Time

Due to the nature of long chain organic molecules, deposition on the crystal surface is more difficult and time consuming than for most inorganic chemicals. This must be taken into consideration and additional time allowed between cycles so that the supersaturation can be relieved. Another effective method for handling this potential problem is to limit the supersaturation generated during each cycle.

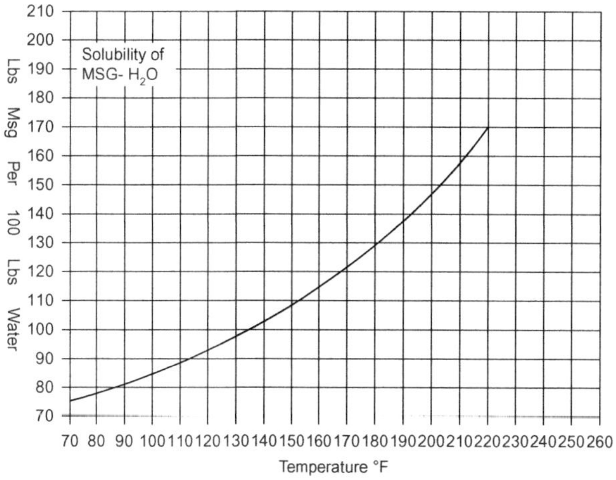

FIGURE 15.7 Solubility curve for MSG.

5.4 Slow Crystal Growth Rate

The problem of deposition is compounded by an increased film resistance, due to high viscosity and to the long chains of the molecules. The result is a decrease in the average growth rate; however, solutions of organic compounds can, as a general rule, support a higher supersaturation than those of inorganic chemicals. This allows the designer to use a higher driving force, but usually, a longer retention time is also required.

6.0 METHOD OF CALCULATION

Now, having discussed the theory, the equipment, the required design data, and the special considerations, an actual design will be considered. Crystallization of monosodium glutamate shall be used for this example. The first step is to gather the design data.

1. Solubility curve—See Figure 15.7
2. Physical properties of solutions

lio	Feed	Mother Liquor at 55°C
Specific gravity	1.23	1.254
Specific heat	0.67	0.64*
Temperature, °C	60	55
Viscosity, cp	4	6
Composition		
MSG, %	40	48
H_2O, %	60	52
Boiling point elevation, °C	—	9

*Estimated

3. Physical properties of crystals

Specific gravity	1.65
Heat of crystallization	38 Btu/lb (exothermic)
Heat of concentration	None

4. Plant parameters

Cooling water	38°C
Steam	30 psig
Electricity	440 V/3 phase/60 Hz
Production rate	300 T/day
Crystal size	100–125 U.S. standard mesh
Material of construction	316 L stainless steel, food grade

5. Operating data

Retention time	8 hours
Supersaturation	10 lb/100 gal
Suspension density	25% crystal, by wt.
Volumetric production rate	2.5 lb/ft^3-hr
Solution upward velocity	50 gpm/ft^2

The second step is to select an operating mode and to calculate the heat and material balances. For this example, an evaporative crystallizer will be used. Complete evaporation and crystallization of the feed will be assumed.

Material Balance (values in lb/hr)

	Liquid Temperature (°C)	Vapor Temperature (°C)	MSG	H_2O	Total
Feed	60	–	25,000	37,500	62,500
Remove	–	–	25,000	37,500	62,500
Mother Liquor	55	46	(48)	(52)	(100)
Heat Balance (Btu/hr)					
Sensible heat	(62,500) (55–60) (1.8) (0.67)			= – 376.875	
Heat of crystallization	(25,000) (– 38)			= – 950,000	
Heat of vaporization	(37,500) (1,028.4)			= 38,565,000	
Total heat required by system				= 37,238,125	

From the preceding information, the nozzles connecting the crystallizer with the remainder of the process can be sized and a heat exchanger design prepared. While these are important to the design of the crystallization, they will not be pursued further at this time.

The final step to be examined is the sizing of the crystal growth container.

1. Volume
   - Quantity of crystals of fluidized bed
     8 hr retention time × 25,000 lb/hr = 200,000 lb
   - Quality of slurry in container
     200,000 lb/0.25 = 800,000 lb
   - Apparent specific gravity of slurry

$$\frac{(1.254)(1.65)}{(0.25)(1.254) + (0.75)(1.65)} = 1.33$$

   - Volume
     800,000/(62.4)/(1.33) = 9,640 ft^3
   - Check-volumetric production rate
     25,000/9,640 = 2.59 lb/hr-ft^3
     Target value 2.5 lb/hr-ft^3
     This is acceptable
2. Diameter
   - Circulation rate

$$\frac{25,000\ \text{lb/hr}}{(60\ \text{min/hr})(10\ \text{lb})} X \times 1,000 = 42,000\ \text{gpm}$$

   - Cross-sectional area
     42,000 gpm/50 gpm/ft^2 = 840 ft^2
   - Diameter-use 33'-0" (855.3 ft^2)
3. Vessel straight side
   - Head volume-4,000 ft^3
   - Volume of cylinder–5,640 ft^3
   - Straight side
     5,640 ft^3/855.3 ft^2 = 6.6 ft

7.0 TROUBLESHOOTING

At some point during operation of a crystallizer, difficulties are going to occur. A list of some of the more common difficulties along with probable causes and recommended remedies is given below.

7.1 Deposits

1. *Local cooling due to lack of insulation.* This causes an increase in local supersaturation into the labile region. To remedy, insulate all areas of the crystallizer and piping carrying saturated liquors, particularly protruding points such as reinforcing rings.
2. *Low suspension density.* Since the solution cannot return to saturation before being *resupersaturated* in the circulation loop, the residual supersaturation builds up to the point that the solution is in the labile region. To remedy, increase the crystals in suspension to maintain the design density.
3. *Protruding gaskets, rough areas on process surface.* These areas provide a place for crystal nucleation and growth. Since the object is to grow the crystals in the solution, not on the vessel walls, it is necessary to remove protrusions and polish the rough areas.

7.2 Crystal Size Too Small

1. *Low suspension density.* This decreases average retention time, hence, the amount of time for crystal growth is insufficient. To correct, increase crystals in suspension to maintain design density.
2. *High circulation rate.* This causes the fluidized crystal bed to become overextended resulting in too many void areas. This will result in improper release of supersaturation. To correct, maintain the circulation rate at the design rate.
3. *Solids in feed.* This introduces nuclei into the crystallizer in excess of the number required for size control. To correct, make sure the feed solution is free of solids, especially crystals.
4. *Design feed rate exceeded.* At design suspension density, this results in a reduced average retention time. Within limits, this can be corrected by increasing the suspension density. An increase in production rate usually requires an increase in circulation rate to handle the additional supersaturation and the heavier fluidized bed of crystal
5. *Excessive nucleation.* In addition to points 1 to 3 above, this is caused by excessive turbulence, local cold spots and subsurface boiling. To correct, maintain level at design point and maintain pump or agitator speed at design point.

6. In some cases it is very difficult to prevent excess nucleation. Excess nucleation results in high surface area to weight ratio, which prevents proper growth. In some cases it becomes necessary to remove fine salt (nuclei) from the system by dissolving or settling. A portion of liquor which contains fine salt is pumped from the crystallizer to a settler or heat exchanger where either all or a portion of the fine salt is removed. This is referred to as a *fines removal system.*

7.3 Insufficient Vacuum

1. *Obstruction in vapor system.* This causes excessive pressure drop. The obstruction is usually a deposit in the noncondensible take off nozzle of the condenser. The obstruction must be removed to correct the problem.
2. *Insufficient cooling water or cooling water above design temperature.* This results in overloading the vacuum system because of insufficient subcooling of the noncondensible gases resulting in excess water of saturation in the noncondensible stream. Cooling water at the design flow rate and at or below the design temperature must be provided to correct the problem.
3. *Air leaks in system.* This results in overloading the vacuum system because of excess noncondensibles and water of saturation in noncondensible stream. Air leakage must be stopped to correct the problem.
4. *Excessive backpressure on vacuum system.* This is caused by an obstruction in the noncondensible discharge pipe or the discharge pipe sealed too deeply in the hot well. The obstruction must be removed or the depth of the seal in the hot well reduced to correct the problem.
5. *Flooded intercondenser.* This is usually caused by a blockage in the discharge line or by using an excess amount of cooling water. The flooded condenser causes excessive pressure drop in the vacuum system. To correct, remove blockage or reduce cooling water flow to design rate.
6. *Low steam pressure.* This applies to steam ejectors only. The cause is low line pressure, wet steam or blockage in the steam line. This reduces the driving force of the ejector and reduces its air handling capacity. By removing the cause of the low steam pressure, the problem of insufficient vacuum is corrected.
7. *Low seal water flow.* This applies to mechanical vacuum pumps only. This reduces the subcooling of the noncondensible, increasing the loading to the system. Seal water must be maintained at design flow to correct the problem.
8. *Low rpm for vacuum pump.* This usually is caused by V-belt slippage or low voltage to the motor. To correct, tighten V-belts or reduce load on electric circuit to motor.

7.4 Instrument Malfunction

1. *Air leaks.* This causes erroneous reading at the instrument. To remedy, seal air leak.
2. *Plugged purge line.* If low pressure purge line is plugged, the instrument will give the minimum reading; conversely, if the high pressure side is plugged, the maximum reading will be indicated. To prevent purge lines from plugging, they should be given a good flushing at least twice a shift.
3. *Purge liquor boiling in purge line.* This occurs when vapor pressure of purge liquor is higher than vapor pressure in crystallizer vessel. To prevent this problem use purge liquor (usually water) which is at or below the maximum operating temperature of the crystallizer.
4. *Improper adjustment.* Proportioning band and reset should be adjusted to give smooth control. Damping must not be so great that sensitivity is lost. Consult manufacturer's manual for instrument adjustment procedures.

7.5 Foaming

1. If foaming is not inherent to the solution, it can usually be traced to air entering the circulating piping via the feed stream, leakage at the flanges or by leakage through the pump packing. By eliminating the air leakage the problem is corrected.
2. If foaming is inherent to the solution, a suitable antifoam agent may be used. Selection of a suitable antifoam must include the effects upon the crystal habit and growth rate as well as the amount required, availability, and cost.

7.6 Pump Performance

1. *Loss of capacity.* This is usually caused by loose V-belts or blockage in line. Check pump rpm and tighten V-belts if below design speed. If pump speed is correct, check for blockage in piping.
2. *Leaks in packing.* Care must be taken to keep packing in good condition. When pump is repacked, wash out the packing housing thoroughly with clean water before installing new packing.
3. *Cavitation.* This can be detected by a popping, gravel-rolling-around sound in the pump. It is caused by air entering into suction or insufficient net positive suction head (NPSH). If the pump is operating at design condition, check for air leaks or blockage in the pump suction piping. Before pump speed is increased above design point, consult the pump curve for rpm and NPSH data.
4. *Low solids content in product slurry.* The cause of this is probably a restriction in the slurry pump suction line. Lumps can cause such restrictions and act as

partial filters. When the problem occurs, it can usually be corrected by flushing the slurry line.

5. *Slurry settling in line.* This is usually caused by a heavy slurry or low slurry pump speed. Check pump rpm and tighten V-belts if necessary. If a heavy slurry is causing the problem, dilute the slurry with mother liquor before pumping to the dewatering equipment.

8.0 SUMMARY

In this chapter, crystallization technology and how it can be applied to fermentation processes have been examined.

The main steps in the unit operation of crystallization are:

1. Formation of a supersaturated solution
2. Appearance of crystalline nuclei
3. Growth of nuclei to size

The selection of crystallization equipment depends mainly upon the solubility characteristics of the solute. Several types of equipment have been described:

1. Evaporative crystallizer
2. Vacuum cooling crystallizer
3. Cooling crystallizer
4. Batch crystallizer

Data required for proper crystallizer design are:

1. Solubility curve
2. Physical properties of the solution, heat of crystallization and of concentration
3. Utilities available; production required
4. Pilot plant test or operating data

Liquors from fermentation processes have special considerations, e.g.:

1. Temperature limitation
2. High viscosity
3. Long desupersaturation time
4. Slow crystal growth rate

A sample calculation was shown to illustrate the basic approach to sizing a crystal growth container.

The author hopes that this chapter will enable the fermentation engineer to decide when crystallization may be useful in his process and what basic information he will have to provide the crystallizer designer.

9.0 AMERICAN MANUFACTURERS

1. Swenson Technology Inc.
   26000 Whiting Way
   Monee, Illinois 60449-8060
   www.swensontechnology.com
   (708) 587 – 2300
2. HPD Evaporation and Crystallization
   23563 West Main Street, IL Route 126
   Plainfield, Illinois 60544
   www.veoliawaterstna.com
   (815) 609 – 2000

FURTHER READING

[1] J.W. Mullin, Crystallization, second ed., Butterworth & Co., London, 1972.

[2] D.B. Wilson, Crystallization, Chem. Eng., 119–138 (Dec. 6, 1965).

[3] J.W. Mullin, Crystallization, in: Kirk, Othmer (Eds.), Encyclopedia of Chemical Technology, vol. 6, John Wiley and Sons, New York.

[4] H. Svanoe, Solids recovery by crystallization, Chem. Eng. Progr. 55 (1959) 47–54.

[5] H. Svanoe, Crystallization of organic compounds from solution, J. Chem. Educ. 27 (1950) 549–553.

[6] H. Svanoe, "Krystal", classifying crystallizer, Ind. Eng. Chem. 32 (1960) 636–639.

[7] A. Mersmann, Crystallization Technology Handbook, second ed, CRC, 2001.

[8] N.S. Tavare, Industrial Crystallization, Plenum Press, New York, 1995.

[9] McCabe, Smith, Unit Operations of Chemical Engineering', McGraw-Hill, New York, 2000.

Chapter 16

Chromatography

Frederick J. Dechow

1.0 INTRODUCTION

In 1850 Thompson [1] reported the first ion exchange applications which used naturally occurring clays. However, ion exchange resins have only been used in biochemical and fermentation product recovery since the 1930's [2,3]. In these early studies, biochemicals such as adenosine triphosphate [4], alcohols [5], alkaloids [6], amino acids [7], growth regulators [8], hormones [9], penicillin [10] and vitamin B12 [11] were purified using ion exchange resins.

Ion exchange applications intensified following the work of Moore and Stein [12], which showed that very complex mixtures of biochemicals, in this case, amino acids and amino acid residues could be isolated from each other using the ion exchange resin as a column chromatographic separator. In biotechnology applications today, ion exchangers are important in preparing water of the necessary quality to enhance the desired microorganism activity during fermentation. Downstream of the fermentation, ion exchange resins may be used to convert, isolate, purify or concentrate the desired product or by-products. This chapter discusses ion exchange resins and their use in commercial fermentation and protein purification operations.

1.1 Chromatographic Processes with Ion Exchange Resins

Processes involving ion exchange resins usually make use of ion interchange with the resin. Examples of these processes are demineralization, conversion, purification and concentration. Chromatographic processes with ion exchange resins merely make use of the ionic environment that the resins provide in separating solutes.

Demineralization is the process in which the salts in the feed stream are removed by passing the stream through a cation exchange column in the hydrogen ion form, followed by an anion exchange column in the hydroxide or "free-base" form. Water is the most common feed stream in demineralization. It may also be necessary to remove the salts from a feed stream before fermentation.

High metallic ion concentrations and high total salt content in the carbohydrate feed has been found to decrease the yield in citric acid fermentation [13]. These ions can be removed by passing the carbohydrate solution through cation and anion exchange resin beds. The salts required for optimum microorganism activity can be added in the desired concentration prior to fermentation.

Conversion or metathesis is a process in which salts of acids are converted to the corresponding free acids by reaction with the hydrogen form of a strong acid cation resin. One such example would be the conversion of calcium citrate to citric acid.

The terms may also be used to describe a process in which the acid salt is converted to a different salt of that acid by interaction with an ion exchange resin regenerated to the desired ionic form.

Many fermentation products may be purified by adsorbing them on ion exchange resins to separate them from the rest of the fermentation broth. Once the resin is loaded, the product is eluted from the column for further purification or crystallization.

Adsorbing lysine on ion exchange resin is probably the most widely used industrial method of purifying lysine. The fermented broth is adjusted to pH 2.0 with hydrochloric acid and then passed through a column of strong acid cation resin in the NH_4^+ form. Dilute aqueous ammonia may be used to elute the lysine from the resin [14].

Gordienko [15] has reported that treating the resin with a citrate buffer solution of pH 3.2 and rinsing with distilled water before elution results in an 83–90% yield of lysine, with a purity of 93–96%.

Ion exchange can be used to concentrate valuable or toxic products of fermentation reactions in a manner similar to purification. The difference between the two processes is in the lower concentration of the desired product in the feed solution of concentration processes.

Shirato [16] reported the concentration process for the antibiotic tubercidan produced from fermented rice grain using the microorganism, *Streptomyces tubercidicus*. Macroporous strong acid cation resin was used to concentrate the antibiotic from 700 μg/ml in the fermentation broth to 13 mg/ml when eluted with 0.25 N HCl. The yield of the antibiotic was about 83%.

1.2 Chromatographic Separation

In most ion exchange operations, an ion in solution is replaced with an ion from the resin and the former solution ion remains with the resin. In contrast, ion exchange chromatography uses the ion exchange resin as an adsorption or separation media, which provides an ionic environment, allowing two or more solutes in the feed stream to be separated. The feed solution is added to the chromatographic column filled with the separation beads and is eluted with solvent, often water in the case of fermentation products. The resin beads selectively slow some solutes while others are eluted down the column (Figure 16.1). As the solutes move down the column, they separate and their individual purity increases. Eventually, the solutes appear at different times at the column outlet where each can be drawn off separately.

Chromatographic separations can be classed according to four types depending on the type of materials being separated: affinity difference, ion exclusion, size exclusion and ion retardation chromatography. These types of separations may be described in terms of the distribution of the materials to be separated between the phases involved.

Figure 16.2 shows a representation of the resin-solvent-solute components of a column chromatographic system. The column is filled with resin beads of the solid stationary phase packed together with the voids between the beads filled with solvent. The phases of interest are *(i)* the liquid phase between the resin beads, *(ii)* the liquid phase held within the resin beads and *(iii)* the solid phase of the polymeric matrix of the resin beads. When the feed solution is placed in contact with the hydrated resin in the chromatographic column, the solutes distribute themselves between the liquid inside the resin and that between the resin beads. The distribution for component i is defined by the distribution coefficient, K_{di}:

$$K_{di} = C_{ri}/C_{li} \tag{16.1}$$

Where C_{ri} is the concentration of component i in the liquid within the resin bead and $C_{\ell i}$ is the concentration of component i in the interstitial liquid. The distribution coefficient for a given ion or molecule will depend upon that component's structure and concentration, the type and ionic form of the resin and the other components in the feed solution. The distribution coefficients for several organic compounds are given in Table 16.1 [17].

The ratio of individual distribution coefficients is often used as a measure of the possibility of separating two solutes and is called the separation factor, α, or relative retention factor.

$$\alpha = K_{d1}/K_{d2} \tag{16.2}$$

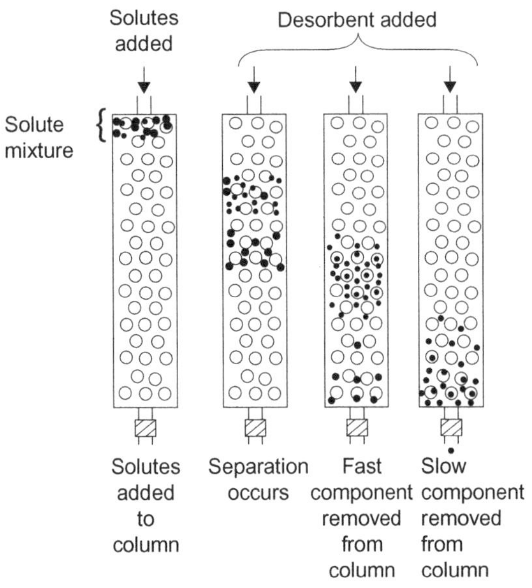

FIGURE 16.1 The steps of chromatographic separation are: addition of the mixed solutes to the column, elution to effect separations, and removal of the separated solutes.

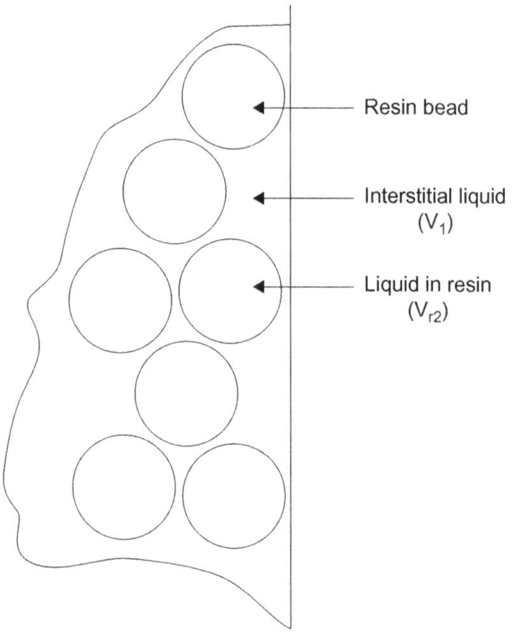

FIGURE 16.2 Representation of the three phases involved in chromatographic separation.

TABLE 16.1 Distribution Coefficients [17]

Solute	Resin	K_d
Ethylene Glycol	Dowex 50-X8, H^+	.67
Sucrose	Dowex 50-X8, H^+	.24
d-Glucose	Dowex 50-X8, H^+	.22
Glycerine	Dowex 50-X8, H^+	.49
Triethylene Glycol	Dowex 50-X8, H^+	.74
Phenol	Dowex 50-X8, H^+	3.08
Acetic Acid	Dowex 50-X8, H^+	.71
Acetone	Dowex 50-X8, H^+	1.20
Formaldehyde	Dowex 50-X8, H^+	.59
Methanol	Dowex 50-X8, H^+	.61
Formaldehyde	Dowex 1-X7.5, Cl^-	1.06
Acetone	Dowex 1-X7.5, Cl^-	1.08
Glycerine	Dowex 1-X7.5, Cl^-	1.12
Methanol	Dowex 1-X7.5, Cl^-	.61
Phenol	Dowex 1-X7.5, Cl^-	17.70
Formaldehyde	Dowex 1-X8, $SO_4=$, 50–100	1.02
Acetone	Dowex 1-X8, $SO_4=$, 50–100	.66
Xylose	Dowex 50-X8, Na^+	.45
Glycerine	Dowex 50-X8, Na^+	.56
Pentaerythritol	Dowex 50-X8, Na^+	.39
Ethylene Glycol	Dowex 50-X8, Na^+	.63
Diethylene Glycol	Dowex 50-X8, Na^+	.67
Triethylene Glycol	Dowex 50-X8, Na^+	.61
Ethylene Diamine	Dowex 50-X8, Na^+	.57
Diethylene Triamine	Dowex 50-X8, Na^+	.57
Triethylene Tetramine	Dowex 50-X8, Na^+	.64
Tetraethylene Pentamine	Dowex 50-X8, Na^+	.66

From Table 16.1, the separation factors for acetone-formaldehyde separability are 0.49, 0.98 and 1.54 for Dowex 50WX8 (H^+), Dowex $1 \times 8(Cl^-)$ and Dowex $1 \times 8(SO_4^{-2})$ resins, respectively. For comparison purposes, it may be necessary to use the inverse of a, so that the values would be 2.03 and 1.02 for Dowex 50WX8 (H^+) and Dowex $1 \times 8(Cl^-)$, respectively. When α is less than 1, the solute in the numerator will exit the column first. When α is greater than 1, the solute in the denominator will exit the column first.

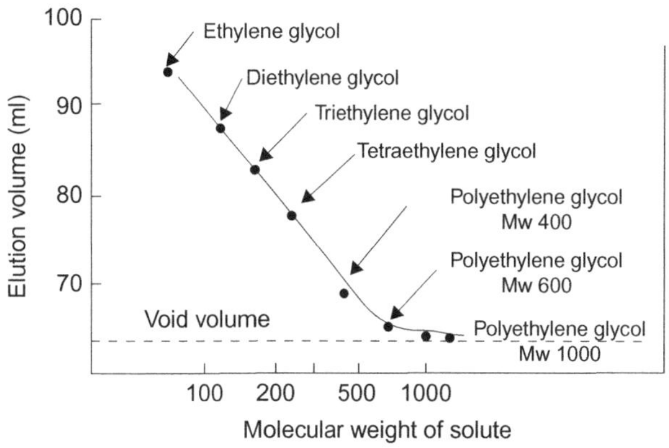

FIGURE 16.3 Effect of molecular weight on the elution volume required for glycol compounds [8].

The acetone-formaldehyde separation would be an example of affinity difference chromatography in which molecules of similar molecular weight or isomers of compounds are separated on the basis of differing attractions or distribution coefficients for the resin. The largest industrial chromatography application of this type is the separation of fructose from glucose to produce 55% or 90% fructose corn sweetener.

Ion exclusion chromatography involves the separation of an ionic component from a nonionic component. The ionic component is excluded from the resin beads by ionic repulsion, while the nonionic component will be distributed into the liquid phase inside the resin beads. Since the ionic solute travels only in the interstitial volume, it will reach the end of the column before the nonionic solute which must travel a more tortuous path through the ion exchange beads. A major industrial chromatography application of this type is the recovery of sucrose from the ionic components of molasses.

In size exclusion chromatography, the resin beads act as molecular sieves, allowing the smaller molecules to enter the beads while the larger molecules are excluded. Figure 16.3 [18] shows the effect of molecular size on the elution volume required for a given resin. The ion exclusion technique has been used for the separation of monosodium glutamate from other neutral amino acids [19].

Ion retardation chromatography involves the separation of two ionic solutes with a common counter ion. Unless a specific complexing resin is used, the resin must be placed in the form of the common counter ion. The other solute ions are separated on the basis of different affinities for the resin. Ion retardation chromatography is starting to see use in the recovery of acids from waste salts following the regeneration of ion exchange columns.

2.0 THEORY

The important features of ion exchange reactions are that they are stoichiometric, reversible and possible with any ionizable compound. The reaction that occurs in a specific length of time depends on the selectivity of the resin for the ions or molecules involved and the kinetics of that reaction.

The stoichiometric nature of the reaction allows resin requirements to be predicted and equipment to be sized. The reversible nature of the reaction, illustrated as follows:

$$R - H^+ + Na^+Cl^- \rightleftharpoons R - Na^+ + H^+Cl^- \quad (16.3)$$

allows for the repeated reuse of the resin since there is no substantial change in its structure.

The equilibrium constant, K, for Eq. (16.1), is defined for such monovalent exchange by the equation:

$$K = \frac{[R - Na^+][H^+Cl^-]}{[R - H^+][Na^+Cl^-]} \quad (16.4)$$

In general, if K is a large number, the reverse reaction is much less efficient and requires a large excess of regenerant chemical, HCl in this instance, for moderate regeneration levels.

With proper processing and regenerants, the ion exchange resins may be selectively and repeatedly converted from one ionic form to another. The definition of the proper processing requirements is based upon the selectivity and kinetic theories of ion exchange reactions.

2.1 Selectivity

When ion B, which is initially in the resin, is exchanged for ion A in solution, the selectivity is represented by:

$$\ln K_B^A = \frac{\pi(|Z_A|V_B - |Z_B|V_A)}{RT} \quad (16.5)$$

where Z_i is the charge and V_i is the partial volume of ion i. The selectivity which a resin has for various ions is affected by many factors. The factors include the valence and size of the exchange ion, the ionic form of the resin, the total ionic strength of the solution, the cross-linkage of the resin, the type of functional group and the nature of the non-exchanging ions.

The *ionic hydration theory* has been used to explain the effect of some of these factors on selectivity [20]. According to this theory, the ions in aqueous solution are hydrated and the degree of hydration for cations increases with increasing charge and decreasing crystallographic radius, as shown in Table 16.2 [21]. It is the high dielectric constant of water molecules that is responsible for the hydration of ions in aqueous solutions. The hydration potential of an ion depends on the intensity of the change on its surface. The degree of hydration of an ion increases as its valence increases and decreases as its hydrated radius increases. Therefore, it is expected that the selectivity of a resin for an ion is inversely proportional to the ratio of the valence/ionic radius for ions of a given radius. In dilute solution, the following selectivity series are followed:

TABLE 16.2 Ionic Size of Cations [21]

Ion	Crystallographic Radius (Å)	Hydrated Radius (Å)	Ionization Potential
Li	0.68	10.00	1.30
Na	0.98	7.90	1.00
K	1.33	5.30	0.75
NH_4	1.43	5.37	—
Rb	1.49	5.09	0.67
Cs	1.65	5.05	0.61
Mg	0.89	10.80	2.60
Ca	1.17	9.60	1.90
Sr	1.34	9.60	1.60
Ba	1.49	8.80	1.40

$$\mathrm{Li} < \mathrm{Na} < \mathrm{K} < \mathrm{Rb} < \mathrm{Cs}$$
$$\mathrm{Mg} < \mathrm{Ca} < \mathrm{Sr} < \mathrm{Ba}$$
$$\mathrm{F} < \mathrm{Cl} < \mathrm{Br} < \mathrm{I}$$

The selectivity of resins in the hydrogen ion or hydroxide ion form, however, depends on the strength of the acid or base formed between the functional group and the ion. The stronger the acid or base formed, the lower is the selectivity coefficient. It should be noted that these series are not followed in nonaqueous solutions, at high solute concentrations or at high temperature.

The dependence of selectivity on the ionic strength of the solution has been related through the mean activity coefficient to be inversely proportional to the Debye-Huckel parameter, a^o [22]:

$$\log\gamma_{\pm} = \frac{-A\sqrt{\mu}}{1 + B\alpha^o\sqrt{\mu}} \quad (16.6)$$

Where $\gamma_{\pm}$ is the mean activity coefficient, A and B are constants, and μ is the ionic strength of the solution. The mean activity coefficient in this instance represents the standard free energy of formation $(-\Delta F^{\circ})$ for the salt formed by the ion exchange resin and the exchanged ion. Figure 16.4 [23] shows this dependence as the ionic

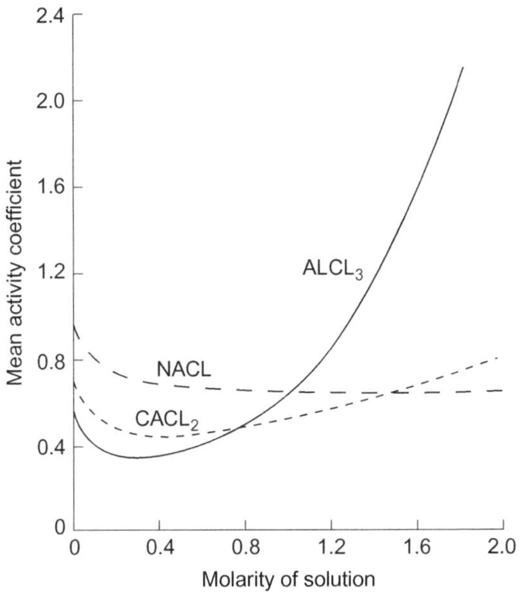

FIGURE 16.4 Dependence of the activity coefficient on the ionic concentration of aqueous solutions [23].

TABLE 16.3 Selectivity and Hydration of Cation Resins With Different Degrees of Crosslinking [25]

	4% DVB		8% DVB		16% DVB	
Cation	*K*	*H*	*K*	*H*	*K*	*H*
Li	1.00	418	1.00	211	1.00	130
H	1.30	431	1.26	200	1.45	136
Na	1.49	372	1.88	183	2.23	113
NH_4	1.75	360	2.22	172	3.07	106
K	2.09	341	2.63	163	4.15	106
Cs	2.37	342	2.91	159	4.15	102
Ag	4.00	289	7.36	163	19.4	102
Tl	5.20	229	9.66	113	22.2	85

K = Selectivity compared to Li
H = Hydration (g H_2O/eq resin)
DVB = divinylbenzene

concentration of the solution is changed. As the concentration increases, the differences in the selectivity of the resin for ions of different valence decreases and, beyond certain concentrations, the affinity is seen to be greater for the lower valence ion.

The selectivity of an ion exchange resin will also depend on its cross-linking. The polymer structure of the ion exchange resin can be thought of as collections of coiled springs, which can swell, or contract during the exchange of ions [24]. The cross-linking of the polymer limits the extent to which the resin may swell—the higher the degree of cross-linking, the lower the extent to which the resin can be hydrated. This limit on resin hydration determines the relative equivalent volumes of hydrated ions, which the cross-linked polymer network can accommodate. This is shown in Table 16.3 [25]. As the resin cross-linking or the fixed ion concentration is lowered, the selectivity of the resin decreases.

The degree of cross-linking can affect the equilibrium level obtained, particularly as the molecular weight of the organic ion becomes large. With highly cross-linked resins and large organic ions, the concentration of the organic ions in the outer layers of the resin particles is much higher than in the center of the particle.

The selectivity of the resin for a given ion is also influenced by the dissociation constants of the functional group covalently attached to the resin (the fixed ion) and of the counter-ions in solutions. Since the charge per unit volume within the resin particle is high, a significant percentage of the functional groups may not be ionized. This

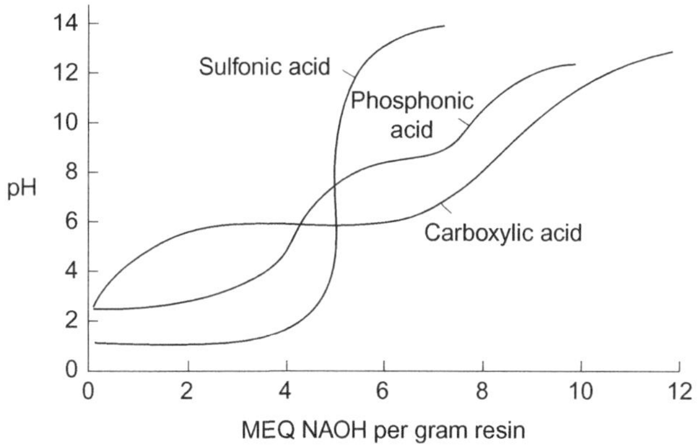

FIGURE 16.5 Titration curves of typical cation exchange resins [26].

is particularly true if the functional group is a weak acid or base. For cation exchange, the degree of dissociation for the functional group increases as the pH is increased; however, the degree of dissociation for the ions in solution decreases with increasing pH. Therefore, if a cation resin had weak acid functionality, it would exhibit little affinity at any pH for a weak base solute. Similarly, an anion resin with weak base functionality exhibits little affinity at any pH for a weak acid solute.

The influence of pH on the dissociation constants for resin with a given functionality can be obtained by titration in the presence of an electrolyte. Typical titration curves are shown in Figure 16.5 for cation resins and in

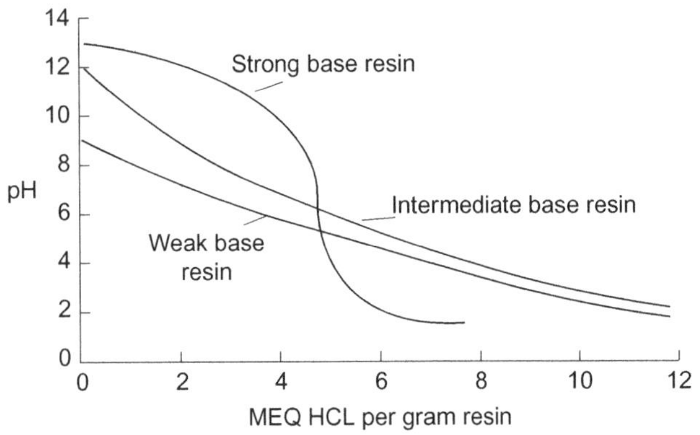

FIGURE 16.6 Titration curves of typical anion exchange resins [26].

TABLE 16.4 Selectivity Coefficients for Strongly Basic Anion Resin [27,28]

Type I Anion		Type II Anion	
Anion	K^{x}_{Cl}	Anion	K^{x}_{Cl}
Salicylate	32.2	Salicylate	28
I^-	8.7	$C_6H_5O^-$	8.7
$C_6H_5O^-$	5.2	I^-	7.3
HSO_4^-	4.1	HSO_4^-	6.1
NO_3^-	3.8	NO_3^-	3.3
Br^-	2.8	Br^-	2.3
CN^-	1.6	CN^-	1.3
HSO_3^-	1.3	HSO_3^-	1.3
NO_2^-	1.2	NO_2^-	1.3
Cl^-	1.00	Cl^-	1.00
HCO_3^-	0.32	OH^-	0.65
H_2PO_4	0.25	HCO_3^-	0.53
$HCOO^-$	0.22	H_2PO_4	0.34
CH_3COO^-	0.17	$HCOO^-$	0.22
$H_2NCH_2COO^-$	0.10	CH_3COO^-	0.18
OH^-	0.09	F^-	0.13
F^-	0.09	$H_2NCH_2COO^-$	0.10

Figure 16.6 for anion resins [26]. For sulfonic acid functional groups, the hydrogen ion is a very weak replacing ion and is similar to the lithium ion in its replacing power. However, for resin with carboxylic acid functionality, the hydrogen ion exhibits the highest exchanging power. Table 16.4 [27,28] summarizes the effect different anion exchange resin functionalities have on the equilibrium exchange constants for a wide series of organic and inorganic anions.

The selectivity can also be influenced by the non-exchanging ions (co-ions) in solution even though these ions are not directly involved in the exchange reaction. An example of this influence would be the exchange of calcium ascorbate with an anion resin in the citrate form. Although calcium does not take part in the exchange reaction, sequestering of citrate will provide an additional driving force for the exchange. This effect, of course, would have been diminished had a portion of the ascorbate been added as the sodium ascorbate rather than the calcium ascorbate.

For nonpolar organic solutes, association into aggregates, perhaps even micelles, may depress solution activity. These associations may be influenced by the co-ions present.

2.2 Kinetics

The overall exchange process may be divided into five sequential steps:

1. The diffusion of ions through the solution to the surface of the ion exchange particles
2. The diffusion of these ions through the ion exchange particle
3. The exchange of these ions with the ions attached to the functional group
4. The diffusion of these displaced ions through the particle
5. The diffusion of these displaced ions through the solution

Each step of the diffusion, whether in the resin or solution phase, must be accompanied by an ion of the opposite charge to satisfy the law of electro neutrality.

Kinetics of ion exchange is usually considered to be controlled by mass transfer in ion exchange particles or in the immediately surrounding liquid phase. The theory used to describe mass transfer in the particle is based on the Nernst-Planck equations developed by Helfferich [29] which accounted for the effect of the electric field generated by ionic diffusion, but excluded convection.

It is recognized that the Nernst-Planck theory fails to take into account the effect of swelling and particle size changes which accompany ion exchange or to take into account the slow relaxation of the resin network which causes the diffusion coefficient to vary with time. However, the approximations, which these equations provide, are a reasonable starting point and will most likely be found to be sufficient for the biotechnology engineer. Any further refinements would lead rapidly to diminished returns. Likewise, the mass transfer in the liquid phase is usually described according to the Nernst film concept

TABLE 16.5 Dependence of Ion Exchange Rates on Experimental Conditions [32]

Factor	Model 1	Model 2	Model 3	Model 4	Model 5*
Particle size (r)	$\propto {}^1/r$	$\propto {}^1/r^2$	independent	$\propto {}^1/r^2$	${}^1/r^2$
Solution (concentration) (c)	$\propto c$	independent	$\propto c$	$\propto c$	$\propto c^{\dagger}$
Separation factor (α)	independent up to a specific time when ≥ 1; $\propto \alpha$ when $\alpha << 1$.	independent‡	independent	independent	independent‡
Stirring rate	sensitive	independent	independent	independent	independent
Resin exchange capacity (c)	$\propto {}^1/c$	independent	independent	independent	$\propto {}^1/c$
Temperature (T)	$\approx 4\%/°K$	$\approx 6\%/°K$	function of E_{Act}	function of E_{Act}	$\approx 6\%/°K$

*Applicable to forward exchange only.
†Provided partition coefficient is independent of solution concentration.
‡For complete conversion and constant solution composition.

using a version [30] of the Nernst-Planck equation or Glueckauf's [31] simpler linear driving force approximation.

There are five models [32] which can be used to represent the kinetics in ion exchange systems which involve liquid exchange phase mass transfer, solid phase mass transfer, and chemical reaction at the exchange group.

Model 1. The liquid phase mass transfer with a linear driving force is the controlling element. This model assumes that there are no concentration gradients in the particle, that there is a quasi-stationary state of liquid phase mass transfer, that there is a linear driving force and that there is a constant separation factor at a given solution concentration.

Model 2. The rate-controlling step is diffusion within the ion exchange particles. This model assumes that there are no concentration gradients in the liquid phase and that there is no convection, either through solvent uptake or release, in the solid phase.

Model 3. This model is controlled by the exchange reaction at the fixed ionic groups. This model assumes that the slowness of the exchange reaction allows for sufficient time for mass transfer to establish and maintain equilibrium so that no concentration gradient exists in either the ion exchange particles or in the liquid phase.

Model 4. This is a variation of Model 3 in which the counter-ion from the solution does not permeate beyond the portion of particle which has been converted to the exchanging ionic form. The boundary of the unreacted core reduces the time such that this is called the *shrinking core model.* It is this sharp boundary between the reacted and unreacted portion of the particles that distinguishes Model 4 from Model 3.

Model 5. The rate-controlling step is the diffusion of the counter ion across the converted portion of the particle. Since the exchange groups undergo a fast and essentially irreversible reaction with the counter ions, their type of reaction affects the rate of reaction and the geometry of the diffusing zone.

Table 16.5 [32] summarizes the effect of operating parameters (particle size, solution concentration, separation factor, stirring rate, resin exchange capacity, and temperature) on ion exchange kinetics described by these different models in batch reactors.

For the cases of interest, the rate of ion exchange is usually controlled by diffusion, either through a hydrostatic boundary layer, called *film diffusion* control or through the pores of the resin matrix, called *particle diffusion* control.

In the case of film diffusion control, the rate of ion exchange is determined by the effective thickness of the film and by the diffusivity of ions through the film. When resin particle size is small, the feed stream is dilute or when a batch system has mild stirring, the kinetics of exchange is controlled by film diffusion.

In the case of particle diffusion control, the rate of ion exchange depends on the charge, spacing and size of the diffusing ion and on the micropore environment. When the resin particle size is large, the feed stream is concentrated, or when a batch system has vigorous stirring, the kinetics is controlled by particle diffusion.

The limits at which one or the other type of diffusion is controlling have been determined by Tsai [33]. When $Kk^2\delta > 50$, the rate is controlled by film diffusion. When $Kk^2\delta < 0.005$, the rate is controlled by particle diffusion. In these relationships, K is the distribution coefficient, k^2 is the diffusivity ratio (D_p/D_f), δ is the relative film thickness on a resin particle with a radius of a. Between these two limits, the kinetic description of ion exchange processes must include both phenomenon.

The characteristic Nernst parameter δ, the thickness of the film around the ion exchange particle, may be converted to the mass transfer coefficient and dimensionless numbers (Reynolds, Schmidt and Sherwood) that engineers normally employ [34].

In terms of the solute concentration in the liquid, between 0.1 to 0.01 mol/L, the rate-limiting factor is the transport to the ion exchange bead. Above this concentration, the rate-limiting factor is the transport inside the resin beads [35]. During the loading phase of the operating cycle, the solute concentration is in the low range. During regeneration however, in which the equilibrium is forced back by addition of a large excess of regenerant ions, the solute is above the 0.1 mol/L limit.

One of the important factors in the kinetic modeling of organic ions is their slow diffusion into the ion exchange resin. The mean diffusion time is listed in Table 16.6 as a function of resin particle size for different size classifications of substances [36]. With the larger organic ions, the contact time for the feed solution and the resin must be increased to have the ion exchange take place as a well-defined process such as occurs with the rapidly diffusing ions of mineral salts.

In principle, fluidized ion exchange beds are similar to stirred tank chemical reactors. The general equations of kinetics and mass transfer can be applied to the individual fluidized units in an identical manner to those for chemical reactors. The primary difference lies in accounting for the behavior of suspended particles in the turbulent fluid [37].

The operation of these fluidized ion exchange beds is identical to that of the fixed beds, with the exception that the resin of each stage is confined by perforated plates and maintained in a fluidized suspension using liquid flow or impellers.

The critical design parameter for fluidized beds is the loss or leakage of the solute through a given stage. The design equation for a single stage bed has been described by Marchello and Davis [38].

TABLE 16.6 Characteristic Diffusion into Spherical Resin Particles for Various Substances [36]

Coefficient of Diffusion (Order of Magnitude) (cm^2/sec)	Type of Sorbed Substance (Ion)	Particle Radius (cm)	Mean Time of Intraparticle Diffusion
10^{-6}	Ions of mineral salts bearing a single charge	0.05	3 min
		0.01	7 sec
		0.005	1.8 sec
10^{-7}	Ions of mineral salts bearing several charges, amino acids	0.05	30 min
		0.01	1.2 min
		0.005	18 sec
10^{-8}	Tetraalkylammonium ions, antibiotic ions on macroporous resins	0.05	5 hr
		0.01	12 min
		0.005	3 min
10^{-9}	Dyes, alkaloids, antibiotics in standard ion exchange resins	0.05	over 2 days
		0.01	2 hr
		0.005	0.5 hr
10^{-10}	Some dyes, polypeptides and proteins	0.05	over 20 days
		0.01	over 20 hr
		0.005	over 5 hr

2.3 Chromatographic Theory

Mathematical theories for ion exchange chromatography were developed in the 1940's by Wilson [39], DeVault [40] and Glueckauf [41,42]. These theoretical developments were based on adsorption considerations and are useful in calculating adsorption isotherms from column elution data. Of more interest for understanding preparative chromatography is the theory of column processes originally proposed by Martin and Synge [43] and augmented by Mayer and Thompkins [44], which was developed analogous to fractional distillation so that plate theory could be applied.

One of the equations developed merely expressed mathematically that the least adsorbed solute would be eluted first and that if data on the resin and the column dimensions were known, the solvent volume required to elute the peak solute concentration could be calculated. Simpson and Wheaton [45] expressed this equation as:

$$V_{MAX} = K_d V_{rl} + V_l \tag{16.7}$$

where V_{MAX} is the volume of liquid that has passed through the column when the concentration of the solute is maximum (the midpoint of the elution of the solute). K_d, defined in Eq. (16.1), is the distribution coefficient of the solute in a *plate* of the column; V_{rl} is the volume of liquid solution inside the resin and V_l is the volume of interstitial liquid.

The mathematical derivation of Eq. (16.7) assumes that complete equilibrium has been achieved and that no forward mixing occurs. Glueckauf [46] pointed out that equilibrium is practically obtained only with very small diameter resin beads and low flow rates. Such restricting conditions may be acceptable for analytical applications, but would severely limit preparative and industrial chromatography. However, column processing conditions and solute purity requirements are often such that any deviations from these assumptions are slight enough that the equation still serves as an adequate first approximation for scaled-up chromatography applications.

Theoretical Plate Height. A second important equation for chromatography processes is that used for the calculation of the number of theoretical plates, i.e., the length of column required for equilibration between the solute in the resin liquid and the solute in the interstitial liquid. If the elution curve approximates a Gaussian distribution curve, the equation may be written as:

$$P = \frac{2c(c+1)}{W^2} \tag{16.8}$$

where P is the number of theoretical plates; $c(= K_d V_{rl}/V_l)$ is the equilibrium constant; W is the half-width of the elution curve at an ordinate value of $1/e$ of the maximum solute concentration. For a Gaussian distribution, $W = 4\sigma$, where σ is the standard deviation of the Gaussian distribution. The equilibrium constant is sometimes called the *partition ratio.*

An alternate form of this equation is:

$$P = \frac{2V_{MAX}(V_{MAX} - V_1)}{W^2} \tag{16.9}$$

Here W is measured in the same units as V_{MAX}. This form of the equation is probably the easiest to calculate from experimental data. Once the number of theoretical plates has been calculated, the height equivalent to one theoretical plate (H.E.T.P.) can be obtained by dividing the resin bed height by the value of P.

The column height required for a specific separation of two solutes can be approximated by [47]:

$$\sqrt{H} = \frac{3.29}{c_2 - c_1}\frac{c_2 + 0.5}{\sqrt{P_2}} + \frac{c_1 + 0.5}{\sqrt{P_1}} \tag{16.10}$$

where H is the height of the column, P is the number of plates per unit of resin bed height and c is the equilibrium constant defined above. Note that the number of plates in a column will be different for each solute. While this equation may be used to calculate the column height needed to separate 99.9% of solute 1 from 99.9% of solute 2, industrial and preparative chromatography applications typically make more efficient use of the separation resin by selectively removing a narrow portion of the eluted solutes, as illustrated in Figure 16.7 [48].

Table 16.7 shows how the theoretical plate number for a chromatographic system may be calculated from various combinations of experimental data. The band variance, σ_t^2, is calculated from the experimental data and combined with the retention time, t_R, for a given solute. Figure 16.8 shows the different experimental values which may be used to calculate σ_t.

Zone Spreading. The net forward progress of each solute is an average value with a normal dispersion about the mean value. The increased band or zone width that

TABLE 16.7 Calculation of Plate Number from Chromatogram

Measurements	Conversion to Variance	Plate Number
t_R *and* σ_T	–	$N = (t_R/\sigma_t)^2$
t_R *and baseline width* W_b	$\sigma_t = W_b/4$	$N = 16\ (t_R/W_b)^2$
t_R and width at half height $W_{0.5}$	$\sigma_t = W_{0.5}/\sqrt{8 \ln 2}$	$N = 5.54\ (t_R/W_{0.5})^2$
t_R and width at inflection points (0.607 *h*) W_1	$\sigma_t = W_i/^2$	$N = 4\ (t_R/W_i)^2$
t_R and band area An and height h	$\sigma_t = A/h\sqrt{2\pi}$	$N = 2\pi\ (t_R\ h/A)^2$

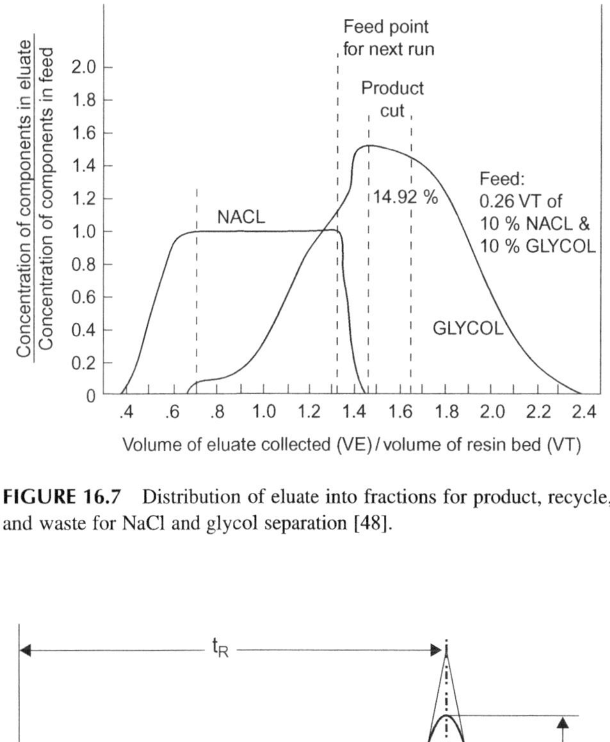

FIGURE 16.7 Distribution of eluate into fractions for product, recycle, and waste for NaCl and glycol separation [48].

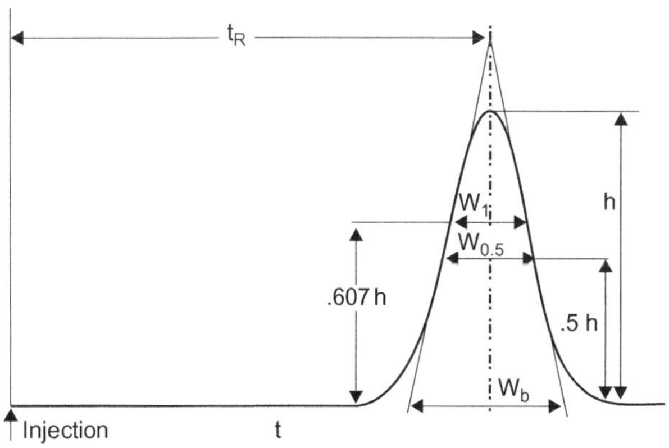

FIGURE 16.8 Identification of chromatographic peak segments for the calculation of column performance.

results from a series of molecular diffusion and non-equilibrium factors is known as zone spreading.

The plate height as a function of the mobile phase velocity may be written as a linear combination of contributions from eddy diffusion, mass transfer and a coupling term:

$$H = \frac{B}{V} + E_s v + \frac{1}{1/A + E_M/v} \quad (16.11)$$

A plot of Eq. (16.11) for any type of linear elution chromatography describes a hyperbola, as shown in Figure 16.9 [49]. There is an optimum velocity of the mobile phase for carrying out a separation at which the plate height is a minimum, and thus, the chromatographic separation is most efficient:

$$v_{\text{optimum}} = \sqrt{D_M/[R_t(1 - R_t)d_p^2/D_s]} \quad (16.12)$$

where D_M is the diffusion coefficient of the solute molecule in the mobile phase, D_s is the diffusion coefficient in the stationary phase, d_p is the diameter of the resin bead and $R_t = L/vt$, where L is the distance the zone has migrated in time t.

Resolution. A variation on calculating the required column height is to calculate the resolution or degree of separation of two components. Resolution is the ratio of peak separation to average peak width:

$$R = \frac{V_{\text{MAX2}} - V_{\text{MAX1}}}{0.5(W_1 + W_2)} \quad (16.13)$$

The numerator of Eq. (16.13) is the separation of the two solutes' peak concentrations and the denominator is the average bandwidth of the two peaks. This form of the equation is evaluating the resolution when the peaks are separated by four standard deviations, σ. If $R = 1$ and the two solutes have the same peak concentration, this means that the adjacent tail of each peak beyond 2σ from the v_{MAX} would overlap with the other solute peak. In this instance there would be 2% contamination of each solute in the other.

Resolution can also be represented [50] by:

$$R = \frac{\sqrt{P_2}}{4} \frac{\alpha - 1}{1} \frac{c_2}{1 + c_2} \quad (16.14)$$

Resolution can be seen to depend on the number of plates for solute 2, the separation factor for the two solutes and the equilibrium constant for solute 2.

In general, the larger the number of plates, the better the resolution. There are practical limits to the column lengths that are economically feasible in industrial and preparative chromatography. It is possible to change P also by altering the flow rate, the mean resin bead size or the bead size distribution since P is determined by the rate processes occurring during separation. As the separation factor increases, resolution becomes greater since the peak-to-peak separation is becoming larger. Increases in the equilibrium constant will usually improve the resolution since the ratio $c_2/(1 + c_2)$ will increase. It should be noted that this is actually only true when c_2 is small since the ratio approaches unity asymptotically as c_2 gets larger. The separation factor and the equilibrium factor can be adjusted for temperature changes or other changes which would alter the equilibrium properties of the column operations.

Equation (16.14) is only applicable when the two solutes are of equal concentration. When that is not the case, a correction factor must be used

$$(A_1^2 + A_2^2)/2A_1A_2$$

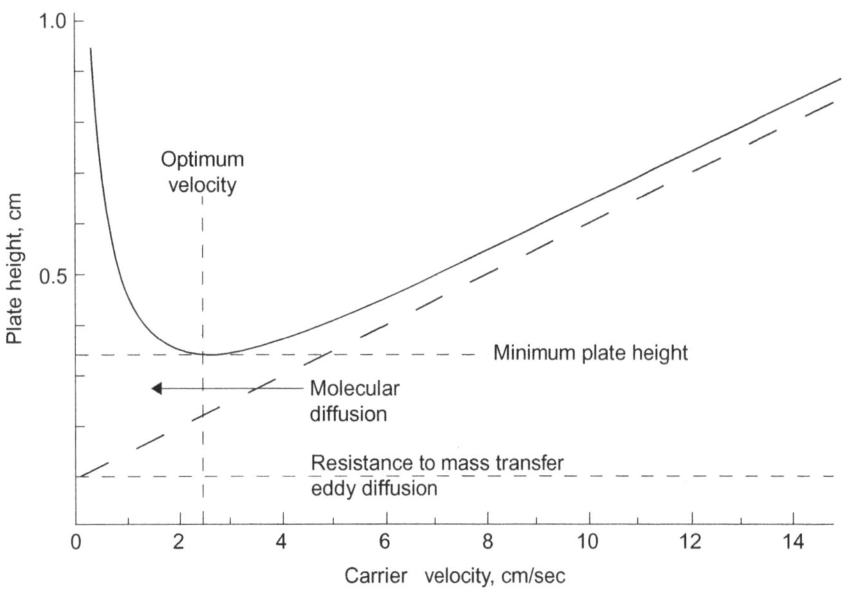

FIGURE 16.9 Relationship between late height and velocity of the mobile phase [49].

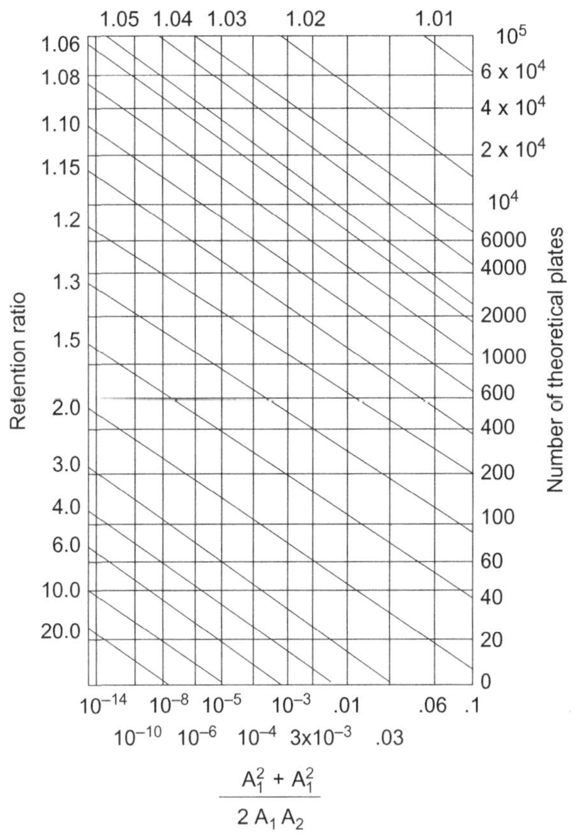

FIGURE 16.10 Relationship between relative retention ratio number of theoretical plates, and product purity [46].

where A_1 and A_2 are the areas under the elution curve for solutes 1 and 2, respectively. Figure 16.10 shows the relationship between product purity (η), the separation ratio and the number of theoretical plates. This graph can be used to estimate the number of theoretical plates required to attain the desired purity of the products.

For example, when the product purity must be 98.0%, then $\eta = \Delta m/m = 0.01$, when the amount of the two solutes is equal. If the retention ratio, a, is equal to 1.2, then the number of theoretical plates from Figure 16.10 is about 650. With a plate height of 0.1 cm, the minimum bed height would be 65 cm. In practice, a longer column is used to account for any deviation from equilibrium conditions.

3.0 ION EXCHANGE MATERIALS AND THEIR PROPERTIES

Ion exchange materials are a special class of polyelectrolytes. The chemical and physical properties of an ion exchange material play a more important role in determining its suitability for a biochemical application than for other types of applications. The chemical properties to be considered are the matrix and the ionic functionality

TABLE 16.8 Producers of Synthetic Ion Exchange Resins

Company	Country	Tradename
Bayer	Germany	Lewatit
Chemolimfex	Hungary	Varion
Dow	United States	Dowex
Ionac	United States	Ionac
Mitsubishi	Japan	Diaion
Montecatini-Edison	Italy	Kastel
Ostion	Czechoslovakia	Ostion
Permutit	United Kingdom	Zeocarb, Deacidite, Zerolit
Permutit, AG	Germany	Orzelith, Permutit
Resindion	Italy	Relite
Rohm & Haas	United States	Amberlite, Duolite
	Russia	AW-, AV-, KB-, KU-

attached to the matrix. The important physical properties are the pore size, the pore volume, the surface area, the density and the particle size. A list of commercial producers of granular or bead ion exchange materials is given in Table 16.8.

3.1 Ion Exchange Matrix

These materials can be broadly categorized into those which are totally inorganic in nature and those that are synthetic organic resins.

Inorganic ion exchangers [51] include both naturally occurring materials such as mineral zeolites (sodalite and clinoptilolite), the greensands, and clays (themontmorillonite group) and synthetic materials such as gel zeolites, the hydrous oxides of polyvalent metal (hydrated zirconium oxide) and the insoluble salts of polybasic acids with polyvalent metals (zirconium phosphate).

The synthetic organic resins consist of cross-linked polymer matrix which is functionalized to provide their ion exchange capacity. The matrix usually must undergo additional reactions to provide the strong acid cation, strong base anion, weak acid cation or weak base anion functionality.

Cross-linked polystyrene, epoxy-polyamine, phenol-formaldehyde, and cross-linked acrylic methacrylic acid resins are the most commonly used ion exchanges in industrial applications and have been used in biochemical applications, such as protein purifications and enzyme immobilizations. However, the hydrophobic matrices have the disadvantages that they might denature the desired biological material or that the high charge density may give such strong binding that only a fraction of the absorbed material might be recovered.

Resins with cellulosic matrices are much more hydrophilic and these do not tend to denature proteins. Cellulosic resins have been used extensively in the laboratory analyses of biological materials, enzyme immobilizations and small-scale preparations. The low capacity and poor flow characteristics have limited the usefulness of these matrices for larger applications.

Recently, diethylaminoethyl (DEAE) silica gel was shown [52] to be an improvement over typical cellulosic-matrices resins for the separation of acidic and neutral lipids from complex ganglioside mixtures. The specific advantages claimed were:

1. An increase in flow rate was possible through the DEAE-silica gel.
2. The DEAE-silica gel was able to be equilibrated much more rapidly with the starter buffer.
3. The DEAE-silica gel was more easily regenerated.
4. The DEAE-silica gel was less susceptible to microbial attack.
5. The preparation of DEAE-silica gel from inexpensive silica gel was described as a simple method that could be carried out in any laboratory.

3.2 Functional Groups

The strong acid cation exchange resins are made by the sulfonation of the matrix copolymer. Strong acid cation resins are characterized by their ability to exchange cations or split neutral salts. They will function throughout the entire pH range.

The synthesis of weak acid cation resins has been described above. The ability of this type of resin to split neutral salts is very limited. The resin has the greatest affinity for alkaline earth metal ions in the presence of alkalinity. Only limited capacities for the alkali metals are obtained when alkalinity other than hydroxide is present. Effective use is limited to solutions above pH 4.0.

The anion exchange resins require the synthesis of an active intermediate. This is usually performed in the process called *chloromethylation.* The subsequent intermediate is reactive with a wide variety of amines which form different functional groups.

The Type I resin is a quaternized amine resin resulting from the reaction of trimethylamine with the chloromethylated copolymer. This functionalized resin has the most strongly basic functional group available and has the greatest affinity for weak acids. However, the efficiency of regenerating the resin to the hydroxide form is somewhat lower than Type II resins, particularly when the resin is exhausted with monovalent anions.

The Type II resin results when dimethylethanolamine is reacted with the chloromethylated copolymer. This quaternary amine has lower basicity than that of the Type I resin, yet it is high enough to remove the anions of weak acids in most applications. While the caustic regeneration efficiency is significantly greater with Type II resins, their thermal and chemical stability is not as good as Type I resins.

Weak base resins may be formed by reacting primary or secondary amines or ammonia with the chloromethylated copolymer. Dimethylamine is commonly used. The ability of the weak base resins to absorb acids depends on the basicity of the resin and the pK of the acid involved. These resins are capable of absorbing strong acids in good capacity, but are limited by kinetics. The kinetics may be improved by incorporating about 10% strong base capacity. While strong base anion resins function throughout the entire pH range, weak base resins are limited to solutions below pH 7.

The desired functionality on the selected matrix will be determined by the nature of the biochemical solute that is to be removed from solution. Its isoelectric point, the pH restrictions on the separation and the ease of eventually eluting the absorbed species from the resin play important roles in the selection process.

Some resins have been developed with functional groups specifically to absorb certain types of ions. The resins shown in Table 16.9 are commercially available.

The selectivity of these resins depends more on the complex that is formed rather than on the size or charge of the ions. Generally they are effective in polar and nonpolar solvents. However, the capacity for various ions is pH sensitive so that adsorption and elution can be accomplished by pH changes in the solution.

TABLE 16.9 Commercial Resins with Special Functional Groups

Functionality	Structure
Iminodiacetate	$R\text{-}CH_2N(CH_2COOH)_2$
Polyethylene Polyamine	$R\text{-}(NC_2H_4)_mH$
Thiol	R-SH
Aminophosphate	$R\text{-}CH_2NHCH_2PO_3H_2$
Amidoxime	$R\text{-}C(=N\text{-}OH)\text{-}NH_2$
Phosphate	$R\text{-}PO_3H$

These chelating resins have found most of their use in metal ion recovery processes in the chemical and waste recovery industries. They may find use in fermentation applications where the cultured organism requires the use of metal ion cofactors. Specific ion exchange resins have also been used in laboratory applications that may find eventual use in biotechnology product recovery applications [53].

A review of selective ion exchange resins has been compiled by Warshawsky [54]. A diaminotetratacetic polymer developed by Mitsubishi [55] was developed for the purification of amino acid feed solutions. The conversions of chloromethylpolystyrene into thiolated derivatives for peptide synthesis have been described by Warshawsky and coworkers [56].

3.3 Porosity and Surface Area

The porosity of an ion exchange resin determines the size of the molecules or ions that may enter an ion exchange particle and determines their rate of diffusion and exchange. Porosity is inversely related to the cross-linking of the resin. However, for gel-type or microporous resins, the ion exchange particle has no appreciable porosity until it is swollen in a solvating medium such as water.

The pore size for microporous resins is determined by the distance between polymer chains or cross-linking subunits. If it is assumed that the cross-linking is uniform throughout each ion exchange particle, the average pore diameter of these resins can be approximated from the water contained in the fully swollen resin. The moisture content of cation resins as a function of the degree of cross-linking is shown in Figure 16.11 and, of anion resins, in Figure 16.12. The calculated average pore size for sulfonic cation resins ranges from 16 to 20 Å as the concentration of the resin cross-linking agent (divinylbenzene) decreases from 20 to 2%. The calculated average pore size of the anion resins ranges from 18 to 14 Å as the cross-linking is decreased from 12 to 2%. Even at low cross-linking and full hydration, microporous resins have average pore diameters of less than 20 Å. The dependence of pore size on the percent of the cross-linking is shown in Table 16.10 for swollen microporous resins of styrene-divinylbenzene [57].

Sulfonic acid and carboxylic acid resins have also been equilibrated with a series of quaternary ammonium ions of different molecular weights to measure the average pore size when these resins have increasing degrees of cross-linking. These results are shown in Figures 16.13 and 16.14.

Figure 16.15 shows the change in the ionic diffusion coefficients of tetra-alkylammonium ions in strong acid cation resins as a function of mean effective pore diameter of the resins [58]. As the pore diameter is increased, the penetrability of the resins with respect to the large ions also increased.

If an inert diluent (porogen) is incorporated into the monomer mixture before the copolymer is formed, it is possible to form a structure containing varying degrees of

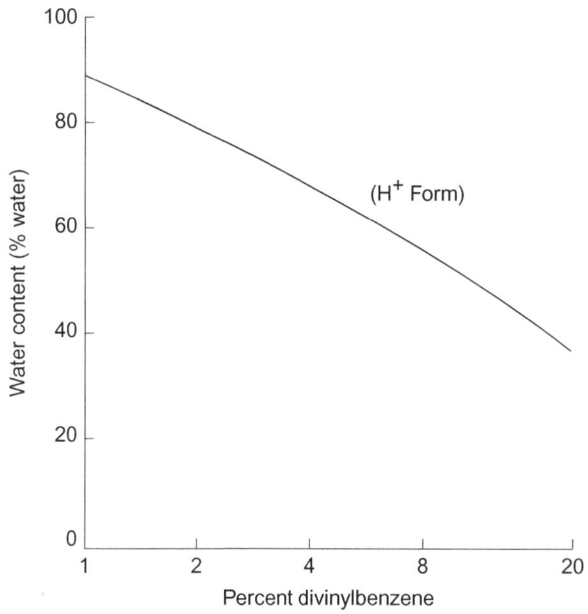

FIGURE 16.11 Moisture content of strong acid cation resins as a function of divinylbenzene content.

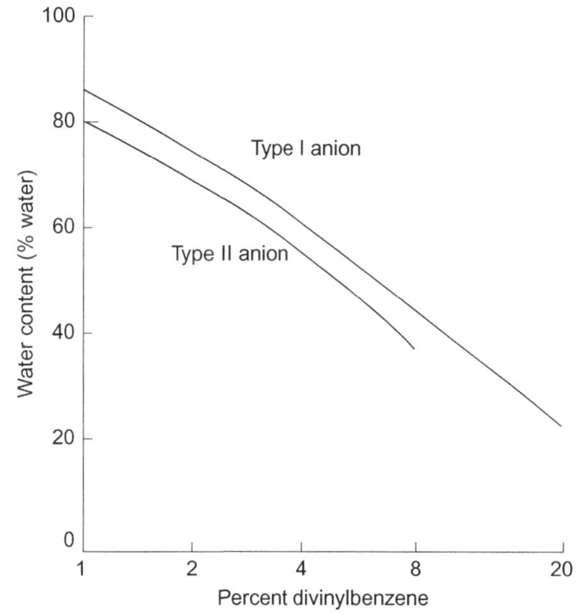

FIGURE 16.12 Moisture content of strong acid anion resins as a function of divinylbenzene content.

TABLE 16.10 Average Swollen Diameter of Cross-linked Polystyrene Beads in Tetrahydrofuran [57]

Divinylbenzene Concentration (Cross-linking) (%)	Swollen Pore Diameter (%)
1	77
2	54
4	37
8	14
16	13

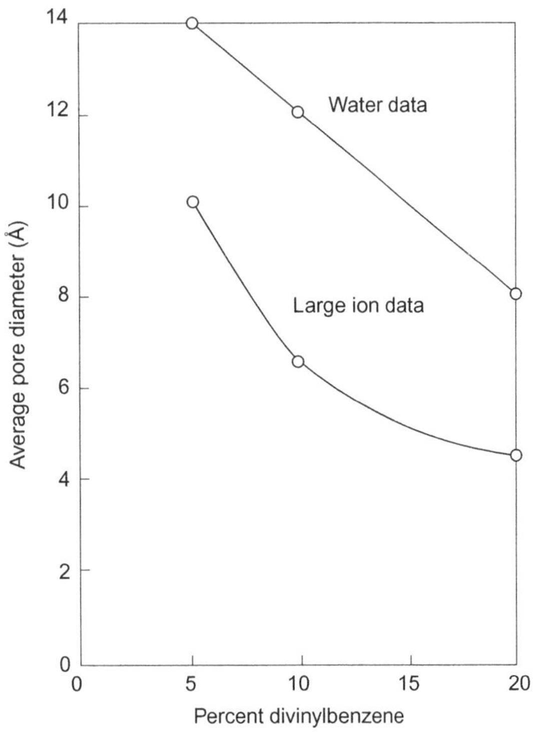

FIGURE 16.13 Average pore diameter of sulfonic acid cation exchange resin as a function of degree of cross-linking. *(Ref. [20], page 46.)*

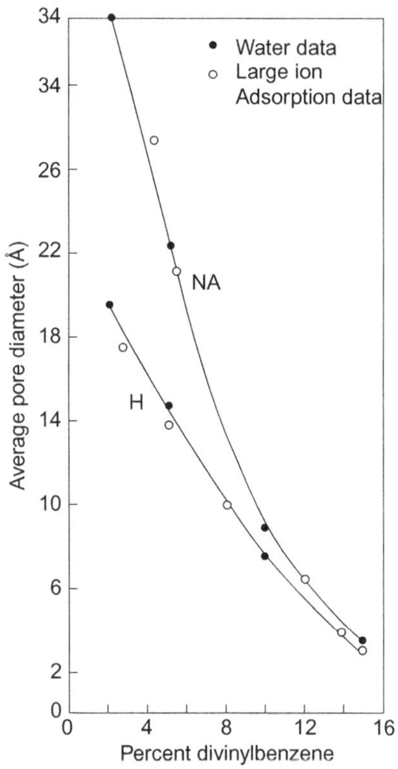

FIGURE 16.14 Average pore diameter of carboxylic acid cation exchange resin as a function of degree of cross-linking. *(Ref. [20], page 47.)*

true porosity or void volume [59,60]. Variations in the amount of divinylbenzene cross-linking and diluent allow for a range of particle strengths and porosity to be made. Subsequent reactions with the appropriate chemicals result in the introduction of the same functional groups as discussed above. These are called *macroporous* or *macroreticular resins.*

Since each bead of a given external diameter that is made by the inert diluent process will contain some void volume, there is actually less polymer available per unit volume for the introduction of functional groups.

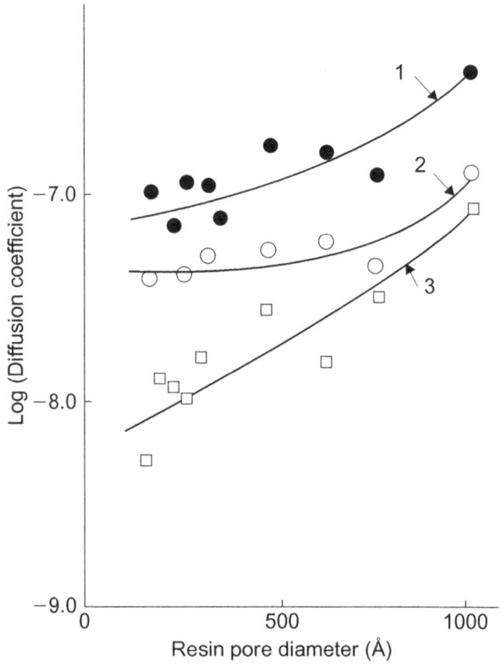

FIGURE 16.15 Ion diffusion coefficients in macroporous sulfonated cation exchange resins [58]. *(1)* Tetramethyl; *(2)* tetraethyl; *(3)* tetrabutyl ammonium ions.

Therefore, these macroporous resins are inherently of lower total exchange capacity than gel-type resins of the same composition.

Macroporous resins are most useful when extremely rigorous osmotic shock conditions are encountered, when the very high porosity is desirable from the standpoint of the molecular weight of the material being treated or when nonpolar media are involved. The drawbacks of using macroporous resins are poorer regeneration efficiencies, lower total exchange capacities and higher regeneration costs.

Until the advent of macroporous resins, the synthetic organic ion exchange resins were of such low porosity that large proteins and other macromolecules would be adsorbed or interact only with the exterior exchange sites on the resins. Therefore, although the microporous resins may have higher total exchange capacities than macroporous resins, the effective capacity of macroporous resins for protein or macromolecule adsorption may often times be greater than that of microporous resins.

Typical macroporous ion exchange resins may have average pore diameters ranging from 100 Å to 4000 Å. Table 16.11 shows the pore sizes of several resins of different matrices that have been used in enzyme immobilization [61]. Pore volumes for macroporous resins may range from 0.1 to 2.0 ml/g.

Normally, as the mean pore diameter increases, the surface area of the resin decreases. These surface areas can be as low as $2\ m^2/g$ to as high as $300\ m^2/g$. Table 16.11 also points out that the total exchange capacity is not utilized in these biochemical fluid processes. Whereas, in water treatment applications, one can expect to utilize 95% of the total exchange capacity, in biotechnology applications it is often possible to use only 10 to 20% of the total exchange capacity of gel resins. Macroporous resins have increased the utilization to close to 90% for the immobilization of enzymes, but biochemical fluid processing applications where the fluid flows through an ion exchange resin bed still are limited to about 35% utilization even with macroporous resins.

Table 16.12 shows the molecular size of some biological macromolecules for comparison to the mean pore size of the resins. When selecting the pore size of a resin for the recovery or immobilization of a specific protein, a general rule is that the optimum resin pore diameter should be about 4 to 5 times the length of the major axis of the protein. Increasing the pore size of the resin beyond that point will result in decreases in the amount of protein adsorbed because the surface area available for adsorption is being decreased as the pore size is increased. An example of this optimal adsorption of glucose oxidase, as defined by enzyme activity, is shown in Figure 16.16 [62]. Enzyme activity is a measure of the amount of enzyme adsorbed and accessible to substrate.

3.4 Particle Density

The typical resin densities may range from 0.6 g/cc to 1.3 g/cc for organic polymers. Silicate materials may be more dense up to 6 g/cc. Since the fermentation broth or other biochemical fluid may be more dense than water,

TABLE 16.12 Molecular Size of Biopolymers

Biopolymer	Molecular Weight	Maximum Length of Biopolymer
Catalase	250,000	183 Å
Glucose Isomerase	100,000–250,000	75–100 Å
Glucose Oxidase	15,000	84 Å
Lysozyme	14,000	40 Å
Papain	21,000	42 Å

TABLE 16.11 Physical Properties and Capacities for Ion Exchange Resins [61]

Resin Matrix	Functionality	Pore Size	Surface Area	Resin Capacity	Adsorption Capacity for Enzyme
phenolic	3° polyethylene polyamine	250 Å	$68.1\ m^2/g$	4.38 meq/g	3.78 meq/g
phenolic	partially 3° polyethylene polyamine	290 Å	$95.3\ m^2/g$	4.24 meq/g	3.57 meq/g
polystyrene	polyethylene polyamine	330 Å	$4.6\ m^2/g$	4.20 meq/g	3.92 meq/g
polystyrene	polyethylene polyamine	560 Å	$5.1\ m^2/g$	4.75 meq/g	4.32 meq/g
polyvinyl chloride	polyethylene polyamine	1400 Å	$15.1\ m^2/g$	4.12 meq/g	3.72 meq/g

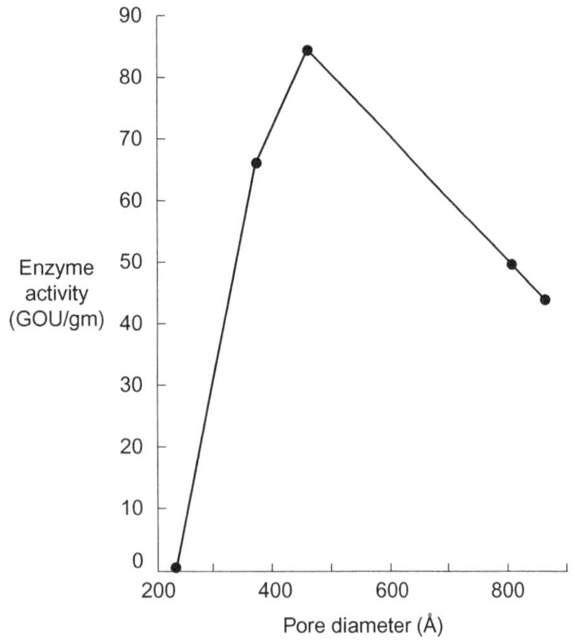

FIGURE 16.16 Effect of resin pore diameter on the enzyme activity of glucose oxidase [62].

TABLE 16.13 Water Soluble Standard Samples for Pore Measurements [64]

Sample	Mean Pore Diameter (Å)
D_2O	3.5
Ribose	8
Xylose	9
Lactose	10.5
Raffinose	15
Stachyose	19
T-4[a]	51
T-10	140
T-40	270
T-70	415
T-500	830
T-2000	1500

[a]*The T-Standards are Dextrans from Pharmacia*

the slow flow rates that are usually involved may require resins that have a greater density than water. A minimum flow rate may be necessary to maintain a packed bed when a fluid denser than water is being processed by a medium density resin. If this is not possible, an up-flow operation or batch process may be necessary. This is discussed in more detail in Sec. 6.

The lower density resins are usually associated with a highly porous structure which has less mechanical strength than the typical gel or macroporous resins. When the mean pore diameter of a resin is greater than 2000 Å, the resin would be subject to attrition in a stirred tank or may collapse in a tall column.

3.5 Particle Size

Many of the resins used in the early biochemical separations were quite small (75–300 microns). With the development of macroporous resins, protein purifications were performed with resins of the 400–1000 micron size since the macroporous structure allowed sufficient surface area for adsorption almost independent of particle size.

4.0 LABORATORY EVALUATION OF RESIN

The total exchange capacity, the porosity, the operating capacity and the efficiency of regeneration need to be evaluated in the laboratory when comparing resins for a given application.

The total exchange capacity is usually determined by titrating the resin with a solution of acid or base to a specific end point. This type of information is readily available from the manufacturers of commercial ion exchange resins.

The pore size of a microporous resin can be determined using water soluble standards, such as those listed in Table 16.13 [63]. If the resin is made with an inert, extractable diluent to generate the macroporous structure, it is easier to determine the mean pore size and pore size distribution. Care must be taken so that the pores are not collapsed during the removal of the water from the resin. Martinola and Meyer [64] have devised a method of preparing a macroporous resin for BET surface analysis or pore size analysis by mercury porosimetry.

1. Convert the ion exchange resin to the desired ionic form.
2. Add 500 ml of water-moist resin to a round bottom flask with an aspirator. Add one liter of anhydrous isopropyl alcohol and boil under reflux at atmospheric pressure for one hour, then remove the liquid. Repeat the isopropyl addition, boiling and aspirating four times. After this procedure the resin will contain less than 0.1% water.
3. After drying to constant weight at 10^{-3} torr and 50°C, the resin sample is ready for pore size analysis.

The design of an ion exchange unit requires knowledge of the capacity of the resin bed and the efficiency of the exchange process. The "theoretical" capacity of a resin is the number of ionic groups (equivalent number of exchangeable ions) contained per unit weight or unit volume of resin. This capacity may be expressed as milliequivalents (meq) per ml or per gram of resin.

When deciding which resin to use for a given operation, batch testing in a small beaker or flask will allow resin selection and an approximation of its loading capacity. A useful procedure is to measure out 1, 3, 10 and 30 milliliter volumes of resin and add them to a specific volume of the feedstream. These volumes were chosen to have even spacing on a subsequent log-log plot of the data.

After the resin bed feed solution has been mixed for at least one half hour, the resin is separated from the liquid phase. The solute concentration remaining in the solution is then determined. The residual concentration is subtracted from the original concentration and the difference is divided by the volume of the resin. These numbers and the residual concentration are plotted on log-log paper and frequently give a straight line.

A vertical line drawn at the feed concentration intersects at a point extrapolated from the data points to give an estimate of the loading of the solute on the resin. Figure 16.17 shows such a plot for glutamic acid absorbed on an 8% cross-linked strong acid cation microporous resin in a fermentation broth with 11 mg/ml of amino acid. The resin was placed in a beaker with 250 ml of broth. The extrapolation of the line for the 1, 3, 10, and 30 ml resin adsorption data indicates that the loading of glutamic acid on this resin is expected to be 60 g/liter resin.

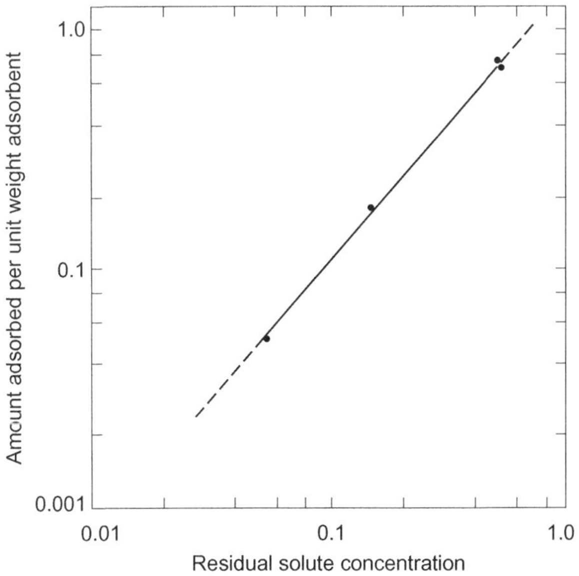

FIGURE 16.17 Plot of glutamic acid adsorbed on an 8% cross-linked strong cation resin.

After several resins have been tested in this manner, the resin is selected for column evaluation which has a high loading per ml of resin or a low residual with larger resin quantities.

In practice, the ion exchange resin is generally operated at a level considerably below its theoretical capacity. Since the ion exchange reactions are equilibrium reactions, an impracticably large quantity of regenerant would be required to drive the reaction to completion. The "operating" capacity of a resin is the number of ionic groups actually utilized per unit weight or volume of resin under a given set of operating conditions.

The operating capacity of a resin is not directly proportional to the amount of regenerant used. "Efficiency" is the concept used to designate the degree of utilization of the regenerant. Column efficiency is the ratio of the operating exchange capacity of a unit to the exchange that theoretically could be derived from a specific weight of applied regenerant.

It is recommended that operating capacity and column efficiency be run initially on a small, laboratory scale to determine if the reaction desired can be made to proceed in the desired direction and manner. The column should be at least 2.5 cm in diameter to minimize wall effects. The preferential flow in a resin column is along the wall of the column. The percentage of the total flow along the wall of the column decreases as the column diameter increases and as the resin particle size decreases.

The bed depth should be at least 0.5 m and the flow rate should be about 0.5 bed volumes per hour for the initial trial. These conditions are good starting points since it is desirable that the transition zone not exceed the length of the column. Using much larger columns would require quantities of the feedstream which are larger than may be readily available.

A suggested operating procedure is outlined below.

1. Soak the resin before adding it to the column to allow it to reach its hydrated volume.
2. After the resin has been added to the column, backwash the resin with distilled water and allow the resin to settle.
3. Rinse the column of resin with distilled water for ten minutes at a flow rate of 50 ml/min.
4. Start the treatment cycle. Monitor the effluent to develop a breakthrough curve, such as shown in Figure 16.18, until the ion concentration in the effluent reaches the concentration in the feed solution.
5. Backwash the resin with distilled water to 50–100% bed expansion for 5 to 10 minutes.
6. Regenerate the resin at a flow rate that allows at least forty-five minutes of contact time. Measure the ion

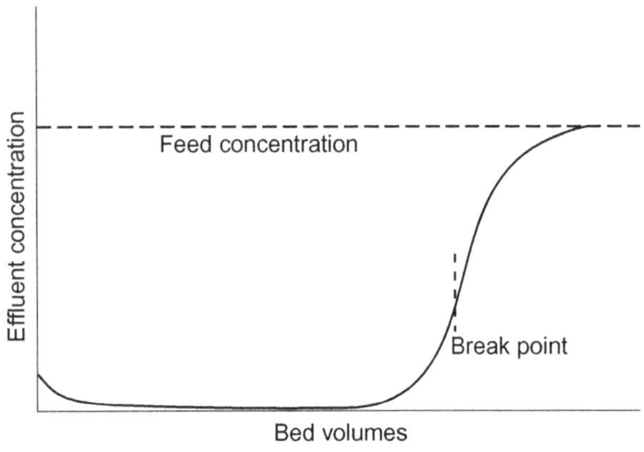

FIGURE 16.18 Concentration of adsorbed species in column effluent during column loading.

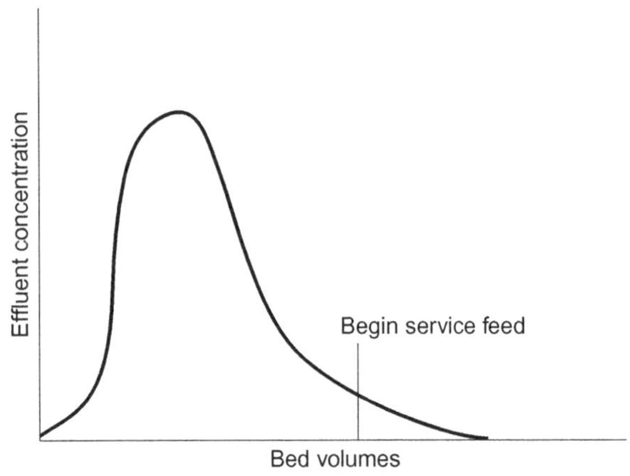

FIGURE 16.19 Elution curve showing concentration of adsorbed species eluted during resin regeneration.

concentration of the spent regenerate to determine the elution curve (Figure 16.19) and the amount of regenerant actually used.

7. Rinse column with distilled water until the effluent has reached pH 7.

Feed concentrations, flow rates, and regenerant dosages may be varied to develop the relationship between resin utilization and regenerant efficiency so that the optimum operating conditions can be selected for the system.

The first portion of the breakthrough curve in Figure 16.18 shows the quality of product that can be obtained under the processing conditions. An integration of the area up to the breakthrough point provides an estimate for commercial column capacities for the space velocities used in the experiment. The velocity at which the mass transfer zone is moving through the column is given by dividing the length of the column by the time it takes to detect the solute in the column effluent. The difference between that time and the time at which the selected breakthrough concentration appears in the effluent, when multiplied by the velocity of the mass transfer zone, results in an approximation of the mass transfer zone.

For simple molecules with large differences in distribution coefficients, a single eluting solution may be used to develop the chromatogram. However, more complex materials, such as peptides and proteins, require a shift in the ionic strength of the eluent. This can be done stepwise or as a gradient. Sememza [65] has proposed the following rules for the proper choice of eluent:

1. Use cationic buffers (Tris-HCI, piperazine-$HClO_4$, etc.) with anion resins and anionic buffers (phosphate, acetate, etc.) with cation exchange resins.
2. With anion resins use decreasing pH gradients and with cation resins use rising pH gradients.
3. Avoid using buffers whose pH lies near the pK of the adsorbent.

If the chromatographed solutes are to be isolated by solvent evaporation, the use of volatile buffers, such as carbonic acid, carbonates, acetates and formates of ammonium should be used.

If better resolution is required, it may be obtained by changing the type of gradient applied. A convex gradient may be useful in improving the resolution during the last portion of a chromatogram or to speed up separation when the first peaks are well separated and the last few are not adequately spaced. A concave gradient can be used if it is necessary to improve resolution in the first part of the chromatogram or to shorten the separation time when peaks in the latter portion are more than adequately spaced.

During the column test, the starting volume and the final volume of the resin should be measured. If there is a change of more than 5%, progressive volume changes as the resin is operated through several cycles should be recorded. These changes may be significant enough to affect the placement of laterals or distributors in the design of commercial equipment. For instance, carboxylic resins may expand by 90% when going from the hydrogen form to the sodium form. This type of volume change may dictate how the resin must be regenerated to prevent the breakage of glass columns due to the pressure from the swelling resin.

Gassing, the formation of air pockets, within the resin bed is to be avoided. Gassing may occur because of heat released during the exchange reaction. It will also occur if a cold solution is placed in a warm bed or if the liquid level falls below the resin level. Keeping the feed solution 5°C warmer than the column temperature should prevent the gassing due to thermal differences.

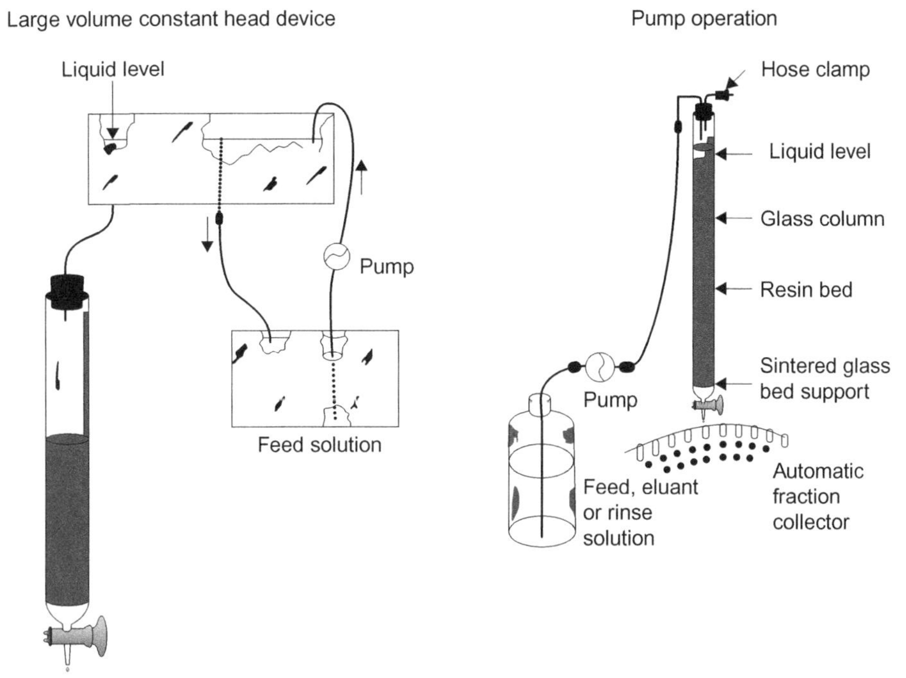

FIGURE 16.20 Equipment for laboratory evaluation of ion exchange resins.

It is necessary to configure the experimental apparatus to insure that the feedstream moves through the column at a steady rate to maintain a well-defined mass transfer zone. Possible methods of maintaining constant flow are shown in Figure 16.20.

Once it is determined that the action will proceed as desired, subsequent optimization of the system in the laboratory calls for setting a packed resin column of approximately the bed depth to be used in the final equipment, typically one to three meters.

5.0 PROCESS CONSIDERATIONS

5.1 Design Factors

The engineer designing an ion exchange column operation usually will prefer to work with the simplest kinetic model and linear driving force approximations. The weakness of this approach is that any driving force law only regards the momentary exchange rate as a function of the solute concentration in the bulk solution and the average concentration in the particle, neglecting the effect of concentration profiles in the particle. Nevertheless, the linear driving force approach provides an approximation that is sufficiently accurate for the engineer.

5.2 Scaling-up Fixed Bed Operations

Rodrigues [66] has presented empirical and semi-empirical approaches which may be used to design ion exchange columns when the solute in the feedstream is c_0 and the flow-rate is u_0. The breakthrough point is usually set at the point where the effluent concentration increases to 5% of c_0. The design equations relate the total equilibrium ion exchange capacity (Q) to the volume of resin required (V_r) to the time of breakthrough (t_B).

In the empirical approach, the overall mass balance is given by the equation:

$$V_r = c_0 \xi t_s/(1+\xi)Q \qquad (16.15)$$

where

$$t_s = \tau(1+\xi) \quad \text{(the stoichiometric time)} \qquad (16.16)$$

$$\xi = (1-\varepsilon)Q/\varepsilon Q_0 \quad \text{(the mass capacity factor)} \qquad (16.17)$$

$$\tau = \varepsilon V/u_0 \quad \text{(the space time)} \qquad (16.18)$$

and V is the bed volume with void space ε.

It is usually necessary to modify this resin amount by a safety factor (1.2 to 1.5) to adjust for the portion of the total equilibrium capacity that can actually be used at flow rate u and to adjust for any dispersive effects that might occur during operation.

The semi-empirical approach involves the use of the mass transfer zones. This approach has been described in detail specifically for ion exchange resins by Passino [67]. He referred to the method as the operating line and regenerating line process design and used a graphical description to solve the mass transfer problems.

For the removal of Ca^{++} from a feedstream, the mass transfer can be modeled using Figure 16.21. The upper part shows an element of ion exchange column containing a volume v of resin to which is added a volume V_{ex} of the feedstream containing Ca^{++}. It is added at a flow rate (F_L) for an exhaustion time t_{ex}. The concentration of Ca^{++} as it passes through the column element is reduced from x_{ex1} to x_{ex2}. Therefore, the resin, which has an equilibrium ion exchange capacity C, increases its concentration of Ca^{++} from y_{ex2} to y_{ex1}. In this model, fresh resin elements are continuously available at a flow rate $(F_s) = v/t_0$, which is another way of saying the mass transfer zone passes down through the column.

The lower part of Figure 16.21 shows the operating lines for this process. The ion exchange equilibrium line describes the selectivity in terms of a Freundlich, Langmuir or other appropriate model.

The equations for the points in the lower part are given by:

$$x_{ex1} = x_0 \quad (Ca^{++} \text{ in the feedstream}) \tag{16.19}$$

$$y_{ex1} = y_{ex2} + (x_{ex1} - x_{ex2})\frac{c_0 V_{ex}}{Cv} \quad (Ca^{++} \text{ in the exhausted streamstream}) \tag{16.20}$$

$$X_{ex2} = \frac{x\,dv}{V_{ex}} \quad (\text{average } Ca^{++} \text{ in the effluent}) \tag{16.21}$$

$$y_{ex2} = 0 \quad (Ca^{++} \text{ in the regenerated resin}) \tag{16.22}$$

and the slope of the operating line:

$$\frac{DV}{Dx_{ex}} = \frac{y_{ex1} - y_{ex2}}{x_{ex1} - x_{ex2}} = \frac{c_0 V_{ex}}{Cv} = \frac{c_0 (F_L)_{ex}}{CF_s} \tag{16.23}$$

The value of c_0, C and v are known so that for any V_{ex} value, the slope of the operating line can be calculated from Eq. (16.23). The specific points: x_0 is given, x_{ex2} is obtained by graphic integration from the breakthrough curves. After operation and regeneration, the value of y_{ex2} may not be zero but may be between 0.02 and 0.05 if the regeneration is not complete. The application of this technique has been described in terms of basic design parameters, such as number of transfer units, the height of a transfer unit and mass transfer coefficient [68,69].

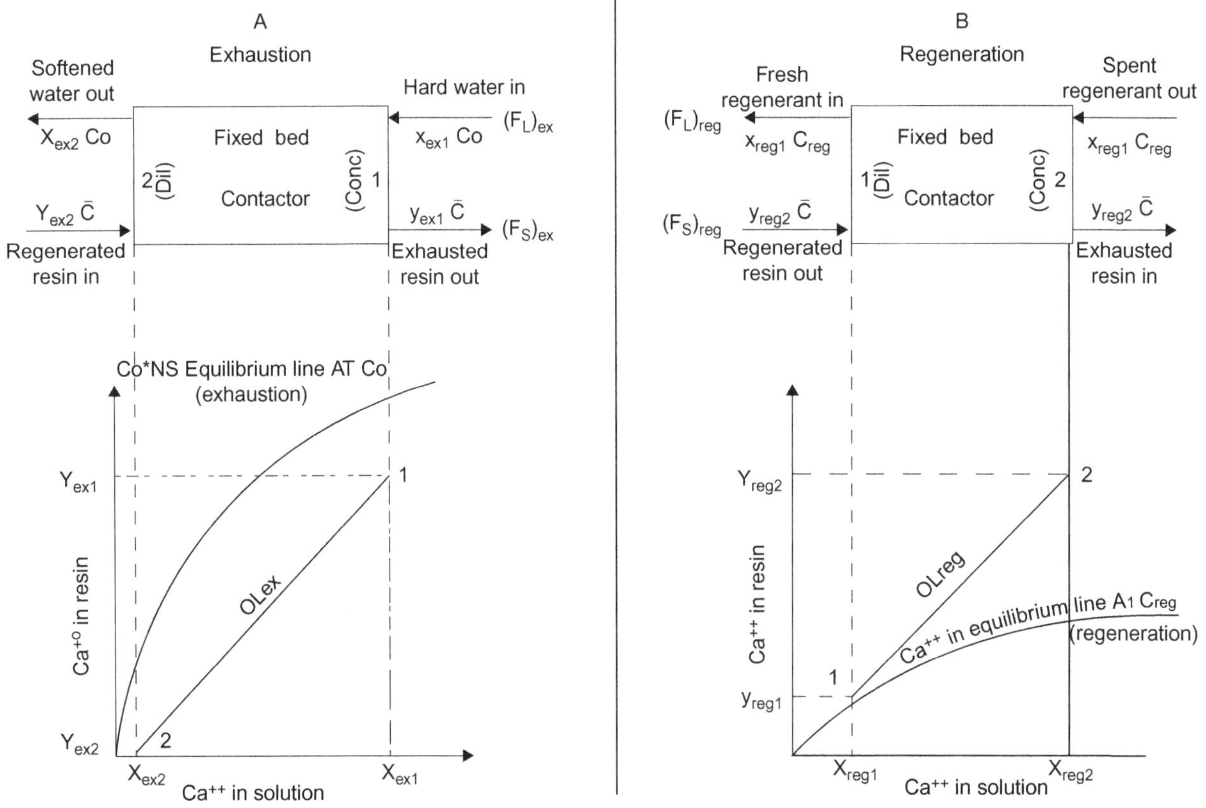

FIGURE 16.21 Schematic representation of the softening of a hard water feedstream on an ideally continuous bed of strong cation resin in the sodium form [67].

The data generated with the laboratory column may be scaled-up to commercial size equipment. Using the same flow rate (on a mass basis) as used in the laboratory experiments, the appropriate increase in column size over that used in the laboratory is a direct ratio of the volumes to be treated compared to that treated in the laboratory equipment.

If a reasonable height to diameter ratio (approximately 1:1) is obtained in the scaleup using the bed depth involved in the laboratory procedure, then that bed depth is maintained and the cross-sectional area of the column is increased. However, if the sizing is such that the column is much larger in diameter than the bed depth, scaleup should be done to maintain a height to diameter ratio of approximately 1. The required resin volume is determined by maintaining the same mass flow conditions (liters of feed solution per minute per cubic meter of installed resin) as was used in the laboratory operation.

Appropriate tank space must be left to accommodate the backwash operation. This is typically 50% of bed depth for cation exchange resins and 100% expansion in the case of anion exchange resins.

5.3 Sample Calculation

The purification of lysine-HCl from a fermentation broth will be used to illustrate the calculations involved in scaling-up laboratory data.

The laboratory fermentation broth, which is similar to the commercial broth, contained 2.0 g/0.11 lysine, much smaller amounts of Ca^{++}, K^{+} and other amino acids. The broth was passed through 500 ml of strong acid cation resin, Dowex® HCR-S, in the NH_4^{+} form. The flow rate was 9 ml/min or 1.77 ml/min per cm^2 of resin. It was determined that the resin capacity averaged 115 g of lysine-HCl per liter of resin. It may be noted that since the equivalent molecular weight of lysine-HCl is 109.6 g and the "theoretical" capacity of Dowex® HCR-S is 2.0 meq/ml, the operating capacity is 52% of theoretical capacity.

The commercial operation must be capable of producing 9,000 metric tons of lysine (as lysine-dihydrochloride-H_2O) per year. With a 2.0 g/0.11 concentration of lysine in the fermentation broth, the number of liters of broth to be treated each year are:

$$\frac{9{,}000 \text{ m tons}}{\text{yr}} \times \frac{146.19(\text{MW of lysine})}{237.12(\text{MW of L-HCl-H}_2O)} \times \frac{0.11}{2.0 \text{ g}} \times \frac{10^6 \text{ g}}{\text{M ton}} = 27.7 \times 10^7 \text{ l/yr} \quad (16.24)$$

If the plant operates 85% of the time, the flow rate would have to be:

$$27.7 \times 10^7 \; l/\text{yr} \times \frac{1}{0.85} \times \frac{1 \text{ yr}}{365 \text{ days}} \times \frac{1 \text{ day}}{24 \text{ hr}} = 3.73 \times 10^4 \; l/\text{hr} \quad (16.25)$$

At a resin capacity of 115 g/l of resin, the amount of resin that must be available is:

$$\frac{1(\text{resin})}{115 \text{ g}} \times \frac{219.12(\text{MW of L}-\text{HCl})}{146.19(\text{MW of L})} \times \frac{2.0 \text{ g}}{0.11} \times \frac{3.73 \times 10^4 \text{ l/hr}}{\text{hr}} = 9.71 \times 10^3 \text{ l/hr} \quad (16.26)$$

To obtain the maximum utilization of the resin in this operation, series bed operation (Carrousel) operation is recommended. This operation uses three beds of resin in a method having two beds operating in series while the product is being eluted from the third. The freshly regenerated resin is placed in the polishing position when the totally loaded lead bed is removed for regeneration.

The elution/regeneration step, which includes backwashing, eluting, and rinsing the resin, might take up to four hours. Therefore, enough resin must be supplied to take up the lysine-HCl presented during that time. The resin requirement for the commercial scaleup operation would be:

$$\frac{9.71 \times 10^3 \text{ 1(resin)}}{\text{hr}} \times \frac{4 \text{ hr}}{\text{bed}} = 3.88 \times 10^4 \text{ 1(resin)/bed} \quad (16.27)$$

Thus, three beds of 39 m^3 resin each are required to produce 9,000 metric tons of lysine/year.

5.4 Comparison of Packed and Fluidized Beds

Belter and co-workers [70] developed a periodic countercurrent process for treating a fermentation broth to recover novobiocin. They found that they were able to scaleup the laboratory results to production operations if the two systems have similar mixing patterns and the same distribution of residence times in the respective columns. The mixing patterns are the same when the space velocity (F/V_e) and the volume ration (V_R/c) are the same. This is shown in Figure 16.22 for the effluent concentration of novobiocin from laboratory and production columns.

The scaling-up of packed beds is subject to the difficulties of maintaining even flow distributions. Removal of solution through screens on sidewalls is not

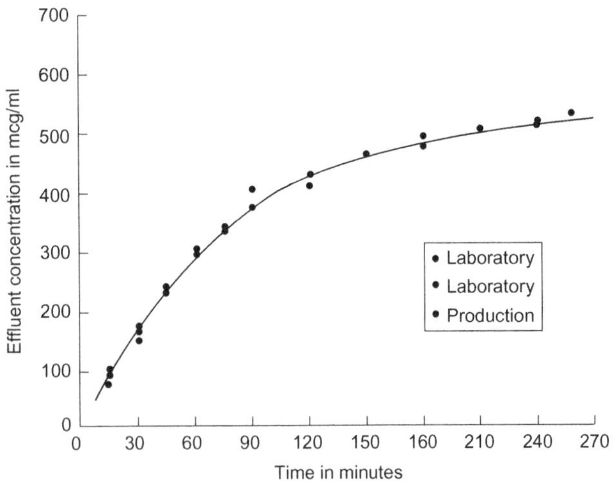

FIGURE 16.22 Comparison of experimental curves for laboratory and production columns [70].

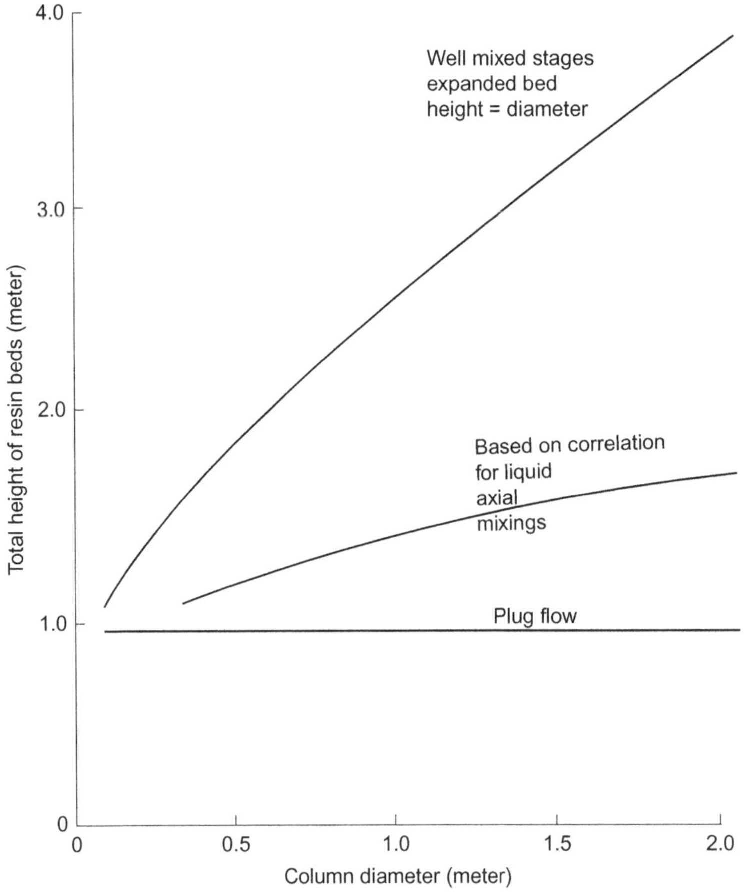

FIGURE 16.23 Scaleup relationships for fluidized beds [71].

recommended and the flow of resin from one section into another of much greater area could distort the resin flow profile.

The problems of scaling-up fluidized bed operations are more difficult in terms of design calculations, but flow distribution is more easily designed because of the mobility of the resin. The degree of axial mixing of the liquid and the resin has to be taken into account when calculating the changes necessary in the bed diameter and bed height. Figure 16.23 shows the increases in bed height necessary when scaling-up packed and fluidized beds with bed diameter increases [71].

Mass transfer coefficients have been correlated for packed and fluidized beds [72]. The mass transfer coefficients for packed beds are 50 to 100% greater than for fluidized beds.

The volume of resin in a packed bed is about half that in a fluidized bed, but the packed bed column may be up to eight times smaller. Despite this, a complete fluidized bed operation may still be smaller than a fixed bed operation. Also, fluidized bed columns do not operate at high pressures so they can be constructed more economically.

5.5 Chromatographic Scale-Up Procedures

The aim of scaling-up a chromatographic process is to obtain the same yield and product quality in the same time period on laboratory, preparative and industrial scale. Laboratory analytical purifications tend to be optimized only for resolution of individual solutes, however, at the preparative and production scale, it is necessary also to maximize throughput. Table 16.14 [73] shows the effect of chromatographic operating parameters on resolution and throughput. While column length is a critical factor for resolution with gel filtration and isocratic elutions, it has little effect on resolution with gradient elution in adsorption chromatography. The wall effects on resolution are very noticeable with small radius columns, but decrease as the column length is increased.

Voser and Walliser [74] viewed scaleup as a three step process involving selection of a process strategy, evaluation of the maximum and optimal bed height and finally column design. The selection of a process strategy involves choosing the direction of flow, frequency of backwashing, the operation with positive head pressure or only hydrostatic pressure, and the use of a single column or a series of columns in batch or semi-continuous operation. The maximum feasible bed height is determined by keeping the optimal laboratory-scale specific volume velocity (bed volume/hour) constant. The limiting factors will be either pressure drop or unfavorable adsorption/desorption kinetics since the linear velocity also increases with increasing bed height. Bed heights from 15 cm to as high as 12 m have been reported. The column diameter is then selected to give the required bed volume. The column design combines these column dimensions with the practical considerations of available space, needed flexibility, construction difficulty and flow distribution and dilution for the columns.

The scaleup considerations of column chromatography for protein isolation has been described by Charm and Matteo [75]. When several hundred liters of a protein feedstream must be treated, the resin may be suspended in the solution, removed after equilibration by filtration and loaded into a column from which the desired proteins may be eluted. Adsorption onto a previously packed column was not recommended by them since they feared suspended particles would clog the interstices of the column, causing reduced flow rates and increased pressure drops across the column. The reduced flow rate may lead to loss of enzyme activity because of the increased time the protein is adsorbed on the resin.

It is important that all of the resin slurry be added to the column in one operation to obtain uniform packing and to avoid the formation of air pockets in the column. An acceptable alternative, described by Whatman [76], allows the addition of the adsorbent slurry in increments. When the resin has settled to a packed bed of approximately 5 cm, the outlet is opened. The next increment of the slurry is added after the liquid level in the column has dropped. It is important that the suspended adsorbent particles do not completely settle between each addition.

Stacey and coworkers [77] have used the relationships shown in Table 16.15 to scaleup the purification from an 8 mg protein sample to a 400 mg sample. The adsorbent used in both columns was a Delta-Pak wide-pore C-18 material. When eluting the protein, the flow rate should change so that the linear velocity of the solvent through the column stays the same. The flow rate is proportional to the cross-sectional area of the column. The gradient duration must be adjusted so that the total number of column volumes delivered during the gradient remain the same. As with size exclusion chromatography, the mass load on the preparative column is proportional to the ratio of the column volumes. Figure 16.24 shows that the chromatograms from the 8 mg separation is very similar to that obtained for the 400 mg sample.

TABLE 16.14 Effects of Process Parameters on Resolution and Throughput [73]

Parameter	Resolution Varies with	Throughput Varies with
Column Length (L)	L	I/L
Column Radius (r)	Some effect	r^2
Temperature (T)	Positive effect	T
Viscosity (η)	Negative effect	$1/\eta$
Sample Volume (V)	$1/\|V - V_{optimum}\|$	V
Flow Rate (J)	$1/\|J - J_{optimum}\|$	J

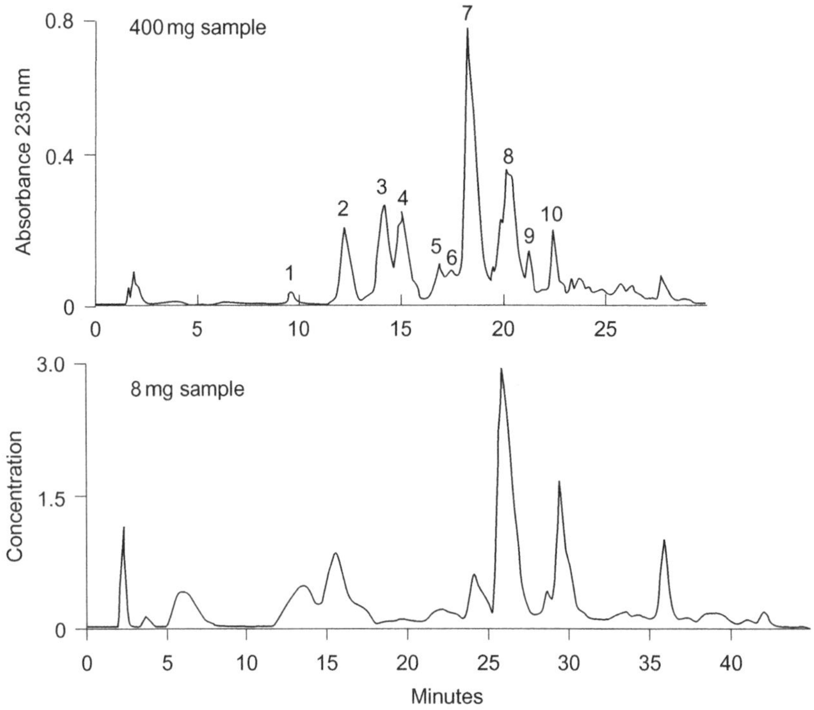

FIGURE 16.24 Laboratory and preparative scale separation of cytochrome C digest [77].

TABLE 16.15 Scale-Up Calculations [77]

	Small Scale	Preparative Scale
Column Dimensions	0.39×30 cm	3×25 cm
Flow Rate Scale Factor		
$\frac{(3.0)^2}{(0.39)^2} = 59$	1.5 ml/min	90 ml/min
Sample Load Scale Factor		
$\frac{(3.0)^2 \times 25}{(0.39)^2 \times 25} = 49$	8 mg	400 mg
Gradient Duration Calculation	40 min	33 min
$\frac{1.5 \times 40}{(0.195)^2 \pi \times 30} = 16.7$	Column Volume $\frac{90 \times (\text{Gradient Duration})}{(1.5)^2 \pi \times 25} = 16.7$	
	Grad. Duration = 33 min.	

Ladisch [78] has worked with a variety of column sizes ranging from 2 to 16 mm in diameter and 10 to 600 cm in length. His experience is that published semi-empirical scaleup correlations are useful in obtaining a first estimate on large scale column performance.

When scaling-up a chromatographic process, it may be necessary to change the order of certain steps from that used in the laboratory. Gel filtration, though a frequent first step at the laboratory scale, is not suitable for handling large scale feedstream volumes [79]. When gel filtration is used to separate molecules of similar molecular weights, sample sizes may range from 1% to 5% of the total gel volume. Thus, a 100 liter feedstream would require a gel filtration column of 2,000 to 10,000 liters. When one is separating a large molecule from small molecules, as in desalting operations, the applied volume may be up to 30% of the gel volume.

On the other hand, ion exchange chromatography is a very good first step because its capacity is approximately 30 mg of protein per ml of resin. This capacity is relatively independent of feed volume. For the same 100 liter feedstream, only a 20 liter ion exchange column would be required.

5.6 Pressure Drop

The pressure drop across an ion exchange bed has been represented by an equation [80] which depends on the average particle diameter, the void fraction in the bed, an exponent and a friction factor dependent on the Reynolds number, a shape factor, the density of the fluid, the viscosity of the fluid and the flow rate.

While that equation has internally consistent units (English system), the variables are not normally measured in those units. Another disadvantage is that one must check graphs of the exponent and friction factor versus the Reynolds number to use the equation.

For laminar flow with spherical particles, the equation can be simplified to:

$$\frac{\Delta P}{L}(\text{bar/cm}) = \frac{0.0738\mu(cp)V_0(\text{l/min})}{D_p^2(\text{mm}^2)}\frac{(1-\varepsilon)^3}{\varepsilon^3} \quad (16.28)$$

For most ion exchange resins, the void volume is about 0.38, so that $(1 - \varepsilon^3)/\varepsilon^3 = 4.34$ and:

$$\frac{\Delta P}{L}(\text{bar/cm}) = \frac{0.32\mu(cp)V_0(\text{l/min})}{D_p^2(\text{mm}^2)} \quad (16.29)$$

Table 16.16 shows the agreement between results from experiments and those calculated with this equation for several ion exchange resins.

Figure 16.25 shows the use of a size factor for resins which is used to develop a pressure factor. This pressure factor can then be used to calculate the pressure drop under conditions of different solution viscosities, flow rates or particle size distributions once the pressure drop is known at one viscosity, flow rate and particle size distribution.

5.7 Ion Exchange Resin Limitations

When ion exchange resins are used for an extended period of time, the exchange capacities are gradually decreased. Possible causes for these decreases include organic contamination due to the irreversible adsorption of organics dissolved in the feedstream, the oxidative

TABLE 16.16 Pressure Drop for Commercial Ion Exchange Resins

Resin	Flow Rate (L/min)	Mean Bead Diameter (mm)	Pressure Drop (bar/cm) Calculated	Pressure Drop (bar/cm) Measured
Dowex SBR-P	151.4	0.750	5.28	5.08
Dowex WGR-2	22.7	0.675	1.00	1.06
Dowex MWA-1	15.1	0.675	0.67	0.77
DowexMSA-1	37.8	0.650	1.79	2.13
DowexMSC-1	30.3	0.740	1.08	1.22

Wet mesh size	Factor
12	0.5
16	1.0
20	2.0
30	4.0
35	6.0
40	8.0
50	16.0
70	32.0
100	64.0

Example:

Retained on	%	Factor
12 mesh	–	0.5 = 0.0
16 mesh	0.6	1 = 0.6
20 mesh	37.0	2 = 74.0
30 mesh	46.3	4 = 185.2
35 mesh	13.0	6 = 78.0
40 mesh	2.0	8 = 16.0
50 mesh	0.9	16 = 14.4
70 mesh	0.2	32 = 6.4
		374.6

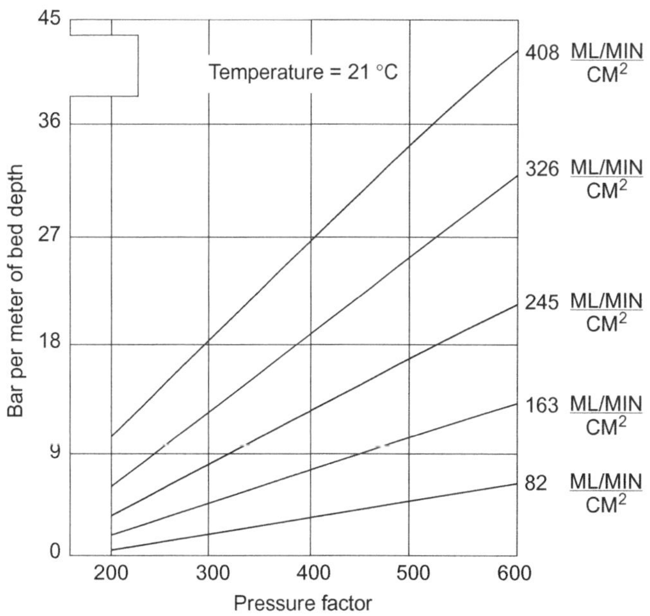

FIGURE 16.25 Size factor calculations for use in determining estimations for pressure drops in ion exchange columns.

decomposition due to the cleaving of the polymer cross-links by their contact with oxidants, the thermal decomposition of functional groups due to the use of the resin at high temperature and the inorganic contaminations due to the adsorption of inorganic ions.

When a resin bed has been contaminated with organic foulants, a procedure is available that can help to restore the resin [81]. Three bed volumes of 10% NaCl − 2% NaOH solution are passed through the resin bed at 50°C. The first bed volume is passed through at the same flow rate, followed by a through rinsing and two regenerations with the standard regenerate.

If the fouling is due to microbial contamination, the same authors recommend backwashing the bed and then filling the entire vessel with a dilute solution (< 0.05%) of organic chlorine. This solution should be circulated through the bed for 8 hours at a warm (50°C) temperature. After this treatment, the resin should be backwashed, regenerated and rinsed before returning it to service. This procedure may cause some oxidation of the polymeric resin, thereby reducing its effective cross-linking and strength. Therefore, treating the resin in this fashion should not be a part of the normal resin maintenance program.

Physical stability of a properly made cation or anion exchange resin is more than adequate for any of the typical conditions of operation. These resins can be made to have a physical crush strength in excess of 300 grams per bead.

Perhaps more important are the limitations inherent in the structure of certain polymers or functional groups due to thermal and chemical degradation. Thermally, styrene-based cation exchange resins can maintain their chemical and physical characteristics at temperatures in excess of 125°C. At temperatures higher than this, the rate of degradation increases. Operating temperatures as high as 150°C might be used, depending on the required life for a particular operation to be economically attractive.

Strong base anion exchange resins, on the other hand, are thermally degraded at the amine functional group. Operating in the chloride form, this is not a severe limitation, with temperatures quite similar to those for cation exchange resins being tolerable. However, most strong base anion exchange resins used involve either the hydroxide form, the carbonate or bicarbonate form. In these ionic forms, the amine functionality degrades to form lower amines and alcohols. Operating temperatures in excess of 50°C should be avoided for Type I strong base anion exchange resins in the hydroxide form. Type II strong base resins in the hydroxide form are more susceptible to thermal degradation and temperatures in excess of 35°C should be avoided.

The amine functionality of weak base resins is more stable in the free-base form than that of strong base resins. Styrene-divinylbenzene weak base resins may be used at temperatures up to 100°C with no adverse effects.

Chemical attack most frequently involves degradation due to oxidation. This occurs primarily at the cross-linking sites with cation exchange resins and primarily on the amine sites of the anion exchange resins. From an operating standpoint and, more importantly, from a safety standpoint, severe oxidizing conditions are to be avoided in ion exchange columns.

Oxidizing agents, whether peroxide or chlorine, will degrade ion exchange resins [82]. On cation resins, it is the tertiary hydrogen attached to a carbon involved in a double bond that is most vulnerable to oxidative degradation. In the presence of oxygen, this tertiary hydrogen is transformed first to the hydroperoxide and then to the ketone, resulting in chain scission. The small chains become soluble and are leached from the resin. This chain scission may also be positioned such that the cross-linking of the resin is decreased, as evidenced by the gradual increase in water retention values.

The degradation of anion resins occurs not only by chain scission, but also at the more vulnerable nitrogen on the amine functionality. As an example, the quaternary nitrogen on Type I strong base anion resins is progressively transformed to tertiary, secondary, primary nitrogen and finally to a nonbasic product.

Oxidative studies on resins with different polymer backbones and functionalities have been performed as accelerated tests [83]. The data is shown in Table 16.17 for polystyrene and polydiallylamine resins. Although the polydiallylamine resins have a higher initial capacity, they are much more susceptible to oxidative degradation. When the polystyrene resin has a mixture of primary and secondary amino groups or when a hydroxy-containing

TABLE 16.17 Oxidation of Polystyrene and Polylamine Resins in a One-Week Accelerated Test at 90°C [83]

Resin Backbone	Functional Group	Initial Base Capacity (meq/g)	Base Capacity Lost During Test (%)
Polystyrene	$R\text{-}CH_2N(CH_3)_2$	4.4	1
Polystyrene	$R\text{-}CH_2N(CH_2CH_3)_2$	4.5	2
Polystyrene	$R\text{-}CH_2N(C_2H_4OH)_2$	3.6	1 7
Polydiallylamine	N-H	8.3	37
Polydiallylamine	$N\text{-}CH_2CH_3$	7.3	37
Polydiallylamine	$N\text{-}C_3H_7$	6.7	31

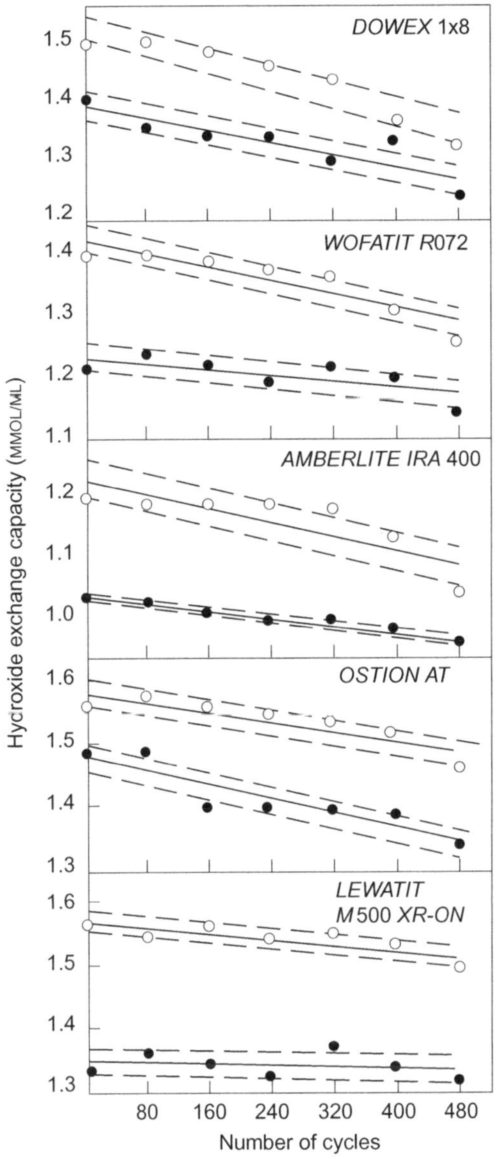

FIGURE 16.26 Dependence of the strong base capacity on the number of cycles between 20°C and 80°C [84].

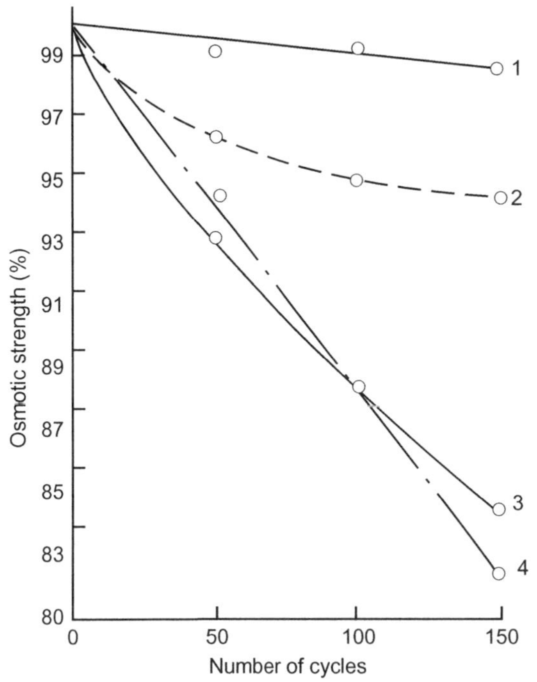

FIGURE 16.27 The effect of thermal and osmotic cycling on the osmotic strength of a strong base anion resin, AV-17. *(1)* Effect of thermal cycling; *(2)* effect of osmotic cycling; *(3)* effect of thermal and osmotic cycling (cold alkali to hot acid); *(4)* effect of thermal and osmotic cycling (hot alkali to cold acid) [85].

group is attached to the amine of the functional group, the susceptibility to oxidation is enhanced. Thus one can understand the lower thermal limit for Type II anion resins compared to Type I resins.

The effect of thermal cycling on strong base anion resins has been studied by Kysela and Brabec [84]. The average drop per cycle in strong base capacity was 2.1×10^{-4} mmol (OH^-)/ml over the 480 cycles between 20°C and 80°C. Figure 16.26 shows the decrease in total exchange capacity (*open circles*) and in strong base (salt splitting) capacity (*solid circles*) for each of the individual resins included in the study.

As would be expected, the combination of thermal and osmotic shocks has been shown to have a large effect on the osmotic strength of anion resins [85]. This is illustrated in Figure 16.27, where the decrease is shown to be over 15% in just 150 cycles. The important processing point to remember from this is that if the regeneration of a resin is performed at a temperature more than 20°C different from the temperature of the operating (loading) stream, it is advisable first to adjust the temperature of the resin with distilled water until it is at the temperature of the regenerant.

5.8 Safety Considerations

Safety considerations in the use of ion exchange systems include understanding the reactions which will take place during the contact of resin and solution being treated. The possibility of reactions being catalyzed in the presence of the hydrogen form of a strong acid cation or of the hydroxide form of the strong anion must be studied. These resins produce acid or base, respectively, during treatment of salt solutions.

If reactants or products of the ion exchange reaction are hazardous, appropriate protective equipment should be included in the system design. For example, during the demineralization of metal cyanide by-products, hydrogen cyanide solution is an intermediate formed in the cation exchange unit. The processing equipment must be designed to ensure containment of this solution until it has passed through the anion exchange unit which picks up the cyanide ions.

Since the ion exchange resins are synthetic organic compounds, they are susceptible to decomposition reactions with strong oxidizing agents, such as nitric acid. Close control of reactant or regenerant concentrations, temperature, and contact time may be required.

When the treated solution is to be used as food, food additives, or in pharmaceutical applications, resins must be used which are acceptable for such processing.

6.0 ION EXCHANGE OPERATIONS

The typical cycle of operations involving ion exchange resins include pretreatment of the resin and possibly of the feed solution; loading the resin with the solutes to be adsorbed by contacting the resin with the feed solution; and elution of the desired material from the resin. The scale of operation has ranged from analytical applications with a few milligrams of resin and microgram quantities of material to commercial production units containing several cubic meters of resin to produce metric tons of material. The loading may be applied batchwise, to a semi-continuous batch slurry, or to a column filled with resin which may be operated in a semi-continuous, continuous, or chromatographic manner.

A list of designers and manufacturers of ion exchange equipment is given in Table 16.18. The list is weighted toward those companies that operate in the United States.

TABLE 16.18 Ion Exchange Process Designers and Manufacturers

Company	Location
Belco Pollution Control Corp.	Parsippany, NJ
Chem Nuclear Systems	Columbia, SC
Cochrane Divison of Crane Corp.	King of Prussia, PA
Downey Welding & Manufacturing Co.	Downey, CA
Envirex Corp.	Waukesha, WI
Epicor	Linden, NJ
Ermco Water Conditioners	St. Louis, MO
Hittman Nuclear & Development	Columbia, MD
Hungerford & Terry, Inc.	Clayton, NJ
Industrial Filter and Pump	Cicero, IL
Infilco Degremont, Inc.	Richmond, VA
Intensa	Mexico City, Mexico
Kinetico, Inc.	Newbury, OH
Kurita	Japan
Liquitech, Div. of Thermotics, Inc.	Houston, TX
Mannesmann	Germany
Mitco Water Labs, Inc.	Winter Haven, FL
Permutit, Division of Zum	Warren, NJ
Rock Valley Water Conditioning	Rockford, IL
Techni-Chem, Inc.	Belvidere, IL
Unitech, Divsion of Ecodyne	Union, NJ
United States Filter, Fluid System Corp.	Whittier, CA
U. S. Filter/IWT	Rockford, IL
Wynhausen Water Softener Company	Los Angeles, CA

6.1 Pretreatment

Filtration, oil flotation and chemical clarification are the pretreatment operations commonly employed with the feed stream. This pretreatment is primarily concerned with the removal of excess amounts of suspended solids, oils, greases and oxidative compounds. Suspended solids, including bacteria and yeasts, in amounts exceeding approximately 50 mg/L should be removed prior to applying the fluid to the column to prevent excessive pressure buildup and short operating cycle times. The presence of oils and greases in excess amounts would coat the resin particles, thereby dramatically reducing their effectiveness. Oils and greases in concentrations above 10 mg/L should not be applied to resins in either column or batch operations. Synthetic resins are subject to decross-linking if oxidative materials are present in the feed solution or eluant.

For many biochemical recovery applications, it is necessary to pretreat the resin to ensure that the extractable level of the resin complies with Food Additive Regulation 21 CFR 173.25 of the Federal Food, Drug and Cosmetic Act. The pretreatment recommended for a column of resin in the backwashed, settled and drained condition is:

1. Add three bed volumes of 4% NaOH at a rate sufficient to allow 45 minutes of contact time.
2. Rinse with five bed volumes of potable water at the same flow rate.
3. Add three bed volumes of 10% H_2SO_4 or 5% HCl at a flow rate that allows 45 minutes of contact time.
4. Rinse with five bed volumes of potable water.
5. Convert the resin to the desired ionic form by applying the regenerant that will be used in subsequent cycles.

If the column equipment has not been designed to handle acid solutions, a 0.5% $CaCl_2$ solution or tap water may be used in place of H_2SO_4 or HCl for cation resins. Similarly, for anion resins, a 10% NaCl solution could be used in place of acids.

6.2 Batch Operations

The batch contactor is essentially a single stage stirred reactor with a strainer or filter separating the resin from the reaction mass once the reaction is complete. This type of contactor has an advantage in some fermentation operations because of its ability to handle slurries. Additional advantages include the low capital cost and simplicity of operation.

In a batch operation, an ion exchange resin in the desired ionic form is placed into a stirred reaction vessel containing the solution to be treated. The mixture is stirred until equilibrium is reached (about 0.5 to 3 hours). Then the resin is separated from the liquid phase by rinsing with the eluting solution. An additional step may be required to reconvert the resin to the regenerated form if this is not done by the eluting solvent. The cycle may then be repeated.

The batch system is basically inefficient since the establishment of a single equilibrium will give incomplete removal of the solute in the feed solution. When the affinity of the resin for this solute is very high, it is possible that the removal is sufficiently complete in one stage. The batch process has the advantage over fixed bed processes that solutions containing suspended solids may be treated. In these cases, the resin particles, loaded with the adsorbate and rinsed from the suspended solids, may be placed in a column for recovery of the adsorbate and for regeneration.

The batch contactor is limited to use with reactions that go to completion in a single stage or in relatively few stages. Difficulties may also arise with batch contactors if resin regeneration requires a greater number of equilibrium stages than the service portion of the cycle.

6.3 Column Operations

Fixed Bed Columns. Column contactors allow multiple equilibrium stages to be obtained in a single unit. This contactor provides for reactions to be driven to the desired level of completion in a single pass by adjusting the resin bed depth and the flow conditions. The main components of a column contactor are shown in Figure 16.28. At the end of its useful work cycle, the resin is backwashed, regenerated and rinsed for subsequent repetition of the work cycle. Typically, this nonproductive portion of the cycle is a small fraction of the total operating cycle.

Column contactors may be operated in cocurrent, countercurrent or fluidized bed modes of operation. The *cocurrent* mode means that the regenerant solution flows

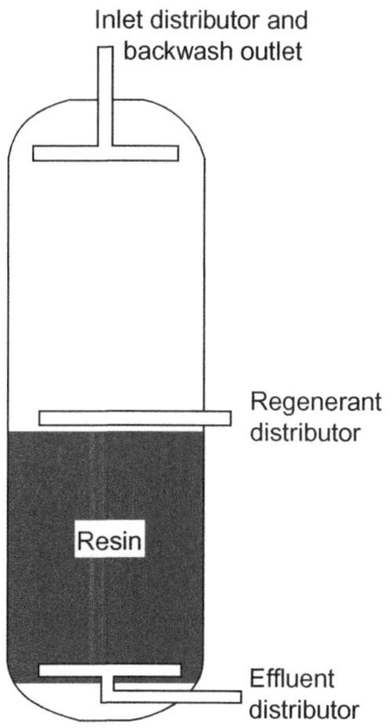

FIGURE 16.28 Ion exchange column contactor.

through the column in the same direction as the feed solution. The *countercurrent* mode has the regenerant flowing in the opposite direction as the feed solution.

Countercurrent operation of a column may be preferred to reduce the ion leakage from a column. Ion leakage is defined as the amount of ion being removed from solution which appears in the column effluent during the course of the subsequent exhaustion phase. The leakage caused by re-exchange of non-regenerated ions during the working phase of concurrently regenerated resin is substantially reduced with countercurrent regeneration.

The fixed bed column is essentially a simple pressure vessel. Each vessel requires a complexity of ancillary equipment. Each column in a cascade will require several automatic control valves and associated equipment involving process computer controls to sequence the proper flow of different influent streams to the resin bed.

Combinations of column reactors may sometimes be necessary to carry out subsequent exchange processes, such as in the case of demineralization (Figure 16.29). As this figure shows, a column of cation exchange resin in the hydrogen form is followed by a column of anion exchange resin in the hydroxide form. A mixed bed, such as shown in Figure 16.30, may also be used for demineralization. Mixed bed operation has the advantage of producing a significantly higher quality effluent than the cocurrently regenerated beds of Figure 16.29, but has the added difficulty of requiring separation of the two types of resin prior to regeneration.

An important requirement for the successful operation of a mixed bed is the careful separation of the strong base

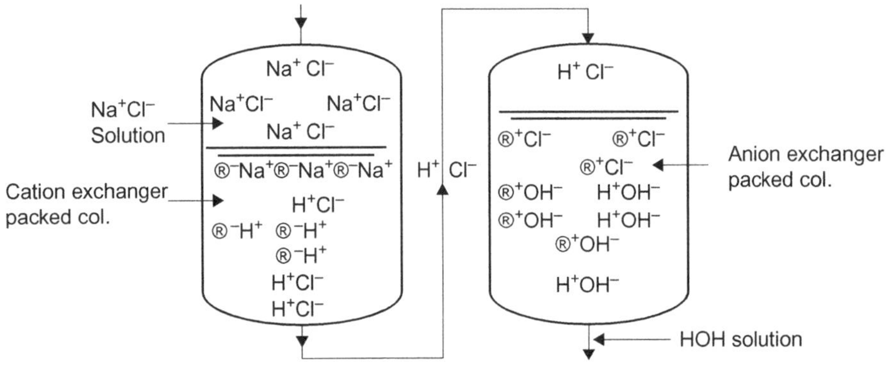

FIGURE 16.29 Demineralization ion exchange column scheme.

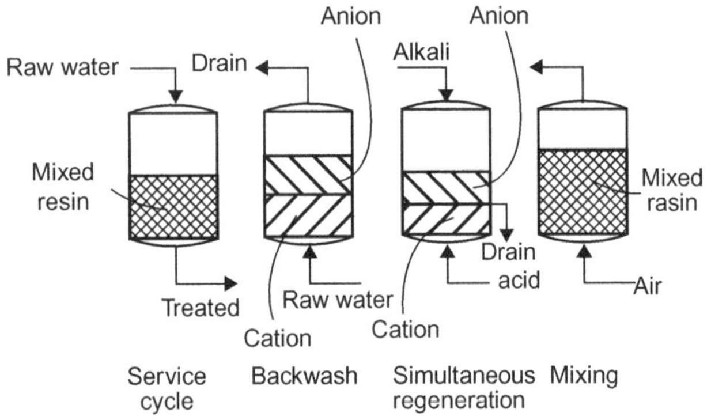

FIGURE 16.30 Operation of a mixed bed demineralization ion exchange column.

anion resin from the strong acid cation resin by backwash fluidization. This is followed by contact of each type of resin with its respective regenerant in a manner which minimizes the cross contamination of the resins with the alternate regenerant. This requires that the quantity of resin, particularly the cation resin, be precisely maintained so that the anion-cation interface will always be at the effluent distributor level. Typically, matched pairs of resins are used so that an ideal separation can be repeatedly achieved during this process. Inert resins are marketed which enhance the distance between the anion-cation interface and allow less cross contamination during regeneration [86].

The air mixing of the anion and cation must also be performed in such a manner that complete mixing of the resin and minimum air entrapment are obtained at the end of the regeneration cycle.

Ion exchange is usually in a fixed bed process. However, a fixed bed process has the disadvantage that it is cyclic in operation, that at any one instant only a relatively small part of the resin in the bed is doing useful work and that it cannot process fluids with suspended solids. Continuous ion exchange processes and fluid bed systems have been designed to overcome these shortcomings.

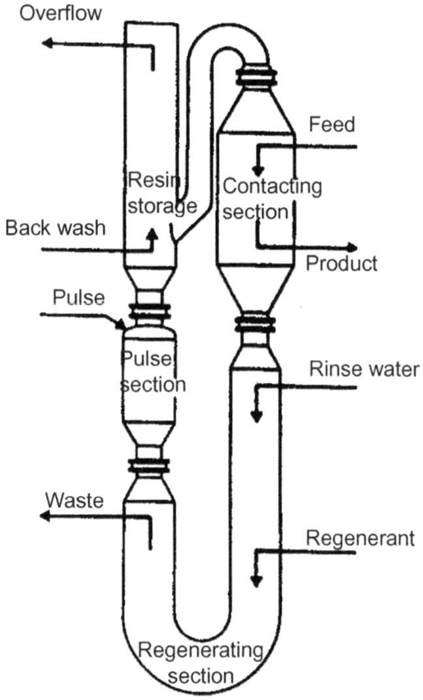

FIGURE 16.31 The Higgins contactor for continuous operation [87].

Continuous Column Operations. When the ionic load of the feed solution is such that the regeneration/elution portion of the operating cycle is nearly as great or greater than the working portion, continuous contactors are recommended instead of column contactors.

Continuous contactors operate as intermittently moving packed beds, as illustrated by the Higgins contactor [87] (Figure 16.31), or as fluidized staged (compartmented) columns, as shown in Figure 16.32, by the Himsley contactor [88–90].

In the Higgins contactor, the resin is moved hydraulically up through the contacting zone. The movement of resin is intermittent and opposite the direction of solution flow except for the brief period of resin advancement when both flows are cocurrent. This type of operation

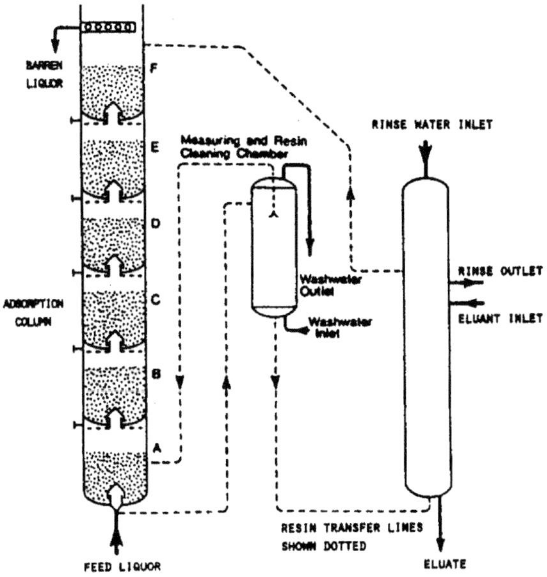

FIGURE 16.32 The Himsley Continuous Fluid-Solid contactor [88].

results in a close approach to steady-state operations within the contactor.

Elegant slide valves are used to separate the adsorption, regeneration and resin backwash stages. The contactor operates in predetermined cycles and is an ideal process for feedstreams with no suspended solids.

The Higgins type of contactor is able to handle a certain amount of slurry due to the continued introduction of fresh resin material to act as a filter media during the operation. A lower resin inventory should result with continuous contactors than with column reactors handling the same ionic load feedstream.

The major disadvantage of the Higgins contactor is the lifetime of the resin. Estimates range from as high as 30% resin inventory replacement per year due to attrition and breakage of the resin as it passes through the valves [91].

In the Himsley type of contactor, the resin is moved from one compartment or stage to the next countercurrent to the feed solution flow on a timed basis that allows for the rate of equilibrium resin loading in each stage. Thus each compartment or tray is designed to accommodate specific feed compositions and effluent requirements. Equipment from different commercial suppliers differ in the manner of resin transfer.

Fluidized Column Operations. Column contactors, when operated with the feedstream in a downflow mode, are poorly suited to handle fermentation slurries because of the excellent filtration characteristics of packed resin beds. For such slurries it is preferable to use a fluidized bed of resin such as is shown in Figure 16.33 [92]. While most of the commercial applications discussed in the literature pertain to slurries of uranium tailings and paper mill

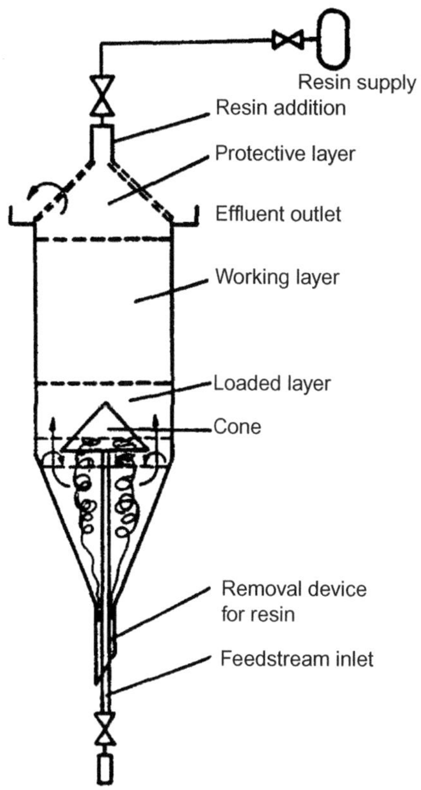

FIGURE 16.33 Fluid bed ion exchange column [92].

effluent, the equipment may be adapted for use in treating fermentation broths.

The shape of the fluidized bed is important in controlling the position of the resin in the column. The effluent from the column should pass over a vibrating screen, e.g., as SWECO, to retain entrained resin, but allow mycellia to pass through.

The Ashai contactor, shown in Figure 16.34 [91], uses conventional pressure vessels as the resin column. These vessels have a resin support grid at the base and a resin screen at the top. The feedstream is fed in up-flow through the packed bed. Periodically the liquid contents of the column are allowed to drain rapidly which causes the resin to flow from the bottom of the bed to a similar vessel for regeneration. At the same time, fresh resin is added to the top of the active column from a resin feed hopper. This hopper contains a ball valve which passes the resin in during downflow operation and seals itself during the up-flow portion of the adsorption cycle.

The Cloete-Street ion exchange equipment is a multistage fluidized bed containing perforated distributor plates (Figure 16.35) [93]. The hole size in the plates is greater than the maximum resin particle. The countercurrent movement of resin occurs due to the controlled cycling of the feedstream. With each cycle, the entire amount of resin in one chamber is transferred to the next chamber.

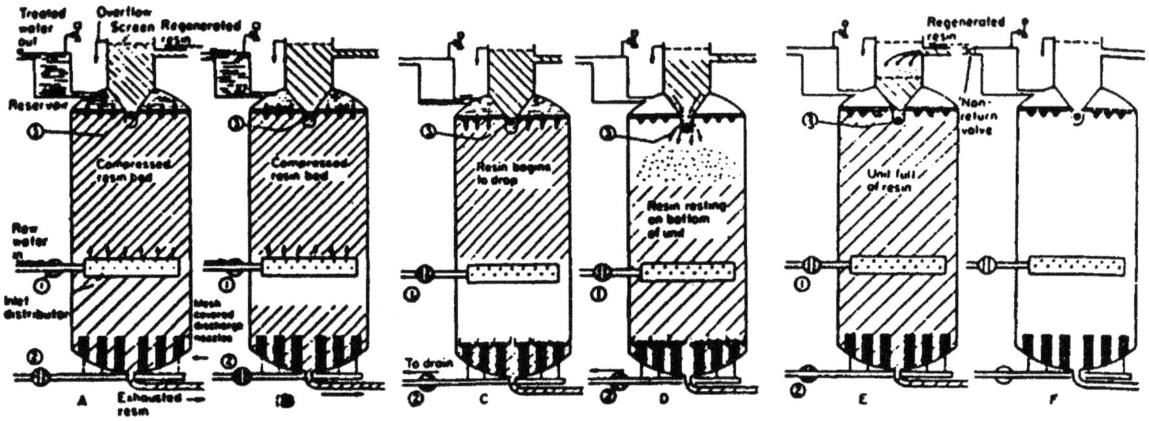

FIGURE 16.34 Operating sequence of the Asahi type countercurrent moving bed contactor [91].

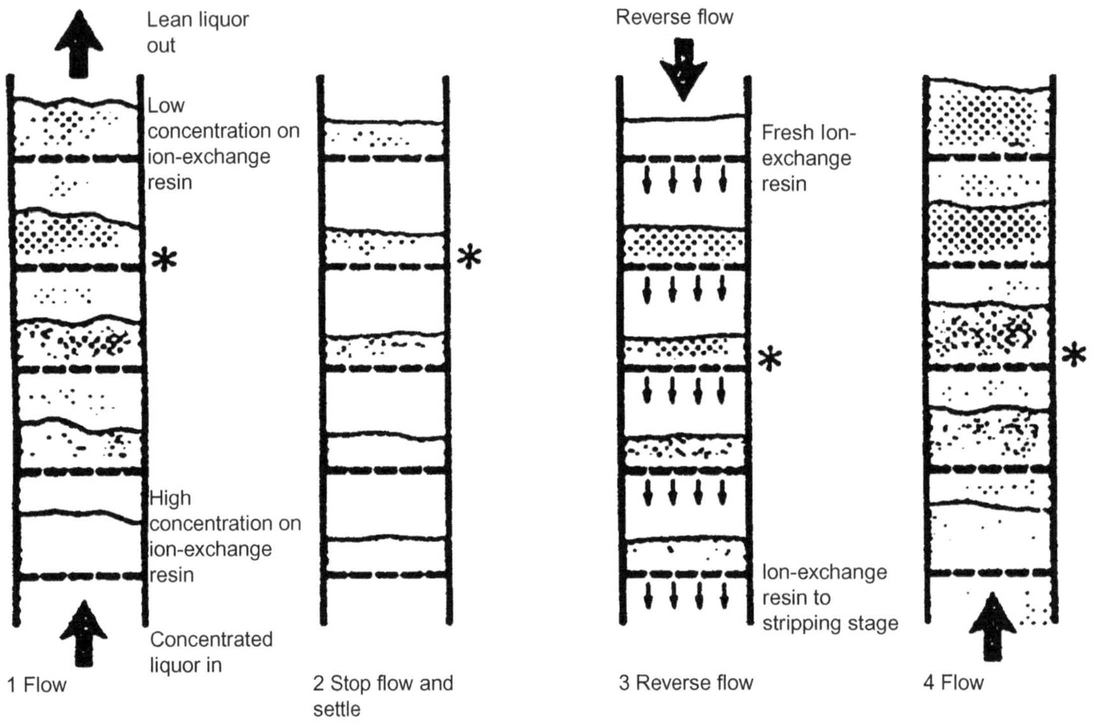

FIGURE 16.35 Sequence of operation of the Cloete-Streat countercurrent ion exchange process [93].

Equipment with 4.5 m diameter columns and eight stages for a total height of about 20 m are in commercial operation.

The USBM equipment, shown in Figure 16.36 [94], is very similar to the Cloete-Street system. The differences are in the plate design, the method of transferring solids and in the method of removing the resin. This system and the Cloete-Street system are able to handle slurry feeds with up to 15% by weight solids.

The advantage of these fluidized bed columns is the relatively low capital cost, low operating cost, small space requirement, simple instrumentation and control compactability with conventional solvent extraction equipment [91]. Only those systems that can accommodate slurries with suspended solids are commercially feasible for biotechnology and fermentation operations. Otherwise, the small volume of fermentation feedstreams which need to be processed are not the scale of operations necessary to make continuous ion exchange processes cost effective.

The principal disadvantage of fluidized bed columns is the mixing of resin in various stages of utilization. This

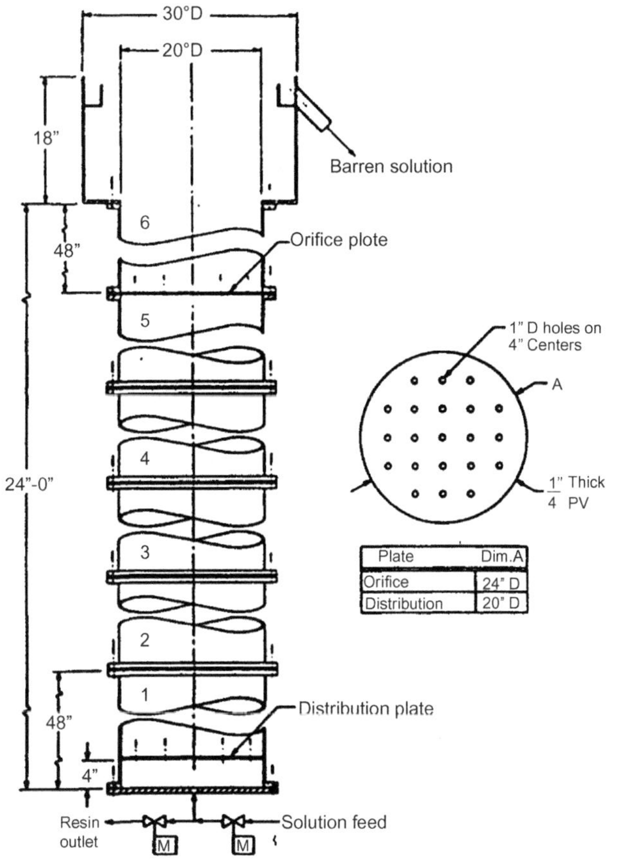

FIGURE 16.36 Schematic diagram of the USBM ion exchange column [94].

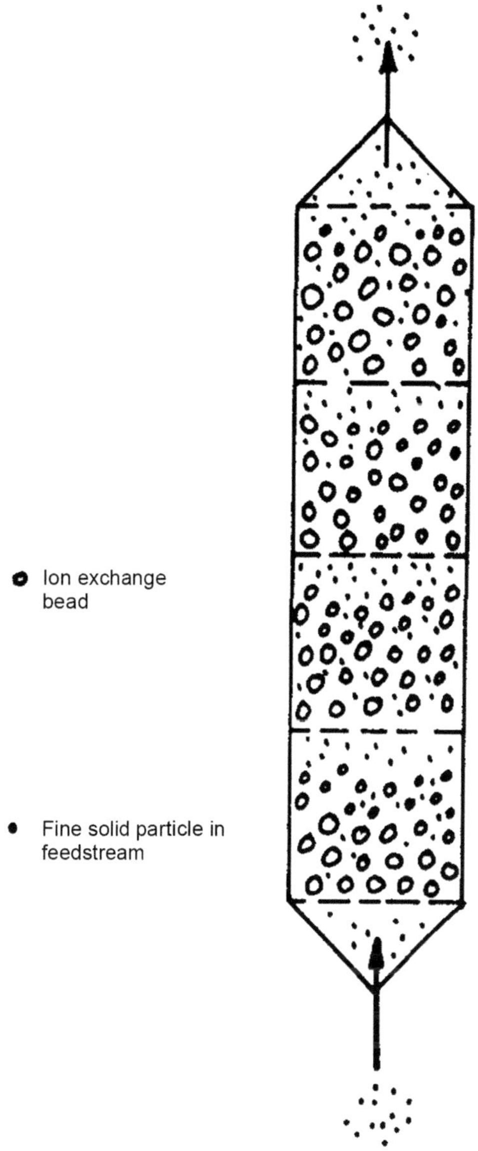

FIGURE 16.37 The perforated plates in this fluidized bed column allow fine solid particles to flow through the column while the ion exchange beads are, for the most part, confined to individual compartments [95].

mixing means that breakthrough occurs sooner and the degree of resin capacity utilization is much lower than in packed bed columns. By placing perforated plates in a column (Figure 16.37) [95], the resin beads only mix within a restricted area allowing more complete utilization of the ion exchange resin's capacity.

Smaller continuous fluid bed systems, like the one shown in Figure 16.38 [96], have been developed which operate with a high concentration of ion exchange resin and suspended solids. These units are 80% smaller than the conventional resin-in-pulp plants of the type which are used in the treatment of uranium ore slurries [97]. The pilot plant unit, which would probably be the size needed for processing commercial fermentation broths, had dimensions per contact chamber of 0.82 m × 0.82 m with a fluid bed height of 0.82 m and an additional 0.16 m for free board. The unit has been successfully operated with 25 to 50% resin and up to 45% suspended solids.

The effect of the degree of regeneration of the resin on the degree of extraction of a solute was measured by Slater [98] using a seven stage unit. The results are shown in Figure 16.39 for the extraction of uranium using a fluidized bed slurry of a strong base resin in a 10% uranium ore leach slurry.

However, the fluidized bed column, even with perforated plates separating it into as many as 25 compartments, may not be appropriate for applications in which there are strict requirements (<1% of influent concentration) on the effluent. For readily exchangeable ions, the optimum utilization of this technique occurs when an actual effluent concentration of 5% of the influent is the

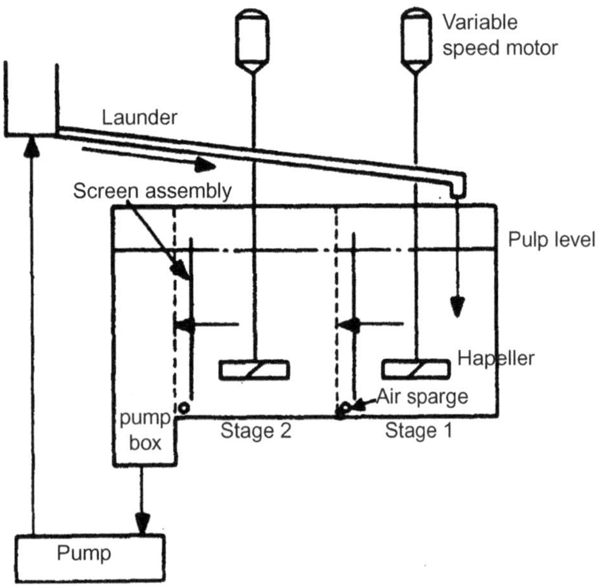

FIGURE 16.38 Pilot plant for resin-in-pulp contactor unit [96].

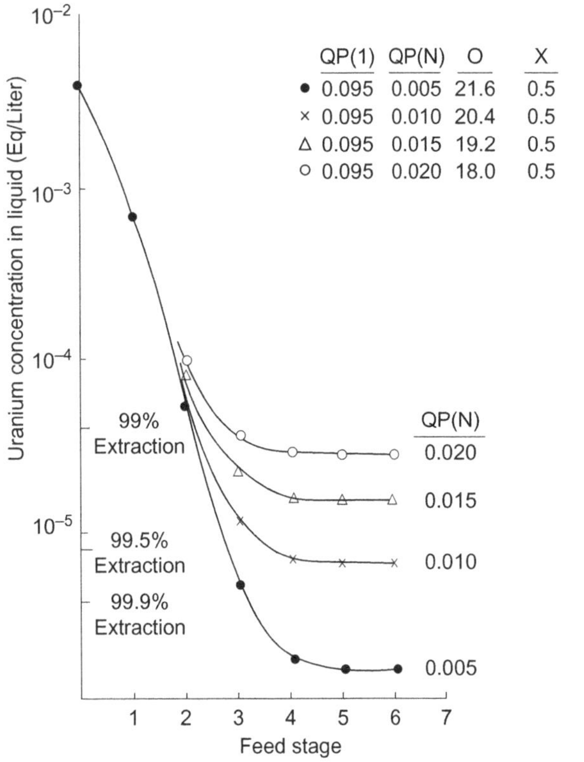

FIGURE 16.39 Effect of residual feedstream ions on the resin on the efficiency of a staged system. *QP(1)* is the concentration of the feedstream ions on the resin at the end of the cycle; *QP(N)* is the concentration of the feedstream ions on the resin at the beginning of the cycle; *Q* is the cycle time; *X* is the fraction of the resin bed volume in a stage which is removed per cycle [98].

breakthrough point. At such times, the ion exchange resin capacity would be 70% utilized [95]. Should it be acceptable that the breakthrough point occur when the average effluent reaches a concentration of 10% of the influent, the utilization of the ion exchange capacity is 90%. If the ions are not readily exchangeable (low selectivity), the resin utilization would be significantly less and fluidized bed operations should not be used.

6.4 Elution/Regeneration

Elution of proteins from ion exchangers can be achieved with buffers containing salts such as sodium chloride or ammonium acetate or by an appropriate pH change, provided that the pH change does not result in denaturation of the eluted protein [99]. The elution may be performed with a series of stepped changes or with a continuous gradient change in the eluting power of the eluant. With such changes, it is possible to separate different proteins or protein fractions from each other based on their different affinities for the ion exchange resin.

Elution of compounds such as penicillin with either acids or bases will render the penicillin inactive. Although aqueous salt solutions can elute penicillin without inactivating it, the large volumes required make this option impractical. Wolf and co-workers [100] developed an elution solvent combination of organic solvent, water and salt that can elute the penicillin with a minimum volume and no inactivation. The mixture of organic solvent and salt is chosen so that the salt is soluble in the resulting organic solvent-water mixture and the organic substance eluted from the resin is soluble in the elution mixture. Table 16.19 shows the elution volume required to recover the indicated amount of antibiotic when the elution solvent is 70% methanol and 5% or 7.5% ammonium chloride in water.

Regeneration alone is not sufficient to prevent fouling or microbial growth on ion exchange resins. If the resin is left standing in the regenerant during nonoperating times, it is possible to suppress the microbial growth [101]. The regenerant in this instance was 10 or 20% NaCl. When the initial microbe count was 10 per milliliter, at the end of three weeks in 20% NaCl, the count had risen to just 800/ml compared to 200,000/ml for the resin stored in water. When an alternate regenerant is used (NaOH or HCl), it is preferable to change the storage medium to 20% NaCl since extended time in an acid or base media can adversely affect the resin matrix.

Since most fluids contain some suspended matter, it is necessary to backwash the resin in the fixed bed column on a regular basis to remove any accumulation of these substances. To carry out a backwashing operation,

TABLE 16.19 Amount of Antibiotic Recovered with Increasing Volumes of Methanol in Aqueous Ammonium Chloride [100]

Antibiotic	Eluant	Total Volume of Eluant (Bed Volumes)	Amount of Antibiotic Recovered (%)
Dihydronovobiocin	70% MeOH with 5% NH_4Cl	1	50.0
		2	83.3
		3	93.3
		4	96.6
Novobiocin	70% MeOH with 5% NH_4Cl	3.5	99.8
Penicillin	70% MeOH with 7.5% NH_4Cl	0.5	11.0
		1.0	69.0
		1.5	94.0
		2.0	97.1
		2.5	99.7

a flow of water is introduced at the base of the column. The flow is increased to a specific rate to classify the resin hydraulically and remove the collection of suspended matter. Figures 16.40 and 16.41 show the types of flow rate which provide certain degrees of expansion of cation and anion resins, respectively. Since the anion resins are significantly less dense than the cation resin shown, it would be necessary to have different amounts of freeboard above the normal resin bed height so that backwashing may be accomplished with only a negligible loss of ion exchange resin. Typically, an anion resin bed may be expanded by 100% during backwashing, while a cation resin bed will only be expanded by 50%.

It is also necessary that the water used for the backwashing be degassed prior to use. Otherwise, resin particles will attach themselves to gas bubbles and be carried out of the top of the column to give an unacceptable increase in resin losses.

When treating fermentation broth filtrates, frequent backwashing of the resin bed is necessary to prevent accumulation of suspended matter. In such cases, the column height should be designed of such a size that the bed is regenerated at least every 10 hours. Shorter columns have been designed to be regenerated at least every hour

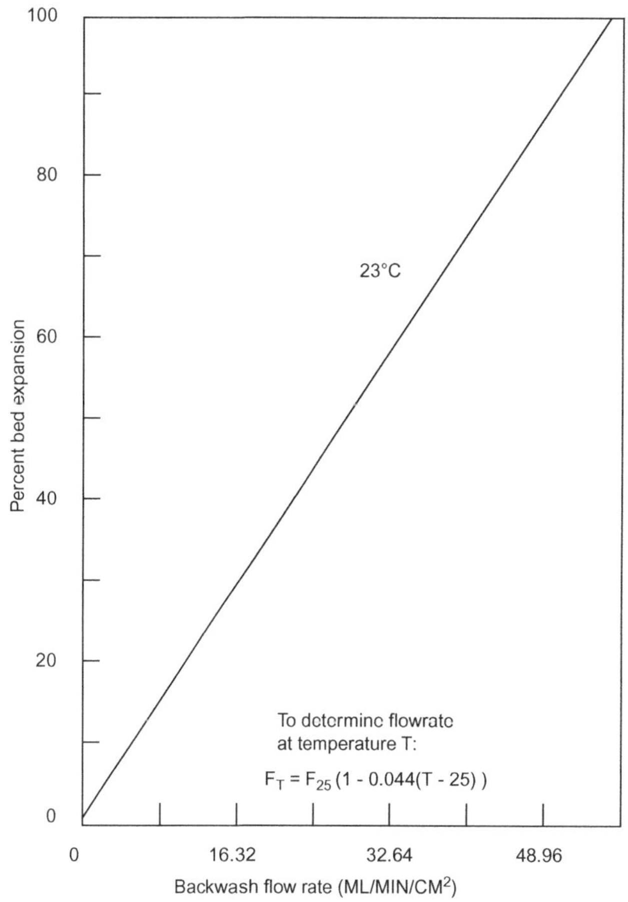

FIGURE 16.40 Backwash expansion characteristics of a macroporous strong acid cation resin, Dowex 88.

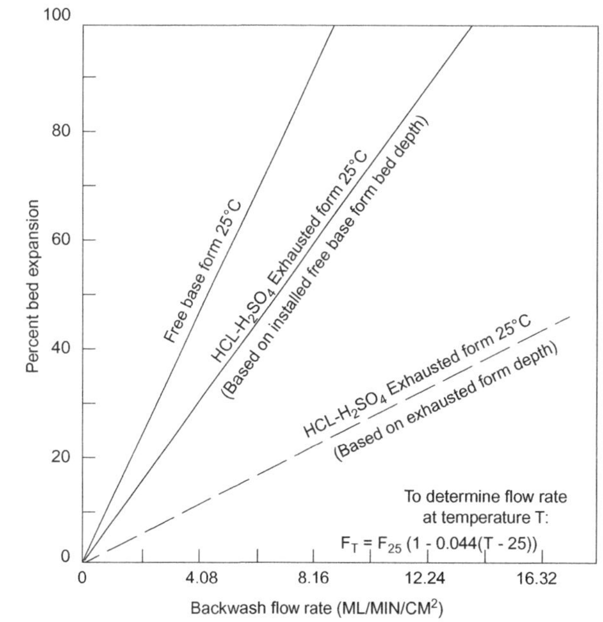

FIGURE 16.41 Backwash expansion characteristics of regenerated and exhausted macroporous weak base anion resin, Dowex 66.

[102]. These shorter beds can then use finer resins and achieve a high level of efficiency with lower capital costs. This may be taken to the point of using very fine resins, as with the Powdex® system [103] which discards the powdered resin after a single use.

After a bed is backwashed, unless it is air-mixed as the level of water is drained down to the surface, the beads or particles classify according to size. The fine beads end up on top and the large beads on the bottom of the column. In cocurrent operations, the regenerant first contacts the top of the bed. The fast kinetics of the fine particles gives a high regeneration efficiency. However, the large beads on the bottom will regenerate more slowly and may end up only partially regenerated. Thus, when the feedstream is next passed through the resin bed, leakage of undesirable ions may occur from the large beads in the bottom of the column. This may be overcome by using a countercurrent flow arrangement described earlier or by using an air-mixing system during the post-backwash draining [104].

7.0 INDUSTRIAL CHROMATOGRAPHIC OPERATIONS

Packed bed, process scale high performance liquid chromatography (HPLC) equipment was first introduced by Millipore and Elf Aquitaine in 1982 [105]. Since then, several companies, listed in Table 16.20, have entered the large scale HPLC market. The systems use packed beds at moderate pressures (30–140 bar). While there are substantial time-savings in using these systems compared to other purification techniques, the short life of the packing material and its high cost continue to restrict this technique to applications that warrant the $100/kg separation cost.

The Waters Kiloprep Chromatography pilot plant is one example of the successful extension of an analytical chromatography process to the process scale. The ability to control the various operation parameters to scaleup directly from the laboratory to the pilot plant and beyond to commercial production has been developed [106]. Figure 16.42 illustrates how the performance of this larger system can be predicted from the data generated in an equivalent laboratory apparatus.

Voser and Walliser [74] have described the approaches different companies have devised so that fine and soft adsorbents may be utilized in large scale chromatography operations. The scaling-up has usually been achieved by increasing the column diameter and using stack columns. The problem then becomes one of achieving uniform distribution of the feed solution over the entire resin bed surface, particularly when gradient elution is involved.

TABLE 16.20 Manufacturers of Process Scale HPLC Equipment

Amicon	Danvers, MA
FldSmith (Dorr-Oliver)	Midvale, UT
Elf Aquitaine (Varex in U.S.)	Rockville, MD
Millipore Corporation	Bedford, MA
GE Healthcare (Pharmacia)	Piscataway, NJ
Separations Technology	Wakefield, RI
Y M C	Morris Plains, NJ

In the Pharmacia approach, the fluid input is split and distributed through six ports on the column end plates [107]. At the entrance of each stream, an anti-jetting device spread the liquid over a fine mesh net. A coarse net between the net and the end plate acts as both a support and a spacer. For the soft Sephadex gels G-50 to G-200, the maximum feasible bed height is 15 cm. Scaleup operations have used as many as six such squat columns in series [108]. The drawbacks of this approach are the cumbersome adsorbent in the column.

In Amicon/Wright columns, the flow is distributed through a carefully designed system of radial ribs cut out of the end plate. There is a single central port with a suitable anti-jetting device to reduce and divert the high velocity of the entering stream. The adsorbent bed is covered with a sintered plate. Sintered plates are claimed to be more efficient than most nets to achieve an even distribution. The larger the pore size of the sintered plate, however, the less efficient the system. Drawbacks of sintered plates are their tendency to adsorb substances on their very large surfaces and the possibility of fouling. These columns may also be stacked.

Whatman has developed a column with a new flow distribution system specifically designed for Whatman's cellulosic ion exchange resins. The bed height is 18 cm. The diameter of the first commercial unit was 40 cm and contained 25 liters of adsorbent. The resin bed is covered with a perforated plate with a high free surface. The slightly conical head plate covers an empty space and has a steep cone in the middle. The total empty head space is about 5% of the bed volume. The feed solution enters the steep cone tangentially in its upper part. The resulting rotary movement efficiently mixes the supernatant liquid and allows gradient elution. It is claimed that filling and equilibration take only one hour.

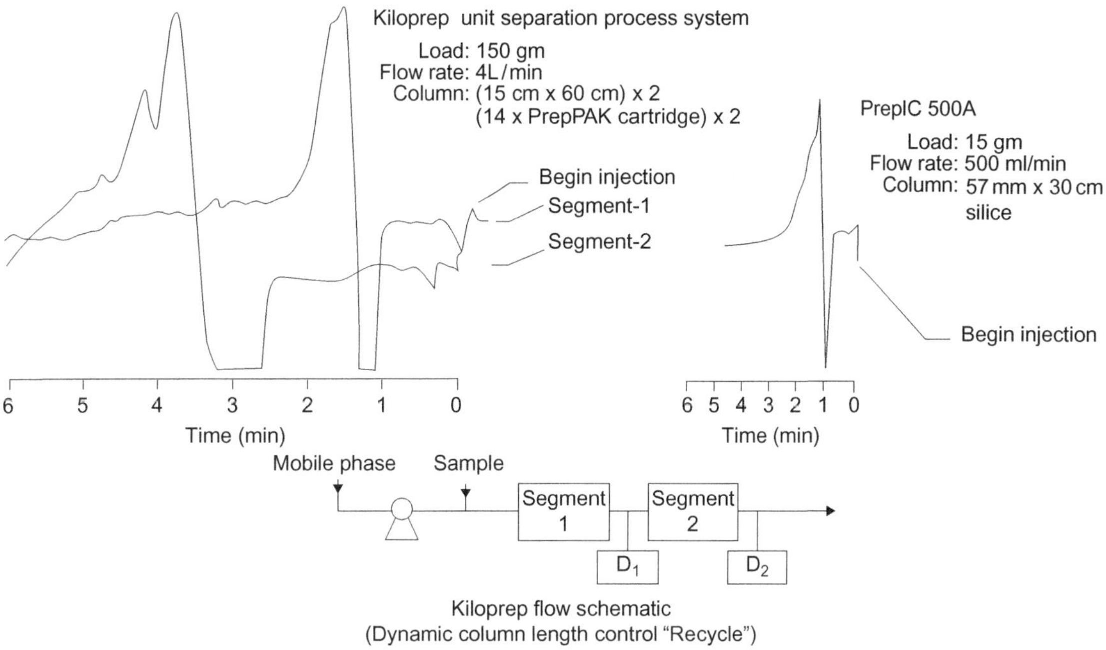

FIGURE 16.42 Comparison of column performance at equivalent loading [106].

AMF has developed an unconventional new approach with its ZETA-PREP cartridge. The cartridges consist of concentric polymer screens which bear the ionic groups and are supported by cellulosic sheets. The flow is radial from the outer rim toward a perforated central pipe. The available nominal cartridge lengths are from 3 cm to 72 cm with a constant diameter of about 7 cm throughout. Scaling-up with this approach is quite straightforward. Single cartridges, each mounted in a housing, can be combined to a multi-cartridge system. For such a system, flow rates up to 12 liter/min and bovine serum albumin capacities of 1400 g are claimed. The present ion exchange functionalities available are DEAE, QAE and SP.

The step-wise transition from high pressure liquid chromatography to medium pressure chromatography, such as described for the preparation of pectic enzyme [109], illustrate the progression toward large scale industrial application of the techniques developed in analytical laboratories. The pressure in these medium pressure chromatography applications is only 6 bar instead of the 100 to 150 bar associated with HPLC. The lower pressure results in longer processing times (about one hour) compared to the 5 to 20 minutes required for an analytical determination with HPLC.

Studies, such as the one by Frolik and coworkers [110], which examine the effect and optimization of variables in HPLC of proteins, can be expected to contribute to the implementation of this type of protein resolution technique into future commercial biotechnology processes.

The first chromatographic systems capable of handling more than 100 kg/day were merely scaled up versions of laboratory chromatography [111,112]. Even with some of these systems it was necessary to recycle a portion of the overlap region to have an economical process. A typical example of such a system would be the Techni-Sweet System of Technichem [113] used for the separation of fructose from glucose. The unique distributors and recycle system are designed to maximize the ratio of sugar volume feed solution per unit volume of resin per cycle while at the same time minimizing the ratio of volume of water required per unit volume of resin per cycle.

The flow through the Technichem system is 0.56 m^3/(hr-m^3) with a column height of 3.05 m. The feed solution contains 45% dissolved solids, and a feed volume equal to 22% of the volume of the resin is added to the column each cycle. The rinse water added per cycle is equal to 36% of the volume of the resin. This is much less rinse water than the 60% volume that was required by the earlier systems.

This technique is known as the *stationary port* technique since the feed solution and the desorbent solution are always added at the same port and the product streams and the recycle stream are always removed from another port. Technichem and Finn Sugar manufacture chromatography systems which utilize the stationary port technique.

One of the earlier attempts [114] at industrial chromatography used an adaptation of the Higgins contactor for the ion exclusion purification of sugar juices. The physical movement of the low cross-linked resin caused attrition as it was moved around the contactor. It was also difficult to maintain the precise control needed on flow rates because of the pressure drop changes and volume changes of the resin as it cycled from the mostly water zone to the mostly sugar solution zone.

An alternate approach [115] utilizes *moving port* or *pseudo-moving bed* techniques. With these techniques, the positions on the column where the feed solution is added and where the product streams are removed are periodically moved to simulate the countercurrent movement of the adsorbent material. At any given time the resin column can be segmented into four zones (Figure 16.43). Zone 1 is called the adsorption zone and is located between the point where the feed solution is added and the point where the fast or less strongly adsorbed component is removed. In this zone the slow or more strongly adsorbed component is completely adsorbed onto the ion exchange resin. The fast component may also be adsorbed, but to a much smaller extent. The second zone, Zone 2, is the purification zone and is located between the point where the fast component is removed and the point where the desorbent solution is added. Zone 3 is called the desorbent zone and is between the point where the desorbent is added and the point where the slow component is removed. In this zone the slow component is removed from the resin and exits the column. The final zone, Zone 4, is called the buffer zone and is located between the point where the slow component is removed and the point where the feed solution is added. There is a circulating pump which unites the different zones into a continuous cycle.

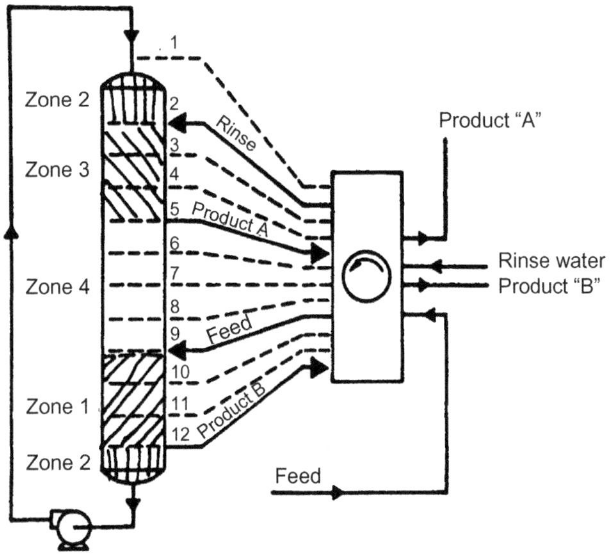

FIGURE 16.43 Moving port chromatographic column with four zones for continuous chromatographic separation [115].

Different sections of the column serve as a specific zone during the cycle operation. Unlike the stationary port technique, the liquid flow is not uniform throughout the column. Because of the variations in the additions and withdrawals of the different fluid streams, the liquid flow rate in each of the zones will be different.

With such a system, one must slowly develop the chromatographic distribution pattern through the different zones. It may take from 8 to 36 hours for the pattern to be established. Other practical considerations are that the recirculation system must represent a small ($< 10\%$) portion of a single zone to prevent unacceptable back-mixing which would alter the established chromatographic pattern.

The flow rate and the pressure drop per unit length of the chromatographic column are much lower for the stationary port compared to the moving port system. Also, the moving port system is much less capital intensive. The moving port technique, however, is calculated to require only one-third of the column volume and ion exchange volume and two-thirds of the desorbent volume compared to the stationary port technique.

After the expiration of the UOP patent covering the rotary valve, there have been several modifications to the moving port technique by Amalgamated Sugar [116], Illinois Water Treatment [117], and Mitsubishi [118]. Each manufacturer has its own proprietary approach for the establishment and control of the chromatographic pattern.

Wankat [119] proposed a hybrid system which has some of the characteristics of both elution chromatography and the pseudo-moving bed system. During the feed pulse, the feed position was moved continuously up into the column at a velocity that lies between the two solute velocities. The eluting solvent was continuously fed into the bottom of the column. Elution development with solvent was used when the feed pulse was over. This method reduces irreversible mixing of solutes near the feed point. Wankat and Ortiz [120] have used this system for gel permeation chromatography and claim improved resolution, narrower bands and higher feed throughputs compared to conventional systems. McGary and Wankat [121] have had similar results applying it to preparative HPLC. This technique uses less adsorbent and produces more concentrated products compared to normal preparative chromatography, but more adsorbent and less concentrated products than pseudo-moving bed systems. Wankat [122] has proposed that his system will be of most value for intermediate size applications or when only one product is desired.

The key items to identify when considering an industrial chromatographic project are the capital for the equipment, yield and purity of the product, the amount of dilution of the product and waste stream, the degree of flexibility the computer controls allow, the expected life of the ion exchange material and whether the equipment allows for periodic expansion of the resin.

New techniques are continuing to be developed which can be expected to be used in future specialized industrial applications. Multi-segmented columns have been demonstrated for the preparative purification of urokinase [123]. Begovich and coworkers [124,125] have developed a technique for continuous spiral cylinder purifications which allow separation of the basis of electropotential in addition to the selective affinity of the adsorbent resin for the components in solution. A schematic of this device is shown in Figure 16.44.

Another new technology that offers promise for commercial biotechnology purifications is the use of parametric pumping with cyclic variations of pH and electric field. This has been described by Hollein and coworkers [126]. They worked with human hemoglobin and human serum albumin protein mixtures on a CM-Sepharose cation exchanger. The extensive equations they reported for parametric separations allow analysis of other systems of two or more proteins which may be candidates for this type of separation.

Applications of ion exchange and column chromatography techniques have been incorporated into the commercial purification scheme for fermentation products, biomaterials and organic chemicals. While the majority of these applications are on the small scale (less than 500 kg/month but greater than 10 g/month), several large industrial scale applications have arisen in the last decade. The extraction of sugar from molasses, the separation of glucose from fructose, the separation of polyhydric alcohols, the separation of xylene isomers and the separation of amino acids are carried out in industrial scale operations preparing thousands of metric tons of purified material each year. Two recent books [127,128] provide extensive examples of these applications. Additional examples, mostly of laboratory studies, are available in books specifically on the HPLC of peptides and proteins [129,130]. *LC-GC* and *Chromatography* are two periodicals with helpful operational suggestions.

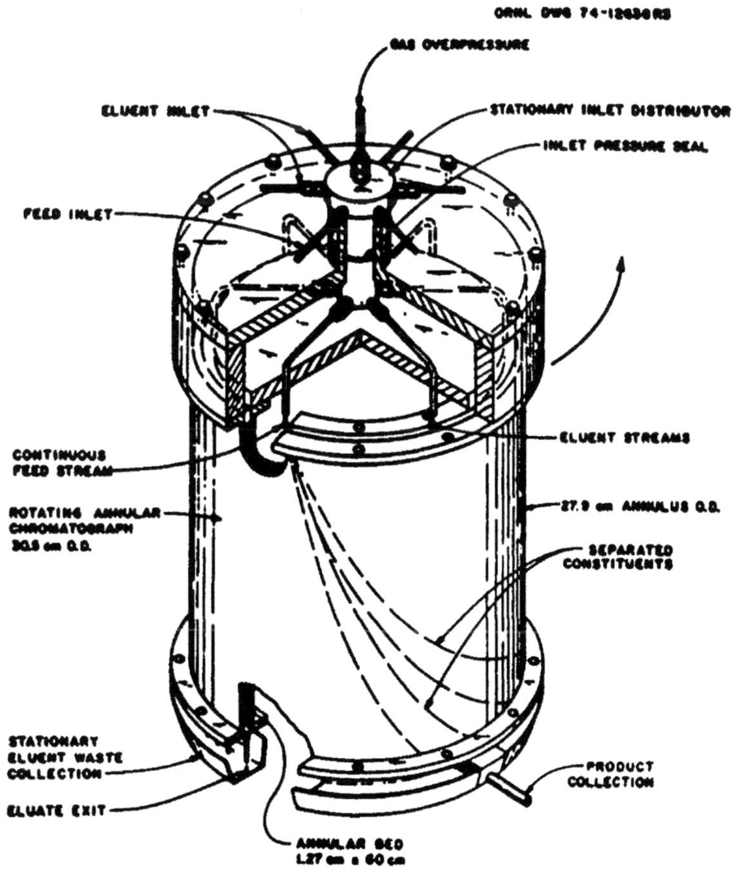

FIGURE 16.44 Schematic of the pressurized continuous annular chromatograph [124].

REFERENCES

[1] H.S. Thompson, J. Roy, Agr. Soc., Eng. 11 (1850) 68.
[2] T. Bersin, Naturwissenschaften 33 (1946) 108.
[3] J.C. Winters, R. Kunin, Ind. Eng. Chem. 41 (1949) 460.
[4] B.D. Polis, O. Meyerhoff, J. Biol. Chem. 169 (1947) 389.
[5] J.F. Carson, W.D. Maclay, J. Am. Chem. Soc. 67 (1945) 1808.
[6] S. Nagai, K. Murakami, J. Soc. Chem. Ind., Jpn. 44 (1941) 709.
[7] T. Wieland, Berichte 77 (1944) 539.
[8] M.S. Bergdoll, D.M. Doty, Ind. Eng. Chem., Ancl. Ed., 18, 1946.
[9] A Lejwa, Biochem. Z. 256 (1939) 236.
[10] E.C. Cruz, F. Gonzales, W. Hulsen, Science 101 (1945) 340.
[11] W.G. Jackson, G. Whitefield, W. DeVries, H. Nelson, J. Evans, J. Am. Chem. Soc. 73 (1951) 337.
[12] S. Moore, W.H. Stein, J. Biol. Chem. 211 (1954) 893.
[13] A Puente, Microbiol. Espan. 14 (1961) 209.
[14] M. Ishida, Y. Sugita, T. Hori, K. Sato, U. S. Patent No. 3,565,951 (Feb. 23, 1971).
[15] S.V. Gordienko, J. Appl. Chem. 39 (1966) 10 USSR
[16] S. Shirato, Y. Miyazaki, I. Suzuki, Ferment. Ind. 45 (1) (1967) 60 Japan
[17] R.M. Wheaton, W.C. Bauman, Ann. N.Y. Acad. Sci. 57 (1953) 159.
[18] J.A Dean, Chemical Separation Methods, Van Nostrand, New York, 1969. p. 295.
[20] R. Kunin, Ion Exchange Resins, Robert E. Krieger Publ. Co., Huntington, NY, 1972. p. 26.
[21] L. Wiklander, Ann. Roy. Agr. Coll., Sweden 14 (1946) 1.
[22] G.E. Boyd, J. Schubert, A.W. Adamson, J. Am. Chem. Soc. 69 (1947) 2818.
[23] H.S. Harned, B.B. Owen, The Physical Chemistry of Electrolytic Solutions, Reinhold Publishing Corp, 1943.
[24] H.P. Gregor, J. Am. Chem. Soc. 70 (1948) 1293.
[25] W. Argersinger, A Davidson, O. Bonner, Trans. Kansas Acad. Sci. 53 (1950) 404.
[26] F.J. Wolf, Separation Methods in Organic Chemistry and Biochemistry, Academic Press, New York, 1969. p. 144.
[27] S. Peterson, Ann. N. Y. Acad. Sci. 57 (1953) 144.
[28] T. Iko, Kino Zairyo 2 (5) (1983) 8.
[29] F.G. Helfferich, M.S. Plesset, J. Chem. Phys. 28 (1958) 418.
[30] R. Schogl, F.G. Helfferich, J. Chem. Phys. 26 (1957) 5.
[31] E. Glueckauf, J.I. Coates, J. Chem. Soc. (Lond.) (1947) 1315.
[32] F.G. Helfferich, L. Liberti, D. Petruzzelli, R. Passino, Israel J. Chem. 26 (1985) 3.
[33] F.N. Tsai, J. Phys. Chem. 86 (1982) 2339.
[34] T. Vermeulen, G. Klein, N.K. Hiester, in: R.H. Perry, C.H. Chilton (Eds.), Perry's Chemical Engineers' Handbook, Sec. 16, McGraw-Hill, New York, 1973.
[35] F.G. Helfferich, Ion Exchange, McGraw-Hill, New York, 1962. p. 255.
[36] G.V. Samsonov, G.E. Elkin, Ion Exch. Solv. Extr. 9 (1985) 211.
[37] S.K. Friedlander, A. I. Ch. E. J. 3 (1957) 381.
[38] J.M. Marchello, M.W. Davis Jr., I EC Fund. 2 (1) (1963) 27.
[39] J.N. Wilson, J. Am. Chem. Soc. 62 (1940) 1583.
[40] D. DeVault, J. Am. Chem. Soc. 65 (1943) 532.
[41] E. Glueckauf, J. Chem. Soc. 69 (1947) 1321.
[42] E. Glueckauf, Discuss. Faraday Soc. 7 (1949) 42.
[43] A.J.P. Martin, R.L.M. Synge, Biochem. J. 35 (1941) 1358.
[44] S.W. Mayer, E.R. Thompkins, J. Am. Chem. Soc. 69 (1947) 2866.
[45] D.W. Simpson, R.M. Wheaton, Chem. Eng. Prog. 50 (1) (1954) 45.
[46] E. Glueckauf, Trans. Faraday Soc. 51 (1955) 34.
[47] J. Beukenkamp, W. Rieman III, S. Lindenbaum, Anal. Chem. 26 (1954) 505.
[48] D.W. Simpson, W.C. Bauman, Ind. Eng. Chem. 46 (1954) 1958.
[49] J.A. Dean, Chemical Separation Methods, Van Nostrand Co., New York, 1969. p. 60.
[50] B.L. Karger, J. Chem. Educ. 43 (1966) 47.
[51] A. Clearfield, G.H. Nanocollas, R.H. Blessing, Ion Exchange and Solvent Extraction, in: J.A. Marinsky, Y. Marcus (Eds.), vol. 5, 1973, p. 1.
[52] S.L. Daniels, in: G. Britton, K.C. Marshall (Eds.), Adsorption of Microorganisms to Surfaces, John Wiley and Sons, New York, 1980.
[53] H. Gold, C. Calmon, A. I. Ch. E. Symp. Ser. 76 (192) (1980) 60.
[54] A Warshawsky, Die Angewandte Makromol. Chem. 109/110 (1982) 171.
[55] K. Yotsumoto, M. Hinoura, M. Goto, Ger. Offen. 2 (718) (1977) 649.
[56] A Warshawsky, M. Fidkin, M. Stern, J. Polym. Sci. Chem. 20 (6) (1982) 1469.
[57] D.H. Freeman, S.B. Schram, Anal. Chem. 53 (1981) 1235.
[58] K.M. Sladaze, T.K. Brutskus, Teploenergetika 23 (9) (1976) 6.
[59] J.R. Millar, D.G. Smith, W.E. Merr, T.R.E. Kressman, J. Chem. Soc. 183 (1963).
[60] M. Mindick, J. Svarz, U. S. Patent No. 3,549,562 (Dec. 22, 1970).
[61] R.A Messing, Immobilized Enzymes for Industrial Reactors, Academic Press, New York, 1975.
[62] T. Okuda, S. Awataguchi, U. S. Patent No. 3,718,742 (Feb. 27, 1973).
[63] T. Crispin, I. Halasz, J. Chromatogr. 239 (1982) 351.
[64] F. Martinola, A Meyer, Ion Exch. Membr. 2 (1975) 111.
[65] G. Sememza, Chimica 14 (1960) 325.
[66] A.E. Rodrigues, in: L. Liberti, F.G. Helfferich (Eds.), Mass Transfer and Kinetics of Ion Exchange, Martinus Nijhoff, The Hague, 1983, p. 259.
[67] R. Passino, in: L. Liberti, F.G. Helfferich (Eds.), Mass Transfer and Kinetics of Ion Exchange, Martinus Nijhoff, The Hague, 1983, p. 313.
[68] G. Boari, L. Liberti, C. Merli, R. Passino, Env. Prot. Eng. 6 (1980) 251.
[69] L. Liberti, R. Passino, Ind. Eng. Chem., Pros. Des. Dev. 21 (2) (1982) 197.
[70] M.J. Slater, Effluent Water Treat. J. 461 (1981).
[72] C. Snowdon, J. Turner, Proceedings of the International Symposium on Fluidization, Netherlands University Press, Amsterdam, 1967. p. 599.
[73] G. Sofer, C. Mason, Biotechnology 5 (3) (1987) 239.
[74] W. Voser, H.P. Walliser, in: M.S. Verrall (Ed.), Discovery and Isolation of Microbial Products, Ellis Horwood Ltd., Chichester, UK, 1985, p. 116.
[75] S.E. Charm, C.C. Matteo, R. Carlson, Anal. Biochem. 30 (1969) 1.
[76] Laboratory Methods of Column Packing, Whatman Data Sheet, Reeve-Angel Co., Clifton, New Jersey 07014 (1968).
[77] C. Stacy, R. Brooks, M. Merion, J. Anal. Purif. 2 (1) (1987) 52.

[78] S.R. Rudge, M.R. Ladisch, in: J.A. Asenjo, J. Hong (Eds.), Separation, Recovery and Purification in Biotechnology, American Chemical Society, Washington, DC, 1986. p. 122.
[79] G. Sofer, C. Mason, Biotechnology 5 (3) (1987) 239.
[80] S. Ergun, Chem. Eng. Progr. 48 (1952) 89.
[81] P. Pelosi, J. McCarthy, Chem. Eng. (1982) 125.
[82] L.F. Wirth Jr., C.A. Feldt, K. Odland, Ind. Eng. Chem. 58 (1961) 639.
[83] B.A Bolto, R.J. Eldrige, K.H. Eppringer, M.B. Jackson, React. Polym. 2 (1984) 5.
[84] J. Kysela, J. Brabec, Jad. Energ. 27 (12) (1981) 445.
[85] B.N. Khodyrev, A.F. Prokhorov, Teploenergetika 25 (3) (1978) 60.
[86] Commercial inert resins for this application include Dowex Buffer Beads (XFS-43179), Ambersep 359 and Duolite S-3TR.
[87] I.R. Higgins, U. S. Patent No. 3,580,842 (May 25, 1971).
[88] A. Himsley, Canadian Patent No. 980,467 (Dec. 23, 1970).
[89] H. Brown, U. S. Patent No. 3,549,526 (Dec. 22, 1970).
[90] F.L.D. Cloete, M. Streat, U. S. Patent No. 3,551,118 (Dec. 29, 1970).
[91] M.J. Streat, Separ. Proc. Technol. 1 (3) (1980) 10.
[92] C. Mallon, M. Richter, Zfl-Mitteilungen, Leipzig 86 (1984) 39.
[93] F.L.D. Cloete, M. Streat, British Patent No. 1,070,251 (1972).
[94] D.R. George, J.R. Ross, J.D. Prater, Min. Eng. 1 (1968) 73.
[95] A. Buijs, J.A Wesselingh, Polytech. Tijdschrift Procestechniek 36 (2) (1981) 70.
[96] D. Naden, G. Willey, E. Bicker, D. Lunt, in: D. Nader, M. Streat (Eds.), Ion Exchange Technology, Horwood, Chichester, UK, 1984, p. 690.
[97] D. Naden, M.R. Bandy, Presented at SCI Meeting, Impact of SX and IX on Hydrometallurgy, University of Slaford, UK, 1978.
[98] M.J. Slater, Can. J. Chem. Eng. 52 (1974) 43.
[99] H.A. Chase, in: D. Nader, M. Streat (Eds.), Ion Exchange Technology, Horwood, Chichester, UK, 1984.
[100] F.J. Wolf, I. Putter, G.V. Downing Jr., J. Gillin, U. S. Patent No. 3,221,008 (Nov. 30, 1965).
[101] H.C. Flemming, Vom Wasser 56 (1981) 215.
[102] Powdex® is a registered tradename of the Graver Water Treatment Company.
[103] I.H. Spinnerr, P.J. Simmons, C.J. Brown, Proceedings of the 40th International Water Conference, Pittsburgh, 1979.
[104] C. Calmon, A. I. Ch. E. Symp. Ser. 80 (233) (1984) 84.
[105] A.H. Heckendorf, Chem. Proc (1987) 33.
[106] A.H. Heckendorf, E. Ashare, C. Rausch, in: D. LeRoith, J. Shiloach, T.J. Leahy (Eds.), Purification of Fermentation Products, American Chemical Society, Washington, DC, 1985, p. 91.
[107] J.C. Janson, P. Dunnill, Industrial Aspects of Biochemistry, Americal Elsevier, New York, 1974. p. 81.
[108] G.J. Strobel, Chemie Technik 11 (1982) 1354.
[109] L. Rexova-Benkova, J. Omelkova, O. Mikes, J. Sedlackova, J. Chromatogr. 238 (1982) 183.
[110] C.A. Frolik, L.L. Dart, M.B. Sporn, Anal. Biochem. 125 (1982) 203.
[111] D. Gross, Proceedings of the 14th General Assembly CITS, 445 (1971).
[112] F.F. Sutthoff, W.J. Nelson, U. S. Patent No. 4,022,637 (May 10, 1977).
[113] H.W. Keller, A.C. Reents, J.W. Laraway, Starch/Staerke 33 (1981) 55.
[114] L. Norman, G. Rorabaugh, H. Keller, J. Am. Soc. Sugar Beet Tech. 12 (5) (1963) 363.
[115] D.B. Broughton, C.G. Gerhold, U. S. Patent No. 2,985,589 (May 23, 1961).
[116] For more information contact Schoenrock, K., Amalgamated Sugar, Ogden, Utah.
[117] D.J. Burke, Presented at A. I. CH. E. 23rd Annual Symposium (May 12, 1983).
[118] H. Ishikawa, H. Tanabe, K. Usui, Japanese Patent No. 102,288 (Aug. 11, 1979).
[119] P.C. Wankat, Ind. Eng. Chem. Fundam. 16 (1977) 468.
[120] P.C. Wankat, P.M. Ortiz, Ind. Eng. Chem. Process Des. Dev. 21 (1982) 416.
[121] R.S. McGary, P.C. Wankat, Ind. Eng. Chem. Fundam. 22 (1983) 10.
[122] P.C. Wankat, Ind. Eng. Chem. Fundam. 23 (1984) 256.
[123] L.J. Novak, P.H. Bowdle, U. S. Patent No. 4,155,846 (May 22, 1979).
[124] J.M. Begovich, C.H. Byers, W.G. Sisson, Sep. Sci. Technol. 18 (12 &13) (1983) 1167.
[125] J.M. Begovich, W.G. Sisson, A. I. Ch. E. J. 30 (1984) 705.
[126] H.C. Hollein, J. Ma, C. Huang, H.T. Chen, Ind. Eng. Chem. Fundam. 21 (1982) 205.
[127] F. Dechow, Separation and Purification Techniques in Biotechnology, Noyes Publications, Park Ridge, New Jersey, 1989.
[128] F. Dechow, in: K. Dorfner (Ed.), Ion Exchangers, Walter de Greyter, Berlin/New York, 1991. Ch. 2.12 and Ch. 3.2.
[129] C. Horvath (Ed.), High Performance Liquid Chromatography, vol. 3, Academic Press, New York, 1983.
[130] R.J. Hamilton, P.A Sewell, Introduction to High Performance Liquid Chromatography, second ed., Chapman and Hall, London, 1982.

Part V

Plant Operations

Chapter 17

Water Systems for Pharmaceutical Facilities

Mark Keyashian

1.0 INTRODUCTION

Common, everyday water is a major consideration in a pharmaceutical plant. The final product or any of its intermediate materials can only be as contaminant-free as the water available at that stage. Water may be an ingredient or used principally to wash and rinse product contact components and equipment. Water is also used to humidify the air, to generate clean steam for sterilization, to cool or heat, as a solvent, for drinking and sanitary uses, etc. To better control this critical media, the pharmaceutical industry has defined two additional types of water: *purified water* and *water for injection*, both of which are highly regulated. Special attention to a good understanding of the water systems in a pharmaceutical facility is essential.

2.0 SCOPE

This chapter is an overview of the various water systems used in a pharmaceutical facility. It will help bring about a better understanding of how they are generated, stored and distributed and what equipment is involved. Starting with raw water as it is sourced, this chapter will:

1. Take the reader step-by-step through various treatments to generate different types of water.
2. Outline applicable cGMPs (current Good Manufacturing Practices)
3. Point out some potential pitfalls to watch for during installation and start-up.

In addition, for a better all around understanding, an overview of how these systems are designed and some of the more important design parameters will be discussed.

3.0 SOURCE OF WATER

Water supply to the plant is ground water (wells), surface water (lakes, rivers), or city water. Raw water is typically contaminated with salts, oils, various organic substances, calcium, clay, silica, magnesium, manganese, aluminum, sulfate, fertilizers, ammonia, insecticides, carbon dioxide and, of course, bacteria and pyrogens. A city water treatment plant removes most of these impurities, but adds chlorine or chloramines and fluoride. Table 17.1 summarizes the level of contaminants by type of raw water.

Regardless of the source, the first step in knowing the water supply or designing a system is to obtain a complete analysis of the supply water. Table 17.2 is an example water analysis. Please note that a water analysis on a sample obtained at the city treatment plant may be significantly different from one obtained at the site.

Usually, immediately upon entering the plant, supply water is split into potable water and process water. This is done by using an air break or back flow preventers. This is a precaution against process contaminants backing up into potable or city water and vice versa. Often a break tank is used as the air break since it also provides storage capacity for demand surges at the use points.

4.0 POTABLE WATER

Potable water, also called drinking or tap water, is used for sanitary purposes such as drinking fountains, showers, toilets, hand-wash basins, cooking, etc. If the water supply to the facility is from a public system such as city water, the Environmental Protection Agency (EPA) Standards, Title 40 CFR, Part 141 set the maximum contaminant levels. Table 17.3 is a highlight of a typical water supply standard. Primary drinking water regulations, Appendix I outlines the existing and proposed U.S. EPA drinking water maximum contaminant levels.

Please note that the proposed EPA drinking water standards reduces the coliform count from 1 to 0 per 100 ml. All types of water discussed from this point on will fall under the category process water.

Fermentation and Biochemical Engineering Handbook.

TABLE 17.1 Contaminants by Type of Source Water

	Tap Water	Surface Water	Ground Water
Particulates	3–5	3–7	4–9
Dissolved Solids	2–5	1–5	5–10
Dissolved Gases	3–5	7–10	5–8
Organics	1–4	3–8	0–5
Colloids	0–5	3–8	0–4
Bacteria	1–2	6–9	2–5
Pyrogens	7–9	6–9	2–5

0 = None
10 = Very High

TABLE 17.2 Typical Water Supply Analysis

Item	Plant Feed
Turbidity	0
Color	0
pH	8.8
Alkalinity	16 mg/L
Hardness (as $CaCO_3$)	38 mg/L
Calcium	10 mg/L
Magnesium	3.2 mg/L
Sodium	23 mg/L
Potassium	3.1 mg/L
Iron	0.04 mg/L
Manganese	0.03 mg/L
Sulfate	27 mg/L
Chloride	49 mg/L
Nitrogen (ammonia)	0.05 mg/L
Nitrogen (nitrite)	0.30 mg/L
Nitrogen (nitrate)	0.002 mg/L
Copper	0.002 mg/L
SDI (fouling index)	25

TABLE 17.3 Minimum Potable Water Standard

Item	Specification
Appearance	1 Turbidity Unit
Chloride	250 ppm
Fluoride	1.4 to 2.4 mg/L
Sulfate	250 ppm
Lead	0.05 mg/L
Fecal Coliforms	1/100 ml (Proposed: 0/100 ml)
Pyrogens	Not Specified
Other Microbes	Not Specified
Total Dissolved Solids	500 mg/L
Arsenic	0.05 mg/L
Barium	1.0 mg/L
Cadmium	0.010 mg/L
Chromium Hexavalent	0.05 mg/L
Chloroform	0.7 mg/L
Cyanide	0.2 mg/L
Mercury	0.002 mg/L
Nitrate	10 mg/L
Selenium	0.01 mg/L
Silver	0.05 mg/L
Pesticides	
Chlorodane	0.003 mg/L
Endrin	0.0002 mg/L
Heptachlor	0.0001 mg/L
Heptachlor Epoxide	0.0001 mg/L
Lindane	0.004 mg/L
Methoxychlor	0.1 mg/L
Toxaphene	0.005 mg/L
2, 4-D	0.1 mg/L
2, 4, 5-TP (Silvex)	0.01 mg/L
Specific Resistance	10,000 ohms/cm (typically)
pH	6.5–8.5

5.0 WATER PRETREATMENT

After the break tank, process water is treated using various equipment and technologies depending on its intended use and the water analysis. Some of the technologies are: multimedia filtration, water softening, activated carbon adsorption, UV treatment, deionization, ultrafiltration, reverse osmosis, final filtration and distillation.

Figures 17.1 and 17.2 depict two alternative equipment trains for treating water. However, these diagrams

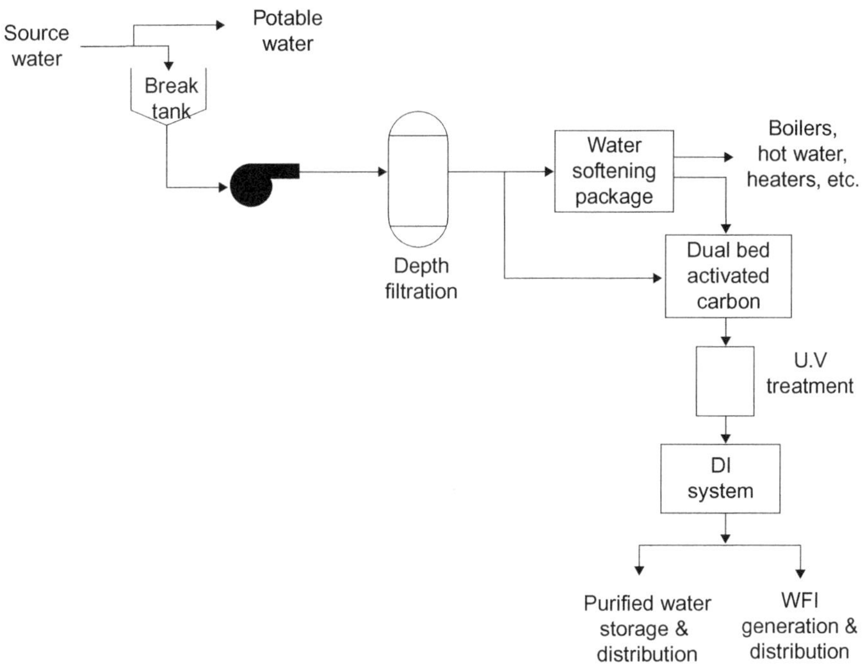

FIGURE 17.1 Water pretreatment.

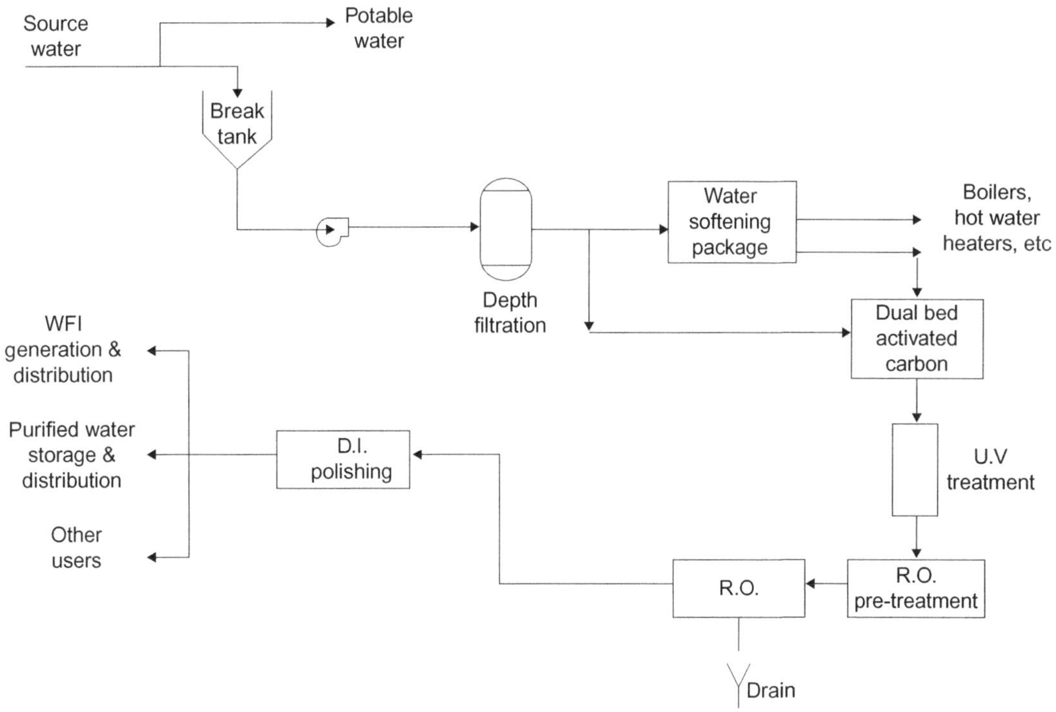

FIGURE 17.2 Water pretreatment.

are not all inclusive. For example, if the water analysis shows a high concentration of insoluble iron oxides, the first-step would be to inject a flocculent agent and then filter. The softener or the Figure 17.1 system can remove clear water iron.

6.0 MULTIMEDIA FILTRATION

Multimedia filtration (also called prefiltration, sand filtration or multilayered filtration) is mainly aimed at removing sediments and suspended matter. Suspended contaminants are trapped in small crevices and, as a result, water turbidity is improved. A number of media are distinctly layered with the coarsest on top so the suspended matter is collected throughout the depth of the filter according to size.

The filter beds need to be backwashed periodically as the backpressure increases; however, backwashing removes the filter from use. To avoid downtime, often a dual filter bed system is installed.

During construction, the filtration unit should be installed before all the walls are erected so it can be kept upright, in which case the vendor before shipping can charge the filters. This would reduce chances of damage to the internals during loading. The unit, of course, should be inspected thoroughly upon receiving. Before shipping, the vendor will often disconnect controls to minimize potential damage. Sufficient time should be allowed to reconnect all of these. Finally, to avoid bacteria building up, start-up should be delayed until a constant water flow is assured.

7.0 WATER SOFTENING

Water is softened to remove the scale-forming hardness elements. Soft water is required for boilers, water heaters, cooling towers, reverse osmosis systems, etc. Softening is an ion-exchange process that replaces almost all of the metallic or cations by sodium ions and sometimes, the anions with chlorine ions. Therefore, a constant supply of salt is required.

A softener may be used in conjunction with a deionizer on certain water supplies to provide softened water for use in regeneration. This will prevent the formation of insoluble precipitates within the deionizer resin bed.

It is important to note that softening does not remove silica, which forms a very hard scale that is not easily removed. In addition, softening does not remove chloride that can cause stress corrosion cracking in stainless steel.

A freshly regenerated resin bed is in the sodium (Na^{+}) form. When in service, sodium cations are exchanged for undesirable quantities of calcium (Ca^{++}), magnesium (Mg^{++}), and iron (Fe^{++}) ions. Sodium ions already present in the raw water pass through the process unchanged. Upon exhaustion of the resin, as indicated by unacceptable hardness leakage, most systems are designed to go automatically into regeneration. It should be noted that although the water is softened, the total dissolved solids content remains unchanged. Further, the effluent contains the same anions as the supply water.

Softeners can be a microbial concern. A dark and moist column interior can provide a growth environment. The regeneration cycle which uses concentrated brine solution and a backwash cycle aids in reducing the bio burden. Softeners should be regenerated based on a time clock set for twice weekly regenerations and on a volumetric flow of water, whichever is shorter. Since the regeneration cycle removes the softener bed from operation, a dual bed system is often specified.

8.0 ACTIVATED CARBON

Activated carbon has long been used as an effective means of removing organics, chlorine, chlorates, other chlorine compounds and objectionable tastes and odors. The organics removed include pesticides, herbicides and industrial solvents for which activated carbon has diverse capacity. Typically, carbon filters are operated at a flow rate of 1–2 gpm/ft^3 of activated carbon.

Since chlorine is removed from water by the carbon, extra care is required from here on to protect against bio burden growth. Carbon beds themselves are good breeding grounds for bacteria. To keep the system in check, a recirculation system as depicted in Figure 17.3 is recommended. The constant recirculation avoids water stagnation and reduces viable bio burden growth.

Activated carbon is manufactured by heating selected grades of coal or other higher carbonaceous material in the absence of oxygen. This "activation" process burns out impurities and produces a honeycomb-like structure containing millions of tiny pores. The structure provides a large total surface area that enables the carbon to adsorb (attract and hold to the surface) large quantities of contaminants. Chlorine, or its related elements, is first adsorbed on the surface of the pores where they react with the carbon to liberate chloride. Because of this reaction and deterioration of chlorine, the capacity of activated carbon for chlorine removal is exceedingly high. In addition to chlorine removal and adsorption of organics, the granular carbon is an effective filter. Although removal of turbidity will shorten the carbon life by blocking pores, the carbon will function as an excellent filter. Particle removal down to 40 microns can be achieved with freshly backwashed beds of carbon.

Carbon beds are backwashed to remove carbon fines and suspended matter that have been filtered by the bed. Backwashing does not regenerate the carbon. Sanitizing and some degree of regeneration can be affected by passing low-pressure steam or hot water through the carbon bed. The degree of regeneration is limited and the carbon must be replaced periodically (once every 1–2 years). Steam is of course more effective than hot water for sanitization, but it does cause some carbon degradation.

9.0 ULTRAVIOLET PURIFICATION

In high purity water systems, UV light is often used inline to control microorganism contamination. Use of UV as a disinfectant is somewhat controversial. In the

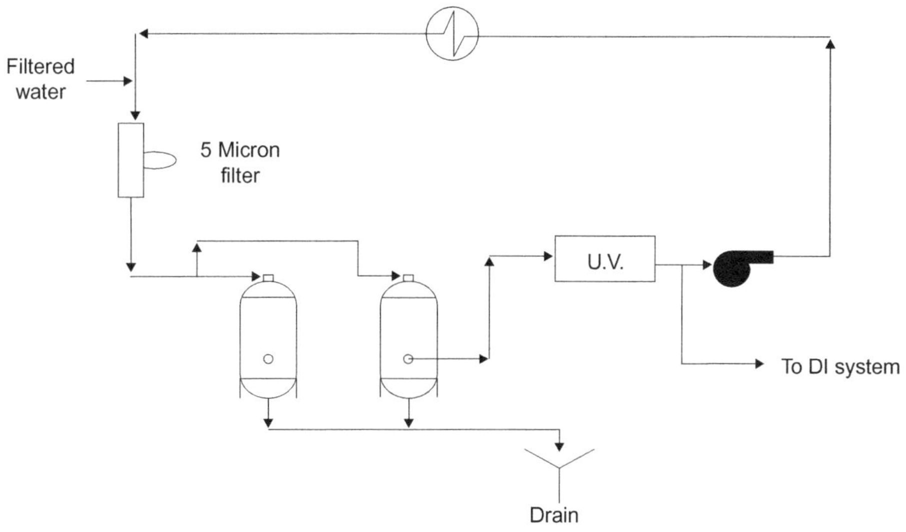

FIGURE 17.3 Activated carbon.

author's opinion, UV as an added measure is worthwhile; however, it should not be totally relied on to keep the water clear of bacterial contaminants. UV systems cannot correct for a poorly designed water system. Also, note that UV kills microorganisms and hence generates pyrogens [1]. In most cases, microorganisms can be filtered out, while pyrogens cannot be.

To be effective, UV radiation at a wavelength of 2537 Å must be applied. A minimum dosage of 16,000 microwatt-seconds per cm^2 must be reached at all points throughout the water chamber. Appendix II is a summary statement by the Department of Health, Education and Welfare on the use of UV as a disinfectant.

During construction and installation, extra care should be taken in handling the UV unit. The UV lamp sleeves are made of quartz, since glass filters UV radiation, and are very fragile. The same is true in the start-up; the lamps can break when the unit is first pressurized. It is recommended that spare lamps be kept on hand. Lamps also get broken during start-up if they are turned on when there is no flow. They get hot before the flow is established and then cold water causes them to break. Finally, avoid looking directly at the lamps while they are on. UV radiation can cause eye damage. A port equipped with a thick glass cover is provided to visually check the lamps.

10.0 DEIONIZATION

Deionization is the process of removing the dissolved ionized solids from water by ion exchange. Ion exchange can be defined as a reversible exchange of ions between a solid (resin) and a liquid (water). The major portion of total dissolved solids is mineral salt, such as calcium bicarbonate, magnesium sulfate, and sodium chloride. Since deionization requires the removal of all ions, both the negatively charged anions and the positively charged cations, materials capable of altering both are required. These materials are known as cation exchange resins and anion exchange resins. The ion exchange resins are contained in pressure tanks, and the water to be deionized is forced through the resins. Typically, deionizers are either dual bed or mixed bed systems.

Dual-bed models have two separate resin vessels, the first being a cation unit followed by an anion unit. Cation resin collects the positively charged cations such as calcium, magnesium or sodium and exchanges them for hydrogen. The discharge from the cation tank is very acidic.

There are two types of anion units. Strong base anion resin units remove all anions including silica and carbon dioxide. Removal of silica and CO_2 are especially important prior to distillation in a unit such as a WFI still. They typically produce a deionized water with a pH greater than 7. Weak base anion units are used when removal of silica and carbon dioxide are not required. Mixed bed units contain both the anion and the cation resins in one vessel. Mixed bed discharge pH is typically around 7.0, neutral.

After a time, the resins are exhausted and must be regenerated. This is done with a strong acid and a strong base. Cation resin is typically regenerated with hydrochloric or sulfuric acid. Anion resin is normally regenerated with sodium hydroxide. A neutralization tank is generally necessary to adjust the pH before waste effluent from regeneration can be discharged into the sewer. The neutralization tank and system should be placed close to the DI (deionization) system, this is due to the fact that strong acid and base solutions will have to be piped between the two systems. Before hookup, all lines should be flushed. For obvious reasons, mixed bed deionizers are more difficult to regenerate.

The quality or degree of deionization is generally expressed in terms of specific resistance (ohms) or specific conductance (mhos). Ionized material in water will conduct electricity. The more ions, the more conductivity and the less resistance. When ions are removed, resistance goes up, and therefore the water quality is improved. Completely deionized water has a specific resistance of 18.3 megohms centimeter.

During construction, the DI system should preferably be positioned before all the walls are erected so the skid can be kept upright, in which case the vessels can be charged with the resins by the vendor before they are shipped. This would reduce chances of damage to the internals during loading. Again, sufficient time should be allowed to reconnect all the control air (or water) lines that are disconnected before shipping. To avoid bacteria buildup, start-up should be delayed until the system is ready to be placed in use with constant water flow.

If the vessels are going to be charged with the resins at the site, it is better to pump a slurry solution of the resin into the vessels instead of physically dumping it through the man way. Make sure to backwash for fines after loading. Upon completion, test for resin and other leaks.

As usual, good planning is important. Make sure sufficient amounts of all the necessary chemicals are on hand and are of the right grade. Along with the skid, materials will include a number of loose boxes containing plastic pipes and fittings, remote items, and perhaps the resin, all of which should be identified and kept safe for the installation. Do not put chlorinated city water directly into the resin beds even for washing. Also, do not recirculate DI water directly to the carbon unit as it will leach out organics. Here, also, a recirculation system is recommended to keep a constant flow through the unit at all times.

To minimize down time, alternating deionization systems are specified so that one DI unit is on line while the second unit regenerates or recirculates in a standby mode. A regeneration cycle is usually 3 to 4 hours long. The frequency of regeneration is governed by both operating cost and potential bacterial buildup. The regeneration with acid and caustic serves to sanitize the resin bed. Deionizers, in the pharmaceutical industry are generally regenerated every one to three days.

Off-site regenerated DI canisters are available as a service to the industry. Due to the difficulty associated with handling, storage and sewer discharge of the caustic and acid chemicals needed for the regeneration, many users choose this alternative. Service exchange DI (SDI) is economically justified when the quantities of DI needed are relatively small (0.5 to 25 gpm). SDI systems are also used downstream to polish water that has been treated before. When the resins are exhausted, a service technician exchanges them for fully regenerated units.

Another alternative for water deionization is continuous deionization (CDI). This technologically innovative deionization process was developed by Millipore Corporation and is currently marketed by Ionpure. It uses electricity across ion exchange membranes and resins to remove ions from a continuous water stream. No chemical regeneration is required. A waste stream, carrying the rejected ions, of less than 10% of the feed water is required.

11.0 PURIFIED WATER

Purified water is typically prepared by ion exchange, reverse osmosis or a combination of the two treatment processes. Purified water is intended for use as an ingredient in the preparation of compedial dosage forms. It contains no added substances, and is not intended for use in parenteral products. It contains no chloride, calcium, or sulfate, and is essentially free of ammonia, carbon dioxide, heavy metals, and oxidizable substances. Total solids content will be no more than 10 ppm, pH will be 5–7, and the water will contain no coliforms. The United States Pharmacopoeia National Formulary (USP) requires that purified water comply with EPA regulations for bacteriological purity of drinking water (40 CFR 141.14, 141.21). Table 17.4 is a quantitative interpretation of United States Pharmacopoeia XXI standards for purified water [2].

TABLE 17.4 USP Purified Water [2]

Constituent	Purified Water
pH	5.0–7.0
Chloride	0.5 mg/L
Sulfate	<1.0 mg/L
Ammonia	<0.1 mg/L
Calcium	<1.0 mg/L
Carbon Dioxide	<5.0 mg/L
Heavy Metals	<0.1 mg/L as Cu
Oxidizable Substances	Passes USP Permanganate Test
Total Solids	<10 mg/L
Total Bacterial Count	<50 cfu/ml
Pyrogens	None Specified

USP XXII (published 1990) purified water standards remain the same as USP XXI. Purified water is essentially equal to deionized water, at least chemically (not necessarily biologically). Figures 17.1 and 17.2 outline the most common methods of purified water generation. After the deionization process, water is collected in a storage tank. A distribution loop takes water from the storage tank to all use points and then back to the storage tanks.

The purified water temperature is typically maintained at 60 to 80°C (hot), ambient, or 4°C (cold). A number of heat exchangers are located around the loop and after the DI system to achieve and maintain the desired temperature. If the system is hot, point-of-use heat exchangers should be used to obtain ambient water. A design engineer would need to evaluate a given system, and strategically locate and size heat exchangers to both maintain the temperature in the loop and to provide water to the use points at the desired temperatures.

Regardless of the system temperature selected, the storage tank and the loop must be sanitized periodically. For the stainless steel system outlined above, sanitization implies raising the water temperature to 80°C (at a minimum) at the cold point and maintaining it for the validated time interval. This is often done automatically off shift.

Another commonly used approach to purified water generation, storage and distribution is RO/DI. Figure 17.4 is a schematic of an RO/DI approach. The components in this type of system are usually all plastic, therefore, sanitization is done chemically. Use of a sterilizing 0.2-micron filter in addition to, or instead of, the resin filter should be resisted. This practice may appear beneficial, but it is specifically prohibited by the proposed LVP GMP's (proposed CFR 212.49) and it is not recommended.

12.0 REVERSE OSMOSIS

The only component of RO/DI purified water not yet discussed is the reverse osmosis system (RO). Reverse osmosis operates at a pressure in the range of 200 to 400 psig or higher, forcing water through membranes. The reverse osmosis process should reject about 95 to 97% of the ionizable salts and 99% of organics with

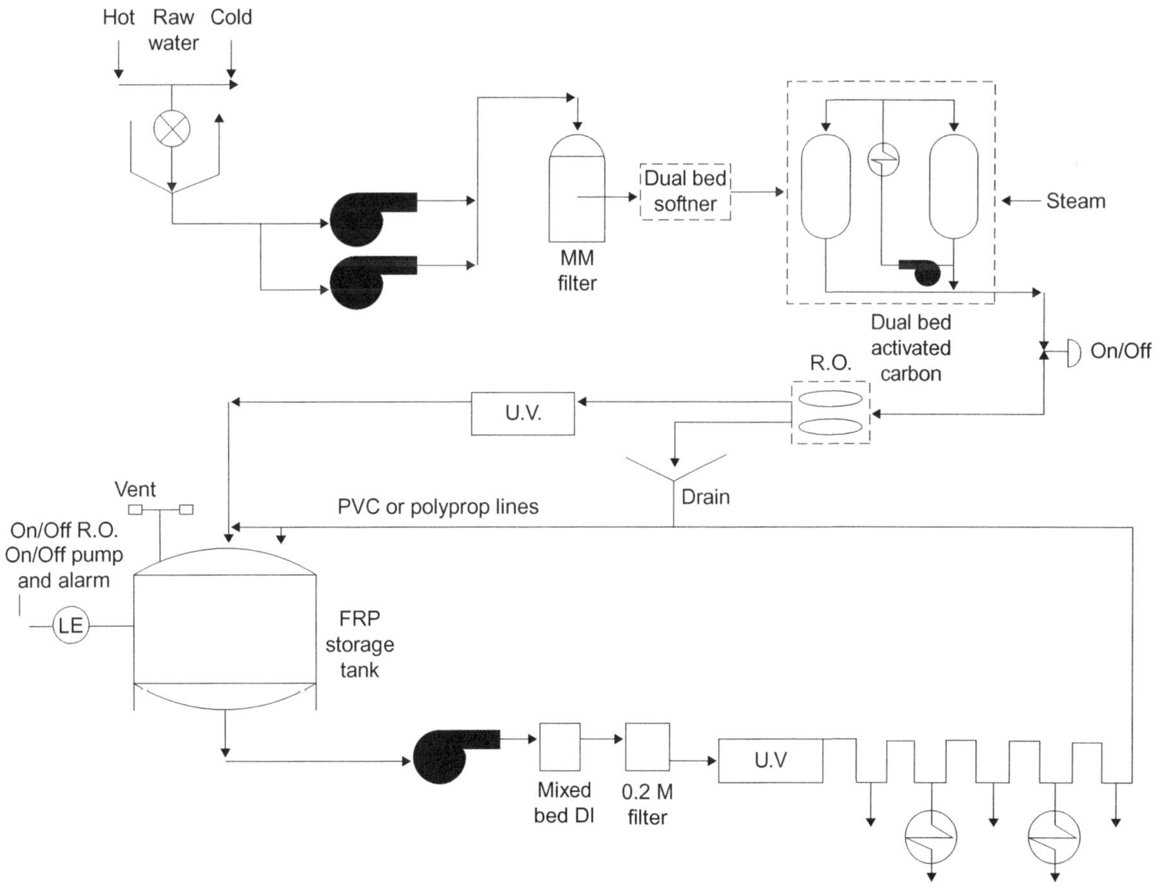

FIGURE 17.4 USP purified water (RO/DI water).

molecular weight over 300. It is extremely effective in rejecting bacteria and pyrogens.

Due to the significant reduction in ionizable salt concentrations, RO systems are often used as a pretreatment method before a DI system. An RO before a DI reduces the size of the deionizer, reduces the consumption of regenerate chemicals and may reduce the length of the deionizer required service cycle.

Osmosis is the procedure by which two solutions separated by semipermeable membrane interchange a solvent. The solvent moves from the solution that is low in solute to the solution that is high in solute through the semipermeable membrane in order to equalize the concentration on both sides of the membrane. By applying water, under pressure, to this semipermeable membrane, the process of osmosis is reversed, forcing pure water through the membrane and leaving a concentrated solution behind. The concentrated side is continuously removed to prevent fouling. A typical RO used in water systems is designed to reject 25 to 50% of the feed water continuously.

Even at these rejection rates, the dirty side of the membranes rapidly builds up undesirable bacterial concentrations. To alleviate this potential problem, the membranes are normally automatically flushed on a continued cycle basis, say 3 to 8 minutes every four hours. Full sanitization with a sanitization chemical like phosphoric acid is required periodically based on continual monitoring of pressure drop, conductivity and bacterial count. To further reduce bacterial count, RO systems should be sized for 24 hours per day operation to minimize water stagnation.

The two most common RO membrane configurations used in water treatment today are spiral-wound and hollow fiber. The spiral-wound elements can operate at a higher pressure and at a higher silt density index (SDI) than the hollow fiber type, and thus may require less pretreatment (and are more tolerant of pretreatment upsets). They also are easier to clean than the hollow fiber type. The main advantage of the hollow fiber configuration is that it has the highest amount of membrane area per unit volume, thus requiring less space. Since there is only one hollow fiber element per pressure vessel, it is easier to troubleshoot, and it is easier to replace membrane modules.

Each membrane configuration is available in different materials, the most common being cellulose acetate and polyamide. Cellulose acetate type membranes have a tight feed water pH specification (5.0–6.5) usually requiring acidification of the feed water and are subject to bacterial degradation, requiring some (up to 1.0 ppm) free chlorine in the feed water. Polyamide membranes can operate continuously over a broader pH range (4.0–11.0) and thus may utilize softening instead of acidification in order to prevent formation of insoluble precipitates at the membrane interface. They are subject to oxidation by even trace amounts of free chlorine, thus requiring activated carbon prefiltration and/or sodium bisulfite addition. The operating temperature range is typically 32–104°F (0–40°C), but the membrane productivity usually is rated at 77°F (25°C), thus equipment is often used to regulate the feed water temperature to the 77°F design point.

Should the RO system outlet conductivity be unsatisfactory, the outlet water should be diverted automatically to drain until the problem is resolved.

13.0 WATER FOR INJECTION

USP requires water for injection (WFI) to be produced by distillation or by reverse osmosis. In the pharmaceutical industry however, distillation is currently the preferred method for WFI generation. A double-pass reverse osmosis unit is sometimes used. A single pass RO is not recommended for WFI generation. Water for injection is intended for use as a solvent for the preparation of parenteral solutions and the final rinse of all parenteral product contact surfaces.

Water for injection must meet the USP purified water requirement discussed and contain no added substances. Table 17.5 is a quantitative interpretation of United States Pharmacopoeia XXI, Standards For Water For Injection [2].

Note that WFI is essentially the same as purified water with the exception of endotoxins and bacteriological purity. USP requires WFI to contain less than 0.25 USP Endotoxin Unit per ml. The USP has no bacteriological purity requirements for WFI at all. However, the proposed large volume parental GMPs (CFR 212.49) requires counts less than 10 CFU/100 ml.

TABLE 17.5 USP Water for Injection [2]

Constituent	Water For Injection
pH	5.0–7.0
Chloride	<0.5 mg/L
Sulfate	<1.0 mg/L
Ammonia	<0.1 mg/L
Calcium	<1.0 mg/L
Carbon Dioxide	<5.0 mg/L
Heavy Metals	<0.1 mg/L as Cu
Oxidizable Substances	Passes USP Permanganate Test
Total Solids	<10 mg/L
Total Bacteria Count	<10 cfu/100 ml
Pyrogens	0.25 EU/ml

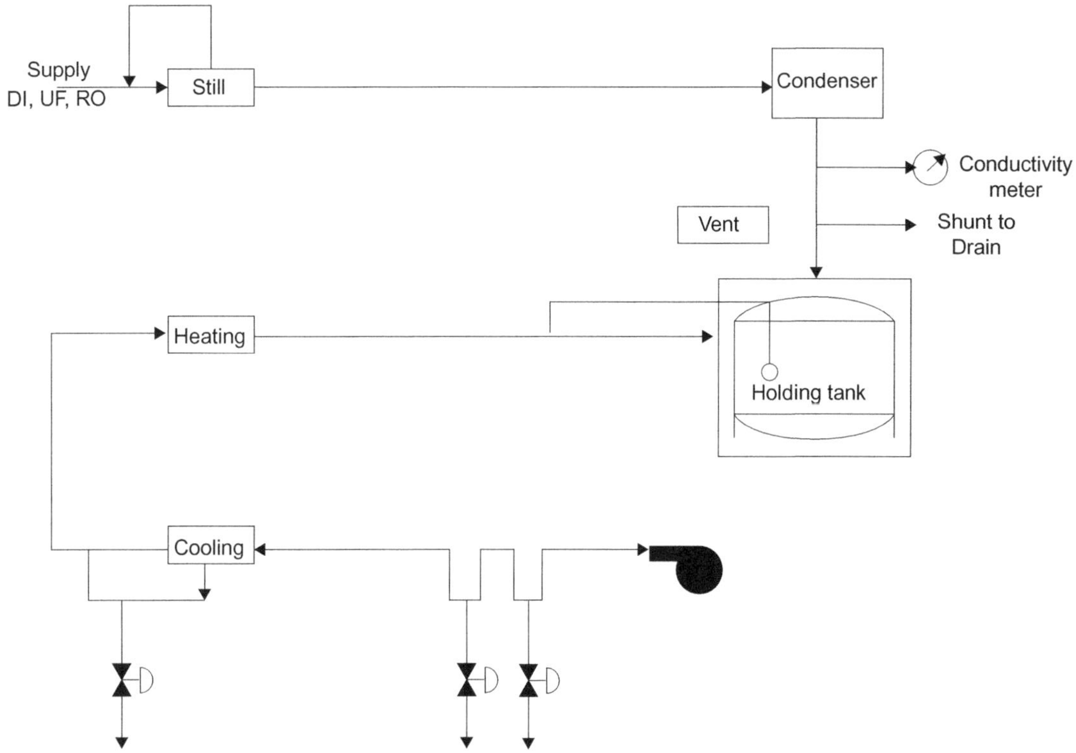

FIGURE 17.5 Water for injection.

Figure 17.5 summarizes a typical WFI storage and distribution system. Pretreated water is fed to the WFI still preheater by level control and on to an evaporator heated with the plant process steam. The evaporated water should go through filters/separators to remove entrained droplets. The steam is condensed with cooling water, and then partially reboiled to remove dissolved gases. The distillate is fed to a WFI storage tank. A conductivity monitor diverts under specification distillate to drain. WFI production is controlled by an on/off level control in the WFI storage tank. Boiler controls are incorporated in the WFI still. The still is vented automatically when on standby waiting for level control to request water.

The WFI recirculating loop velocity is designed to ensure turbulent flow and is generally 5–10 ft/sec. The WFI recirculation loop is designed to run continuously. At peak use rate, the water velocity in the pipes should be 2 ft/sec or more.

Standard control methods are used for the WFI tank. A water high level switch (LSH) turns the WFI still off or on depending on whether the water level is at or below the LSH point. A water low level switch (LSL) is interlocked to the recirculation pump to shut it off should the level reach the LSL point.

The WFI storage tank water is usually maintained at about 80°C. This water may be temperature controlled (heated) at the return end of the WFI loop via a WFI heat exchanger (shell and tube double tube sheet) or a hot jacket may be used. A heat exchanger is the preferred method. Where cool WFI is required, point of use coolers (double tube sheet) or a cool WFI loop is provided. Considerable control effort is needed for the point of use cooler design to meet the continuous flow non-stagnancy standards.

All pipelines used in a WFI generation, storage or distribution must be sloped to provide for complete drainage. No pipe segment not in regular use can be greater in length than six diameters of the unused pipe measured from the axis of the pipe in use.

A WFI system must be sampled and tested at least once a day. All sampling ports or points of use in the distribution system shall be sampled at least weekly.

It must be kept in mind that WFI is an extremely aggressive solvent, especially at 80°C. Therefore, the still, storage tank(s), and the distribution system are generally 300 series stainless steel with welded joints wherever possible. All surfaces that come in contact with the water are, at a minimum, smooth and manually polished to a #4 finish (150 grit) and passivated to prevent corrosion. Welds are made with automatic arc welders under inert atmosphere to prevent chromium migration, carbide and oxide formation, inclusions, or incomplete penetration of the joint. All connections that are not welded should be sanitary in design to eliminate crevices where corrosion can occur and bacteria can grow.

GMPs do allow storage at ambient temperatures, but if this option is chosen, the water must be tested on a batch basis, and can only be held for 24 hours before it must be discarded. Therefore, a hot loop may turn out to be less expensive than a system without the heated loop in the long term.

All maintenance on the WFI system must be performed by trained personnel and carefully documented. Maintenance personnel must be fully aware of any impact that their activities may have on the system and on the facility. All maintenance will require careful planning and coordination with manufacturing and quality control personnel.

14.0 WATER SYSTEM DOCUMENTATION

It is necessary to maintain accurate blueprints of all water systems for FDA review and to comply with cGMPs. It is also critical to the integrity of the system that the validation is kept current. In order to accomplish these objectives, a change control procedure must be implemented that ensures that all changes to the system are fully documented, and all anticipated changes are evaluated by appropriate personnel for potential adverse effects on the system prior to implementation. Based on this evaluation, decisions are made about the need for revalidation to guarantee that the system remains under control.

With this procedure in place, it is much less likely that the status of the system will be altered haphazardly, and that changes will not occur without the review and consent of appropriate personnel.

APPENDIX I: EXISTING AND PROPOSED U.S. EPA DRINKING WATER STANDARDS

In addition to the eight regulated volatile organics, there are 51 unregulated VOCs which may require an initial monitoring once during a four-year period.

PRIMARY REGULATIONS

Contaminants	Existing MCL mg/l[1]	Proposed MCL mg/l[2]	Best Available Technologies (BAT)[3]
INORGANICS			
Arsenic	0.05	0.05	C/F, LS, RO
Asbestos	—	7 MFL[4]	UF, SF, C/E
Barium	1.0	5.0	LS, CS, RO
Cadmium	0.01	0.005	C/F, LS, CS, RO
Chromium	0.05	0.1	C/F, LS, CS, AX, RO
Fluoride	4.0	4.0	AX, RO
Lead	0.05	0.005	C/F, LS, CS, SF, RO
Mercury	0.002	0.002	C/F, GAC, CS, RO
Nitrate	10	10	AX, RO
Nitrite		1.0	LS, RO
Selenium	0.01	0.05	LS, RO, C/F
MICROBIALS			
Coliforms	< 1/100 ml	0	C/F, CL, SF
Giardia Lamblia		0	C/F, CL, SF
Legionella	—	0	C/F, CL, SF
Viruses	—	0	C/F, CL, SF
Standard Plate Count	—	—	C/F, CL, SF
Turbidity	1–5 mtu	0.5–5 mtu	C/F, CS, SF

(Continued)

(Continued)

Contaminants	Existing MCL mg/l[1]	Proposed MCL mg/l[2]	Best Available Technologies (BAT)[3]
ORGANICS			
Acrylamide	—	0.0005	GAC, OX
Alachlor	—	0.002	GAC
Aldicarb	—	0.01	GAC, OX
Aldicarb Sulfoxide	—	0.01	GAC
Aldicarb			
Sulfone	—	0.04	GAC
Atrazine	—	0.003	GAC
Carbofuran	—	0.04	GAC, RO, OX
Chlordane	—	0.002	GAC, PTA, RO
cis-1,2,-Dichloroethylene	—	0.07	GAC, PTA
Dibromochloropropane (DBCP)	—	0.0002	GAC, PTA
1,2-Dichloropropane		0.005	GAC, PTA
0-Dichlorobenzene	—	0.6	GAC, PTA
2,4-D	0.1	0.07	GAC, RO
Endrin	0.0002	0.0002	GAC, PTA, RO
Ethylenedibromide (EDB)	—	0.00005PTA, GAC	
Epichlorohydrin	—	0.005	not known
ORGANICS			
Ethylbenzene	—	0.7	PTA
Heptachlor	—	0.0004	GAC, OX
Heptachlor epoxide	—	0.0002	GAC
Lindane	0.004	0.0002	GAC, RO, OX
Methoxychlor	0.1	0.4	GAC, RO, C/F
Monochlorobenzene	—	0.1	GAC
PCB's Polychlorinated			
Biphenyls	—	0.0005	GAC, OX, RO
Pentachlorophenol	—	0.2	GAC
Styrene	—	0.005	GAC, PTA, OX
Tetrachloroethylene	—	0.005	GAC, PTA
Toluene	—	2	GAC, PTA
2,4,4-TP	0.02	0.05	GAC
Toxaphene	0.005	0.005	GAC, PTA
trans-1,2-Dichloro-ethylene		0.1	GAC, PTA
Xylenes (Total)		10	GAC, PTA
Trihalomethanes	0.1	0.2	GAC

(Continued)

(Continued)

Contaminants	Existing MCL mg/l[1]	Proposed MCL mg/l[2]	Best Available Technologies (BAT)[3]
RADIOISOTOPES			
Beta particles	4 mrem	4 mrem	RO
Gross alpha particles	15 pCi/l	15 pCi/l	CS, LS, RO
Radium 226 & 228	5 pCi/l	5 pCi/l	CS, LS, RO
RADIOISOTOPES			
Radon 222	—	200–2000 pCi/l	PTA
Uranium	—	20–40 pCi/l	CS, LS, RO
VOLATILE ORGANIC CHEMICALS			
Benzene	0.005	0.005	GAC, PTA
Carbon Tetrachloride	0.005	0.005	GAC, PTA
1,1-Dichloroethylene	0.007	0.007	GAC, PTA
1,2-Dichloroethane para-Dichlorobenzene	0.005 0.075	0.005 0.075	GAC, PTA GAC, PTA
1,1,1-Trichloroethane	0.2	0.2	GAC, PTA
Trichloroethylene	0.005	0.005	GAC, PTA
Vinyl Chloride	0.002	0.002	PTA
Aluminum	–	0.05	CS, RO, LS
Chloride	250	250	RO
Copper	1	1	LS, CS, RO
Fluoride	2	2	AX, RO
Iron	0.3	0.3	C/F, LS, CS, SF
Manganese	0.05	0.05	C/F, LS, CS, RO
Silver	—	0.09	C/F, LS, CS, RO
Sulfate	250	250	C/F, AX, RO
TDS	500	500	C/F, RO
Zinc	5	5	C/F, LS, CS, RO

KEY TO BEST AVAILABLE TECHNOLOGIES (BAT)

AX–Anion exchange
C/F–Coagulation/flocculation (i.e., addition of alum or ferric sulfate followed by settling and filtration)
CL–Disinfection by chlorine
CS–Cation softening with salt
GAC–Granular activated carbon
LS–Lime softening
OX–Oxidation by ozone
PTA–Packed tower aeration
RO–Reverse osmosis
SF–Sand filtration or similar media
US–Ultra filtration

[1]*Existing maximum contaminant levels, National Drinking Water Standards.*
[2]*Proposed or likely maximum contaminant levels under current development per revisions of the Safe Drinking Water Act.*
[3]*The stated best available technologies are a guideline only for general approaches to treatment of the listed contaminants. See keys below.*
[4]*Million fibers per liter (fibers over 10 micron).*

APPENDIX II: DEPARTMENT OF HEALTH, EDUCATION AND WELFARE PUBLIC HEALTH SERVICE

Division of Environmental Engineering and Food Protection Policy Statement on Use of the Ultraviolet Process for Disinfection of Water

The use of the ultraviolet process as a means of disinfecting water to meet the bacteriological requirements of the Public Health Service Drinking Water Standards is acceptable provided the equipment used meets the criteria described herein.

In the design of a water treatment system, care must be exercised to insure that all other requirements of the Drinking Water Standards relating to the Source and Protection, Chemical and Physical Characteristics, and Radioactivity are met. (In the case of an individual water supply, the system should meet the criteria contained in the *Manual of Individual Water Supply Systems*, Public Health Service Publication No. 24.) The ultraviolet process of disinfecting water will not change the chemical and physical characteristics of the water. Additional treatment, if otherwise dictated, will still be required, including possible need for residual disinfectant in the distribution system.

Color, turbidity, and organic impurities interfere with the transmission of ultraviolet energy and it may be necessary to pretreat some supplies to remove excess turbidity and color. In general, units of color and turbidity are not adequate measures of the decrease that may occur in ultraviolet energy transmission. The organic nature of materials present in waters can give rise to significant transmission difficulties. As a result, an ultraviolet intensity meter is required to measure the energy levels to which the water is subjected.

Ultraviolet treatment does not provide residual bactericidal action, therefore, the need for periodic flushing and disinfection of the water distribution system must be recognized. Some supplies may require routine chemical disinfection, including the maintenance of a residual bactericidal agent throughout the distribution system.

Criteria for the Acceptability of an Ultraviolet Disinfecting Unit

1. Ultraviolet radiation at a level of 2,537 Angstrom units must be applied at a minimum dosage of 16,000 microwatt-seconds per square centimeter at all points throughout the water disinfection chamber.
2. Maximum water depth in the chamber, measured from the tube surface to the chamber wall, shall not exceed three inches.
3. The ultraviolet tubes shall be:
   - **a.** Jacketed so that a proper operating tube temperature of about 150°F is maintained.
   - **b.** The jacket shall be of quartz or high silica glass with similar optical characteristics.
4. A flow or time delay mechanism shall be provided to permit a two-minute tube warm-up period before water flows from the unit.
5. The unit shall be designed to permit frequent mechanical cleaning of the water contact surface of the jacket without disassembly of the unit.
6. An automatic flow control valve, accurate within the expected pressure range, shall be installed to restrict flow to the maximum design flow of the treatment unit.
7. An accurately calibrated ultraviolet intensity meter, properly filtered to restrict its sensitivity to the disinfection spectrum, shall be installed in the wall of the disinfection chamber at the point of greatest water depth from the tube or tubes.
8. A flow diversion valve or automatic shut-off valve shall be installed which will permit flow into the potable water system only when at least the minimum ultraviolet dosage is applied. When power is not being supplied to the unit, the valve should be in a closed (fail-safe) position that prevents the flow of water into the potable water system.
9. An automatic, audible alarm, shall be installed to warn of malfunction or impending shutdown if considered necessary by the Control or Regulatory Agency.
10. The materials of construction shall not impart toxic materials into the water either as a result of the presence of toxic constituents in materials of construction or as a result of physical or chemical changes resulting from exposure to ultraviolet emergency.
11. The unit shall be designed to protect the operator against electrical shock or excessive radiation.

As with any potable water treatment process, due consideration must be given to the reliability, economics, and competent operation of the disinfection process and related equipment, including:

1. Installation of the unit in a protected enclosure not subject to extremes of temperature that cause malfunctions.
2. Provision of a spare ultraviolet tube and other necessary equipment to effect prompt repair of qualified personnel properly instructed in the operation and maintenance of the equipment.
3. Frequent inspection of the unit and keeping a record of all operations, including maintenance problems.

Special Note

This criterion was established after numerous tests were conducted on an *Ultra dynamics* Ultraviolet Water Purifier System by the U. S. P. H. S. Ultra dynamics Purifiers meet and surpass the above criteria.

REFERENCES

[1] P.L. Parise, B.S. Panekh, G. Waddington, Ultrapure Water (1990).
[2] J. Brown, N. Jayawardena, Y. Zelmanovich, Water systems for pharmaceutical facilities, Pharm. Eng. 11 (4) (1991) 15–22.

Chapter 18

Sterile Formulation

Michael J. Akers, Curtis S. Strother and Mark R. Walden

1.0 INTRODUCTION

Historically, sterile bulk pharmaceutical manufacturing processes, prior to filling operations, have followed general bulk pharmaceutical guidelines. As technology and equipment have improved, the requirements for aseptic manufacture have increased. It is important to understand that product quality often is realized in the manufacturing phase and should be maintained throughout the remaining filling/packaging processes. It is the Food and Drug Administration's current opinion that Current Good Manufacturing Practice for Finished Pharmaceuticals [1] apply to sterile bulk operations [2]. Adherence to the Guideline on Sterile Drug Products Produced by Aseptic Processing [3] is considered essential for non-terminally sterilized products as is the case for sterile bulk pharmaceutical dry powders. The facility design and manufacturing process should be integrated with current regulatory guidelines, the interpretation and application of which can be found in several publications [4–9].

This chapter focuses on the preparing and filling of injectable solid bulk pharmaceutical formulations. The material presented is general in nature but with references to direct the reader to more in-depth treatment of the subject matter. Coverage includes sterile bulk product preparation, filtration, isolation, filling, and environmental conditions required for aseptic processing.

2.0 STERILE BULK PREPARATION

The solutions used for the dissolution of injectable products are prepared by using Water for Injection (WFI) USP that has been made as described in Ch. 19 of this handbook. In some cases, solutions are prepared using organic solvents (e.g., acetone, methanol, ethanol, isopropanol) alone or in combination with WFI. The potential for preventing microbial contamination should dominate the delivery and storage systems for water and solvents.

A typical solution system will consist of a dissolution vessel, a sterile filtration transfer line, and a vessel to hold the sterile filtered solution prior to further processing. Dissolution areas tend to have Class 100,000* air quality with smooth, easy-to-clean surfaces. The sterile side of the system should have the capability of being cleaned and steam sterilized in place or easily dismantled for cleaning and sterilization [10]. Normally, type 316 stainless steel can be used throughout the facility unless process conditions dictate otherwise. Passivation of welds will minimize the potential for microbial growth at rough edges. Metal particulates should be a concern when welding into the processing system. Computer automated systems tend to be the method of choice for validated cleaning and sterilizing operations.

The solution filtration system should have a prefilter and final sterilization filter. The selection of filters is dependent on the type of solutions to be filtered. The sterile filters should be validated for the intended use with the product/solution systems. Sterile filters for gases (air or nitrogen) need to be discussed with filter manufacturers to ensure that pressure ratings are appropriate with the intended use. Appropriate pressure regulation of ancillary systems should always be a design consideration. Vent filters will be needed in the processing system to maintain sterility during transfer operations. Filter integrity testing (e.g., bubble point or diffusion testing) is required to ensure that filters remain functional after their usage. Redundancy of filters will provide a greater safety factor for product during manufacturing operations. Sterilization of diaphragm valves tends to present fewer concerns with microbial penetration compared to ball type valves. The number of connections should be kept to a minimum. Thread-fitted piping connections are not recommended and should be replaced with soldered, passivated or sanitary clamp connections. The transport of liquid streams can be accomplished using either pressure or pumps. For pressure transfer with organic solvent, nitrogen is preferred due to its noncombustible properties; however, appropriate safety precautions need to be considered in the system design. A flow diagram illustrating solution preparation is shown in Figure 18.1. The location of the sterile filter traditionally has been on the non-sterile side primarily for ease of changing and to

*Class 100,000 means no more than 100,000 particles per cubic foot greater than or equal to 0.5 micrometers.

Fermentation and Biochemical Engineering Handbook.

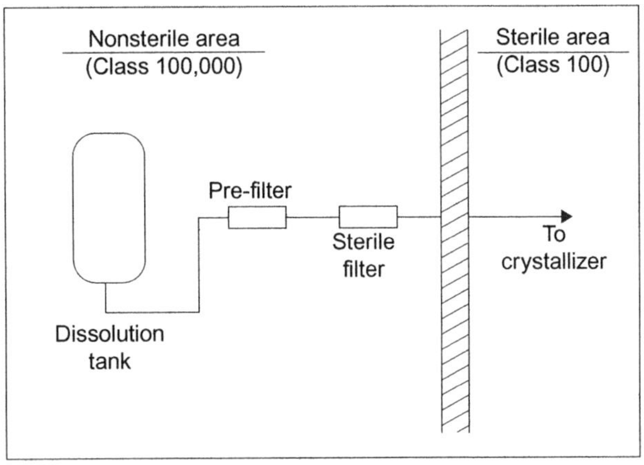

FIGURE 18.1 Bulk solution preparation.

minimize contamination of sterile area if leakages occur. However, new designs have the filter on the sterile side.

3.0 ISOLATION OF STERILE BULK PRODUCT

3.1 General Considerations

All equipment should be easy to clean and steam sterilizable and have a sanitary finish. If the facility is not dedicated to one product, computer automated "recipes" provide the greatest control and flexibility for processing. The overall operation must be designed so as to minimize the personnel required to operate the equipment and thus minimize the exposure of product to people. One of the most important facility design factors is in the isolation of product from its surrounding environment. Within the constraints of product quality, prevention of bacterial and particulate matter contamination should dominate the design concept and selection of equipment.

When product is exposed, air quality should be Class 100* or better, which can be achieved by High Efficiency Particulate Air (HEPA) filtration. Documentation of initial HEPA certification and periodic test results should be available at all times. Air pressure balancing should provide airflow from clean to less clean areas. Temperature and humidity are properties important to control in order to minimize the potential for microbial growth within the constraints of impact on product. Frequent rotation of sanitizing agents reduces the potential development of resistant organisms. Environmental monitoring is required to verify that product protection systems are working as expected. Environmental and safety concerns have reduced the practicality of ethylene oxide sterilization while other methods such as peracetic acid and VPHP (vapor pressure hydrogen peroxide) are also used as sterilants although more widely used in Europe than the U.S.

*Class 100 means no more than 100 particles per cubic foot greater than or equal to 0.5 micrometers.

4.0 CRYSTALLIZATION

Crystallizers should have variable speed agitators, temperature control, and sterilizable vent filters. As many controls as possible should be located outside of the sterile area. The crystallization vessel should be located as close to the filtration unit as possible. Time, temperature, and agitation speed are critical variables that may need strict control during the crystallization process. The crystallization vessel should be part of a closed system and often is jacketed for glycol temperature control.

5.0 FILTERING/DRYING

The filtration unit can be a centrifuge or closed filter that is either a pressure or vacuum unit. Some processes may require solution washing of the crystalline product. Facility design should therefore be optimized for flexibility. Recent pressure/vacuum filtration units can perform several functions such as collection washing with appropriate solvents, solution washing, and drying of a crystalline product. These filter/dryer units offer the advantage of a closed system that protects product from people and vice versa. The unit's agitator can resuspend and smooth product cake. After washing the product cake, the filter/dryer can be rotated to facilitate drying. The filter dryer should be readily sterilizable and allow continuous flow of product to the next operation. Drying can be done in vacuum dryers, fluid bed dryers, continuous or manual tray dryers; the latter is least preferable. Solvent emissions and recovery will be an important consideration for any solvent drying system.

6.0 MILLING/BLENDING

The dried product is aseptically discharged into suitable bulk containers or, alternately, to the milling unit. Bulk containers need to be designed for cleanability/sterilization. Milling and blending can be done as separate steps or in series by feeding the milled product directly to a blender. Mill parts are generally sterilized in place and blenders must be capable of cleaning and sterilizing in place. The working size of the blender should dictate batch size for the crystallization process. Blending is normally achieved in a tumbler type blender such as drum, double cone, twin, or a cube, or in a stationary shell type blender such as a ribbon or vertical screw mixer. Aseptic filling and sampling of the final bulk container should be part of the design considerations in order to minimize product exposure. If possible, the final bulk product should be filled into its final marketed container at the same facility as manufactured. However, if the final bulk container must be transported, the container must be designed and tested for container-closure integrity and product compatibility. A flow diagram illustrating a typical isolation process for a filter/dryer or spray dryer process is shown in Figure 18.2.

7.0 BULK FREEZE DRYING

A suitably sized solution preparation system similar to that mentioned under the previous sections can be used to provide material for bulk freeze-drying. (Since product solutions can be sterile-filtered directly into the final container, microbial and particulate exposure will be minimized.) The sterile solution is subdivided into trays and placed into a sterilized freeze dryer. Aseptic transfer of sterile product in trays to the freeze dryer must be validated. After tray drying, the sterile product is aseptically transferred through a mill into suitably designed sterile containers. The preparation of sterile bulk material is usually reserved for those cases where the product cannot be isolated by more common and relatively less expensive crystallization methods. Due to recent advances in this field, a freeze-drying process should be considered as a viable option [11].

8.0 SPRAY DRYING

Spray drying processes can be batch or continuous depending on production needs and the stability of the solutions to be spray dried. Because of reduced product manipulation, microbial and particulate burden can be reduced. Normally there is a solution vessel, a filtration system with prefilters and sterile filters, and a pressure vessel to feed the spray dryer at a controlled rate, the spray dryer itself, and bulk containers.

The air used for product drying should be HEPA filtered. When designed with silicone gaskets, the system will withstand sterilization temperatures. The atomizing device can be either a spray nozzle or a high-speed centrifugal device.

Spray dried products are typically temperature sensitive, therefore, air temperature should be controlled and as low as possible. Design of the atomizing device should ensure that product will not adhere to vessel walls. Surface drying and depyrogenation can be done in a continuous operated tunnel or batch oven. The former method is preferred since it minimizes the potential of particulate contamination during loading.

The spray dryer is normally dry heat sterilized by a hot air system that is used for drying the product. All lines entering the spray dryer must be sterilizable. The selection of spray dryer size and solution atomizing device is best determined by trial runs on sized pilot equipment. As with freeze drying, operational expense may limit spray drying operations to specific product applications. A flow diagram illustrating the spray drying process is shown in Figure 18.3.

9.0 EQUIPMENT PREPARATION

All portable equipment and tools used in a sterile area must be thoroughly pre-washed with proper cleaning agents, finally rinsed with WFI, and wrapped if required. These items are usually passed into the sterile area through a double door autoclave. Elimination of all particulate matter from any object entering the sterile area should be a major design consideration. All product-contact equipment, especially large mixers, should be electro polished. When stability is a concern, product should be cooled as soon as possible after leaving the cyclone separator.

Materials that cannot be sterilized should be transferred into the sterile area through an isolated area in which an outer wrapping is removed. The object is then wiped down with a sanitizing agent such as isopropanol or hydrogen peroxide.

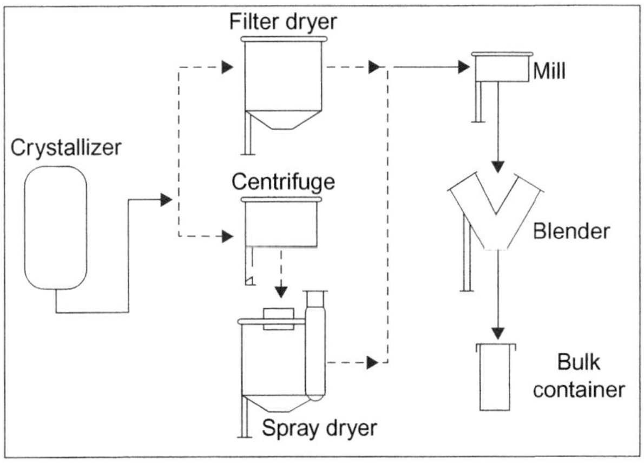

FIGURE 18.2 Typical isolation process for a filter/dryer or spray dryer.

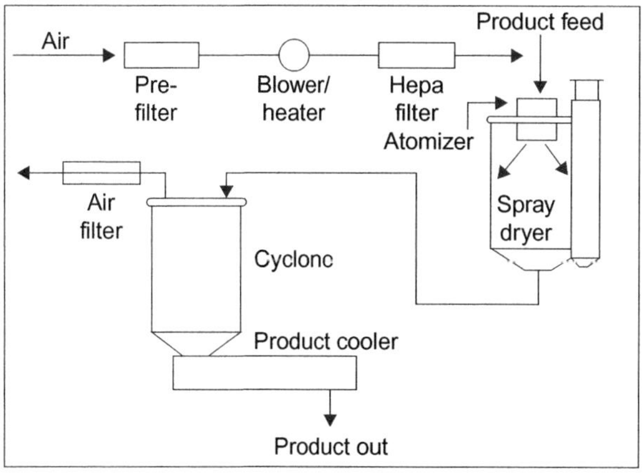

FIGURE 18.3 Spray drying process.

Stationary equipment such as conveyors and filling equipment must be sanitized at some specified frequency. This can be accomplished by wiping down with a sanitizing agent or fogging the sterile area with formaldehyde. All product contact parts such as powder hoppers, filling wheels, and stopper bowls are removed from the sterile area, cleaned and sterilized as previously described.

Freeze dryers are usually steam sterilized or sterilized using VPHP (vapor phase hydrogen peroxide). Trays used in a freeze dryer are usually cleaned and sterilized separately.

10.0 VALIDATION

Procedures must be developed and staffing provided for the collection of data that proves that the processes and equipment meet all parameters claimed [12]. Systems should be in place for equipment qualifications, validation, changes, and replacement. The manufacturing process validation could be invalidated without proper documentation of equipment maintenance. A minimum of three consecutive manufacturing lots should be evaluated for process validation. Parameters involved in process validation include in-process and final bulk product test, deviation analysis of the process, stability testing of final product and equipment qualification and validation. Sawyer and Stats discuss other validation requirements [13].

11.0 FILLING VIALS WITH STERILE BULK MATERIALS

11.1 Vial and Stopper Preparation

Vials must be thoroughly washed, dried, sterilized, and depyrogenated. They should be handled in a clean room to minimize contamination by particulate matter. Washing is normally done in automated vial washers using purified water, filtered oil-free air, and a final rinse of WFI.

Rubber closures for vials are also washed and depyrogenated in an automatic washer. The final rinse of the stoppers should be WFI. The use of detergent is optional. These operations should occur in a clean room to minimize contamination. After washing, stoppers are batched and autoclaved prior to entering the sterile area.

Depending on stoppering equipment and tendency of stoppers to clump during sterilization, a silicone lubricant may be added to the stoppers prior to sterilization. Several manufacturers offer equipment, which is capable of all these operations washing, silicone addition, and sterilization.

Vial and stopper washers are available that will allow processing from the clean room area into the sterile area in one operation. This equipment eliminates the transfer of vials and stoppers into the sterile area through ovens or autoclaves, thereby minimizing the potential for viable or nonviable particulate contamination. A typical flow sheet for the handling of vials, stoppers, and miscellaneous equipment is shown in Figure 18.4.

11.2 Filling of Vials

Vials used in a filling operation are fed into the system automatically by a conveyor from a vial sterilizer or manually from trays that have been processed through a batch oven. Because of the increased risk of contamination, the former method is preferred.

Powder fills are made by aseptically transferring the sterile bulk powder from its containers into the hopper of the filling machine. The transfer is usually done from a container that is mechanically positioned over the hopper with a solid aseptic connection to the hopper.

The type of filling machine to be used is best determined from trial runs of various supplier machines. All filling lines and equipment should be designed to prevent contamination by people and particulate matter. A typical vial filling operation is shown in Figure 18.5. More recent

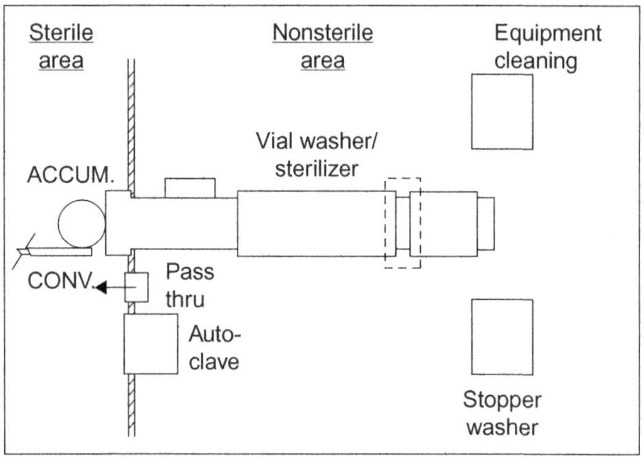

FIGURE 18.4 Sterile vial preparation.

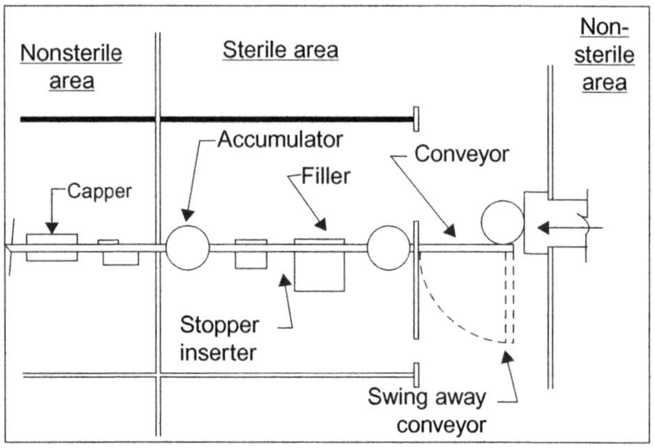

FIGURE 18.5 Sterile filling line.

designs incorporate barrier technology to accomplish this objective [14].

Once a vial has been filled with powder, it is stoppered and transported out of the sterile area, and is capped. The current regulatory trend is to perform the capping operation in a sterile area using sterilized caps. After capping, vials are usually visually inspected, labeled, and packaged.

A liquid fill operation is delivered to a pump through lines that have been sterilized in place or sterilized and assembled aseptically.

Freeze dried vials are usually partially stoppered just before entering the dryer. Closures are seated into the vials mechanically at the end of the drying cycle. A typical freeze-drying flow diagram is shown in Figure 18.6.

12.0 ENVIRONMENT

The term *environment* in sterile product manufacture means more than air quality and surface cleanliness. Other environmental concerns include water systems, compressed air and gas systems, temperature and humidity control, and the monitoring of personnel.

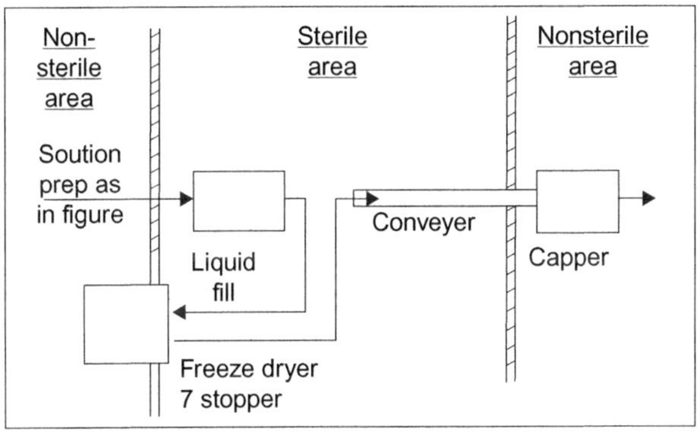

FIGURE 18.6 Sterile freeze drying operation.

There are at least four classified areas in sterile bulk manufacturing, each with different requirements for cleanliness: aseptic, controlled clean, clean, and general.

Good Manufacturing Practice regulations (1) (211.42(c) 10) require floors, walls, and ceilings to be smooth and easily cleanable. Temperature and humidity should be controlled. The air supply is filtered through high-efficiency particulate air (HEPA) filters, and systems are used to monitor environmental conditions, cleaning and disinfecting the room, and equipment product aseptic conditions. Federal Standard 209E [15] and The European Community Guide to Good Manufacturing Practice for Medicinal Products (EC-CGMP) [16] provide air classifications for the manufacture of sterile products. A recent information section in the USP has proposed similar microbial quality standards for clean room (see Table 18.1).

12.1 Aseptic Areas

Aseptic areas maintain air cleanliness at no more than 100 particles per cubic foot greater than or equal to 0.5 pm. This is achieved by using HEPA filtration of air over areas where product is exposed to the environment.

For aseptic air systems, the static pressure of the innermost room of a series of rooms should have pressure higher than the adjacent room leading towards the non-sterile room(s). Air pressure differentials should be monitored on a periodic basis to assure that air from the most critical manufacturing areas is always sufficiently positive and meets predetermined values.

Personnel must wear garments which shed virtually no fibers or particulate matter and, of course, retain particles shed by the body. Strict procedures must exist for the use of the following sterilized garments and protective coverings:

- Headgear, which totally encloses the hair and beard
- Eye covering such as goggles– Non-linting facemask

TABLE 18.1 Recommended Classification of Clean Rooms and Clean Zones for Aseptic Processing [17]

	CFU	CFU	Surfaces (2 in^2) (12.9 cm^2)	Personnel (2 in^2) (12.9 cm^2)	
	ft^3*	m^3		mask, boots, gloves	gown
Critical Processing Area (Class M-1)	0.03	1	3	3	5
Less Critical Processing Area (Class M-2)	0.15	5	5 10 (floor)	5	10
Preparation Areas/Change Rooms (Class M-3)	2.5	87	20 30 (floor)	15	30

*Determined by use of slit-to-agar sampler. Other types of samplers can be used but must be calibrated against slit-to-air sampler with use of correction factors, if necessary.

- Powder-free gloves
- Footwear that totally encloses the feet
- Single or two-piece trouser suits

No cosmetics or jewelry should be worn in the aseptic areas as these are sources of particulate matter and bacterial contamination.

12.2 Controlled Areas

Personnel change rooms and non-sterile manufacturing or preparation areas are common examples of controlled areas. Particulate matter in the air should be no greater than 100,000 particles greater than or equal to 0.5 μm per-cubic foot. Air locks must be provided to entrances and exits, surfaces must be easily cleanable, and air supply should be filtered and conditioned. The number of air changes should be at least 20 per hour. For explosion-proof areas where solvents are used, the air supply operates on a once-through basis.

Dress requirements in controlled areas should include hair covering, beard covering, and a long-sleeved protective over garment. Garments should be free from the shedding of particles and fibers.

12.3 Monitoring the Environment

To assure a consistently acceptable high quality production environment the following microbiological programs should be in place [18]:

1. Sound facility design and maintenance
2. Documentation systems
3. Validated/qualified decontamination procedures
4. Reliable process controls
5. Good housekeeping practices
6. Effective area access controls
7. Effective training and performance programs
8. Quality assurance of materials and equipment

The environmental monitoring program will confirm the effectiveness of these controls in the manufacturing environment.

12.4 Evaluation of the Air

There are at least eight methods used in the pharmaceutical industry for air quality. Seven measure microbiological contamination and include slit-to-agar impact samplers, sieve samplers, rotary centrifugal air samplers, cascade impactors, liquid impingement, membrane filtration and settling plates. The eighth method, air particle counters, measures both viable and nonviable particulates in the air. The most commonly used of these methods are settling plates, slit-to-agar samplers, and the particle counters. Settling plates are the simplest, but also the most unreliable or inaccurate method. The slit-to-agar sampler is probably the preferred method for monitoring microbiological air quality, while air particle counters are essential to monitor the overall quality of air. In areas where sterile solids are manufactured, particulate counts of the air are monitored prior to the start of manufacture to evaluate and benchmark the performance of air quality. Alert and action limits should be established based on historical and achievable low-level particulate and microbial counts. Procedures should clearly describe what actions are to be taken when these limits are exceeded. Alert limits typical are 2 σ and action limits 3 σ above mean. For example if the mean particle count is 20 particles per ft^3 then 1 σ (standard deviation) is 8, the alert limit will be 36, action limit, 44 and reject limit, 100.

12.5 Evaluation of Surfaces

There are three basic methods that have been employed for evaluation of microbiological content on surfaces. These include RODAC (Replicate Organism Detection and Counting) plates, swab testing, and agar overlay or rinse techniques. RODAC plates are the most commonly used of the surface monitoring methods. However, they are not suitable for irregular surfaces, in which case swab techniques are used.

12.6 Evaluation of Water

Water is used in sterile bulk operations for final rinsing of equipment, tanks and other items used in final compounding, processing and filling of sterile drug products. The quality of water must meet the requirements of the USP Water for Injection. Among the most important of these requirements are extremely low (e.g. 0–2 CFU) coliform bacterial counts. Water for Injection outlets are sampled daily in large amounts (>500 ml). Appropriate culture media, temperatures and times for incubation of water samples are selected for enumeration of bacteria [18].

12.7 Evaluation of Compressed Gases

Compressed air, nitrogen, or other inert gases are monitored for microbial content, oil content and other potential contaminants, e.g., moisture. In most instances, membrane filters are used to collect contaminants and incubated in culture media to permit microbial growth.

12.8 Evaluation of Personnel

A normal healthy person sheds about ten million *skin scales* daily. Such scales potentially carry *microorganisms* such as *Staphylococcus* and *Propionibacterium.* Microorganisms are present in noses and throats, wounds and skin infections. Poor personal *hygiene* will result in microorganisms contaminating our hands, therefore, before personnel are allowed to work in aseptic environments, they must pass medical examinations, be adequately trained on aseptic techniques and correct gowning procedures, and periodically be evaluated for their ability to maintain aseptic conditions in the manufacturing environment. Several good references are available for more in-depth treatment of training and evaluation of personnel working in aseptic manufacturing environments [19–21].

In finished product manufacturing areas, production personnel should be evaluated twice a year for their ability to maintain the sterility of the product by undergoing media fills where each employee manipulates sterile filling equipment and fills 300 + vials aseptically with sterile culture media. Additionally, personnel should be monitored daily for levels of contamination by RODAC contact plates on fingers and other parts of the sterile gown. This requirement is becoming standard practice for bulk manufacturing personnel.

13.0 EQUIPMENT LIST

Vessels (316L Stainless Steel or Hastelloy, electro polished)
- DCI; St. Cloud, MN
- Mueller; Springfield, MO
- Precision Stainless; Springfield, MO

Vessels (316L Stainless Steel or Hastelloy, mechanically polished)
- Enerfab; Cincinnati, OH
- Mann Welding; Chattanooga, TN
- Northland Stainless; Tomahawk, WI
- Walker Stainless; New Lisbon, WI

Filters or Filter-Dryers Product
- Cogeim, Charlotte, NC
- DeDietrich (Rosenmund, Guedu); Union, NJ
- Jaygo, Mahwah, NJ

Andritz (Krauss-Maffei), Florence, KY
- Micro Powder Systems; Summit, NJ
- Rosenmund; Charlotte, NC
- Sparkler Filter; Conroe, TX
- Steri-Technologies (Zwag); Bohemia, NY

Dryers, Spray
- APV/Crepaco; Tonawanda, NY
- Niro Atomizer; Columbia, MD

Dryer/Blenders
- GEMCO; Middlesex, NJ
- J. H. Day; Cincinnati, OH
- Micron Powder Systems; Summit, NJ
- Niro-Fielder; Columbia, MD
- Patterson-Kelly; East Stroudsburg, PA
- Processall; Cincinnati, OH

Dryers, Freeze
- Edward High Vacuum; Grand Island, NY
- Finn-Aqua; Windsor Locks, CT
- Hull; Hatboro, PA
- Stokes; Warminster, PA
- Virtis; Gardner, NY

Clean Stem Generators
- AMSCO; Erie, PA
- Mueller; Springfield, MO

Sanitary Pumps
- Cherry Burrell; Cedar Rapids, IA
- Ladish Co; Kenosha, WI
- Waukesha; Waukesha, WI

Filters, Sterilizing
- AMF-Cuno; Meriden, CT
- Gelman; Ann Arbor, MI
- Millipore Corp.; Bedford, MA
- Pall Corp.; Glen Cove, NY

Sterilizers
- AMSCO; Erie, PA
- Finn Aqua; Windsor Locks, CT
- Getinge-Sterilizer Corp.; Secaucus, NJ

Vial Washers and Sterilizers
- Bausch & Stroebel; Clinton, CT
- Cazzoli; Plainfield, NJ
- Despatch Inc.; Wheeling, IL
- Gilowy; Hicksville, LI, NY
- Strunck-Bosch Packaging Inc.; Piscataway, NJ

Stopper Washers
- Huber; Hicksville, LI, NY
- Industrial Washing Machine Co.; Matawan, NJ

Filling Equipment
- Bausch & Stroebel; Clinton, CT
- Bosch Packaging Inc.; Piscataway, NJ
- Cozzoli; Plainfield, CT
- TL Systems; Minneapolis, MN

Stoppering Equipment
- Cozzoli; Plainfield, NJ
- TL Systems; Minneapolis, MN
- West Co.; Phoenixville, PA

HEPA Filters and Systems
- American Air Filter; Louisville, KY
- Envirco; Alburquerque, NM
- Flanders Filters; Washington, NC
- Farr Co.; Los Angeles, CA
- Lunaire Environmental; Williamsport, PA

Valves, Sanitary
Hill McCanna, Carpenterville, IL
Page ITT Corp.; Lancaster, PA
Saunders; Houston, TX

REFERENCES

[1] Food and Drug Administration, Current good manufacturing practices for finished pharmaceuticals, Fed. Reg. 43 (1978) 45076.

[2] Guide to Inspection of Bulk Pharmaceutical Chemicals, Food and Drug Administration, Rockville, MD (September, 1991).

[3] Food and Drug Administration, Guideline on Sterile Drug Products Produced by Aseptic Processing, Government Printing Office, Washington, DC, 1987.

[4] H.L. Avallone, Aseptic processing of non-preserved parenterals, J. Parenteral Sci. Tech. 43 (3) (1989) 113.

[5] H.L. Avallone, CGMP inspection of new drug products, Pharm. Tech. 13 (10) (1989) 60–68.

[6] S. Khan, Automatic flexible aseptic filling and freeze-drying of parenteral drugs, Pharm. Tech. 13 (10) (1989) 24–28.

[7] H.L. Avallone, Manufacture of clinical products, Pharm. Tech. 14 (9) (1990) 150–158.

[8] R.M. Thornton, Pharmaceutically sterile clean rooms, Pharm. Tech. 14 (2) (1990) 44–48.

[9] D.H. Gold, GMP issues in bulk pharmaceutical chemical manufacturing, Pharm. Tech. 16 (4) (1992) 74–83.

[10] J. Agalloco, Steam sterilization-in-place technology, J. Parenteral Sci. Tech. 44 (5) (1990) 253–256.

[11] J.W. Snowman, Freeze drying of sterile products, in: M.J Groves, W.P Olson, M Anisfeld (Eds.), Sterile Pharmaceutical Manufacturing, vol. 1, Interpharm Press, USA, 1991.

[12] Committee on Microbial Purity, Validation and environmental monitoring of aseptic processing, J. Parenteral Sci. Tech. 44 (5) (1990) 272–277.

[13] C.J. Sawyer, R.W. Stats, Validation requirements for bulk pharmaceutical chemical facilities, Pharm. Eng. (1992) 44–52.

[14] P. Oles, B. St. Martin, J. Teillon, D. Meyer, C. Picard, Isolator technology for manufacturing and quality control, in: W.P. Olson, M.J. Groves (Eds.), Aseptic Pharmaceutical Manufacturing, Interpharm Press, IL, 1987, pp. 219–245. J.P. Lysfjord, P.J. Haas, H.L. Melgaard, I.J. Pflug, Barrier isolation technoloogy: a systems approach, in: M.J. Groves, R. Murty (Eds.), Aseptic Pharmaceutical Manufacturing II, Interpharm Press, IL, 1995, pp. 369–414.

[15] Federal Standard 209, Airborne Particulate Cleanliness Classes in Clean Rooms and Clean Zones, GSA Specifications Section, Washington, D.C, 1992.

[16] Commission of the European Community, Guide to Good Manufacturing a practice for Medicinal Products, vol. IV of The Rules Governing Medicinal Products in the European Community, ECSC-EEC-EAEC, Brussels, 1989.

[17] Microbial Evaluation and Classification of Clean Rooms and Clean Zones, Pharm. Forum, United States Pharmacopeial Convention, Inc., 1991, pp. 2399–2404.

[18] PDA Environmental Task Force, Fundamentals of a microbiological environmental monitoring program, J. Parenteral Sci. Tech. 44 (1990) S3–S16.

[19] C.J. Luna, Personnel: the key factor in clean room operations in pharmaceutical dosage forms, in: K.E. Avis, L.L. Lachman, H.A. Hieberman (Eds.), Parenteral Medications, Vol. 1, Marcel Dekker, New York, 1986.

[20] J.W. Levchuk, A.G. Lord, Personnel issues in aseptic processing, Biopharm (1989) 34–40.

[21] M.J. Akers, Good aseptic practices: education and training of personnel involved in aseptic processing, in: M.J. Groves, R. Murty (Eds.), Aseptic Pharmaceutical Manufacturing II, Interpharm Press, Buffalo Grove, IL, 1995, pp. 181–222.

Chapter 19

Environmental Concerns

Elliott Goldberg and Maung K. Min

1.0 ENVIRONMENTAL REGULATIONS AND TECHNOLOGY

1.1 Regulatory Concerns

Environmental laws and regulations including permits are reviewed in this chapter. Included are the Federal Clean Air Act Amendment (CAAA), the Federal Clean Water Act (CWA) regulations, the Resource Conservation and Recovery Act (RCRA) or, as it is also known, the Solid Waste Disposal Act. Along with the regulations under OSHA the National Institute for Occupational Safety and Health (NIOSH) and the Hazardous Waste Operations and Emergency Response (HAZWOPER) are also discussed.

The environmental regulations covered here are not intended to be all-inclusive but to provide a basic understanding of the important environmental laws and regulations.

1.2 Technology

The environmental technology section includes reviews of waste water treatment and air and waste minimization/pollution prevention. Waste water treatment procedures discussed include biological treatment, activated carbon adsorption, air and steam stripping, chemical precipitation, ion exchange, and membrane separation.

Air pollution control technology includes thermal incineration, catalytic incineration, carbon adsorption, absorption, condensation, baghouse filtration, wet scrubbing, and electrostatic precipitation.

The range of technology will provide the engineer with a sufficient background to understand the important air control measures.

2.0 LAWS, REGULATIONS AND PERMITS

2.1 Air

The Federal Clean Air Act Amendments (CAA) were initially enacted in 1963 and modified in 1970 and 1977. The Clean Air Act Amendments of 1990 involved major changes to environmental regulations. These included a national permitting system to regulate air pollution emissions. Its purpose was to protect the environment public health by indicating how and when the various industries involved must control a list of air toxics. The regulatory authority was given to the states and local governments. Congress, through the CAA, authorized the EPA to develop the necessary regulations to carry out the provisions of the act.

The EPA established the National Ambient Air Quality Standards (NAAQS), which included allowable ceilings for specific pollutants. However, the states have the option to make any or all parts of the Clean Air Act requirements more stringent than the minimums set by EPA. The EPA is required to regularly evaluate the compliance status of all geographic areas with respect to pollutants, that is, whether the NAAQS is being met for each criteria pollutant. An area where the NAAQS are not met is designated as a *non-attainment area* (N.A.) for that pollutant.

Areas where the Federal Ambient Air Quality Standards are being met are designated as *attainment areas*, are subject to Prevention of Significant Deterioration (PSD) requirements and are required to identify that are attaining or not attaining the standards.

Compliance and noncompliance can be costly. It has been estimated that the installed cost of equipment and systems to control emissions could range from $20 to $50 billion or higher. The technologies expected to be used include wet scrubbing, thermal incineration, catalytic incineration, carbon absorption, and solvent recovery. New sources and modifications of existing sources of air pollution in an attainment area are regulated under the Prevention of Significant Deterioration Program (PSD). PSD review is required if the new source or modifications result in a net emission increase above specified levels.

The specific pollutants referred to include carbon monoxide, nitrogen dioxide, lead, ozone, inhalable particulates, and sulfur dioxide.

Primary and secondary standards also were set by EPA, with secondary standards reflecting levels necessary to protect welfare in addition to health.

An area may be in an attainment status for one pollutant and in a non-attainment status for another pollutant. In most areas, PSD authority has been assigned to either the state or local jurisdiction. The use of the Best Available Control Technology (BACT) is required for each pollutant and is based on the emission level and capital and operating costs. Regulations in non-attainment areas are required to meet the EPA's New Source Review (NSR) regulations.

The Clean Air Act of 1990 included a list of 189 toxic chemicals to be controlled and such emissions are to be reduced 90% by the year 2000. It also included the phasing out of chlorinated fluorocarbons (CFC's) and carbon tetrachloroflurocarbons (HCFC's) by 2030.

All new and modified emission sources must meet the New Source Performance Standards (NSPS). These standards are generally less stringent than either the Best Available Control Technology (BACT) or the Lowest Available Emission Rate (LAER).

The National Emission Standards for Hazardous Air Pollutants (NESHAPS) specify emission standards for various hazardous air pollutants and cover asbestos, arsenic, benzene, beryllium, mercury, vinyl chloride, PVC, etc.

The CAAA was promulgated to strengthen the federal air protection program and concerns about air toxics by including an expanded National Emission Standards for Hazardous Air Pollutants (NESHAP) program. Concerns over the effects of hazardous air pollutants (HAP) or air toxics resulted in the Title V operating permit program. The relationship of the Title V program to other CAA titles is shown in Figure 19.1.

2.2 Water

In 1972 Congress enacted the Federal Water Pollution Control Act known as the Clean Water Act (CWA). In 1977 further amendments were enacted which strengthened the provisions of the Clean Water Act. Further refinements

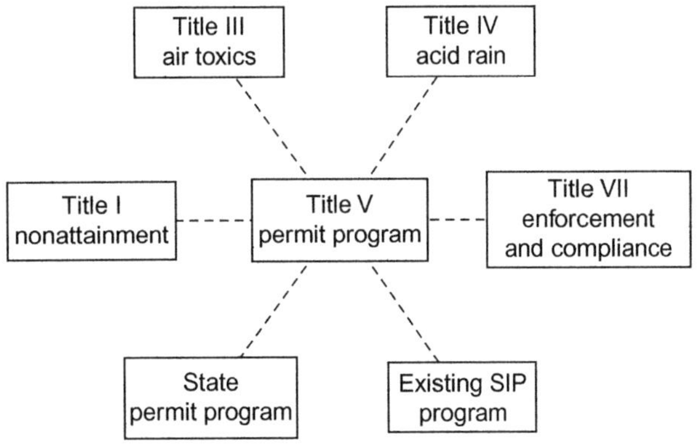

FIGURE 19.1 The relationship of the Title V program to other CAA titles.

were enacted by Congress with the Water Quality Act Amendments of 1987. The purpose of the CWA was to restore and maintain the chemical, physical, and biological integrity of our country's waters. It set up specific effluent guidelines for SIC industry categories with BOD and suspended solids continuing to serve as the primary parameters. The National Pollutant Discharge Elimination System (NPDES) was set up and authorized EPA to establish and enforce effluent limitations on waste water discharges.

Designated priority pollutants were introduced with a permit program and aquatic toxicity also became a permit requirement. Volatile Organic Compounds (VOC) emissions from waste water treatment plants were severely restricted and control of nutrients such as nitrogen and phosphorous were required.

The Clean Water Act established standards that area-wide waste water treatment plants be developed and implemented to assume adequate control of the quality of the effluent for industrial discharges of toxic pollutants into Publicly Owned Treatment Works (POTW). It also stated that federal financial assistance would be provided to construct publicly owned waste water treatment works.

It also established that the federal agencies, the state water pollution control agencies, interstate agencies, and the municipalities and industries involved, prepare or develop comprehensive programs for preventing, reducing or eliminating the pollution of navigable waters and ground waters and improving the sanitary condition of surface and underground waters. Under this legislation, due regard shall be given improvements which are necessary to conserve such waters for the protection and propagation of fish and aquatic life.

The Clean Water Act lays the basis for technology based effluent standards of conventional pollutants such as Biochemical Oxygen Demand (BOD), Total Suspended Solids (TSS), fecal coliform, oil and grease, pH, toxic pollutants, and non-conventional pollutants such as active pesticides, ingredients used in the pesticide manufacturing industry, etc.

A complete list of toxic pollutants can be found in the Code of Federal Regulations, 40 CFR, Part 401.15.

The CWA established requirements for setting standards for discharges from new sources for specific industries. It also lists requirements for preventing and responding to accidental discharges of oil or hazardous substances into navigable waters with notification requirements for releases, removal requirements, liability standards and civil penalties. Furthermore, the CWA established permitting programs to control discharges and severe civil and criminal enforcement provisions for failure to comply with the law.

Compliance with the CWA must be incorporated into the design and operation of every chemical process plant.

To summarize, the focus of the CWA is the control of pollutants in effluent discharged from a facility through any conveyance to virtually any stream or significant body of water. These discharges are primarily controlled through the National Pollution Discharge Elimination System (NPDES).

If the discharge is to a Publicly Owned Treatment Works, the plant needs to meet pretreatment standards to limit pollutants that cannot be readily removed by the POTW.

Discharges from the POTW are required to be in accordance with the effluent limitations contained in the NPDES permit for the POTW. If the facility discharges directly into receiving waters, the facility must file for and obtain its own NPDES permit.

2.3 Solid Waste

Resource Conservation and Recovery Act (RCRA) (Solid Waste Disposal Act) was originally enacted by Congress in 1976 and amended several times subsequently. The 1984 amendments set deadlines for enforcing the regulations. They also placed restrictions on disposal of wastes on land and forced tighter regulation of hazardous wastes.

In effect, Congress gave EPA the authority to control hazardous wastes from their generation to their ultimate disposal.

Congress also sought to encourage the recycling of recoverable material. The RCRA included the statements that:

- Millions of tons of recoverable material which could be used are needlessly buried each year.
- Methods are available to separate usable materials from solid waste.
- The recovery and conservation of such materials can reduce the dependence of the United States on foreign resources and reduce the deficit in its balance of payments.

Principally, however, Congress aimed at the environment and health:

- Disposal of solid waste and hazardous waste in or on the land without careful planning and management can present a danger to human health and the environment.
- As a result of the Clean Air Act, the Water Pollution Control Act, and other federal and state laws respecting public health and the environment, greater amounts of solid waste, in the form of sludge and other pollution treatment residues, have been created. Similarly, inadequate and environmentally unsound practices for the disposal or use of solid waste have created increased amounts of air and water pollution and other problems for the environment and health.
- Open dumping is particularly harmful to health since it can contaminate drinking water from underground and surface supplies and pollutes the air and land.
- Alternatives to existing methods of land disposal must be developed, since many of the cities in the United States will be running out of suitable solid waste disposal sites within five years, unless immediate action is taken. Objectives of the Act are to promote the protection of health and the environment and to conserve valuable material and energy resources by providing technical and financial support to state and local governments and interstate agencies for the development of solid waste management plans, including resource recovery and resource conservation systems.

Furthermore the act proposed to prohibit future open dumping on the land and required the conversion of existing open dumps to facilities which do not pose a danger to the environment or health. The Act requires that hazardous waste be properly managed, thereby reducing the need for corrective action at a future date. An important consideration of RCRA was that it required the promulgation of guidelines for solid waste collection, transport, separation, recovery, and disposal practices and systems.

The Act established specific procedures for creating standards. Enforcement of job safety and health standards were also written into the Act.

2.4 Occupational Safety and Health Act (OSHA)

In 1970, Congress enacted the Occupational Safety and Health Act, which requires employers to provide safe and healthful working conditions for their employees. It authorized the Secretary of Labor to set mandatory occupational safety and health standards to protect employees.

As a result, the Occupational Safety and Health Review Commission was created to review the enforcement actions taken by OSHA. The National Institute for Occupational Safety and Health (NIOSH) was set up to research work place safety and health and to recommend standards to OSHA for controlling exposure to harmful and toxic substances.

The OSHA act of 1970 is comprehensive in scope and covers enforcement of standards, penalties, research activities, state programs, financial assistance, employee duties and rights, and OSHA's effect on other laws.

The Hazardous Waste Operations and Emergency Response Standard (HAZWOPER; 29 CFR 1910.120) were issued by OSHA in March 1990. These regulations serve as a guide to a safety and health plan for hazardous waste operations.

The HAZWOPER regulations include the following:

- RCRA Corrective Actions
- Clean up operations for uncontrolled hazardous waste sites, including voluntary operations, and routine operations
- Emergency responses where there is a release of hazardous substances or a potential release of hazardous substances exists

The formulation of safety and health plans for hazardous waste operations, which include the following:

- A preliminary site characterization analysis and hazard assignment before entering a site known to be contaminated
- A site specific safety and health program to control safety and health hazards
- A training program for all employees and contractors employed on the site who may be involved in hazardous waste activities
- A medical surveillance program
- Proper work practices, including appropriate personnel protective equipment.
- A site control system to prevent contamination of personnel and equipment
- A monitoring program to establish the appropriate levels of personnel protective equipment
- Decontamination procedures before entering a site
- Observation of applicable regulations issued by the Department of Transportation, Environmental Protection Agency, and the Occupational Safety and Health Act, in handling, labelling, moving and disposing of containers containing contaminated material
- An emergency response plan for emergencies which may occur on site

The preliminary site plan characteristic analysis and hazard assessment should be performed by an experienced and trained technician before entering the site. A more detailed site evaluation and analysis must be done to establish the necessary engineering controls and personnel protective equipment. All potential hazards must be identified and evaluated and an air monitoring program must be set up to ascertain that it is safe for work to begin and proceed. In summary, the hazards must be identified, a health risk assessment performed, a medical surveillance program instituted, potential sources of fire and explosion identified, and other possible risks and hazards evaluated. Site specific safety and health rules must be set up, distributed and posted. Such a plan should include the applicable items from the following list. Workers at the site must be informed of the potential hazards and must be cognizant of the site specific safety and health plan.

Safety and Health Plan:

Safety and Health Procedures
Personnel Responsibilities
Decontamination Procedures
Required Monitoring
Training Requirements
Hazards Identification
Personnel Protective Equipment
Emergency Procedures
Hazardous and Toxic Materials On Site
Medical Surveillance
Hospital Location
Emergency Response Personnel

2.5 Environmental Auditing

Environmental auditing can cover a wide range of objectives. The approach can focus on how well a manufacturing facility is complying with the various environmental regulations, such as the Clean Air Act Amendments (CAAA), the Clean Water Act (CWA), the Resource Conservation and Recovery Act (RCRA), Title III of the Superfund Amendments and Reauthorization Act (SARA), the Comprehensive Environmental Response, Compensation and Liability Act (CERCLA), the Toxic Substances Control Act (TSCA), various aspects of the Occupational Safety and Health Act (OSHA), and can also cover property acquisition. It can also cover the various State regulations, for instance, in New Jersey an environmental audit can cover the Toxic Catastrophe Prevention Act (TCPA), the Spill Act, and the State Permit Compliance. If desired, it can also cover waste minimization and pollution prevention.

The environmental audit offers a comprehensive assessment of a facility's compliance with applicable federal, state and local regulations. It also can identify problems before the local or state regulator can be made aware of them and allows time to correct the inadequacies.

Other advantages of the environmental audit is to allow time to properly assess the problem, plan its solution and allow for funding the capital cost required. There are potential problems in having an environmental audit performed. The results of recent court decisions indicate that the environmental audit results may not be able to be kept in confidence and furthermore, they may be used as evidence of noncompliance in civil or criminal court actions. It is conceivable that an audit can increase the potential liabilities. Consequently, management should be prepared to satisfy any negative findings before the audit is undertaken.

The Department of Justice has developed guidelines for compliance and reporting that will be taken into consideration before assessing penalties for environmental regulations. Consequently, the scope and purpose of the audit should be fully understood and agreed to by both management and the consultant prior to undertaking the audit, and questions such as who does the consultant report, who will be given copies of the environmental audit, who will be in charge of document control, who should meet with the consultant, etc., should be decided.

Audit—Environmental auditing consultants have developed forms for collecting/developing information on various aspects of the site. The more important of these will include the following:

- Site information data
- Types of adjacent land use
- Primary site use
- Site configuration
- Buildings on site, number and size
- For manufacturing on site:
  - Products
  - Intermediates
  - Waste materials disposal
    - Solid
    - Liquid
    - Gaseous
- Chemical storage on site
  - Solid
  - Liquid
  - Gas
- Are underground storage tanks used?
  - Number
  - How long in service
  - Volume
  - Material stored
  - Material of construction
  - Leak tested
- Above ground storage tanks
  - Number
  - How long in service
  - Volume
  - Material stored
  - Material of construction
  - Leak tested
- Waste piles
  - Size/volume
  - How long used
  - Material
  - How contained
- Hazardous/toxic wastes generated on site
- How are wastes handled on site/off site treatment
- Disposal—How
- Site ownership history
- Regulatory/environmental history
- Spills history
- Accident history
- Environmental problems
- List of permits
  - Air emissions
  - Waste water
  - Solid wastes/hazardous wastes

2.6 National Environmental Policy Act

The purpose of the Act as stated in a Congressional Declaration of National Environmental Policy is:

To declare a National Policy which will encourage productive and enjoyable harmony between man and his environment; to promote efforts which will prevent or eliminate damage to the environment and biosphere and stimulate the health and welfare of man; to enrich the understanding of the ecological systems and natural resources to the Nation; and to establish a Council on Environmental Quality.

Congress further agreed that to carry out the policy set forth in this Act; it is the continuing responsibility of the federal government to use all practicable means to improve and coordinate federal plans, function programs, and resources that the nation may:

- Fulfill the responsibilities of each generation as trustee of the environment for succeeding generations.
- To assure for all Americans safe, healthful, productive area, aesthetically and culturally pleasing surroundings.
- Attain the widest range of beneficial uses of the environment without degradation, risk to health or safety, or other undesirable and unintended consequences.
- Preserve important historic cultural and natural aspects of our national heritage and maintain, whenever possible, an environment which supports diversity and variety of individual choice.
- Achieve a balance between population, and resource use which will permit high standards of living and a wide sharing of life's amenities.
- Enhance the quality of renewable resources and approach the maximum attainable recycling of depletable resources.

2.7 Storm Water Regulations

An overview of storm water regulations is included in this section. As a result of the 1987 amendments to the Federal Clean Water Act, the United States Environmental Protection Agency (EPA) adopted rules in 1990 which require permit applications for a number of storm water discharges. The intent of storm water regulations is to reduce and prevent pollution due to storm water.

A primary approach is source reduction and pollution minimization. A number of different regulatory programs cover storm water, which may be treated as either a point or a non-point source discharge. The new federal storm water permitting regulations require permit applications to be submitted for all large and medium municipal separate storm sewer systems.

Storm water discharges from residential or commercial sites, except for construction activities, are not subject to current federal storm water permit application regulations; however, such storm water discharges may be subject to existing state regulations and may be subject to future federal regulations.

The discharge of contaminated storm water to surface water or ground water, including discharges through separate storm sewers, requires an NJPDES permit in the State of New Jersey and other states. Traditionally, discharges of storm water in ground water have not been controlled by the NPDES program.

3.0 TECHNOLOGY (WASTE WATER)

3.1 NPDES

Under the NPDES program, all industrial and municipal facilities that discharge waste water directly into United States waters must obtain a permit. Specifically, the water act requires NPDES permits for discharges from point sources such as municipal waste water treatment plants, industries, animal feed lots, aquatic animal production facilities, and mining operations. NPDES permits specify effluent limitations for each individual industrial and municipal discharge, a compliance schedule, monitoring and reporting requirements, and other terms and conditions necessary to protect water quality. NPDES permits are valid for five years, although EPA may issue them for shorter terms. NPDES permits may be revoked, transferred, or modified.

NPDES permits are available from the EPA or from a state authorized to issue NPDES permits. Upon authorization of a state NPDES program, the state is primarily responsible for issuing permits and administrating the NPDES permit program. State NPDES programs must be consistent with minimum federal requirements.

Under the Federal Clean Water Act's National Pollutant Discharge Elimination System (NPDES) permitting program, two approaches exist for controlling pollutant discharges from individual and municipal waste water treatment facilities: the technology-based approach and the water-quality based approach.

Technology-based controls consist of uniform EPA established standards of treatment that apply to direct industrial dischargers and publicly owned waste water treatment works. These uniform standards, known as *effluent limitations*, generally are in the heart of NPDES permits and place numeric limits on the amount of effluent pollutant concentrations permitted at the point of discharge (end-of-pipe).

Industrial effluent limitations are derived from technologies that are available for treating the effluent and removing pollutants, and also are based on considerations of cost and economic achievability. The water quality based approach is used to develop stricter effluent limitations where technology based controls will not be stringent enough to ensure that waters can support their intended uses.

3.2 Effluent Limitations

EPA and the states issue waste water discharge permits to individual factories, power plants, refineries, and other private companies, based on national effluent limitation guidelines. These are based on chemical, physical and biological characteristics of effluent that industry may dump into water ways. An effluent limitation guideline sets the degree of reduction of a pollutant that can be achieved through the application of various levels of technology. An effluent limitation is a restriction on the amount of a pollutant that can be released from a point source into a water body. The discharge of waste water containing metals has effluent limitations, standards, or prohibitions, expressed in terms of the total metal, that is, the sum of the dissolved and suspended fractions of the metal.

3.3 Continuous Discharger

All permit effluent limitations, standards, and prohibitions, including those necessary to achieve water quality standards, will be stated as maximum daily and average monthly discharge limitations for all dischargers.

3.4 Non-Continuous Discharger

A permittee's noncontinuous discharge is limited and described as follows:

- Frequency
- Total mass
- Maximum rate of discharge of pollutants during the discharge
- Prohibition or limitation of specified pollutants by mass, concentration, or other appropriate measure

3.5 Mass Limitations

All pollutants limited in a discharger's permit will have limitations, standards, or prohibitions expressed in terms of mass except:

- pH, temperature, radiation, or other pollutants which cannot be expressed by mass
- when standards and limitations are expressed in terms of other units of measurement
- if the permit limitations were issued on a case-by-case basis, limitations expressed in terms of mass are infeasible because the mass of the pollutant discharged cannot be related to a measure of operation
- permit conditions to ensure that dilution will not be used as a substitute for treatment

A permittee must comply with pollutants limited in terms of mass. Additionally, pollutants may be limited in terms of other units of measurement, in which case a permittee must comply with both limitations.

3.6 Waste Water Characterization

An understanding of the nature of the physical, chemical, and biological characteristics of waste water is essential in the design and operation of collection, treatment, and disposal facilities, and in the engineering management of environmental quality.

The analyses performed on waste waters may be classified as physical, chemical, and biological. These analyses vary from precise quantitative chemical determinations to the more qualitative biological and physical determinations.

Physical Characteristics. The most important physical characteristic of waste water is its total solids content, which is composed of floating matter, matter in suspension, colloidal matter, and matter in solution. Other physical characteristics include temperature, color, and odor.

Chemical Characteristics. These consist of organic matter, the measurement of organic content, the inorganic matter, and the gases found in waste water. The measurement of organic content is very important because of its importance in both the design and operation of waste water treatment plants and the management of water quality.

Biological Characteristics. Biological aspects with which the sanitary engineer must be familiar include knowledge of the principal groups of microorganisms found in surface and waste waters, as well as those responsible for biological treatment, knowledge of the organisms used as indicators of pollution and their significance, and knowledge of the methods used to evaluate the toxicity of treated waste waters.

3.7 Common Pollutants

Generally, under NPDES program, the following pollutants are required to be monitored and reported.

Oxygen Demand
- Biochemical Oxygen Demand
- Total Oxygen Demand
- Total Organic Carbon

Solids
- Total Suspended Solids
- Total Dissolved Solids

Nutrients
- Inorganic Phosphorus Compounds
- Inorganic Nitrogen Compounds

Detergents and Grease
- MBAS (Methylene Blue Active Substances)
- Oil and Grease

Minerals
- Calcium
- Chloride
- Fluoride
- Magnesium
- Sodium
- Potassium
- Sulfur
- Sulfate
- Total Alkalinity
- Total Hardness

Metals
- Aluminum
- Cobalt
- Iron
- Vanadium

Inorganics
- Cyanide
- Total Residual Chlorine

4.0 WASTE WATER TREATMENT STRATEGY

Different types of waste water streams are generated from various processes in pharmaceutical and biotechnology industries. Treatment of these streams is not only ethically required for not polluting the waters of the nation, but USEPA and the states have mandated strict discharge standards that their plants must meet with their NPDES (or equivalent) permitting requirements. Waste water treatment can no longer be considered as a secondary issue in the plant's scheme of things, but must be considered an integral issue. The cost of operation and the capital cost are considerable and, therefore, should attract management's attention.

Waste water treatment process is generally divided into two types:

- Biological Treatment
- Physical Treatment

Biological treatment utilizes microbial organisms to reduce pollutant loadings of process waste streams to EPA (and/or state) acceptable limits. Physical treatment involves reduction of pollutant of process waste stream utilizing physical procedures, such as stripping, ion exchange, membrane separation, etc.

4.1 Activated Carbon

Activated carbon is a recommended and established process used in separating organic and certain inorganic species from aqueous waste streams. Normally, concentration of the waste species should be 1% or less so that carbon regeneration is less frequent. Since this process is generally cost-effective, it has been applied in numerous industrial municipal and pharmaceutical waste water treatment facilities.

Adsorption is based on physical/chemical interaction between the organic pollutant and the carbon surface. In this process, a filter bed of activated carbon is placed in a vessel and used to adsorb certain components. A large adsorptive surface area is generally used for the process. Different raw materials, including coal, wood, coconut shells, peat, and coke, are used to produce activated carbon. These products may impact carbon effectiveness for a given application. Carbons for waste water adsorption generally have a large adsorptive surface area, about 2.5 to 7.5 million ft./lb. Pore sizes range within 50 Å and 1000 Å. Activated carbon is available in granular and powdered form. Granular carbons are used in the treatment of continuous and semi-continuous waste water streams. It can be packed in canisters or beds which then constitute a process unit. Powdered carbon is used for batch adsorption applications and in conjunction with biological treatment.

Favorable results are achieved when the pollutant is slightly soluble in water, has a high molecular weight, high polarity, and low ionization capability, and when the concentration of any suspended solids is less than 50 ppm.

Once the carbon has become saturated, regeneration of the carbon is required. Regeneration methods include the following:

- Thermal reactivation in a multiple hearth furnace or rotary kiln at temperatures of 1200°F to 1700°F
- Steam stripping
- Ozone medicated on-site oxidation

The method used will vary according to characteristics of the plant site and properties of a given system. Energy cost from carbon adsorption ranges from 5 to 25% of the total operating cost. If carbon is thermally regenerated, equipment cost increases since a furnace, after burner, and a scrubber are needed. If carbon is not regenerated, a means of carbon disposal is needed.

4.2 Air Stripping

For pharmaceutical waste water streams containing volatile organic compounds (VOCs), air stripping can be used to reduce containment discharge. Air stripping can be used for treating waste water containing less than 100 ppm VOC and insoluble organics, such as methylene chloride and toluene. Treatment efficiency using air stripping is dependent on Henry's law, which is as follows:

$$P_A = H_A X_A$$

where,

P_A = Vapor pressure of compound A (atm)
H_A = Henry's law constant of compound A (atm)
X_A = Liquid phase mole fraction of compound A

Henry's constant is a function of temperature and a weak function of composition and pressure, however, the air stripping process is normally run at ambient conditions and published data within 20° and 30°C is available.

Maximum pollutant removal by air stripping can be predicted by the following:

$$\%\ \text{Removal} = \frac{C_{Ai} - C_{Ao}}{C_{Ai}} \times 100 = 1 - \frac{1}{1 + H_A R_v R\, T} \times 100$$

where:

C_{Ai} = initial concentration
C_{Ao} = final concentration
H_A = Henry's constant atm m^3/mole
R_v = volumetric air/water ratio = V/L
V = air fed in cfm
L = water fed in cfm
R = gas constant = 8.206×10^{-5} atm m^3/mol K
T = temperature K

The packed tower is compact and efficient for air stripping. The waste water, in a lime slurry (for phosphate removal and pH control), is first sent to a mixing tank and then to a shell settling tank to settle out calcium phosphate and calcium carbonate. The clarified waste water enters near the top of the packed tower, while air is introduced countercurrently at the bottom of the tower. Waste water product is then sent to a recarbonation basin so that calcium carbonate may be precipitated, removed, and reused.

Air strippers are typically 15 to 60 feet in height, and have diameters ranging from 1 to 10 feet. Water flow rates are typically in the range of 40 to 250 lb/cfm air and air/water volumetric ratios are typically in the range of 10 to 300. Generally, contaminated air from the stripper may be routed to an incubator or vapor phase carbon adsorption.

4.3 Steam Stripping

Steam stripping is used to remove dilute concentrations of ammonia, hydrogen sulfide, and other volatile components from pharmaceutical waste stream. Steam stripping can typically achieve contaminant removal of 99% or better and is effective for the removal of organics having boiling points of less than 150°C. The steam stripping process is carried out in a distillation column, which may be either a packed or tray tower. Steam enters at the bottom of the column while waste water is countercurrently supplied from the top of the distillation column. The product stream, rich in volatile components, may further be treated to recover these components.

Steam strippers are generally designed by the use of computer simulation programs, although preliminary estimates can be prepared by modifications of McCabe-Thiele or Fenske-Underwood-Gilliland methods. The development of reliable equilibrium data is critical to the completion of a successful design. In general, such data should be obtained through pilot scale testing on the actual waste water stream to be treated.

4.4 Heavy Metals Removal

While heavy metals (i.e., chromium, copper, nickel) are not typical pollutants in a pharmaceutical waste water stream, removal becomes an issue in some segments of the industry, namely chemical intermediates. These streams are generally treated at the process source in order to minimize the waste water volume. Also, heavy metal streams must be treated prior to any biological treatment that the waste water also requires. Since heavy metals are toxic to microorganisms (even at very low concentrations), their presence reduces biological treatment efficiency.

4.5 Chemical Precipitation

Waste stream metal loading can be reduced by hydroxide precipitation. Hydroxide precipitation uses lime or liquid sodium hydroxide as reactants to form insoluble metal hydroxide. Solids are settled, filtered, and removed as sludge. Liquid sodium hydroxide or quick lime is commonly used as a precipitation reagent. Generally, lime is cheaper than other reagents, however, it has a higher operating cost because it is difficult to handle.

Chrome bearing waste water requires pretreatment, since hexavalent chromium will not react with hydroxide. Hexavalent chrome must be reduced to the trivalent form by reaction with ferrous sulfate, sodium meta-bisulfite or sulfur dioxide. The reduction must be carried out at a pH below 3.0. Achieving complete reduction is important, since any remaining hexavalent chromium will remain in solution in the effluent. The pH is then raised to pH 4.5 to 8.0 and hydroxide precipitation is carried out.

Sludge generated from this process requires careful handling as this waste is considered hazardous. Water content of the sludge can be reduced, generally using plate and frame presses or clarifiers. Plate and frame presses, generally require more operation and attention, however, they can achieve higher solids concentration in the cake, typically 50% to 60%.

4.6 Electrolysis

Electrolysis is the reaction of either oxidation or reduction taking place at the surface of conductive electrodes immersed in an electrolyte, under the influence of an applied potential. This process is used for reclaiming heavy metals from concentrated aqueous solutions. Application to waste water treatment may be limited because of cost factors. A frequent application is the recovery for recycle or reuse of metals, like copper, from waste streams. Pilot applications include oxidation of cyanide waste and separation of oil-water mixtures. Gaseous emissions may occur and, if they are hazardous and cannot be vented to the atmosphere, further treatment, such as scrubbing, is required. Waste water from the process may also require further treatment.

The most common waste water treatment application of electrolysis is the partial removal of heavy metals from spent pickling solutions. When the typical concentration of the spent pickling solution is a 2 to 7% copper, the system design is similar to that of a conventional electroplating bath. When recovering from more dilute streams, mixing and stirring are necessary to increase the rate of diffusion; it may also be necessary to use a large electrode surface area and a short distance between electrodes.

Another concern with this method is the removal of collected ions from the electrodes. This removal may or may not be difficult, but must be addressed. The material collected must be ultimately disposed of if it is not suitable for reuse.

4.7 Ion Exchange

Ion exchange can be useful for heavy metal removal, particularly for nickel, zinc, copper, or chrome, where the metals can be recovered from the regenerating solution and recycled to the process or sold. Ion exchange has also been applied to treatment of streams containing complexing agents or their compounds that would interfere with a precipitation process.

Ion exchange is a two-step process. First, a solid material, the ion exchanger, collects specific ions after coming into contact with the aqueous waste stream. The exchanger is then exposed to another aqueous solution of a different composition that picks up the ions originally removed by the exchanger. The process is usually accomplished by sending the two aqueous streams through one or more fixed beds of exchangers. The ion-rich product stream may be recovered or disposed and the ion-poor stream is usually dilute enough for discharge to sewers.

The chemistry of the ion exchange process may be represented by the following equilibrium equations:

Reaction:

$$RH + Na^{+} \leftrightarrows RNa + H^{+}$$
$$RNa_2 + Ca^{++} \leftrightarrows RCa + 2Na^{+}$$

Regeneration:

$$RNa + HCl \leftrightarrows RH + NaCl$$
$$RCa + 2NaCl \leftrightarrows RNa_2 + CaCl_2$$

where R represents the resin.

These equations represent the reactions involved in the removal of sodium and calcium ions from water, using a synthetic cationic-exchange resin. The extent of completion of the removal reactions shown depends on the equilibrium that is established between the ions in the aqueous phase and those in the solid phase. For the removal of sodium, this equilibrium is defined by the following expression:

$$\frac{[H]X_{RNa}}{[Na]X_{RH}} = K_{H} \rightarrow Na$$

Where

$K_{H} \rightarrow Na$ = selectivity coefficient
[ ] = concentration in solution phase
X_{RH} = mole fraction of hydrogen on exchange resin
X_{RNa} = mole fraction of sodium on exchange resin

The selectivity coefficient depends primarily on the nature and valence of the ion, the type of resin and its saturation, and the ion concentration in waste water. Although both natural and synthetic ion exchange resins are available, synthetic resins are used more widely because of their durability.

To make ion exchange economical for advanced waste water treatment, it would be desirable to use regenerants and restorants that would remove both the inorganic anions and the organic material from the spent resin. Chemicals successful in the removal of organic material from resins include sodium hydroxide, hydrochloric acid, and methanol.

4.8 Membrane Technology

Reverse osmosis uses semipermeable membranes and high pressure to produce a clean permeate and a retentate solution containing salts and ions, including heavy metals. The technique is effective if the retentate solution can be reused in the process. The equipment tends to be expensive, and fouling of the membranes has been a common problem. Considerable research effort is being carried out on membrane processes, however, and they are likely to be more commonly applied in the future. Concentrations of dissolved components are usually about 34,000 ppm or less.

Reverse osmosis employs a semipermeable membrane that allows passage of the solvent molecules, but not those of the dissolved organic and inorganic material. A pressure gradient is applied to cause separation of the solvent and solute. Any components that may damage or restrict the function of the membrane must be removed before the process is performed. Capital investment and operating costs depend on the waste stream composition.

4.9 Organic Removal

Sequencing Batch Reactor (SBR). SBRs are used in batch modes. They consist of one or more reactors used for equalization, aeration and clarification in sequence.

Operation cycle of a typical SBR system consists of five steps, (*A*) fill, (*B*) react, (*C*) settle, (*D*) draw and (*E*) idle.

A. Fill—the waste water is pumped into reactor tank, controlled either by volume or time.
B. React—in this step, the aeration equipment is energized to supply air to the batch, where oxidation of the organic occurs.
C. Settle—solids are separated from the liquid in this step and the clarified liquid is then discharged from the system.
D. Draw—in this step, clarified water is discharged from the reactor.
E. Idle—this step is used where two or more reactors are connected in an SBR system. One reactor remains idle while others (one or more) are filled.

SBR systems are generally advantageous to use over conventional flow systems. They include:

- One SBR tank can serve as an equalization tank, reactor tank and clarifier.
- SBR can tolerate high peak hourly flows, since it serves as an equalization tank. Effluent quality is not compromised.
- Operating level in SBR tank is adjustable. Hence, during low flow conditions, a smaller batch can be collected and treated as desired.
- Solids in tank do not get washed out by hydraulic surges as solids can be held in the tank as long as necessary.

4.10 Activated Sludge Systems

Different types of activated sludge systems are used in treating pharmaceutical waste water. Some sludge systems include conventional, complete mix, contact stabilization, extended aeration, and step aeration.

- *Conventional*—This type includes an aeration basin, a clarifier, and a sludge recycle system. The recycle sludge (RAS) and influent enter the aeration basin at the inlet and leave the basin in the opposite or outlet end. The solids in the mixed liquor get separated out in the clarifier and are recycled back to the aeration basin. Aeration in the basin can be achieved by diffused air or mechanical aerations.

- *Complete Mix*—The influent and the RAS are introduced at several points in the center of the aeration basin from a header or central channel. Effluent from the aeration basin is collected from both sides of the basin by means of two effluent channels. A clarifier is used to separate the solids in the basin effluent before the sludge is recycled back to the head of the system. Excess sludge is sent to the sludge handling facility of the treatment plant. Aeration can be accomplished by diffused air or mechanical aeration.
- *Contact Stabilization*—The influent is first mixed with the return activated sludge for approximately 30 minutes or long enough for the organics to be absorbed in a contact basin. Effluent from the contact basin is fed to a clarifier, where the activated sludge is settled out and returned to another aeration/stabilization basin. Inside the stabilization basin the activated sludge is aerated for 3–6 hours before it is mixed with the influent in the contact basin. The absorbed organics are converted to carbon dioxide, water, and new cells, by means of biochemical oxidation. Excess sludge, generated from the process, can be wasted from the bottom of the clarifier or from the stabilization basin outlet.
- *Extended Aeration*—In this process, aeration time is about 24 hours. The activated sludge system operates in an endogenous respiration phase in which there is inadequate organic material in the system to support all the microorganisms present, due to a low BOD loading. Under this type of operating condition, extended aeration process can produce low sludge and a highly treated effluent.
- *Step Aeration*—Influent to the aeration basin is split up into four or more equal streams and are then fed into four parallel channels separated by baffles. Each channel is equivalent to a separate step and these steps are connected together in series, very similar to having four small plug-flow systems arranged in series. The first step is commonly used to reaerate the return activated sludge, assuming the sludge is not oxygen starved when it comes in contact with the influent. With the return activated sludge aerated, the organic material in the influent can be readily absorbed and broken down within a relatively short contact time.

5.0 AIR (EMISSIONS OF CONCERN)

A major concern associated with chemical processing is the emission of air pollutants. The greatest mass of air contaminants consists primarily of the following pollutants:

- Volatile Organic Compounds (VOC)
- Inorganics
- Particulates

5.1 Volatile Organic Compounds (VOC)

VOC's are emitted from chemical processes either controlled or uncontrolled. Control techniques are provided in the following sections. Emission can also result from the incomplete combination of organic constituents and conversion of certain constituents present in the raw material, auxiliary fuel, and/or combustion air.

5.2 Inorganics

Inorganic pollutants, such as HCL, HF, NO_x, SO_x, are also formed as a result of incomplete combustion. Inorganics include:

- Hydrogen chloride and small amounts of chlorine from the combustion of chlorinated hydrocarbons
- Sulfur oxides, mostly as sulfur dioxide (but also SO_3) formed from sulfur or sulfur compounds present in the products and/or fuel mixture
- Nitrogen oxides from the nitrogen in the combustion air and/or from organic nitrogen present in the product.

5.3 Particulates

Particulates emissions are strongly influenced by the chemical composition of the raw material and the auxiliary fuel, type of combustion process, the operating parameters, and the air pollution control system. Most of the pollutants of concern, other than VOC and inorganics, are collected as particulates.

6.0 SELECTING A CONTROL TECHNOLOGY

There are a number of control options for air toxics, particulates, SO_x and NO_x. It is up to the chemical engineer to choose the most cost-effective control equipment for the source application.

Selection of control equipment begins with gathering relevant data on the emissions and key process parameters. The properties of the exhaust-gas stream and the pollutants need to be characterized. These include:

6.1 Exhaust Stream

- Pollutant concentration
- Flowrate
- Temperature
- Pressure

- Moisture content
- Oxygen content
- Heat content
- Corrosiveness
- Explosivity

6.2 Pollutant characteristics

- Particle size distribution
- Molecular weight
- Vapor pressure
- Solubility
- Adsorptive properties
- Lower explosive limits
- Reactivity

This information can be obtained from vent testing, a mass balance, engineering calculations, or simple engineering estimates (using AP-42). Vent testing, though expensive, is the most accurate method. Other important required information is as follows:

- The level of control required by the regulatory agency must be known. This will allow the minimum control efficiency to be established.
- Site-specific issues impacting the selection of the control equipment, must be quantified. These include availability of utilities, pace constraints, disposal options, and cost of residue generated by emissions control.
- Cost effectiveness has to be judged. Cost effectiveness is defined as the annual operating expense required to control each zone of emissions. The cost effectiveness calculation provides a gauge for ranking various control combinations within a facility so that the greatest emission reduction can be selected for the least cost.
- Secondary environmental impacts, as well as energy impact, also must be considered.
- Consideration must be given to potential for waste minimization at the facility. If feasible, this could reduce or even eliminate the need for emission control equipment.

7.0 VOLATILE ORGANIC COMPOUND (VOC) EMISSIONS CONTROL

The control of VOCs is the single largest environmental challenge facing CPI companies, especially with the enactment of CAAA of 1990. About 80% of the annual air toxic emissions (per SARA Title III, Sec. 313, Form R Reporting) are VOC's.

Most commonly employed control technology for VOC emissions control are as follows:

- Thermal Incineration
- Catalytic Incineration
- Carbon Adsorption
- Condensation
- Absorption

The choice is often dependent on VOC concentration of the stream being controlled, because control efficiency depends on VOC content.

7.1 Thermal Incineration

In incineration, gaseous organic-vapor emissions are converted to carbon dioxide and water through combustion. There are two types of thermal incinerator, based on heat recovery employed, regenerative and recuperative.

Thermal incinerators depend upon contact between the contaminant and the high-temperature combustion flame to oxidize the pollutants. The incinerator, generally consists of refractory-lined chamber, one or more burners, a temperature-control system, and heat-recovery equipment.

Contaminated gases are collected by a capture system and delivered to the preheater inlet, where they are heated by indirect contact with the hot incinerator exhaust. Gases are mixed thoroughly with the burner flame in the upstream portion of the unit, and then passed through the combustion zone, where combustion process is completed. An efficient thermal incinerator design must provide:

- Adequate residence time for complete combustion
- Sufficiently high temperature for VOC destruction
- Adequate velocities to ensure proper mixing

The residence times for incinerators are on the order of 0.5 to 1 seconds, at a temperature ranging from 1200°F to 1600°F. Destruction efficiencies in excess of 95% can be commonly achieved.

Advantages:

- Simple operating concept
- Nearly complete (>95%) destruction of VOC's
- No liquid or solid residual waste generation
- Low maintenance requirements
- Low initial capital costs

Disadvantages:

- High fuel cost

The fuel costs can be minimized by utilizing air preheaters incorporated into the incinerator design. With a recuperative heat exchanger, efficiency of 60% is typical. Thermal incinerators with regenerative heat exchangers can recover 80–95% of the systems energy demands. Regenerative incinerators can initially cost roughly 80% more than recuperative designs. With annual fuel costs of about 10–30% those of recuperative units, the savings can be significant if contaminated air has a low VOC concentration (<25 ppmv) and auxiliary fuel costs are high (>$5/mmBTU).

7.2 Catalytic Incineration

The operation of catalytic incineration is similar to thermal incineration in that heat is used to convert VOC to carbon dioxide and water. The presence of a catalyst lowers the oxidation activation energy, allowing the combustion to occur at about 600°F.

Operation. The preheated gas stream is passed through a catalyst bed, where the catalyst initiates and promotes the oxidation of the organic without being permanently altered. The catalyst is normally an active material, such as platinum, copper chromite, chromium, or nickel, on an inert substrate, such as honeycomb-shaped ceramic. For the catalyst to be effective, the active sites upon which the organic gas molecules react must be accessible. The buildup of polymerized material or reaction with certain metal particulates will prevent contact between active sites and the gas. A catalyst can be reactivated by removing such a coating.

Catalyst cleaning methods are as follows:

- Air blowing
- Steam blowing
- Operating at elevated temperature (about 100°F above operating temperature)

Advantages:

- Nearly complete destruction of VOC (> 95%)
- No residual waste generation
- Low maintenance cost

Disadvantages:

- High capital costs
- Catalyst deactivation over time
- Inability to handle halogenated organics
- Supplement fuel cost

7.3 Carbon Adsorption

Adsorption is a process by which organics are retained on the surface of granulated solids. The solid adsorbent particles are highly porous and have very large surface-to-volume ratios. Gas molecules penetrate the pores of the adsorbent and contact the large surface area available for adsorption. Activated carbon is the most common adsorbent for organic removals.

The amount of VOC retained on the carbon may be represented by adsorption isotherm, which relates the amount of VOC adsorbed to the equilibrium pressure (or VOC concentration) at a constant temperature. The adsorptive capacity of the carbon (expressed as VOClb/Clb) depends not only on properties on the carbon, but also on the properties of the organic. Generally, the adsorptive capacity increases with:

- Increased molecular weight of the VOC
- Polarity
- Degree of cyclization (ringed compound more easily adsorbed than straight chain hydrocarbons)

Regenerative carbon adsorption systems operate in two modes—adsorption and desorption. Adsorption is rapid and removes essentially all the VOCs in the stream. Eventually, the adsorbent becomes saturated with the VOC and system efficiency drops. At this *breakthrough* point, the contaminated stream is directed to another bed containing regenerated adsorbent and the saturated bed is then regenerated. The adsorption cycle typically lasts two hours to many days, depending on the inlet VOC concentration, the variability of organic loading, and the design parameters of the carbon bed. The regenerative cycle typically lasts from one to two hours, including the time needed for drying and cooling the bed.

One important consideration of this system is the operating temperature of the process gas stream. Operating temperature must be less than 100°F. This is because the adsorption capacity decreases with the increase in temperature. The efficiency of carbon adsorption depends on both the concentration of VOC in the gas stream and its composition. Generally, efficiencies of over 95% can be achieved when the organic concentrations are greater than 1,000 ppmv.

Advantages:

- Recovery of relatively pure product for recycle
- High removal efficiency (> 95%)
- Low fuel costs

Disadvantages:

- Potential generation of hazardous organic wastes
- Generation of potentially contaminated waste water
- Higher operating and maintenance cost for disposal of these waters

7.4 Adsorption and Incineration

This process involves a combination of activated carbon adsorption with incineration. The adsorber concentrates the organic laden air before treatment by incineration. This approach is particularly useful for organic streams with a low concentration and higher volumes (<100 ppmv and flowrates over 20,000 cfm), such as paint spray booths. This process has many advantages. These include:

- High destruction efficiency
- Little or no generation of liquid or solid waste of incineration
- Low fuel consumption

7.5 Condensation

Condensation is a basic separation technique where a contaminated gas stream is first brought to saturation and then the contaminants are condensed to a liquid. The conversion of vapor to liquid phase can be accomplished either by increasing the pressure at constant temperature, or reducing the temperature, keeping the pressure constant. Generally, condensation systems are operated at a constant pressure.

The design and operation of the system is affected by the concentration and type of VOC's in the emission stream. Before condensation can occur, the dew point of the system (where the partial pressure of the organic is the same as the system pressure) must be reached. As condensation continues, VOC concentration in the vapor decreases, and the temperature must be lowered even further.

The removal efficiency of the condenser ranges from 50% to 95% or more and depends on the partial pressure of the organic in the gas stream, which is a function of the concentration of the organic and the condenser temperature. For a given temperature, the greatest potential removal efficiencies are achieved with the largest initial concentrations. VOC removal efficiencies via condensation may reach 95% or more for concentrations in excess of 5,000 ppmv.

Plots of vapor pressure versus temperature (Cox charts) are used to determine the temperature required to achieve the desired removal efficiency. Generally, the condenser outlet organic concentration will be greater than 10,000 ppmv for a water-cooled system. For higher removal efficiencies, other coolants, such as a brine solution (−30 °F to 40 °F) may be used.

Condensation offers the advantage of:

- Product recovery
- No disposal problems
- Modest space requirements

Disadvantages include:

- Limited applicability to streams
- Limited applicability to streams with single components if the product is to be recycled and reused.

7.6 Absorption

Absorption is the mass transfer of selected components from a gas stream into a nonvolatile liquid. Such systems are typically classified by the absorbent used. The choice of absorbent depends on the solubility of the gaseous VOC and the cost of the absorbent.

Absorption is a function of both the physical properties of the system and the operating parameters of the absorber. The best absorption systems are characterized by low operating temperatures, large contacting surface areas, high liquid/gas (L/G) ratios, and high VOC concentration in the gas stream. For inlet concentration of 5,000 ppmv, removal efficiencies of greater than 98% may be achieved. Absorption may also be efficient for dilute streams provided the organic is highly soluble in the absorbent, removals of 90% may be attained for concentrations as low as 300 ppmv. Packed towers, venturi scrubbers, and spray chambers may be used for absorption.

The efficiency of an absorber depends on:

- Solubility of VOC in solvent
- Concentration of VOC in the gas stream
- Temperature
- L/G ratio
- Contact surface area

Higher gas solubilities and inlet gas concentrations provide a greater driving force and hence, a higher efficiency. Also, lower temperature causes higher solubility, absorption as enhanced at lower temperatures. Generally, the most economical absorption factor is 1.25 to 2 times the minimum L/G. Absorption increases with contact surface are, thereby, removal efficiency, however, this also raises overall pressure drop through the packed bed, hence increasing energy costs.

8.0 PARTICULATE CONTROL

Most commonly used particulate control technologies are:

- Fabric filters (*baghouses*)
- Cyclones/mechanical collectors
- Electrostatic precipitators

8.1 Fabric Filters (Baghouses)

The basic components of a baghouse are:

- Filter medium in form of fabric bags
- Tube sheet to support the bags
- Gas-tight enclosures
- Mechanism to dislodge accumulated dust from the bags

The particulates laden gas normally enters the lower portion of the baghouse near the collection hoppers, then passing upward through the device, either on the outside or the inside of the bags, depending on the specific design.

Commercially available baghouses employ either felted or woven fabric. A fabric is selected based on its mechanical, chemical, and thermal characteristics. Some fabrics (like nomex) are better suited than others (like polyester) for high temperature operations, some perform well in the presence of acid gases, while others are especially good at collecting sticky particulates because of good release characteristics.

The two design and operational parameters that determine fabric filter performance are air-to-cloth ratio and pressure drop. The air-to-cloth ratio is the volumetric flowrate of the gas stream divided by the surface area of the fabric. The higher the ratio, the smaller the baghouse and higher the pressure drop. Shaker and reverse-air baghouses with woven fabrics generally have a lower air-to-cloth ratio—ranging from 2.0 to 3.5, depending on the dust type being collected. Pulse-jet collectors with felted fabrics have higher air-to-cloth ratios, ranging from 5 to 12.

The pressure drop across the filter medium is a function of the velocity of the gas stream through the filter and the combined resistance of the fabric and accumulated dust layer. Pressure drop across the filter medium is usually limited to 6–8 in. H_2O.

Advantages:

- Performance
- Uniform collection efficiency independent of particle resistivity
- Collection efficiencies exceeding 99 wt%

Disadvantages:

- Clogging of the filter medium, due to condensation in the gas stream
- Cementation of the filter cake in humid, low-temperature gases (especially in the presence of lime from a scrubber)
- Excursions of high particulate concentrations during bag breaks

8.2 Cyclones/Mechanical Collectors

Cyclones are seldom used as the primary means of particulate collection, but often serve as *first-stage* air cleaning devices that are followed by other methods of particulate collection. Cyclone collection efficiency is probably more susceptible to changes in particulate characteristics than are other types of devices, therefore, care should be taken if used. Cyclone operation is dependent, generally, on physical parameters such as particle size, density, and velocity, as opposed to the chemical nature or properties of the material being collected.

8.3 Electrostatic Precipitators

ESPs are generally used to remove particulates from gas streams that can be easily ionized. A typical ESP consists of charged wires or grids and positively grounded collection plates. A high voltage is applied between the negative electrodes and the positive collection plates, producing an electrostatic field between the two elements. In the space between the electrodes, a corona is established around the negatively charged electrode. As the particulate-laden gas passes through this space, the corona ionizes molecules of the electronegative gases present in the stream. These particles get charged and migrate to the oppositely polarized collection plates.

ESPs can be designed for virtually any control efficiency, with most units operating in the 95–99% range. ESP control factors include the Specific Collection Area (SCA), which is the area of the collecting electrodes or plates divided by the volumetric flowrate of the gas. The higher the SCA, the greater the collection efficiency.

Particles with resistivities in the range of 10^4–10^{10} ohm-cm are the most suitable for control by ESP's and most common industrial particulates will exhibit resistivities in this range. Less resistive particles will give up their charge too easily when they contact the collection plates and may be re-entrained in the flue gas. More resistive particles will adhere to the collection plates and be difficult to dislodge, acting as an insulator and reducing the ability of the electrode to further collect particulate matter. Since resistivity changes with temperature, efficient particulate collection requires selection of an optimum ESP operating temperature.

Entrained water droplets in the flue-gas can encapsulate the particles, thus lowering resistivity. High flowrates decrease the residence time of the particles in the ESP, reducing the number of charged particles that migrate to the collection plates.

The particle migration velocity, the rates at which charged particles travel toward the collection plates, also affects ESP efficiency and the unit's design specifications. A slow migration velocity indicates less particle capture per unit of collection plate area. The surface area of the collecting electrode would, therefore, have to be increased for applications involving large quantities of small particles.

Advantages:

- Reliability and low maintenance requirements
- Relatively low power requirements, due to low pressure drops
- High collection efficiencies over a wide ranges of particle sizes
- Ability to treat relatively humid gas streams

A 50,000 acfm ESP operating on a coal fired boiler, with an estimated mean particle diameter of 7 micrometers, could achieve 99.9% control with a total ESP pressure drop of 0.38 in H_2O.

Disadvantages:

- Collection efficiency unreliable if gas property or particle size distribution changes.
- Inorganic particulates difficult to collect
- Requires heating during start-up and shutdown to avoid corrosion because of condensation of acid gases

9.0 INORGANICS

With the passage of the CAAA, emissions control for numerous inorganics have become mandatory. Technologies that can be used effectively to control emissions include:

- Wet scrubbing
- Adsorption
- Incineration

Adsorption and incineration are discussed in Sec. 7.

9.1 Wet Scrubbing

The ionic nature of acids, bases and salts are removed from flue gases by wet scrubbing because the ionic separation that occurs in water creates advantageous equilibrium conditions. Removal may often be enhanced by manipulation of the chemistry of the scrubbing solution.

Both spray towers and packed-bed towers operate based on common principles of absorption. Pollutants in the form of gases are transferred from the gas stream to the scrubbing liquid as long as the gas is not equilibrium in the liquid stream. An important consideration in design of the spray or packed-bed towers is *flooding*. Flooding, where liquid is carried back up the column by the gas stream, occurs when the gas stream velocity approaches the flooding velocity. Tower diameter is established based on superficial gas velocity ranging from 50% to 75% of the flooding velocity.

Spray towers operate by delivering liquid droplets through a spray-distribution system. Generally, the droplets fall through countercurrent gas stream by gravity. A mist eliminator removes liquid entrained in the gas stream prior to its discharge to the exhaust stack. Typical pressure drops in a spray tower are 1–2 in H_2O, and design L/G ratios are, generally, about 20–100 gal/1000 ft^3. Spray towers have relatively low energy requirements (about 3×10^{-4} kW/actm airflow), however, water usage is high. Economics of spray tower operation is influenced by waste water disposal costs. Capital costs consist mainly of cost of the vessel, chemical treatment system, and waste water treatment system.

In packed-bed scrubbers, liquid is flown from the top of the tower and it flows over a random or structured packing. Generally, in the industry, countercurrent flow with high L/G ratio packed-bed scrubber are prevalent, when particulate loadings are higher. These provide the highest theoretical removal efficiencies, because gas with the lowest pollutant concentration contacts liquid with the lowest pollutant concentration, thus maximizing the absorption driving force. Pressure drops of 1–8 in H_2O are typical, while an L/G ratio range of 10–20 gal/1000 ft^3 is generally employed.

Packed-bed towers can achieve removal efficiencies of over 99% and have relatively lower water consumption requirements. They also offer design and retrofit flexibility. Disadvantages include high system pressure drops, relatively high clogging and fouling potential, potentially high maintenance costs, and waste water disposal requirements. Packed-bed scrubbers are also more expensive to install and operate than spray towers.

FURTHER READING

[1] E. Kirschner, Prevention takes priority over cure, Chem. Week (1993) 2732.
[2] Clean Air Act Amendments of 1990 Overview – Policy Guide 1993, Bureau of National Affairs, Inc., 100:101–110.
[3] Ambient Air Quality Standards Overview-Policy Guide 1993, Bureau of National Affairs, Inc., pp. 100:101–1003.
[4] Clean Air Act, 71:1101–1182, Bureau of National Affairs, Inc., 1984.
[5] P. Zahodiakin, Puzzling out the new clean air act, Chem. Eng. (1990) 2427.
[6] K. Hiller, Clean air: a fresh challenge, Chem. Week (1991) 2223.
[7] W.W. Eckenfelder, Industrial wastewater management, Ind. Wastewater (1993) 7072.
[8] G.B. Davenport, Understanding the water pollution, laws governing CPI plants, Chem. Eng. Prog. (1992) 3033.
[9] Environmental Statutes, 1990 Edition, Government Institutes, Inc.
[10] J.E. Roughten, A guide to safety and health plan development for hazardous waste operations, HAZMAT World (1993) 40–43.
[11] Occupational Safety and Health Act – Policy Guide, The Bureau of National Affairs, Inc., 1988, 291:301–306.
[12] Resource Conservation and Recovery Act, RCRA – Policy Guide, The Bureau of National Affairs, Inc., 1991, 291:401–408.
[13] J.J. Balco, Avoiding environmental audit pitfalls, Nat. Environ. J. (1993) 1214.
[14] L. Theodore, J. Reynolds, Introduction to Hazardous Waste Incineration, Wiley-Interscience, New York, 1987.
[15] National Technical Information Services, Physical, Chemical, and Biological Treatment Techniques for Industrial Wastes, vol. 1, PB-275–054, 1977.
[16] Metcalf & Eddy, Wastewater Engineering – Treatment, Disposal, and Reuse, third ed., McGraw-Hill Book Company, New York, 1991.
[17] Office of Research & Development, USEPA, Treatability Manual: vol. III – Technologies for Control/Removal of Pollutants, EPA-600/8-80-042C, July 1980.
[18] K. Noll, et al., Recovery, Recycle, and Reuse of Industrial Wastes, Lewis Chelsea, Ann-Arbor, MI, 1985.
[19] L.L. Tavlarides, Process Modification for Industrial Source Reduction, Lewis Clark, Ann-Arbor, MI, 1985.
[20] USEPA, Control of Volatile Organic Compound Emissions from Air Oxidation Processes in Synthetic Organic Chemical Manufacturing Industry, PB85-164275.

Chapter 20

Instrumentation and Control Systems

John P. King

1.0 INTRODUCTION

The widespread use of advanced control and process automation for biochemical applications has been lagging as compared with industries such as refining and petrochemicals whose feedstocks are relatively easy to characterize and whose chemistry is well understood and whose measurements are relatively straightforward.

Biological processes are extraordinarily complex and subject to considerable variability. The reaction kinetics cannot be completely determined in advance in a fermentation process because of variations in the biological properties of the inoculant. Therefore, information regarding the activity of the process must be gathered as the fermentation progresses. Directly measuring all the necessary variables which characterize and govern the competing biochemical reactions, even under optimum laboratory conditions, is not yet achievable. Developing mathematical models which can be utilized to infer the biological processes underway from the measurements available, although useful, is still not sufficiently accurate. Add to this the constraints and compromises imposed by the manufacturing process and the task of accurately predicting and controlling the behavior of biological production processes is formidable indeed.

The knowledge base in fermentation and biotechnology has expanded at an explosive rate in recent years aided in part by the development of sophisticated measurement, analysis and control technology. Much of this research and technology development has progressed to the point where commercialization of many of these products is currently underway.

The intent of this chapter is to survey some of the more innovative measurement and control instrumentation and systems available as well as to review the more traditional measurement, control and information analysis technologies currently in use.

2.0 MEASUREMENT TECHNOLOGY

Measurements are the key to understanding and therefore controlling any process. As it relates to biochemical engineering, measurement technology can be separated into three broad categories. These are biological, such as cell growth rate, florescence, and protein synthesis rate; chemical, such as glucose concentration, dissolved oxygen, pH and off gas concentrations of CO_2, O_2, N_2, ethanol, ammonia and various other organic substances; and physical, such as temperature, level, pressure, flow rate and mass. The most prevalent are the physical sensors while the most promising for the field of biotechnology are the biological sensors.

One concern when considering measuring biological processes is the maintenance of a sterile environment. This is necessary to prevent foreign organisms from contaminating the process. In-line measurement devices must conform to the AAA Sanitary Standards specifying the exterior surface and materials of construction for the "wetted parts." Instruments must also be able to withstand steam sterilization which is needed periodically to prevent bacterial buildup. Devices located in process lines should be fitted with sanitary connections to facilitate their removal during extensive clean-in-place and sterilize-in-place operations. Sample ports, used for the removal of a small portion of the contents from the bioreactor for analysis in a laboratory, must be equipped with sterilization systems to ensure organisms are not inadvertently introduced during the removal of a sample.

3.0 BIOSENSORS

Biosensors are literally the fusing of biological substrates onto electric circuits. These have long been envisioned as the next generation of analytical sensors measuring specific biomolecular interactions. The basic principle is first to immobilize one of the interacting molecules, the ligand, onto an inert substrate such as a dextran matrix which is bonded (covalently bound) to a metal surface such as gold or platinum. This reaction must then be converted into a measurable signal typically by taking advantage of some transducing phenomenon. Four popular transducing techniques are:

Potentiometric or amperometric, where a chemical or biological reaction produces a potential difference or current flow across a pair of electrodes.

Fermentation and Biochemical Engineering Handbook.

Enzyme thermistors, where the thermal effect of the chemical or biological reaction is transduced into an electrical resistance change.

Optoelectronic, where a chemical or biological reaction evokes a change in light transmission.

Electrochemically sensitive transistors whose signal depends upon the chemical reactions underway.

One example is the research [1] to produce a biomedical device which can be implanted into a diabetic to control the flow of insulin by monitoring the glucose level in the blood via an electrochemical reaction. One implantable glucose sensor, designed by Leland Clark of the Childrens Hospital Research Center in Cleveland, utilizes a microprobe where the outside wall is constructed of glucose-permeable membrane such as cuprophan. Inside, an enzyme which breaks the glucose down to hydrogen peroxide is affixed to an inert substrate. The hydrogen peroxide then passes through an inner membrane, constructed of a material such as cellulose acetate, where it reacts with platinum producing a current which is used to monitor the glucose concentration.

A commercial example of a biosensor, introduced by Pharmacia Biosensor AB[2], is utilizing a photoelectric principle called *surface plasmon resonance* (SPR) for detection of changes in concentration of macromolecular reactants. This principle relates the energy transferred from photons bombarding a thin gold film at the resonant angle of incidence to electrons in the surface of the gold. This loss of energy results in a loss of reflected light at the resonant angle.

The resonant angle is affected by changes in the mass concentration in the vicinity of the metal's surface which is directly correlated to the binding and dissociation of interacting molecules.

Pharmacia claims its BIAcore system can provide information on the affinity, specificity, kinetics, multiple binding patterns, and cooperativity of a biochemical interaction on line without the need of washing, sample dilution or labeling of a secondary interactant. Their scientists have mapped the epitope specificity patterns of thirty monoclonal antibodies (Mabs) against recombinant core HIV-1 core protein.

4.0 CELL MASS MEASUREMENT

The on-line direct measurement of cell mass concentration by using optical density principles promises to dramatically improve the knowledge of the metabolic processes underway within a bioreactor. This measurement is most effective on spherical cells such as *E. coli.* The measurement technology is packaged in a sterilizable stainless steel probe which is inserted directly into the bioreactor itself via a flange or quick-disconnect mounting (Fig. 20.1).

FIGURE 20.1 Photo of MAX Cell Mass Sensor. *(Courtesy of CEREX Ijamsville, Maryland.)*

By comparing the mass over time, cell growth rate can be determined. This measurement can be used in conjunction with metabolic models which employ such physiological parameters as oxygen uptake rate (OUR), carbon dioxide evolution rate (CER) and respiratory quotient (RQ) along with direct measurements such as dissolved oxygen concentration, pH, temperature, and off gas analysis to more precisely control nutrient addition, aeration rate and agitation. Harvest time can be directly determined as can shifts in metabolic pathways possibly indicating the production of an undesirable by-product.

Cell mass concentrations of up to 100 grams per liter are directly measured using the optical density probe. In this probe, light of a specific wavelength, created by laser diode or passing normal light through a sapphire crystal, enters a sample chamber containing a representative sample of the bioreactor broth and then passes to optical detection electronics. The density is determined by measuring the amount of light absorbed, compensating for backscatter. Commercial versions such as those manufactured by Cerex, Wedgewood, and Monitec are packaged as stainless steel probes that can be mounted directly into bioreactors ten liters or greater, and offer features such as sample debubblers to eliminate interference from entrained air.

Another technique used to determine cell density is spectrophotometric titration which is a laboratory procedure which employs the same basic principles as the probes discussed above. This requires a sample to be withdrawn from the broth during reaction and therefore exposes the batch to contamination.

5.0 CHEMICAL COMPOSITION

The most widely used method for determining chemical composition is chromatography. Several categories have been developed depending upon the species being separated. These include gas chromatography and several varieties of liquid chromatography including low pressure (gel permeation) and high pressure liquid chromatography and thin layer chromatography. The basic principle behind these is the separation of the constituents traveling through a porous, sorptive material such as a silica gel. The degree of retardation of each molecular species is based on its particular affinity for the sorbent. Proper selection of the sorbent is the most critical factor in determining separation. Other environmental factors such as temperature and pressure also play a key role. The chemical basis for separation may include adsorption, covalent bonding or pore size of the material.

Gas chromatography is used for gases and for liquids with relatively low boiling points. Since many of the constituents in a biochemical reaction are of considerable molecular weight, high pressure liquid chromatography is the most commonly used. Specialized apparatus is needed for performing this analysis since chromatograph pressures can range as high as 10,000 psi.

Thin layer chromatography requires no pressure but instead relies on the capillary action of a solvent through a paper-like sheet of sorbent. Each constituent travels a different distance and the constituents are thus separated. Analysis is done manually, typically using various coloring or fluorescing reagents.

Gel permeation chromatography utilizes a sorbent bed and depends on gravity to provide the driving force but usually requires a considerable time to effect a separation.

All of these analyses are typically performed in a laboratory; therefore they require the removal of samples. As the reaction is conducted in a sterile environment, special precautions and sample removal procedures must be utilized to prevent contaminating the contents of the reactor.

6.0 DISSOLVED OXYGEN

Dissolved oxygen is one of the most important indicators in a fermentation or bioreactor process. It determines the potential for growth. A sterilizable probe inserted directly into the aqueous solution of the reactor makes the measurement of dissolved oxygen. Two principles of operation are used for this measurement: the first is an electrochemical reaction while the second employs an amperometric (polarographic) principle.

The electrochemical approach uses a sterilizable stainless steel probe with a cell face constructed of a material which will enable oxygen to permeate across it and enter the electrochemical chamber which contains two electrodes of dissimilar reactants (forming the anode and cathode) immersed in a basic aqueous solution (Fig. 20.2). The entering oxygen initiates an oxidation reduction reaction which in turn produces an EMF which is amplified into a signal representing the concentration of oxygen in the solution.

In the amperometric (polarographic) approach, oxygen again permeates a diffusion barrier and encounters an electrochemical cell immersed in basic aqueous solution. A potential difference of approximately 1.3 V is maintained between the anode and cathode. As the oxygen encounters the cathode, an electrochemical reaction occurs:

$$O_2 + 2H_2O + 4e^- \rightarrow 4OH^- \quad \text{(at cathode)}$$

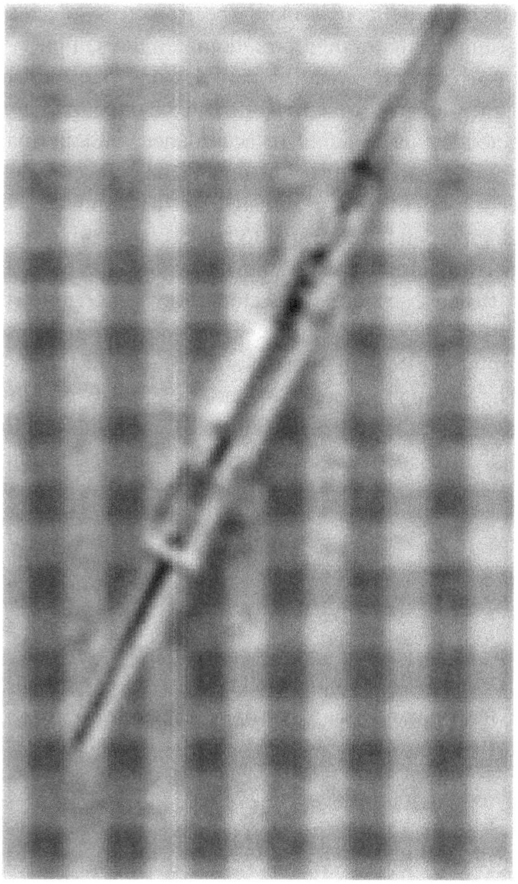

FIGURE 20.2 Sterilizable polarimetric dissolved oxygen probe. *(Courtesy of Ingold Electrodes, Inc., Wilmington, Mass.)*

The hydroxyl ion then travels to the anode where it completes the electrochemical reaction process:

$$4OH^- \rightarrow O_2 + 2H_2O + 4e^- \quad \text{(at anode)}$$

The concentration of oxygen is directly proportional to the amount of current passed through the cell.

7.0 EXHAUST GAS ANALYSIS

Much can be learned from the exchange of gases in the metabolic process such as O_2, CO_2, N_2, NH_3, and ethanol. In fact, most of the predictive analysis is based upon such calculations as oxygen uptake rate, carbon dioxide exchange rate or respiratory quotient. This information is best obtained by a component material balance across the reactor. A key factor in determining this is the analysis of the bioreactor off gas and the best method for measuring this is with a mass spectrometer because of its high resolution. Two methods of operation are utilized. These are magnetic deflection and quadrapole. The quadrapole has become the primary commercial system because of its enhanced sensitivity and its ability to filter out all gases but the one being analyzed.

Magnetic deflection mass spectrometers inject a gaseous sample into an inlet port, bombard the sample with an electron beam to ionize the particles and pass the sample through a magnetic separator. The charged particles are deflected by the magnet in accordance with its mass-to-energy (or charge) ratio—the greater this ratio, the less the deflection. Detectors are located on the opposing wall of the chamber and are located to correspond to the trajectory of specific components as shown in Fig. 20.3. As the ionized particles strike the detectors, they generate a voltage proportional to their charge. This information is used to determine the percent concentration of each of the gases.

The quadrapole mass spectrometer also employs an electron beam to ionize the particles using the quadrapole instead of a magnet to deflect the path of the particles and filter out all but the specific component to be analyzed. The quadrapole is a set of four similar and parallel rods (see Fig. 20.4) with opposite rods electrically connected. A radio frequency and dc charge of equal potential, but opposite charge, is applied to each set of the rods. By varying the absolute potential applied to the rods, it is possible to eliminate all ions except those of a specific mass-to-energy ratio. Those ions which successfully travel the length of the rods strike a Faraday plate which releases electrons to the ions thereby generating a measurable change in EMF. For a given component the strength of the signal can be compared to references to determine the concentration.

The quadrapole, when used in conjunction with a gas chromatograph to separate the components, can measure a wide range of gases, typically from 50 to 1000 atomic mass units (amu).

As mass spectrometers are relatively expensive, the exhaust gas of three or more bioreactors is typically directed to a single analyzer. This is possible because the off gas analysis is done outside the bioreactors themselves. However, the multiplexing of the streams results in added complexity with regard to sample handling and routing, particularly if concerns of cross contamination need be addressed. The contamination issue is usually

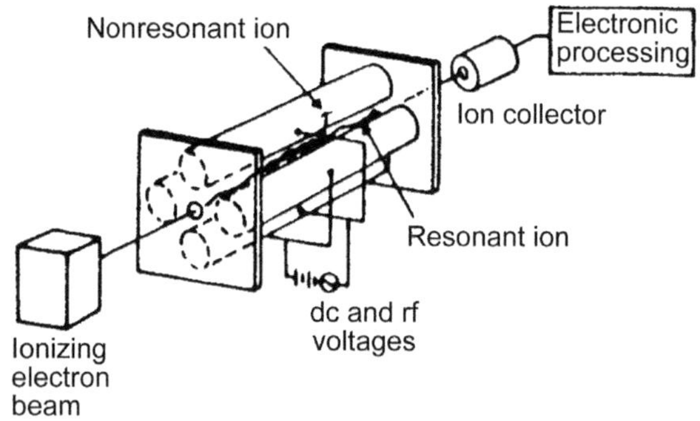

FIGURE 20.4 Quadrapole principle.

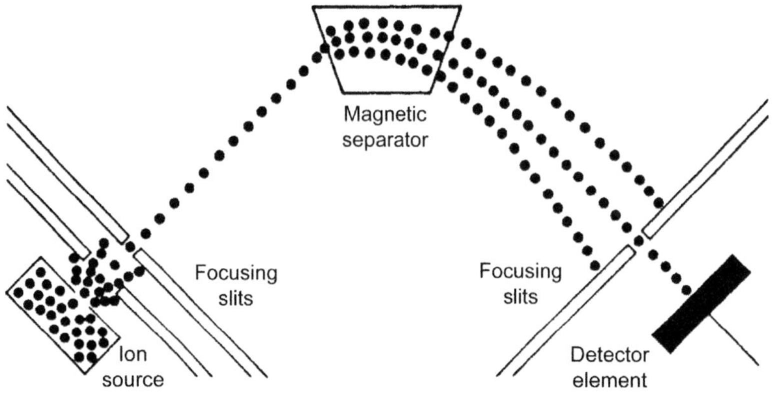

FIGURE 20.3 Magnetic deflection principle.

handled by placing ultrafilters in the exhaust lines. Care, however, must be taken to ensure that these filters don't plug resulting in excessive backpressure. Periodic measurement calibration utilizing reference standards must be sent to the spectrometer to check its calibration.

8.0 MEASUREMENT OF PH

Metabolic processes are typically highly susceptible to even slight changes in pH, and therefore, proper control of this parameter is critical. Precise manipulation of pH can determine the relative yield of the desired species over competing by-products. Deviations of as little as 0.2 to 0.3 may adversely affect a batch in some cases. Like the cell mass probe and dissolved oxygen probes described earlier, the pH probe (see Fig. 20.5) is packaged in a sterilizible inert casing with permeable electrode facings for direct insertion into the bioreactor. The measurement principle is the oxidation reduction potential of the hydrogen ion and the electrode materials are selected for that purpose.

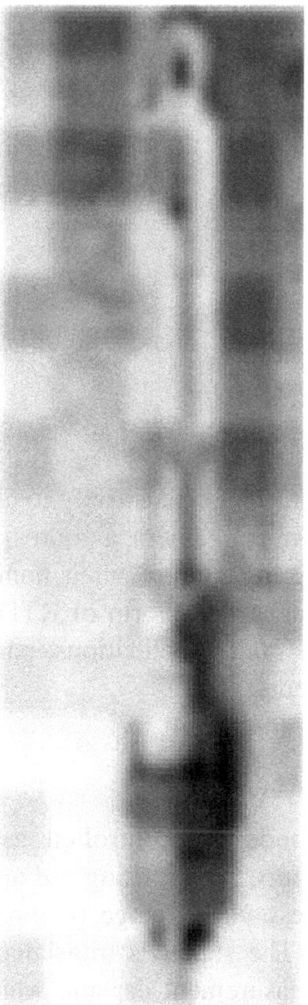

FIGURE 20.5 Ingold sterilizable pH probe. *(Courtesy of Ingold Electrodes, Inc., Wilmington, Mass.)*

9.0 WATER PURITY

Water purity is often very important in biochemical processes. One of the best methods to detect the presence of salts or other electrolytic materials is to measure its resistivity. Conductivity or resistivity probes are capable of measuring conductivities as high as 20,000 microsiemens per centimeter and resistivities as high as 20 megohms per centimeter.

10.0 TEMPERATURE

Precise temperature control and profiling are key factors in promoting biomass growth and controlling yield. Temperature is one of the more traditional measurements in bioreactors so there is quite a variety of techniques.

Filled thermal systems, Fig. 20.6, are among the more traditional temperature measuring devices. Their operating principle is to take advantage of the coefficient of thermal expansion of a sealed fluid to transduce temperature into pressure or movement. This has the advantage of requiring essentially no power and therefore is very popular in mechanical or pneumatic control loops. Although the trend in control is toward digital electronic, pneumatic and mechanical systems are still very popular in areas where solvent or other combustible gases may be present and therefore represent a potential safety hazard. The primary constraint in these types of systems is that the receiver (indicator, recorder, controller) must be in close proximity to the sensor.

Thermocouple assemblies, Fig. 20.7, are a popular measurement choice in electronic systems or in pneumatics where the sensor must be remote. The thermoelectric principle, referred to as the *Seebeck Effect*, is that two dissimilar metals, when formed into a closed circuit, generate an electromotive force when the junction points of the metals are at different temperatures. This conversion of thermal energy to electric energy generates an electric current. Therefore, if the temperature of one juncture point (the cold junction) is known, the temperature of the hot juncture point is determined by the current flow through the circuit. Depending upon the alloys chosen, thermocouples can measure a wide temperature range (−200 to +350°C for copper, constantan) and are quite fast acting assuming the assembly doesn't contribute too much lag in its absorbance and dissipation of heat. Its primary disadvantages are its lack of sensitivity (copper, constantan generates only 40.5 microvolts per °C) and requirement for a precise cold junction temperature reading.

Resistance temperature detectors, RTD's, are more sensitive than thermocouples especially when measuring small temperature ranges. As a result, they are preferred for accurate and precise measurements. The principle behind these devices is based on the use of materials,

FIGURE 20.6 Filled thermal system assembly for temperature measurement. *(Courtesy of the Foxboro Co., Foxboro, Mass.)*

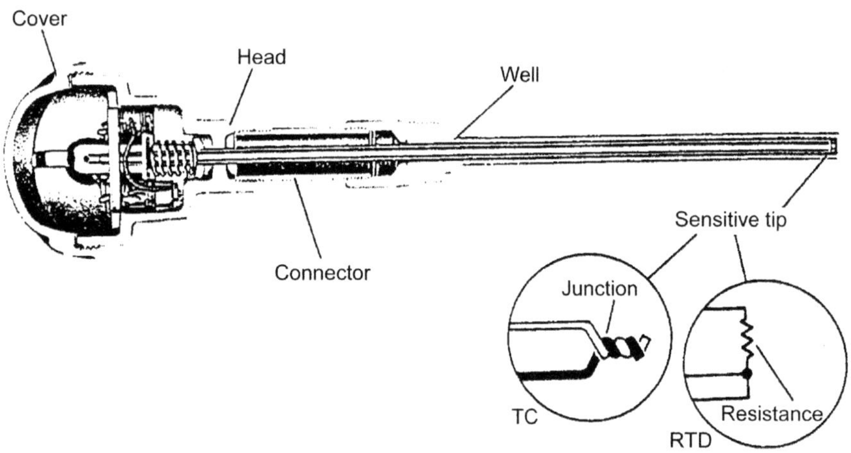

FIGURE 20.7 Cutaway view of thermocouple or resistance temperature detector probe for temperature measurement. *(Courtesy of the Foxboro Co., Foxboro, Mass.)*

such as platinum or nickel, whose resistance to current flow changes with temperature. These materials are used as one leg in a wheatstone bridge circuit with the other legs being known precision resistors. A voltage is applied across the bridge and the voltage drop midway through each path of the circuit is compared. The potential difference at the midway point is directly related to the ratio of each set of resistances in series. Since three of these are known, the resistance of the RTD can be calculated and the temperature inferred. If the RTD is remote from the bridge circuit, the resistance of lead wires can affect the measurement. Therefore, for highly precise measurements, compensating circuits are included which require increasing the wiring for this measuring device from two to as many as four leads.

Thermistors are a special class of RTD's and are constructed from semiconductor material. Their primary advantage is their greater sensitivity to changes in temperature, therefore making them a more precise measuring method. Their disadvantage is their nonlinear response to temperature changes. This form of RTD is gaining popularity for narrow range applications, particularly in laboratory environments.

11.0 PRESSURE

Pressure is an important controlled variable. The measurement is obtained by exposing a diaphragm surface or seal to the process via a flange or threaded tap through the vessel wall. The signal is translated through a filled capillary to a measurement capsule which will transduce the signal to one measurable by an electronic circuit by one of several methods. One method is to employ a piezoelectric phenomenon whereby the pressure exerted on

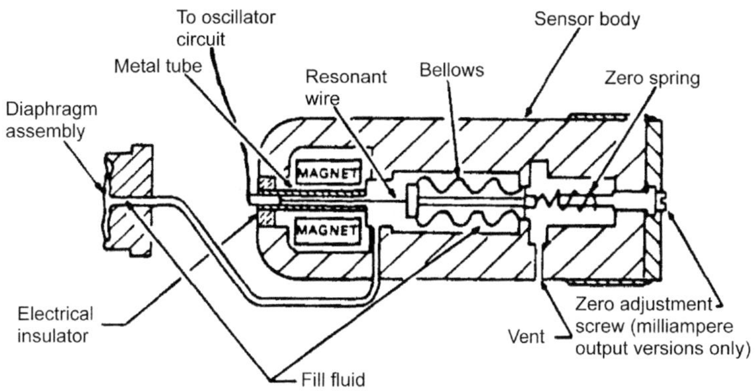

FIGURE 20.8 Diagram of resonant wire technology pressure measurement. *(Courtesy of the Foxboro Co., Foxboro, Mass.)*

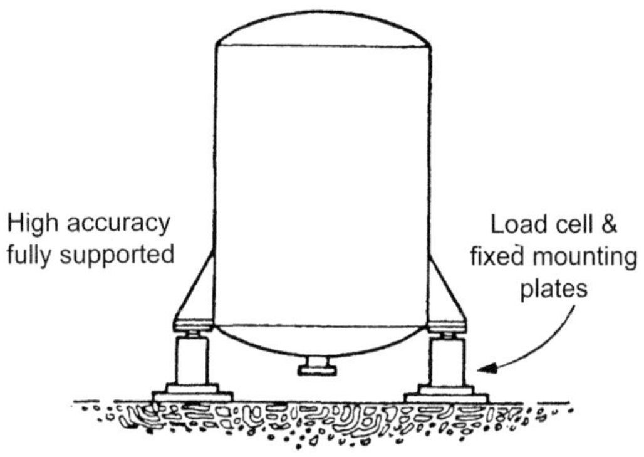

FIGURE 20.9 Schematic of the installation of a load cell.

an asymmetric crystal creates an elastic deformity which in turn causes the flow of an electric charge. A second technology is variable resistance whereby flexure on a semiconductive wafer affects its resistivity which is measured in a similar fashion to RTD's. The third, shown in Fig. 20.8, is the use of a vibrating wire where changes in the tension of the wire changes its resonant frequency which is measured as a change in pulse rate.

Several types of pressure measurements can be taken. These include absolute pressure, where one side of the capsule is exposed to 0 psia in a sealed chamber. Gauge pressure is measured with one side of the capsule vented to atmosphere. Vapor pressure transmitter seals one side of the capsule, filling it with the chemical composition of the vapor to be measured. The vapor pressure in the sealed chamber is compared with the process pressure (at the same temperature). If equal, the compositions are inferred to be equal. This technique is used primarily for binary mixtures as multicomponent compositions have too many degrees of freedom.

12.0 MASS

Weigh cells or load cells are typically used to measure the mass of the contents of a vessel. These are electromechanical devices which convert force or weight into an electrical signal. The technique is to construct a wheatstone bridge similar to that used in the RTD circuit with one resistor being a rheostat which changes resistance based on load.

Three configurations are popular. These are the column, where the cell is interposed between one leg of the vessel and the ground (see Fig. 20.9) and is typically used for weights exceeding 5000 pounds. The second is the cantilever design, where the weight is applied to a bending bar and is used for weights under 500 pounds. The third is the shear design, where the weight is applied to the center of a dual strain gage arrangement.

13.0 MASS FLOW RATE

A Coriolis meter utilizes a measurement technology which is capable of directly measuring mass flow (instead of inferring mass flow from volumetric flow and density). The Coriolis effect is the subtle correction to the path of moving objects to compensate for the rotation of the earth. This appears as a force exerted perpendicular to the direction of motion and creates a counterclockwise rotation in the Northern Hemisphere and a clockwise rotation in the Southern Hemisphere. This phenomenon is used by the mass flow meter to create a vibration whose frequency is proportional to the mass of the fluid flowing through the meter. This is accomplished via the geometry of the meter (Fig. 20.10), specifically the bends to which the fluid is subjected as it travels through the meter.

14.0 VOLUMETRIC FLOW RATE

Quite a number of technologies are available for measuring volumetric flow rates. These include differential pressure transmitters, vortex meters and magnetic flow meters. Each has its advantages and disadvantages.

The differential pressure transmitter is the most popular and has been in use the longest. Its measurement principle is quite simple. Create a restriction in the line with an orifice plate and measure the pressure drop across the restriction. The measurement takes advantage of the physical relationship between pressure drop and flow. That is, the fluid velocity is proportional to the square root of the pressure drop, and in turbulent flow, the volumetric flow rate is essentially the velocity of the fluid multiplied by the cross-sectional area of the pipe (Fig. 20.11).

Inaccuracies with regard to transmitting the pressures between the sensor and transducer occur at very low flow rates, therefore closely coupled units have been designed for this purpose. Using this approach and small bore orifice plates, extremely low flows can be measured. A 0.38 millimeter diameter bore can accurately measure flows in the 0.02 liters per minute range for liquids and 0.03 cubic meters per hour for gases. Jeweled orifice plates can have a bore as small as 0.05 millimeters in diameter. The primary disadvantages of the differential pressure producing flow measurements are the permanent pressure drop caused by the restriction in the line; sediment buildup behind the orifice plate (which could be a source of bacterial buildup) and loss of accuracy over time as the edge of the plate is worn by passing fluid and sediment. This type of transmitter typically has a limited range (*turndown*)—usually a 4 to 1 ratio between its maximum and minimum accurate flow rates.

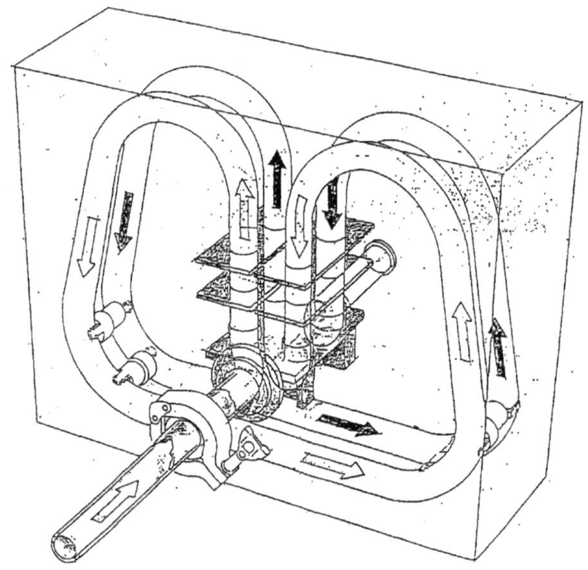

FIGURE 20.10 Schematic of Coriolis Meter flowpath. *(Courtesy of the Foxboro Co., Foxboro, Mass.)*

Vortex meters utilize a precision constructed bar or bluff through the diameter of the flow path to create a disruption in flow which manifests itself as eddy currents or vortices being generated, starting at the downstream side of the bar (Fig. 20.12). The frequency at which the vortices are created are directly proportional to velocity of the fluid. Although these devices contain a line obstruction, the turbulence created by the vortices make the bluff self-cleaning and they are available for sanitary applications. Also, their linear nature makes them a wide-range device with a ratio of as much as 20:1 between the maximum and minimum flow rate. Line sizes as small as 1" are available which are capable of reading flow rates as low as 0.135 liters per minute.

Magnetic flowmeters take advantage of the electrolytes in an aqueous solution to induce a magnetic field in the coils surrounding the meter's flow tube, see Fig. 20.13. The faster the flow rate, the greater the induced field. Interestingly, the ionic strength of the electrolytes has only negligible effect on the induced field so long as it is above the threshold value of 2 microsiemens per centimeter. Because these meters create no obstructions to the flow path they are the preferred meter for sanitary applications.

15.0 BROTH LEVEL

As the broth in a fermenter or bioreactor becomes more viscous and is subjected to agitation from *sparging* (the

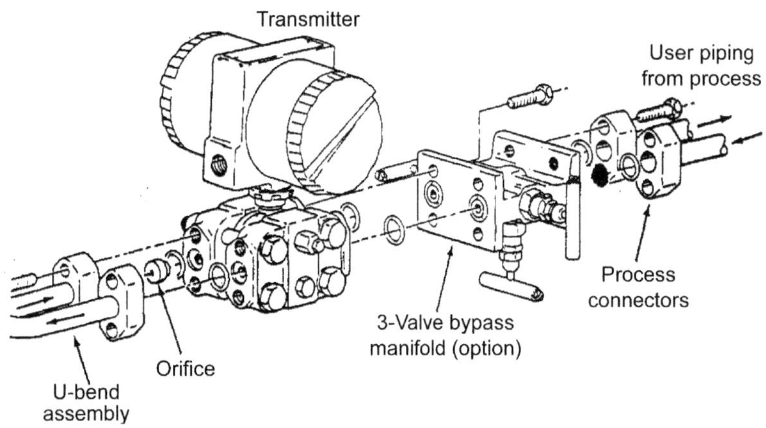

FIGURE 20.11 Integral flow orifice assembly, U-bend configuration. *(Courtesy of Foxboro Co., Foxboro, Mass.)*

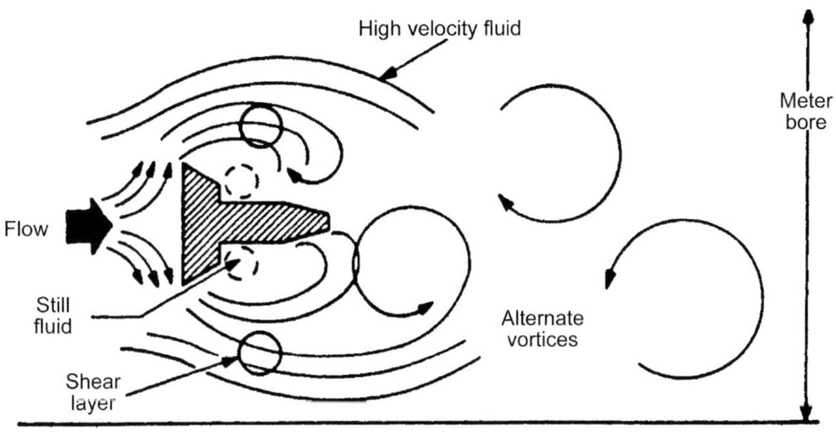

FIGURE 20.12 Vortex creation via shedding bluff. *(Courtesy of Foxboro Co., Foxboro, Mass.)*

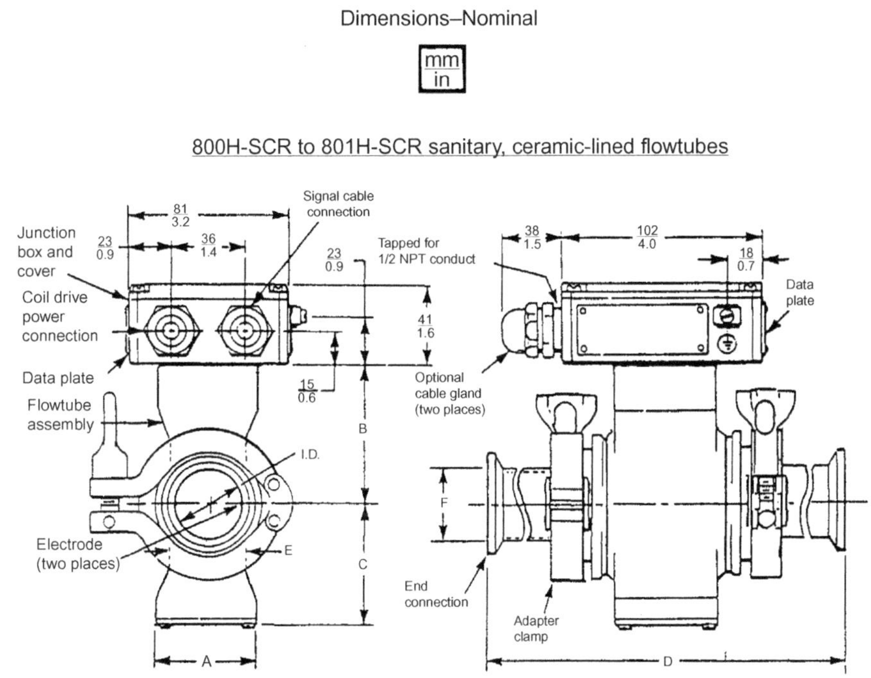

FIGURE 20.13 Cutaway schematic of a sanitary magnetic flowmeter. *(Courtesy of Foxboro Co., Foxboro, Mass.)*

introduction of tiny sterilized air bubbles at the bottom of the liquid) and from mixing by the impeller, it has a tendency to foam. This can be a serious problem as the level may rise to the point where it enters the exhaust gas lines clogging the ultrafilters and possibly jeopardizing the sterile environment within the reactor. Various antifoam strategies can be employed to correct this situation, however, detection of the condition is first required.

Capacitance probes (Fig. 20.14) are one means to accomplish this. The basic principle is to measure the charge between two conductive surfaces maintained at different voltage potentials and separated by a dielectric material. The construction of the probe provides an electrode in the center surrounded by an insulator, air, and a conductive shell. The length of the probe is from the top of the reactor to the lowest level measuring point. As the level in the reactor rises the broth displaces the air between the capacitance plates and thereby changes the dielectric constant between the plates to the level of the broth. The result is a change in the charge on the plate. If the vessel wall can act as a plate (is sufficiently conductive), the preferred approach would be to use an unshielded probe (inner electrode with insulator) to prevent erroneous readings resulting from fouling of the probe. Because of the

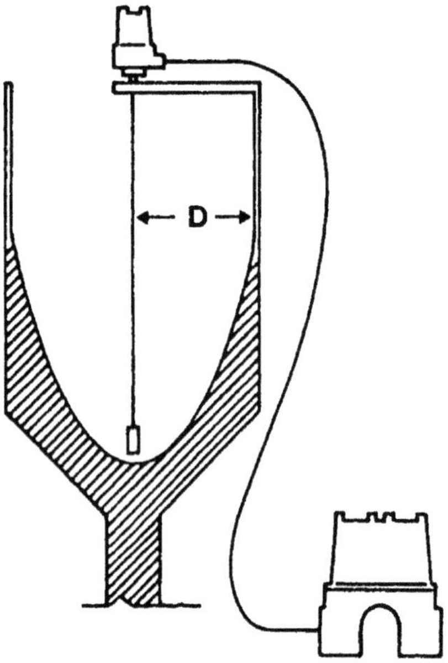

FIGURE 20.14 Installation schematic of a capacitance probe in a vessel.

uncertain dielectric character of the broth, this measurement should only be used as a gross approximation of level for instituting antifoaming strategies.

Several other forms of level measurement technologies are available. One is the float and cable system, where the buoyancy of the float determines the air-broth interface boundary and the length of the cable determines the level. The density of the broth may render this measurement questionable.

A second is hydrostatic tank gauging, where level is inferred from pressure. Again, density, particularly if two phases exist (aqueous and foam), may render this approach questionable.

A third is sonic, which computes the distance from the device to the broth surface based on the time it takes for the sound wave initiating from the device to reflect off the surface of the air-liquid boundary and return.

Several other ingenious variations of these basic approaches are commercially available as well.

16.0 REGULATORY CONTROL

Automatic regulatory control systems (Fig. 20.15) have been in use in the process industries for over seventy years.

Utilizing simple feedback principles, measurements were driven toward their setpoints by manipulating a controlled variable such as flow rate through actuators like throttling control valves. Through successive refinements in first mechanical, then pneumatic, then electronic and finally digital electronic systems, control theory and practice has progressed to a highly sophisticated state.

16.1 Single Stage Control

The fundamental building block has been the proportional plus integral plus derivative (PID) controller whereby the proportional term would adjust the manipulated variable to correct for a deviation between measurement and target or setpoint; the integral term would continue the action of the proportional term over time until the measurement reached the setpoint and the derivative term would compensate for lags in the action in the measurement in responding to actions of the manipulated variable. The classic equation is:

$$m = 100/PB(e + 1/R\, e\mathrm{d}t - D\, \mathrm{d}e/\mathrm{d}t)$$

Judicious application of this control strategy on essentially linear single variable control systems which don't exhibit a prolonged delay (dead time) between action by the manipulated variable and measured response by the controlled variable has proven quite effective. Fortunately most single loop control systems exhibit this behavior.

In highly nonlinear applications such as pH control, or in situations where the dynamics of the process change over time as occurs in many chemical reactions, adjustments to the tuning coefficients are needed to adequately control the modified process dynamics. Self-tuning controllers employing expert rule sets for dynamic retuning the PID settings are available for this class of problem. Many users also use self-tuning controllers to determine the optimum settings for the linear systems described above. One such rule system is the EXACT controller by Invensys (formerly Foxboro) (Fig. 20.16), which automatically adjusts the controller tuning parameters based on the pattern of the measurement signal received.

When the process under control exhibits significant dead time, the problem is considerably more difficult. One approach is to use a simple model-based predictor corrector algorithm such as the Smith predictor [2] which is interposed between the manipulated and controlled variable in parallel with a conventional controller and conditions the measurement signal to the controller based on time conditioned changes to the manipulated variable made by the controller. This works exceedingly well if properly tuned, but is sensitive to changes in process dynamics. Another scheme, introduced by Shinsky [3], utilizes a standard PID controller with a dead time function added to the external reset feedback portion of the loop. This appears to be less sensitive to changes in process conditions.

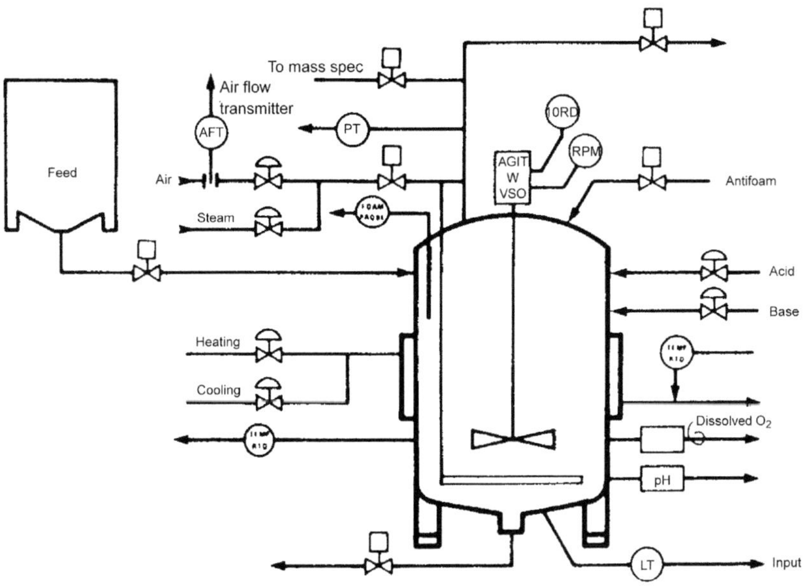

FIGURE 20.15 Typical instrument configuration around a fermenter.

FIGURE 20.16 Model 761 Controller with EXACT tuning. *(Courtesy of the Foxboro Co., Foxboro, Mass.)*

17.0 DYNAMIC MODELING

A control system which anticipates adjustments to the manipulated variables based on changes to one or more controlled variables can be constructed by combining single station controllers with signal characterizers, dynamic compensators and computational elements such as summers and multipliers. Simpler implementations, such as cascade control, will minimize the effect of a deviation of a controlled variable from its target value while dynamic models will anticipate changes to process conditions and adjust the control strategy to compensate based on a leading indicator. A simple example would be the effect on the draw rate and energy input to a distillation column based on a change to its feed rate. The dynamic model in this case would be a material and energy balance around the column compensating for the time delays encountered on each tray as the increased flow rate works its way through the column.

18.0 MULTIVARIABLE CONTROL

Characterizing a process as a set of nonlinear time dependent equations and then developing a strategy which manipulates sets of outputs based on changes to the inputs is another approach gaining momentum in other industries such as petroleum refining. One approach is called Dynamic Matrix Control [4] (DMC) which first automates the process of determining the coefficients for the set of nonlinear equations based on sets of controlled and manipulated variables declared. The method perturbs each of the manipulated variables and determines the corresponding response of the controlled variables. Once the model is constructed, the information is represented in a relative gain matrix to predict the control actions necessary to correct for changing process conditions. Once the DMC is correctly tuned, including dynamic compensations, a predictor corrector algorithm is applied to compensate to changes in the process dynamics over time.

This technique has been applied quite successfully to reaction processes in the petroleum industry including fluid catalytic cracking units and catalytic reformers.

18.1 Batch Control

Batch is a general term given to a diverse set of time dependent control strategies including:

State variable control, such as the opening and closing of a solenoid or the starting and stopping of a motor, including the use of any timing circuits which may be used for alarming in the event the action doesn't achieve its specified results in the allotted time.

The interlocking, sequencing or coordinating of systems of devices to ensure their proper and coordinated operation. Examples include interlocking a discharge pump to the opening of the discharge valve and the alignment of pumps and valves to transfer materials from one vessel to another. This may include actions such as the resetting and starting of totalizers to ensure the proper amount of material was successfully transferred.

The modification of selected process variables in accordance with a prespecified time-variable profile. Two examples are the changing of the reactor temperature over time to conform with a specified profile or the timed periodic addition of nutrient into the bioreactor.

Conducting event driven actions such as adding antifoam upon the detection of excess foam or invoking an emergency shut down routine if an exothermic reaction goes beyond controllable limits.

Performing a sequence of operations in a coordinated manner to produce the desired changes to the contents of a process unit. This would typically include combinations of the above mentioned activities on various sets of equipment associated with the unit.

The Instrument Society of America Committee Group SP88, Batch Control Structure drafted a specification which decomposes batch control into a hierarchal set of activities each with their own purview and problem definition. The objective is to define the properties of the control problem at each level and identify conceptually the appropriate control and information management tools needed for each level. Once defined, a building block approach is taken whereby successively higher levels rely on the foundation established by the controls implemented at the lower-levels. A strategy directed at the operation of a reflux condenser would rely on the definitions already in place for throttling flow to achieve proper temperature control and would merely direct the devices (such as PID controllers) as to the actions required.

This hierarchy is currently depicted [5] as:

Loop/Device, Element Level, *which deals with the real-time devices which interface directly with the process.*

Equipment Module Level, *which utilizes combinations of loops and devices to manage an equipment function such as a reflux condenser within a reactor.*

Unit Level, *which coordinates the equipment modules to manage the process unit.*

Train/Line Level, *which coordinates a set of units to manufacture a batch of specified product.*

Area Level, *which coordinates the manufacture of sets of products being made at the train/line level so as to ensure adequate availability of resources and the optimum utilization of capital equipment.*

Plant Level *is the integration of the manufacturing process with other plant functions such as accounting, quality control, inventory management, purchasing, etc.*

Corporate Level *is the coordination of various plants to ensure a proper manufacturing balance with market needs and financial goals.*

19.0 ARTIFICIAL INTELLIGENCE

A considerable amount of attention is being given to the use of various forms of artificial intelligence for the control of bioreactor systems. Two forms of systems are currently being explored. These are expert systems and neural networks. Expert systems combine stored knowledge and rules about a process with inference engines (forward and backward chaining algorithms) to choose a best or most reasonable approach among a large number of choices when no correct answer can be deduced and in some situations the information may appear to be contradictory.

Neural networks are also being seriously explored for certain classes of optimization applications. These employ parallel solution techniques which are patterned after the way the human brain functions. Statistical routines and back propagation algorithms are used to force closure on a set of cross linked circuits (equations). Weighting functions are applied at each of the intersections.

The primary advantage for using neural networks is that no model of the problem is required (some tuning of the weighting functions may facilitate "learning", however). The user merely furnishes the system with cause and effect data which the program uses to learn the relationships and thereby model the process from the data. Given an objective function, it can assist in the selection of changes to the causes (manipulated variables) to achieve the optimum results or effects (controlled variables).

At BPEC, the Engineering Research Center of Excellence at MIT, advanced computer control of bioprocesses was researched with an eye toward industrial commercialization. Professor Charles Cooney directed the effort to develop expert systems and artificial neural networks to achieve this goal. One of the products resulting from this effort is the Bioprocess Expert developed by Dr. Gregory O'Connor, President of Bioprocess Automation, Inc. in Cambridge. This uses an expert system called G2 from Gensym Corporation, also located in Cambridge.

20.0 DISTRIBUTED CONTROL SYSTEMS

As the knowledge of the physiology and reaction kinetics of biochemical processes has progressed and the measurement systems for monitoring their activity has improved, the need for sophisticated systems able to execute coordinated control strategies including batch has increased. Fortunately the state of the art of control systems has rapidly evolved to the point where all of the control strategies described above can be embodied in a Distributed Control System (DCS), see Fig. 20.17. This transformation has been facilitated to a great extent by the technology breakthroughs in computer, communications, and software technology.

Distributed control systems are organized into five subsystems.

Process interface, which is responsible for the collection of process data from measurement instruments and the issuing of signals to actuating devices such as pumps, motors and valves.

Process control, which is responsible for translating the information collected from the process interface subsystem and determining the signals to be sent to the process interface subsystem based on preprogrammed algorithms and rules set in its memory.

Process operations, which is responsible for communicating with operations personnel at all levels including operator displays, alarms, trends of process variables and activities, summary reports, and operational instructions and guidelines. It also tracks process operations and product batch lots.

Applications engines, which are the repository for all of the programs and packages for the system from control, display and report configuration tools to program language compilers and program libraries to specialized packages such as database managers, spreadsheets and optimization or expert system packages to repositories for archived process information.

Communications subsystems, which enable information flow between the various DCS subsystems as well as to other computerized systems such as laboratory information management systems (LIMS); plant inventory management and scheduling systems such as MRP II; plant maintenance systems and business systems.

The integration of these systems into a cohesive whole has dramatically increased the level of automation possible to improve the quality, productivity and economics of manufacturing.

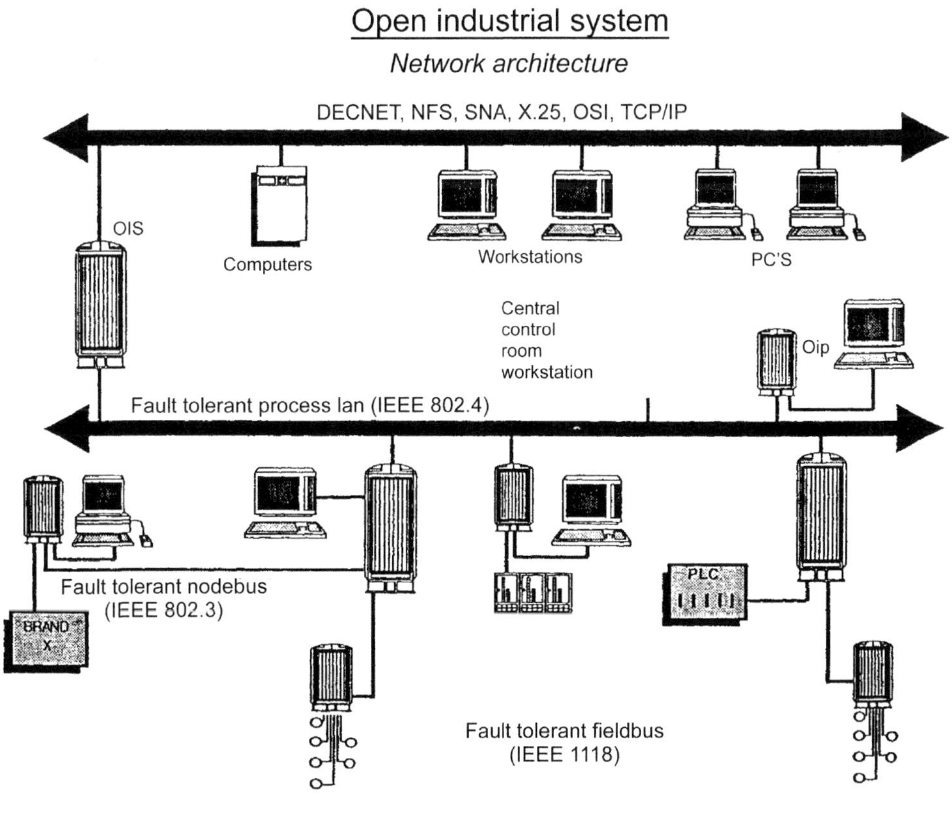

FIGURE 20.17 Schematic drawing of Foxboro's Distributed Control System called Intelligent Automation. *(Courtesy of Foxboro, Foxboro, Mass.)*

REFERENCES

[1] H.G. de Young, Biosensors, High Technol. Mag. (1983) 41–49.

[2] O.J.M. Smith, A controller to overcome dead time, ISA J. (1959).

[3] F.G. Shinsky, Model predictors: the first smart controllers, Instrum. Control Syst. (1991).

[4] C.R. Cutler, B.L. Ramaker, Dynamic Matrix Control - A Computer Control Algorithm, Paper #51b, AIChE 86[to] Meeting (April, 1979).

[5] Gensym Real Times, Gensymk Joins MIT's Biotechnology Consortium, vol. 1, Gensym Corporation, 1991.

FURTHER READING

[1] S. Evans, Genet. Eng. News (1991) 2–20.

[2] CEREX Brochure, Cerex announces MAXimum Accuracy in cell mass measurement, CEREX Corporation, Gaithersburg, MD.

[3] Wedgewood Technology Specification Sheet, Model 650 Cell Growth Probe, Wedgewood Technology, San Carlos, CA.

[4] Foxboro Bulletin F-20B, Temperature Measurement and Control Systems, (May, 1982).

[5] D. Considine, Process Instruments and Control Systems, third ed., McGraw Hill, 1985.

[6] R.W. Miller, Flow Measurement Engineering Handbook, McGraw Hill, 1983.

[7] Foxboro Instrument Catalog 583E (October, 1991).

[8] H.C. Vogel, Fermentation and Biochemical Engineering Handbook, first ed., Noyes Publications, Park Ridge, NJ, 1983.

[9] Meeting Minutes. ISA/SP88/WG1–1990–2. Draft Specification ISAdS88.01, Attachment 3, Batch Control Systems Models and Terminology, Draft 1, (October, 1990).

Chapter 21

Statistical Methods for Fermentation Optimization

Edwin O. Geiger

1.0 INTRODUCTION

A common problem for a biochemical engineer is to be handed a microorganism and be told he has six months to design a plant to produce the new fermentation product. Although this seems to be a formidable task, with the proper approach this task can be reduced to a manageable level. There are many ways to approach the problem of optimization and design of a fermentation process. One could determine the nutritional requirements of the organism and design a medium based upon the optimum combination of each nutrient, i.e., glucose, amino acids, vitamins, minerals, etc. This approach has two drawbacks. First, it is very time-consuming to study each nutrient and determine its optimum level, let alone its interaction with other nutrients. Secondly, although knowledge of the optimal nutritional requirements is useful in designing a media, this knowledge is difficult to apply when economics dictate the use of commercial substrates such as corn steep liquor, soy bean meal, etc., which are complex mixtures of many nutrients.

Production data can be used but it is often ambiguous. Multiple factors and their relevance and interdependence are obscured by patterns that may or may not be significant.

2.0 TRADITIONAL ONE-VARIABLE-AT-A-TIME METHOD

The traditional approach to the optimization problem is the one-variable-at-a-time method. In this process, all variables but one are held constant and the optimum level for this variable is determined. Using this optimum, the second variable's optimum is found, etc. This process works if, and only if, there is no interaction between variables. In the case shown in Figure 21.1, the optimum found using the one-variable-at-a-time approach was 85%, far from the real optimum of 90%. Because of the interaction between the two nutrients, the one-variable-at-a-time approach failed to find the true optimum. In order to find the optimum conditions, it would have been necessary to repeat the one-variable-at-a-time process at each step to verify that the true optimum was reached. This requires numerous sequential experimental runs, a time-consuming and ineffective strategy, especially when many variables need to be optimized. Because of the complexity of microbial metabolism, interaction between the variables is inevitable, especially when using commercial substrates which are a complex mixture of many nutrients. Therefore, since it is both time-consuming and inefficient, the one-variable-at-a-time approach is not satisfactory for fermentation development. Fortunately, there are a number of statistical methods which will find the optimum quickly and efficiently.

3.0 DESIGN OF EXPERIMENT (DOE)

A statistical method for finding the significant variables from large sets of variable through a set of planned experiments is called Design of Experiment (DOE). Two significant methods of DOE, Evolutionary Optimization (EVOP) and Response Surface Methodology (RSM) are discussed herein.

3.1 Evolutionary Optimization

An alternative to the one-variable-at-a-time approach is the technique of *evolutionary optimization*, a DOE. Evolutionary optimization (EVOP), also known as *method of steepest ascent*, is based upon the techniques developed by Spindley et al. [1]. The method is an iterative process in which a *simplex figure* is generated by running one more experiment than the number of variables to be optimized. It gets its name from the fact that the process slowly evolves toward the optimum. A simplex process is designed to find the optimum by ascending the reaction surface along the lines of the steepest slope, i.e., path with greatest increase in yield.

Fermentation and Biochemical Engineering Handbook.

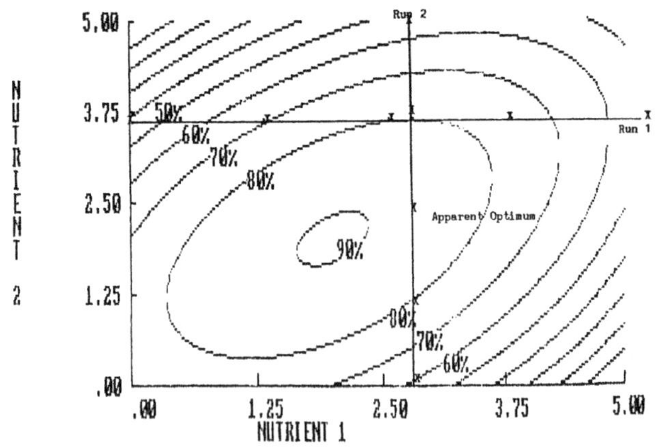

FIGURE 21.1 Example of one-variable-at-a-time approach. Contour plot of yield.

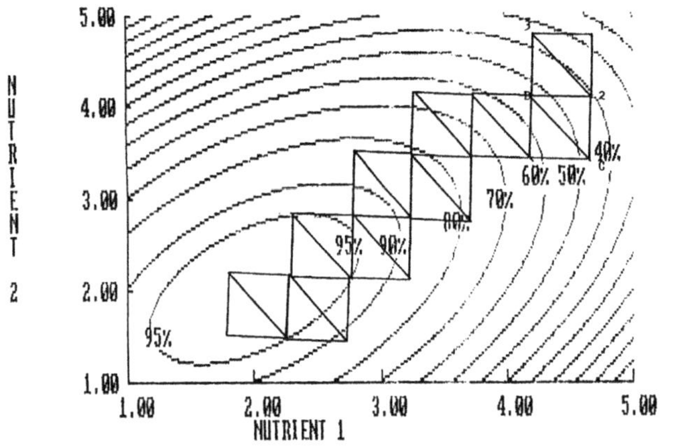

FIGURE 21.2 Example of evolutionary optimization contour plot of yield.

The procedure starts by the generation of a simplex figure. The simplex figure is a triangle when two variables are optimized, a tetrahedron when three variables are optimized, increasing to an $n+1$ polyhedron, where n is the number of variables to be optimized. The experimental point with the poorest response is eliminated and a new point generated by reflection of the eliminated point through the centroid of the simplex figure. This process is continued until an optimum is reached. In Figure 21.2, experimental points 1, 2, and 3 form the vertices of the original simplex figure. Point 1 was found to have the poorest yield, and therefore was eliminated from the simplex figure and a new point (B) generated. Point 3 was then eliminated and the new point (C) generated. The process was continued until the optimum was reached. The EVOP process is a systematic method of adjusting the variables until an optimum is reached.

Numerous modifications have been made to the original simplex method. One of the more important modifications was made by Nelder and Mead [2] who modified the method to allow expansions in directions which are favorable and contractions in directions which are unfavorable. This modification increased the rate at which the optimum is found. Other important modifications were made by Brissey [3] who describes a high-speed algorithm, and Keefer [4] who describes a high-speed algorithm and methods dealing with bounds on the independent variables.

Additional modifications were reported by Nelson [5], Bruley [6], Deming [7], and Ryan [8]. For reviews on the simplex methods see papers by Deming et al. [7,9,10].

EVOP does have its limitations. First, because of its iterative nature, it is a slow process which can require many steps. Secondly, it provides only limited information about the effects of the variables. Upon completion of the EVOP process only a limited region of the reaction surface will have been explored and therefore, minimal information will be available about the effects of the variables and their interactions. This information is necessary to determine the ranges within which the variables must be controlled to insure optimal operation. Further, EVOP approaches the nearest optimum. It is unknown whether this optimum is a local optimum or the optimum for the entire process.

Despite the limitations, EVOP is an extremely useful optimization technique. EVOP is robust, can handle many variables at the same time, and will always lead to an optimum. Also, because of its iterative nature, little needs to be known about the system before beginning the process and non-conforming runs are usually avoided. Most important, however, is the fact that it can be useful in plant optimization where the cost of running experiments using conditions that result in low yields or unusable product cannot be tolerated. In theory, the process improves at each step of the optimization scheme, making it ideal for a production situation. For application of EVOP to plant scale operations, see Refs. [11–13].

The main difficulty with using EVOP in a plant environment is performing the initial experimental runs. Plant managers are reluctant to run at less than optimal conditions. Attempts to use process data as the initial experiments in the simplex are, in general, not successful because of confounding. Confounding occurs because critical variables are closely controlled, and therefore, the error in measuring the conditions and results tend to be greater than the effect of the variables. Because of this, operating data usually gives a false perspective as to which variables are important and the changes to be made for the next step.

The successful use of EVOP depends heavily upon the choice for the initial experimental runs. If the initial points are far from the optimum and relatively close to one another, many iterations will be required. Reasonable step sizes must be chosen to insure that a significant effect of the variable is observed between the points, however, the step size should not be so great as to encompass the optimum. A second factor to consider is magnitude effects. If one variable is measured over a range of 0.1 to 1.0 while another is measured over a range of 1 to 100 the magnitude difference between the variables can effect the simplex. Scaling factors should be used to keep all variables within the same order of magnitude.

3.2 Response Surface Methodology

The best method for process optimization is response surface methodology (RSM). This process will not only determine optimum conditions, but also give the information necessary to design a process.

Response surface methodology (RSM) is a method of optimization using statistical techniques based upon the special factorial designs of Box and Behenkin [14] and Box and Wilson [14]. It is a scientific approach to determining optimum conditions which combines special experimental designs with Taylor first and second order equations. The RSM process determines the *surface* of the Taylor expansion curve which describes the *response* (yield, impurity level, etc.) The Taylor equation, which is the heart of the RSM method, has the form:

$$\text{Response} = A + B \cdot X1 + C \cdot X2 + \cdots H \cdot X1^2 + I \cdot X2^2 + \cdots M \cdot X1 \cdot X2 + N \cdot X1 \cdot X3 + \cdots$$

where A,B,C,... are the coefficients of the terms of the equation, and

X1 = linear term for variable 1
X2 = linear term for variable 2
.
.
.
$X1^2$ = nonlinear squared term for variable 1
$X2^2$ = nonlinear squared term for variable 2
.
.
.
$X1 \cdot X2$ = interaction term for variable 1 and variable 2
$X1 \cdot X3$ = interaction term for variable 1 and variable 3
.
.
.

The Taylor equation is named after the English mathematician Brook Taylor who proposed that any continuous function can be approximated by a power series. It is used in mathematics for approximating a wide variety of continuous functions. The RSM protocol, therefore, uses the Taylor equation to approximate the function which describes the response in nature, coupled with the special experimental designs for determining the coefficients of the Taylor equation.

The use of RSM requires that certain criteria must be met. These are:

1. The factors which are critical for the process are known. RSM programs are limited in the number of variables that they are designed to handle. As the number of variables increases the number of experiments required by the designs increases exponentially. Therefore, most RSM programs are limited to 4 to 5 variables. Fortunately for the scale up of most fermentations the number of variables to be optimized are limited. Some of the more important variables are listed in Table 21.1.
2. The factors must vary continuously over the experimental range tested. For example, the variables of pH, aeration rate, and agitation rate are continuous and can be used in an RSM model. Variables such as carbon source (potato starch vs corn syrup) or nitrogen source (cotton seed meal vs soy bean meal) are noncontinuous and cannot be optimized by RSM. However, level of corn syrup or level of soybean meal are continuous and can be optimized.
3. There exists a mathematical function which relates the response to the factors.

For reviews on the RSM process see Henika [15] or Giovanni [16]. For details on the calculation methods see Cochran and Cox [17], or Box [18]. The difficult and time-consuming nature of these calculations have inhibited the wide spread use of RSM. Fortunately, numerous computer programs are available to perform this chore. They range from the expensive and sophisticated, such as SAS™, to inexpensive, PC based programs, SPSS-X™, E-Chip™, and X STAT™ [19]. The availability of these programs, however, has led to a "black box" approach to RSM. This approach can lead to many problems if the

TABLE 21.1 Typical Variables in a Fermentation

Aeration rate	Agitation rate
Temperature	Carbon/Nitrogen ratio
Phosphate level	Magnesium level
Back pressure	Sulfur level
Carbon Source	Nitrogen source
pH	Dissolved oxygen level
Power input	

user does not have a thorough understanding of the process or the meaning of the results.

4.0 ADVANTAGES OF RSM

The response surface methodology approach has many advantages over other optimization procedures. These are listed in Table 21.2.

4.1 Maximum Information from Experiments

RSM yields the maximum amount of information from the minimum amount of work. For example, in the one-variable-at-a-time approach, shown in Figure 21.1, ten experiments were run only to find the suboptimum conditions. However, using RSM and thirteen properly designed experiments not only would the true optimum have been found, but also the information necessary to design the process would have been made available. Secondly, since all of the experiments can be run simultaneously, the results could be obtained quickly. This is the power of response surface methodology.

RSM is a very efficient procedure. It utilizes partial factorial designs, such as central composite or star designs, and therefore, the number of experimental points required is a minimum (Table 21.3). A full factorial three level design would require n^3 experiments; while a full factorial five level design would require n^5 experiments, where n is the number of variables to be optimized. Response surface protocols, being a partial factorial design, require fewer experiments. For example, if one were to examine five variables at five different levels, a full factorial design approach would require 3125 experiments. Response Surface Methodology, on the other hand, requires only 48 experiments, clearly a large savings in time, effort, and expense.

TABLE 21.2 Advantages and Disadvantages of RSM

Advantages of RSM

1. Greatest amount of information from experiments.
2. Forces you to plan.
3. Know how long project will take.
4. Gives information about the interaction between variables.
5. Multiple responses at the same time.
6. Gives information necessary for design and optimization of a process.

Disadvantages of RSM

1. Tells what happens, not why.
2. Notoriously poor for predicting outside the range of study.

4.2 Forces One to Plan

The successful use of an RSM protocol requires careful planning on the part of the experimenter before beginning the protocol. The ranges over which the variables are to be tested must be chosen with care. Choosing a range which is too narrow can result in a variable being discarded as not significant, not because the variable did not have an effect, but rather because the effect of the variable over the range evaluated was small in comparison to the experimental error. The range must be large enough so that the variable has a significant effect over the range evaluated. On the other hand, choosing a range which is too large can also result in a variable being discarded as not significant, not because the variable did not have an effect, but rather because the Taylor equation could not adequately explain the effect of the variable. It must be remembered that RSM does not determine the function which describes the results, but rather determines the Taylor expansion equation which best fits the data. Over a limited range, the Taylor equation will approximate the function which describes the results. The wider the range chosen the less likely a Taylor expansion equation which meaningfully explains the data will be obtained. Therefore, ranges which include extreme minimums and maximums for a variable should be avoided. Further, the experimenter needs to have an approximation as to where the optima exist. It is a sad state of affairs to have completed the RSM protocol only to find that the optimum conditions were outside of the range evaluated. RSM is notorious for its inability to predict outside the range evaluated. It is strongly advised that preliminary experiments be done to determine the ranges over which the variables are to be evaluated.

TABLE 21.3 Experimental Efficiency of RSM

Number of Variables	Number of Combinations	Number of Actual Experiments
Narrow Three Level Design		
2	9	13
3	27	15
4	81	27
5	234	46
Broad Five Level Exploratory Design		
2	25	13
3	125	20
4	625	31
5	3125	48

4.3 Know How Long Project Will Take

A distinct advantage of the RSM procedure is that one knows how many experiments and the time frame needed to complete the process. This is especially helpful for budgetary purposes and the allocation of scarce scientific resources. Using RSM, the experimenter has the information necessary to determine whether a project is worth undertaking.

4.4 Interaction Between Variables

With the one-variable-at-a-time approach, it is difficult to determine the amount of interaction between variables. Response surface methodology, since it looks at all the variables at the same time, can calculate the interaction between them. This information is essential for optimizing conditions and determining what control limits are needed for the variables.

4.5 Multiple Responses

RSM has the ability to model as many responses as one wishes to measure. For example, one may not only be interested in optimum yield, but also the level of a difficult to remove impurity. Both the yield and impurity levels could be modeled using data from the same set of experiments. Decisions could then be made between the cost to remove an impurity and changes in yield.

4.6 Design Data

Last, but most important, RSM gives the information necessary to design the process. For example, Figure 21.3 shows the effect of temperature and degree of saccharification on alcohol yield. This plot not only shows the conditions necessary for optimum yield, it also indicates the sensitivity of the process to changes in temperature and degree of saccharification. It shows the range over which these variables must be controlled for optimum yield. Temperature needs to be controlled within a 5-degree range and the degree of saccharification within a 10% range. This information can now be used in designing control loops for these variables.

In any industrial process, the cost-effective conditions are influenced by factors other than optimum reaction conditions. There exists a compromise between optimum reaction conditions and economic factors such as capital and purification costs. In addition to determining optimum conditions and the ranges within which the variables need to be controlled, the regression equations generated by the RSM procedure allow the process to be modeled for a wide variety of operating parameters. The regression equations, therefore, are an ideal tool for evaluating various economic trade-offs. For example, in Figure 21.4, 98% yields are obtained at low carbohydrate levels and long fermentation times. Although this is a high yield, both capital costs for the fermentation capacity and distillation cost for the resulting low alcohol beer makes this an uneconomical operating condition. Using the model developed by the RSM process, the trade-off between capital and purification costs can be weighed against lower yields to determine the best process.

5.0 DISADVANTAGES OF RSM

There are two major disadvantages of RSM. First, it tells what happened, not why it happened. Aesthetically, this is not appealing to many scientists. This perhaps explains why, with the exception of analytical method

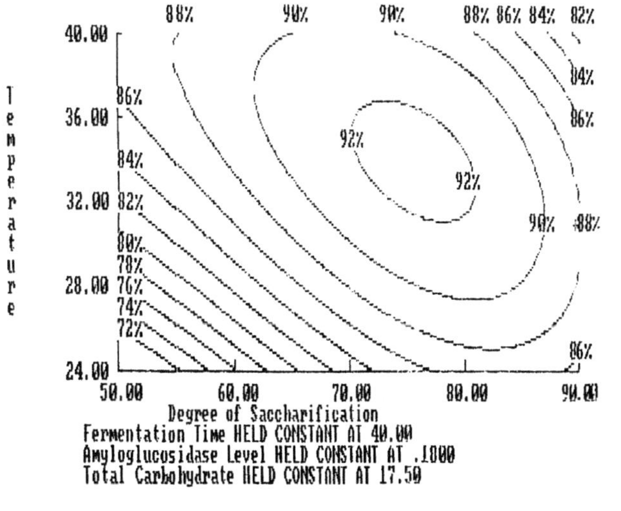

FIGURE 21.3 Contour plot of alcohol yield. Degree of saccharification.

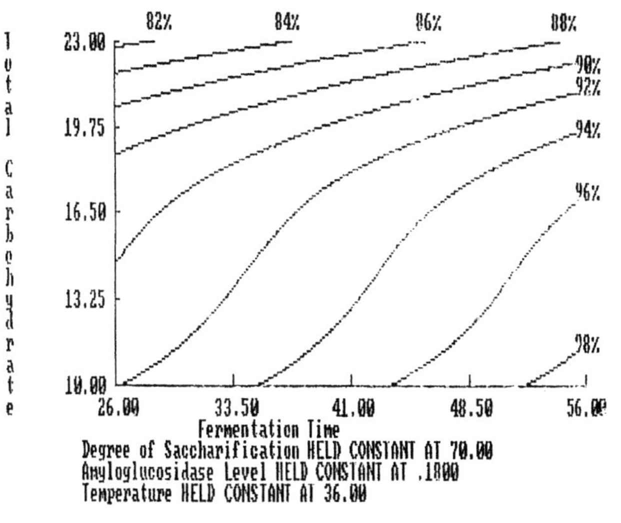

FIGURE 21.4 Contour plot of alcohol yield. Fermentation time.

development, few papers appear in the literature using RSM. This is an unfortunate circumstance since RSM is such a powerful and timesaving tool. In many cases, knowing what happens can lead to an explanation of the why or point to alternative directions for future research. For example, in Figure 21.5 there is a definite optimum for the degree of saccharification. Hypotheses to explain this phenomenon are slow substrate production at low saccharification levels and substrate inhibition at high saccharification levels. Having seen the effect of saccharification, one can readily design experiments to determine the cause.

6.0 POTENTIAL DIFFICULTIES WITH RSM

It must be remembered that RSM uses multiple regression techniques to determine the coefficients for the Taylor expansion equation which best fits the data. The RSM does not determine the function which describes the data. The Taylor equation only approximates the true function. The RSM process fits one of a series of curves to the data. Most RSM programs use only the first and second order terms of the Taylor equation to the data, which limits the number of curves available to fit the data. The first order Taylor equation is a linear model. Therefore, the only curves available are a series of straight lines. The second order Taylor equation is a nonlinear model where two types of curves are available; a peak or a saddle surface. Over a narrow range, these curves will approximate the true function that exists in nature; but they are not necessarily the function that describes the response.

Although RSM is a rapid method for determining optimum conditions for a process, caution must be used when interpreting the results. Always remember the quote by

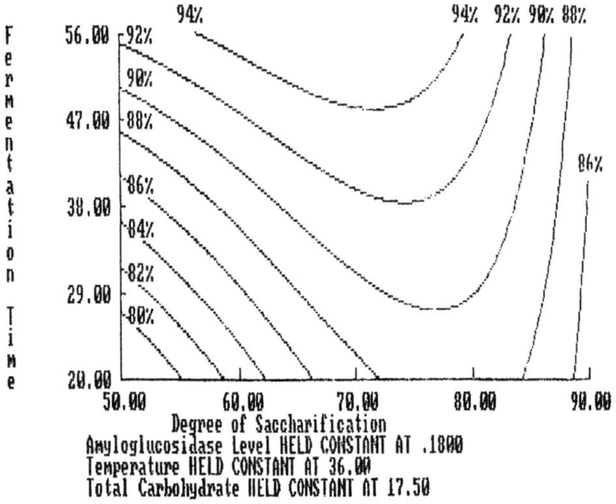

FIGURE 21.5 Contour plot of alcohol yield. Degree of saccarification.

Mark Twain, "There are liars, damn liars, and statisticians." Unless the RSM output is used properly, it is easy to make this quote true. RSM will always give the user a number. The question remains as to how good is that number and what does it mean? Some of the important statistical values which should be considered in evaluating the RSM output are listed below.

6.1 Correlation Coefficient

The correlation coefficient is a measure of the relationship between the Taylor expansion term and the response obtained. The correlation coefficient can vary from 0 (absolutely no correlation) to 1 or -1 (perfect correlation). A correlation coefficient of 0.5 shows a weak but useful correlation. A positive sign for the correlation coefficient indicates that the response increases as the variable increases while a negative sign indicates that the response decreases as the variable increases.

6.2 Regression Coefficients

The regression coefficients are the coefficients for the terms of the Taylor expansion equation. These coefficients can be determined either by using the actual values for the independent variables or coded values. Using the actual values makes it easy to calculate the response from the coefficients since it is not necessary to go through the coding process. However, there is a loss of important information. The reason for coding the variables is to eliminate the effect that the magnitude of the variable has upon the regression coefficient. When coded values are used in determining the regression coefficients, the importance of the variable in predicting the results can be determined from the absolute value of the coefficient. Using coded values for the independent variables, those variables which are important and must be closely controlled can readily be determined. The formula for coding values is:

Coded Value = (Value minus Midpoint value)/Step value

where:

Value = the level of the variable used
Midpoint Value = Level of variable at the mid point of the range
Step Value = Midpoint value minus next lowest value

6.3 Standard Error of the Regression Coefficient

RSM determines the best estimate of the coefficients for the Taylor equation which explains the response. The estimated regression coefficient is not necessarily the exact value but rather an estimate for the coefficient. The advantage of statistical techniques is that from the standard error one has information about how valid is the estimate for the coefficient (the range within which the exact value for the coefficient may be found). The greater the standard error, the larger the range within which the exact value for the coefficient may be, i.e., the larger the possible error in the value for the coefficient. The standard error of the regression coefficient should be as small as possible. A standard error which is 50% of the coefficient indicates a coefficient which is useful in predicting the response. Designing a process using coefficients with a large standard error can lead to serious difficulties.

6.4 Computed T Value

The T test value is a measure of the regression coefficient's significance, i.e., does the coefficient have a real meaning or should it be zero. The larger the absolute value of T the greater the probability that the coefficient is real and should be used for predictions. A T test value 1.7 or higher indicates that there is a high probability that the coefficient is real and the variable has an important effect upon the response.

6.5 Standard Error of the Estimate

The standard error of the estimate yields information concerning the reliability of the values predicted by the regression equation. The greater the standard error of the estimate, the less reliable the predicted values.

6.6 Analysis of Variance

Three other statistical numbers which should be closely examined relate to the source of variation in the data. The variation attributable to the regression reflects the amount of variation in the data explained by the regression equation. The deviation from regression is a measure of the scatter in the data which is not explained, i.e., the experimental error. Ideally the deviation from the regression should be very small in comparison to the amount of variation explained by the regression. If this is not the case, it means that the Taylor equation does not explain the data and the regression equation should not be used as a design basis. The third important factor is the relationship between the explained and unexplained variation. The greater the amount of variation explained by the regression equation, the greater the probability that the equation meaningfully explains the results. The F value is a measure of this relationship. The larger the F values the greater significance the regression equation has in explaining the data. The F value is also helpful in comparing different models. Models with the larger F value are better in explaining the response data.

7.0 METHODS TO IMPROVE THE RSM MODEL

The output from an RSM program is only as good as the data entered. The cliche GIGO (garbage in garbage out) applies especially to the RSM process. Since the minimum amount of experiments is being used, any inaccuracies in the data can have a large effect upon the results. One acceptable method to increase the accuracy of the results is to perform replicate experiments and use the averages as the input data. Care must be taken, however, to avoid confounding the results by performing replicates of only a portion of the experimental design. This will result in the experimental error being understated in some areas of the response surface and over stated in others. All experimental points must be treated in a similar manner in order to insure that a meaningful response surface is obtained. A common error, especially when using multiple regression programs, is to use all the data available. Performing the regression analysis with missing data points or the addition of data points to the design leads to misleading results unless special care is taken. The design used must be symmetrical to prevent the uneven weighting of specific areas of the response surface from distorting the final model. Although adding the extra data points may improve the statistics of the model, it can also reduce its reliability. RSM users are strongly cautioned to resist the temptation to add extra data points to the model simply because they are available.

Another method to improve the reliability of the RSM model is the use of backward elimination, i.e., the removal of those variables whose T test value is below the 95% confidence limit. This process, however, must be used with care. There are two types of statistical errors. A Type I error is saying a variable is significant when it is not. A Type II error is saying a variable is not significant when it is. Statistical procedures are designed to minimize the chances of committing a Type I error. The statistical process determines the probability that a variable is indeed important. Elimination of those variables not significant at the 95% confidence limit reduces the chances for making a Type I error. This does not mean that the variables eliminated were not important. Lack of statistical significance means the variable was not proven to be important. There is a large difference between unimportant and not proven important. While elimination of the variables not significant at the 95% confidence limit decreases the probability of making a Type I error, it increases the chances of making a Type II error; disregarding a variable which was important.

Some mathematical considerations also need to be taken into account when eliminating variables from the equation. An equation where the linear term was eliminated while the nonlinear term was retained can mathematically produce only a curve with the maxima, or minima, centered in the region evaluated. It is necessary to retain the linear term in order to move the maxima or minima to the appropriate area on the plot. Similarly, an equation containing only an interaction term, can mathematically produce only a saddle surface centered on the region evaluated. The other terms for the variables are necessary to move the optimum to the appropriate area of the response surface. When eliminating terms, it is best to eliminate the entire variable and not just selected terms for the variable. Failure to heed these warnings will result in a process being designed for conditions which are not optimum.

8.0 SUMMARY

The problem of designing and optimizing fermentation processes can be handled quickly using a number of statistical techniques. It has been our experience that the best technique is response surface methodology, a Design of Experiment (DOE). Although not reported widely in the literature, most pharmaceutical companies for the optimization of their antibiotic fermentations use this process. RSM is a highly efficient procedure for determining not only the optimum conditions, but also the data necessary to design the entire process. In cases where RSM cannot be applied, evolutionary optimization (EVOP) is an alternative method for optimization of a process. These methods are systematic procedures that guarantee optimum conditions will be found.

REFERENCES

[1] W. Spindely, G.R. Hext, F.R. Himsworth, Technometrics 4 (1962) 411.

[2] J.A. Nelder, R. Mead, Comput. J. 7 (1965) 308.

[3] G.W. Brissey, R.B. Spencer, C.L. Wilkins, Anal. Chem. 51 (1979) 2295.

[4] D. Keefer, Ind Eng. Chem. Process Des. Dev. 12 (1) (1973) 92.

[5] L. Nelson, Annual Conference Transactions of the American Society for Quality Control, May 1973, pp. 107–117.

[6] R.W. Glass, D.F. Bruley, Ind. Eng. Chem. Process Des. Dev. 12 (1) (1973) 6.

[7] S.N. Deming, LR Parker, Crit. Rev. Anal. Chem. 7 (1978) 187.

[8] P.B. Ryan, R.L. Barr, H.D. Tood, Anal. Chem. 52 (1980) 1460.

[9] S.N. Deming, S.L. Morgan, M.R. Willcott, Am. Lab. 8 (10) (1976) 13.

[10] C.L. Shavers, M.L. Parsons, S.N. Deming, J. Chem. Educ. 56 (1976) 307.

[11] B.H. Carpenter, H C Sweeny, Chem. Eng. 72 (1965) 117.

[12] T. Umeda, A. Ichikawa, Ind. Eng. Chem. Process Des. 10 (1971) 229.

[13] W.D. Basel, Chem. Eng. 72 (1965) 147.

[14] G.E.P. Box, K.B. Wilson, J. R. Stat. Soc. B. 13 (1951) 1.

[15] R.G. Henika, Ceral Sci. Today 17 (1972) 309.

[16] M. Giovanni, Food Technol. (1983) 41.

[17] W.G. Cochran, G.M. Cox, Experimental Designs, John Wiley & Sons, New York City, 1957, pp. 335.

[18] G.EP Box, W.G. Hunter, J.S. Hunter, Statistics for Experimenters, John Wiley & Sons, New York City, 1978, pp. 510.

[19] SAS is a trademark of SAS Institute, Cary, NC; SPSS-X is a trademark of SPSS, Chicago, IL; E Chip is a trademark of E-CHIP, Inc., Hockessin, DE; X STAT is a trademark of Wiley and Sons, New York, NY.

FURTHER READING

[1] P.G. King, S.N. Deming, Anal. Chem. 46 (1974) 1476.

[2] W.J. Hill, W.G. Hunter, Technometrics 8 (1966) 571.

Index

Note: Page numbers followed by "*f*" and "*t*" refer to figures and tables, respectively.

D

E

F

G

Q

R

Dedication

For my parents,
Ernest and Charlotte Todaro,
whose pursuit of knowledge
inspired me and continues
to do so.

9781455725533